# CAMBRIDGE MONOGRAPHS ON MATHEMATICAL PHYSICS

General editors: P. V. Landshoff, W. H. McCrea, D. W. Sciama, S. Weinberg

## SPINORS AND SPACE–TIME

Volume 1: Two-spinor calculus and relativistic fields

# SPINORS AND SPACE–TIME

## Volume 1
## Two-spinor calculus and relativistic fields

ROGER PENROSE

*Rouse Ball Professor of Mathematics, University of Oxford*

WOLFGANG RINDLER

*Professor of Physics, University of Texas at Dallas*

CAMBRIDGE UNIVERSITY PRESS

*Cambridge*

*New York New Rochelle*

*Melbourne Sydney*

Published by the Press Syndicate of the University of Cambridge
The Pitt Building, Trumpington Street, Cambridge CB2 1RP
32 East 57th Street, New York, NY 10022, USA
10 Stamford Road, Oakleigh, Melbourne 3166, Australia



First published 1984

Reprinted with corrections 1986
First paperback edition 1987
Reprinted 1987

Printed in Great Britain at the University Press, Cambridge

Library of Congress catalogue card number: 82–19861

*British Library cataloguing in publication data*

Penrose, Roger
Spinors and space–time. – (Cambridge monographs
on mathematical physics)
Vol. 1: Two-spinor calculus and relativistic fields
1. Spinor analysis
I. Title II. Rindler, Wolfgang
512′.57 QA433

ISBN 0 521 24527 3 hard covers
ISBN 0 521 33707 0 paperback

TP

# Contents


| | | |
|---|---|---|
| | Preface | vii |
| **1** | **The geometry of world-vectors and spin-vectors** | 1 |
| 1.1 | Minkowski vector space | 1 |
| 1.2 | Null directions and spin transformations | 8 |
| 1.3 | Some properties of Lorentz transformations | 24 |
| 1.4 | Null flags and spin-vectors | 32 |
| 1.5 | Spinorial objects and spin structure | 41 |
| 1.6 | The geometry of spinor operations | 56 |
| **2** | **Abstract indices and spinor algebra** | 68 |
| 2.1 | Motivation for abstract-index approach | 68 |
| 2.2 | The abstract-index formalism for tensor algebra | 76 |
| 2.3 | Bases | 91 |
| 2.4 | The total reflexivity of $\mathfrak{S}^{\cdot}$ on a manifold | 98 |
| 2.5 | Spinor algebra | 103 |
| **3** | **Spinors and world-tensors** | 116 |
| 3.1 | World-tensors as spinors | 116 |
| 3.2 | Null flags and complex null vectors | 125 |
| 3.3 | Symmetry operations | 132 |
| 3.4 | Tensor representation of spinor operations | 147 |
| 3.5 | Simple propositions about tensors and spinors at a point | 159 |
| 3.6 | Lorentz transformations | 167 |
| **4** | **Differentiation and curvature** | 179 |
| 4.1 | Manifolds | 179 |
| 4.2 | Covariant derivative | 190 |
| 4.3 | Connection-independent derivatives | 201 |
| 4.4 | Differentiation of spinors | 210 |
| 4.5 | Differentiation of spinor components | 223 |
| 4.6 | The curvature spinors | 231 |
| 4.7 | Spinor formulation of the Einstein–Cartan–Sciama–Kibble theory | 237 |

4.8 The Weyl tensor and the Bel–Robinson tensor 240
4.9 Spinor form of commutators 242
4.10 Spinor form of the Bianchi identity 245
4.11 Curvature spinors and spin-coefficients 246
4.12 Compacted spin-coefficient formalism 250
4.13 Cartan's method 262
4.14 Applications to 2-surfaces 267
4.15 Spin-weighted spherical harmonics 285

**5 Fields in space–time** 312
5.1 The electromagnetic field and its derivative operator 312
5.2 Einstein–Maxwell equations in spinor form 325
5.3 The Rainich conditions 328
5.4 Vector bundles 332
5.5 Yang–Mills fields 342
5.6 Conformal rescalings 352
5.7 Massless fields 362
5.8 Consistency conditions 366
5.9 Conformal invariance of various field quantities 371
5.10 Exact sets of fields 373
5.11 Initial data on a light cone 385
5.12 Explicit field integrals 393
Appendix: diagrammatic notation 424
References 435
Subject and author index 445
Index of symbols 457

# Preface

To a very high degree of accuracy, the space–time we inhabit can be taken to be a smooth four-dimensional manifold, endowed with the smooth Lorentzian metric of Einstein's special or general relativity. The formalism most commonly used for the mathematical treatment of manifolds and their metrics is, of course, the tensor calculus (or such essentially equivalent alternatives as Cartan's calculus of moving frames). But in the specific case of four dimensions and Lorentzian metric there happens to exist – by accident or providence – another formalism which is in many ways more appropriate, and that is the formalism of 2-spinors. Yet 2-spinor calculus is still comparatively unfamiliar even now – some seventy years after Cartan first introduced the general spinor concept, and over fifty years since Dirac, in his equation for the electron, revealed a fundamentally important role for spinors in relativistic physics and van der Waerden provided the basic 2-spinor algebra and notation.

The present work was written in the hope of giving greater currency to these ideas. We develop the 2-spinor calculus in considerable detail, assuming no prior knowledge of the subject, and show how it may be viewed either as a useful supplement or as a practical alternative to the more familiar world-tensor calculus. We shall concentrate, here, entirely on 2-spinors, rather than the 4-spinors that have become the more familiar tools of theoretical physicists. The reason for this is that only with 2-spinors does one obtain a practical alternative to the standard vector–tensor calculus, 2-spinors being the more primitive elements out of which 4-spinors (as well as world-tensors) can be readily built.

Spinor calculus may be regarded as applying at a deeper level of structure of space–time than that described by the standard world-tensor calculus. By comparison, world-tensors are less refined, fail to make transparent some of the subtler properties of space–time brought particularly to light by quantum mechanics and, not least, make certain types of mathematical calculations inordinately heavy. (*Their* strength lies in a general applicability to manifolds of arbitrary dimension, rather than in supplying a specific space–time calculus.)

In fact any world-tensor calculation can, by an obvious prescription, be translated entirely into a 2-spinor form. The reverse is also, in a sense, true – and we shall give a comprehensive treatment of such translations later in this book – though the tensor translations of simple spinor manipulations can turn out to be extremely complicated. This effective equivalence may have led some 'sceptics' to believe that spinors are 'unnecessary'. We hope that this book will help to convince the reader that there are many classes of spinorial results about space–time which would have lain undiscovered if only tensor methods had been available, and others whose antecedents and interrelations would be totally obscured by tensor descriptions.

When appropriately viewed, the 2-spinor calculus is also simpler than that of world-tensors. The essential reason is that the basic spin-space is two-complex-dimensional rather than four-real-dimensional. Not only are two dimensions easier to handle than four, but complex algebra and complex geometry have many simple, elegant and uniform properties not possessed by their real counterparts.

Additionally, spinors seem to have profound links with the complex numbers that appear in quantum mechanics.* Though in this work we shall not be concerned with quantum mechanics as such, many of the techniques we describe are in fact extremely valuable in a quantum context. While our discussion will be given entirely classically, the formalism can, without essential difficulty, be adapted to quantum (or quantum-field-theoretic) problems.

As far as we are aware, this book is the first to present a comprehensive development of space–time geometry using the 2-spinor formalism. There are also several other new features in our presentation. One of these is the systematic and consistent use of the *abstract index* approach to tensor and spinor calculus. We hope that the purist differential geometer who casually leafs through the book will not automatically be put off by the appearance of numerous indices. Except for the occasional bold-face upright ones, our indices differ from the more usual ones in being abstract markers without reference to any basis or coordinate system. Our use of abstract indices leads to a number of simplifications over conventional treatments. The use of some sort of index notation seems, indeed, to be virtually essential in order that the necessary detailed manipulations can

* The view that space–time geometry, as well as quantum theory, may be governed by an underlying complex rather than real structure is further developed in the theory of twistors, which is just one of the several topics discussed in the companion volume to the present work: *Spinors and space–time, Vol. 2: Spinor and twistor methods in space–time geometry*, (Cambridge University Press 1985).

be presented in a transparent form. (In an appendix we outline an alternative and equivalent diagrammatic notation which is very valuable for use in private calculations.)

This book appears also to be breaking some new ground in its presentation of several other topics. We provide explicit geometric realizations not only of 2-spinors themselves but also of their various algebraic operations and some of the related topology. We give a host of useful lemmas for both spinor and general tensor algebra. We provide the first comprehensive treatment of (not necessarily normalized) spin-coefficients which includes the compacted spin- and boost-weighted operators ð and þ and their conformally invariant modifications $ð_{\mathscr{C}}$ and $þ_{\mathscr{C}}$. We present a general treatment of conformal invariance; and also an abstract-index-operator approach to the electromagnetic and Yang–Mills fields (in which the somewhat ungainly appearance of the latter is, we hope, compensated by the comprehensiveness of our scheme). Our spinorial treatment of (spin-weighted) spherical harmonics we believe to be new. Our presentation of exact sets of fields as the systems which propagate uniquely away from arbitrarily chosen null-data on a light cone has not previously appeared in book form; nor has the related explicit integral spinor formula (the generalized Kirchhoff–d'Adhémar expression) for representing massless free fields in terms of such data. The development we give for the interacting Maxwell–Dirac theory in terms of sums of integrals described by zig-zag and forked null paths appears here for the first time.

As for the genesis of this work, it goes back to the spring of 1962 when one of us (R.P.) gave a series of seminars on the then-emerging subject of 2-spinors in relativity, and the other (W.R.) took notes and became more and more convinced that these notes might usefully become a book. A duplicated draft of the early chapters was distributed to colleagues that summer. Our efforts on successive drafts have waxed and waned over the succeeding years as the subject grew and grew. Finally during the last three years we made a concerted effort and re-wrote and almost doubled the entire work, and hope to have brought it fully up to date. In its style we have tried to preserve the somewhat informal and unhurried manner of the original seminars, clearly stating our motivations, not shunning heuristic justifications of some of the mathematical results that are needed, and occasionally going off on tangents or indulging in asides. There exist many more rapid and condensed ways of arriving at the required formalisms, but we preferred a more leisurely pace, partly to facilitate the progress of students working on their own, and partly to underline the down-to-earth utility of the subject.

Fortunately our rather lengthy manuscript allowed a natural division into two volumes, which can now be read independently. The essential content of Vol. 1 is summarized in an introductory section to Vol. 2. References in Vol. 1 to Chapters 6–9 refer to Vol. 2.

We owe our thanks to a great many people. Those whom we mention are the ones whose specific contributions have come most readily to mind, and it is inevitable that in the period of over twenty years in which we have been engaged in writing this work, some names will have escaped our memories. For a variety of different kinds of assistance we thank Nikos Batakis, Klaus Bichteler, Raoul Bott, Nick Buchdahl, Subrahmanyan Chandrasekhar, Jürgen Ehlers, Leon Ehrenpreis, Robert Geroch, Stephen Hawking, Alan Held, Nigel Hitchin, Jim Isenberg, Ben Jeffryes, Saunders Mac Lane, Ted Newman, Don Page, Felix Pirani, Ivor Robinson, Ray Sachs, Engelbert Schücking, William Shaw, Takeshi Shirafuji, Peter Szekeres, Paul Tod, Nick Woodhouse, and particularly, Dennis Sciama for his continued and unfailing encouragement. Our thanks go also to Markus Fierz for a remark leading to the footnote on p. 321. Especially warm thanks go to Judith Daniels for her encouragement and detailed criticisms of the manuscript when the writing was going through a difficult period. We are also greatly indebted to Tsou Sheung Tsun for her caring assistance with the references and related matters. Finally, to those people whose contributions we can no longer quite recall we offer both our thanks and our apologies.

Roger Penrose

1984 Wolfgang Rindler

# 1
# The geometry of world-vectors and spin-vectors

## 1.1 Minkowski vector space

In this chapter we are concerned with geometry relating to the space of *world-vectors*. This space is called *Minkowski vector space*. It consists of the set of 'position vectors' in the space–time of special relativity, originating from an arbitrarily chosen origin-event. In the curved space–time of general relativity, Minkowski vector spaces occur as the tangent spaces of space–time points (events). Other examples are the space spanned by four-velocities and by four-momenta.

A Minkowski vector space is a four-dimensional vector space $\mathbb{V}$ over the field $\mathbb{R}$ of real numbers, $\mathbb{V}$ being endowed with an orientation, a (bilinear) inner product of signature $(+---)$, and a time-orientation. (The precise meanings of these terms will be given shortly.) Thus, as for any vector space, we have operations of addition, and multiplication by scalars, satisfying

$$\begin{aligned} \boldsymbol{U}+\boldsymbol{V}=\boldsymbol{V}+\boldsymbol{U}, \quad \boldsymbol{U}+(\boldsymbol{V}+\boldsymbol{W})=(\boldsymbol{U}+\boldsymbol{V})+\boldsymbol{W}, \\ a(\boldsymbol{U}+\boldsymbol{V})=a\boldsymbol{U}+a\boldsymbol{V}, \quad (a+b)\boldsymbol{U}=a\boldsymbol{U}+b\boldsymbol{U}, \\ a(b\boldsymbol{U})=(ab)\boldsymbol{U}, \quad 1\boldsymbol{U}=\boldsymbol{U}, \quad 0\boldsymbol{U}=0\boldsymbol{V}=:\boldsymbol{0} \end{aligned} \tag{1.1.1}$$

for all $\boldsymbol{U}, \boldsymbol{V}, \boldsymbol{W} \in \mathbb{V}$, $a, b \in \mathbb{R}$. $\boldsymbol{0}$ is the neutral element of addition. As is usual, we write $-\boldsymbol{U}$ for $(-1)\boldsymbol{U}$, and we adopt the usual conventions about brackets and minus signs, e.g., $\boldsymbol{U}+\boldsymbol{V}-\boldsymbol{W}=(\boldsymbol{U}+\boldsymbol{V})+(-\boldsymbol{W})$, etc.

The four-dimensionality of $\mathbb{V}$ is equivalent to the existence of a *basis* consisting of four linearly independent vectors $\boldsymbol{t}, \boldsymbol{x}, \boldsymbol{y}, \boldsymbol{z} \in \mathbb{V}$. That is to say, any $\boldsymbol{U} \in \mathbb{V}$ is uniquely expressible in the form

$$\boldsymbol{U}=U^0\boldsymbol{t}+U^1\boldsymbol{x}+U^2\boldsymbol{y}+U^3\boldsymbol{z} \tag{1.1.2}$$

with the *coordinates* $U^0, U^1, U^2, U^3 \in \mathbb{R}$; and only $\boldsymbol{0}$ has all coordinates zero. Any other basis for $\mathbb{V}$ must also have four elements, and *any* set of four linearly independent elements of $\mathbb{V}$ constitutes a basis. We often refer to a basis for $\mathbb{V}$ as a *tetrad*, and often denote a tetrad $(\boldsymbol{t}, \boldsymbol{x}, \boldsymbol{y}, \boldsymbol{z})$ by $\boldsymbol{g}_{\mathbf{i}}$, where

$$\boldsymbol{t}=\boldsymbol{g}_0, \boldsymbol{x}=\boldsymbol{g}_1, \boldsymbol{y}=\boldsymbol{g}_2, \boldsymbol{z}=\boldsymbol{g}_3. \tag{1.1.3}$$

Then (1.1.2) becomes

$$\boldsymbol{U} = U^0\boldsymbol{g}_0 + U^1\boldsymbol{g}_1 + U^2\boldsymbol{g}_2 + U^3\boldsymbol{g}_3 = U^{\mathbf{i}}\boldsymbol{g}_{\mathbf{i}}. \tag{1.1.4}$$

Here we are using the Einstein *summation convention*, as we shall henceforth: it implies a summation whenever a *numerical* index occurs twice in a term, once up, once down. Bold-face upright lower-case latin indices $\mathbf{a}, \mathbf{i}, \mathbf{a}_0, \mathbf{a}_1, \hat{\mathbf{a}}$, etc., will always be understood to range over the four values 0, 1, 2, 3. Later we shall also use bold-face upright capital latin letters $\mathbf{A}, \mathbf{I}, \mathbf{A}_0, \mathbf{A}_1, \hat{\mathbf{A}}$, etc., for numerical indices which will range only over the two values 0, 1. Again the summation convention will apply.

Consider two bases for $\mathbb{V}$, say $(\boldsymbol{g}_0, \boldsymbol{g}_1, \boldsymbol{g}_2, \boldsymbol{g}_3)$ and $(\boldsymbol{g}_{\hat{0}}, \boldsymbol{g}_{\hat{1}}, \boldsymbol{g}_{\hat{2}}, \boldsymbol{g}_{\hat{3}})$. Note that we use the 'marked index' notation, in which indices rather than kernel letters of different bases, etc., carry the distinguishing marks (hats, etc.). And indices like $\mathbf{a}, \hat{\mathbf{a}}, \hat{\hat{\mathbf{a}}}$, etc., are as unrelated numerically as $\mathbf{a}, \mathbf{b}, \mathbf{c}$. The reader may feel at first that this notation is unaesthetic but it pays to get used to it; its advantages will becomes apparent later. Now, each vector $\boldsymbol{g}_{\mathbf{i}}$ of the first basis will be a linear combination of the vectors $\boldsymbol{g}_{\hat{\mathbf{i}}}$ of the second:

$$\begin{aligned} \boldsymbol{g}_{\mathbf{i}} &= g_{\mathbf{i}}{}^{\hat{0}}\boldsymbol{g}_{\hat{0}} + g_{\mathbf{i}}{}^{\hat{1}}\boldsymbol{g}_{\hat{1}} + g_{\mathbf{i}}{}^{\hat{2}}\boldsymbol{g}_{\hat{2}} + g_{\mathbf{i}}{}^{\hat{3}}\boldsymbol{g}_{\hat{3}} \\ &= g_{\mathbf{i}}{}^{\hat{\mathbf{j}}}\boldsymbol{g}_{\hat{\mathbf{j}}}. \end{aligned} \tag{1.1.5}$$

The 16 numbers $g_{\mathbf{i}}{}^{\hat{\mathbf{j}}}$ form a $(4 \times 4)$ real non-singular matrix. Thus $\det(g_{\mathbf{i}}{}^{\hat{\mathbf{j}}})$ is non-zero. If it is *positive*, we say that the tetrads $\boldsymbol{g}_{\mathbf{i}}$ and $\boldsymbol{g}_{\hat{\mathbf{i}}}$ have the *same orientation*; if *negative*, the tetrads are said to have *opposite orientation*. Note that the relation of 'having the same orientation' is an equivalence relation. For if $\boldsymbol{g}_{\hat{\mathbf{i}}} = g_{\hat{\mathbf{i}}}{}^{\mathbf{j}}\boldsymbol{g}_{\mathbf{j}}$, then $(g_{\hat{\mathbf{i}}}{}^{\mathbf{j}})$ and $(g_{\mathbf{i}}{}^{\hat{\mathbf{j}}})$ are inverse matrices, so their determinants have the same sign; if $\boldsymbol{g}_{\mathbf{i}} = g_{\mathbf{i}}{}^{\hat{\mathbf{j}}}\boldsymbol{g}_{\hat{\mathbf{j}}}$ and $\boldsymbol{g}_{\hat{\mathbf{i}}} = g_{\hat{\mathbf{i}}}{}^{\underset{\cdot}{\mathbf{j}}}\boldsymbol{g}_{\underset{\cdot}{\mathbf{j}}}$, then the matrix $(g_{\mathbf{i}}{}^{\underset{\cdot}{\mathbf{j}}})$ is the product of $(g_{\mathbf{i}}{}^{\hat{\mathbf{j}}})$ with $(g_{\hat{\mathbf{i}}}{}^{\underset{\cdot}{\mathbf{j}}})$ and so has positive determinant if both the others have. Thus the tetrads fall into two disjoint equivalence classes. Let us call the tetrads of one class *proper* tetrads and those of the other class *improper* tetrads. It is this selection that gives $\mathbb{V}$ its *orientation*.

The *inner product operation on* $\mathbb{V}$ assigns to any pair $\boldsymbol{U}, \boldsymbol{V}$ of $\mathbb{V}$ a real number, denoted, by $\boldsymbol{U}\cdot\boldsymbol{V}$, such that

$$\boldsymbol{U}\cdot\boldsymbol{V} = \boldsymbol{V}\cdot\boldsymbol{U}, \quad (a\boldsymbol{U})\cdot\boldsymbol{V} = a(\boldsymbol{U}\cdot\boldsymbol{V}), \quad (\boldsymbol{U}+\boldsymbol{V})\cdot\boldsymbol{W} = \boldsymbol{U}\cdot\boldsymbol{W} + \boldsymbol{V}\cdot\boldsymbol{W}, \tag{1.1.6}$$

i.e., the operation is symmetric and bilinear. We also require the inner product to have signature $(+ - - -)$. This means that there exists a tetrad $(\boldsymbol{t}, \boldsymbol{x}, \boldsymbol{y}, \boldsymbol{z})$ such that

$$\boldsymbol{t}\cdot\boldsymbol{t} = 1, \quad \boldsymbol{x}\cdot\boldsymbol{x} = \boldsymbol{y}\cdot\boldsymbol{y} = \boldsymbol{z}\cdot\boldsymbol{z} = -1 \tag{1.1.7}$$

$$\boldsymbol{t}\cdot\boldsymbol{x} = \boldsymbol{t}\cdot\boldsymbol{y} = \boldsymbol{t}\cdot\boldsymbol{z} = \boldsymbol{x}\cdot\boldsymbol{y} = \boldsymbol{x}\cdot\boldsymbol{z} = \boldsymbol{y}\cdot\boldsymbol{z} = 0. \tag{1.1.8}$$

If we denote this tetrad by $\boldsymbol{g}_{\mathbf{i}}$ according to the scheme (1.1.3), then we can rewrite (1.1.7) and (1.1.8) succinctly as

$$\boldsymbol{g}_{\mathbf{i}} \cdot \boldsymbol{g}_{\mathbf{j}} = \eta_{\mathbf{ij}}, \tag{1.1.9}$$

where the matrix $(\eta_{\mathbf{ij}})$ is given by

$$(\eta_{\mathbf{ij}}) = (\eta^{\mathbf{ij}}) = \begin{pmatrix} 1 & 0 & 0 & 0 \\ 0 & -1 & 0 & 0 \\ 0 & 0 & -1 & 0 \\ 0 & 0 & 0 & -1 \end{pmatrix}. \tag{1.1.10}$$

(The raised-index version $\eta^{\mathbf{ij}}$ will be required later for notational consistency.) We shall call a tetrad satisfying (1.1.9) a *Minkowski tetrad*. For a given vector space over the real numbers, it is well known (Sylvester's 'inertia of signature' theorem) that for *all* orthogonal tetrads (or 'ennuples' in the *n*-dimensional case), i.e., those satisfying (1.1.8), the number of positive self-products (1.1.7) is invariant.

Given any Minkowski tetrad $\boldsymbol{g}_{\mathbf{i}}$, we can, in accordance with (1.1.4), represent any vector $\boldsymbol{U} \in \mathbb{V}$ by its corresponding Minkowski coordinates $U^{\mathbf{i}}$; then the inner product takes the form

$$\begin{aligned} \boldsymbol{U} \cdot \boldsymbol{V} &= (U^{\mathbf{i}}\boldsymbol{g}_{\mathbf{i}}) \cdot (V^{\mathbf{j}}\boldsymbol{g}_{\mathbf{j}}) = U^{\mathbf{i}}V^{\mathbf{j}}(\boldsymbol{g}_{\mathbf{i}} \cdot \boldsymbol{g}_{\mathbf{j}}) \\ &= U^{\mathbf{i}}V^{\mathbf{j}}\eta_{\mathbf{ij}} \\ &= U^0V^0 - U^1V^1 - U^2V^2 - U^3V^3. \end{aligned} \tag{1.1.11}$$

Note that $\boldsymbol{U} \cdot \boldsymbol{g}_{\mathbf{j}} = U^{\mathbf{i}}\eta_{\mathbf{ij}}$. Thus,

$$U^0 = \boldsymbol{U} \cdot \boldsymbol{g}_0,\ U^1 = -\boldsymbol{U} \cdot \boldsymbol{g}_1,\ U^2 = -\boldsymbol{U} \cdot \boldsymbol{g}_2,\ U^3 = -\boldsymbol{U} \cdot \boldsymbol{g}_3. \tag{1.1.12}$$

A particular case of inner product is the *Lorentz norm*

$$\|\boldsymbol{U}\| = \boldsymbol{U} \cdot \boldsymbol{U} = U^{\mathbf{i}}U^{\mathbf{j}}\eta_{\mathbf{ij}} = (U^0)^2 - (U^1)^2 - (U^2)^2 - (U^3)^2. \tag{1.1.13}$$

We may remark that the inner product can be defined in terms of the Lorentz norm by

$$\boldsymbol{U} \cdot \boldsymbol{V} = \tfrac{1}{2}\{\|\boldsymbol{U} + \boldsymbol{V}\| - \|\boldsymbol{U}\| - \|\boldsymbol{V}\|\}. \tag{1.1.14}$$

The vector $\boldsymbol{U} \in \mathbb{V}$ is called

$$\left.\begin{array}{ll} \textit{timelike} \text{ if} & \|\boldsymbol{U}\| > 0 \\ \textit{spacelike} \text{ if} & \|\boldsymbol{U}\| < 0 \\ \textit{null} \text{ if} & \|\boldsymbol{U}\| = 0. \end{array}\right\} \tag{1.1.15}$$

In terms of its Minkowski coordinates, $\boldsymbol{U}$ is *causal* (i.e., timelike or null) if

$$(U^0)^2 \geqslant (U^1)^2 + (U^2)^2 + (U^3)^2, \tag{1.1.16}$$

with equality holding if $\boldsymbol{U}$ is null. If each of $\boldsymbol{U}$ and $\boldsymbol{V}$ is causal, then applying in succession (1.1.16) and the Schwarz inequality, we obtain

$$|U^0V^0| \geqslant \{(U^1)^2+(U^2)^2+(U^3)^2\}^{\frac{1}{2}}\{(V^1)^2+(V^2)^2+(V^3)^2\}^{\frac{1}{2}}$$
$$\geqslant U^1V^1+U^2V^2+U^3V^3, \qquad (1.1.17)$$

Hence unless $\boldsymbol{U}$ and $\boldsymbol{V}$ are both null and proportional to one another, or unless one of them is zero (the only cases in which both inequalities reduce to equalities), then by (1.1.11), the sign of $\boldsymbol{U}\cdot\boldsymbol{V}$ is the same as the sign of $U^0V^0$. Thus, in particular, no two non-zero causal vectors can be orthogonal unless they are null and proportional.

As a consequence the causal vectors fall into two disjoint classes, such that the inner product of any two non-proportional members of the same classes is *positive* while the inner product of non-proportional members of different classes is *negative*. These two classes are distinguished according to the sign of $U^0$, the class for which $U^0$ is positive being the class to which the timelike tetrad vector $\boldsymbol{t}=\boldsymbol{g}_0$ belongs. The *time-orientation* of $\mathbb{V}$ consists in calling *future-pointing* the elements of one of these classes, and *past-pointing* the elements of the other. We often call a future-pointing timelike [null, causal] vector simply a future-timelike [-null, -causal] vector. If $\boldsymbol{t}$ is a future-timelike vector, then the Minkowski tetrad $(\boldsymbol{t}, \boldsymbol{x}, \boldsymbol{y}, \boldsymbol{z})$ is called *orthochronous*. When referred to an orthochronous Minkowski tetrad, the future-causal vectors are simply those for which $U^0>0$. The zero vector, though null, is neither future-null nor past-null. The negative of any future-causal vector is past-causal.

The *space-orientation* of $\mathbb{V}$ consists in assigning 'right-handedness' or 'left-handedness' to the three spacelike vectors of each Minkowski tetrad. This can be done in terms of the orientation and time-orientation of $\mathbb{V}$. Thus the triad $(\boldsymbol{x}, \boldsymbol{y}, \boldsymbol{z})$ is called *right-handed* if the Minkowski tetrad $(\boldsymbol{t}, \boldsymbol{x}, \boldsymbol{y}, \boldsymbol{z})$ is both proper and orthochronous, or neither. Otherwise the triad $(\boldsymbol{x}, \boldsymbol{y}, \boldsymbol{z})$ is *left-handed*. A Minkowski tetrad which is both proper and orthochronous is called *restricted*. Any two of the orientation, time-orientation, and space-orientation of $\mathbb{V}$ determine the third, and if any two are reversed, the third must remain unchanged. When making these choices in the space–time we inhabit, it may be preferable to begin by choosing a triad $(\boldsymbol{x}, \boldsymbol{y}, \boldsymbol{z})$ and calling it right- or left-handed according to that well-known criterion which physicists use and which is based on the structure of the hand with which most people write.* Similarly statistical physics determines a unique future sense.

* In view of the observed non-invariance of weak interactions under space-reflection ($P$) and of $K^0$-decay under combined space-reflection and particle–antiparticle interchange ($CP$) it is now possible to specify the space-orientation of physical space–time independently of such cultural or physiological considerations: *cf.* Lee and Yang (1956), Wu, Ambler, Hayward Hoppes and Hudson (1957), Lee, Oehme and Yang (1957), Christenson, Cronin, Fitch and Turlay (1964), Wu and Yang (1964); also Gardner (1967) for a popular account.

## *Minkowski space–time*

As we mentioned earlier, Minkowski vector space $\mathbb{V}$ can be regarded as the space of position vectors, relative to an arbitrarily chosen origin, of the points (events) which constitute *Minkowski space–time* $\mathbb{M}$. That space–time is the stage for special relativity theory. None of its points is preferred, and specifically it has no preferred origin: it is invariant under translations, i.e., it is an affine space. The relation between $\mathbb{M}$ and $\mathbb{V}$ can be characterized by the map

$$\mathrm{vec}\colon \mathbb{M} \times \mathbb{M} \to \mathbb{V} \tag{1.1.18}$$

for which

$$\mathrm{vec}(P, Q) + \mathrm{vec}(Q, R) = \mathrm{vec}(P, R), \tag{1.1.19}$$

whence $\mathrm{vec}(P, P) = 0$ and $\mathrm{vec}(P, Q) = -\mathrm{vec}(Q, P)$. We can regard $\mathrm{vec}(P, Q)$ as the position vector $\overrightarrow{PQ} \in \mathbb{V}$ of $Q$ relative to $P$, where $P, Q \in \mathbb{M}$. Evidently $\mathbb{V}$ induces by this map a norm, here called the *squared interval* $\Phi$, on any pair of points $P, Q \in \mathbb{M}$:

$$\Phi(P, Q) := \| \mathrm{vec}(P, Q) \| \tag{1.1.20}$$

The standard coordinatization of $\mathbb{M}$, $\mathbb{M} \leftrightarrow \mathbb{R}^4$, where $\mathbb{R}^4$ is the space of quadruples of real numbers, consists of a choice of origin $O \in \mathbb{M}$ and a choice of Minkowski tetrad $\boldsymbol{g}_{\mathbf{i}} = \overrightarrow{OQ_{\mathbf{i}}}$ for $Q_0, Q_1, Q_2, Q_3 \in \mathbb{M}$. Then the coordinates $P^0, P^1, P^2, P^3$ of any point $P \in \mathbb{M}$ are the coordinates of the vector $\overrightarrow{OP}$ relative to $\boldsymbol{g}_{\mathbf{i}}$, i.e. $\overrightarrow{OP} = P^{\mathbf{i}}\boldsymbol{g}_{\mathbf{i}}$. From (1.1.19) we find, by putting $O$ for $Q$, the following coordinates of $\overrightarrow{PR}$ relative to $\boldsymbol{g}_{\mathbf{i}}$:

$$(\overrightarrow{PR})^{\mathbf{i}} = R^{\mathbf{i}} - P^{\mathbf{i}}, \tag{1.1.21}$$

clearly independently of the choice of origin. Substituting this and (1.1.20) into (1.1.13) yields

$$\Phi(P, Q) = (Q^0 - P^0)^2 - (Q^1 - P^1)^2 - (Q^2 - P^2)^2 - (Q^3 - P^3)^2. \tag{1.1.22}$$

A linear self-transformation of $\mathbb{V}$ which preserves the Lorentz norm – and therefore, by (1.1.14), also the inner product – is called an (*active*) *Lorentz transformation.* If such a transformation preserves both the orientation and time-orientation of $\mathbb{V}$, it is called a *restricted* Lorentz transformation. Clearly the [restricted] Lorentz transformations form a group, and this group is called the [*restricted*] *Lorentz group.* Similarly a self-transformation of $\mathbb{M}$ which preserves the squared interval (no linearity assumption being here needed) is called an (*active*) *Poincaré transformation.* Any such transformation induces a Lorentz transformation on $\mathbb{V}$, and can accordingly also be classified as restricted or not. Again, the restricted Poincaré transformations clearly form a group.*

Any physical experiment going on in the Minkowski space–time of our experience may be subjected to a Poincaré transformation – i.e., rotated in space, translated in space and time, and given a uniform motion – without altering its intrinsic outcome. This is the basis of special relativity theory, and it can be stated without reference to coordinates or to the other laws of physics.

### Coordinate change

If not further qualified, Lorentz and Poincaré transformations in this book will be understood to be *active*. But it is sometimes useful to consider 'passive' Lorentz [and Poincaré] transformations. These are transformations of the *coordinate space* $\mathbb{R}^4$, i.e. re-coordinatizations of $\mathbb{V}$ [*or* $\mathbb{M}$]. Any Minkowski tetrad $\boldsymbol{g}_{\mathbf{i}}$ in $\mathbb{V}$ [*or* tetrad $\boldsymbol{g}_{\mathbf{i}}$ and origin $O$ in $\mathbb{M}$] defines a quadruple of coordinates $U^{\mathbf{i}}$ for each $\boldsymbol{U}$ of $\mathbb{V}$ [*or* $\boldsymbol{U}=\overrightarrow{OP}$ of $\mathbb{M}$], with $\boldsymbol{U}=U^{\mathbf{i}}\boldsymbol{g}_{\mathbf{i}}$. A change in this *reference tetrad*, $\boldsymbol{g}_{\mathbf{i}}\mapsto\boldsymbol{g}_{\hat{\mathbf{i}}}$ in $\mathbb{V}$ [*or* of tetrad and origin in $\mathbb{M}$] induces a change in the coordinates for $\mathbb{V}[\mathbb{M}]$. The resulting correspondence

$$G: U^{\mathbf{i}}\mapsto U^{\hat{\mathbf{i}}} \tag{1.1.23}$$

$$[\textit{or}\ U^{\mathbf{i}}\mapsto U^{\hat{\mathbf{i}}}+K^{\hat{\mathbf{i}}}\ \text{with}\ K^{\hat{\mathbf{i}}}\ \text{const.}]$$

is called a *passive Lorentz* [*Poincaré*] *transformation*. It is called *restricted if it can be generated by two restricted Minkowski tetrads* $\boldsymbol{g}_{\mathbf{i}}$ *and* $\boldsymbol{g}_{\hat{\mathbf{i}}}$. For the sake of conciseness, we shall now concentrate on Lorentz transformations, obvious generalizations being applicable to Poincaré transformations.

It the two reference tetrads are related by

$$\boldsymbol{g}_{\mathbf{i}}=g_{\mathbf{i}}{}^{\hat{\mathbf{i}}}\boldsymbol{g}_{\hat{\mathbf{i}}}, \tag{1.1.24}$$

then

$$\boldsymbol{U}=U^{\hat{\mathbf{i}}}\boldsymbol{g}_{\hat{\mathbf{i}}}=U^{\mathbf{i}}\boldsymbol{g}_{\mathbf{i}}=U^{\mathbf{i}}g_{\mathbf{i}}{}^{\hat{\mathbf{i}}}\boldsymbol{g}_{\hat{\mathbf{i}}},$$

and thus the passive transformation (1.1.23) is given explicitly by

$$U^{\hat{\mathbf{i}}}=U^{\mathbf{i}}g_{\mathbf{i}}{}^{\hat{\mathbf{i}}}, \tag{1.1.25}$$

which is evidently linear. It is fully characterized by the matrix $g_{\mathbf{i}}{}^{\hat{\mathbf{i}}}$.

It is often convenient, though slightly misleading, to describe even an

* Note that we use the term 'Lorentz group' here only for the six-parameter *homogeneous* group on Minkowski vector space, while referring to the corresponding ten-parameter inhomogeneous group on Minkowski space–time as the Poincaré group.

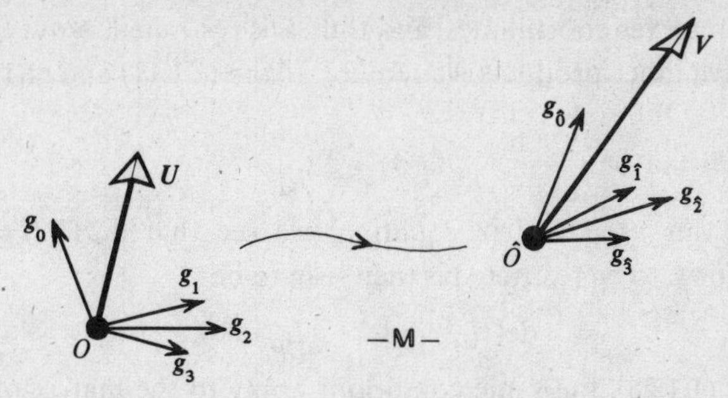


Fig. 1-1. An active Poincaré transformation sends the world vector $\boldsymbol{U}$ at $O$ to a world vector $\boldsymbol{V}$ at $\hat{O}$. If it also sends the tetrad $\boldsymbol{g}_i$ at $O$ to $\boldsymbol{g}_{\hat i}$ at $\hat{O}$, then the coordinates $U^i$, of $\boldsymbol{U}$ in $\boldsymbol{g}_i$, are the same as those, $V^{\hat i}$, of $\boldsymbol{V}$ in $\boldsymbol{g}_{\hat i}$, (i.e. $U^i = V^{\hat i}$). Hence the (reversed) passive transformation induced by $\{\boldsymbol{g}_{\hat i}$ at $\hat{O}\} \mapsto \{\boldsymbol{g}_i$ at $O\}$ takes the original coordinates $U^i (= V^{\hat i})$ of $U$ to the original coordinates $V^i$ of $\boldsymbol{V}$.

*active* Lorentz transformation by means of coordinates. (It is slightly misleading because an active Lorentz transformation exists independently of all coordinates, whereas a passive Lorentz transformation does not.) Thus, for a given active Lorentz transformation $L: \boldsymbol{U} \mapsto \boldsymbol{V}$, we can refer both $\boldsymbol{U}$ and its image $\boldsymbol{V}$ to one (arbitrary) Minkowski tetrad $\boldsymbol{g}_{\hat{\mathbf{i}}}$, whose pre-image under $L$, let us say, is $\boldsymbol{g}_{\mathbf{i}}$ as in (1.1.24). Since by the assumed linearity of $L$ the expression of $\boldsymbol{V}$ in terms of $\boldsymbol{g}_{\hat{\mathbf{i}}}$ must be identical with the expression of $\boldsymbol{U}$ in terms of $\boldsymbol{g}_{\mathbf{i}}$, we then have, from (1.1.25), (see also Fig. 1-1)

$$U^{\hat{\mathbf{i}}} = V^{\hat{\mathbf{j}}} g_{\mathbf{j}}{}^{\hat{\mathbf{i}}}, \qquad (1.1.26)$$

where, in violation of the general rule, we here for once understand summation over the unlike index pair $\mathbf{j}$ and $\hat{\mathbf{j}}$. We therefore have the following explicit form of the transformation,

$$V^{\mathbf{j}} = U^{\hat{\mathbf{i}}} L_{\hat{\mathbf{i}}}{}^{\mathbf{j}}, \qquad (1.1.27)$$

where

$$(L_{\hat{\mathbf{i}}}{}^{\mathbf{j}}) = (g_{\mathbf{j}}{}^{\hat{\mathbf{i}}})^{-1}. \qquad (1.1.28)$$

Thus the active Lorentz transformation $L$ that carries $\boldsymbol{g}_{\mathbf{i}}$ into $\boldsymbol{g}_{\hat{\mathbf{i}}}$ is *formally* equivalent, in its effect on the coordinates of a vector, to the passive Lorentz transformation $G^{-1}$ induced by the passage from $\boldsymbol{g}_{\hat{\mathbf{i}}}$ to $\boldsymbol{g}_{\mathbf{i}}$ as reference tetrad.

If $L$ is a restricted Lorentz transformation, it clearly carries a restricted Minkowski tetrad into a restricted Minkowski tetrad, and thus the corresponding passive transformation $G$ is restricted also. If, conversely, $G$ is restricted, suppose it is generated by the restricted tetrads $\boldsymbol{g}_{\mathbf{i}}$ and $\boldsymbol{g}_{\hat{\mathbf{i}}}$; then the corresponding $L$ preserves norms, products, and orientation since, in fact, it preserves coordinates, and thus $L$ is restricted. Now in order for $L$ to preserve inner products we require – from (1.1.11) and (1.1.27), dropping hats –

$$\eta_{\mathrm{ij}} L_{\mathrm{k}}{}^{\mathrm{i}} L_{\mathrm{l}}{}^{\mathrm{j}} = \eta_{\mathrm{kl}}. \tag{1.1.29}$$

Regarding this as a matrix equation, we see that $\det(L_{\mathrm{i}}{}^{\mathrm{j}}) = \pm 1$. The condition for $L$ to be restricted is then seen to be

$$\det(L_{\mathrm{i}}{}^{\mathrm{j}}) = 1, \qquad L_0{}^0 > 0. \tag{1.1.30}$$

Because of (1.1.28), the same conditions apply to the matrix of a passive restricted Lorentz transformation. They can, of course, also be derived directly from the definitions:

$$\eta_{\mathrm{ij}} g_{\hat{\mathrm{i}}}{}^{\mathrm{i}} g_{\hat{\mathrm{j}}}{}^{\mathrm{j}} = \eta_{\hat{\mathrm{i}}\hat{\mathrm{j}}}, \quad \det(g_{\hat{\mathrm{i}}}{}^{\mathrm{i}}) = 1, \quad g_{\hat{0}}{}^0 > 0. \tag{1.1.31}$$

## 1.2 Null directions and spin transformations

In §1.1 the conventional representation of a world-vector $\boldsymbol{U}$ in terms of *Minkowski* coordinates was considered. Now we examine another way of representing world-vectors by coordinates. In particular, we shall obtain a coordinatization of the null cone (i.e., the set of null vectors) in terms of complex numbers. This will lead us to the concept of a spin-vector.

To avoid unnecessary indices, we write $T, X, Y, Z$ for the coordinates $U^0, U^1, U^2, U^3$ of $\boldsymbol{U}$ with respect to a restricted Minkowski tetrad $(\boldsymbol{t}, \boldsymbol{x}, \boldsymbol{y}, \boldsymbol{z})$:

$$\boldsymbol{U} = T\boldsymbol{t} + X\boldsymbol{x} + Y\boldsymbol{y} + Z\boldsymbol{z}. \tag{1.2.1}$$

For *null* vectors the coordinates satisfy

$$T^2 - X^2 - Y^2 - Z^2 = 0. \tag{1.2.2}$$

Often we wish to consider just the null *directions*, say at the origin $O$ of (Minkowski) space–time. Note that $\pm\boldsymbol{U}$ will be considered to have unequal (namely, opposite) directions. The abstract space whose elements are the future [past] null directions we call $\mathscr{S}^+$ [$\mathscr{S}^-$]. These two spaces can be *represented* in any given coordinate system $(T, X, Y, Z)$ by the intersections $S^+$ [$S^-$] of the future [past] null cone (1.2.2) with the hyperplanes $T = 1$ [$T = -1$]. In the Euclidean $(X, Y, Z)$-space $T = 1$ [$T = -1$], $S^+$ [$S^-$] is a sphere with equation*

$$x^2 + y^2 + z^2 = 1. \tag{1.2.3}$$

(See Fig. 1-2) Of course, the direction of *any* vector (1.2.1) through $O$

* We here reserve lower case letters $x, y, z$ for coordinates on $S^+$ and $S^-$

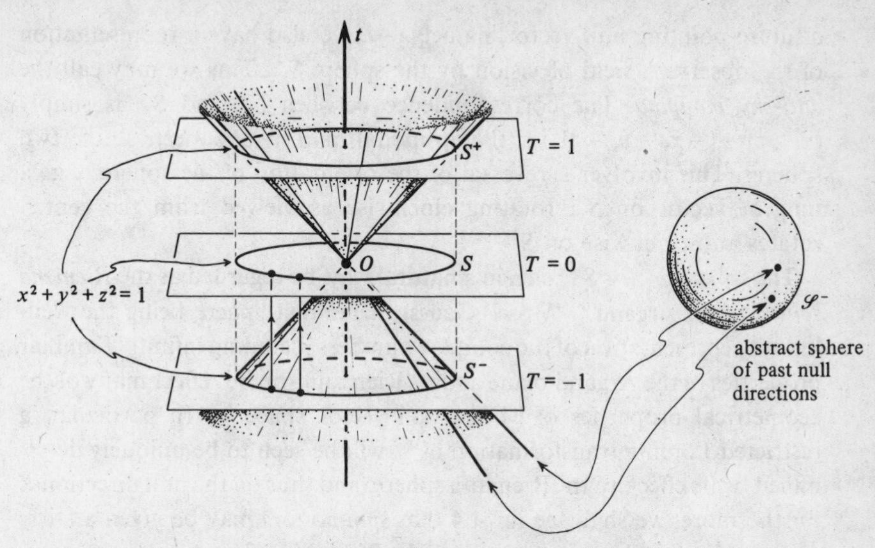


Fig. 1-2. The abstract sphere $\mathscr{S}^-$ naturally represents the observer's celestial sphere while $S^-$, or its projection to $S$, gives a more concrete (though somewhat less invariant) realization.

(whether null or not), unless it lies in the hyperplane $T = 0$, can be represented by a point of $T = 1$ or $T = -1$. The direction of $\boldsymbol{U}$ from the origin is represented, on the appropriate hyperplane, by the point $(X/|T|, Y/|T|, Z/|T|)$. The interior of $S^-$ represents the set of past timelike directions and the interior of $S^+$ the set of future timelike directions. The exteriors of these spheres represent spacelike directions.

Let us consider the significance of $S^-$ and $S^+$ in physical terms. We imagine an observer situated at the event $O$ in space–time. Now, light rays through his eye correspond to null straight lines through $O$, whose past directions constitute the field of vision of the observer. This is $\mathscr{S}^-$ and is represented by the sphere $S^-$. In fact, $S^-$ is an accurate geometrical representation of what the observer actually 'sees' provided he is stationary relative to the frame $(t, x, y, z)$, i.e., his world velocity is $\boldsymbol{t}$. For he can imagine himself permanently situated at the centre of a unit sphere $S$ (his sphere of vision) onto which he maps all he sees at any instant. The lines from his eye to these image points on $S$ are the projections of the world lines of the incoming rays to his instantaneous space $T = 0$. Hence these images are congruent with those on $S^-$ (*cf.* Fig. 1-2), and we can refer to $\mathscr{S}^-$ or $S^-$ as the *celestial sphere* of $O$. The mapping of the past null directions at $O$ to the points of $S^-$ we shall call the *sky mapping*. Since any past-pointing null vector $\boldsymbol{L}$ is uniquely (and invariantly) associated with a future-pointing null vector, namely $-\boldsymbol{L}$ we also have a representation of the observer's field of vision by the sphere $S^+$. This we may call the *anti-sky mapping*. The correspondence between $S^+$ and $S^-$ is simply $(x, y, z) \leftrightarrow (-x, -y, -z)$, i.e. the antipodal map if we superpose the two spheres. This involves a reversal of the orientation of the sphere: e.g., a tangent vector on $S^-$ rotating clockwise as viewed from the center, rotates anti-clockwise on $S^+$.

The sphere $S^+$ (or $S^-$) can, in a natural way, be regarded as the *Riemann sphere* of an Argand (–Wessel–Gauss) plane, that sphere being the well-known representation of the complex numbers including infinity. Familiar properties of the Argand plane and its Riemann sphere reflect many of the geometrical properties of Minkowski vector space $\mathbb{V}$. In particular, a restricted Lorentz transformation of $\mathbb{V}$ will be seen to be uniquely determined by its effect on the Riemann sphere (and thus on the null directions). Furthermore, we shall see in §1.4 that *spin-vectors* may be given a fairly direct geometrical interpretation on the Riemann sphere.

We can replace the coordinates $x, y, z$ on $S^+$ by a single complex number, obtained by means of the 'stereographic' correspondence between a sphere and a plane (See Fig. 1-3.) Draw the plane $\Sigma$ with equation $z = 0$ in the Euclidean 3-space $T = 1$, and map the points of $S^+$ to this plane by projecting from the north* pole $N(1, 0, 0, 1)$. Let $P(1, x, y, z)$ and $P'$ $(1, X', Y', O)$ denote corresponding points on $S^+$ and $\Sigma$. Let the points $A$ and $B$ denote the feet of the perpendiculars drawn from $P$ to $CP'$ and $CN$, respectively. Labelling the points of $\Sigma$ by the single complex parameter

$$\zeta = X' + \mathrm{i}Y', \tag{1.2.4}$$

we have

$$x + \mathrm{i}y = h\zeta, \tag{1.2.5}$$

where

$$h = \frac{CA}{CP'} = \frac{NP}{NP'} = \frac{NB}{NC} = 1 - z.$$

Hence the expression for $\zeta$ in terms of the coordinates $(1, x, y, z)$ of the point $P$ becomes

$$\zeta = \frac{x + \mathrm{i}y}{1 - z}. \tag{1.2.6}$$

* We choose the north pole rather than the south pole to be consistent with most of the directly relevant spinor literature, and also because it leads to a right-handed rotation of a spin-vector's flag plane under positive phase change (cf. §§1.4, 3.2). It may be noted, however, that this assigns $S^+$ the opposite orientation, when viewed from the outside, to that of $\Sigma$, when viewed from above. This has the effect that, in relation to the conventions of §§4.14, 4.15, $\zeta$ turns out to be an *anti*-holomorphic coordinate on $S^+$ (and a holomorphic coordinate on $S^-$) (cf. Figs. 4-2, 4-6).

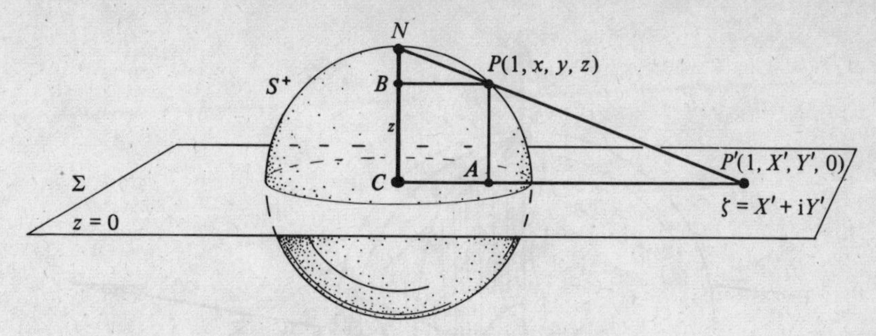


Fig. 1-3. Stereographic projection of $S^+$ to the Argand plane.

To obtain the inverse relations, we first eliminate $x$ and $y$ from (1.2.6) by means of (1.2.3):

$$\zeta\bar{\zeta} = \frac{x^2 + y^2}{(1-z)^2} = \frac{1+z}{1-z}. \tag{1.2.7}$$

Solving (1.2.7) for $z$ and substituting in (1.2.6), we obtain

$$x = \frac{\zeta + \bar{\zeta}}{\zeta\bar{\zeta} + 1}, \; y = \frac{\zeta - \bar{\zeta}}{\mathrm{i}(\zeta\bar{\zeta} + 1)}, \; z = \frac{\zeta\bar{\zeta} - 1}{\zeta\bar{\zeta} + 1}. \tag{1.2.8}$$

Equations (1.2.6) and (1.2.8) are the algebraic expressions for a standard stereographic correspondence between the Argand plane of $\zeta$ and the unit sphere in $(x, y, z)$-space centered on $(0, 0, 0)$. The correspondence is one-to-one provided we regard $\zeta = \infty$ as one 'point' added to the Argand plane, and associate this point with the north pole of the sphere. In this way the sphere $S^+$ gives a standard realization of the Argand plane of $\zeta$ with $\zeta = \infty$ adjoined: it is the *Riemann sphere* of $\zeta$.

As an alternative coordinatization of $S^+$ we can use standard spherical polar coordinates, related to $x, y, z$ by the equations

$$x = \sin\theta\cos\phi, \quad y = \sin\theta\sin\phi, \quad z = \cos\theta. \tag{1.2.9}$$

To find the relation between $\zeta$ and $(\theta, \phi)$, we substitute (1.2.9) into (1.2.6) and find

$$\zeta = \mathrm{e}^{\mathrm{i}\phi}\cot\frac{\theta}{2}. \tag{1.2.10}$$

This relation can also be obtained directly, by reference to Fig. 1-4 and use of simple trigonometry.

Formulae (1.2.6), (1.2.7), (1.2.8), (1.2.10) apply to the anti-sky mapping

$$\text{future null cone} \to S^+ \to \Sigma.$$

We shall also be interested in the corresponding formulae for the sky

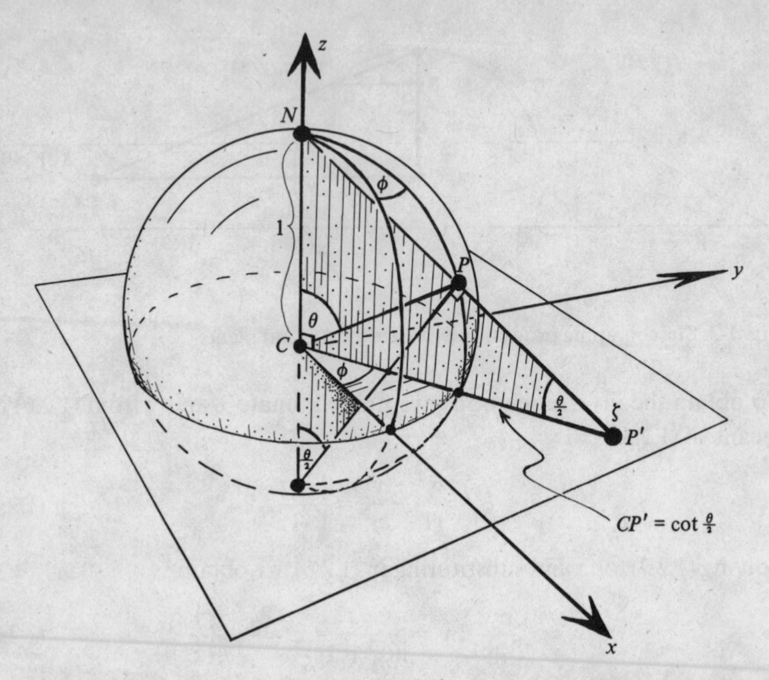


Fig. 1-4. The geometry of the equation $\zeta = e^{i\phi} \cot \theta/2$ which relates the spherical polar angles $\theta, \phi$ to the complex stereographic coordinate $\zeta$. (The angle $CP'N$ is equal to that subtented at the south pole by $PN$, since each is complementary to the angle $PNC$.)

mapping, in which each null direction at $O$ is represented by a typical *past* event $-(1, x, y, z)$ rather than by a future event $+(1, x, y, z)$. If we require $\zeta$ in both cases to represent the same null *line*, then on $S^+$ and $S^-$ it must correspond to antipodal points $\pm(x, y, z)$. The relevant formulae are therefore obtained from (1.2.6), (1.2.7), (1.2.8), (1.2.10) by the antipodal transformation $(x, y, z) \mapsto -(x, y, z)$ or, equivalently, $(\theta, \phi) \mapsto (\pi - \theta, \pi + \phi)$. Equation (1.2.10), in particular, becomes

$$\zeta = -e^{i\phi} \tan \frac{\theta}{2}. \tag{1.2.11}$$

(Note that the effect of the antipodal map is $\zeta \mapsto -1/\bar{\zeta}$.)

The above correspondence between the set of future [past] null directions at $O$ and the complex $\zeta$-plane could have been obtained more directly than via a stereographic projection. To achieve this direct correspondence (see Fig. 1-5), we slice the $(T, X, Y, Z)$-space with the null hyperplane $\Pi$ whose equation is

$$T - Z = 1 \tag{1.2.12}$$

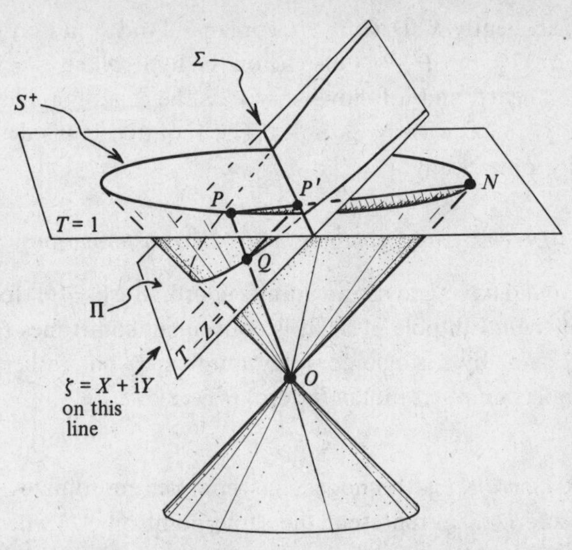


Fig. 1-5. As the plane through $ON$ varies it provides the stereographic projection $P \mapsto P'$ and also the correspondences $P \mapsto Q$ and $Q \mapsto P'$. (The "parabolic" intersection of Π with the cone has the same intrinsic Euclidean metric as the plane Σ – the Argand plane of ζ.)

rather than with the spacelike hyperplane $T = 1$. Consider a null straight line through $O$ which meets $S^+$ at the point $P = (1, x, y, z)$. This null straight line clearly also contains the point

$$Q = \left(\frac{1}{1-z}, \frac{x}{1-z}, \frac{y}{1-z}, \frac{z}{1-z}\right)$$

lying on Π. Now the '$x$' and '$y$' coordinates of $Q$ are precisely

$$X' = \frac{x}{1-z} \quad \text{and} \quad Y' = \frac{y}{1-z}$$

with

$$\zeta = X' + \mathrm{i}Y'$$

as in (1.2.4), (1.2.6), and so $\zeta$ is obtained by simple orthogonal projection from Π to Σ. In the exceptional case $\zeta = \infty$ $(z = 1)$, the null line through $O$ is parallel to Π and so meets it in no finite point.

We refer to Fig. 1-5 to elucidate the geometric relation between our two different constructions. Let $N = (1, 0, 0, 1)$ be the north pole of $S^+$, as before. Let $OPQ$ be the null straight line under consideration, with $P \in S^+$ and $Q \in \Pi$. Let $P'$ be the orthogonal projection of $Q$ to the plane $\Sigma(T = 1, Z = 0)$. Then the direction of $QP'$ is $1:0:0:1$, the same as that of $ON$. Consequently $P'$, $Q$, $O$, $N$ are coplanar. And $P$ lies on their plane since it lies on $OQ$. But $P'$, $P$, $N$ also lie on the hyperplane $T = 1$. They are therefore collinear*, and it follows that $P'$ is the stereographic projection of $P$ (from $S^+$ to $\Sigma$ with $N$ as pole). The required equivalence is thus established geometrically.

### *Lorentz transformations and spin transformations*

In order to avoid having to use an infinite coordinate ($\zeta = \infty$) for the point $(1, 0, 0, 1)$ at the north pole of $S^+$ it is convenient sometimes to label the points of $S^+$ not by a single complex number, $\zeta$, but rather by a pair $(\xi, \eta)$ of complex numbers (not *both* zero), where

$$\zeta = \xi/\eta. \tag{1.2.13}$$

These are to be *projective* (homogeneous) complex coordinates, so that the pairs $(\xi, \eta)$ and $(\lambda\xi, \lambda\eta)$ represent the same point on $S^+$, where $\lambda$ is any non-zero complex number. With these coordinates, the additional point at infinity, $\zeta = \infty$, is given a finite label, e.g., $(1, 0)$. Thus, we now regard $S^+$ as a realization of a *complex projective line*. Written in terms of these complex homogeneous coordinates, equations (1.2.8) become

$$x = \frac{\xi\bar{\eta} + \eta\bar{\xi}}{\xi\bar{\xi} + \eta\bar{\eta}}, \quad y = \frac{\xi\bar{\eta} - \eta\bar{\xi}}{\mathrm{i}(\xi\bar{\xi} + \eta\bar{\eta})}, \quad z = \frac{\xi\bar{\xi} - \eta\bar{\eta}}{\xi\bar{\xi} + \eta\bar{\eta}}. \tag{1.2.14}$$

Note that $x$, $y$, and $z$ are homogeneous of degree zero in $\xi, \eta$, and so are invariant under a rescaling of $\xi, \eta$.

Recall that the role of the point $P(1, x, y, z)$ on $S^+$ was simply that of representing a future null direction at $O$. We could, if desired, choose any other point on the line $OP$ to represent the same null direction. In particular, we could choose the point $R$ on $OP$ whose coordinates $(T, X, Y, Z)$ are obtained from those of $P$ by multiplying by the factor $(\xi\bar{\xi} + \eta\bar{\eta})/\sqrt{2}$. This will eliminate the denominators in (1.2.14). (The factor $1/\sqrt{2}$ is included for later convenience.) Then $\boldsymbol{K} := \overrightarrow{OR}$ has coordinates

$$T = \frac{1}{\sqrt{2}}(\xi\bar{\xi} + \eta\bar{\eta}), \qquad X = \frac{1}{\sqrt{2}}(\xi\bar{\eta} + \eta\bar{\xi}),$$
$$Y = \frac{1}{\mathrm{i}\sqrt{2}}(\xi\bar{\eta} - \eta\bar{\xi}), \qquad Z = \frac{1}{\sqrt{2}}(\xi\bar{\xi} - \eta\bar{\eta}). \tag{1.2.15}$$

Unlike the point $P$, however, $R$ is not independent of the real scaling of

* In *four* dimensions, the intersection of a plane (two linear equations) with a hyperplane (one linear equation) is a straight line.

$(\xi, \eta)$, i.e. $(\xi, \eta) \to (r\xi, r\eta)$, $r \in \mathbb{R}$, although it is independent of the phase 'rescaling' $(\xi, \eta) \to (e^{i\theta}\xi, e^{i\theta}\eta)$, $\theta \in \mathbb{R}$. Thus, the position of $R$ is not just a function of $\zeta$ alone, although the direction $OQ$ depends only on $\zeta$.

Now it is not difficult to see from (1.2.15) that any complex linear transformation of $\xi$ and $\eta$ will result in a real linear transformation of $(T, X, Y, Z)$ (given explicitly in (1.2.26) below). Since the null vectors span the whole space $\mathbb{V}$, a linear transformation of the null vectors induces a linear transformation of $\mathbb{V}$, which is given formally by the same equation (namely (1.2.26)) on the general coordinates $(T, X, Y, Z)$. Under such a transformation the property (1.2.2) will be preserved. Thus we get a Lorentz transformation together, possibly, with a dilation. In any case, the effect on the null *directions* at $O$ will be the same as that of a Lorentz transformation, since dilations produce no effect on directions.

Consider, then, a complex linear (non-singular) transformation of $\xi$ and $\eta$:

$$\begin{aligned} \xi \mapsto \tilde{\xi} &= \alpha\xi + \beta\eta \\ \eta \mapsto \tilde{\eta} &= \gamma\xi + \delta\eta. \end{aligned} \tag{1.2.16}$$

Here, $\alpha$, $\beta$, $\gamma$, and $\delta$ are arbitrary complex numbers subject only to the condition $\alpha\delta - \beta\gamma \neq 0$ (non-singularity). Expressed in terms of $\zeta$, the transformation (1.2.16) becomes*

$$\zeta \mapsto \tilde{\zeta} = \frac{\alpha\zeta + \beta}{\gamma\zeta + \delta}. \tag{1.2.17}$$

We may, without loss of generality as regards the transformation of $\zeta$, normalize (1.2.16) by imposing the 'unimodular' condition

$$\alpha\delta - \beta\gamma = 1. \tag{1.2.18}$$

The transformations (1.2.16) (or (1.2.17)), subject to (1.2.18), are called *spin transformations* in the context where $\zeta$ is related to the Minkowski null vectors through equations (1.2.13) and (1.2.15). We note that these equations imply

$$\zeta = \frac{X + iY}{T - Z} = \frac{T + Z}{X - iY}. \tag{1.2.19}$$

In the same context, we define the *spin-matrix* $\mathbf{A}$ by

$$\mathbf{A} := \begin{pmatrix} \alpha & \beta \\ \gamma & \delta \end{pmatrix}, \qquad \det \mathbf{A} = 1. \tag{1.2.20}$$

* A bilinear transformation of this kind is in fact the most general global holomorphic (i.e., complex-analytic, i.e., conformal and orientation preserving) transformation of the Riemann sphere to itself.

The last condition is simply the normalization condition (1.2.18). In terms of **A**, (1.2.16) takes the form

$$\begin{pmatrix}\tilde{\xi}\\ \tilde{\eta}\end{pmatrix}=\mathbf{A}\begin{pmatrix}\xi\\ \eta\end{pmatrix}. \tag{1.2.21}$$

We see from (1.2.21) that the composition of two successive spin transformations is again a spin transformation: the spin-matrix of the composition is given by the product of the spin-matrices of the factors. Also, any spin-matrix has an inverse,

$$\mathbf{A}^{-1}=\begin{pmatrix}\delta & -\beta\\ -\gamma & \alpha\end{pmatrix}, \tag{1.2.22}$$

which is also a spin-matrix. Thus, the spin transformations form a group – referred to as $SL(2,\mathbb{C})$.

Note that the two spin-matrices **A** and $-\mathbf{A}$ give rise to the *same* transformation of $\zeta$ even though they define different spin transformations. Conversely, suppose **A** and **B** are spin-matrices each of which defines the same transformation of $\zeta$. Then $\mathbf{B}^{-1}\mathbf{A}$ is a spin-matrix which defines the identity transformation on $\zeta$. We see from (1.2.17) that this implies $\beta=\gamma=0$, $\alpha=\delta$. The normalization (1.2.18) implies $\alpha=\delta=\pm1$. Thus $\mathbf{B}^{-1}\mathbf{A}=\pm\mathbf{I}$ (identity matrix), whence $\mathbf{A}=\pm\mathbf{B}$. A spin transformation is therefore defined *uniquely up to sign* by its effect on the Riemann sphere of $\zeta$.

Let us examine the effect of the spin transformation (1.2.21) on the coordinates $(T, X, Y, Z)$. We observe that (1.2.15) can be inverted and re-expressed as:

$$\frac{1}{\sqrt{2}}\begin{pmatrix}T+Z & X+\mathrm{i}Y\\ X-\mathrm{i}Y & T-Z\end{pmatrix}=\begin{pmatrix}\xi\bar{\xi} & \xi\bar{\eta}\\ \eta\bar{\xi} & \eta\bar{\eta}\end{pmatrix}=\begin{pmatrix}\xi\\ \eta\end{pmatrix}(\bar{\xi}\quad\bar{\eta}). \tag{1.2.23}$$

From this, we see that the spin transformation (1.2.21) effects:

$$\begin{pmatrix}T+Z & X+\mathrm{i}Y\\ X-\mathrm{i}Y & T-Z\end{pmatrix}\mapsto\begin{pmatrix}\tilde{T}+\tilde{Z} & \tilde{X}+\mathrm{i}\tilde{Y}\\ \tilde{X}-\mathrm{i}\tilde{Y} & \tilde{T}-\tilde{Z}\end{pmatrix}=\mathbf{A}\begin{pmatrix}T+Z & X+\mathrm{i}Y\\ X-\mathrm{i}Y & T-Z\end{pmatrix}\mathbf{A}^*, \tag{1.2.24}$$

where $\mathbf{A}^*$ denotes the conjugate transpose of **A**. As we remarked earlier, this is a linear transformation of $(T, X, Y, Z)$, it is real (since Hermiticity is preserved in (1.2.24)), and it preserves the condition $T^2-X^2-Y^2-Z^2=0$. Also, if

$$\boldsymbol{U}=T\boldsymbol{t}+X\boldsymbol{x}+Y\boldsymbol{y}+Z\boldsymbol{z} \tag{1.2.25}$$

is *any* world-vector (i.e. *not* necessarily null), then the spin-matrix **A** still defines a transformation of $\boldsymbol{U}$ according to (1.2.24). We note that this transformation is not only linear and real, but that in addition it leaves the form $T^2 - X^2 - Y^2 - Z^2$ actually *invariant*. For this form is just the determinant of the left-hand matrix in (1.2.24), and the determinant of the right-hand side is simply this form multiplied by $\det \mathbf{A} \det \mathbf{A}^*$, which is 1 (*cf.* (1.2.20)). Thus (1.2.24) defines a *Lorentz transformation*. Regarded as a transformation on $(T, X, Y, Z)$, its explicit form is

$$\begin{pmatrix} T \\ X \\ Y \\ Z \end{pmatrix} \mapsto \begin{pmatrix} \tilde{T} \\ \tilde{X} \\ \tilde{Y} \\ \tilde{Z} \end{pmatrix} = \tfrac{1}{2}\mathbf{B} \begin{pmatrix} T \\ X \\ Y \\ Z \end{pmatrix},$$

$$\mathbf{B} = \begin{pmatrix} \alpha\bar{\alpha} + \beta\bar{\beta} + \gamma\bar{\gamma} + \delta\bar{\delta} & \alpha\bar{\beta} + \beta\bar{\alpha} + \gamma\bar{\delta} + \delta\bar{\gamma} & \mathrm{i}(\alpha\bar{\beta} - \beta\bar{\alpha} + \gamma\bar{\delta} - \delta\bar{\gamma}) & \alpha\bar{\alpha} - \beta\bar{\beta} + \gamma\bar{\gamma} - \delta\bar{\delta} \\ \alpha\bar{\gamma} + \gamma\bar{\alpha} + \beta\bar{\delta} + \delta\bar{\beta} & \alpha\bar{\delta} + \delta\bar{\alpha} + \beta\bar{\gamma} + \gamma\bar{\beta} & \mathrm{i}(\alpha\bar{\delta} - \delta\bar{\alpha} + \gamma\bar{\beta} - \beta\bar{\gamma}) & \alpha\bar{\gamma} + \gamma\bar{\alpha} - \beta\bar{\delta} - \delta\bar{\beta} \\ \mathrm{i}(\gamma\bar{\alpha} - \alpha\bar{\gamma} + \delta\bar{\beta} - \beta\bar{\delta}) & \mathrm{i}(\delta\bar{\alpha} - \alpha\bar{\delta} + \gamma\bar{\beta} - \beta\bar{\gamma}) & \alpha\bar{\delta} + \delta\bar{\alpha} - \beta\bar{\gamma} - \gamma\bar{\beta} & \mathrm{i}(\gamma\bar{\alpha} - \alpha\bar{\gamma} + \beta\bar{\delta} - \delta\bar{\beta}) \\ \alpha\bar{\alpha} + \beta\bar{\beta} - \gamma\bar{\gamma} - \delta\bar{\delta} & \alpha\bar{\beta} + \beta\bar{\alpha} - \gamma\bar{\delta} - \delta\bar{\gamma} & \mathrm{i}(\alpha\bar{\beta} - \beta\bar{\alpha} + \delta\bar{\gamma} - \gamma\bar{\delta}) & \alpha\bar{\alpha} - \beta\bar{\beta} - \gamma\bar{\gamma} + \delta\bar{\delta} \end{pmatrix}. \tag{1.2.26}$$

In fact, this must be a *restricted* Lorentz transformation. For: (i) a Lorentz transformation continuous with the identity must be restricted, since no continuous Lorentz motion can transfer the positive time axis from inside the future null cone to inside the past null cone, or achieve a space reflection; (ii) the transformation (1.2.24) is evidently continuous with the identity if $\mathbf{A}$ is; (iii) and $\mathbf{A}$, like every spin-matrix, *is* continuous with the identity. For consider the matrix $\mathbf{B} := \lambda\mathbf{I} + (1 - \lambda)\mathbf{A}$. Since it is singular for at most two values of $\lambda$, we can find a path in the complex $\lambda$-plane from 0 to 1 which avoids these values. Then $(\det \mathbf{B})^{-\frac{1}{2}}\,\mathbf{B}$ defines a continuous succession of spin transformations from $\mathbf{A}$ to $\mathbf{I}$ or $-\mathbf{I}$. (The latter occurs if the path is such that $(\det \mathbf{B})^{-\frac{1}{2}}$ changes from 1 to $-1$, as is inevitable, for example, if $\mathbf{A} = -\mathbf{I}$. But $-\mathbf{I}$ is continuous with $\mathbf{I}$, for example by the spin transformation $\operatorname{diag}(e^{i\theta}, e^{-i\theta})$, $0 \leqslant \theta \leqslant \pi$, and thus (iii) is established.

We shall presently give a constructive proof showing that, conversely, any restricted Lorentz transformation is expressible in the form (1.2.24), with $\mathbf{A}$ a spin-matrix. Then we shall have established the following basic result:

(1.2.27) PROPOSITION

*Every spin transformation corresponds* (*via* (1.2.24) *to a unique restricted Lorentz transformation; conversely every restricted Lorentz transformation so corresponds to precisely two spin transformations, one being the negative of the other.*

Actually the required converse part of this result is a simple consequence of a general property of Lie groups. For the subgroup of the Lorentz group which arises in the form (1.2.24) must have the full dimensionality *six*. This is because spin-matrices form a six-real-dimensional (i.e. three-complex-dimensional) system and because only a discrete number (namely two) of spin-matrices define a single Lorentz transformation. This full-dimensional subgroup must contain the entire connected component of the identity in the Lorentz group.

It is instructive, however, to give an alternative demonstration of the converse part of (1.2.27) by simply constructing those spin-matrices explicitly which correspond to certain basic Lorentz transformations sufficient for generating the whole group. These basic transformations are space rotations, and 'boosts' (i.e. pure velocity transformations) such as in the well-known equations

$$\tilde{T} = (1 - v^2)^{-\frac{1}{2}}(T + vZ), \quad \tilde{X} = X, \quad \tilde{Y} = Y, \quad \tilde{Z} = (1 - v^2)^{-\frac{1}{2}}(Z + vT), \quad (1.2.28)$$

in which $v$ is the *velocity parameter*. Any restricted (active) Lorentz transformation can be compounded of a proper space rotation, followed by a boost in the z-direction, followed, finally, by a second space rotation. For the transformation is characterized by its effect on a Minkowski tetrad. Choose the first rotation so as to bring $\boldsymbol{z}$ into the space–time plane containing both the initial and final $\boldsymbol{t}$ directions. The boost (1.2.28) now sends $\boldsymbol{t}$ into its final direction, and the second rotation appropriately orients $\boldsymbol{x}, \boldsymbol{y}$, and $\boldsymbol{z}$. Thus we have only to show that space rotations and $\boldsymbol{z}$-boosts can be obtained from spin transformations. We shall consider *rotations* first and, in fact, establish the following result:

(1.2.29) PROPOSITION:

*Every unitary spin transformation corresponds to a unique proper rotation of $S^+$; conversely every proper rotation of $S^+$ corresponds to precisely two unitary spin transformations, one being the negative of the other.* (A unitary spin transformation is one given by a spin-matrix which is unitary: $\mathbf{A}^{-1} = \mathbf{A}^*$.)

First, let us fix our ideas a little more as to the geometrical significance of the transformations. The Lorentz transformations are here regarded as *active*. The spheres $S^+$ and $S^-$ are regarded as part of the coordinate frame, and do *not* partake of the transformation: as each future [past] null direction gets shifted, its representation on $S^+$ [$S^-$] shifts. For example, a rotation of $(\boldsymbol{x}, \boldsymbol{y}, \boldsymbol{z})$ which leaves $\boldsymbol{t}$ unchanged corresponds to a rotation of the images on $S^+$ $[S^-]$, which we may loosely call a rotation 'of' $S^+$ $[S^-]$. The plane $\Sigma$ also is part of the coordinate structure, and remains fixed while the images $\zeta$ of the null lines shift *on* it. Again, loosely, we may speak of motions 'of' $\Sigma$. (Of course, $S^+$, $S^-$, $\Sigma$ are no more invariant than the various coordinate hyperplanes: vectors that terminate on them generally will not do so after a Lorentz transformation has been applied.) It is of importance to remember that, while we here have a representation only of the null directions of $\mathbb{V}$, their transformations uniquely determine the transformation of *all* vectors of $\mathbb{V}$.

Now it is clear from (1.2.24) that $T$ is invariant under a unitary spin transformation, since the trace $(=2T)$ is always invariant under unitary transformations. (Equivalently we may see this from the invariance of the expression $\xi\bar{\xi} + \eta\bar{\eta}$, which is the Hermitian norm of $(\xi, \eta)$.) Restricted Lorentz transformations for which $T$ is invariant are simply proper rotations of $S^+$ (since they leave $X^2 + Y^2 + Z^2$ invariant) as required. To demonstrate the converse explicitly, we begin by noting that any proper rotation $(\boldsymbol{x}, \boldsymbol{y}, \boldsymbol{z}) \mapsto (\boldsymbol{x}', \boldsymbol{y}', \boldsymbol{z}')$ of $S^+$ may be compounded of successive rotations about the $Y$- and $Z$-axes. For the triad $(\boldsymbol{x}', \boldsymbol{y}', \boldsymbol{z}')$ is determined by the polar coordinates $\theta, \phi$ of $\boldsymbol{z}'$ relative to $(\boldsymbol{x}, \boldsymbol{y}, \boldsymbol{z})$ and by the angle $\psi$ subtended by the plane of $\boldsymbol{x}', \boldsymbol{z}'$ with that of $\boldsymbol{z}, \boldsymbol{z}'$. (These three angles are essentially the well-known Euler angles of mechanics: see Goldstein 1980, Arnold 1978). Thus a rotation through $\psi$ about $\boldsymbol{z}$, followed by a rotation through $\theta$ about the original $\boldsymbol{y}$, followed, finally, by a rotation through $\phi$ about the original $\boldsymbol{z}$, will achieve the required transformation. We shall show how these elementary rotations may be represented by unitary spin transformations. It will then follow that any proper rotation of $S^+$ can be so represented, since a product of unitary matrices is unitary.

A rotation of $S^+$ about the $z$-axis, through an angle $\psi$, evidently arises from a rotation of the Argand plane about the origin, through an angle $\psi$. This is given by

$$\tilde{\zeta} = e^{i\psi}\zeta, \tag{1.2.30}$$

i.e. by the spin transformations

$$\begin{pmatrix} \tilde{\xi} \\ \tilde{\eta} \end{pmatrix} = \pm \begin{pmatrix} e^{i\psi/2} & 0 \\ 0 & e^{-i\psi/2} \end{pmatrix} \begin{pmatrix} \xi \\ \eta \end{pmatrix}. \tag{1.2.31}$$

Next, we assert that a rotation of $S^+$ through an angle $\theta$ about the $y$-axis is represented by the following unitary spin transformations:

$$\begin{pmatrix} \tilde{\xi} \\ \tilde{\eta} \end{pmatrix} = \pm \begin{pmatrix} \cos\theta/2 & -\sin\theta/2 \\ \sin\theta/2 & \cos\theta/2 \end{pmatrix} \begin{pmatrix} \xi \\ \eta \end{pmatrix}. \tag{1.2.32}$$

Since (1.2.32) is unitary, it certainly represents *some* rotation. Furthermore, since $\xi\bar{\eta} - \eta\bar{\xi}$ is invariant, as well as $\xi\bar{\xi} + \eta\bar{\eta}$, it follows from (1.2.14) that the $y$-coordinates of points on $S^+$ are invariant under (1.2.32). Hence the rotation is about the $y$-axis. Finally, the transformation (1.2.32) sends the point (1, 0, 0, 1) into (1, sin $\theta$, 0, cos $\theta$), so the angle of rotation is indeed $\theta$. (By a similar argument one verifies that the unitary spin transformations

$$\begin{pmatrix}\tilde{\xi}\\ \tilde{\eta}\end{pmatrix} = \pm\begin{pmatrix}\cos\chi/2 & \mathrm{i}\sin\chi/2\\ \mathrm{i}\sin\chi/2 & \cos\chi/2\end{pmatrix}\begin{pmatrix}\xi\\ \eta\end{pmatrix} \tag{1.2.33}$$

correspond to a rotation through an angle $\chi$ about the $x$-axis.) Proposition (1.2.29) is now established. For reference we exhibit the resultant spin-matrix corresponding to the (general) rotation through the Euler angles $\theta, \phi, \psi$:

$$\pm\begin{pmatrix}\cos\frac{\theta}{2}\,\mathrm{e}^{\mathrm{i}(\phi+\psi)/2} & -\sin\frac{\theta}{2}\,\mathrm{e}^{\mathrm{i}(\phi-\psi)/2}\\ \sin\frac{\theta}{2}\,\mathrm{e}^{-\mathrm{i}(\phi-\psi)/2} & \cos\frac{\theta}{2}\,\mathrm{e}^{-\mathrm{i}(\phi+\psi)/2}\end{pmatrix} \tag{1.2.34}$$

Its elements are in fact the Cayley–Klein rotation parameters of mechanics. (Goldstein 1980).

We now complete the proof of Proposition (1.2.27) by showing that every $z$-boost (1.2.28) can be obtained from a spin transformation. To do this, we rewrite (1.2.28) in the form

$$\tilde{T} + \tilde{Z} = w(T + Z),\ \tilde{T} - \tilde{Z} = w^{-1}(T - Z),\ \tilde{X} = X,\ \tilde{Y} = Y, \tag{1.2.35}$$

where

$$w = \left(\frac{1+v}{1-v}\right)^{\frac{1}{2}}. \tag{1.2.36}$$

(Here $w$ is the relativistic Doppler factor and $\log w = \tanh^{-1} v$ is the 'rapidity' corresponding to $v$.) By reference to (1.2.24) we see at once that (1.2.35) is achieved by the spin transformation

$$\begin{pmatrix}\tilde{\xi}\\ \tilde{\eta}\end{pmatrix} = \pm\begin{pmatrix}w^{\frac{1}{2}} & 0\\ 0 & w^{-\frac{1}{2}}\end{pmatrix}\begin{pmatrix}\xi\\ \eta\end{pmatrix}, \tag{1.2.37}$$

or, in terms of the Argand plane of $\zeta$, by the simple expansion

$$\tilde{\zeta} = w\zeta. \tag{1.2.38}$$

Thus (1.2.27) is established.

We finally remark that *any* pure boost (two hyperplanes orthogonal to $t$ kept invariant, e.g. $X = 0$, $Y = 0$ above) corresponds to a positive- [ *or* negative-] definite Hermitian spin-matrix, and vice versa. For the $z$-

boost (1.2.37) is of this form, and to obtain a boost in any other direction we need merely rotate that direction into the $z$-direction, apply a $z$-boost, and rotate back. This corresponds to the spin-matrix $\mathbf{A}^{-1}\mathbf{B}\mathbf{A}$, where $\mathbf{A}$ is the required rotation and $\mathbf{B}$ the $z$-boost; by elementary matrix theory, $\mathbf{A}^*\mathbf{B}\mathbf{A}$ is still positive- [negative-] definite Hermitian. Conversely, any positive- [negative-] definite Hermitian matrix $\tilde{\mathbf{B}}$ may be diagonalized by a unitary matrix $\mathbf{A}$: $\mathbf{A}\tilde{\mathbf{B}}\mathbf{A}^{-1} = \mathrm{diag}(\alpha, \delta)$, which must be of the form $\pm\,\mathrm{diag}(w^{\frac{1}{2}}, w^{-\frac{1}{2}}) = \pm\,\mathbf{B}$, since Hermiticity, definiteness, and unit determinant are preserved. Consequently $\tilde{\mathbf{B}}$ is of the form $\pm\,\mathbf{A}^{-1}\mathbf{B}\mathbf{A}$ and our result is established.

It is easy to see from the preceding work that *any* restricted Lorentz transformation $L$ is uniquely the composition of *one* boost followed by *one* proper space rotation, and also the other way around. For we need merely determine the spatial direction $\boldsymbol{w}$ orthogonal to $\boldsymbol{t}$ in the plane containing the original and final $\boldsymbol{t}$, apply a '$\boldsymbol{w}$-boost' to send $\boldsymbol{t}$ into its final position, and then apply a space rotation to re-orient $\boldsymbol{x}, \boldsymbol{y}, \boldsymbol{z}$ suitably.* Evidently, if we perform these transformations in reverse, we have a decomposition of $L^{-1}$. Again in the light of what has been shown before, the reader will recognize in this result an example of the mathematical theorem according to which any non-singular complex matrix is uniquely expressible as the product of a unitary matrix with a positive definite Hermitian matrix, and also the other way around.

*Relation to quaternions*

We conclude this section with some remarks on quaternions. Several of our spin-matrix results in connection with rotations will thereby be illuminated. In fact, the representation of proper rotations by unitary spin-matrices is effectively the same as their more familiar representation in terms of quaternions. Set

$$\mathbf{I} = \begin{pmatrix} 1 & 0 \\ 0 & 1 \end{pmatrix}, \ \mathbf{i} := \begin{pmatrix} 0 & \mathrm{i} \\ \mathrm{i} & 0 \end{pmatrix}, \ \mathbf{j} := \begin{pmatrix} 0 & -1 \\ 1 & 0 \end{pmatrix}, \ \mathbf{k} := \begin{pmatrix} \mathrm{i} & 0 \\ 0 & -\mathrm{i} \end{pmatrix}. \qquad (1.2.39)$$

Then these matrices have the following multiplication table

$$\begin{array}{c|cccc|} & \mathbf{I} & \mathbf{i} & \mathbf{j} & \mathbf{k} \\ \hline \mathbf{I} & \mathbf{I} & \mathbf{i} & \mathbf{j} & \mathbf{k} \\ \mathbf{i} & \mathbf{i} & -\mathbf{I} & \mathbf{k} & -\mathbf{j} \\ \mathbf{j} & \mathbf{j} & -\mathbf{k} & -\mathbf{I} & \mathbf{i} \\ \mathbf{k} & \mathbf{k} & \mathbf{j} & -\mathbf{i} & -\mathbf{I} \\ \hline \end{array}\ , \qquad (1.2.40)$$

* It follows that the topology of the restricted Lorentz group is the topological product of the rotation group with $\mathbb{R}^3$.

which defines $\mathbf{I}, \mathbf{i}, \mathbf{j}$ and $\mathbf{k}$ as the elementary quaternions. The general quaternion will then be represented by the matrix

$$\mathbf{A} = \mathbf{I}a + \mathbf{i}b + \mathbf{j}c + \mathbf{k}d = \begin{pmatrix} a + \mathrm{i}d & -c + \mathrm{i}b \\ c + \mathrm{i}b & a - \mathrm{i}d \end{pmatrix}, \tag{1.2.41}$$

where $a, b, c, d \in \mathbb{R}$. The sum or product of two quaternions is obtained simply as the matrix sum or product. Again, $\mathbf{A}^*$ is defined via the corresponding matrix operation, and we note that

$$\mathbf{A}^* = \mathbf{I}a - (\mathbf{i}b + \mathbf{j}c + \mathbf{k}d). \tag{1.2.42}$$

The matrix $\mathbf{A}$ in (1.2.41) will be a unitary spin-matrix if it is unimodular and unitary. But from (1.2.41),

$$\det \mathbf{A} = a^2 + b^2 + c^2 + d^2, \tag{1.2.43}$$

$$\mathbf{A}\mathbf{A}^* = \mathbf{I}(a^2 + b^2 + c^2 + d^2), \tag{1.2.44}$$

so that both conditions are satisfied if the quaternion has unit 'norm':

$$N(\mathbf{A}) := a^2 + b^2 + c^2 + d^2 = 1. \tag{1.2.45}$$

Thus, unit quaternions can be represented by unitary spin-matrices. Particular examples of unit quaternions are the elementary quaternions $\mathbf{I}, \mathbf{i}, \mathbf{j}, \mathbf{k}$. We see from (1.2.31), (1.2.32), (1.2.33), and (1.2.39) that $\mathbf{i}, \mathbf{j}$, and $\mathbf{k}$ define, respectively, rotations through $\pi$ about the $X$, $Y$, and $Z$ axes.

If we write

$$\mathbf{A} = \mathbf{I}a + \mathbf{i}b + \mathbf{j}c + \mathbf{k}d = a + \mathbf{v} \tag{1.2.46}$$

and regard

$$\mathbf{v} = (b, c, d)$$

as a vector having components $(b, c, d)$ relative to *some* basis, and if, similarly, $\mathbf{A}' = \mathbf{I}a' + \cdots = a' + \mathbf{v}'$, it is easy to verify that

$$\mathbf{A} + \mathbf{A}' = a + a' + \mathbf{v} + \mathbf{v}'.$$

$$\mathbf{A}\mathbf{A}' = aa' - \mathbf{v}\cdot\mathbf{v}' + a'\mathbf{v} + a\mathbf{v}' + \mathbf{v} \times \mathbf{v}', \tag{1.2.47}$$

where vectorial sums and products are formed in the usual way from components. It is therefore clear that a valid equation of quaternions, involving sums and products, remains valid when a rotational transformation is applied to the 'vector' components $(b, c, d)$, $(b', c', d')$, etc.; for the 'vectorial' part of the equation will be form-invariant under such transformations.

We saw that some quaternions can be represented by spin-matrices. In that sense, they may be regarded as transformations. But quaternions play a dual role, in that they can also function as 'transformands', e.g., as three-vectors *being* transformed. As we saw in (1.2.24), it is sometimes useful to combine the components $(T, X, Y, Z)$ of a four-vector into a certain Hermitian matrix. In the particular case when $T = 0$, and after multiplying that matrix by i, it can be identified with a 'vectorial' quaternion **Q** (*cf.* (1.2.41))*:

$$\mathbf{Q} = \begin{pmatrix} \mathrm{i}Z & \mathrm{i}X - Y \\ \mathrm{i}X + Y & -\mathrm{i}Z \end{pmatrix} = \mathbf{i}X + \mathbf{j}Y + \mathbf{k}Z. \tag{1.2.48}$$

Equation (1.2.24) then reads

$$\tilde{\mathbf{Q}} = \mathbf{AQA}^*. \tag{1.2.49}$$

Now, from the spin-matrix interpretation of this equation we know that *any* unit quaternion **A** will, via this equation, effect a certain proper space rotation on the vector **Q**. The most general unit quaternion can clearly be written in the form

$$\begin{aligned} \mathbf{A} &= \mathbf{I}\cos\frac{\psi}{2} + (\mathbf{i}l + \mathbf{j}m + \mathbf{k}n)\sin\frac{\psi}{2} \\ &= \cos\frac{\psi}{2} + \mathbf{v}\sin\frac{\psi}{2}, \end{aligned} \tag{1.2.50}$$

where $\mathbf{v} = (l, m, n)$ and $l^2 + m^2 + n^2 = 1$. We assert that this **A** effects a rotation through $\psi$ about **v**. For proof we need merely note that (1.2.49) is a quaternion equation and as such unaffected by a change of (quaternion-) vector basis: rotate that basis so that **v** becomes (0, 1, 0). Then our result is immediate by comparison with (1.2.32), **A** having reduced to the spin-matrix which effects a rotation through $\psi$ about the $y$-axis, and **v** having reduced to that axis. A corollary of this result is the important fact that *any* proper space rotation (unit quaternion) is a rotation about some axis **v**, through some angle $\psi$.

Writing (1.2.50) in matrix notation,

$$\mathbf{A} = \begin{pmatrix} \cos\frac{\psi}{2} + \mathrm{i}n\sin\frac{\psi}{2} & (-m + \mathrm{i}l)\sin\frac{\psi}{2} \\ (m + \mathrm{i}l)\sin\frac{\psi}{2} & \cos\frac{\psi}{2} - \mathrm{i}n\sin\frac{\psi}{2} \end{pmatrix}, \tag{1.2.51}$$

we obtain the most general unitary spin-matrix in a form which allows us

* No such 'trick' works to relate the full four-vector $(T, X, Y, Z)$ with real quaternions.

to read off its transformational effect by inspection. Note, incidentally, that **A** changes sign under $\psi \mapsto \psi + 2\pi$.

Although unitary spin-matrices and unit quaternions are effectively the same thing, there is no such close relationship between spin-matrices in general and quaternions. The underlying reason for this is that quaternions are associated with quadratic forms of positive definite signature (*cf.* (1.2.45)) whereas spin-matrices and Lorentz transformations are concerned with the Lorentzian signature $(+, -, -, -)$. Of course, one can avoid this difficulty by introducing 'quaternions' with suitably complex coefficients. Such objects do not share with the real quaternions their fundamental property of constituting a division algebra. Nevertheless, the mere use of quaternion *notation* (especially (1.2.47)) can bring considerable advantages to certain manipulations of general spin-matrices (see, e.g., Ehlers, Rindler and Robinson 1966).

## 1.3 Some properties of Lorentz transformations

As a consequence of the correspondence between the restricted Lorentz group and the group of spin transformations, it is possible to give simple derivations of many of the standard properties of rotations and Lorentz transformations This we shall now do.

It is easily seen that when the spin transformation (1.2.16) is unitary then (1.2.17) becomes

$$\tilde{\zeta} = \frac{\alpha\zeta - \bar{\gamma}}{\gamma\zeta + \bar{\alpha}}. \tag{1.3.1}$$

The fixed points (i.e. $\tilde{\zeta} = \zeta$) are then given by

$$\gamma\zeta^2 + (\bar{\alpha} - \alpha)\zeta + \bar{\gamma} = 0.$$

Clearly if $\zeta$ is one root of this quadratic equation, then $-1/\bar{\zeta}$ is the other. Consequently the fixed points have the form $\zeta, -1/\bar{\zeta}$, which correspond to antipodes on the sphere $S^+$ (*cf.* after (1.2.11)). This constitutes yet another proof of the fact that every rotation of the sphere is equivalent to a rotation about a single axis.

A *circle* on the sphere $S^+$ is defined as the intersection of $S^+$ with some plane in the Euclidean 3-space $T = 1$, given by a real linear equation $lX + mY + nZ = p\,(p^2 < l^2 + m^2 + n^2)$. Substituting (1.2.8) into this, we get (provided $p \neq n$) an equation of the form $\zeta\bar{\zeta} - \kappa\bar{\zeta} - \bar{\kappa}\zeta + \kappa\bar{\kappa} = r^2$ $(r > 0, \kappa$ complex), i.e. $|\zeta - \kappa| = r$. This is the equation of a circle in the Argand plane, with centre $\kappa$ and radius $r$. When $p = n$ (i.e., when the original circle passes through the north pole on $S^+$), we get the equation of a straight line on the Argand plane. So we have established the well-known fact that under stereographic projection circles on the sphere project to circles or straight lines on the plane – and vice versa, since the above argument is reversible.

Now, in the previous section we showed that every spin transformation can be compounded of transformations which induce either rotations of $S^+$ or simple expansions of the Argand plane. The first type clearly preserves circles on $S^+$ while the second clearly preserves circles or straight lines on the Argand plane. It therefore follows from the above discussion that every spin transformation induces a transformation on $S^+$ which sends circles into circles. (This is actually a familiar property of bilinear transformations (1.2.17) of the Riemann sphere and is not hard to establish directly.)

Any circle-preserving transformation must necessarily be *conformal* (i.e. angle-preserving). This is basically because infinitesimal circles must transform to infinitesimal circles rather than ellipses. Alternatively, we may verify the conformal nature of stereographic projection directly by observing that the squared interval $\mathrm{d}\sigma^2$ on the sphere is related to that, $\mathrm{d}\zeta\,\mathrm{d}\bar{\zeta}$, on the Argand plane by

$$\mathrm{d}\sigma^2 = \mathrm{d}x^2 + \mathrm{d}y^2 + \mathrm{d}z^2 = \frac{4\mathrm{d}\zeta\,\mathrm{d}\bar{\zeta}}{(\zeta\bar{\zeta}+1)^2}, \tag{1.3.2}$$

as follows from (1.2.8). And the conformal nature of the bilinear transformation can be deduced from the mere fact that it is holomorphic (i.e., complex-analytic)*; for then $\tilde{\zeta} = f(\zeta)$ implies $\mathrm{d}\tilde{\zeta} = f'(\zeta)\mathrm{d}\zeta$. In consequence of all this it is seen that a Lorentz transformation effects an *isotropic* expansion and rotation of the neighbourhood of each point of $S^+$.

The above conformal and circle-preserving properties have as a corol-

* Conversely, *every* local proper (i.e. orientation preserving) conformal transformation of the Riemann sphere arises from a holomorphic mapping of $\zeta$. (For conformality means $\mathrm{d}\tilde{\zeta} = a\,\mathrm{d}\zeta$ for *some* complex $a$, i.e. $(\mathrm{d}\tilde{x} + \mathrm{i}\mathrm{d}\tilde{y}) = a(\mathrm{d}x + \mathrm{i}\mathrm{d}y)$, from which follow the Cauchy–Riemann equations $\partial\tilde{x}/\partial x = \mathrm{Re}(a) = \partial\tilde{y}/\partial y$, $-\partial\tilde{x}/\partial y = \mathrm{Im}(a) = \partial\tilde{y}/\partial x$, hence holomorphicity.) Since the only global holomorphic mappings of the Riemann sphere to itself are bilinear transformations, it follows that the only proper global conformal maps of $\mathscr{S}^+$ (to be understood via $S^+$) to itself are those induced by spin transformations. The group of proper conformal self-transformations of $\mathscr{S}^+$ is thus the restricted Lorentz group. The structure of $\mathscr{S}^+$ which is significant, therefore, is *precisely* its *conformal structure and orientation.*

In fact *any* 2-surface $T$ with the topology of a sphere $S^2$ and with a (positive definite) conformal structure is conformally identical with a metric sphere (say $S^+$). Thus, the proper conformal self-transformations of $T$ also form a group isomorphic with the restricted Lorentz group. This result will have importance for us in Chapter 9 (*cf.* (9.6.31)).

lary the familiar but surprising special-relativity effects sometimes known as 'the invisibility of the Lorentz contraction'. Let us envisage an observer at $O$. As remarked earlier, his field of vision or *celestial sphere* may be conveniently represented by $S^-$. Each light ray entering his eye is represented by a null straight line through $O$ and hence by a single point of his celestial sphere $S^-$ (sky mapping). Now $S^-$ is related to $S^+$ simply by the antipodal map. Thus, any restricted Lorentz transformation of $\mathbb{V}$ induces a conformal circle-preserving map of the celestial sphere onto itself. It follows from the conformal property that an object subtending a *small* angle at a given observer will present a similar shape visually to any other observer momentarily coincident with the first, no matter what his velocity relative to the first observer may be (Terrell 1959). Only the apparent angular size and direction will, in general, be different for two such observers. Moreover, from the circle-preserving property it follows that if one inertial observer perceives an object of *any* size to have a *circular* outline, then all inertial observers momentarily coincident with the first will perceive circular outlines (or, in special cases, 'straight' outlines, if we consider a great circle on the celestial sphere as appearing 'straight'). Hence, in particular, uniformly moving spheres, in spite of the Lorentz contraction, present circular outlines to all observers (Penrose 1959).

A bilinear transformation of the Riemann sphere is fully determined if we specify any three distinct points as images of any other three distinct points of the sphere. This well-known fact is a simple consequence of (1.2.17). (The three complex ratios $\alpha:\beta:\gamma:\delta$ define the transformation and these are fixed by three complex equations.) It follows that every restricted Lorentz transformation is fully determined if we specify the (distinct) maps of three distinct null directions. (Thus, by a unique adjustment of his velocity and orientation an observer can make any three given stars take up three specified positions on his celestial sphere.)

Again, each bilinear transformation (1.2.17) – not merely the special case (1.3.1) – (apart from the identity transformation) possesses just *two* (possibly coincident) fixed points on the Riemann sphere, as is readily seen if we set $\tilde{\zeta} = \zeta$ in (1.2.17) and solve the resulting quadratic equation. Consequently every (non-trivial) Lorentz transformation leaves invariant just two (possibly coincident) null directions.*

* In fact, according to a theorem of topology, every continuous orientation-preserving map of the sphere onto itself must possess at least *one* fixed point and, 'properly' counted, precisely two fixed points, since the Euler characteristic of the sphere is 2. Compare the discussion of 'fingerprints' in §8.7.

### *The kinds of Lorentz transformation, in terms of $S^+$*

Let us examine the structure of Lorentz transformations in the light of this fact. Consider first the case when the two fixed null directions are distinct. We can obtain a canonical form for such a Lorentz transformation by choosing our reference Minkowski tetrad so that $\boldsymbol{t}$ and $\boldsymbol{z}$ lie in the 2-plane spanned by these null directions. The latter must then have components $(1, 0, 0, \pm 1)$ whence the fixed points lie at the north and south poles of $S^+$ ($\zeta = \infty, 0$). The most general bilinear transformation (1.2.17) which leaves both poles invariant is of the form

$$\tilde{\zeta} = w e^{i\psi}\zeta \tag{1.3.3}$$

where $w$ and $\psi$ are real numbers. This is the composition (in either order) of a rotation through an angle $\psi$ about the $z$-axis, with a boost with rapidity $\phi = \log w$ along the $z$-axis (*cf.* Figs. 1-6, 1-7, 1-8). In terms of Minkowski

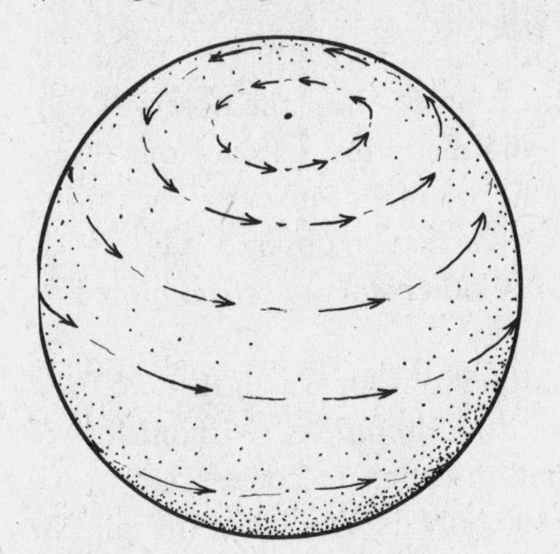

Fig. 1-6. The effect of a rotation on $S^+$.

Fig. 1-7. The effect of a boost on $S^+$.

Fig. 1-8. The effect of a four-screw on $S^+$.

coordinates (1.2.15) we have

$$\tilde{X} = X\cos\psi - Y\sin\psi, \quad \tilde{Y} = X\sin\psi + Y\cos\psi,$$
$$\tilde{Z} = Z\cosh\phi + T\sinh\phi, \tilde{T} = Z\sinh\phi + T\cosh\phi. \qquad (1.3.4)$$

This is what Synge (Synge 1955, p. 86) calls a 'four-screw'.

We have already seen that a pure boost in the $z$-direction corresponds to an expansion $\tilde{\zeta} = w\zeta$ of the Argand plane. In terms of the sky mapping ($S^-$), and with spherical polar coordinates for the celestial sphere (*cf.* (1.2.11)), this leads to the aberration formula for incoming light rays in the useful form:

$$\tan\frac{\tilde{\theta}}{2} = w\tan\frac{\theta}{2},$$

with (1.3.5)

$$w = \left(\frac{1-V}{1+V}\right)^{1/2}.$$

Here $V = -v$ is the velocity in the direction $\theta = 0$ of the observer who measures angle $\tilde{\theta}$, relative to the one who measures $\theta$. (Since our transformations are active, we must think of the rest of the universe as acquiring a velocity $-V = v$ in the $z$-direction.) We see that an observer who travels at high speed toward a star $P$ perceives all other stars to crowd more and more around $P$ an his speed increases.

Next we examine the Lorentz transformations for which the two fixed null directions coincide. These are called *null rotations*. Without loss of generality we may choose the fixed null direction to correspond to the north pole of $S^+$. Thus $\zeta = \infty$ is to be the only fixed point of the bilinear transformation (1.2.17), and so

$$\tilde{\zeta} = \zeta + \beta \qquad (1.3.6)$$

where $\beta$ is some complex number. This is simply a translation in the Argand plane. (A bilinear transformation of the Argand plane, for which $\zeta = \infty$ is a fixed point, must be of the form $\zeta = \alpha\zeta + \beta$, but if $\alpha \neq 1$ it has also a *finite* fixed point.) The spin transformations giving (1.3.6) are

$$\begin{pmatrix}\tilde{\xi}\\ \tilde{\eta}\end{pmatrix} = \pm\begin{pmatrix}1 & \beta\\ 0 & 1\end{pmatrix}\begin{pmatrix}\xi\\ \eta\end{pmatrix}. \qquad (1.3.7)$$

Without loss of generality we can take, say, $\beta = ia$ with $a$ real. Then in terms of Minkowski coordinates we get

$$\tilde{X} = X, \quad \tilde{Y} = Y + a(T - Z),$$
$$\tilde{Z} = Z + aY + \tfrac{1}{2}a^2(T - Z), \tilde{T} = T + aY + \tfrac{1}{2}a^2(T - Z). \qquad (1.3.8)$$

Note that the null vector $\boldsymbol{z} + \boldsymbol{t}$ is itself invariant, not merely its direction.

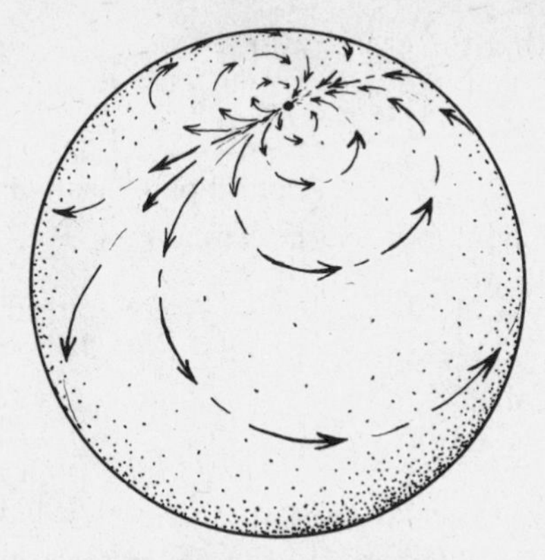

Fig. 1-9. The effect of a null rotation on $S^+$.

To visualize the effect of this null rotation of the Riemann sphere, refer to Fig. 1-9. The rigid translation of the Argand plane projects to a transformation on the Riemann sphere for which the points are displaced along circles through the north pole tangent to the $y$-direction there. The displacements become less and less as the north pole is approached, leaving this as the only fixed point.

As mentioned above, the most general bilinear transformation for which $\zeta = \infty$ is a fixed point is of the form $\tilde{\zeta} = \alpha\zeta + \beta$. This can be broken up into a translation, a rotation, and a dilation of the Argand plane (in any order). Thus the most general restricted Lorentz transformation which leaves invariant a given null direction $\boldsymbol{K}$ (in the plane of $z$ and $t$, say) is the product of a null rotation about $\boldsymbol{K}$, a space rotation about $z$, and a $z$-boost. The first two of these transformations leave the entire vector $\boldsymbol{K}$ invariant, the last only its direction.

We may remark that the Lorentz transformations leaving two *given* null directions invariant and the null rotations leaving one *given* null direction invariant each form a two-dimensional Abelian subgroup of the Lorentz group. In the first case we have the additive group on the complex number $\phi + \mathrm{i}\psi$ (modulo $2\pi\mathrm{i}$) and in the second case, the additive group on $\beta$. The groups are not isomorphic since they have a different topology ($S^1 \times \mathbb{R}$ and $\mathbb{R}^2$, respectively). This is because $\phi + \mathrm{i}\psi + 2\pi\mathrm{i}n$ ($n = \ldots, -1, 0, 1, 2, \ldots$) all give the same transformation, whereas for different $\beta$'s the null rotations are all distinct.

### *Cross-ratios of null directions*

For future use (in Vol. 2 Chapter 8), we conclude this section with some rather specialized results concerning cross-ratios. It is well known (and can easily be verified) that the cross-ratio

$$\chi = \{\zeta_1, \zeta_2, \zeta_3, \zeta_4\} = \frac{(\zeta_1 - \zeta_2)(\zeta_3 - \zeta_4)}{(\zeta_1 - \zeta_4)(\zeta_3 - \zeta_2)} \tag{1.3.9}$$

of four points $\zeta_1, \zeta_2, \zeta_3, \zeta_4$ in the Argand plane is invariant under bilinear transformations. In homogeneous coordinates,

$$\chi = \frac{(\xi_1\eta_2 - \xi_2\eta_1)(\xi_3\eta_4 - \xi_4\eta_3)}{(\xi_1\eta_4 - \xi_4\eta_1)(\xi_3\eta_2 - \xi_2\eta_3)} \; . \tag{1.3.10}$$

Now it is easily seen that

$$\{\alpha, \beta, \gamma, \delta\} = \{\beta, \alpha, \delta, \gamma\} = \{\gamma, \delta, \alpha, \beta\} = \{\delta, \gamma, \beta, \alpha\}. \tag{1.3.11}$$

Consequently there can be at most six different values of the cross-ratio of four points taken in all possible orders, and these are seen to be

$$\chi, \quad 1 - \chi, \quad \frac{1}{\chi}, \quad \frac{1}{1-\chi}, \quad \frac{\chi - 1}{\chi}, \quad \frac{\chi}{\chi - 1}. \tag{1.3.12}$$

When just two $\zeta$'s coincide, $\chi$ degenerates into 1, 0 or $\infty$. With triple or quadruple coincidences, $\chi$ becomes indeterminate.

The cross-ratio of four real null directions is defined to be the cross-ratio (1.3.9) of the four corresponding points in the Argand plane. It is easily seen, e.g. by an interchange of the $\xi$s and $\eta$s in (1.3.10), that

$$\{1/\alpha, 1/\beta, 1/\gamma, 1/\delta\} = \{\alpha, \beta, \gamma, \delta\} = \{-\alpha, -\beta, -\gamma, -\delta\}, \tag{1.3.13}$$

and that consequently the sky mapping and the anti-sky mapping yield cross-ratios which are complex conjugates of each other, since the maps on $S^+$ and $S^-$ of given null directions are as $-1/\bar{\zeta}$ and $\zeta$ respectively.

A knowledge of any three of the complex numbers $\zeta_1, \zeta_2, \zeta_3, \zeta_4$ (without coincidences), and an arbitrarily assigned value of the cross-ratio (1.3.9), determines the fourth number uniquely (counting $\infty$ as a number). In consequence, any four distinct null directions (points of $S^+$) can be transformed by a suitable restricted Lorentz transformation (bilinear transformation) into any other four null directions (points) having the same cross-ratio; for any three of the null directions can be so mapped into any other three noncoincident null directions and then the fourth is determined uniquely by the invariant cross-ratio.

The *reality* of the cross-ratio (1.3.9) is the condition for the four relevant points to be *concyclic* (or collinear) in the Argand plane. This is equivalent to saying that the four corresponding points on the Riemann sphere are concyclic and hence coplanar. Consequently the condition that four null lines lie in a real hyperplane

$$aT + bX + cY + dZ = 0$$

is that their cross-ratio be real. A particular instance of this is a *harmonic* set for which the cross ratio is $-1, 2$, or $1/2$. One harmonic set in the Argand plane is given by the vertices of the square $1, \mathrm{i}, -1, -\mathrm{i}$; these same points on the equator of the Riemann sphere therefore correspond to a set of harmonic null directions. By the remark of the preceding paragraph, any four harmonic null directions can be transformed into these by a suitable restricted Lorentz transformation.

Also of interest is an *equianharmonic* set which possesses even greater intrinsic symmetry. The cross-ratio in this case is $-\omega$ or $-\omega^2$, where $\omega = \mathrm{e}^{2\mathrm{i}\pi/3}$. By a suitable restricted Lorentz transformation four such points on the Riemann sphere can be made the vertices of a regular tetrahedron. This follows from the fact that for all $\lambda$ the set of points $0, \lambda, \lambda\omega, \lambda\omega^2$ is equianharmonic in the plane, and for a suitable real $\lambda$ it evidently projects into the vertices of a regular tetrahedron.

A geometric interpretation of the cross-ratio can be elucidated as follows.* Consider any four distinct real null vectors $\boldsymbol{A}, \boldsymbol{B}, \boldsymbol{C}, \boldsymbol{D}$, and write $[\boldsymbol{AB}]$ for the plane of $\boldsymbol{A}$ and $\boldsymbol{B}$. Let $\Omega$ be that unique timelike plane which contains one vector from each of $[\boldsymbol{AB}]$ and $[\boldsymbol{CD}]$, and one normal vector to each of $[\boldsymbol{AB}], [\boldsymbol{CD}]$. For purposes of computation, assume that $\Omega$ is the plane of $\boldsymbol{z}$ and $\boldsymbol{t}$, and that $\boldsymbol{A} \propto (1, p, q, r)$, $\boldsymbol{B} \propto (1, p', q', r')$. The only normals to $\boldsymbol{A}$ and $\boldsymbol{B}$ contained in $\Omega$ are $(r, 0, 0, 1)$ and $(r', 0, 0, 1)$, respectively; hence $r' = r$. If $[\boldsymbol{AB}]$ is to contain a vector of $\Omega$, $(p', q') \propto (p, q)$. Since $\boldsymbol{A}$ and $\boldsymbol{B}$ are null and distinct, this implies $A \propto (1, p, q, r)$, $B \propto (1, -p, -q, r)$; and similarly for $\boldsymbol{C}, \boldsymbol{D}$. Hence on the Riemann sphere each of these pairs is represented by a pair of points on the same latitude but opposite meridians. There can be only one restricted Lorentz transformation which transforms $\boldsymbol{A} \mapsto \boldsymbol{C}, \boldsymbol{B} \mapsto \boldsymbol{D}$, and preserves $\Omega$ (and thus its null directions, corresponding to the north and south poles). Evidently it is a rotation (1.2.30) about the $z$-axis, followed by a boost (1.2.38) along the $z$-axis:

$$\tilde{\zeta} = \mathrm{e}^{\phi + \mathrm{i}\psi}\zeta = \mathrm{e}^{\rho}\zeta, \tag{1.3.14}$$

where $\mathrm{e}^{\phi} = w$ and $\rho = \phi + \mathrm{i}\psi$. If $\alpha, \beta$ are the points in the complex plane corresponding to $\boldsymbol{A}, \boldsymbol{B}$, then $\beta = -\alpha$; and if $\gamma, \delta$ correspond to $\boldsymbol{C}, \boldsymbol{D}$, then $\delta = -\gamma$ and $\gamma = \mathrm{e}^{\rho}\alpha$. Consequently

$$\chi = \{\alpha, \gamma, \beta, \delta\} = \frac{(1 - \mathrm{e}^{\rho})^2}{(1 + \mathrm{e}^{\rho})^2} = \tanh^2 \frac{\rho}{2}. \tag{1.3.15}$$

* We are indebted to Ivor Robinson for discussion on this point.
The unique existence of $\Omega$ is not entirely obvious. But note that the unique Lorentz transformation sending $\boldsymbol{A}, \boldsymbol{B}, \boldsymbol{C}$ into $\boldsymbol{B}, \boldsymbol{A}, \boldsymbol{D}$ also sends $\boldsymbol{D}$ to $\boldsymbol{C}$ (since the cross-ratios of $\boldsymbol{A}, \boldsymbol{B}, \boldsymbol{C}, \boldsymbol{D}$ and $\boldsymbol{B}, \boldsymbol{A}, \boldsymbol{D}, \boldsymbol{C}$ are equal by (1.3.11)) and hence has period 2. This is a reflection whose invariant planes are $\Omega$ and its spacelike orthogonal complement.

We may call

$$\rho = \phi + \mathrm{i}\psi = 2\tanh^{-1}\sqrt{\chi} \qquad (1.3.16)$$

the *complex angle* between the real planes $[\boldsymbol{AB}]$ and $[\boldsymbol{CD}]$, and we see that it is uniquely determined by the cross-ratio $\{\alpha, \gamma, \beta, \delta\}$ apart from sign, and modulo $\mathrm{i}\pi$ depending on whether $\boldsymbol{A}\mapsto\boldsymbol{C}, \boldsymbol{B}\mapsto\boldsymbol{D}$ or the other way around. Its geometric meaning is that $[\boldsymbol{AB}]$, $[\boldsymbol{CD}]$ 'differ' by a Lorentz transformation compounded of a rotation through an angle $\psi$ about the plane $\Omega$, and a boost through a rapidity $\phi$ in the plane $\Omega$, $\Omega$ being the unique 'normal' to $[\boldsymbol{AB}]$, $[\boldsymbol{CD}]$, as defined above.

## 1.4 Null flags and spin-vectors

The purpose of this section is to lead up to the geometrical concept of a spin-vector (the simplest type of spinor). On this will rest the geometrical content of spinor algebra, which will be treated formally in succeeding chapters. The geometry of the elementary algebraic operations between spin-vectors will be discussed in §1.5.

Our aim is to find some geometrical structure in Minkowski vector space $\mathbb{V}$ – this will be our geometrical picture of a *spin-vector* $\boldsymbol{\kappa}$ – of which the pair $(\xi, \eta)$ of complex numbers introduced in §1.2 can be regarded as a coordinate representation. We have already seen how to associate with $(\xi, \eta)$ a future-pointing null vector $\boldsymbol{K}$ provided a Minkowski coordinate system is given. The pair $(\xi, \eta)$ serve as coordinates for $\boldsymbol{K}$; but for $\boldsymbol{K}$ these coordinates are redundant to the extent that the phase transformation $\xi \mapsto e^{i\theta}\xi, \eta \mapsto e^{i\theta}\eta$ leaves $\boldsymbol{K}$ unchanged (*cf.* (1.2.15)). We now propose to associate a richer geometrical structure with $(\xi, \eta)$, one that reduces the redundancy to a single (essential) sign ambiguity. This structure will, in fact, be a 'null flag', i.e., the previous null vector $\boldsymbol{K}$ representing $\xi$ and $\eta$ up to phase, together with a 'flag plane', or half null plane, attached to $\boldsymbol{K}$, which represents the phase.* However, when the phase angle changes by $\theta$ the flag rotates by $2\theta$, which leads to the above-mentioned sign ambiguity. This sign ambiguity cannot be removed by any local or canonical geometric interpretation in $\mathbb{V}$; it will be discussed later in this section.

An essential requirement on any geometrical picture of $(\xi, \eta)$ is that it be independent of the coordinates used. If $(\tilde{\xi}, \tilde{\eta})$ is obtained from $(\xi, \eta)$ by a spin transformation corresponding – via (1.2.24) – to a *passive* Lorentz transformation of the Minkowski coordinates, then the abstract spin-

* We do not discuss the possible different ways in which the relationship between a spinor and its geometric realization may be affected by a space (or time) reflection. See Staruszkiewicz (1976).

vector $\boldsymbol{\kappa}$ represented by $(\xi, \eta)$ must remain unchanged, as must its geometrical representation. Thus if $(\xi, \eta)$ determines the geometrical representation of $\boldsymbol{\kappa}$ in the first Minkowski coordinate system, $(\tilde{\xi}, \tilde{\eta})$ must determine precisely the same structure in the second system. Note that we are here concerned with invariance under *passive* transformations. In the last two sections we examined the 2-1 local isomorphism between the spin group $SL(2, \mathbb{C})$ and the restricted Lorentz group, taking the groups as active; but the same isomorphism holds for the passive transformations, by the remarks at the end of §1.1, and so the detailed results apply also to passive transformations, *mutatis mutandis.*

### *Description on $\mathscr{S}^+$*

We start by showing how a geometrical picture of $(\xi, \eta)$ can be obtained in $\mathscr{S}^+$, the space of future null directions, and then we represent this in $\mathbb{V}$. As before, we label the points of $\mathscr{S}^+$ by the complex numbers $\zeta = \xi/\eta$ (with $\zeta = \infty$ for $\eta = 0$). We shall show that not merely the ratio $\xi : \eta$ but also $\xi$ and $\eta$ individually (up to a common sign) can be represented, in a natural way, by picking, in addition to the null direction $P$ (labelled by $\zeta$), a real tangent vector $\mathbf{L}$ of $\mathscr{S}^+$ at $P$ (*cf.* Fig. 1-10). To span the space of derivatives of real functions on $\mathscr{S}^+$ (which is an image of the Argand plane), we need the real and imaginary parts of $\partial/\partial\zeta$. Now a real vector $\mathbf{L}$ on $\mathscr{S}^+$ (except at the coordinate singularity $\zeta = \infty$ – for that point we need to replace $\zeta$ by another coordinate, e.g., $1/\zeta$) can be represented by a linear differential operator*

$$\mathbf{L} = \lambda\partial/\partial\zeta + \bar{\lambda}\partial/\partial\bar{\zeta}, \qquad (1.4.1),$$

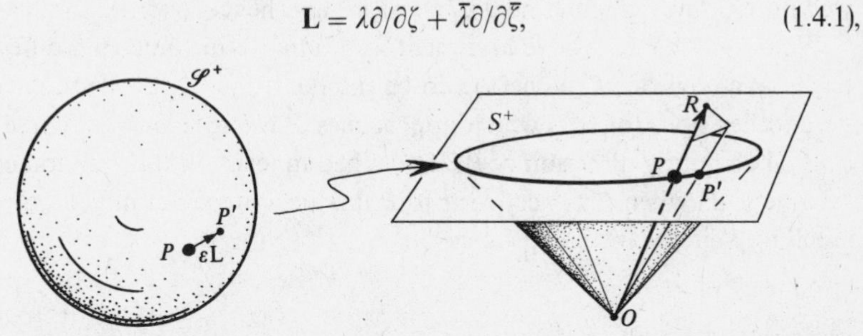


Fig. 1-10. The representation of a spin-vector in terms of a pair of infinitesimally separated points on $\mathscr{S}^+$ or, equivalently, by a null flag.

* A differential operator is a standard representation of a vector, which has become increasingly popular in differential geometry. It automatically incorporates all transformation properties. This representation will play an important role in Chapter 4.

the coefficients being chosen so as to make **L** real. We shall require $\lambda$ to be some definite expression in $\xi$ and $\eta$ so that after applying the (passive) spin transformation

$$\tilde{\xi} = \alpha\xi + \beta\eta, \quad \tilde{\eta} = \gamma\xi + \delta\eta, \quad \tilde{\zeta} = \frac{\alpha\zeta + \beta}{\gamma\zeta + \delta}, \tag{1.4.2}$$

we get

$$\tilde{\lambda}\frac{\partial}{\partial\tilde{\zeta}} + \bar{\tilde{\lambda}}\frac{\partial}{\partial\bar{\tilde{\zeta}}} = \lambda\frac{\partial}{\partial\zeta} + \bar{\lambda}\frac{\partial}{\partial\bar{\zeta}}, \tag{1.4.3}$$

where $\tilde{\lambda}$ is the same expression in $\tilde{\xi}, \tilde{\eta}$ that $\lambda$ is in $\xi, \eta$. From (1.4.2) we get

$$\begin{aligned}\frac{\partial}{\partial\zeta} &= \left\{\frac{\alpha(\gamma\zeta + \delta) - \gamma(\alpha\zeta + \beta)}{(\gamma\zeta + \delta)^2}\right\}\frac{\partial}{\partial\tilde{\zeta}} \\ &= (\gamma\zeta + \delta)^{-2}\frac{\partial}{\partial\tilde{\zeta}} = \eta^2\tilde{\eta}^{-2}\frac{\partial}{\partial\tilde{\zeta}},\end{aligned} \tag{1.4.4}$$

since $\alpha\delta - \beta\gamma = 1$ (by (1.2.18)). Substituting in (1.4.3), we find

$$\tilde{\lambda}\tilde{\eta}^2 = \lambda\eta^2, \tag{1.4.5}$$

and so we must choose $\lambda$ to be some numerical multiple of $\eta^{-2}$. For later convenience we make the choice $\lambda = -(1/\sqrt{2})\eta^{-2}$, which gives

$$\mathbf{L} = -\frac{1}{\sqrt{2}}\left(\eta^{-2}\frac{\partial}{\partial\zeta} + \bar{\eta}^{-2}\frac{\partial}{\partial\bar{\zeta}}\right). \tag{1.4.6}$$

Conversely, if we know **L** at $P$ (as an operator), we know the pair $(\xi, \eta)$ completely up to an overall sign. This is seen from the expression (1.4.6): knowing **L** and comparing coefficients, we can find $\eta^2$; and knowing $P$, we know $\zeta$. Thus we can find $\xi^2$, $\xi\eta$ and $\eta^2$, and hence $\pm(\xi, \eta)$.

We can couch the above argument for finding **L** in somewhat different terms. We consider $P$, as before, to be the point of $\mathscr{S}^+$ labelled $\zeta$. Let $P'$ be another point of $\mathscr{S}^+$ which approaches $P$ along a smooth curve on $\mathscr{S}^+$. The limiting direction of $PP'$ is defined in terms of the location of $P'$ relative to $P$, when $P'$ is very near to $P$. Let us write the complex number labelling $P'$ in the form

$$\zeta' = \zeta - \frac{1}{\sqrt{2}}\frac{\varepsilon}{\eta^2} \tag{1.4.7}$$

when $P'$ is near to $P$, $\varepsilon$ being a small positive quantity whose square is to be neglected. By a simple calculation parallel to (1.4.4) one verifies that under the spin transformation (1.4.2),

$$\tilde{\zeta}' = \tilde{\zeta} - \frac{1}{\sqrt{2}}\frac{\varepsilon}{\tilde{\eta}^2} \tag{1.4.8}$$

is required for the invariance of the construction. Taking $\varepsilon > 0$ as arbitrarily 'given', $P$ and $P'$ define $\zeta$ and $\eta^{-2}$. Hence (as before), $P$ and $P'$ define $\pm(\xi, \eta)$ (where $\zeta = \infty$ may be treated as a limiting case). The ordered pair of 'neighbouring points' $P$, $P'$ on $\mathscr{S}^+$ defines essentially the same situation as the point $P$ together with the tangent vector $\mathbf{L}$ at $P$ (*cf.* Fig. 1-10). For, the vector $\mathbf{L}$ is

$$\mathbf{L} = \lim_{\varepsilon \to 0} \frac{1}{\varepsilon} \overrightarrow{PP'}. \tag{1.4.9}$$

We can see this by observing that, for any $f(\zeta, \bar{\zeta})$,

$$\lim_{\varepsilon \to 0} \left\{ \frac{1}{\varepsilon}(f_{P'} - f_P) \right\} = \lim_{\varepsilon \to 0} \left\{ \frac{1}{\varepsilon} \left[ f\left( \zeta - \frac{\varepsilon}{\sqrt{2}\eta^2}, \bar{\zeta} - \frac{\varepsilon}{\sqrt{2}\bar{\eta}^2} \right) - f(\zeta, \bar{\zeta}) \right] \right\}$$
$$= -\frac{1}{\sqrt{2}} \left( \eta^{-2} \frac{\partial}{\partial \zeta} + \bar{\eta}^{-2} \frac{\partial}{\partial \bar{\zeta}} \right) f = \mathbf{L} f,$$

by (1.4.6).*

### *Description in* $\mathbb{V}$

The tangent vector $\mathbf{L}$ in the abstract space $\mathscr{S}^+$ corresponds to a tangent vector $\boldsymbol{L}$ in the coordinate-dependent representation $S^+$ of $\mathscr{S}^+$. The operator expression for the (coordinate-dependent) vector $\boldsymbol{L}$ is formally the same as that (1.4.6) for the (coordinate-independent) vector $\mathbf{L}$:

$$\boldsymbol{L} = -\frac{1}{\sqrt{2}} \left( \eta^{-2} \frac{\partial}{\partial \zeta} + \bar{\eta}^{-2} \frac{\partial}{\partial \bar{\zeta}} \right) = L^{\mathbf{a}} \frac{\partial}{\partial x^{\mathbf{a}}}, \tag{1.4.10}$$

where $L^{\mathbf{a}}$ are the components** of $\boldsymbol{L}$ relative to the coordinates $x^{\mathbf{a}}$ of $\mathbb{V}$.

* Another invariant way of representing the pair $(\xi, \eta)$ on $\mathscr{S}^+$ is to use the differential form

$$\eta^2 \mathrm{d}\zeta = \eta \mathrm{d}\xi - \xi \mathrm{d}\eta$$

at $P$, since under spin transformations (*cf.* (1.4.4))

$$\tilde{\eta}^2 \mathrm{d}\tilde{\zeta} = \eta^2 \mathrm{d}\zeta.$$

The real part of this differential form (times $-\sqrt{2}$) gives a description essentially equivalent to the one in terms of $\mathbf{L}$ just given, but since the geometrical interpretation of forms is not quite so immediate as that of tangent vectors (*cf.* Chapter 4), we shall not pursue this further here.

We may remark that various possible ways of representing a spin-vector which are equivalent under restricted Lorentz transformations are not necessarily equivalent when the group is widened to include reflections or conformal re-scalings (*cf.* §§3.6 and 5.6). In each case a choice may have to be made as to the behaviour of the spin vector under these additional transformations (e.g., conformal weight, signs, or possible factors of i).

** Henceforth we conform to the usual practice and refer to the coordinates of a vector as 'components'.

The difference arises because in (1.4.10) we interpret the operators $\partial/\partial\zeta$ and $\partial/\partial\bar{\zeta}$ as acting on functions defined on $\mathbb{V}$ instead of on $\mathscr{S}^+$, and two extra constraints are being imposed on the coordinates to define the subspace $S^+$ of $\mathbb{V}$ (say $T = 1 = X^2 + Y^2 + Z^2$).

For later reference we calculate the components $L^{\mathbf{a}}$ explicitly, using (1.2.8):

$$x^0 = 1, x^1 = \frac{\zeta + \bar{\zeta}}{\zeta\bar{\zeta} + 1}, x^2 = \frac{\zeta - \bar{\zeta}}{\mathrm{i}(\zeta\bar{\zeta} + 1)}, x^3 = \frac{\zeta\bar{\zeta} - 1}{\zeta\bar{\zeta} + 1}, \tag{1.4.11}$$

whence

$$\frac{\partial x^0}{\partial \zeta} = 0, \frac{\partial x^1}{\partial \zeta} = \frac{1 - \bar{\zeta}^2}{(\zeta\bar{\zeta} + 1)^2}, \frac{\partial x^2}{\partial \zeta} = \frac{1 + \bar{\zeta}^2}{\mathrm{i}(\zeta\bar{\zeta} + 1)^2}, \frac{\partial x^3}{\partial \zeta} = \frac{2\bar{\zeta}}{(\zeta\bar{\zeta} + 1)^2}; \tag{1.4.12}$$

and since, by (1.4.10),

$$L^{\mathbf{a}}\frac{\partial}{\partial x^{\mathbf{a}}} = -\frac{1}{\sqrt{2}}\left(\eta^{-2}\frac{\partial x^{\mathbf{a}}}{\partial \zeta} + \bar{\eta}^{-2}\frac{\partial x^{\mathbf{a}}}{\partial \bar{\zeta}}\right)\frac{\partial}{\partial x^{\mathbf{a}}}, \tag{1.4.13}$$

we find

$$L^0 = 0, \quad L^1 = \frac{\xi^2 + \bar{\xi}^2 - \eta^2 - \bar{\eta}^2}{\sqrt{2}(\xi\bar{\xi} + \eta\bar{\eta})^2},$$

$$L^2 = \frac{\xi^2 - \bar{\xi}^2 + \eta^2 - \bar{\eta}^2}{\sqrt{2}\mathrm{i}(\xi\bar{\xi} + \eta\bar{\eta})^2}, \quad L^3 = \frac{-\sqrt{2}(\xi\eta + \bar{\xi}\bar{\eta})}{(\xi\bar{\xi} + \eta\bar{\eta})^2}. \tag{1.4.14}$$

From this we can calculate the norm of the (evidently spacelike) vector $\boldsymbol{L}$ (*cf.* (1.1.13)):

$$\| \boldsymbol{L} \| = L^{\mathbf{a}}L^{\mathbf{b}}\eta_{\mathbf{ab}} = \frac{-2}{(\xi\bar{\xi} + \eta\bar{\eta})^2}, \tag{1.4.15}$$

so that $\boldsymbol{L}$ is a unit vector if and only if $\boldsymbol{K}$ (the null vector corresponding to $\xi, \eta$ via (1.2.15)) defines a point $P$ actually on $S^+$ (i.e. $K^0 = T = 1$). In fact, the 'length' $(-\|\boldsymbol{L}\|)^{\frac{1}{2}}$ of $\boldsymbol{L}$ varies inversely as the extent* of $\boldsymbol{K}$, i.e. inversely as the ratio of $\boldsymbol{K}$ to $\overrightarrow{OP}$ (*cf.* (1.2.14), (1.2.15)). We shall write

$$\boldsymbol{K} = (-\|\boldsymbol{L}\|)^{-\frac{1}{4}}\,\boldsymbol{k}, \tag{1.4.16}$$

where $\boldsymbol{k}$ defines a point on $S^+$. Note also that whereas we might have envisaged difficulty with the definition (1.4.10) for $\boldsymbol{L}$ at $\eta = 0, \zeta = \infty$, we

* The term 'extent' will frequently be used for null vectors, since their 'length' is always zero. The extent of a null vector cannot be characterized in an invariant way by a number, nor can null vectors of different directions be compared with respect to extent. The ratio of the extents of null vectors of the *same* direction is meaningful, being just the ratio of the vectors.

see from (1.4.14) that no such problem in fact arises since $\boldsymbol{L}$ is still well-defined at $\eta = 0$.

If $(\xi, \eta)$ and $(\tilde{\xi}, \tilde{\eta})$ are related by a (passive) spin transformation and we calculate $L^{\mathbf{a}}$ relative to the two corresponding Minkowski coordinate systems (using (1.4.14)) we shall in general find that they are *not* related by a Lorentz transformation. (This is clear, for example, from the fact that $\boldsymbol{L}$, being tangent to $S^+$, is necessarily orthogonal to the coordinate $t$-axis, a relation that is *not* generally preserved by a Lorentz transformation.) Nevertheless the plane $\hat{\Pi}$ of $\boldsymbol{K}$ and $\boldsymbol{L}$ *is* invariant, i.e., coordinate-independent, and thus a geometric structure in $\mathbb{V}$. This becomes clear when we remember that $\boldsymbol{L}$ corresponds to a tangent vector in $\mathscr{S}^+$ and so to two infinitesimally close null directions, one of which is $\boldsymbol{K}$. The plane of these null directions is evidently $\hat{\Pi}$.

Now $\hat{\Pi}$ is given by the set of vectors

$$a\boldsymbol{K} + b\boldsymbol{L}, (a, b \in \mathbb{R}), \tag{1.4.17}$$

and therefore has the required invariance. To give significance to the *sense* of $L$ we shall stipulate

$$b > 0, \tag{1.4.18}$$

which makes (1.4.17) into a *half*-plane, say $\Pi$, bounded by $\boldsymbol{K}$. This is the flag we have been looking for. Together with $\boldsymbol{K}$ it determines $(\xi, \eta)$ up to sign. For, knowing $\boldsymbol{K}$ we know $\xi$ and $\eta$ up to common phase, and knowing the direction of $\boldsymbol{L}$ we can get the phase of $\eta$ (and thus of $\xi$) from (1.4.6). Note that $\boldsymbol{L}$ is spacelike and orthogonal to $\boldsymbol{K}$ (being tangent to $S^+$ it has zero time component, and its spatial part is evidently orthogonal to that of $\boldsymbol{K}$). Hence $\Pi$ is one-half of a *null* 2-plane (see next paragraph), i.e. it is tangent to the null cone. It must touch the null cone along the line through $\boldsymbol{K}$. All directions in $\Pi$, other than $\boldsymbol{K}$, are spacelike and orthogonal to $\boldsymbol{K}$. We shall refer to $\Pi$ and $\boldsymbol{K}$ as a *null flag* or simply as a flag. The vector $\boldsymbol{K}$ will be called the *flagpole*, its direction the *flagpole direction*, and the half-plane $\Pi$ the *flag plane*.

We digress briefly to recall some general properties of null planes. Any real plane

$$a\boldsymbol{U} + b\boldsymbol{V} \tag{1.4.19}$$

spanned by two 4-vectors $\boldsymbol{U}$ and $\boldsymbol{V}$ contains at most two real null directions, given by

$$a^2 \| \boldsymbol{U} \| + b^2 \| \boldsymbol{V} \| + 2ab\boldsymbol{U} \cdot \boldsymbol{V} = 0. \tag{1.4.20}$$

When these null directions coincide, the plane is called null. In that case

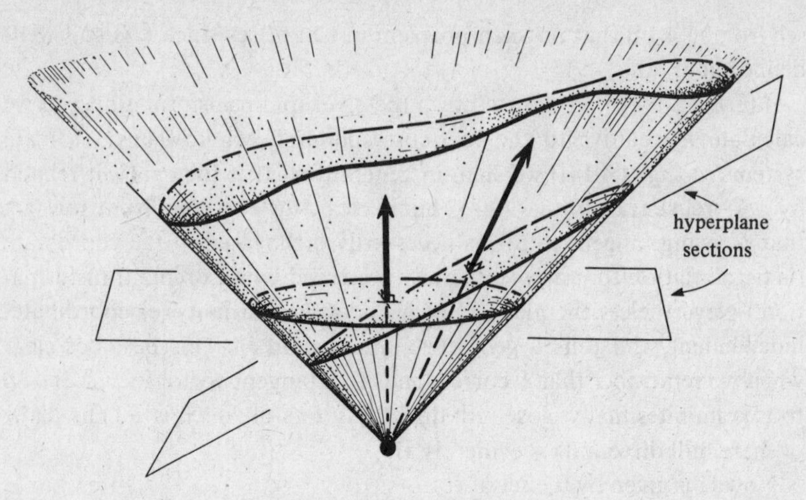


Fig. 1-11. Cross-sections of the null cone are mapped conformally to one another by the generators of the cone. This provides $\mathscr{S}^+$ with a conformal structure. Sphere metrics arise as cross-sections by spacelike hyperplanes. The various unit sphere metrics on $\mathscr{S}^+$ compatible with its conformal structure correspond to the different choices of unit timelike vector (i.e., normal to the hyperplane).

suppose $\boldsymbol{U}$ is the unique null direction in it; (1.4.20) then shows that there is no other null direction in the plane only if $\boldsymbol{U}\cdot\boldsymbol{V}=0$, i.e. *every* other vector $\boldsymbol{V}$ in the plane must be orthogonal to $\boldsymbol{U}$. And since no two distinct causal directions can be orthogonal (*cf.* after (1.1.17)), every such vector $\boldsymbol{V}$ must be spacelike. The angle $\theta$ between two null planes with common null vector, say $\boldsymbol{U}$, can be defined as that between *any* two non-null vectors, one in each plane; for suppose $\boldsymbol{V}$, $\boldsymbol{W}$ are two such vectors, then $(a\boldsymbol{U}+b\boldsymbol{V})\cdot(c\boldsymbol{U}+\mathrm{d}\boldsymbol{W})=b\mathrm{d}\boldsymbol{V}\cdot\boldsymbol{W}$, whence $\cos\theta=\pm\boldsymbol{V}\cdot\boldsymbol{W}/(\|\boldsymbol{V}\|\,\|\boldsymbol{W}\|)^{-\frac{1}{2}}$, independently of $a, b, c, d$. An interesting corollary is that *any* cross-section $T$ of the null cone in $\mathbb{V}$, even one achieved by cutting the null cone with a *curved* hypersurface, is conformally identical to every other, and therefore to $S^+$, where corresponding points lie on the same generator of the null cone (*cf.* Fig. 1-11). This result can be seen in many ways but, in particular, by considering the infinitesimal triangle obtained by cutting three given neighbouring generators of the null cone by *any* hyperplane element. Since neighbouring generators lie on a null plane, our preceding result shows that all these infinitesimal triangles will be similar. The corollary is therefore established. We have already seen an example of it in the use of the hyperplane (1.2.12).

We now examine a little more closely the geometric role of the 'magnitude' of the vector $\boldsymbol{L}$. We cannot attach significance to the norm $\|\boldsymbol{L}\|$, since this would attach unwanted significance to the metric of $S^+$. (We saw above that $\|\boldsymbol{L}\| = -1$ was the condition for $\boldsymbol{K} = \overrightarrow{OP}$ with $P$ on $S^+$.) We have to envisage $\boldsymbol{L}$, at $P$, as 'joined' to the origin $O$. Thus, if we replace $P$ by some other point $R$ on $OP$, to the future of $O$, with must rescale $\boldsymbol{L}$ by the factor $OR/OP$ (Fig. 1-10). So $\boldsymbol{L}$ at $P$ is 'equivalent' to $(OR/OP)\boldsymbol{L}$ at $R$. We can choose $R$ so that $(OR/OP)\boldsymbol{L}$ is a unit vector*. Then $\overrightarrow{OR} = \boldsymbol{K}$, since this gives $\boldsymbol{K} = (-\|\boldsymbol{L}\|)^{-\frac{1}{2}}\,\overrightarrow{OP}$, which is the same as (1.4.16). Thus, the magnitude of $\boldsymbol{L}$ simply locates $\boldsymbol{K}$ (in the direction $OP$). In terms of $P$ and $P'$, we can picture this, intuitively, in the following way. We consider neighbouring null lines $OP$, $OP'$. Then we locate $R$ on $OP$ by proceeding along the line $OP$ until the distance to the neighbouring line attains the value $\varepsilon$. Note that the 'closer' together are $OP$ and $OP'$, the greater will be the extent of $OR$.

The labelling of flags by pairs $(\xi, \eta)$ does not specifically depend on the choice of a *Minkowski* coordinate system for $\mathbb{V}$. (In fact, certain other types of coordinate systems would have led to rather simpler formulae than the ones we have used.) The assignment of a pair $(\xi, \eta)$ to a null flag can be made much more directly once we have the concept of a *spin-frame* available. The coordinate system for $\mathbb{V}$ will then be seen as an irrelevance. Essentially a spin-frame is defined when the flags corresponding to the pairs $(1, 0)$ and $(0, 1)$ are known, but there is a slight difficulty concerning the sign ambiguity. This is removed when the concept of a *spin-vector* is introduced, which we shall do in a moment. The details of the necessary geometric operations will then be given in §1.6.

## *Spin-vectors*

To understand the passage from the concept of flag to that of spin-vector, we must appreciate the essential nature of the sign ambiguity in the representation of a null flag by a pair $(\xi, \eta)$. Let us, for this purpose, examine the effect on a null flag of transformations of the form

$$(\xi, \eta) \mapsto (\lambda\xi, \lambda\eta), \tag{1.4.21}$$

where $\lambda$ is some non-zero complex number. These are the transformations which leave the flagpole direction invariant but may alter the extent of the flagpole or the direction of the flag plane. Set

$$\lambda = r\mathrm{e}^{\mathrm{i}\theta}, \tag{1.4.22}$$

where $r, \theta \in \mathbb{R}$ and $r > 0$. Then in the particular case $\theta = 0$ (i.e., $\lambda$ real),

* Strictly speaking, we must allow that $\boldsymbol{L}$ may also have multiples of $\boldsymbol{K}$ added into it. But this makes no difference to the norm since $\boldsymbol{K}$ is null and orthogonal to $\boldsymbol{L}$.

(1.4.21) gives no change in the flag plane, while the extent of the flagpole is increased by a factor $r^2$; on the other hand, if $r = 1$ (i.e. $\lambda$ of unit modulus), (1.4.21) gives no change in the flagpole but the flag plane rotates through an angle $2\theta$ in the positive sense. This may be seen perhaps most easily if we use the representation in terms of the infinitesimally separated points $P, P'$ on $S^+$. Then $P$ is given by $\zeta$ and $P'$ by $\zeta - 2^{-\frac{1}{2}}\varepsilon\eta^{-2}$. Under (1.4.21) we have $\eta \to \lambda\eta$, hence $\eta^{-2} \to r^{-2}e^{-2i\theta}\eta^{-2}$. Since the extent of the flagpole varies inversely as the infinitesimal separation $PP'$, we have the first part of the above assertion. The second part is seen to follow when we recall that $S^+$ is obtained from the Argand plane of $\zeta$ by a conformal stereographic projection.

Now let us apply a continuous rotation $(\xi, \eta) \to (e^{i\theta}\xi, e^{i\theta}\eta)$ where $\theta$ varies from 0 to $\pi$. We end up with

$$(\xi, \eta) \mapsto (-\xi, -\eta), \tag{1.4.23}$$

but the *flag* is returned to its original position, the flag plane having been rotated through $2\pi$ (i.e., once completely about the flagpole). If we continue the rotation, so that $\theta$ further varies from $\pi$ to $2\pi$, then the original pair $(\xi, \eta)$ is obtained once more. Thus, a rotation of the flag plane through $4\pi$ is required in order to restore $(\xi, \eta)$ to its original state. Such considerations imply that a complete local geometrical representation, in $\mathbb{V}$, of $(\xi, \eta)$, which takes into account its overall sign, is *not possible*. Every local structure in the Minkowski space $\mathbb{V}$ that we might adjoin to the null flag would also be rotated through $2\pi$ and hence returned to its original state, while $(\xi, \eta)$ undergoes (1.4.23). To see this most clearly, we observe first that for any *particular* pair $(\xi, \eta)$ we can achieve $(\xi, \eta) \mapsto (e^{i\theta}\xi, e^{i\theta}\eta)$ by a spin transformation which corresponds to a rotation for which the flagpole direction is an invariant null direction. (For simplicity we could choose $(\xi, \eta) = (0, 1) \mapsto (0, e^{i\theta})$ and use (1.2.31).) As $\theta$ varies continuously from 0 to $\pi$, the spin transformation varies continuously (provided the rotation axis is kept fixed) and takes the final value $-\mathbf{I}$. The corresponding Lorentz transformations also vary continuously, but end up at the identity Lorentz transformation. Thus, *any* geometrical structure in $\mathbb{V}$ would be rotated into its original state by this succession of Lorentz transformations, even though $(\xi, \eta)$ is 'rotated' into $(-\xi, -\eta)$.

Once it is accepted that a complete local geometrical representation in $\mathbb{V}$ is not possible, it becomes clearer what attitude should be adopted. Essentially we must *widen* the concept of geometry in $\mathbb{V}$, so that quantities can be admitted as 'geometrical' which are not returned to their original state when rotated through an angle $2\pi$ about some axis; when rotated through $4\pi$, however, they *must* be returned to their original state. Such quantities will be referred to as *spinorial objects*. A spin-vector differs from a null flag only in that it is a spinorial object, and to each null flag there correspond exactly *two* spin-vectors. The next section is devoted to an elaboration of these ideas.

## 1.5 Spinorial objects and spin structure

To define a spinorial object in more precise terms, we must first consider some properties of ordinary rotations. The manifold of proper rotations in Euclidean 3-space is denoted by $SO(3)$. $SO(3)$ can also be used to represent the different orientations* of an object in space. If one such orientation is designated as the 'original' orientation – and represented by the identity element of $SO(3)$ – each other element represents the orientation which is obtained by acting on the original orientation with the rotation in question.

Any rotation is defined by an axis $\boldsymbol{k}$ and a right-handed turning through an angle $\theta$. It can therefore be represented by a vector of length $\theta$ in the direction of $\boldsymbol{k}$. Since we need only consider the range $0 \leqslant \theta \leqslant \pi$, every point of $SO(3)$ thus corresponds to a point of the closed ball $B$ of radius $\pi$. This correspondence is not unique, however, since a rotation about $\boldsymbol{k}$ through an angle $\pi$ is the same as a rotation about $-\boldsymbol{k}$ through an angle $\pi$. Thus opposite points of the boundary $S$ of $B$ must be identified, giving us a space $\hat{B}$ which represents rotations uniquely, and also continuously (i.e., points close in $\hat{B}$ represent rotations that differ little from each other).

We are interested in the topology, and especially in the connectivity properties of $\hat{B}$. A space is said to be *simply-connected* if every closed loop in it can be deformed continuously to a point. (This property obviously holds for an entire Euclidean space, for an ordinary spherical surface, or for a Euclidean 3-space with a point removed. It does not hold for the surface of a torus, for a circle, for a Euclidean 3-space with a closed curve removed, or for a Euclidean 2-space with a point removed.) A simply-connected space is alternatively characterized by the property that if $c_1$ and $c_2$ are two *open* curves connecting two points in the space, then $c_1$ can be continuously deformed into $c_2$.

The space $\hat{B}$ is in fact *not* simply-connected. The closed loops in $\hat{B}$ fall into two disjoint classes, I and II, according as they have an odd or even

* Not to be confused with our previous use of the word 'orientation' which means 'handedness'.

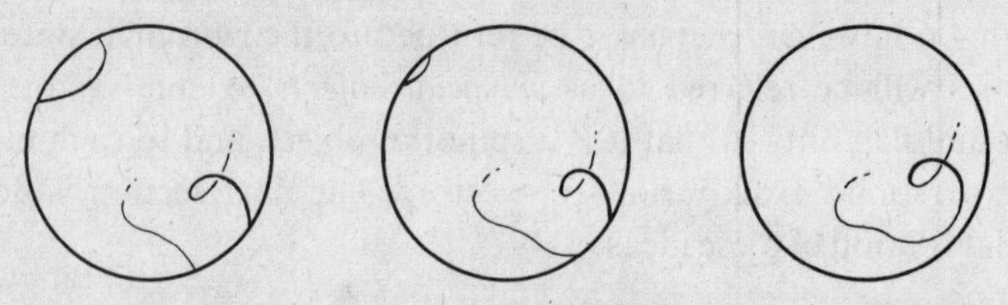

Fig. 1-12. $SO(3)$ is a closed 3-ball with opposite points of its boundary identified. Continuous deformation of a curve in $SO(3)$ can eliminate pairs of its intersections with this boundary.

number, respectively, of 'intersections' with the boundary $S$. An intersection occurs when a curve approaches $S$ and, by the identification of points, reappears diametrically opposite. Class I contains, for example, all diameters of $\hat{B}$. Class II contains all internal loops, and in particular the 'trivial' loops consisting of a single point. No loop of class I can be continuously deformed into a loop of class II, since intersection points with $S$ can appear or disappear in pairs only. On the other hand, all class I loops *can* be deformed into each other*, and the same holds for class II loops. This is because intersections with $S$ can be eliminated in pairs (see Fig. 1-12, which shows how to deal with one strand at a time), and internal loops can clearly be deformed into each other, as can loops which intersect the boundary once.

Now consider a continuous rotation of an object in Euclidean 3-space which takes that object back to its original orientation. This corresponds to a closed loop in $SO(3)$ (hence also in $\hat{B}$), which may be of class I or of class II. In the case of a simple rotation through $2\pi$ we evidently get a class I loop, whereas for a rotation through $4\pi$ we get a class II loop. It is clear from the above discussion that the rotation through $2\pi$ (where the *whole* motion must be considered, not just the initial and final orientations) cannot be continuously deformed into no motion at all, whereas the rotation through $4\pi$ *can*. That this is the case is by no means obvious without some such discussion as that given above in terms of $\hat{B}$.

There are numerous ways of illustrating this result. One way of performing a continuous deformation of a $4\pi$ rotation into no rotation is the following (due to H. Weyl). Consider a pair of right circular cones in space, of equal semi-angle $\alpha$, one cone being fixed while the other is free to roll on the fixed one, the vertices remaining in contact. Start with $\alpha$ very small and roll the moving cone once around the fixed cone: the moving cone executes

* One of the rules for such deformations is that different portions of a loop may be moved through one another. Hence the question of 'knots' does not arise.

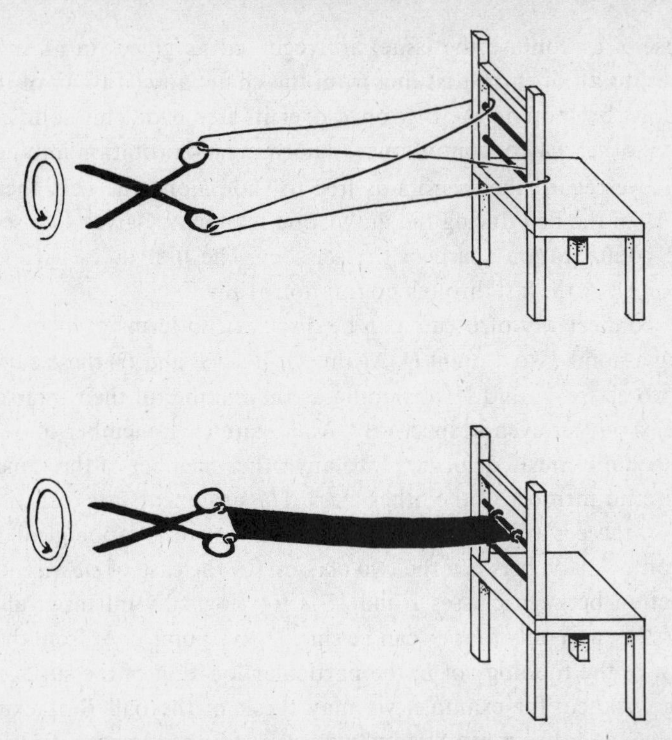

Fig. 1-13. Dirac's scissors problem: rotate the scissors through 720° and then untangle the string without moving the chair or rotating the scissors. With a belt it is easier.

essentially a rotation through $4\pi$. Now increase $\alpha$ gradually from 0 to $\pi/2$. For each fixed $\alpha$ we have a closed motion as the moving cone rolls once about the fixed cone. But when $\alpha$ nears $\pi/2$ the cones become almost flat and the motion becomes a mere wobble. So *at* $\alpha = \pi/2$ we get a 'trivial' loop in $SO(3)$, and the rotation through $4\pi$ has been continuously deformed into no motion at all.

In Dirac's well-known *scissors problem* a piece of string is passed through a finger hole of the scissors, then around a back strut of a chair, then through the other finger hole and around the other back strut, and then tied (see Fig. 1-13). The scissors are rotated through $4\pi$ about their axis of symmetry, and the problem is to disentangle the string without rotating the scissors or moving the chair. The fact that this problem can be solved for $4\pi$ but not* for $2\pi$ is a consequence of the above properties of $SO(3)$. The solution is made trivially easy if the four strands of string (whose main

* That it *cannot* be solved for $2\pi$ requires, strictly speaking, a more involved topological argument (see Newman 1942).

purpose is to confuse the issue) are regarded as glued (in an arbitrary manner) to an open belt issuing from the chair: a twist of $4\pi$ of the belt is undone by looping the belt once over its free end. This solution also yields another way of continuously deforming a $4\pi$ rotation into no rotation. If we regard the scissors as free to slide along the belt, then each position of the belt during the untwisting manoeuvre gives a closed path in the configuration space of the scissors. The first takes it through a rotation of $4\pi$, the last through no rotation at all.

The connectivity of $\hat{B}$ can also be discussed in terms of 'open' curves, joining a point $P$ to a point $Q$. Again (for fixed $P$ and $Q$) these curves fall into two classes, I and II, according as the number of their intersections with $S$ is odd or even, respectively. And again each member of one class can be continuously deformed into any other member of the same class, but into no member of the other class. The arguments are just as above. But now there is a slight difference in that no intrinsic topological distinction can be made between the two classes. (In the case of *closed* loops the distinction between classes I and II *is* topologically intrinsic: all loops of class II – and only those – can be shrunk to a point.) For from the point of view of the topology of $\hat{B}$, the particular location of the surface $S$ has no significance; for example, we may think of the ball $B$ as extending beyond $S$, and then move $S$ radially outward in one direction and radially inward in the opposite direction. If a curve from $P$ to $Q$ intersects the initial location of $S$ once, it need not intersect the final location of $S$ at all. We note that two curves from $P$ to $Q$ belong to the *same* class if and only if the first followed by the second in reverse direction constitute a closed loop of class II.

In terms of the original Euclidean 3-space, the points $P$ and $Q$ correspond to two orientations $\mathscr{P}$ and $\mathscr{Q}$ of an object, and a path from $P$ to $Q$ in $\hat{B}$ corresponds to a continuous motion beginning with $\mathscr{P}$ and ending with $\mathscr{Q}$. We now see that there are two essentially different classes of continuous motions from $\mathscr{P}$ to $\mathscr{Q}$. The motions of each class can be continuously deformed into one another, but not into any motion of the other class. Nevertheless there need be nothing intrinsically to distinguish between one class and the other.

The particular feature of the topology of $SO(3)$ that we have been discussing concerns its *fundamental group* (or first homotopy group) $\pi_1(SO(3))$ (roughly: the group of topologically equivalent loops, in this sense). Here, this group has just *two* elements, so we have $\pi_1(SO(3)) = \mathbb{Z}_2$ (the group of integers mod 2).

### *Universal covering spaces*

An important concept is that of a *universal covering space* $\tilde{T}$ for a connected (but not necessarily simply-connected) topological space $T$. Choose a base point $O$ in $T$ and consider paths in $T$ from $O$ to some other point $X$. There may be several different classes of paths from $O$ to $X$, with the property that each path of one class can be continuously deformed into every other path of the same class but into no path of another class. To construct $\tilde{T}$ we, in effect, allow $X$ to acquire as many different 'identities' as there are such distinct classes. More precisely, the points of $\tilde{T}$ *are* these different classes associated with $X$, where $X$ itself varies over the whole of $T$, $O$ remaining fixed. The continuity properties of $\tilde{T}$ are defined in the obvious way from those of $T$. It is readily seen that $\tilde{T}$ is simply-connected and that – as a topological space – it is essentially independent of the choice of $O$. If $T$ is simply-connected then $\tilde{T}$ is identical with $T$.

For example, take $T$ to be a circle, whose points may be assigned coordinates $\theta$ in the usual way, modulo $2\pi$. The inequivalent paths from $O$ (the base point) to $\theta$ are distinguished by the different number of times that they 'wrap around' the circle. Evidently the space $\tilde{T}$ is parametrized by $\theta$ *without* the equivalence modulo $2\pi$, i.e., $\tilde{T}$ is topologically the real line. Similarly, if $T$ is an infinite cylinder, parametrized by $z$ and by $\theta$ modulo $2\pi$, $\tilde{T}$ is the entire plane parametrized by $z$ and an unrestricted $\theta$. Thus, to pass from $T$ to $\tilde{T}$, we maximally 'unwrap' the space $T$ – and this, indeed, is how we may view the situation in the general case.

Consider, in particular, the space $SO(3)$. There are now just *two* classes for each point of the space, so the universal covering space $\widetilde{SO(3)}$ is just a *twofold* unwrapping of $SO(3)$. A concrete realization of $\widetilde{SO(3)}$ has, in fact, already been obtained in §1.2, namely the space $SU(2)$ of $2 \times 2$ unimodular unitary matrices.* The correspondence referred to in Proposition (1.2.29) establishes precisely the required 1-2 relation between the spaces $SO(3)$ and $SU(2)$. In just the same way, Proposition (1.2.27) establishes the relation between the restricted Lorentz group $O_+^{\uparrow}(1, 3)$ and its (twofold) universal covering space $SL(2, \mathbb{C})$ of spin-matrices. The fact that the topology of $O_+^{\uparrow}(1, 3)$ is essentially no more complicated than that of $SO(3)$ follows from a property of restricted Lorentz transformations established in §1.3: every such transformation is uniquely

* From our discussion of quaternions in §1.2 we see that the topology of $SU(2)$ is the same as that of the space of unit quaternions. This is a 3-sphere $S^3$ (topologically a closed 3-ball whose boundary is identified as *one* point) – which, indeed, is simply-connected.

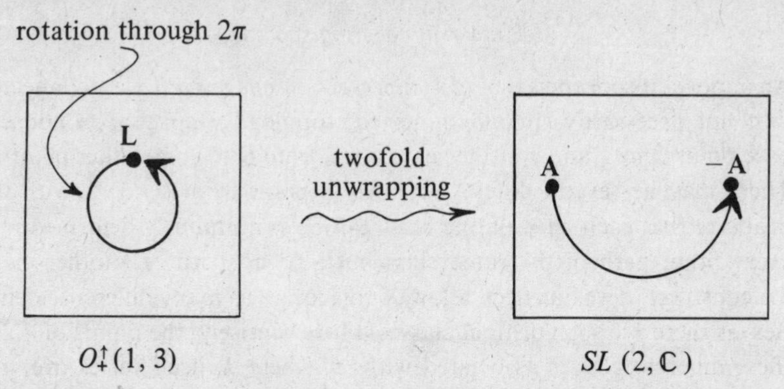


Fig. 1-14. The 1-2 relation between the restricted Lorentz group and $SL(2, \mathbb{C})$.

the product of a rotation with a boost. The topology of the boost is 'trivial' (i.e., that of Euclidean space $\mathbb{R}^3$, in which the rapidities are mapped as lengths in the relevant directions), so the topological properties of $O^{\uparrow}_{+}(1, 3)$ are essentially the same (except for dimension) as those of $SO(3)$.* The 1-2 relation between $O^{\uparrow}_{+}(1, 3)$ and $SL(2, \mathbb{C})$ is indicated in Fig. 1-14.

Consider now a geometrical structure (e.g., a rigid body, a flag, etc.) in Euclidean 3-space $E^3$, space–time, or Minkowski vector space $\mathbb{V}$. Let $\mathscr{C}$ denote the space of orientations $Q$ of that structure. We have shown how to construct a 'spinorized' version $Q_1$ of $Q$, provided the space $\mathscr{C}$ is such that it possesses a *twofold* universal covering space $\tilde{\mathscr{C}}$, and provided the two different images $Q_1, Q_2 \in \tilde{\mathscr{C}}$ of an element $Q \in \mathscr{C}$ are interchanged after a continuous rotation through $2\pi$ is applied to $Q$. For example, if the $Q$s are the orientations of a rigid (asymmetrical) body in $E^3$, $\mathscr{C}$ has the topology of $SO(3)$ and we have seen that $\widetilde{SO(3)}$ has the required properties. In the general case, the elements of $\tilde{\mathscr{C}}$ may be pictured as representing ordinary geometrical structures in space (–time), but with this additional feature, that a rotation through $2\pi$ about any axis (or any other motion continuously equivalent to this) will send the structure into something distinct, and a further rotation through $2\pi$ is needed to send the structure back to its original state. The elements of $\tilde{\mathscr{C}}$ will then be *spinorial objects*.** They are the required 'spinorized' version of the original $Q$s.

* Technically, $O^{\uparrow}_{+}(1, 3)$ is of the same homotopy type as $SO(3)$.

** Although from the point of view of conventional geometry, or of everyday experience, the concept of a spinorial object may seem strange, such objects do appear to have a physical reality. The states of electrons, protons or neutrons are examples of spinorial objects. According to Aharonov and Susskind (1967), genuine spinorial objects on a macroscopic scale are also possible in principle. Their (somewhat idealized) theoretical

*Definition of a spin-vector*

We are now in a position to give the geometrical definition of a spin-vector. For that, we take the $Q$s to be null flags in $\mathbb{V}$, and $\mathscr{C}$ the space of null flags. We must verify that this $\mathscr{C}$ does indeed have the topological properties required of it. Since it is four-dimensional, it cannot be topologically identical with $SO(3)$ (three-dimensional) or $O^{\uparrow}_{+}(1, 3)$ (six-dimensional). Nevertheless, as was the case with $O^{\uparrow}_{+}(1, 3)$, the *essential* part of its topology is the same as that of $SO(3)$ (We have $\mathscr{C} \cong SO(3) \times \mathbb{R}$, and consequently $\pi_1(\mathscr{C}) = \mathbb{Z}_2$). To see this, we can think in terms of the representation on $S^+$. Each element $Q$ of $\mathscr{C}$ is represented by a point $P$ of $S^+$ and a non-zero tangent vector $\boldsymbol{L}$ to $S^+$ at $P$. We can associate a Cartesian frame continuously (but not invariantly) with $Q$ by choosing the $z$-axis from the origin to $P$, the $x$-axis parallel to $\boldsymbol{L}$, and the $y$-axis so as to complete the frame. Such a frame uniquely corresponds to points of $SO(3)$. The only remaining parameter defining $Q$ is $\|\boldsymbol{L}\|$ which is a positive real number and topologically trivial, so it is seen that $\mathscr{C}$ has the properties required of it.

The elements of the space $\tilde{\mathscr{C}}$ are therefore the spinorized null flags, which we identify with the (non-zero) *spin-vectors* of $\mathbb{V}$. Each null flag $Q$ defines *two* associated spin-vectors, which we label $\boldsymbol{\kappa}$ and $-\boldsymbol{\kappa}$. A continuous rotation through $2\pi$ will carry $\boldsymbol{\kappa}$ into $-\boldsymbol{\kappa}$, and since on repeating the process $-\boldsymbol{\kappa}$ is carried back to $\boldsymbol{\kappa}$, we have

$$-(-\boldsymbol{\kappa}) = \boldsymbol{\kappa},$$

as suggested by the notation. In addition, there is a unique *zero spin-vector*, written $\mathbf{0}$, which does not correspond to a flag. It is associated with the zero world-vector, which is its 'flagpole', but no 'flag plane' is defined.

The pair $(\xi, \eta)$ may genuinely be thought of as components for $\boldsymbol{\kappa}$. Spin transformations applied to $(\xi, \eta)$ will correspond to active motions which carry $\boldsymbol{\kappa}$ about the space $\mathbb{V}$. A continuous rotation of $\boldsymbol{\kappa}$ through $2\pi$ corresponds to a succession of spin transformations acting on $(\xi, \eta)$ which end up with $(-\xi, -\eta)$. Thus $(-\xi, -\eta)$ will in fact be the components of $-\boldsymbol{\kappa}$, as the notation implies.

---

apparatus involves splitting wave functions of a large number of electrons between two containers $A$, $B$ which fit together in a unique relative orientation. When together, a current flows from $A$ to $B$, but after they are separated, and $B$ rotated relative to $A$ through $2\pi$, the current then flows from $B$ to $A$ when they are re-united. After a further relative rotation through $2\pi$ the original direction of the current from $A$ to $B$ is restored. A corresponding effect on the scale of single neutrons in a split beam has actually been observed experimentally. (Bernstein 1967, Klein and Opat 1975, Rauch *et al.* 1975, Werner *et al.*, 1975.)

### *Spinor structure*

Before proceeding to a detailed discussion of the basic operations on spin-vectors, we briefly consider how the above arguments are affected by the *global* topology of a general curved space–time manifold $\mathcal{M}$. The topology we have been concerned with until now arose from *local* considerations in space–time. (For example, we discussed the topology of the space of null flags at *one* space–time point.) But certain space–time manifolds themselves have a non-trivial (i.e. non-Euclidean) topology, which must be considered together with local topological properties. Indeed, the question arises of what restrictions to put on a manifold for it to allow objects like spin-vectors to be defined globally.

We shall not, at this stage, enter into the precise definition of a space–time manifold, beyond saying that locally its structure is that of Minkowski space – i.e. it has a Lorentzian metric – and that it is an ordinary (i.e., Hausdorff, paracompact, connected) $C^\infty$ 4-manifold (see Chapter 4 for a definition of these concepts).

To fix ideas, let us consider a space $\mathcal{F}$, each point of which represents a null flag at a point of $\mathcal{M}$. This space is called the *null-flag bundle* of $\mathcal{M}$.

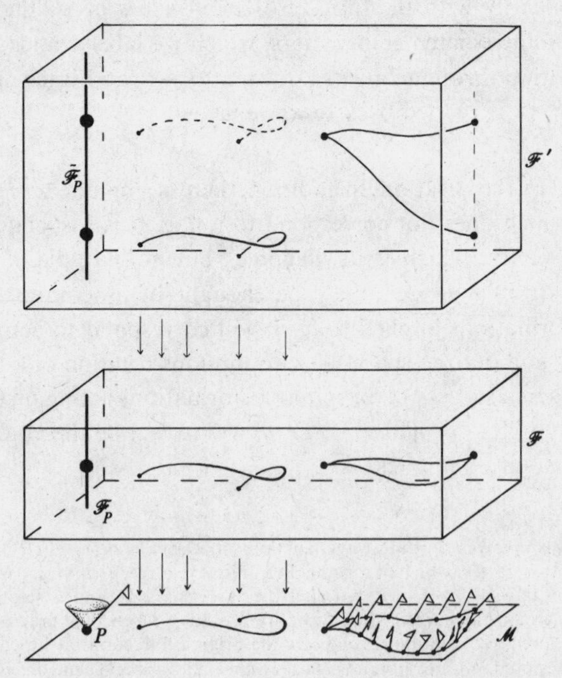


Fig. 1-15. The null-flag bundle $\mathcal{F}$ of $\mathcal{M}$, and its twofold covering space, the spin-vector bundle $\mathcal{F}'$.

(See Fig. 1-15 and *cf.* §5.4.) It is an 8-dimensional space, since $\mathscr{M}$ itself is 4-dimensional, and the space $\mathscr{F}_P$ of null flags at any one point $P$ of $\mathscr{M}$ (previously denoted by $\mathscr{C}$) is also 4-dimensional. The null flags at $P$ are thought of as structures in the *tangent space* (*cf.* Chapter 4) at $P$ – which is a Minkowski vector space. Now, already for the existence of $\mathscr{F}$, there are two global restrictions on $\mathscr{M}$. In the first place, null flags are associated with only one of the two null half-cones in the tangent space at $P$, namely that labelled *future*-pointing. Thus it is necessary to be able to make a consistent continuous choice, over the whole of $\mathscr{M}$, of null half-cones. In other words,

$$\mathscr{M} \textit{ must be time-orientable.} \tag{1.5.1}$$

In the second place, the algebra for spin-vectors requires a choice of space–time orientation at each point, since multiplication by $e^{i\theta}$ must rotate the null flags in a particular sense. That this requires a space–time orientation, rather than a space orientation, can be seen from the fact that the positive rotation of the null flags assigns a positive orientation to $S^+$, and, correspondingly, a negative orientation to $S^-$ (*cf.* §§1.2, 1.4, 3.2) We must therefore be able to make a consistent continuous choice, over the whole of $\mathscr{M}$, of a space–time orientation. Thus,

$$\mathscr{M} \textit{ must be space–time-orientable.} \tag{1.5.2}$$

But if we want to pass from the concept of null flag to that of spin-vector, these two global requirements on $\mathscr{M}$ are not sufficient. $\mathscr{M}$ must also permit a *spin structure** to be defined on it, which means, roughly speaking, a prescription for keeping track of the sign of a spin-vector not only if we move it around at a fixed point of $\mathscr{M}$, but also if we move it around from point to point within $\mathscr{M}$. If $\mathscr{M}$ is topologically trivial, this spin structure exists and is unique. But if topologically non-trivial, $\mathscr{M}$ may or may not permit a consistent spin structure, and if it does, the possible spin structures may or may not be unique. It turns out, generally, that (assuming (1.5.1) and (1.5.2) hold) the conditions on $\mathscr{M}$ for existence and uniqueness of spin structure depend only on the *topology* of $\mathscr{M}$ and not on the nature of its (Lorentzian) metric. We shall see shortly (*cf.* (1.5.4); also (1.5.6)) the precise topological condition on $\mathscr{M}$ that is required.

In accordance with our earlier discussion, we require $\mathscr{F}$ to possess

* It should be emphasized that the question of the existence of spin structure on a manifold $\mathscr{M}$ is not the same question as that of the existence of certain (e.g. non-vanishing) spinor fields on $\mathscr{M}$. The latter is analogous to the question of whether there is a non-vanishing vector field on a 2-sphere. But without spin structure, the very concept of a global spinor field does not exist.

an appropriate twofold covering space $\mathscr{F}'$, which will in fact be the space of spin-vectors on $\mathscr{M}$. (Unlike the universal covering space, a general covering space need merely be connected and map to the original space so that the local topology is preserved and the inverse map of a point is a discrete set of points.) $\mathscr{F}'$ must be 'appropriate' in the sense of reducing to $\tilde{\mathscr{F}}_P$, the universal covering space of $\mathscr{F}_P$, above each point $P$ of $\mathscr{M}$. We might expect that the universal covering space $\tilde{\mathscr{F}}$ of $\mathscr{F}$ will normally achieve this (i.e., $\mathscr{F}' = \tilde{\mathscr{F}}$), but since the complete 'unwrapping' of $\mathscr{F}$ also involves unwrapping $\mathscr{M}$, this may not be what is required. Moreover the situation is more complicated than this. We shall find that in fact there are two somewhat different *obstructions* to the existence of $\mathscr{F}'$ which can occur. The first of these can arise whether or not $\mathscr{M}$ is simply-connected, while the second occurs only for non-simply-connected $\mathscr{M}$.

To investigate this, let us examine closed loops in $\mathscr{F}$, and their projections to $\mathscr{M}$. A projection from $\mathscr{F}$ to $\mathscr{M}$ maps each flag at $P$ to the point $P$; thus the whole of each $\mathscr{F}_P$ maps to the single point $P$. (See Fig. 1-15.) Any path in $\mathscr{F}$ projects to a path in $\mathscr{M}$; a closed loop in $\mathscr{F}$ evidently projects to a closed loop in $\mathscr{M}$. Each path in $\mathscr{F}$ corresponds to a motion which carries a null flag around in $\mathscr{M}$, and which finally returns it, in the case of a *closed* path, to the starting configuration. The projection simply describes the motion of the base point in $\mathscr{M}$.

A loop in $\mathscr{F}$ that lies entirely in $\mathscr{F}_P$, for some fixed $P$, projects to a 'trivial' loop (the point $P$) in $\mathscr{M}$. As we have seen earlier, there are precisely two classes (I and II) of closed loops in $\mathscr{F}_P$. The *first* kind of obstruction that can arise when the topology of $\mathscr{M}$ is suitably non-trivial is that these two classes can become united into one, which means that class I (i.e., nonshrinkable) loops in any one $\mathscr{F}_P$, after being deformed within $\mathscr{F}$, can return to $\mathscr{F}_P$ as class II (i.e., shrinkable) loops. In that case spin-vectors could not exist on $\mathscr{M}$. For suppose they exist, and consider a class I loop $\lambda$ in a given $\mathscr{F}_P$ which consists simply of a $2\pi$-rotation of the flag plane of some given null flag, and which therefore sends the corresponding spin-vector $\boldsymbol{\kappa}$ into $-\boldsymbol{\kappa}$. Every closed loop in $\mathscr{F}$ into which $\lambda$ can be transformed continuously, carries a nonzero spin-vector continuously into its negative. But if $\lambda$ can be continuously moved to become a single point on $\mathscr{F}$, we find that the spin-vector must equal its negative. Hence $\mathscr{M}$ does not admit a spin-vector concept.

Now assume that this first kind of obstruction is absent. Then a second kind of difficulty can sometimes arise when $\mathscr{M}$ contains an unshrinkable loop $\gamma$, i.e., when $\mathscr{M}$ is not simply-connected. If a null flag is carried around $\gamma$ to its starting position $P$, a corresponding spin-vector $\boldsymbol{\kappa}$ would have to be returned either to its starting value or to $-\boldsymbol{\kappa}$. Accordingly we must choose between these two alternatives. If $\gamma$ is such that no multiple $m\gamma$ (i.e., $\gamma$ traversed $m$ times) is shrinkable to a point, then both choices are equally valid, but lead to *different spin structures* for $\mathcal{M}$ (assuming that spin structure is not ruled out for other loops). The choice between the two alternatives is then part of the *definition* of what shall be meant by spin-vector. Once the choice is made for $\gamma$, it determines the choice for all loops in $\mathcal{F}$ which can be projected to $\gamma$ or to loops deformable to $\gamma$ through $\mathcal{M}$.

Next suppose that $\gamma$ is such that some *odd* multiple $m\gamma$ of it is shrinkable to a point within $\mathcal{M}$. Then for any loop $\lambda$ in $\mathcal{F}$ that projects to $\gamma$, $m\lambda$ is deformable within $\mathcal{F}$ to a loop in one $\mathcal{F}_P$ – following the deformation of $m\gamma$ within $\mathcal{M}$ to a point $P$. If that final loop in $\mathcal{F}_P$ is of class I, then, by continuity, a spin-vector $\boldsymbol{\kappa}$ taken around $m\lambda$ must go to $-\boldsymbol{\kappa}$; if of class II, such a $\boldsymbol{\kappa}$ must go to $\boldsymbol{\kappa}$. Since $m$ is odd, this fixes $\lambda$ as taking $\boldsymbol{\kappa}$ into $-\boldsymbol{\kappa}$ or $\boldsymbol{\kappa}$ respectively, without ambiguity.

Finally, it may happen that while no odd multiple of $\gamma$ is shrinkable, some (smallest possible) *even* multiple $2n\gamma$ *can* be shrunk to a point. Then one of two things can happen. The corresponding loops $2n\lambda$ in $\mathcal{F}$, following the deformation of $2n\gamma$ to a point $P$ in $\mathcal{M}$, may all end up as loops of class II in $\mathcal{F}_P$, or else some (and then in fact all) end up as of class I. In the former case, by continuity, a spin-vector $\boldsymbol{\kappa}$ taken around $2n\gamma$ must go into itself. Hence *either* choice of $\boldsymbol{\kappa} \mapsto \pm\boldsymbol{\kappa}$ for *one* traversal of $\lambda$ is valid, and we end up with different possible spin structures for $\mathcal{M}$ as before (unless spin structure is ruled out for other loops). But suppose a loop $2n\lambda$ ends up as a class I loop in $\mathcal{F}_P$, requiring $\boldsymbol{\kappa} \mapsto -\boldsymbol{\kappa}$ around $2n\lambda$. *Neither* choice $\boldsymbol{\kappa} \mapsto \pm\boldsymbol{\kappa}$ around $\lambda$ is now consistent, and this is the second obstruction to $\mathcal{M}$ having spin structure. Unlike the first, in can only occur if $\mathcal{M}$ is not simply-connected, and, also unlike the first, it obviously disappears when we pass to the universal covering space of $\mathcal{M}$.

Examples of space–times can be constructed (*cf.* Penrose 1968, p. 155, Geroch 1968, 1970, Hitchin 1974) in which one or the other or both of the above-mentioned difficulties occur, but which nevertheless satisfy (1.5.1) and (1.5.2) and which in no obvious other way seem to be physically unacceptable. The phenomenon is actually an instance of a more general one arising in manifolds of arbitrary dimension. There is a topological invariant, known as the *second Stieffel–Whitney class*, $w_2$, whose vanishing, in the case of an orientable* manifold $\mathcal{M}$, is necessary and sufficient

* The concept of orientability, for an $n$-manifold, is the same as has been referred to here (and in §1.1) as space–time orientability.

for the property

$$\mathscr{M} \textit{ has spin structure,} \tag{1.5.3}$$

i.e., for the existence of general (but still two-valued) spinorial objects in $\mathscr{M}$ (Milnor 1963, Lichnerowicz 1968; *cf.* Milnor and Stasheff 1974). The condition $w_2 = 0$ can be roughly stated as follows:

(1.5.4) CONDITION:

*On any closed 2-surface $\mathscr{S}$ in the manifold $\mathscr{M}$ (of dimension $n \geqslant 3$) there exists a set of $n - 1$ continuous fields of tangent vectors to $\mathscr{M}$, linearly independent at every point of $\mathscr{S}$. If $\mathscr{M}$ is orientable (which, in fact, is the condition $w_1 = 0$), we can replace '$n - 1$' by '$n$'.*

We shall show that when (1.5.4) is satisfied, for a space–time manifold $\mathscr{M}$ satisfying (1.5.1) and (1.5.2) (so now $n = 4$), neither of the above-mentioned obstructions to spin structure can occur.

As a preliminary, we consider the rotation group $SO(4)$ (that is, the identity-connected component of the group of rotations in four Euclidean dimensions) and show that, like $SO(3)$, closed paths within it fall into two classes I and II, non-shrinkable and shrinkable respectively, such that double a class I path is a class II path (i.e., $\pi_1(SO(4)) = \mathbb{Z}_2$). (In fact the same holds for $SO(n)$, for all $n \geqslant 3$, but we shall not need this more general result here.) We recall the quaternions of §1.2 and remark that any given element of $SO(4)$ may be obtained as an action on unit quaternions $\boldsymbol{q}$:

$$\boldsymbol{q} \mapsto \tilde{\boldsymbol{q}} = \boldsymbol{aqb} \tag{1.5.5}$$

where $\boldsymbol{a}$ and $\boldsymbol{b}$ are fixed unit quaternions. (This follows because $\tilde{\boldsymbol{q}}\tilde{\boldsymbol{q}}^* = \boldsymbol{qq}^*$ the 4-dimensional Euclidean norm, and the full dimensionality, 6, of $SO(4)$ is obtainable in the above way.) We have the one ambiguity

$$(\boldsymbol{a}, \boldsymbol{b}) = (-\boldsymbol{a}, -\boldsymbol{b}),$$

but apart from this, the pair $(\boldsymbol{a}, \boldsymbol{b})$ is uniquely determined by the $SO(4)$ element it represents.

Next, suppose that (1.5.4) and the orientability conditions (1.5.1), (1.5.2) hold for the 4-dimensional space–time manifold $\mathscr{M}$. Let us imagine that the tangent space* $T_P$ at each point $P$ of a closed 2-surface $\mathscr{S}$, in $\mathscr{M}$, is mapped linearly to $\mathbb{R}^4$ in such a way that the four linearly independent vectors at $P$ provided by (1.5.4) are mapped, respectively, to the four coordinate basis vectors in $\mathbb{R}^4$ (i.e., we use the four vector fields of (1.5.4) as coordinate

* See §4.1 for a precise definition of this concept.

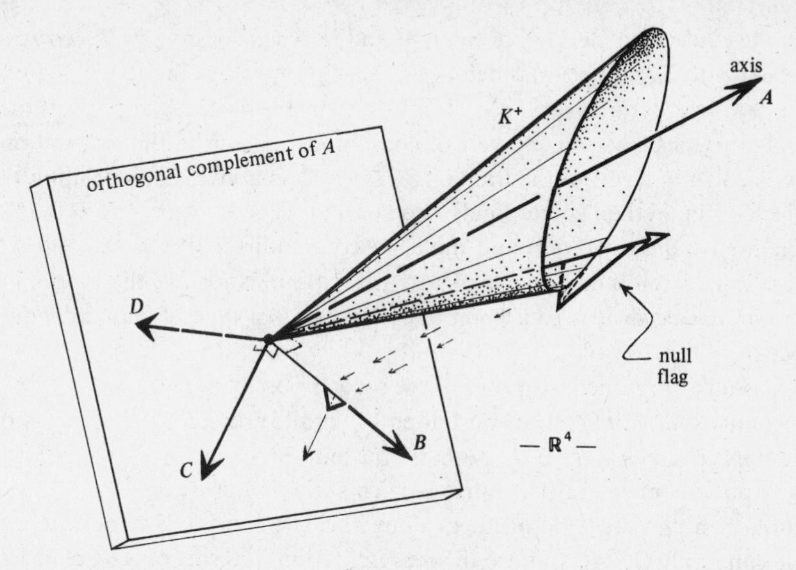


Fig. 1-16. The map to $\mathbb{R}^4$, of future null cone and null flag, provides, in continuous fashion, a unique frame ***ABCD***, right-handed and orthonormal with respect to the Euclidean metric of $\mathbb{R}^4$.

axes at each point of $\mathscr{S}$). The future null cone at $P$ will be mapped to a half-cone $K^+$ in $\mathbb{R}^4$. (See Fig. 1-16.) Just one of the principal semi-axes of $K^+$ (with respect to the standard Euclidean geometry of $\mathbb{R}^4$) will be the image, $\boldsymbol{A}$, in $\mathbb{R}^4$, of a future-timelike vector in $T_P$ (namely the axis *within* $K^+$). As the point $P$ moves about $\mathscr{S}$, the vector $\boldsymbol{A} \in \mathbb{R}^4$ moves continuously with $P$. Now consider a null flag at $P$. Its image in $\mathbb{R}^4$ will be a 'flag' whose flagpole points along a generator of $K^+$ and whose flag plane is tangent to $K^+$. Let $\boldsymbol{B}$ be the projection orthogonal to $\boldsymbol{A}$ (with respect to the Euclidean metric of $\mathbb{R}^4$) of this flagpole. The projection orthogonal to $\boldsymbol{A}$ of the flag *plane* contains just one direction, $\boldsymbol{C}$, perpendicular to $\boldsymbol{B}$ (as well as to $\boldsymbol{A}$). Take $\boldsymbol{D}$ to complete a right-handed frame with $\boldsymbol{A}, \boldsymbol{B}, \boldsymbol{C}$, and finally normalize all of $\boldsymbol{A}, \boldsymbol{B}, \boldsymbol{C}, \boldsymbol{D}$ to be unit vectors (in the metric of $\mathbb{R}^4$). We thus have a continuous way of assigning a right-handed orthonormal frame ***ABCD*** to any null flag at any point of $\mathscr{S}$, i.e. to any point of $\mathscr{F}$ which lies above $\mathscr{S}$. We note that in this correspondence if the null flag executes a class I [*or* II] path, keeping the point $P$ fixed, then the corresponding frame ***ABCD*** executes a class I [*or* II] continuous rotation in $SO(4)$. (Consider a $2\pi$-rotation of the flag plane and then argue by continuity.)

Let us now examine our two types of possible obstruction to the existence of spinors in a space- and time-oriented space–time manifold $\mathscr{M}$.

In the case of the merging of classes I and II, when a loop $\lambda$ in $\mathscr{F}_P$ corresponding to a $2\pi$-rotation is deformed through $\mathscr{F}$ to a point, its projection to $\mathscr{M}$ traces out a closed surface $\mathscr{S}$ to which (1.5.4) can be applied. If the stated frames exist on $\mathscr{S}$, we can continuously describe the orientation of our flag in terms of the frame $\boldsymbol{ABCD}$ in $\mathbb{R}^4$ as above. Each position of the loop in $\mathscr{F}$ then corresponds to a continuous motion of $\boldsymbol{ABCD}$ in $\mathbb{R}^4$, the first to a $2\pi$-rotation and the last – continuous with the first – to no rotation, which is impossible. Thus, if (1.5.4) holds for $\mathscr{M}$, the loop $\lambda$ in $\mathscr{F}$ can *not* be shrunk to a point in $\mathscr{F}$, and so *that* type of failure cannot occur.

A similar argument disposes of the second possible reason for failure – incompatibility of flag transport after $2n$ circuits around a loop $\gamma$, i.e. two circuits around $\eta = n\gamma$. By hypothesis, the loop $2n\gamma = 2\eta$ is to be shrinkable to a point $P$ in $\mathscr{M}$. In the course of such shrinking, it traces out a closed surface in $\mathscr{M}$, 'welded together' along the one loop $\eta$. To this surface we can apply (1.5.4), and use the representation in terms of $\boldsymbol{ABCD}$ in $\mathbb{R}^4$ as before to map the flags taken around $2\eta$ at various stages of its deformation to a point. Now the obstruction in question arises if a loop $2\zeta$ in $\mathscr{F}$, where $\zeta$ projects to $\eta$, is deformable to a loop of class I in $\mathscr{F}_P$. But a flag taken around $2\zeta$ in $\mathscr{F}$ is represented by a *double* motion of $\boldsymbol{ABCD}$ and thus a class II loop in $SO(4)$. If the final loop in $\mathscr{F}_P$ were of class I, the original class II path in $SO(4)$ would be continuously deformable to a class I path in $SO(4)$, which is impossible, and so this obstruction cannot arise either.

We shall use the more specific term *spinor structure* (rather than the general term spin structure) to mean that *all three* of the properties (1.5.1), (1.5.2), and (1.5.3) hold. Thus, if $\mathscr{M}$ has spinor structure, a spinor system (based on null flags and spin-vectors) of the type that concerns us in this book will exist for $\mathscr{M}$. In other words, the space $\mathscr{F}'$, defined above, will exist. If $\mathscr{M}$ is simply-connected, $\mathscr{F}'$ will in fact be $\tilde{\mathscr{F}}$. (In each $\tilde{\mathscr{F}}_P$, a path between the two points which represent a single point in $\mathscr{F}_P$ corresponds to a $2\pi$-rotation, as we saw earlier; this ensures the same property also for $\tilde{\mathscr{F}}$.)

But even when $\mathscr{M}$ possesses spinor structure, that structure will generally not be unique if $\mathscr{M}$ is not simply-connected. In that case $\mathscr{F}' \neq \tilde{\mathscr{F}}$. (For $\mathscr{F}'$ must 'unwrap' each class I loop in each $\mathscr{F}_P$, and no more; yet $\tilde{\mathscr{F}}$ would unwrap also those loops that correspond to unshrinkable loops in $\mathscr{M}$.) In fact, there are then $2^k$ different spinor structures, where $\boldsymbol{k}$ is the number of 'independent' loops in $\mathscr{M}$ of which no odd multiple can be shrunk to a point.

### Spinor structure for non-compact space–times

We end this section by mentioning without proof a very simple criterion for $\mathcal{M}$ to have spinor structure in the case when $\mathcal{M}$ is *non-compact* (i.e., loosely speaking, 'open'). In fact non-compactness follows from the very 'reasonable' physical requirement that $\mathcal{M}$ should possess no closed time-like curves (Bass and Witten 1957; *cf.* also Penrose 1968, Hawking and Ellis 1973 p. 189).

(1.5.6) THEOREM (Geroch 1968)

*If $\mathcal{M}$ is a non-compact space–time, then a necessary and sufficient condition that it should have spinor structure is the existence of four continuous vector fields on $\mathcal{M}$ which constitute a Minkowski tetrad in the tangent space at each point of $\mathcal{M}$.*

The *sufficiency* of the condition in (1.5.6) is, indeed, evident. For the continuity of the Minkowski tetrads implies time- and space-orientability for $\mathcal{M}$ as required by (1.5.1) and (1.5.2) – and without loss of generality we can assume Minkowski tetrads are all *restricted*. Choosing a fixed abstract ('restricted') Minkowski coordinate space (playing the role of $\mathbb{R}^4$ above, but the discussion is now simplier) we may refer each null flag in $\mathcal{M}$ to it by using the flag's representation in the local Minkowski tetrad provided by (1.5.6). This enables us to keep track of the parity of the number of $2\pi$-rotations executed by any null flag on $\mathcal{M}$ and spin structure is assured, as in (1.5.4).

We remark that for a non-simply-connected space–time the selection of a spinor structure is fixed once a Minkowski tetrad field is chosen in accordance with (1.5.6), this choice being normally topologically non-unique. But even when $\mathcal{M}$ is simply-connected, topological non-uniqueness in the Minkowski-tetrad field may occur. An instructive example is the Einstein static universe (to be discussed in more detail in §§9.2, 9.5), for which $\mathcal{M}$ has the topology $S^3 \times \mathbb{R}$ (with $\mathbb{R}$ corresponding to the time-direction). A 3-frame at any point of $S^3$ may, for example, be carried into a continuous field of 3-frames all over $S^3$ either by *right-translation* (given by the motions of $S^3$ to itself defined by the quaternionic transformations (1.5.5) of the special form $\boldsymbol{q} \mapsto \tilde{\boldsymbol{q}} = \boldsymbol{qb}$) or by *left-translation* (given by those of the special form $\boldsymbol{q} \mapsto \tilde{\boldsymbol{q}} = \boldsymbol{aq}$). These are topologically inequivalent even though they give rise to the same (unique) spin structure.

We have remarked that the non-compactness assumption required for $\mathcal{M}$ in Theorem (1.5.6) is very desirable on physical grounds. There are also rather strong reasons for believing that physical space–time should actually possess the global requirements of spinor structure. Time-orientability may be regarded as strongly suggested by the time-asymmetry of statistical physics which apparently assigns, on a reasonably local level, a time-orientation everywhere throughout physical space–time. Space-orientability would likewise follow from the (seemingly universal) reflection-asymmetry of weak interactions ($P$-noninvariance) and of $K^0$-decay ($CP$-noninvariance) which apparently provide a natural space-orientation throughout space–time (see footnote on p. 4). Finally, the existence of spinor fields in physics seems to imply that physical space–time should possess spin structure (see footnote on pp. 46, 47). Hence the existence of globally defined restricted Minkowski tetrad fields would appear to be physically assured by Theorem (1.5.6).

## 1.6 The geometry of spinor operations

As we have seen, every null flag in Minkowski vector space $\mathbb{V}$ defines a pair of spinorial objects, namely the (non-zero) spin-vectors $\boldsymbol{\kappa}$ and $-\boldsymbol{\kappa}$. With the help of a Minkowski coordinate system we assigned two complex components $(\xi, \eta)$ to $\boldsymbol{\kappa}$. Let us now write

$$\xi = \kappa^0, \quad \eta = \kappa^1.$$

Similarly, if $\boldsymbol{\omega}$ is some other spin-vector we can denote its components by $(\omega^0, \omega^1)$, etc. Now, we shall be interested in operations between spin-vectors which have geometrical (and therefore coordinate independent) meanings. But we saw in §1.4 that any passive Lorentz transformation (i.e., change of Minkowski coordinate system) corresponds to a *spin-transformation* applied to the components $(\xi, \eta)$. Thus operations between spin-vectors, when written as relations between components, must be invariant under (passive) spin-transformations.

Let $\mathfrak{S}^{\bullet}$ denote *spin-space*, i.e., the space of spin-vectors, and $\mathbb{C}$ the field of complex numbers. There are three basic operations on $\mathfrak{S}^{\bullet}$ to be considered. These are

(1.6.1) *scalar multiplication*: $\mathbb{C} \times \mathfrak{S}^{\bullet} \to \mathfrak{S}^{\bullet}$,
i.e., given $\lambda \in \mathbb{C}$ and $\boldsymbol{\kappa} \in \mathfrak{S}^{\bullet}$, we have an element $\lambda\boldsymbol{\kappa} \in \mathfrak{S}^{\bullet}$;

(1.6.2) *addition*: $\mathfrak{S}^{\bullet} \times \mathfrak{S}^{\bullet} \to \mathfrak{S}^{\bullet}$,
i.e. given $\boldsymbol{\kappa}, \boldsymbol{\omega} \in \mathfrak{S}^{\bullet}$ we have an element $\boldsymbol{\kappa} + \boldsymbol{\omega} \in \mathfrak{S}^{\bullet}$;

(1.6.3) *inner product*: $\mathfrak{S}^{\bullet} \times \mathfrak{S}^{\bullet} \to \mathbb{C}$
i.e. given $\boldsymbol{\kappa}, \boldsymbol{\omega} \in \mathfrak{S}^{\bullet}$ we have an element $\{\boldsymbol{\kappa}, \boldsymbol{\omega}\} \in \mathbb{C}$.

Representing each spin-vector by its components relative to some given reference system, we can define these three operations, respectively, by:

$$\lambda(\kappa^0, \kappa^1) = (\lambda\kappa^0, \lambda\kappa^1), \quad (1.6.4)$$

$$(\kappa^0, \kappa^1) + (\omega^0, \omega^1) = (\kappa^0 + \omega^0, \kappa^1 + \omega^1), \quad (1.6.5)$$

$$\{(\kappa^0, \kappa^1), (\omega^0, \omega^1)\} = \kappa^0\omega^1 - \kappa^1\omega^0. \quad (1.6.6)$$

The first two of these are obviously invariant under the spin transformation

$$\begin{pmatrix} \kappa^0 \\ \kappa^1 \end{pmatrix} \mapsto \begin{pmatrix} \kappa^{\hat{0}} \\ \kappa^{\hat{1}} \end{pmatrix} = \begin{pmatrix} \alpha & \beta \\ \gamma & \delta \end{pmatrix} \begin{pmatrix} \kappa^0 \\ \kappa^1 \end{pmatrix} \quad (1.6.7)$$

because of its linearity. The invariance of the third is easily verified:

$$\kappa^{\hat{0}}\omega^{\hat{1}} - \kappa^{\hat{1}}\omega^{\hat{0}} = \begin{vmatrix} \kappa^{\hat{0}} & \omega^{\hat{0}} \\ \kappa^{\hat{1}} & \omega^{\hat{1}} \end{vmatrix} = \left| \begin{pmatrix} \alpha & \beta \\ \gamma & \delta \end{pmatrix} \begin{pmatrix} \kappa^0 & \omega^0 \\ \kappa^1 & \omega^1 \end{pmatrix} \right|$$

$$= \begin{vmatrix} \alpha & \beta \\ \gamma & \delta \end{vmatrix} \begin{vmatrix} \kappa^0 & \omega^0 \\ \kappa^1 & \omega^1 \end{vmatrix} = \kappa^0\omega^1 - \kappa^1\omega^0,$$

since $\alpha\delta - \beta\gamma = 1$.

The following relations are immediate consequence of the definitions (1.6.4), (1.6.5) (1.6.6).

$$\lambda(\mu\boldsymbol{\kappa}) = (\lambda\mu)\boldsymbol{\kappa}, \quad (1.6.8)$$

$$1\boldsymbol{\kappa} = \boldsymbol{\kappa}, \quad (1.6.9)$$

$$0\boldsymbol{\kappa} = \mathbf{0}, \quad (1.6.10)$$

$$(-1)\boldsymbol{\kappa} = -\boldsymbol{\kappa}, \quad (1.6.11)$$

$$(\lambda + \mu)\boldsymbol{\kappa} = (\lambda\boldsymbol{\kappa}) + (\mu\boldsymbol{\kappa}), \quad (1.6.12)$$

$$\boldsymbol{\kappa} + \boldsymbol{\omega} = \boldsymbol{\omega} + \boldsymbol{\kappa}, \quad (1.6.13)$$

$$(\boldsymbol{\kappa} + \boldsymbol{\omega}) + \boldsymbol{\tau} = \boldsymbol{\kappa} + (\boldsymbol{\omega} + \boldsymbol{\tau}), \quad (1.6.14)$$

$$\lambda(\boldsymbol{\kappa} + \boldsymbol{\omega}) = (\lambda\boldsymbol{\kappa}) + (\lambda\boldsymbol{\omega}), \quad (1.6.15)$$

$$\{\boldsymbol{\kappa}, \boldsymbol{\omega}\} = -\{\boldsymbol{\omega}, \boldsymbol{\kappa}\}, \quad (1.6.16)$$

$$\lambda\{\boldsymbol{\kappa}, \boldsymbol{\omega}\} = \{\lambda\boldsymbol{\kappa}, \boldsymbol{\omega}\}, \quad (1.6.17)$$

$$\{\boldsymbol{\kappa} + \boldsymbol{\omega}, \boldsymbol{\tau}\} = \{\boldsymbol{\kappa}, \boldsymbol{\tau}\} + \{\boldsymbol{\omega}, \boldsymbol{\tau}\}. \quad (1.6.18)$$

Furthermore,

$$\{\boldsymbol{\kappa}, \boldsymbol{\omega}\}\boldsymbol{\tau} + \{\boldsymbol{\omega}, \boldsymbol{\tau}\}\boldsymbol{\kappa} + \{\boldsymbol{\tau}, \boldsymbol{\kappa}\}\boldsymbol{\omega} = 0, \quad (1.6.19)$$

as follows by Laplace expansion of

$$\begin{vmatrix} \tau^0 & \kappa^0 & \omega^0 \\ \tau^1 & \kappa^1 & \omega^1 \\ \tau^{\mathbf{A}} & \kappa^{\mathbf{A}} & \omega^{\mathbf{A}} \end{vmatrix} = 0 \qquad (\mathbf{A} = 0, 1)$$

with respect to the last row. Note the particular case of (1.6.16):

$$\{\boldsymbol{\kappa}, \boldsymbol{\kappa}\} = 0. \tag{1.6.20}$$

Regarding (1.6.8)–(1.6.15) as (not independent) axioms, we recognize that $\mathfrak{S}^{\bullet}$ *is a vector space over* $\mathbb{C}$. By (1.6.16)–(1.6.18) the inner product is a skew-symmetrical bilinear form on $\mathfrak{S}^{\bullet}$. Relation (1.6.19), together with

$$\{\boldsymbol{\kappa}, \boldsymbol{\omega}\} \neq 0 \quad \text{for some } \boldsymbol{\kappa}, \boldsymbol{\omega} \in \mathfrak{S}^{\bullet}, \tag{1.6.21}$$

implies that the vector space is two-dimensional. For, if $\boldsymbol{\kappa}$ and $\boldsymbol{\omega}$ satisfy (1.6.21), neither can be a multiple of the other (by (1.6.20) and (1.6.17)), so the dimension is at least two; then (1.6.19) shows how to express an arbitrary spin-vector $\boldsymbol{\tau}$ as a linear combination of the two spin-vectors $\boldsymbol{\kappa}$ and $\boldsymbol{\omega}$.

With the aid of the inner product, we can readily obtain the general representation of a spin-vector in terms of components. We choose any pair of spin-vectors $\boldsymbol{o}$ and $\boldsymbol{\iota}$ (omicron and iota) normalized so that their inner product is unity:

$$\{\boldsymbol{o}, \boldsymbol{\iota}\} = 1 = -\{\boldsymbol{\iota}, \boldsymbol{o}\}. \tag{1.6.22}$$

We call the pair $\boldsymbol{o}, \boldsymbol{\iota}$ a *spin-frame*. The components of a spin-vector $\boldsymbol{\kappa}$ in this spin-frame are*

$$\kappa^0 = \{\boldsymbol{\kappa}, \boldsymbol{\iota}\}, \quad \kappa^1 = -\{\boldsymbol{\kappa}, \boldsymbol{o}\}. \tag{1.6.23}$$

Thus, from (1.6.19), (1.6.22) and (1.6.16) we get

$$\boldsymbol{\kappa} = \kappa^0 \boldsymbol{o} + \kappa^1 \boldsymbol{\iota}. \tag{1.6.24}$$

The components of $\boldsymbol{o}$ are (1, 0) and those of $\boldsymbol{\iota}$ are (0, 1). Now, if we *start* from (1.6.24) and (1.6.22), then we can directly re-obtain the expressions (1.6.4), (1.6.5) and (1.6.6) for scalar multiplication, addition and inner product. A passive spin transformation (1.6.7) is achieved when the spin-frame $\boldsymbol{o}, \boldsymbol{\iota}$ is replaced by another spin-frame $\hat{\boldsymbol{o}}, \hat{\boldsymbol{\iota}}$. The particular spin-frame which gave the representation of spin-vectors by components $(\xi, \eta)$ according to §§1.2 and 1.4, is related (*cf.* Fig. 1-17) to the given Minkowski tetrad $(\boldsymbol{t}, \boldsymbol{x}, \boldsymbol{y}, \boldsymbol{z})$ as follows (*cf.* (1.2.15), (1.4.14)): the flagpole of $\boldsymbol{o}$ is $(\boldsymbol{t} + \boldsymbol{z})/\sqrt{2}$ with flag plane extending from this line in the direction of $\boldsymbol{x}$; the flagpole of $\boldsymbol{\iota}$ is $(\boldsymbol{t} - \boldsymbol{z})/\sqrt{2}$ with flag plane extending from this line in the direction of $-\boldsymbol{x}$; the relative signs of the *spin-vectors* are defined by the fact that $\boldsymbol{o}$ is rotated into $\boldsymbol{\iota}$ by a continuous rotation about $\boldsymbol{y}$ in the positive sense through an angle $\pi$ (and hence the same rotation sends $\boldsymbol{\iota}$ into $-\boldsymbol{o}$).

* Expression (1.6.23) looks a little more natural in terms of the 'lowered indices' that we shall introduce later. For, we shall have $\kappa_0 = -\kappa^1$, $\kappa_1 = \kappa^0$.

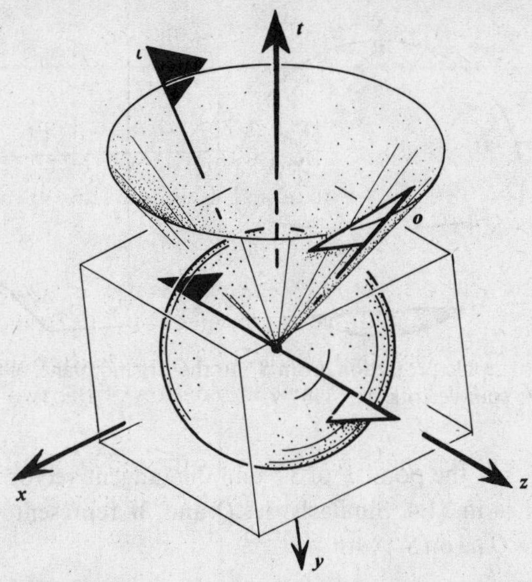


Fig. 1-17. The standard relation between a spin-frame $\boldsymbol{o}$, $\boldsymbol{\iota}$ and a (restricted) Minkowski tetrad $\boldsymbol{t}, \boldsymbol{x}, \boldsymbol{y}, \boldsymbol{z}$.

*The geometry of inner product*

We conclude this chapter by giving geometrical interpretations of each of the three basic operations on spin-vectors. (Readers less interested in this geometry may pass directly on to Chapter 2.) The first operation, namely scalar multiplication, has already been dealt with in §1.4. Let us recapitulate the conclusion here. The spin-vector $\lambda\boldsymbol{\kappa}$ is obtained from $\boldsymbol{\kappa}$ by keeping the flagpole direction fixed, multiplying the extent of the flagpole by the factor $\lambda\bar{\lambda}$ and rotating the flag plane in the positive sense through an angle $2 \arg \lambda$.

Next let us consider the inner product (since this turns out to be rather simpler than addition). To begin with, the *modulus* of $\{\boldsymbol{\kappa}, \boldsymbol{\omega}\}$ is just $2^{-\frac{1}{2}}$ times the spacelike interval between the extremeties of the flagpoles. For if $\boldsymbol{K}$ is the flagpole of $\boldsymbol{\kappa}$ and $\boldsymbol{W}$ that of $\boldsymbol{\omega}$, we have, using coordinates as in §1.2,

$$\begin{aligned}
&\|\boldsymbol{K} - \boldsymbol{W}\| \\
&\quad = \{(K^0 + K^3) - (W^0 + W^3)\}\{(K^0 - K^3) - (W^0 - W^3)\} \\
&\qquad - \{(K^1 + \mathrm{i}K^2) - (W^1 + \mathrm{i}W^2)\}\{(K^1 - \mathrm{i}K^2) - (W^1 - \mathrm{i}W^2)\} \\
&\quad = 2(\kappa^0\overline{\kappa^0} - \omega^0\overline{\omega^0})(\kappa^1\overline{\kappa^1} - \omega^1\overline{\omega^1}) - 2(\kappa^0\overline{\kappa^1} - \omega^0\overline{\omega^1})(\kappa^1\overline{\kappa^0} - \omega^1\overline{\omega^0}) \\
&\quad = -2(\kappa^0\omega^1 - \kappa^1\omega^0)\overline{(\kappa^0\omega^1 - \kappa^1\omega^0)} = -2|\{\boldsymbol{\kappa}, \boldsymbol{\omega}\}|^2. \qquad (1.6.25)
\end{aligned}$$

It remains to interpret $\arg\{\boldsymbol{\kappa}, \boldsymbol{\omega}\}$. This is most easily done in terms of

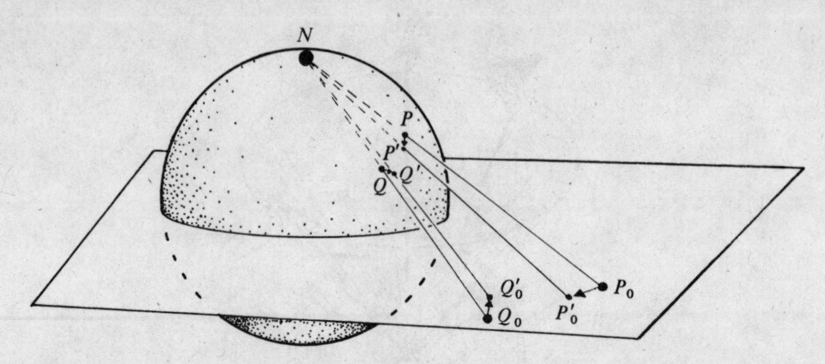


Fig. 1-18. Stereographic projection, from $S^+$ to the Argand plane, of the point pairs representing two spin-vectors.

the sphere $S^+$. Let the point $P$ of $S^+$ and the tangent vector $\boldsymbol{L}$ to $S^+$ at $P$ represent $\pm\boldsymbol{\kappa}$ as in §1.4. Similarly, let $Q$ and $\boldsymbol{M}$ represent $\pm\boldsymbol{\omega}$. Choose $P' = P'(\varepsilon)$, $Q' = Q'(\varepsilon)$ on $S^+$ with

$$\boldsymbol{L} = \lim_{\varepsilon\to 0}\frac{1}{\varepsilon}\overrightarrow{PP'}, \quad \boldsymbol{M} = \lim_{\varepsilon\to 0}\frac{1}{\varepsilon}\overrightarrow{QQ'} \tag{1.6.26}$$

as in (1.4.9). Let $P_0, P'_0, Q_0, Q'_0, \boldsymbol{L}_0, \boldsymbol{M}_0$ be the respective projections of $P, P', Q, Q', \boldsymbol{L}, \boldsymbol{M}$ in the Argand plane (*cf*. Fig. 1-18). Representing vectors in the Argand plane by complex numbers, we have:

$$\overrightarrow{P_0Q_0} \leftrightarrow \frac{\omega^0}{\omega^1} - \frac{\kappa^0}{\kappa^1} = \alpha, \tag{1.6.27}$$

$$\boldsymbol{L}_0 = \lim_{\varepsilon\to 0}\frac{1}{\varepsilon}\overrightarrow{P_0P'_0} \leftrightarrow \frac{-1}{\sqrt{2}(\kappa^1)^2} = \beta, \tag{1.6.28}$$

$$\boldsymbol{M}_0 = \lim_{\varepsilon\to 0}\frac{1}{\varepsilon}\overrightarrow{Q_0Q'_0} \leftrightarrow \frac{-1}{\sqrt{2}(\omega^1)^2} = \gamma. \tag{1.6.29}$$

Hence

$$\{\boldsymbol{\kappa}, \boldsymbol{\omega}\}^2 = \tfrac{1}{2}(\alpha\beta^{-1})(\alpha\gamma^{-1}) \tag{1.6.30}$$

and therefore $2\arg\{\boldsymbol{\kappa}, \boldsymbol{\omega}\}$ is minus the sum of the angles that $\boldsymbol{L}_0$ and $\boldsymbol{M}_0$ make with $\overrightarrow{P_0Q_0}$. Since stereographic projection is conformal, these angles are the same as the corresponding angles measured on the sphere $S^+$ (although the sign of each angle is reversed since the projection reverses the orientation of the surface). Now the straight line $P_0Q_0$ (oriented in the direction $P_0Q_0$) is the projection of the oriented circle $c = NPQ$ ($N$ being the north pole of $S^+$). Thus, $2\arg\{\boldsymbol{\kappa}, \boldsymbol{\omega}\}$ (mod $2\pi$) is the sum of the two angles (measured in the positive sense) which $\boldsymbol{L}$ and $\boldsymbol{M}$

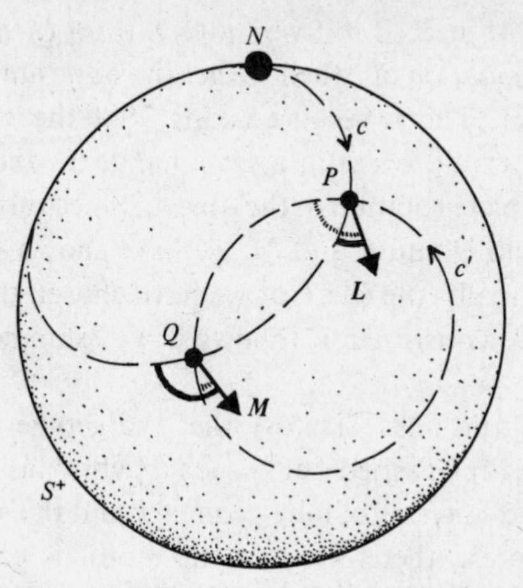


Fig. 1-19. Any circle on $S^+$ through $P$ and $Q$ makes the same sum-of-angles with $\boldsymbol{L}$ and $\boldsymbol{M}$.

make with $c$, respectively. This defines $\arg\{\boldsymbol{\kappa}, \boldsymbol{\omega}\}$ geometrically up to the possible addition of $\pi$. Hence we have $\pm\{\boldsymbol{\kappa}, \boldsymbol{\omega}\}$ defined geometrically. From the invariance of $\{\boldsymbol{\kappa}, \boldsymbol{\omega}\}$ it is clear that $N$ must actually be irrelevant to this construction. Indeed, it follows at once, by elementary geometry (*cf*. Fig. 1-19), that the sum of these angles is the same whichever oriented circle $c$ we choose on $S^+$, through the two points $P$, $Q$.

To obtain the correct sign for $\{\boldsymbol{\kappa}, \boldsymbol{\omega}\}$, we must think of $\boldsymbol{L}$ and $\boldsymbol{M}$ as spinorial objects rather than simply as tangent vectors to $S^+$. Now, imagine $\boldsymbol{L}$ moved continuously along $c$, following the orientation from $P$ to $Q$ and always keeping the same angle with $c$. When $\boldsymbol{L}$ arrives at $Q$ we expand (or contract) $\boldsymbol{L}$ until its length is the same as that of $\boldsymbol{M}$, then we rotate $\boldsymbol{L}$ and $\boldsymbol{M}$ equally in opposite directions (tangentially to $S^+$) until they coincide. The angle (measured in the positive sense) that the now coincident $\boldsymbol{L}$ and $\boldsymbol{M}$ make with $c$ is the required value of $\arg\{\boldsymbol{\kappa}, \boldsymbol{\omega}\}$. The point of this construction is that whereas there are *two* possible directions at $Q$ along which the *vectors* (and so the corresponding flags) finally coincide, only *one* of these is a coincidence between the spinorial objects represented by $\boldsymbol{L}$ and $\boldsymbol{M}$ (so that the corresponding spin-vectors coincide). It is *this* coincidence that we must choose. Hence we have defined $\{\boldsymbol{\kappa}, \boldsymbol{\omega}\}$, *including* its sign, geometrically.

It follows from the continuity of this construction that the resulting value of $\arg\{\boldsymbol{\kappa}, \boldsymbol{\omega}\}$ (mod $2\pi$) is independent of the choice of $c$ through $P$ and $Q$. The same angle also results if instead of moving $\boldsymbol{L}$ forwards from $P$ to $Q$, we move $\boldsymbol{M}$ back (against the orientation of $c$) from $Q$ to $P$ and then rotate $\boldsymbol{L}$ and $\boldsymbol{M}$ at $P$. But if we move $\boldsymbol{L}$ from $Q$ to $P$ along $c$ in the direction of the *orientation* of $c$ then we get the opposite sign for the value of the inner product. (This is because taking $\boldsymbol{M}$ all the way around $c$ once, from $P$ back to $P$, would result in a sign change for the spinorial object, this being one complete rotation of the object.) Since this is an interchange of the roles of $P$ and $Q$ and $\boldsymbol{L}$ and $\boldsymbol{M}$, we have shown $\{\boldsymbol{\kappa}, \boldsymbol{\omega}\} = -\{\boldsymbol{\omega}, \boldsymbol{\kappa}\}$, as required also. Finally, the fact that we have chosen the correct sign for $\{\boldsymbol{\kappa}, \boldsymbol{\omega}\}$ in the above construction follows if we examine the special case $\{\boldsymbol{o}, \boldsymbol{\iota}\} = 1$.

It is perhaps a little unsatisfactory that the geometrical definition of $\{\boldsymbol{\kappa}, \boldsymbol{\omega}\}$ should have been carried out in such a hybrid fashion, the modulus having been defined in terms of four-geometry and the argument in terms of $S^+$. It is of interest, therefore, that the modulus can also be simply interpreted in terms of $S^+$ as follows:

$$|\{\boldsymbol{\kappa}, \boldsymbol{\omega}\}| = \frac{PQ}{2|\boldsymbol{L}|^{\frac{1}{2}}|\boldsymbol{M}|^{\frac{1}{2}}}, \tag{1.6.31}$$

where $|\boldsymbol{L}|$ denote the *Euclidean length* $(= (-\,\|\boldsymbol{L}\|)^{\frac{1}{2}})$ of $\boldsymbol{L}$. To see this, we observe, first, that this is simply (1.6.25) in the case when both flagpole tips lie on $S^+$, so $|\boldsymbol{L}| = 1 = |\boldsymbol{M}|$. For the general case, we simply scale up the flagpoles (recalling (1.6.17)) from the special case above, by the respective factors $|\boldsymbol{L}|^{-\frac{1}{2}}$ and $|\boldsymbol{M}|^{-\frac{1}{2}}$ (*cf.* (1.4.16)). Note that the modulus of (1.6.30) is the limiting case of (1.6.31), when $S^+$ becomes a plane.

As an alternative, let us interpret $\arg\{\boldsymbol{\kappa}, \boldsymbol{\omega}\}$ in terms of 4-geometry. Let $\sigma$ denote the spacelike 2-plane through $O$ which is the orthogonal

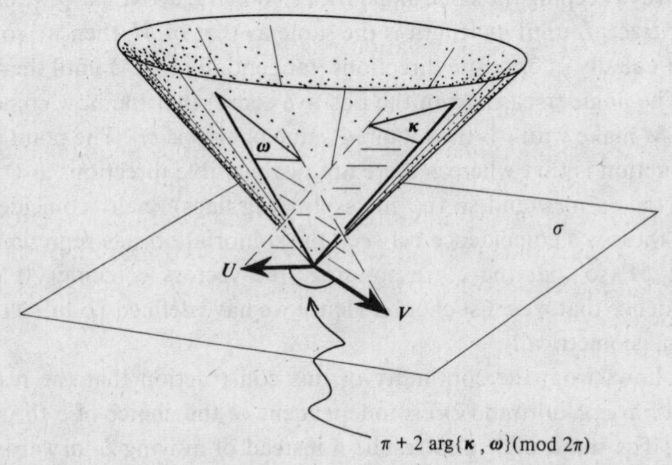


Fig. 1-20. An interpretation of $\arg\{\boldsymbol{\kappa}, \boldsymbol{\omega}\}$ in terms of 4-geometry.

complement of the timelike 2-plane spanned by the two flagpoles. Let $\boldsymbol{U}$ and $\boldsymbol{V}$ be the unit vectors along the intersections of $\sigma$ with the respective flag planes of $\boldsymbol{\kappa}$ and $\boldsymbol{\omega}$ (Fig. 1-20). Then the angle between $\boldsymbol{U}$ and $\boldsymbol{V}$, measured in the appropriate sense, turns out to be $\pi + 2\arg\{\boldsymbol{\kappa}, \boldsymbol{\omega}\}$ (mod $2\pi$). To obtain the sense in which this angle is to be measured and also to obtain the sign of $\{\boldsymbol{\kappa}, \boldsymbol{\omega}\}$, consider a spatial rotation (the direction of the fixed time-axis $\boldsymbol{t}$ being chosen along the sum of the flagpoles) about the line $p$, in $\sigma$, which bisects the angle between $\boldsymbol{U}$ and $\boldsymbol{V}$. If this rotation is built up continuously to $\pi$, then, depending on the sense in which the rotation takes place, $\boldsymbol{\kappa}$ is rotated into $\boldsymbol{\omega}$ or into $-\boldsymbol{\omega}$. (For, $\boldsymbol{U}$ goes to $\boldsymbol{V}$ and the flagpoles go one into the other.) Choose the sense so that $\boldsymbol{\kappa}$ goes to $\boldsymbol{\omega}$. That this rotation be in the *positive* sense about $p$ defines an orientation for $p$. The angle that $\boldsymbol{U}$ makes with the positive direction of $p$ (measured in the sense induced by a positive rotation of the flag plane of $\boldsymbol{\kappa}$ about its flagpole) is then precisely $\arg\{\boldsymbol{\kappa}, \boldsymbol{\omega}\} - (\pi/2)(\text{mod } 2\pi)$.

In order to see the validity of the above prescription, we consider a Minkowski frame with time axis $\boldsymbol{t}$ and associated Riemann sphere $S^+$. The points $P$ and $Q$ represented the flagpoles are now antipodes on the sphere; the vectors $\boldsymbol{U}$ and $\boldsymbol{V}$ are just positive multiplies of $\boldsymbol{L}$ and $\boldsymbol{M}$ transferred to the centre. The result follows by simple geometry.

### *The geometry of addition*

A geometrical interpretation for the *sum* of two spin-vectors is actually implicit in that of the inner product. For, the relation

$$\boldsymbol{\kappa} + \boldsymbol{\omega} = \boldsymbol{\tau} \tag{1.6.32}$$

is evidently equivalent to the relation

$$\{\boldsymbol{\kappa}, \boldsymbol{\rho}\} + \{\boldsymbol{\omega}, \boldsymbol{\rho}\} = \{\boldsymbol{\tau}, \boldsymbol{\rho}\} \tag{1.6.33}$$

holding for all $\boldsymbol{\rho} \in \mathfrak{S}^{\cdot}$. Similarly we could treat any linear combination in place of the simple sum in (1.6.32). Since $\mathfrak{S}^{\cdot}$ is two-dimensional, (1.6.33) needs to hold only for two non-proportional choices of $\boldsymbol{\rho}$. We can take these to be $\boldsymbol{\omega}$ and $\boldsymbol{\kappa}$ themselves (assuming $\boldsymbol{\kappa}$ and $\boldsymbol{\omega}$ are not proportional – otherwise (1.6.32) reduces essentially to scalar multiplication). Thus (1.6.32) is equivalent to

$$\{\boldsymbol{\tau}, \boldsymbol{\omega}\} = \{\boldsymbol{\kappa}, \boldsymbol{\omega}\} = -\{\boldsymbol{\tau}, \boldsymbol{\kappa}\}. \tag{1.6.34}$$

Taking the modulus of (1.6.34) we see, from the interpretation of inner product, that the spatial intervals between the tips of the flagpoles of $\boldsymbol{\omega}$, $\boldsymbol{\kappa}$ and $\boldsymbol{\omega} + \boldsymbol{\kappa}$ must all be equal. That is to say, the tips of the flagpoles are the vertices of an *equilateral triangle* in space-time. As for the flag planes, taking the argument of (1.6.34), we get

$$2\arg\{\boldsymbol{\tau},\boldsymbol{\omega}\} = 2\arg\{\boldsymbol{\kappa},\boldsymbol{\omega}\} = 2\arg\{\boldsymbol{\tau},\boldsymbol{\kappa}\} = \lambda, \tag{1.6.35}$$

say. Consider the representation in terms of $S^+$. Let $P, Q, R$ be points on $S^+$ corresponding to the respective flagpole directions of $\boldsymbol{\kappa}, \boldsymbol{\omega}, \boldsymbol{\tau}$ and let $\boldsymbol{L}, \boldsymbol{M}, \boldsymbol{N}$ be the tangent vectors to $S^+$ at $P, Q, R$, respectively, which complete the representation of the flags defined by $\boldsymbol{\kappa}, \boldsymbol{\omega}, \boldsymbol{\tau}$. By our earlier construction of $2\arg\{\boldsymbol{\kappa},\boldsymbol{\omega}\}$ on $S^+$, the angles that $\boldsymbol{N}$ and $\boldsymbol{M}$ make with the oriented circle $c$ through $P$, $Q$ and $R$, must sum to $\lambda$. The same is true by (1.6.35) for $\boldsymbol{L}$ and $\boldsymbol{M}$, and for $\boldsymbol{N}$ and $\boldsymbol{L}$. Hence, the angles which each of $\boldsymbol{L}, \boldsymbol{M}$ and $\boldsymbol{N}$ make with $c$ must all be equal, being, in fact, either $\frac{1}{2}\lambda$ or $\frac{1}{2}\lambda + \pi \pmod{2\pi}$, according to the choice of orientation of $c$. Thus, in space–time terms, the flag planes of $\boldsymbol{\kappa}, \boldsymbol{\omega}$ and $\boldsymbol{\kappa}+\boldsymbol{\omega}$ must be equally inclined to the circumcircle of the equilateral triangle formed by the flagpole tips (and hence to the triangle itself).

The fact that this configuration of flags is completely symmetrical in $\boldsymbol{\kappa}, \boldsymbol{\omega}$ and $\boldsymbol{\kappa}+\boldsymbol{\omega}$ should not be surprising. For, the flags themselves do not define the *signs* of the spin-vectors, so the flag configuration for $\boldsymbol{\kappa}+\boldsymbol{\omega}=\boldsymbol{\tau}$ is the same as for the symmetrical expression $\boldsymbol{\kappa}+\boldsymbol{\omega}+\boldsymbol{\tau}=0$. Furthermore, both these relations are indistinguishable from $\boldsymbol{\kappa}-\boldsymbol{\omega}=\pm\boldsymbol{\tau}$ if we look at the flags only. Thus, if the flags of $\boldsymbol{\kappa}$ and of $\boldsymbol{\omega}$ are given, the above

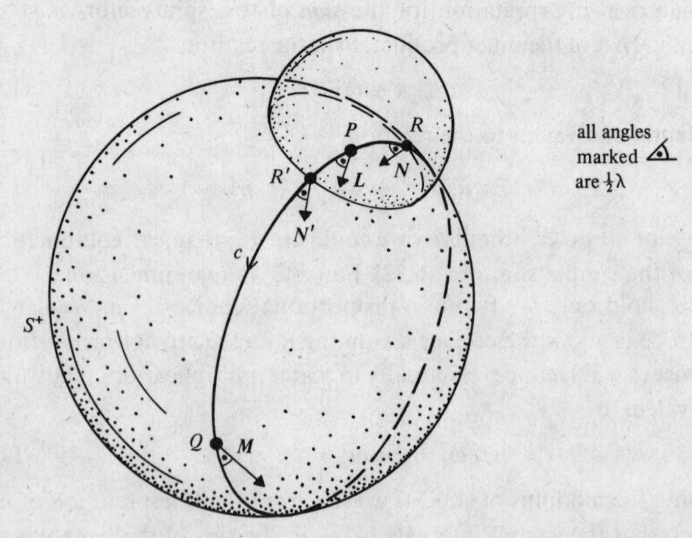


Fig. 1-21. Construction for the sum of two spin-vectors in terms of $S^+$. (Refer to (1.6.36) and (1.6.37).)

characterization for the flag of $\boldsymbol{\kappa}+\boldsymbol{\omega}$ will not fix it uniquely, since it cannot distinguish it from the flag of $\boldsymbol{\kappa}-\boldsymbol{\omega}$. However, these two flags are the *only* possibilities allowed by the construction, as we shall see in a moment.

Let us return to the representation on $S^+$ (Fig. 1-21). We assume $P, Q, \boldsymbol{L}$ and $\boldsymbol{M}$ are given, and try to construct $R$ and $\boldsymbol{N}$. We first construct the circle $c$ as the unique directed circle on $S^+$, through $P$ and $Q$, such that $\boldsymbol{L}$ and $\boldsymbol{M}$ each make an angle $\frac{1}{2}\lambda$ with $c$ (*cf.* (1.6.35)). To locate $R$ on $c$, we make use of (1.6.31) and apply it to (1.6.34):

$$1=\left|\frac{\{\boldsymbol{\kappa},\boldsymbol{\tau}\}}{\{\boldsymbol{\tau},\boldsymbol{\omega}\}}\right|=\frac{PR}{2|\boldsymbol{L}|^{\frac{1}{2}}|\boldsymbol{N}|^{\frac{1}{2}}}\Big/\frac{QR}{2|\boldsymbol{N}|^{\frac{1}{2}}|\boldsymbol{M}|^{\frac{1}{2}}}=\frac{PR}{QR}\cdot\frac{|\boldsymbol{M}|^{\frac{1}{2}}}{|\boldsymbol{L}|^{\frac{1}{2}}}. \tag{1.6.36}$$

By a well-known theorem, the locus of points in Euclidean 3-space, for which the ratio of the distances to two fixed points is constant, is a sphere. Thus, (1.6.36) implies that $R$ lies on a certain sphere defined by $P, Q, \boldsymbol{L}$ and $\boldsymbol{M}$. This sphere meets* $c$ in two points $R, R'$. Since $P$ and $Q$ are separated by the sphere, it follows that $R$ and $R'$ separate $P$ and $Q$ on $c$ (harmonically, in fact). Now one of $R, R'$ will correspond to $\boldsymbol{\kappa}+\boldsymbol{\omega}$ (say $R$) and the other will correspond to $\boldsymbol{\kappa}-\boldsymbol{\omega}$ (say $R'$). In order to be able to pick out which of the intersection points is $R$ and which is $R'$, we need to envisage $\boldsymbol{L}, \boldsymbol{M}$ and $\boldsymbol{N}$ as spinorial objects rather than just as tangent vectors to $S^+$. Let us move $\boldsymbol{L}$ continuously from $P$ to $Q$ along an arc of $c$, keeping the angle which it makes with $c$ a constant $(=\frac{1}{2}\lambda)$. When $\boldsymbol{L}$ has been brought to $Q$ it is expanded by the factor $|\boldsymbol{M}|/|\boldsymbol{L}|$, so that $\boldsymbol{L}$ and $\boldsymbol{M}$ coincide as vectors. If they then also coincide as spinorial objects, $R$ lies *between* the original $\boldsymbol{L}$ and $\boldsymbol{M}$ on the arc of $c$ under consideration and $R'$ lies on the remaining portion of $c$. If they do *not* then coincide as spinorial objects, it is $R'$ that lies between the original $\boldsymbol{L}$ and $\boldsymbol{M}$ on $c$, and $R$ lies on the remaining portion of $c$.

Having located $R$ on $S^+$, we can define $|\boldsymbol{N}|$ by

$$|\boldsymbol{N}|^{\frac{1}{2}}=\left|\frac{\{\boldsymbol{\tau},\boldsymbol{\omega}\}}{\{\boldsymbol{\kappa},\boldsymbol{\omega}\}}\right||\boldsymbol{N}|^{\frac{1}{2}}=\frac{RQ}{PQ}|\boldsymbol{L}|^{\frac{1}{2}}, \tag{1.6.37}$$

while the direction of $\boldsymbol{N}$ is fixed such that $\boldsymbol{N}$ makes an angle $\frac{1}{2}\lambda$ with $c$. Finally, as a spinorial object, $\boldsymbol{N}$ is defined so that: if it is moved continuously along $c$, keeping its angle with $c$ a constant, until it *first* encounters $P$ or $Q$; and if it is then expanded until it coincides, as a vector, with either

* The intersection of the sphere with $S^+$ is, of course, a circle. Indeed, (1.6.36) shows that.

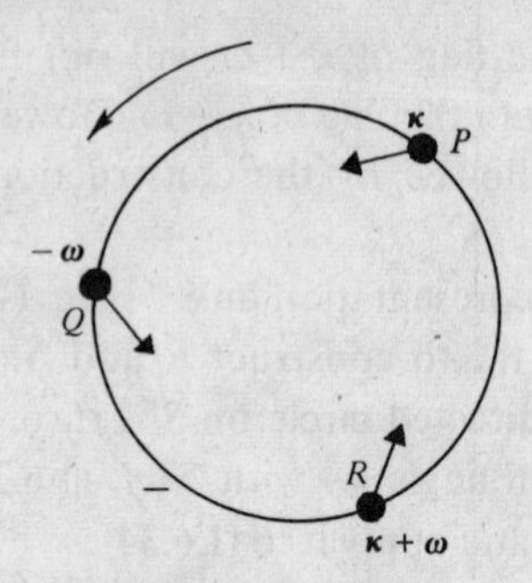


Fig. 1-22. Two spin-vectors and their sum, represented in the Argand plane with respect to a special frame which brings out the symmetry (1.6.38).

***L*** or ***M*** (as the case may be); then this coincidence must also be a coincidence of the spinorial objects. (See Fig. 1-21).

There are two special Minkowski frames which are convenient for visualizing the above situation. Since we can choose any three points of $S^+$ to occupy any three (distinct) pre-assigned positions (*cf.* §1.3) by a suitable choice of frame, let us choose $P$, $Q$ and $R$ to be equally spaced around the equator of $S^+$. Then $P$, $Q$ and $R$ are the vertices of an equilateral triangle. The vectors ***L***, ***M*** and ***N*** are now of equal length (and are inclined equally to the circle $PQR$). It is clear from the symmetry that a rotation through $2\pi/3$ about an axis perpendicular to $PQR$ will send the configuration into itself. It will also have the effect:

$$\boldsymbol{\kappa} \mapsto -\boldsymbol{\omega}, \quad \boldsymbol{\omega} \mapsto \boldsymbol{\kappa} + \boldsymbol{\omega}, \quad \boldsymbol{\kappa} + \boldsymbol{\omega} \mapsto \boldsymbol{\kappa} \tag{1.6.38}$$

(or the inverse of this), by virtue of the above description of the signs of the spinorial objects (Fig. 1-22). The transformation (1.6.38) is readily seen to be obtainable as the result of a (unique) spin transformation. In fact, we may use this to establish the correctness of the above prescription for the signs.

Alternatively, we may choose our ***t***-axis so that the *four* points $P$, $R'$, $Q$, $R$ are equally spaced around the equator of $S^+$, forming the vertices of a square. That is, we pre-assign $P$, $Q$ and $R$; then by symmetry $R'$ takes up the position diametrically opposite to $R$, since $PR = QR$, whence $|\boldsymbol{L}| = |\boldsymbol{M}|$, so that $PR' = QR'$. Now a rotation through $\pi/4$ about an axis perpendicular to $PQR$ achieves (see Fig. 1-23)

$$\boldsymbol{\kappa} \mapsto \frac{1}{\sqrt{2}}(\boldsymbol{\kappa} - \boldsymbol{\omega}), \frac{1}{\sqrt{2}}(\boldsymbol{\kappa} - \boldsymbol{\omega}) \mapsto -\boldsymbol{\omega}, \boldsymbol{\omega} \mapsto \frac{1}{\sqrt{2}}(\boldsymbol{\omega} + \boldsymbol{\kappa}), \frac{1}{\sqrt{2}}(\boldsymbol{\omega} + \boldsymbol{\kappa}) \mapsto \boldsymbol{\kappa},$$

which is again obtainable by a unique spin transformation. The factor

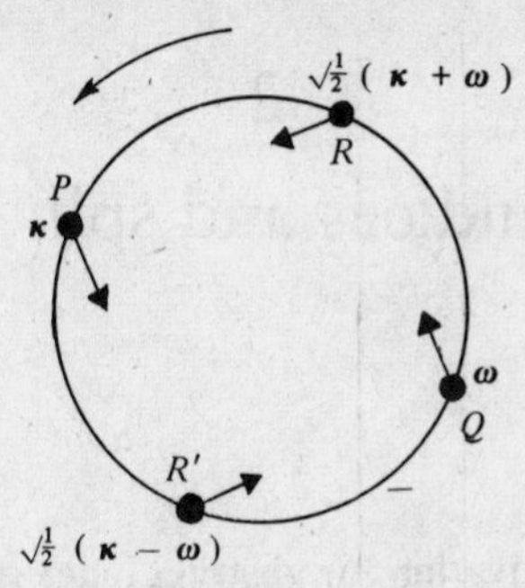


Fig. 1-23. Representation in the Argand plane with respect to a frame bringing out the symmetry (1.6.39).

$1/\sqrt{2}$ arises because now $|\boldsymbol{N}| = \frac{1}{2}|\boldsymbol{L}|$. (That is, the $t$-value of the flagpole for $\boldsymbol{\kappa} + \boldsymbol{\omega}$ is just *twice* that for $\boldsymbol{\kappa}$.)

Now that we have obtained geometrical descriptions of all the three basic spinor operations, the way is opened for a completely synthetic geometrical definition of the basic algebra of spin-vectors. All that would remain to be done is a geometrical verification of the basic properties (1.6.8)–(1.6.19). We do not propose to spell this out here, and merely leave it as an exercise for the interested reader. Some of the properties are trivially verified, but others are rather tedious if tackled directly. It is perhaps worth mentioning that (1.6.19) is almost immediate from the above constructions for inner product and for addition. This property is of help in verifying some of the others.

# 2
# Abstract indices and spinor algebra

## 2.1 Motivation for abstract index approach

In Chapter 1 we introduced the concept of a spin-vector. We saw that a spin-vector is to be pictured essentially as a null flag in Minkowski vector space, but with the additional property that under rotation through $2\pi$ about any axis it is returned, not to its original state, but to another spin-vector associated with the same null flag, called the negative of the original spin-vector. Spin-vectors form a two-dimensional complex vector space, called spin-space, on which a skew-symmetric inner product is also defined. All operations have explicit geometrical, Lorentz invariant interpretations, in space–time terms.

Later in this chapter (in §2.5) we shall develop the *algebra of spinors*. The essential idea is that spinors may be constructed, starting from the basic concept of spin-space, analogously to the way that *tensors* are built up starting from the concept of a vector space. It will emerge, moreover (in §3.1), that the world-tensor algebra of space–time is *contained* in the spinor algebra. Thus spin-space is, in a sense, even more basic than world-vector space. It is, therefore, *conceptually* very valuable that spin-space has a clear-cut geometrical space–time interpretation. For this removes much of the abstractness which has tended to cloud the spinor concept. It shows, furthermore, that while we shall describe spinors and spinor operations in this (and subsequent) chapters in a largely *algebraic* way, nevertheless each such object and operation will have an essential geometrical content in space–time terms.

However, our algebraic development will by no means rest on these geometrical interpretations. Our treatment can be made to be logically independent of the geometrical background suggested in Chapter 1. We shall describe the structures we are interested in by using algebraic formulations. This would actually enable us to turn the logical sequence around the other way. We could *define* the space–time geometry in terms of the algebraic structure that we shall erect – which should seem fairly simple and natural once the main idea is grasped. Thus, ultimately we may tend to regard the algebraic rules defining the spinor system as more primitive than the (somewhat complicated) explicit geometrical constructions of Chapter 1. Even the concept of a spinorial object will be effectively incorporated by our algebraic approach. Thus it will not be *necessary* to base a rigorous development of spinor algebra on the geometry and topology described in Chapter 1. In general, we shall find that the value of the geometrical constructs will be primarily *conceptual*, whereas the algebraic method will be indispensable for detailed manipulations.

The spinor algebra that we shall build up will involve generalizations in two different directions from the concept of spin-space that we developed in §1.5. In the first place, it will be convenient to consider not just spin-vectors at a single point in space–time,* but spin-vector *fields*. This generalization is analogous to that which takes us from the concept of a vector at one point to the concept of a vector field. World vectors at a single point in a space–time are subject to the operations of addition, scalar multiplication and inner product. Such vectors form a Minkowski vector space, called the tangent space at the point, over the division ring** of scalars at that point. Vector fields are subject to just the same operations. For example, to add two vector fields, we simply add the two vectors at each point to obtain the resultant vector field. In scalar multiplication a vector field is multiplied by a scalar field, the value of the scalar field at each point multiplying the vector at that point. The familiar laws (1.1.1) for a vector space still hold for vector and scalar fields. The only new feature which arises here is that the scalar fields do not form a division ring, but only a commutative ring with unit. (For example, if $h$ and $k$ are two infinitely differentiable scalar fields, it may be that $h$ is non-zero only in a region throughout which $k$ vanishes. Then $hk = 0$ but neither $h$ nor $k$ need be zero, so divisors of zero exist. In any case, it is clear that $h^{-1}$ cannot exist if $h$ vanishes at any point.) Because of this feature, vector fields are not said to form a vector space, but a system called, instead, a *module*, over the ring of scalar fields (MacLane and Birkhoff 1967, Herstein 1964).

In the same way, we generalize the concept of spin-space, which is a two-dimensional vector space over the divison ring of complex numbers, to the concept of the *module of spin-vector fields*. The scalars must here be

* The precise meaning of the term 'space–time' will not concern at here, but cf. §§1.5, 3.2, 4.1.

** It is unfortunate that two quite distinct meanings for the word 'field' come into conflict here. For this reason we are using the term 'division ring' for the algebraic notion of 'field'. Our division rings will all be commutative.

suitably smooth *complex* scalar fields. A spin-vector field defines a null flag at each point of the space–time. A null flag at a point is a structure in the tangent space at that point, this tangent space being the Minkowski vector space of world vectors at the point. In addition, there must be suitable smoothness in the way that the null flags vary from point to point, and there will also be the topological requirements arising from the global compatibility of the concept of spinorial object. All this will be implicit in the axioms for the precise system that we shall set up. In essence, then, our first generalization involves passage from a vector space over a division ring to a module over a ring.

The second generalization involves passage from 'univalent' objects (spin-vectors) to 'polyvalent' objects (spinors). This will be analogous to the way in which the concept of a tensor is built up from that of an ordinary vector. The construction is essentially the same whether we start with (spin-) vectors at one point or with (spin-) vector fields. In general, therefore, we shall not be too explicit about which type of system we are working with. We shall be able to develop this second generalization to a considerable extent before worrying about the details of the first generalization.

## *Classical tensor algebra*

Let us motivate our discussion by recalling the basic operations of classical tensor algebra. The latter deals with arrays

$$A_{\rho\ldots\tau}^{\alpha\ldots\gamma}, \tag{2.1.1}$$

where each of the indices $\alpha,\ldots,\gamma,\ \rho,\ldots,\tau$ takes values from $1, 2,\ldots,n$ (the dimension of the space under consideration being $n$). To fix ideas we may imagine these to be arrays of real (or complex) numbers. Alternatively, they could be, say, arrays of functions of $n$ variables. However, as we shall elaborate shortly, we should not really think of the array (2.1.1) actually *as* a tensor, but merely as a set of tensor components. Furthermore, tensors in general need not be globally describable in this way.

The permitted operations are as follows. Given two such arrays $A_{\rho\ldots\tau}^{\alpha\ldots\gamma}$ and $B_{\rho\ldots\tau}^{\alpha\ldots\gamma}$, where the two sets of upper indices are equal in number and the two sets of lower indices are also equal in number, then we can add corresponding elements to obtain the *sum*:

$$A_{\rho\ldots\tau}^{\alpha\ldots\gamma} + B_{\rho\ldots\tau}^{\alpha\ldots\gamma} = C_{\rho\ldots\tau}^{\alpha\ldots\gamma}. \tag{2.1.2}$$

Given any two such arrays (with no restriction now on numbers of indices) $A_{\rho\ldots\tau}^{\alpha\ldots\gamma}, D_{\xi\ldots\zeta}^{\lambda\ldots\nu}$, we can multiply each element of one array with each separate element of the other to obtain the (outer) *product* array:

$$A^{\alpha\ldots\gamma}_{\rho\ldots\tau} D^{\lambda\ldots\nu}_{\xi\ldots\zeta} = E^{\alpha\ldots\gamma\lambda\ldots\nu}_{\rho\ldots\tau\xi\ldots\zeta}. \tag{2.1.3}$$

Given any array $F^{\alpha\ldots\gamma\lambda}_{\rho\ldots\tau\mu}$ with at least one upper index and at least one lower index, we can form the contraction:

$$F^{\alpha\ldots\gamma\lambda}_{\rho\ldots\tau\lambda} = G^{\alpha\ldots\gamma}_{\rho\ldots\tau}. \tag{2.1.4}$$

Here the summation convention is being employed for the repeated index $\lambda$ so each element of the resulting array is a sum of $n$ terms of the original array. Finally, given an array $A^{\alpha\ldots\gamma}_{\rho\ldots\tau}$, with $p$ upper and $q$ lower indices, we can form $p!q!$ (generally) different arrays from it, by simply relabelling the upper indices and the lower indices in their various possible orders. For example, given the array $A^{\alpha\beta\gamma}_{\rho\tau}$ we can form arrays such as

$$H^{\alpha\beta\gamma}_{\rho\tau} = A^{\beta\alpha\gamma}_{\rho\tau}, \quad K^{\alpha\beta\gamma}_{\rho\tau} = A^{\gamma\alpha\beta}_{\tau\rho}. \tag{2.1.5}$$

The *elements* of these arrays are, of course, precisely the same as the elements of the original array, but they are arranged in different orders.

We may ask what is the special significance of these four particular operations. The answer is that they commute with the transformation law for tensor components. Thus, if we consider a replacement of each array by a new one according to the scheme:

$$A^{\alpha\ldots\gamma}_{\rho\ldots\tau} \mapsto A^{\lambda\ldots\nu}_{\varphi\ldots\psi} t^{\alpha}_{\lambda} \cdots t^{\gamma}_{\nu} T^{\varphi}_{\rho} \cdots T^{\psi}_{\tau} \tag{2.1.6}$$

(summation convention!), where the matrices $t^{\alpha}_{\beta}$ and $T^{\alpha}_{\beta}$ are made up of elements of the same type as those appearing in (2.1.1) (e.g., real numbers), and are inverses of one another,

$$t^{\alpha}_{\beta} T^{\beta}_{\gamma} = \delta^{\alpha}_{\gamma} (= T^{\alpha}_{\beta} t^{\beta}_{\gamma}), \tag{2.1.7}$$

then we can verify that each of the equations (2.1.2)–(2.1.5) is preserved unchanged.

In the particular case when there is just a single upper index, we may regard the replacement (2.1.6),

$$V^{\alpha} \mapsto V^{\beta} t^{\alpha}_{\beta}, \tag{2.1.8}$$

as reassigning components to a vector $\boldsymbol{V}$ under a change of basis*. In the same way, we may regard the replacement (2.1.6) as reassigning components to a *tensor* under the change of basis which effects (2.1.8). But what exactly *is* this abstract object we have called a 'tensor'? There are in fact several different ways of defining a tensor. These will emerge in our subsequent discussion. But in the present context, the most immediate

* The notation in (2.1.8) is slightly at variance with our general conventions for a 'passive' transformation (*cf.* (1.1.25)), but this should not cause confusion.

definition of a tensor $\boldsymbol{A}$, once we have the concept of a vector $\boldsymbol{V}$, is simply that $\boldsymbol{A}$ is a *rule* which assigns to every choice of basis for our space of vectors an array of components (2.1.1), this rule being such that whenever one basis is replaced by another, the resulting array is replaced according to the scheme (2.1.6); here $T^{\alpha}_{\beta}$ is defined by (2.1.7) and $t^{\alpha}_{\beta}$ is such that (2.1.8) gives the change of *vector* components under this basis change*. The consistency of the definition follows from the group properties of the $t^{\alpha}_{\beta}$ matrices. Thus, if $t^{\beta}_{\alpha}, \tilde{t}^{\beta}_{\alpha}$ and $\tilde{\tilde{t}}^{\beta}_{\alpha}$ are the transformation matrices corresponding, respectively, to a replacement of a first basis by a second; to a replacement of the second basis by a third; and finally, to a replacement of the first basis by the third, then we have $t^{\beta}_{\alpha}\tilde{t}^{\gamma}_{\beta} = \tilde{\tilde{t}}^{\gamma}_{\alpha}$. If $T^{\beta}_{\alpha}$, $\tilde{T}^{\beta}_{\alpha}$ and $\tilde{\tilde{T}}^{\beta}_{\alpha}$ are the corresponding inverse matrices, then $T^{\gamma}_{\beta}\tilde{T}^{\beta}_{\alpha} = \tilde{\tilde{T}}^{\gamma}_{\alpha}$. This ensures that if a basis change is made in two stages, the resulting array of components will be the same as if the basis change is made directly. The significance, now, of the fact that the transformation law for tensor components commutes with the operations of addition, multiplication, contraction and index permutation as applied to arrays, is that it implies that these operations may be applied to the *tensors* themselves, not merely to the components.

A tensor $\boldsymbol{A}$ is said to have *valence* $\left[{p \atop q}\right]$ if there are $p$ upper indices and $q$ lower indices in (2.1.1). The upper indices are called *contravariant* and the lower indices are called *covariant* (The *total* number $p+q$ of indices is sometimes called the *rank* of $\boldsymbol{A}$. We prefer the term *total valence*.) Each tensor of valence $\left[{1 \atop 0}\right]$ is naturally associated with a unique vector, namely the one whose components are identical (in any basis) with those of the tensor. If it were not for the logical circularity that would result because of our particular tensor definition, we could actually *identify* tensors of valence $\left[{1 \atop 0}\right]$ as vectors. (One trouble that arises with this tensor definition is that a basis is already a set of vectors!) Let us call a tensor of valence $\left[{1 \atop 0}\right]$ a *contravariant vector*. A tensor of valence $\left[{0 \atop 1}\right]$ is also a kind vector. Let us call it a *covariant vector* or *covector*.** For any valence

* This stated definition is adequate only if the space of vectors possesses a (finite) basis. This will actually be the case in the situations of interest to us here, although not *obviously* so in the case of globally defined fields. However, the present definition is provisional in any case, serving mainly to motivate the subsequent discussion.

** The intuitive geometrical picture of a (contravariant) vector field on a manifold is a field of 'arrows' on the manifold (which we may think of as pointing from each point to some neighboring point, defining a 'flow' on the manifold). A covector field, on the other hand, defines a field of oriented hyperplane elements, each of which is assigned a kind of 'strength'. In the case of the gradient of a scalar, the hyperplane elements are tangent to the level hypersurfaces of the scalar, the 'strength' measuring its rate of increase.

$\left[{p \atop q}\right]$, there is a special tensor $\mathbf{0}$ whose components are zero in any basis. Also there is a special tensor $\boldsymbol{\delta}$ of valence $\left[{1 \atop 1}\right]$ whose components, in *any* basis, constitute the 'Kronecker delta'

$$\delta_{\boldsymbol{\alpha}}^{\boldsymbol{\beta}} = \begin{cases} 1 & \text{if } \boldsymbol{\alpha} = \boldsymbol{\beta} \\ 0 & \text{if } \boldsymbol{\alpha} \neq \boldsymbol{\beta} \end{cases}. \tag{2.1.9}$$

A tensor of valence $\left[{0 \atop 0}\right]$ is called a *scalar*.

### *Coordinate-free tensor algebra*

The definition of a tensor that we gave above is reasonably satisfactory from the logical point of view (although there are some contexts in which such a definition cannot be used). However, conceptually it lays rather too much emphasis on bases and components. It leaves us with the impression that the tensor concept is closely bound up with the concept of a vector basis (or of a coordinate system). It tends also to suggest that detailed calculations in terms of tensors will normally require bases to be introduced, so that the calculations can be performed in terms of the tensor components. But these impressions are false. The modern algebraic definition of a tensor (in its various guises) avoids any reference to bases or coordinate systems. (In fact, the tensor concept has relevance also in situations in which bases or coordinate systems need not exist in the ordinary sense.) The emphasis in this book will be very much on such an algebraic, coordinate-free, development. For, particularly in the case of spinor analysis, we feel that there has been a tendency to read too much significance into the apparent need for coordinate bases.

As regards detailed tensorial calculations, on the other hand, the modern algebraic approach to tensors, *in the form in which it has been normally given*, has certain distinct disadvantages. The reason lies essentially in the notation. When calculations are carried out using tensor *components*, the classical tensor index notation, together with the Einstein summation convention, is the foundation of a very powerful and versatile technique, a technique whose utility rests in large measure on the possibility of treating the individual indices separately. But in the usual abstract algebraic approach there are *no* indices, and consequently much of this versatility is lost. In fact, it is often the case with this more abstract approach that, when a detailed computation becomes necessary, one reverts briefly to the description in terms of components, and re-interprets the equations as relating abstract tensors only at the end of the computation.

To be more explicit about the difficulties confronting the abstract approach, let us consider how we might operate algebraically with tensors $\boldsymbol{A}, \boldsymbol{B}, \boldsymbol{C}, \ldots$ directly, while attempting to dispense with any form of index notation. The first of our operations, namely addition, presents no difficulty. If $\boldsymbol{A}$ and $\boldsymbol{B}$ are two tensors of equal valence $\left[{p \atop q}\right]$, then we can form the sum:

$$\boldsymbol{A} + \boldsymbol{B} = \boldsymbol{C}, \tag{2.1.10}$$

also of valence $\left[{p \atop q}\right]$, the components being related by (2.1.2). For each $\left[{p \atop q}\right]$, we have an *Abelian group* structure defined by addition, namely (all tensors in the next four lines having the same valence $\left[{p \atop q}\right]$):

$$\boldsymbol{A} + \boldsymbol{B} = \boldsymbol{B} + \boldsymbol{A}; \tag{2.1.11}$$

$$(\boldsymbol{A} + \boldsymbol{B}) + \boldsymbol{C} = \boldsymbol{A} + (\boldsymbol{B} + \boldsymbol{C}); \tag{2.1.12}$$

$$\text{there exists } \boldsymbol{0} \text{ such that } \boldsymbol{0} + \boldsymbol{A} = \boldsymbol{A}; \tag{2.1.13}$$

$$\text{for each } \boldsymbol{A} \text{ there exists } -\boldsymbol{A} \text{ such that } \boldsymbol{A} + (-\boldsymbol{A}) = \boldsymbol{0}. \tag{2.1.14}$$

Similarly, outer multiplication presents no real problem. If $\boldsymbol{A}$ has valence $\left[{p \atop q}\right]$ and $\boldsymbol{D}$ has valence $\left[{r \atop s}\right]$ then we form the product

$$\boldsymbol{A}\boldsymbol{D} = \boldsymbol{E} \tag{2.1.15}$$

of valence $\left[{p+r \atop q+s}\right]$, the components being related according to (2.1.3). Outer multiplication assigns a (non-commutative) semigroup structure to the entire system of tensors:

$$(\boldsymbol{A}\boldsymbol{B})\boldsymbol{C} = \boldsymbol{A}(\boldsymbol{B}\boldsymbol{C}). \tag{2.1.16}$$

(We have $\boldsymbol{A}\boldsymbol{B} \neq \boldsymbol{B}\boldsymbol{A}$ in general, since, although the sets of components of $\boldsymbol{A}\boldsymbol{B}$ and of $\boldsymbol{B}\boldsymbol{A}$ are the same, they are arranged in different orders.) Multiplication is distributive over addition:

$$\boldsymbol{A}(\boldsymbol{B} + \boldsymbol{C}) = \boldsymbol{A}\boldsymbol{B} + \boldsymbol{A}\boldsymbol{C}, \tag{2.1.17}$$

$$(\boldsymbol{B} + \boldsymbol{C})\boldsymbol{A} = \boldsymbol{B}\boldsymbol{A} + \boldsymbol{C}\boldsymbol{A}, \tag{2.1.18}$$

where $\boldsymbol{B}$ and $\boldsymbol{C}$ have the same valence.

The contraction operation whose component form is defined by (2.1.4) again presents no real problem. If $\boldsymbol{F}$ has valence $\left[{p+1 \atop q+1}\right]$, then, denoting the contraction operation by $\mathscr{C}$, we have (*cf.* (2.1.4))

$$\mathscr{C}\boldsymbol{F} = \boldsymbol{G}, \tag{2.1.19}$$

which is a tensor of valence $\left[{p \atop q}\right]$. Contraction is related to addition and multiplication by the two rules

$$\mathscr{C}(\boldsymbol{A} + \boldsymbol{B}) = \mathscr{C}\boldsymbol{A} + \mathscr{C}\boldsymbol{B}, \tag{2.1.20}$$

$$\mathscr{C}(\boldsymbol{E}\boldsymbol{F}) = \boldsymbol{E}\mathscr{C}\boldsymbol{F}, \tag{2.1.21}$$

where $\boldsymbol{A}$ and $\boldsymbol{B}$ have the same valence and where the valence of $\boldsymbol{F}$ is positive in both upper and lower positions.

On the other hand, the apparently innocuous but *essential* fourth operation of index permutation presents in this approach a very serious problem. We need the operation in order to express symmetries of tensors; we need a special case of it in order to express the relation between $\boldsymbol{AB}$ and $\boldsymbol{BA}$; having defined but the one contraction operation $\mathscr{C}$, we also need it, in conjunction with $\mathscr{C}$, in order to form the general contractions (in which indices other than just the final ones are contracted over). It is possible to devise special index-free notations to cope with certain simple tensorial relations of this kind, but for generality, transparency, and flexibility, such notations compare very unfavourably with the original tensor index notation when expressions of any complexity occur. (The only alternative to the tensor index notation which retains all these virtues seems to be some form of diagrammatic notation: *cf.* Penrose 1971, Cvitanović 1976; *cf.* also t'Hooft and Veltman 1973. Some suggestions along these lines are outlined in the Appendix. Unfortunately, such notations present severe difficulties for the printer and would appear to be appropriate mainly for private calculations.)

Once we have resigned ourselves to this fact, it becomes clearer how we must proceed. The tensor index notation should be retained. However, this does *not* imply abandonment of our ideal of a completely basis-independent approach. The advantages enjoyed by the abstract algebraic and by the tensor component methods are not mutually exclusive. In the approach that we shall adopt in the next section, the full flexibility of the index notation will be retained, together with complete frame-independence from the outset. The key to this formalism will lie in the recognition that an index letter *need not* represent one of a set of integers (e.g. $1, 2, \ldots, n$) over which it is to range. One may, rather, regard a tensor index simply as a *label* whose sole purpose is to keep track of the type of tensor under discussion and of the particular operations to which the tensor is being subjected. The calculations may be performed using indexed quantities exactly as in classical tensor algebra. But now the meanings of the symbols will be quite different. Each indexed symbol will describe an entire tensor, no coordinate system or basis frame being implicitly or explicitly involved. Exactly how this is achieved will be described in the next section.

## 2.2 The abstract-index formalism for tensor algebra

A symbol $V^\alpha$, say, will be taken to mean, *not* an $n$-tuple of components $(V^1, V^2, \ldots, V^n)$ but a single element of some abstract vector space or module. It may, indeed, sometimes be convenient to choose a basis frame and represent the vector $V^\alpha$ in this basis. But then a different type of index letter has to be employed: Bold face upright index letters serve for this purpose. Thus, $V^{\boldsymbol{\alpha}}$ *will* stand for the set of components $(V^1, \ldots, V^n)$. Bold face upright indices will thus be used in the conventional way, but in general we avoid their use as much as possible. The presence of a bold face index in an expression will signify two things. In the first place, it indicates that a (possibly arbitrary) basis frame is implicitly involved in the expression, in addition to all the tensors or scalars which appear explicitly. Secondly, such indices will be subject to the summation convention (*cf.* (1.1.4)). As applied to light face italic indices, the summation convention in the usual sense would be meaningless. However, contraction (as well as the other tensor operations) *will* be defined on abstract tensors. We shall define basis-free operations in such a way that it will be possible to mirror exactly the familiar rules for the manipulation of tensor components.

There is, however, one awkward feature that makes its appearance at the outset, and to which one must become accustomed. In the classical tensor notation, expressions such as $V^{\boldsymbol{\alpha}}V^{\boldsymbol{\beta}}$ or $V^{\boldsymbol{\alpha}}U^{\boldsymbol{\beta}} - V^{\boldsymbol{\beta}}U^{\boldsymbol{\alpha}}$ are frequently considered. If we are to mirror this notation with abstract indices, i.e. $V^\alpha V^\beta$ and $V^\alpha U^\beta - V^\beta U^\alpha$, then we need an object $V^\beta$, in addition to $V^\alpha$, both of which stand for the same vector $\boldsymbol{V}$. Clearly, $V^\alpha$ and $V^\beta$ must be *different* objects. For if $V^\alpha = V^\beta$ and $U^\alpha = U^\beta$ were valid equations, then we should be led to $V^\alpha U^\beta - V^\beta U^\alpha = V^\alpha U^\beta - V^\alpha U^\beta = 0$. Furthermore, $V^\alpha = V^\beta$ would mirror the invalid classical expression $V^{\boldsymbol{\alpha}} = V^{\boldsymbol{\beta}}$. Thus any vector $\boldsymbol{V}$ must have associated with it an infinite collection of distinct copies $V^\alpha, V^\beta, V^\gamma, \ldots, V^{\alpha_0}, \ldots, V^{\alpha_1}, \ldots$ (since arbitrarily long expressions must be allowable). The entire module $\mathfrak{S}^\cdot$ to which $\boldsymbol{V}$ belongs must therefore possess infinitely many completely separate copies of itself. These we shall denote by $\mathfrak{S}^\alpha, \mathfrak{S}^\beta, \mathfrak{S}^\gamma, \ldots, \mathfrak{S}^{\alpha_0}, \ldots, \mathfrak{S}^{\alpha_1}, \ldots$. The modules will be canonically isomorphic to one another and to $\mathfrak{S}^\cdot$, with $\boldsymbol{V} \in \mathfrak{S}^\cdot$ corresponding to $V^\alpha \in \mathfrak{S}^\alpha$ and to $V^\beta \in \mathfrak{S}^\beta$, etc. Thus, $a\boldsymbol{V} + b\boldsymbol{U} = \boldsymbol{W}$ iff* $aV^\alpha + bU^\alpha = W^\alpha$ iff $aV^\beta + bU^\beta = W^\beta$, etc., where $a, b$ are elements of $\mathfrak{S}$, the ring of scalars.

It will, no doubt, seem rather unnatural to need infinitely many separate mathematical objects to represent what is really a single entity.

* We use the term 'iff' in its usual mathematical sense of 'if and only if', or '$\Leftrightarrow$'.

Nevertheless, one must get used to it. But there is a slightly different way of viewing the situation which may seem a little less unnatural. The set of abstract labels

$$\mathscr{L} = \{\alpha, \beta, \gamma, \dots, \omega, \alpha_0, \beta_0, \dots, \alpha_1, \dots\} \tag{2.2.1}$$

has an organizational significance only. Its one necessary property is to be infinite. The vectors (or vector fields) with which we are concerned constitute the module $\mathfrak{S}^{\bullet}$. The elements of the various sets $\mathfrak{S}^{\alpha}, \mathfrak{S}^{\beta}, \dots,$ $\mathfrak{S}^{\alpha_0}, \dots$ are simply the elements of $\mathfrak{S}^{\bullet} \times \mathscr{L}$, with $\mathfrak{S}^{\xi_3}$, for example, being $\mathfrak{S}^{\bullet} \times \{\xi_3\}$. That is to say, $V^{\xi_3}$ is simply a *pair* $(V, \xi_3)$ with $V \in \mathfrak{S}^{\bullet}$ and $\xi_3 \in \mathscr{L}$. Each abstract index such as $\xi_3$ is therefore just a kind of organizational marker used to 'store' the vector $V$ in the 'compartment $\xi_3$'.

*Axiomatic development*

Let us now be rather formal about the rules our system has to satisfy. By setting up axioms we shall be able to avoid tying ourselves down to any one particular interpretation. The generality provided will be ample for application to the particular systems that we shall need, and its should also suffice in other contexts.

In the first place, the axioms for the set $\mathfrak{S}$ of scalars state that it is a commutative ring with unit. That is to say, there exist operations of addition and multiplication on $\mathfrak{S}$, satisfying the following

(2.2.2) AXIOMS:

(i) $a + b = b + a$
(ii) $a + (b + c) = (a + b) + c$
(iii) $ab = ba$
(iv) $a(bc) = (ab)c$
(v) $a(b + c) = ab + ac$

*for all* $a, b, c \in \mathfrak{S}$ *and*

(vi) *there is a zero element* $0 \in \mathfrak{S}$ *such that* $0 + a = a$ *for all* $a \in \mathfrak{S}$
(vii) *there is a unit element* $1 \in \mathfrak{S}$ *such that* $1a = a$ *for all* $a \in \mathfrak{S}$
(viii) *for each* $a \in \mathfrak{S}$ *there exists an element* $-a \in \mathfrak{S}$, *the additive inverse of* $a$, *such that* $a + (-a) = 0$.

In general, we shall not have multiplicative inverses. (For example, as we have remarked earlier, scalar fields on a manifold can have divisors of zero.) If a scalar does have a multiplicative inverse, then this is unique. The additive inverse is always unique, as are the zero and unit elements.

We now require an infinite labelling set $\mathscr{L}$. We shall denote the elements of $\mathscr{L}$ as in (2.2.1). Select an element $\alpha$ from $\mathscr{L}$. Then the system $\mathfrak{S}^\alpha$ is to be an $\mathfrak{S}$-module; that is to say, an operation of addition is defined on $\mathfrak{S}^\alpha$ and an operation of scalar multiplication is defined from $\mathfrak{S} \times \mathfrak{S}^\alpha$ to $\mathfrak{S}^\alpha$, satisfying the following axioms

(2.2.3) AXIOMS

(i) $U^\alpha + V^\alpha = V^\alpha + U^\alpha$
(ii) $U^\alpha + (V^\alpha + W^\alpha) = (U^\alpha + V^\alpha) + W^\alpha$
(iii) $a(U^\alpha + V^\alpha) = aU^\alpha + aV^\alpha$
(iv) $(a + b)U^\alpha = aU^\alpha + bU^\alpha$
(v) $(ab)U^\alpha = a(bU^\alpha)$
(vi) $1U^\alpha = U^\alpha$
(vii) $0U^\alpha = 0V^\alpha$

for all $a, b \in \mathfrak{S}$ and all $U^\alpha, V^\alpha, W^\alpha \in \mathfrak{S}^\alpha$. The unique zero vector $0V^\alpha$ is written $0^\alpha$ or, more frequently, simply 0. Each $V^\alpha \in \mathfrak{S}^\alpha$ has an additive inverse $(-1)V^\alpha$, since $V^\alpha + (-1)V^\alpha = (1 + (-1))V^\alpha = 0V^\alpha = 0$. As is conventional, we write $-V^\alpha$ for $(-1)V^\alpha$ and $U^\alpha - V^\alpha$ for $U^\alpha + (-1)V^\alpha$. (In fact, axiom (i) is a consequence of (ii)–(vii). To prove this, expand $(1 + 1)(U^\alpha + V^\alpha)$, once using (iii) and once using (iv), equate the two expressions and cancel the extra $U^\alpha$ and $V^\alpha$, using the existence of additive inverses. In a similar way, in (2.2.2) (i) is also a consequence of the remaining axioms (2.2.2).)

Now select another label $\beta \in \mathscr{L}$. Define $\mathfrak{S}^\beta$ to be canonically isomorphic to $\mathfrak{S}^\alpha$, so $U^\beta + V^\beta$ and $aV^\beta$ correspond respectively to $U^\alpha + V^\alpha$ and to $aV^\alpha$, where $U^\beta \in \mathfrak{S}^\beta$ corresponds to $U^\alpha \in \mathfrak{S}^\alpha$, and $V^\beta \in \mathfrak{S}^\beta$ to $V^\alpha \in \mathfrak{S}^\alpha$. Thus the same rules (2.2.3) hold for $\mathfrak{S}^\beta$ as for $\mathfrak{S}^\alpha$ (except that $\beta$ replaces $\alpha$ throughout). For $\gamma, \delta, \ldots, \xi_3, \ldots \in \mathscr{L}$, we similarly define $\mathfrak{S}^\gamma, \mathfrak{S}^\delta, \ldots, \mathfrak{S}^{\xi_3}, \ldots$. In any valid equation in which the label $\alpha$ appears (but not $\beta$) we may replace $\alpha$ by $\beta$ throughout and a new valid equation results. The same holds for any other pair of elements of $\mathscr{L}$. The elements of each of the sets $\mathfrak{S}^\alpha, \mathfrak{S}^\beta, \ldots, \mathfrak{S}^{\xi_3}, \ldots$ will be called tensors of valence $\left[\begin{smallmatrix}1\\0\end{smallmatrix}\right]$. The elements of $\mathfrak{S}$ are tensors of valence $\left[\begin{smallmatrix}0\\0\end{smallmatrix}\right]$.

To define tensors of valence $\left[\begin{smallmatrix}0\\1\end{smallmatrix}\right]$, we take the *duals* (strictly, $\mathfrak{S}$-duals) of the modules $\mathfrak{S}^\alpha, \mathfrak{S}^\beta, \ldots$. The dual $\mathfrak{S}_\alpha$ of $\mathfrak{S}^\alpha$ is defined as the collection of all $\mathfrak{S}$-linear mappings of $\mathfrak{S}^\alpha$ into $\mathfrak{S}$. That is to say, each element $Q_\alpha \in \mathfrak{S}_\alpha$ is a map $Q_\alpha : \mathfrak{S}^\alpha \to \mathfrak{S}$ such that

$$Q_\alpha(U^\alpha + V^\alpha) = Q_\alpha(U^\alpha) + Q_\alpha(V^\alpha), \tag{2.2.4}$$

$$Q_\alpha(aV^\alpha) = aQ_\alpha(V^\alpha), \tag{2.2.5}$$

for all $U^\alpha, V^\alpha \in \mathfrak{S}^\alpha$ and all $a \in \mathfrak{S}$. Thus, two elements $Q_\alpha, R_\alpha \in \mathfrak{S}_\alpha$ are equal *by definition* if, for each element $V^\alpha$ of $\mathfrak{S}^\alpha$, $Q_\alpha(V^\alpha)$ and $R_\alpha(V^\alpha)$ are the same element of $\mathfrak{S}$. We shall normally omit the parentheses and write $Q_\alpha(V^\alpha)$ simply as $Q_\alpha V^\alpha$. We shall also sometimes write this $V^\alpha Q_\alpha$. This operation is called *scalar product*.

We define addition of pairs of elements of $\mathfrak{S}_\alpha$, and multiplication of elements of $\mathfrak{S}_\alpha$ by elements of $\mathfrak{S}$, as follows:

$$(Q_\alpha + R_\alpha)V^\alpha = Q_\alpha V^\alpha + R_\alpha V^\alpha \tag{2.2.6}$$

$$(aQ_\alpha)V^\alpha = a(Q_\alpha V^\alpha). \tag{2.2.7}$$

Thus, for example, (2.2.6) means that $Q_\alpha + R_\alpha$ is *defined* as that element of $\mathfrak{S}_\alpha$ whose effect on each $V^\alpha \in \mathfrak{S}^\alpha$ is as given by the right-hand side of (2.2.6). Under these operations, $\mathfrak{S}_\alpha$ is an $\mathfrak{S}$-module, i.e. the seven axioms (2.2.3) are satisfied by elements of $\mathfrak{S}_\alpha$ (as is not hard to verify). The zero element of $\mathfrak{S}_\alpha$, written $0_\alpha$ or simply 0, is defined by $0_\alpha V^\alpha = 0$ for all $V^\alpha \in \mathfrak{S}^\alpha$.

In an exactly similar way, we define $\mathfrak{S}_\beta$ as the dual of $\mathfrak{S}^\beta$, $\mathfrak{S}_\gamma$ as the dual of $\mathfrak{S}^\gamma$, etc. Owing to the canonical isomorphism between $\mathfrak{S}^\alpha, \mathfrak{S}^\beta, \ldots$, it follows that $\mathfrak{S}_\alpha, \mathfrak{S}_\beta, \ldots$ are also canonically isomorphic to one another. Thus the element $Q_\alpha \in \mathfrak{S}_\alpha$ corresponds to elements $Q_\beta \in \mathfrak{S}_\beta, Q_\gamma \in \mathfrak{S}_\gamma, \ldots, Q_{\alpha_0} \in \mathfrak{S}_{\alpha_0}, \ldots$, where, for any $V^\alpha$,

$$Q_\alpha V^\alpha = Q_\beta V^\beta = Q_\gamma V^\gamma = \cdots = Q_{\alpha_0} V^{\alpha_0} = \cdots. \tag{2.2.8}$$

(Note that as yet we cannot write expressions such as $Q_\alpha V^\beta$ because the elements of $\mathfrak{S}_\alpha$ do not operate on those of $\mathfrak{S}^\beta$.) As before, in any valid equation in which just one index label appears, we may replace this label throughout by any other and a new valid equation results.

A natural question to ask at this stage is whether the relation between $\mathfrak{S}^\alpha$ and $\mathfrak{S}_\alpha$ is symmetrical. That is to say, is $\mathfrak{S}^\alpha$ effectively the dual of $\mathfrak{S}_\alpha$? It is clear from (2.2.6) and (2.2.7) that any element $V^\alpha \in \mathfrak{S}^\alpha$ does define an $\mathfrak{S}$-linear map from $\mathfrak{S}_\alpha$ to $\mathfrak{S}$, given by $V^\alpha(Q_\alpha) := Q_\alpha V^\alpha$. However, it is not clear that *all* $\mathfrak{S}$-linear maps from $\mathfrak{S}_\alpha$ to $\mathfrak{S}$ can arise in this way. Nor is it clear that knowledge of the map will fix $V^\alpha$ uniquely (or, equivalently, that $W^\alpha \in \mathfrak{S}^\alpha$ and $W^\alpha Q_\alpha = 0$ for all $Q_\alpha \in \mathfrak{S}_\alpha$ implies $W^\alpha = 0$). In fact, if $\mathfrak{S}^\alpha$ is a general module over a general commutative ring $\mathfrak{S}$ with unit, then neither of these desiderata need be true*. However, the modules

* For example, choose $\mathfrak{S}$ to be the $C^\infty$ functions on a manifold, but choose for $\mathfrak{S}^\alpha$ the $C^0$ (contravariant) vector fields. Then the dual, $\mathfrak{S}_\alpha$, of $\mathfrak{S}^\alpha$ contains only the zero element, for there are no other covector fields which when operating on an arbitrary $C^0$ vector field always yield a $C^\infty$ scalar field. Thus $Q_\alpha V^\alpha = Q_\alpha U^\alpha (=0)$ for all $U^\alpha$, $V^\alpha \in \mathfrak{S}^\alpha$, $Q_\alpha \in \mathfrak{S}_\alpha$, even when $U^\alpha \neq V^\alpha$. On the other hand, if $\mathfrak{S}$ is the ring of constants on the manifold and $\mathfrak{S}^\alpha$ the $C^\infty$ contravariant vector fields, then $\mathfrak{S}_\alpha$ contains various distributional forms (of compact support), and its dual turns out to be much larger than $\mathfrak{S}^\alpha$.

that we are concerned with here all have the property that $\mathfrak{S}^\alpha$ is in fact naturally isomorphic to (and therefore identifiable with) the dual of $\mathfrak{S}_\alpha$. Such a module is called *reflexive*. Later, when we allow ourselves to suppose the existence of a finite basis for $\mathfrak{S}^\alpha$, the reflexive nature of the module $\mathfrak{S}^\alpha$ will be a consequence. However, for the moment we shall simply *assume* that $\mathfrak{S}^\alpha$ is reflexive. In fact, we shall shortly have to assume a stronger restriction on $\mathfrak{S}^\alpha$ – that it be *totally reflexive* (a concept we shall define presently). This again will be a consequence of the existence of a finite basis. It is possible that the type of notation we are introducing here is really useful only in the case of totally reflexive modules.

## *Tensors*

We next define tensors of general valence $\left[\begin{smallmatrix}p\\q\end{smallmatrix}\right]$. In fact, we shall give two different definitions giving rise to two different concepts of a tensor. It is the condition that these two different tensor concepts coincide that fixes $\mathfrak{S}^\alpha$ as being totally reflexive. In §2.3 we shall show the equivalence of these definitions to that given in (2.1.6) when a finite basis is assumed to exist.

The first coordinate-free definition of a tensor is the multilinear map definition (*type* I *tensor*). This is perhaps the most natural extension of what has gone before. Choose any two disjoint finite subsets of the labelling set $\mathscr{L}$, say $\{\alpha, \beta, \ldots, \delta\}$ and $\{\lambda, \ldots, \nu\}$, of respective cardinality $p$ and $q$. Then we define a tensor $A^{\alpha\beta\ldots\delta}_{\lambda\ldots\nu}$ (of valence $\left[\begin{smallmatrix}p\\q\end{smallmatrix}\right]$) as an $\mathfrak{S}$-multilinear map:

$$A^{\alpha\beta\ldots\delta}_{\lambda\ldots\nu} : \mathfrak{S}_\alpha \times \mathfrak{S}_\beta \times \cdots \times \mathfrak{S}_\delta \times \mathfrak{S}^\lambda \times \cdots \times \mathfrak{S}^\nu \to \mathfrak{S}. \tag{2.2.9}$$

This means that to each selection $Q_\alpha \in \mathfrak{S}_\alpha, R_\beta \in \mathfrak{S}_\beta, \ldots, T_\delta \in \mathfrak{S}_\delta, U^\lambda \in \mathfrak{S}^\lambda, \ldots, W^\nu \in \mathfrak{S}^\nu$, the type I tensor $A^{\alpha\beta\ldots\delta}_{\lambda\ldots\nu}$ assigns a scalar

$$A^{\alpha\beta\ldots\delta}_{\lambda\ldots\nu}(Q_\alpha, R_\beta, \ldots, T_\delta, U^\lambda, \ldots, W^\nu) \in \mathfrak{S}, \tag{2.2.10}$$

this function being separately $\mathfrak{S}$-linear in each variable, i.e.,

$$A^{\alpha\beta\ldots\delta}_{\lambda\ldots\nu}(aQ_\alpha, \ldots) = aA^{\alpha\beta\ldots\delta}_{\lambda\ldots\nu}(Q_\alpha, \ldots), \ldots, A^{\alpha\beta\ldots\delta}_{\lambda\ldots\nu}(\ldots, aW^\nu) = aA^{\alpha\beta\ldots\delta}_{\lambda\ldots\nu}(\ldots, W^\nu) \tag{2.2.11}$$

for all $a \in \mathfrak{S}$ and

$$A^{\alpha\beta\ldots\delta}_{\lambda\ldots\nu}(\overset{1}{Q}_\alpha + \overset{2}{Q}_\alpha, \ldots) = A^{\alpha\beta\ldots\delta}_{\lambda\ldots\nu}(\overset{1}{Q}_\alpha, \ldots) + A^{\alpha\beta\ldots\delta}_{\lambda\ldots\nu}(\overset{2}{Q}_\alpha, \ldots), \tag{2.2.12}$$

with corresponding properties holding for each other of the variables $R_\beta, \ldots, W^\nu$. We shall write (2.2.10) simply as

$$A^{\alpha\beta\ldots\delta}_{\lambda\ldots\nu} Q_\alpha R_\beta \ldots T_\delta U^\lambda \ldots W^\nu \in \mathfrak{S}. \tag{2.2.13}$$

The set of all such tensors $A^{\alpha\beta\ldots\delta}_{\lambda\ldots\nu}$ we denote by $\mathfrak{S}^{\alpha\beta\ldots\delta}_{\lambda\ldots\nu}$. Note that this definition coincides, in the case of $\mathfrak{S}_\lambda$ with the one already given, and trivially coincides, in the case of $\mathfrak{S}$, with the definition of a scalar. In the case of $\mathfrak{S}^\alpha$, the definition gives us back, in effect, the original module $\mathfrak{S}^\alpha$, by virtue of the assumed reflexiveness.

We now give the second coordinate-free definition of a tensor (*type* II *tensor*). Again select any two disjoint finite subsets of $\mathscr{L}$, say $\{\alpha, \beta, \ldots, \delta\}$ and $\{\lambda, \ldots, \nu\}$, not both empty. Consider all formal expressions which are finite (commutative associative) sums of formal (commutative associative) products of elements, one from each of $\mathfrak{S}^\alpha, \mathfrak{S}^\beta, \ldots, \mathfrak{S}^\delta, \mathfrak{S}_\lambda, \ldots, \mathfrak{S}_\nu$. We can write such an expression

$$B^{\alpha\beta\ldots\delta}_{\lambda\ldots\nu} = \sum_{\mathrm{i}=1}^{m} \overset{\mathrm{i}}{G}{}^\alpha \overset{\mathrm{i}}{H}{}^\beta \ldots \overset{\mathrm{i}}{J}{}^\delta \overset{\mathrm{i}}{L}_\lambda \ldots \overset{\mathrm{i}}{N}_\nu. \tag{2.2.14}$$

However, not all such formal expressions are to be regarded as distinct even if formally distinct. The criterion for equivalence of two such expressions is that it be possible to convert one into the other by means of relations of the form

$$(X^\xi + Y^\xi)C^\rho \ldots E^\tau = X^\xi C^\rho \ldots E^\tau + Y^\xi C^\rho \ldots E^\tau \tag{2.2.15}$$

and

$$(qX^\xi)Y^\eta C^\rho \ldots E^\tau = X^\xi(qY^\eta)C^\rho \ldots E^\tau, \tag{2.2.16}$$

where the commutative and associative nature of sums and products may, of course, also be made use of. The formal expressions (2.2.14), under this equivalence relation, are the type II tensors.*

Any type II tensor defines a type I tensor by giving a multilinear map as follows:

$$B^{\alpha\beta\ldots\delta}_{\lambda\ldots\nu} Q_\alpha R_\beta \ldots T_\delta U^\lambda \ldots W^\nu = \sum_{\mathrm{i}=1}^{m} (\overset{\mathrm{i}}{G}{}^\alpha Q_\alpha)(\overset{\mathrm{i}}{H}{}^\beta R_\beta) \ldots (\overset{\mathrm{i}}{J}{}^\delta T_\delta)(\overset{\mathrm{i}}{L}_\lambda U^\lambda) \ldots (\overset{\mathrm{i}}{N}_\nu W^\nu), \tag{2.2.17}$$

the right-hand side of which clearly belongs to $\mathfrak{S}$ (the sums and products being now the ordinary operations (2.2.2) defined on $\mathfrak{S}$ and the bracketed factors being the scalar products (2.2.8)). This clearly defines an $\mathfrak{S}$-multilinear map. It is also clear from (2.2.4)–(2.2.7) that any two such expressions, which are equivalent by virtue of (2.2.15) or (2.2.16) or the commutative and associative nature of the sums and products, will give rise to the same multilinear map. Thus it is clear that to any type II tensor there corresponds a unique type I tensor.

* This definition may seem formal and non-intuitive. However, apart from certain differences arising from our use of an abstract labelling system (which allows our tensor products to be formally commutative), this is essentially the modern 'tensor product' definition of a tensor, which can be applied to a completely general module.

It is not clear, on the other hand, whether *all* type I tensors are obtainable in this way, nor is it clear whether any two type II tensors which are distinct (inequivalent under (2.2.15) and (2.2.16)) will necessarily give rise to distinct type I tensors. Indeed, neither of these desiderata holds true for a general (reflexive) module $\mathfrak{S}^{\bullet}$. However, we shall henceforth assume that $\mathfrak{S}^{\alpha}$ is *totally reflexive*. This means that every type I tensor *does* arise in this way from a type II tensor and, secondly, that it arises in this way from exactly *one* type II tensor. To put this another way, our assumption that $\mathfrak{S}^{\alpha}$ is totally reflexive means that the two types of tensor are *equivalent*. We shall see later that a sufficient (but by no means necessary) condition for total reflexiveness is the existence of a finite basis for $\mathfrak{S}^{\alpha}$.

It may be remarked that reflexiveness is a part of total reflexiveness. The elements of '$\mathfrak{S}^{\alpha}$', if regarded as type I tensors, would actually belong not directly to the original module $\mathfrak{S}^{\alpha}$, but to the dual of $\mathfrak{S}_{\alpha}$. On the other hand, the elements of the original module $\mathfrak{S}^{\alpha}$ are directly type II tensors. Thus, for these two types of tensors to agree, $\mathfrak{S}^{\alpha}$ must be reflexive.

The criterion of equivalence between formal expressions (2.2.14) that was adopted for type II tensors is a little awkward to handle directly, especially if we wish to prove that two type II expressions are *not* equivalent. The assumed total reflexiveness of $\mathfrak{S}^{\alpha}$ now gives us an alternative and simple criterion for the equivalence between expressions (2.2.14). This is that two such expressions are equivalent if and only if for each $Q_{\alpha} \in \mathfrak{S}_{\alpha}$, $R_{\beta} \in \mathfrak{S}_{\beta}, \ldots, T_{\delta} \in \mathfrak{S}_{\delta}$, $U^{\lambda} \in \mathfrak{S}^{\lambda}, \ldots, W^{\nu} \in \mathfrak{S}^{\nu}$, the two corresponding right-hand sides of (2.2.17) are equal.

## *Tensor operations*

We now come to the tensor operations of addition, outer multiplication, index substitution and contraction. *Addition* is a map: $\mathfrak{S}^{\alpha\ldots\delta}_{\lambda\ldots\nu} \times \mathfrak{S}^{\alpha\ldots\delta}_{\lambda\ldots\nu} \rightarrow \mathfrak{S}^{\alpha\ldots\delta}_{\lambda\ldots\nu}$, for each pair of disjoint subsets $(\alpha, \ldots, \delta)$ and $(\lambda, \ldots, \nu)$ of $\mathscr{L}$, where the sum $A^{\alpha\ldots\delta}_{\lambda\ldots\nu} + B^{\alpha\ldots\delta}_{\lambda\ldots\nu}$ can be defined in an obvious way using either of our definitions of a tensor. If we use the type I definition, the sum is just that multilinear map: $\mathfrak{S}_{\alpha} \times \cdots \times \mathfrak{S}^{\nu} \rightarrow \mathfrak{S}$, whose value is the sum of those defined by $A^{\cdots}_{\cdots}$ and by $B^{\cdots}_{\cdots}$. If we use the type II definition, we simply express each of $A^{\cdots}_{\cdots}$ and $B^{\cdots}_{\cdots}$ in the form (2.2.14) and formally add the two formal sums. It is clear that these definitions are equivalent to one another, that the sum is the same as the one already defined in the cases $\mathfrak{S}$, $\mathfrak{S}^{\alpha}$, $\mathfrak{S}_{\alpha}$, and that

$$A^{\alpha\ldots}_{\lambda\ldots} + B^{\alpha\ldots}_{\lambda\ldots} = B^{\alpha\ldots}_{\lambda\ldots} + A^{\alpha\ldots}_{\lambda\ldots} \tag{2.2.18}$$

$$A^{\alpha\ldots}_{\lambda\ldots} + (B^{\alpha\ldots}_{\lambda\ldots} + C^{\alpha\ldots}_{\lambda\ldots}) = (A^{\alpha\ldots}_{\lambda\ldots} + B^{\alpha\ldots}_{\lambda\ldots}) + C^{\alpha\ldots}_{\lambda\ldots}. \tag{2.2.19}$$

*Outer multiplication* is a map: $\mathfrak{S}^{\alpha\ldots\delta}_{\lambda\ldots\nu} \times \mathfrak{S}^{\rho\ldots\tau}_{\phi\ldots\psi} \to \mathfrak{S}^{\alpha\ldots\delta\rho\ldots\tau}_{\lambda\ldots\nu\phi\ldots\psi}$ for each quadruple of disjoint subsets $(\alpha,\ldots,\delta), (\lambda,\ldots,\nu), (\rho,\ldots,\tau), (\phi,\ldots,\psi)$ of $\mathscr{L}$. The product $A^{\alpha\ldots\delta}_{\lambda\ldots\nu}D^{\rho\ldots\tau}_{\phi\ldots\psi}$ can again be defined in an obvious way using either definition of a tensor. If we use the type I definition, we can define the product to be that multilinear map: $\mathfrak{S}_{\alpha} \times \cdots \times \mathfrak{S}^{\psi} \to \mathfrak{S}$ whose value is the product of those defined by $A^{\cdots}_{\cdots}$ and by $D^{\cdots}_{\cdots}$. (The reason for requiring the index sets to be disjoint in the case of outer multiplication and identical in the case of addition is, in this context, that otherwise we do not get a multilinear map.) If we use the type II definition of a tensor, then to define the product we simply multiply out the corresponding formal sums (2.2.14) formally, using the distributive law. It is clear from (2.2.17) that these definitions are equivalent to one another. Also, the relations

$$A^{\alpha\cdots}_{\lambda\ldots}D^{\rho\cdots}_{\phi\ldots} = D^{\rho\cdots}_{\phi\ldots}A^{\alpha\cdots}_{\lambda\ldots}, \tag{2.2.20}$$

$$A^{\alpha\cdots}_{\lambda\ldots}(D^{\rho\cdots}_{\phi\ldots}E^{\zeta\cdots}_{\xi\ldots}) = (A^{\alpha\cdots}_{\lambda\ldots}D^{\rho\cdots}_{\phi\ldots})E^{\zeta\cdots}_{\xi\ldots}, \tag{2.2.21}$$

and

$$(A^{\alpha\cdots}_{\lambda\ldots} + B^{\alpha\cdots}_{\lambda\ldots})D^{\rho\cdots}_{\phi\ldots} = A^{\alpha\cdots}_{\lambda\ldots}D^{\rho\cdots}_{\phi\ldots} + B^{\alpha\cdots}_{\lambda\ldots}D^{\rho\cdots}_{\phi\ldots} \tag{2.2.22}$$

follow readily using either definition.*

It may be remarked that the notation used in the formal sum of formal products (2.2.14) is consistent with the above. That is to say, we may regard (2.2.14) as a sum of outer products. This again follows at once using either definition.

A particular case of outer multiplication occurs when one of the factors is a scalar. Then we obtain an operation of scalar multiplication on each set $\mathfrak{S}^{\alpha\ldots\delta}_{\lambda\ldots\nu}$. This, together with the operation of addition, gives *each* $\mathfrak{S}^{\alpha\ldots\delta}_{\lambda\ldots\nu}$ the *structure of an* $\mathfrak{S}$*-module* (*cf.* (2.2.3)); for, the required properties additional to (2.2.18)–(2.2.22), namely

$$1A^{\alpha\cdots}_{\lambda\ldots} = A^{\alpha\cdots}_{\lambda\ldots} \tag{2.2.23}$$

and

$$0A^{\alpha\cdots}_{\lambda\ldots} = 0B^{\alpha\cdots}_{\lambda\ldots}, \tag{2.2.24}$$

are obvious. We denote $0A^{\alpha\ldots\delta}_{\lambda\ldots\nu}$ by $0^{\alpha\ldots\delta}_{\lambda\ldots\nu}$ or, more usually, simply by 0. Note that

$$A^{\alpha\cdots}_{\lambda\ldots}0^{\rho\cdots}_{\phi\ldots} = 0^{\alpha\ldots\rho\cdots}_{\lambda\ldots\phi\ldots},\ A^{\alpha\cdots}_{\lambda\ldots} + 0^{\alpha\cdots}_{\lambda\ldots} = A^{\alpha\cdots}_{\lambda\ldots}. \tag{2.2.25}$$

* The fact that outer multiplication is commutative in the sense (2.2.20) is a particularly pleasant feature of the abstract index approach to tensor algebra. In the standard algebraic 'index-free' formalism, tensor products are non-commutative: $\mathbf{A} \otimes \mathbf{D} \neq \mathbf{D} \otimes \mathbf{A}$; the relation between the two expressions $\mathbf{A} \otimes \mathbf{D}$ and $\mathbf{D} \otimes \mathbf{A}$ being rather difficult to express. In our notation, the non-commutation of tensor products reads: $A^{\alpha\cdots}_{\lambda\ldots}D^{\cdots}_{\cdots} \neq D^{\alpha\cdots}_{\lambda\ldots}A^{\cdots}_{\cdots}$.

In the cases of $\mathfrak{S}$, $\mathfrak{S}^{\alpha}$, and $\mathfrak{S}_{\alpha}$, scalar multiplication agrees with that already given.

*Index substitution* is a map: $\mathfrak{S}^{\alpha\ldots\delta}_{\lambda\ldots\nu} \to \mathfrak{S}^{\pi\ldots\tau}_{\phi\ldots\psi}$, defined on each $\mathfrak{S}^{\cdots}_{\cdots}$ and induced simply by some permutation of the labelling set $\mathscr{L}$ (so $\pi, \ldots, \tau$ and $\alpha, \ldots, \delta$ must be equal in number, and $\phi, \ldots, \psi$ and $\lambda, \ldots, \nu$ must also be equal in number but $\{\alpha, \ldots, \delta, \lambda, \ldots, \nu\}$ and $\{\pi, \ldots, \tau, \phi, \ldots, \psi\}$ need not be disjoint). By itself this operation is quite trivial. Any equation will simply have an equally valid analogue obtained by relabelling the indices. Clearly any index substitution commutes with addition and with outer multiplication.

In the particular case when the $\alpha, \ldots, \delta$ are permuted among themselves, and similarly $\lambda, \ldots, \nu$ are permuted among themselves, we get a map $\mathfrak{S}^{\alpha\ldots\delta}_{\lambda\ldots\nu} \to \mathfrak{S}^{\alpha\ldots\delta}_{\lambda\ldots\nu}$ which is referred to as an *index permutation.** Applying it in conjunction with the operation of addition, we can define *symmetry* operations. For example, given $A_{\alpha\beta} \in \mathfrak{S}_{\alpha\beta}$ we can define another element $B_{\alpha\beta} = A_{\beta\alpha} \in \mathfrak{S}_{\alpha\beta}$. Then the symmetric and anti-symmetric parts of $A_{\alpha\beta}$ are, respectively, $\frac{1}{2}(A_{\alpha\beta} + A_{\beta\alpha})$ and $\frac{1}{2}(A_{\alpha\beta} - A_{\beta\alpha})$. Thus, when combined with addition, for example, index substitution ceases to be trivial.

$\binom{\xi}{\eta}$-*Contraction* is a map: $\mathfrak{S}^{\alpha\ldots\delta\xi}_{\lambda\ldots\nu\eta} \to \mathfrak{S}^{\alpha\ldots\delta}_{\lambda\ldots\nu}$, defined for each pair of disjoint subsets $\{\alpha, \ldots, \delta\}$, $\{\lambda, \ldots, \nu\}$ of $\mathscr{L}$, the two elements $\xi, \eta$ of $\mathscr{L}$ belonging to neither subset. We must use the type II definition** of a tensor. Let

$$A^{\alpha\ldots\delta\xi}_{\lambda\ldots\nu\eta} = \sum_{\mathrm{i}=1}^{m} \overset{\mathrm{i}}{D}{}^{\alpha} \ldots \overset{\mathrm{i}}{G}{}^{\delta} \overset{\mathrm{i}}{H}{}^{\xi} \overset{\mathrm{i}}{L}{}_{\lambda} \ldots \overset{\mathrm{i}}{N}{}_{\nu} \overset{\mathrm{i}}{P}{}_{\eta} \in \mathfrak{S}^{\alpha\ldots\delta\xi}_{\lambda\ldots\nu\eta}. \tag{2.2.26}$$

Then we define the $\binom{\xi}{\eta}$-contraction of $A^{\cdots}_{\cdots}$ by

$$A^{\alpha\ldots\delta\zeta}_{\lambda\ldots\nu\zeta} = \sum_{i=1}^{m} (\overset{i}{P}{}_{\zeta} \overset{i}{H}{}^{\zeta}) \overset{i}{D}{}^{\alpha} \ldots \overset{i}{G}{}^{\delta} \overset{i}{L}{}_{\lambda} \ldots \overset{i}{N}{}_{\nu} \in \mathfrak{S}^{\alpha\ldots\delta}_{\lambda\ldots\nu}. \tag{2.2.27}$$

If we absorb the scalar $\overset{\mathrm{i}}{P}{}_{\zeta} \overset{\mathrm{i}}{H}{}^{\zeta}$ into one of the other vectors in the product, then we have an expression of the required form (2.2.14). It clearly does not matter which vector, because of (2.2.16). It remains to be verified that any

* We do not need to permute the indices on the symbol $\mathfrak{S}^{\alpha\ldots\delta}_{\lambda\ldots\nu}$ since this *set* is invariant under index permutation. It is the pair of *unordered* sets $\{\alpha, \ldots, \delta\}$, $\{\lambda, \ldots, \nu\}$ which fixes $\mathfrak{S}^{\alpha\ldots\delta}_{\lambda\ldots\nu}$. On the other hand, for each tensor symbol $A^{\alpha\ldots\delta}_{\lambda\ldots\nu}$, the ordering of the indices *is* significant. Thus $\mathfrak{S}^{\beta\ldots\delta\alpha}_{\nu\ldots\lambda} = \mathfrak{S}^{\alpha\beta\ldots\delta}_{\lambda\ldots\nu}$, but $A^{\beta\ldots\delta\alpha}_{\nu\ldots\lambda} \neq A^{\alpha\beta\ldots\delta}_{\lambda\ldots\nu}$ in general.

** The reason we cannot use the multilinear map definition directly in order to define contraction is that it would give us a contraction concept also in systems (not totally reflexive) in which such a concept does not exist. For example, if $\mathfrak{S}^{\alpha}$ were an infinite-dimensional vector space over a division ring $\mathfrak{S}$, then a 'Kronecker delta' $\delta^{\beta}_{\alpha}$ would exist as a bilinear map (2.2.41). However, no contraction $\delta^{\alpha}_{\alpha}$ could exist since we would need $\delta^{\alpha}_{\alpha} = \infty$, the dimension of the space.

two expressions (2.2.26) which are equivalent under (2.2.15) and (2.2.16) will give rise to equivalent* expressions (2.2.27). In fact it is readily seen that this is the case by referring back to (2.2.15) and (2.2.16) and using the linearity of the scalar product where required.

There is no significance in the fact that the contracted indices of $A_{\dots}^{\dots}$ have been written as its final upper and lower indices. The contraction operation applies equally well whichever upper and lower index are selected. (We could for example, define $B_{\eta\lambda\dots\nu}^{\delta\xi\alpha\dots\gamma} = A_{\lambda\dots\nu\eta}^{\alpha\dots\gamma\delta\xi}$; then $B_{\zeta\lambda\dots\nu}^{\delta\zeta\alpha\dots\gamma} = A_{\lambda\dots\nu\zeta}^{\alpha\dots\gamma\delta\zeta}$.) Also, by (2.2.8), the 'dummy' index $\zeta$ in (2.2.27) could equally well have been any other element of $\mathscr{L}$ (for example $\eta$ or $\xi$) which does not appear among $\alpha,\dots,\delta,\lambda,\dots,\nu$. Thus

$$A_{\lambda\dots\nu\zeta}^{\alpha\dots\delta\zeta} = A_{\lambda\dots\nu\xi}^{\alpha\dots\delta\xi} = \dots. \tag{2.2.28}$$

It is, furthermore, clear from the definition that the order in which two successive contractions are performed is immaterial. Thus we can unambiguously write $A_{\lambda\dots\mu\theta\zeta}^{\alpha\dots\gamma\theta\zeta}$ for the $\binom{\delta}{\nu}$-contraction of $A_{\lambda\dots\mu\nu\zeta}^{\alpha\dots\gamma\delta\zeta}$ or for the $\binom{\sigma}{\chi}$-contraction of $A_{\lambda\dots\mu\theta\chi}^{\alpha\dots\gamma\theta\sigma}$. It is also immediate from the definition that $\binom{\xi}{\eta}$-contraction commutes with addition:

$$A_{\lambda\dots\nu\eta}^{\alpha\dots\delta\xi} = B_{\lambda\dots\nu\eta}^{\alpha\dots\delta\xi} + C_{\lambda\dots\nu\eta}^{\alpha\dots\delta\xi} \text{ implies } A_{\lambda\dots\nu\zeta}^{\alpha\dots\delta\zeta} = B_{\lambda\dots\nu\zeta}^{\alpha\dots\delta\zeta} + C_{\lambda\dots\nu\zeta}^{\alpha\dots\delta\zeta}; \tag{2.2.29}$$

and, in the appropriate sense, with multiplication:

$$A_{\lambda\dots\nu\phi\dots\psi\eta}^{\alpha\dots\delta\rho\dots\tau\xi} = B_{\lambda\dots\nu}^{\alpha\dots\delta}C_{\phi\dots\psi\eta}^{\rho\dots\tau\xi} \text{ implies } A_{\lambda\dots\psi\zeta}^{\alpha\dots\tau\zeta} = B_{\lambda\dots\dots\nu}^{\alpha\dots\dots\delta}C_{\phi\dots\psi\zeta}^{\rho\dots\tau\zeta}, \tag{2.2.30}$$

of which scalar multiplication is a special case:

$$A_{\phi\dots\psi\eta}^{\rho\dots\tau\xi} = bC_{\phi\dots\psi\eta}^{\rho\dots\tau\xi} \text{ implies } A_{\phi\dots\psi\zeta}^{\rho\dots\tau\zeta} = bC_{\phi\dots\psi\zeta}^{\rho\dots\tau\zeta}. \tag{2.2.31}$$

Also $\binom{\xi}{\eta}$-contraction commutes with any index substitution not involving $\xi$ or $\eta$. Finally, it is clear that any index substitution applied to two indices which are subsequently contracted will not affect the result of the contraction.

Observe that these tensor operations allow us to build up tensor expressions, with indices, which are exactly analogous to the expressions of classical tensor algebra, but now the indexed symbols stand for actual tensors instead of for sets of components of a tensor, no basis frame or other coordinate system being involved. We can tell to which set $\mathfrak{S}_{\dots}^{\dots}$ a tensor belongs by simply examining its indices. As in the classical tensor notation, repeated indices are paired off, one upper and one lower. The

* Strictly speaking we should also have explicitly checked this for the other tensor operations, whenever the type II definition of tensor was employed. However, there we always had the type I definition to fall back on. In each case the verification that equivalence under (2.2.15) and (2.2.16) is preserved is quite trivial and does not require total reflexiveness.

indices which remain unpaired then serve to define the type of the tensor, i.e. the set $\mathfrak{S}^{\cdots}_{\cdots}$ to which it belongs. For this reason we shall frequently omit indicating the particular $\mathfrak{S}^{\cdots}_{\cdots}$ to which a given tensor belongs, the indices themselves adequately serving this purpose.

It should, however, be pointed out that there are certain expressions which it is legitimate to write using the present notation, but which do not correspond to normally employed classical tensor expressions. The simplest of these would be an expression of the form $U^\alpha(Q_\alpha V^\alpha)$; this is the product of $U^\alpha$ with the scalar $Q_\alpha V^\alpha$. The use of parentheses is of course necessary since

$$U^\alpha(Q_\alpha V^\alpha) \neq (U^\alpha Q_\alpha)V^\alpha, \tag{2.2.32}$$

but the notation is consistent (as it would also be on the classical interpretation) as long as the parentheses are retained. However, to avoid possible confusion we shall normally rewrite such expressions in a form more in keeping with the classical usage. Thus (2.2.32) can be rewritten

$$U^\alpha Q_\beta V^\beta \neq U^\beta Q_\beta V^\alpha, \tag{2.2.33}$$

which is more economical in any case. An expression such as $(Q_\alpha V^\alpha)^2$, on the other hand, is more economical then $Q_\alpha V^\alpha Q_\beta V^\beta$ and is perfectly legitimate.

An outer multiplication followed by a contraction (or contractions) across the two elements involved is sometimes thought of as a single operation called *contracted* (or *inner*) *product* or, sometimes, *transvection*. (This last term is normally used when a verb is required: 'transvect through by ...'.) Thus we have a product (outer or inner) defined between any two tensors, provided *only* that the two sets of upper indices have no letter in common and the two sets of lower indices have no letter in common. For example, if $A^{\gamma\alpha}_{\eta\beta\delta} \in \mathfrak{S}^{\alpha\gamma}_{\beta\delta\eta}$ and $B^{\delta\eta}_{\chi\gamma} \in \mathfrak{S}^{\delta\eta}_{\gamma\chi}$, the product is a contracted product, $A^{\gamma\alpha}_{\beta\eta\delta}B^{\delta\eta}_{\chi\gamma}$, an element of $\mathfrak{S}^{\alpha}_{\beta\chi}$. Contracted product is clearly commutative and distributive over addition. One has to be careful about the associative law, however, when considering contracted products of three or more tensors. If no index letter appears more than twice (once as an upper and once as a lower index), then no trouble arises and the product is associative. Otherwise an ambiguity arises, of the type encountered in (2.2.32). For example,

$$(A^{\gamma\alpha}_{\beta\eta\delta}B^{\delta\eta}_{\chi\gamma})C^{\gamma\chi} \neq A^{\gamma\alpha}_{\beta\eta\delta}(B^{\delta\eta}_{\chi\gamma}C^{\gamma\chi}) \tag{2.2.34}$$

in general. We can (and normally would) avoid the use of parentheses by replacing the two $\gamma$s inside the parentheses (on both sides of the equation) by some other letter, say $\zeta$, as in (2.2.33).

It should be remarked that the notation employed in (2.2.13), for the multilinear map defined by a tensor, is consistent with this notation for contracted products. This is readily seen if we refer to (2.2.17): the outer product of $B^{\cdots}_{\cdots}$ with $Q_\pi, \ldots, W^\psi$ may be performed first, if desired, and the contraction afterwards. Thus, the multilinear map defined by a tensor is a special case of a repeated contracted product.

*Some useful properties of a totally reflexive* $\mathfrak{S}^{\bullet}$

The condition that $\mathfrak{S}^\alpha$ is totally reflexive has a number of important and pleasant consequences. In the first place we have:

(2.2.35) PROPOSITION:

*The dual of the* $\mathfrak{S}$*-module* $\mathfrak{S}^{\alpha\ldots\gamma}_{\lambda\ldots\nu}$ *may be identified with* $\mathfrak{S}^{\lambda\ldots\nu}_{\alpha\ldots\gamma}$, *the required scalar product being contracted product.*

*Proof*: It is clear that any element $Q^{\lambda\ldots\nu}_{\alpha\ldots\gamma} \in \mathfrak{S}^{\lambda\ldots\nu}_{\alpha\ldots\gamma}$ defines an $\mathfrak{S}$-linear map from $\mathfrak{S}^{\alpha\ldots\gamma}_{\lambda\ldots\nu}$ to $\mathfrak{S}$. (For, $Q^{\lambda\ldots}_{\alpha\ldots}(U^{\alpha\ldots}_{\lambda\ldots} + V^{\alpha\ldots}_{\lambda\ldots}) = Q^{\lambda\ldots}_{\alpha\ldots}U^{\alpha\ldots}_{\lambda\ldots} + Q^{\lambda\ldots}_{\alpha\ldots}V^{\alpha\ldots}_{\lambda\ldots}$ and $Q^{\lambda\ldots}_{\alpha\ldots}(aV^{\alpha\ldots}_{\lambda\ldots}) = a(Q^{\lambda\ldots}_{\alpha\ldots}V^{\alpha\ldots}_{\lambda\ldots})$, by (2.2.22), (2.2.29), (2.2.30), (2.2.21), (2.2.20).) What has to be shown is that *every* $\mathfrak{S}$-linear map from $\mathfrak{S}^{\alpha\ldots\gamma}_{\lambda\ldots\nu}$ to $\mathfrak{S}$ is obtainable in this way by means of an element of $\mathfrak{S}^{\lambda\ldots\nu}_{\alpha\ldots\gamma}$ which is unique. For this we invoke (2.2.14) for the elements of $\mathfrak{S}^{\alpha\ldots\gamma}_{\lambda\ldots\nu}$, i.e., we express these elements as sums of outer products of vectors. Any $\mathfrak{S}$-linear map from $\mathfrak{S}^{\alpha\ldots\gamma}_{\lambda\ldots\nu}$ to $\mathfrak{S}$ is thus defined in terms of its effect on those elements of $\mathfrak{S}^{\alpha\ldots\gamma}_{\lambda\ldots\nu}$ which happen to be outer products of vectors. This effect must, indeed, be an $\mathfrak{S}$-multilinear map from $\mathfrak{S}^\alpha \times \cdots \times \mathfrak{S}^\gamma \times \mathfrak{S}_\lambda \times \cdots \times \mathfrak{S}_\nu$ to $\mathfrak{S}$. Such a multilinear map is achieved by a unique tensor $Q^{\lambda\ldots\nu}_{\alpha\ldots\gamma} \in \mathfrak{S}^{\lambda\ldots\nu}_{\alpha\ldots\gamma}$, so the result is established.

It is often useful, when stating general propositions about tensors, to be able to 'clump together' a set of indices and write them as a single *composite index*. We shall use script letters to denote a general such clumping. We allow both upper and lower indices to be clumped together as a single composite index, if desired. For example, we might wish to clump together the upper index $\rho$ and the two lower indices $\theta$ and $\eta$ as a single upper composite index $\mathscr{A}$. We write this $\mathscr{A} = \rho\theta^*\eta^*$, where an asterisk indicates that the index so marked is to be in the opposite position from $\mathscr{A}$. Then we can denote an element $Q^\rho{}_{\theta\eta}$, say, of $\mathfrak{S}^\rho_{\theta\eta}$, by $Q^{\mathscr{A}} = Q^\rho{}_{\theta\eta}$. (The staggering of indices now becomes necessary for notational consistency; see $W^{\cdots}_{\cdots}$ below.) An element $U_\rho{}^{\theta\eta}$ of $\mathfrak{S}^{\theta\eta}_\rho$ can then be written $U_{\mathscr{A}} = U_\rho{}^{\theta\eta}$. More generally, $W_\alpha{}^{\lambda\mu\nu}{}_{\mathscr{A}}$ is the element $W_\alpha{}^{\lambda\mu\nu}{}_\rho{}^{\theta\eta}$ of $\mathfrak{S}^{\lambda\mu\nu\theta\eta}_{\alpha\rho}$. Contractions may also be performed between composite indices. Thus, $W_\alpha{}^{\lambda\mu\nu}{}_{\mathscr{A}}Q^{\mathscr{A}}$ stands for

$W_\alpha{}^{\lambda\mu\nu}{}_\rho{}^{\theta\eta}Q^\rho{}_{\theta\eta}$. If $\mathscr{B}$ is another composite index defined by $\mathscr{B} = \alpha^*\lambda\mu\nu$, then we can write this $W^{\mathscr{B}}{}_{\mathscr{A}}Q^{\mathscr{A}}$. We shall normally avoid the use of composite indices in which a repetition of a constituent index is not explicit. For example, a symbol $R^{\theta\mathscr{A}}$ would denote an element $R^{\theta\rho}{}_{\theta\eta}$ of $\mathfrak{S}^\rho_\eta$, but the contraction is hidden in the notation. Thus, if a number of composite or ordinary indices appear in one expression, then it will be assumed (in the absence of any explicit statement or convention to the contrary) that the *only* repetitions of indices which occur are those which appear explicitly.

In Proposition (2.2.35) we can set $\mathscr{A} = \alpha \ldots \gamma\lambda^* \ldots \nu^*$. Then the statement asserts that the dual of $\mathfrak{S}^{\mathscr{A}}$ may be identified with $\mathfrak{S}_{\mathscr{A}}$, for any composite index $\mathscr{A}$. The following three propositions all generalize this result.

(2.2.36) PROPOSITION

*The set of all $\mathfrak{S}$-bilinear maps from $\mathfrak{S}^{\mathscr{A}} \times \mathfrak{S}^{\mathscr{B}}$ to $\mathfrak{S}$ may be identified with $\mathfrak{S}_{\mathscr{A}\mathscr{B}}$, the maps being achieved by means of contracted product.*

*Proof*: The proof is similar to that of (2.2.35). Clearly any element $P_{\mathscr{A}\mathscr{B}} \in \mathfrak{S}_{\mathscr{A}\mathscr{B}}$ effects such an $\mathfrak{S}$-bilinear map, the result being $P_{\mathscr{A}\mathscr{B}}X^{\mathscr{A}}Y^{\mathscr{B}}$ for each $X^{\mathscr{A}} \in \mathfrak{S}^{\mathscr{A}}$ and $Y^{\mathscr{B}} \in \mathfrak{S}^{\mathscr{B}}$. (Note that $P_{\mathscr{A}\mathscr{B}}(\overset{1}{X}{}^{\mathscr{A}} + \overset{2}{X}{}^{\mathscr{A}})Y^{\mathscr{B}} = P_{\mathscr{A}\mathscr{B}}\overset{1}{X}{}^{\mathscr{A}}Y^{\mathscr{B}} + P_{\mathscr{A}\mathscr{B}}\overset{2}{X}{}^{\mathscr{A}}Y^{\mathscr{B}}$ and $P_{\mathscr{A}\mathscr{B}}(aX^{\mathscr{A}})Y^{\mathscr{B}} = aP_{\mathscr{A}\mathscr{B}}X^{\mathscr{A}}Y^{\mathscr{B}}$; and similarly for $Y^{\mathscr{B}}$.) To show that every such $\mathfrak{S}$-bilinear map arises this way from a unique $P_{\mathscr{A}\mathscr{B}}$, we can express $X^{\mathscr{A}}$ and $Y^{\mathscr{B}}$ as sums of outer products of vectors. The bilinear map is uniquely defined by its effect on such outer products of vectors, this effect being an $\mathfrak{S}$-multilinear map of the vectors. Thus $P_{\mathscr{A}\mathscr{B}}$ is uniquely determined as a type I tensor.

(2.2.37) PROPOSITION

*The set of all $\mathfrak{S}$-linear maps from $\mathfrak{S}^{\mathscr{A}}$ to $\mathfrak{S}^{\mathscr{K}}$ may be identified with $\mathfrak{S}^{\mathscr{K}}_{\mathscr{A}}$, where the maps are achieved by means of contracted product.*

*Proof*: Clearly any $Q^{\mathscr{K}}{}_{\mathscr{A}}$ effects an $\mathfrak{S}$-linear map from $\mathfrak{S}^{\mathscr{A}}$ to $\mathfrak{S}^{\mathscr{K}}$, the image of $X^{\mathscr{A}}$ being $Q^{\mathscr{K}}{}_{\mathscr{A}}X^{\mathscr{A}}$. Conversely suppose we have an $\mathfrak{S}$-linear map from $\mathfrak{S}^{\mathscr{A}}$ to $\mathfrak{S}^{\mathscr{K}}$. Denote the image of $X^{\mathscr{A}}$ under this map by $U^{\mathscr{K}}$. Then, for $Z_{\mathscr{K}} \in \mathfrak{S}_{\mathscr{K}}$, the map which sends the pair $(X^{\mathscr{A}}, Z_{\mathscr{K}})$ to $U^{\mathscr{K}}Z_{\mathscr{K}}$ is $\mathfrak{S}$-bilinear from $\mathfrak{S}^{\mathscr{A}} \times \mathfrak{S}_{\mathscr{K}}$ to $\mathfrak{S}$. Thus, by (2.2.36) (with $\mathscr{B} = \mathscr{K}^*$) we have a *unique* element $Q^{\mathscr{K}}{}_{\mathscr{A}} \in \mathfrak{S}^{\mathscr{K}}_{\mathscr{A}}$ with $Q^{\mathscr{K}}{}_{\mathscr{A}}X^{\mathscr{A}}Z_{\mathscr{K}} = U^{\mathscr{K}}Z_{\mathscr{K}}$ for all $Z_{\mathscr{K}} \in \mathfrak{S}_{\mathscr{K}}$, so $Q^{\mathscr{K}}{}_{\mathscr{A}}X^{\mathscr{A}}$ and $U^{\mathscr{K}}$ represent the same element of the dual of $\mathfrak{S}_{\mathscr{K}}$. Thus $Q^{\mathscr{K}}{}_{\mathscr{A}}X^{\mathscr{A}} = U^{\mathscr{K}}$ as required, this map characterizing $Q^{\mathscr{K}}{}_{\mathscr{A}}$ uniquely.

The following proposition incorporates (2.2.35), (2.2.36) and (2.2.37) as special cases:

(2.2.38) PROPOSITION

*The set of all $\mathfrak{S}$-multilinear maps from $\mathfrak{S}^{\mathscr{A}} \times \mathfrak{S}^{\mathscr{B}} \times \cdots \times \mathfrak{S}^{\mathscr{D}}$ to $\mathfrak{S}^{\mathscr{K}}$ may be identified with $\mathfrak{S}^{\mathscr{K}}_{\mathscr{A}\mathscr{B}\ldots\mathscr{D}}$, where the maps are achieved by means of contracted product.*

*Proof*: This is just a repeated application of (2.2.37). If we fix $B^{\mathscr{B}}, \ldots, D^{\mathscr{D}}$, we are concerned with $\mathfrak{S}$-linear maps from $\mathfrak{S}^{\mathscr{A}}$ to $\mathfrak{S}^{\mathscr{K}}$, which, by (2.2.37), are effectively the elements of $\mathfrak{S}^{\mathscr{K}}_{\mathscr{A}}$. Allowing $B^{\mathscr{B}}, \ldots, D^{\mathscr{D}}$ now to vary, we see that an $\mathfrak{S}$-multilinear map from $\mathfrak{S}^{\mathscr{A}} \times \mathfrak{S}^{\mathscr{B}} \times \cdots \mathfrak{S}^{\mathscr{D}}$ to $\mathfrak{S}^{\mathscr{K}}$ is equivalent to an $\mathfrak{S}$-multilinear map from $\mathfrak{S}^{\mathscr{B}} \times \cdots \times \mathfrak{S}^{\mathscr{D}}$ to $\mathfrak{S}^{\mathscr{K}}_{\mathscr{A}}$. Repeating this argument with $\mathscr{A}$ replaced successively by $\mathscr{B}, \ldots, \mathscr{D}$ we obtain the result.

There is one aspect of (2.2.38) which it is worth spelling out. Since no two distinct elements of $\mathfrak{S}^{\mathscr{K}}_{\mathscr{A}\mathscr{B}\ldots\mathscr{D}}$ can give the same map, we have:

$$\text{if } A^{\mathscr{K}}{}_{\mathscr{A}\ldots\mathscr{D}} W^{\mathscr{A}} \ldots Z^{\mathscr{D}} = B^{\mathscr{K}}{}_{\mathscr{A}\ldots\mathscr{D}} W^{\mathscr{A}} \ldots Z^{\mathscr{D}} \text{ for all } W^{\mathscr{A}} \in \mathfrak{S}^{\mathscr{A}}, \ldots, Z^{\mathscr{D}} \in \mathfrak{S}^{\mathscr{D}},$$
$$\text{then } A^{\mathscr{K}}{}_{\mathscr{A}\ldots\mathscr{D}} = B^{\mathscr{K}}{}_{\mathscr{A}\ldots\mathscr{D}}. \tag{2.2.39}$$

More particularly still:

$$\text{if } A^{\mathscr{K}}{}_{\mathscr{A}\ldots\mathscr{D}} W^{\mathscr{A}} \ldots Z^{\mathscr{D}} = 0 \text{ for all } W^{\mathscr{A}} \in \mathfrak{S}^{\mathscr{A}}, \ldots, Z^{\mathscr{D}} \in \mathfrak{S}^{\mathscr{D}}, \text{ then } A^{\mathscr{K}}{}_{\mathscr{A}\ldots\mathscr{D}} = 0. \tag{2.2.40}$$

A tensor of especial utility is the Kronecker delta $\delta_\alpha^\beta$ (*cf.* (2.1.9)). This may be defined abstractly in numerous different ways. For example, the map from $\mathfrak{S}^\alpha \times \mathfrak{S}_\beta$ to $\mathfrak{S}$ which assigns the scalar product $X^\alpha Z_\alpha$ to the pair $(X^\alpha, Z_\beta)$ is clearly $\mathfrak{S}$-bilinear and is therefore achieved by some tensor, which we denote by $\delta_\alpha^\beta$. Thus, $\delta_\alpha^\beta$ is formally defined by

$$\delta_\alpha^\beta X^\alpha Z_\beta = X^\alpha Z_\alpha. \tag{2.2.41}$$

Alternatively, we can define $\delta_\alpha^\beta$ to be that element of $\mathfrak{S}_\alpha^\beta$ which effects a map from $\mathfrak{S}_\beta^\alpha$ to $\mathfrak{S}$ by assigning the scalar $Y_\gamma^\gamma$ to each $Y_\beta^\alpha \in \mathfrak{S}_\beta^\alpha$, i.e.,

$$\delta_\alpha^\beta Y_\beta^\alpha = Y_\gamma^\gamma. \tag{2.2.42}$$

(This must be the same $\delta_\alpha^\beta$ since (2.2.41) is a special case of (2.2.42).) Yet again, the map from $\mathfrak{S}_\beta$ to $\mathfrak{S}_\alpha$ which gives the canonical isomorphism between these sets is trivially $\mathfrak{S}$-linear and is therefore achieved by a tensor – again $\delta_\alpha^\beta$:

$$\delta_\alpha^\beta Z_\beta = Z_\alpha. \tag{2.2.43}$$

(That this is the same $\delta_\alpha^\beta$ is obvious since (2.2.43) yields (2.2.41) again.) Or, we could use the dual version of (2.2.43). The tensor $\delta_\alpha^\beta$ effects the map from $\mathfrak{S}^\alpha$ to $\mathfrak{S}^\beta$ which gives the canonical isomorphism:

$$\delta_\alpha^\beta X^\alpha = X^\beta. \tag{2.2.44}$$

Of course, the same tensor $\delta_\alpha^\beta$ effects many other $\mathfrak{S}$-linear maps in consequence of these; for example, the maps $\mathfrak{S}_\beta^{\alpha\mathscr{A}} \to \mathfrak{S}^{\mathscr{A}}$; $\mathfrak{S}^{\alpha\mathscr{A}} \to \mathfrak{S}^{\beta\mathscr{A}}$; $\mathfrak{S}_\beta^{\mathscr{A}} \to \mathfrak{S}_\alpha^{\mathscr{A}}$ expressed by:

$$\delta_\alpha^\beta U_\beta^{\alpha\mathscr{A}} = U_\gamma^{\gamma\mathscr{A}}, \delta_\alpha^\beta V^{\alpha\mathscr{A}} = V^{\beta\mathscr{A}}, \delta_\alpha^\beta W_\beta{}^{\mathscr{A}} = W_\alpha{}^{\mathscr{A}}. \tag{2.2.45}$$

(These follow at once from (2.2.42)–(2.2.44) since we can transvect by an arbitrary $Q_{\mathscr{A}}$ and then cancel it again as in (2.2.39).)

### *Embedding of one tensor system in another*

There is an aspect of the use of composite indices which will have some considerable significance for us later. We have seen that composite indices can be manipulated in just the same way as the original index labels $\alpha, \beta, \ldots$. (The necessity for staggering composite indices which involve reverse position labels $\alpha^*, \beta^*, \ldots$ can easily be circumvented by use of suitable conventions, e.g. we could consider all upper indices as occurring first and all lower ones afterwards.) In fact, given any tensor algebra of the type considered, we can construct a new tensor algebra ('embedded' in the given one) whose labelling set $\mathscr{L}_{\mathscr{A}}$ consists of suitably clumped subsets of $\mathscr{L}$. For example, we could set $\mathscr{A} = \alpha\beta\gamma^*$, $\mathscr{A}_0 = \alpha_0\beta_0\gamma_0{}^*$, $\mathscr{A}_1 = \alpha_1\beta_1\gamma_1{}^*, \ldots,$ and use $\mathscr{L}_{\mathscr{A}} = (\mathscr{A}, \mathscr{A}_0, \mathscr{A}_1, \ldots)$ as our new labelling set. We then consider the tensor system* built up from $\mathfrak{S}$ and $\mathfrak{S}^{\mathscr{A}}$ in a way exactly analogous to the way in which our original system was built up from $\mathfrak{S}$ and $\mathfrak{S}^\alpha$. For example, by (2.2.35), the dual of $\mathfrak{S}^{\mathscr{A}}$ is $\mathfrak{S}_{\mathscr{A}}$. It is not hard to see that the type I and type II definitions of a tensor each lead to higher valence tensors which are the elements of the sets $\mathfrak{S}_{\mathscr{A}_{j_1}\ldots\mathscr{A}_{j_q}}^{\mathscr{A}_{i_1}\ldots\mathscr{A}_{i_p}}$ (*cf.* (2.2.38) in particular). The system is thus totally reflexive, and is indeed embedded in the original one. The tensor operations of the new system are just those of the original system which can be consistently written using the allowed composite indices only. (These remarks do not, of course, depend on the particular choice $\mathscr{A} = \alpha\beta\gamma^*$ made above.)

If we were to consider different types of clumpings simultaneously (for example $\mathscr{A} = \alpha\beta^*$, $\mathscr{B} = \gamma\delta\varepsilon$, $\mathscr{A}_0 = \alpha_0\beta_0^*$, $\mathscr{B}_0 = \gamma_0\delta_0\varepsilon_0, \ldots$) then we should be led to consider tensor systems of a slightly more general type in which more than one labelling set appears. (In the example considered, we have $\mathscr{L}_{\mathscr{A}} = (\mathscr{A}, \mathscr{A}_0, \mathscr{A}_1, \ldots)$ and $\mathscr{L}_{\mathscr{B}} = (\mathscr{B}, \mathscr{B}_0, \mathscr{B}_1, \ldots)$.) The rules for a tensor system with more than one labelling set are essentially the same as for a system with just one labelling set. The only difference arises from the fact

* In more conventional terminology (*cf.* Herstein 1964 MacLane and Birkhoff 1967). these are the tensors on the $\mathfrak{S}$-module $\mathfrak{S}^\cdot \otimes_{\mathfrak{S}} \mathfrak{S}^\cdot \otimes_{\mathfrak{S}} \mathfrak{S}^{\cdot *}$ in abstract-index form.

that index substitutions are only allowed between members of the same labelling set. The two labelling sets are to have no members in common, so contractions between indices of two different types cannot be performed.

A tensor system with more than one labelling set would naturally arise if we were to consider, *initially*, several different $\mathfrak{S}$-modules simultaneously (the ring of scalars $\mathfrak{S}$ being the same in each case). We could denote the initial modules by $\mathfrak{S}^{\alpha}$, $\mathfrak{S}^{\alpha'}$, $\mathfrak{S}^{\alpha''}$,... and define labelling sets $\mathscr{L} = (\alpha, \beta, \ldots, \alpha_0, \ldots)$, $\mathscr{L}' = (\alpha', \beta', \ldots, \alpha'_0, \ldots)$, etc. The definitions of the general sets $\mathfrak{S}^{\cdots}_{\cdots}$ would be as before. Since we suppose *no* canonical isomorphism between $\mathfrak{S}^{\alpha}$ and $\mathfrak{S}^{\alpha'}$, nor between $\mathfrak{S}^{\alpha}$ and $\mathfrak{S}^{\alpha''}$, etc., then, as remarked above, we have no operation of index substitution between indices possessing different numbers of primes. Apart from this new feature, the development of the tensor algebra proceeds exactly as before. The importance of such systems to us here lies in the fact that the spinor algebra that we introduce in §2.5 – and which we consider for the remainder of this book – is, in fact, a system of this kind. The spinor system is built up from two modules $\mathfrak{S}^{A}$ and $\mathfrak{S}^{A'}$ which are not related to each other 'algebraically', but rather by a ('non-algebraic') relation of complex conjunction.

## 2.3 Bases

In this section we consider the consequences of introducing a basis into $\mathfrak{S}^{\alpha}$. Throughout §2 we refrained from using bases in any way. This we did *partly* to emphasize the fact that our development of tensor algebra is completely coordinate-free (despite the use of indices). But in addition it gives us considerably more generality (at least, in a direct fashion) than would have been obtainable had we had to assume the existence of a finite basis. For there are many totally reflexive modules for which bases do not exist.

A *finite basis* for $\mathfrak{S}^{\alpha}$ is a set of elements $\delta^{\alpha}_1, \delta^{\alpha}_2, \ldots, \delta^{\alpha}_n \in \mathfrak{S}^{\alpha}$ such that any $V^{\alpha} \in \mathfrak{S}^{\alpha}$ has a *unique* expansion

$$V^{\alpha} = V^1\delta^{\alpha}_1 + V^2\delta^{\alpha}_2 + \cdots + V^n\delta^{\alpha}_n. \tag{2.3.1}$$

The scalars $V^1, \ldots, V^n \in \mathfrak{S}$ are called the *components* of $V^{\alpha}$ in this basis. If $\mathfrak{S}^{\alpha}$ possesses a finite basis, then any other basis for $\mathfrak{S}^{\alpha}$ must have the same number $n$ of elements. This is a consequence of the fact that the existence of an $n$-element basis for $\mathfrak{S}^{\alpha}$ implies that $\mathfrak{S}_{\alpha_1 \ldots \alpha_n}$ contains a non-zero antisymmetrical element, whereas $\mathfrak{S}_{\alpha_1 \ldots \alpha_{n+k}} (k > 0)$ contains only the zero antisymmetrical element. (We use only the type I definition of $\mathfrak{S}\ldots$, so total reflexiveness need not be assumed.) This serves to define $n$ independently of the basis.* To prove this property, suppose $A_{\alpha_1\ldots\alpha_{n+k}}$ is anti-symmetrical. (This means that $A_{\ldots}$ changes sign whenever an index permutation is applied which interchanges just two of the $\alpha$s (see §3.3).) Then

$$A_{\ldots\alpha_i\ldots\alpha_j\ldots}X^{\alpha_i}X^{\alpha_j}=0 \tag{2.3.2}$$

for any $X^\alpha$, since interchange of the dummy labels $\alpha_i, \alpha_j$ changes the sign. Consider the multilinear map given by

$$A_{\alpha_1\ldots\alpha_{n+k}}R^{\alpha_1}\ldots W^{\alpha_{n+k}}, \quad (k>0). \tag{2.3.3}$$

Expanding each of $R^{\alpha_1},\ldots,W^{\alpha_{n+k}}$ in terms of the basis (2.3.1), and multiplying out, we see that each term contains at least one repeated basis element, and so vanishes by (2.3.2). Thus $A\ldots=0$. On the other hand, we can define a non-zero anti-symmetrical element $\varepsilon_{\alpha_1\ldots\alpha_n}\in\mathfrak{S}_{\alpha_1\ldots\alpha_n}$ (an *alternating tensor*) by the property

$$\varepsilon_{\alpha_1\ldots\alpha_n}U^{\alpha_1}\ldots W^{\alpha_n}=\begin{vmatrix}U^1\ldots W^1\\ \vdots \quad\quad \vdots\\ U^n\ldots W^n\end{vmatrix}, \tag{2.3.4}$$

where $U^1,\ldots,U^n$ are the components of $U^{\alpha_1}$ in the basis $\delta_1^{\alpha_1},\ldots,\delta_n^{\alpha_1}$ etc. This clearly gives an anti-symmetrical multilinear map as required. Also $\varepsilon_{\alpha_1\ldots\alpha_n}\neq 0$, since if $U^\alpha=\delta_1^\alpha,\ldots,W^\alpha=\delta_n^\alpha$, the result of the map is the unit scalar. The integer $n$ we call the *dimension* of the $\mathfrak{S}$-module $\mathfrak{S}^\alpha$.

If $\mathfrak{S}^\alpha$ is the set of tangent vectors at a single point in a manifold, then $\mathfrak{S}^\alpha$ is a finite-dimensional vector space (the tangent space at that point) and a basis exists. However, when $\mathfrak{S}^\alpha$ refers to smooth vector fields on an $n$-dimensional manifold, a basis will often *not* exist. For, a basis now means a set of $n$ vector fields which are linearly independent at each point of the manifold. In the simple example of an ordinary spherical 2-surface ($S^2$) it is clearly impossible to arrange this. By the well-known fixed-point theorem, each of the two vector fields on the surface would have to vanish at some point, so the two vectors would become linearly dependent there. An $n$-manifold which *does* possess $n$ vector fields which are linearly independent at each point, is called parallelizable. Thus the module of tangent vector fields has a basis if and only if the manifold is parallelizable. As was indicated at the end of §1.5, the 3-sphere ($S^3$) is (perhaps rather surprisingly) parallelizable,** but the 4-sphere ($S^4$) is not. (In fact, every

* In fact, if we allow $n=\infty$ for the case when anti-symmetrical elements of arbitrarily large valence $\left[{0 \atop q}\right]$ exist, then this property defines $n$ for *any* module, independently of the existence of a basis. We may regard $n$ as the dimension of $\mathfrak{S}_\alpha$ in the general case.

** We recall that the points of $S^3$ may be represented by unit quaternions $\boldsymbol{q}$, and that the various right rotations, given by $\boldsymbol{q}\mapsto\boldsymbol{qb}$ for the various fixed choices of unit quaternion $\boldsymbol{b}$, will carry a frame at some given point $\boldsymbol{q}_0$ of $S^3$ uniquely and continuous-

orientable 3-manifold turns out to be parallelizable, but many orientable 4-manifolds are not.) By (1.5.6), all space–times which globally possess the type of spinor structure that we require here and which are also non-compact (a reasonable requirement, physically, since compact space–times contain closed timelike curves) are parallelizable (Geroch 1968). Thus, if $\mathfrak{S}^a$ denotes the module of vector fields on a space–time, we may regard it as *physically reasonable* to assume that $\mathfrak{S}^a$ possesses a basis (here $n = 4$).* Moreover, applying to the module of spin-vector fields on a space–time the same reasoning (*cf.* (1.5.6)), one finds also that a spinor basis globally exists (here $n = 2$). Thus, the global existence of a basis, in the situations of interest to us here, may be *assumed* as a reasonable physical requirement.

Even if we are interested in manifolds which are not parallelizable, the discussion of bases will be relevant, since it can always be applied *locally* (e.g., in a coordinate patch). Of course one would have to be careful about drawing conclusions of a global nature from such arguments. In §2.4 we shall show that whether or not the manifold is parallelizable (but provided it is paracompact – which is a normal assumption, redundant in the case of space–times: *cf.* Kelley 1955, Geroch 1968) the module of $C^\infty$ vector fields over the $C^\infty$ scalar fields is still totally reflexive.

## Components in a basis

Let us suppose, then, that a basis $\delta^\alpha_1, \ldots, \delta^\alpha_n$ exists for $\mathfrak{S}^\alpha$. We can use bold face letters to stand for 1, 2, …, $n$ in the conventional way, and adopt the summation convention for such indices. Thus, the basis elements may be collectively denoted by $\delta^\alpha_{\boldsymbol{\alpha}}$ ($\delta^\alpha_{\boldsymbol{\alpha}} \in \mathfrak{S}^\alpha$) and the relation (2.3.1) for the expression of a vector $V^\alpha$ in terms of its components $V^{\boldsymbol{\alpha}}$ ($\in \mathfrak{S}$) in this basis can be written

$$V^\alpha = V^{\boldsymbol{\alpha}} \delta^\alpha_{\boldsymbol{\alpha}}. \tag{2.3.5}$$

---

ly into frames at all the various other points of $S^3$. (If $\boldsymbol{q}_0 \mapsto \boldsymbol{r}$, this is uniquely achieved by the right rotation for which $\boldsymbol{b} = \boldsymbol{q}_0^{-1}\boldsymbol{r}$.) The same argument applies to $S^7$, where Cayley numbers are used in place of the quaternions. A proof that $S^n$ is parallelizable only if $n = 1, 3, 7$ is given by Eckmann (1968), p. 522. It should be noted that the existence of a global basis for vector fields is a much weaker requirement than the existence of a global coordinate system (*cf.* 4.1.33). This is evident from the parallelizability of $S^3$ (and $S^7$).

* We cannot, on the other hand, reasonably assume the existence of a global *holonomic* basis, that is, one arising naturally from a coordinate system in such a way that the basis vectors $(\partial/\partial x^{\mathbf{a}})$ point along the coordinate lines. However, we may set against this disadvantage the fact that our basis may be chosen to be orthonormal everywhere, with one vector timelike and future-pointing throughout (*cf.* also p. 199).

(No special relation is implied between the index symbols $\alpha$ and $\boldsymbol{\alpha}$ here. We could equally well have written (2.3.5) as $V^\alpha = V^{\boldsymbol{\beta}}\delta^\alpha_{\boldsymbol{\beta}}$.)

We now associate with the basis $\delta^\alpha_{\boldsymbol{\alpha}}$ for $\mathfrak{S}^\alpha$ its *dual* $\delta^1_\alpha, \delta^2_\alpha, \ldots, \delta^n_\alpha \in \mathfrak{S}_\alpha$. By definition, $\delta^{\boldsymbol{\alpha}}_\alpha$ is that map from $\mathfrak{S}^\alpha$ to $\mathfrak{S}$ which assigns to a vector $V^\alpha$ its $\boldsymbol{\alpha}$th component $V^{\boldsymbol{\alpha}}$ in the basis $\delta^\alpha_1, \ldots, \delta^\alpha_n$:

$$\delta^{\boldsymbol{\alpha}}_\alpha V^\alpha = V^{\boldsymbol{\alpha}}. \tag{2.3.6}$$

Owing to the uniqueness and linearity of the expansion (2.3.1), the equation (2.3.6) does indeed define a linear map (for each $\boldsymbol{\alpha} = 1, 2, \ldots, n$) from $\mathfrak{S}^\alpha$ to $\mathfrak{S}$ and so gives us a well-defined element of $\mathfrak{S}_\alpha$. By letting $V^\alpha$ be each of $\delta^\alpha_1, \ldots, \delta^\alpha_n$ in turn, we obtain

$$\delta^{\boldsymbol{\alpha}}_\alpha \delta^\alpha_{\boldsymbol{\beta}} = \delta^{\boldsymbol{\alpha}}_{\boldsymbol{\beta}}, \tag{2.3.7}$$

where $\delta^{\boldsymbol{\alpha}}_{\boldsymbol{\beta}}$ is the $(n \times n)$-matrix of elements of $\mathfrak{S}$ consisting of the unit scalar if $\boldsymbol{\alpha} = \boldsymbol{\beta}$ and the zero scalar otherwise (Kronecker delta symbol).

We next show that the $n$ elements $\delta^{\boldsymbol{\alpha}}_\alpha$ of $\mathfrak{S}_\alpha$ form a basis for $\mathfrak{S}_\alpha$. We must establish that any element $Q_\alpha$ of $\mathfrak{S}_\alpha$ has a unique expansion as a linear combination of the $\delta^{\boldsymbol{\alpha}}_\alpha$. Given $Q_\alpha$, define

$$\tilde{Q}_\alpha = (Q_\beta \delta^\beta_{\boldsymbol{\beta}})\delta^{\boldsymbol{\beta}}_\alpha. \tag{2.3.8}$$

Then, for each $V^\alpha$,

$$\begin{aligned} \tilde{Q}_\alpha V^\alpha &= Q_\beta \delta^\beta_{\boldsymbol{\beta}} \delta^{\boldsymbol{\beta}}_\alpha V^\alpha = Q_\beta \delta^\beta_{\boldsymbol{\beta}} \delta^{\boldsymbol{\beta}}_\alpha (V^{\boldsymbol{\alpha}} \delta^\alpha_{\boldsymbol{\alpha}}) \\ &= Q_\beta \delta^\beta_{\boldsymbol{\beta}} \delta^{\boldsymbol{\beta}}_{\boldsymbol{\alpha}} V^{\boldsymbol{\alpha}} = Q_\beta \delta^\beta_{\boldsymbol{\alpha}} V^{\boldsymbol{\alpha}} \\ &= Q_\beta V^\beta = Q_\alpha V^\alpha. \end{aligned} \tag{2.3.9}$$

Since $\tilde{Q}_\alpha$ and $Q_\alpha$ give the same scalar when acting on an arbitrary element of $\mathfrak{S}^\alpha$, we have $\tilde{Q}_\alpha = Q_\alpha$. Thus (2.3.8) establishes that $Q_\alpha$ can be expanded as a linear combination of the $\delta^{\boldsymbol{\beta}}_\alpha$,

$$Q_\alpha = Q_{\boldsymbol{\beta}} \delta^{\boldsymbol{\beta}}_\alpha, \tag{2.3.10}$$

where

$$Q_{\boldsymbol{\beta}} = Q_\beta \delta^\beta_{\boldsymbol{\beta}}. \tag{2.3.11}$$

To show that this expansion is unique, suppose that $Q_\alpha$ can be expressed in a form (2.3.10) where $Q_{\boldsymbol{\beta}}$ is not necessarily given by (2.3.11). Taking the scalar product of (2.3.10) with $\delta^\alpha_{\boldsymbol{\alpha}}$ and using (2.3.7) we get $Q_\alpha \delta^\alpha_{\boldsymbol{\alpha}} = Q_{\boldsymbol{\beta}} \delta^{\boldsymbol{\beta}}_\alpha \delta^\alpha_{\boldsymbol{\alpha}} = Q_{\boldsymbol{\beta}} \delta^{\boldsymbol{\beta}}_{\boldsymbol{\alpha}} = Q_{\boldsymbol{\alpha}}$, which is (2.3.11) again, establishing the required uniqueness.

Note that the components of $Q_\alpha$ in the dual basis are obtained by taking scalar products with the elements of the original basis. This is analogous to the fact that the components of $V^\alpha$ in the original basis were obtained by taking scalar products with the elements of the dual basis. Note, further, that

$$Q_\alpha V^\alpha = Q_\alpha V^{\boldsymbol{\alpha}} \delta^\alpha_{\boldsymbol{\alpha}} = Q_{\boldsymbol{\alpha}} V^{\boldsymbol{\alpha}}, \tag{2.3.12}$$

by (2.3.5) and (2.3.11), so that the scalar product has the familiar form when written in terms of components.

So far in this section we have not used total reflexivity, i.e., the equivalence of type I and type II tensors. We now introduce components for general tensors and, as a corollary, show that total reflexivity holds whenever there is a basis.

If $A^{\alpha\ldots\gamma}_{\lambda\ldots\nu}$ is a general type I tensor, we can apply this multilinear map to the basis elements to define its *components*:

$$A^{\boldsymbol{\alpha}\ldots\boldsymbol{\gamma}}_{\boldsymbol{\lambda}\ldots\boldsymbol{\nu}} = A^{\alpha\ldots\gamma}_{\lambda\ldots\nu}\delta^{\boldsymbol{\alpha}}_{\alpha}\cdots\delta^{\boldsymbol{\gamma}}_{\gamma}\delta^{\lambda}_{\boldsymbol{\lambda}}\cdots\delta^{\nu}_{\boldsymbol{\nu}}, \tag{2.3.13}$$

generalizing (2.3.6) and (2.3.11). Calling such an array $A^{\boldsymbol{\alpha}\ldots\boldsymbol{\gamma}}_{\boldsymbol{\lambda}\ldots\boldsymbol{\nu}}$ a *type* III *tensor* with respect to the basis $\delta^{\alpha}_{\boldsymbol{\alpha}}$ (provisionally – a more complete definition is given in §2.4), we see that, given a basis $\delta^{\alpha}_{\boldsymbol{\alpha}}$, the formula (2.3.13) defines a map (I → III) which assigns a unique type III tensor to each type I tensor. Moreover, from any such array $A^{\boldsymbol{\alpha}\ldots\boldsymbol{\gamma}}_{\boldsymbol{\lambda}\ldots\boldsymbol{\nu}}$ we can form a type II tensor as the sum of outer products

$$A^{\alpha\ldots\gamma}_{\lambda\ldots\nu} = A^{\boldsymbol{\alpha}\ldots\boldsymbol{\gamma}}_{\boldsymbol{\lambda}\ldots\boldsymbol{\nu}}\delta^{\alpha}_{\boldsymbol{\alpha}}\cdots\delta^{\gamma}_{\boldsymbol{\gamma}}\delta^{\boldsymbol{\lambda}}_{\lambda}\cdots\delta^{\boldsymbol{\nu}}_{\nu}, \tag{2.3.14}$$

generalizing (2.3.5) and (2.3.10). Thus we have a map (III → II) which assigns a unique type II tensor to each type III tensor. Finally, we already have a standard scheme (2.2.17) – irrespective of total reflexivity or the existence of bases – which assigns a unique type I tensor to each type II tensor. Let this be the map (II → I). To establish the equivalence of all three types of tensor (and thus total reflexivity), we shall verify that all three of the cyclic compositions of these maps,

$$\text{I} \to \text{III} \to \text{II} \to \text{I}, \quad \text{III} \to \text{II} \to \text{I} \to \text{III}, \quad \text{II} \to \text{I} \to \text{III} \to \text{II}$$

give the identity. To verify the first of these, we start with the type I tensor $A^{\alpha\ldots\gamma}_{\lambda\ldots\nu}$ and apply (2.3.13), (2.3.14) and (2.2.17) successively, the final multilinear map in (2.2.17) being on $Q_{\alpha}, \ldots S_{\gamma}, U^{\lambda}, \ldots W^{\nu}$; thus we obtain

$$\begin{aligned} &[(A^{\alpha\ldots}_{\ldots\nu}\delta^{\boldsymbol{\alpha}}_{\alpha}\ldots\delta^{\nu}_{\boldsymbol{\nu}})\delta^{\alpha_0}_{\boldsymbol{\alpha}}\ldots\delta^{\boldsymbol{\nu}}_{\nu_0}]Q_{\alpha_0}\cdots W^{\nu_0} \\ &= A^{\alpha\ldots}_{\ldots\nu}\delta^{\boldsymbol{\alpha}}_{\alpha}\ldots\delta^{\nu}_{\boldsymbol{\nu}}Q_{\boldsymbol{\alpha}}\ldots W^{\boldsymbol{\nu}} \\ &= A^{\alpha\ldots}_{\ldots\nu}Q_{\alpha}\ldots W^{\nu}, \end{aligned} \tag{2.3.15}$$

by (2.3.11), (2.3.6), and then (2.3.10), (2.3.5). To verify that the second cyclic composition gives the identity, we start with the array $A^{\boldsymbol{\alpha}\ldots}_{\ldots\boldsymbol{\nu}}$ and apply (2.3.14), then (2.2.17) followed by (2.3.13). In fact, to follow (2.2.17) by (2.3.13), we simply substitute the basis elements $\delta^{\boldsymbol{\alpha}}_{\alpha}, \cdots \delta^{\nu}_{\boldsymbol{\nu}}$ for the elements $Q_{\alpha}, \ldots, W^{\nu}$ on which the multilinear map acts, to obtain

$$(A^{\boldsymbol{\alpha}\ldots}_{\ldots\boldsymbol{\nu}}\delta^{\alpha}_{\boldsymbol{\alpha}}\cdots\delta^{\boldsymbol{\nu}}_{\nu})\delta^{\boldsymbol{\alpha}_0}_{\alpha}\cdots\delta^{\nu}_{\boldsymbol{\nu}_0}, \tag{2.3.16}$$

which is the original array $A^{\alpha_0 \dots}_{\dots \nu_0}$, by (2.3.7). Finally, to verify that the third cyclic composition gives the identity, we start with a type II tensor and apply (2.2.17) followed by (2.3.13) (as above) and then (2.3.14):

$$\left[\left(\sum_{\mathrm{i}=1}^{N} \overset{\mathrm{i}}{F}{}^{\alpha} \dots \overset{\mathrm{i}}{N}_{\nu}\right)\delta^{\boldsymbol{\alpha}}_{\alpha} \dots \delta^{\nu}_{\boldsymbol{\nu}}\right]\delta^{\alpha_0}_{\boldsymbol{\alpha}} \dots \delta^{\boldsymbol{\nu}}_{\nu_0}$$

$$= \left[\left(\sum_{\mathrm{i}=1}^{N} \overset{\mathrm{i}}{F}{}^{\alpha_0} \delta^{\alpha}_{\alpha_0} \dots \overset{\mathrm{i}}{N}_{\nu_0} \delta^{\nu_0}_{\nu}\right)\delta^{\boldsymbol{\alpha}}_{\alpha} \dots \delta^{\nu}_{\boldsymbol{\nu}}\right]\delta^{\alpha_0}_{\boldsymbol{\alpha}} \dots \delta^{\boldsymbol{\nu}}_{\nu_0}$$

$$= \left[\sum_{\mathrm{i}=1}^{N} \overset{\mathrm{i}}{F}{}^{\boldsymbol{\alpha}} \dots \overset{\mathrm{i}}{N}_{\boldsymbol{\nu}}\right]\delta^{\alpha_0}_{\boldsymbol{\alpha}} \dots \delta^{\boldsymbol{\nu}}_{\nu_0}$$

$$= \sum_{\mathrm{i}=1}^{N} \overset{\mathrm{i}}{F}{}^{\alpha_0} \dots \overset{\mathrm{i}}{N}_{\nu_0}, \tag{2.3.17}$$

by (2.3.10), (2.3.5), then (2.3.7), then (2.3.10), (2.3.5) again. Thus total reflexivity is established, when a basis exists: the general tensor of valence $\left[{p \atop q}\right]$ (of either type I or II) is in 1-1 correspondence with its array of $n^{p+q}$ components (type III) by the mutually inverse relations (2.3.13), (2.3.14).

Note that the $n^{p+q}$ tensors

$$\delta^{\alpha}_{\boldsymbol{\alpha}} \dots \delta^{\gamma}_{\boldsymbol{\gamma}} \delta^{\boldsymbol{\lambda}}_{\lambda} \dots \delta^{\boldsymbol{\nu}}_{\nu} \tag{2.3.18}$$

forms a basis for $\mathfrak{S}^{\alpha \dots \gamma}_{\lambda \dots \nu}$, since any element of $\mathfrak{S}^{\alpha \dots \gamma}_{\lambda \dots \nu}$ has a unique expression, via (2.3.14), as a linear combination of the tensors (2.3.18). Consequently, the $\mathfrak{S}$-module $\mathfrak{S}^{\alpha \dots \gamma}_{\lambda \dots \nu}$ has dimension $n^{p+q}$. The particular basis (2.3.18) for $\mathfrak{S}^{\alpha \dots \gamma}_{\lambda \dots \nu}$ is said to be *induced* by the basis $\delta^{\alpha}_{\boldsymbol{\alpha}}$ for $\mathfrak{S}^{\alpha}$.

If we define the tensor $\delta^{\beta}_{\alpha} \in \mathfrak{S}^{\beta}_{\alpha}$ by

$$\delta^{\beta}_{\alpha} = \delta^{\beta}_{\boldsymbol{\beta}} \delta^{\boldsymbol{\beta}}_{\alpha}, \tag{2.3.19}$$

then (2.3.10), (2.3.11), (2.3.5), (2.3.6) give

$$Q_{\alpha} = Q_{\beta} \delta^{\beta}_{\ \alpha}, \quad V^{\alpha} = V^{\beta} \delta^{\alpha}_{\beta}. \tag{2.3.20}$$

By referring back to (2.2.43) and (2.2.44) we see that either of these relations establishes the $\delta^{\beta}_{\alpha}$ defined by (2.3.19) as being actually the same as that defined in (2.2.41)–(2.2.44) in a basis-independent manner. Reverting, then, to this original basis-free definition of $\delta^{\beta}_{\alpha}$, we can assert that equations (2.3.19) and (2.3.7) are together necessary and sufficient for $\delta^{\alpha}_{\boldsymbol{\alpha}}$ to be a basis for $\mathfrak{S}^{\alpha}$ (with dual basis $\delta^{\boldsymbol{\alpha}}_{\alpha}$). For, equation (2.3.19) (together with (2.3.20)) shows that $V^{\alpha}$ is a linear combination of the $\delta^{\alpha}_{\boldsymbol{\alpha}}$, while (2.3.7) establishes these components as uniquely given by (2.3.6). (The formal similarity between (2.3.7) and (2.3.19) should not mislead us: (2.3.7) condenses $n^2$ scalar equations whereas (2.3.19) is a single tensor equation. Although the various quantities $\delta_\alpha^\beta, \delta_{\boldsymbol{\alpha}}^\beta, \delta_\alpha^{\boldsymbol{\beta}}, \delta_{\boldsymbol{\alpha}}^{\boldsymbol{\beta}}$ behave formally very similarly, they are conceptually very different from each other.) It is worth remarking that a way of obtaining (2.3.14) is to use the relation

$$A_{\lambda\ldots\nu}^{\alpha\ldots\gamma} = A_{\lambda_0\ldots\nu_0}^{\alpha_0\ldots\gamma_0}\delta_{\alpha_0}^{\alpha}\ldots\delta_{\gamma_0}^{\gamma}\delta_{\lambda}^{\lambda_0}\ldots\delta_{\nu}^{\nu_0} \tag{2.3.21}$$

(which is a repeated application of (2.3.20)), and to substitute (2.3.19) for each occurrence of a $\delta$.

Let us relate the type III tensors we have been considering here to those discussed in §2.1. The notion of a tensor used in §2.1 depended on behaviour under *change* of basis. Suppose, then, that we have two bases for $\mathfrak{S}^\alpha$, namely $\delta_{\mathbf{1}}^\alpha, \ldots, \delta_{\boldsymbol{n}}^\alpha$ and $\delta_{\hat{\mathbf{1}}}^\alpha, \ldots \delta_{\hat{\boldsymbol{n}}}^\alpha$. The dual basis to $\delta_{\hat{\boldsymbol{\alpha}}}^\alpha$ is $\delta_\alpha^{\hat{\boldsymbol{\alpha}}}$ and satisfies $\delta_\alpha^{\hat{\boldsymbol{\alpha}}}\delta_{\hat{\boldsymbol{\beta}}}^\alpha = \delta_{\hat{\boldsymbol{\beta}}}^{\hat{\boldsymbol{\alpha}}}$, where $\delta_{\hat{\boldsymbol{\beta}}}^{\hat{\boldsymbol{\alpha}}}$ is again an ordinary Kronecker delta symbol. In our notation, when components are taken with respect to $\delta_{\hat{\boldsymbol{\alpha}}}^\alpha$ and $\delta_\alpha^{\hat{\boldsymbol{\alpha}}}$, the component indices must bear a circumflex but the kernel symbol remains unchanged. Thus, $V^{\hat{\boldsymbol{\alpha}}} = V^\alpha\delta_\alpha^{\hat{\boldsymbol{\alpha}}}$, etc. This applies also if we take the components of the basis elements of one basis with respect to the other. In this way we get two $(n \times n)$-matrices of scalars, defined by

$$\delta_{\hat{\boldsymbol{\alpha}}}^{\boldsymbol{\alpha}} = \delta_\alpha^{\boldsymbol{\alpha}}\delta_{\hat{\boldsymbol{\alpha}}}^\alpha, \quad \delta_{\boldsymbol{\alpha}}^{\hat{\boldsymbol{\alpha}}} = \delta_{\boldsymbol{\alpha}}^\alpha\delta_\alpha^{\hat{\boldsymbol{\alpha}}}. \tag{2.3.22}$$

The quantities $\delta_{\hat{\boldsymbol{\alpha}}}^{\boldsymbol{\alpha}}$ and $\delta_{\boldsymbol{\alpha}}^{\hat{\boldsymbol{\alpha}}}$ correspond, respectively to the $t_{\boldsymbol{\beta}}^{\boldsymbol{\alpha}}$ and $T_{\boldsymbol{\alpha}}^{\boldsymbol{\beta}}$ which appear in (2.1.6)–(2.1.8). (The use of Kronecker symbols here should not confuse us: such symbols stand for an actual Kronecker delta *only* when the two bold face indices are both of the same kind.) The matrices $\delta_{\hat{\boldsymbol{\alpha}}}^{\boldsymbol{\alpha}}$ and $\delta_{\boldsymbol{\alpha}}^{\hat{\boldsymbol{\alpha}}}$ are in fact inverses of one another ($\delta_{\hat{\boldsymbol{\alpha}}}^{\boldsymbol{\alpha}}\delta_{\boldsymbol{\beta}}^{\hat{\boldsymbol{\alpha}}} = \delta_{\boldsymbol{\beta}}^{\boldsymbol{\alpha}}, \delta_{\boldsymbol{\alpha}}^{\hat{\boldsymbol{\alpha}}}\delta_{\hat{\boldsymbol{\beta}}}^{\boldsymbol{\alpha}} = \delta_{\hat{\boldsymbol{\beta}}}^{\hat{\boldsymbol{\alpha}}}$), as is readily seen. For any vector $V^\alpha$, the components $V^{\hat{\boldsymbol{\alpha}}}$ with respect to $\delta_{\hat{\boldsymbol{\alpha}}}^\alpha$ are related to those, $V^{\boldsymbol{\alpha}}$, with respect to $\delta_{\boldsymbol{\alpha}}^\alpha$, by

$$V^{\hat{\boldsymbol{\alpha}}} = V^\alpha\delta_\alpha^{\hat{\boldsymbol{\alpha}}} = V^{\boldsymbol{\alpha}}\delta_{\boldsymbol{\alpha}}^\alpha\delta_\alpha^{\hat{\boldsymbol{\alpha}}} = V^{\boldsymbol{\alpha}}\delta_{\boldsymbol{\alpha}}^{\hat{\boldsymbol{\alpha}}}, \tag{2.3.23}$$

which may be compared with (2.1.8). In the same way the components with respect to $\delta_{\hat{\boldsymbol{\alpha}}}^\alpha$ of a general tensor $A_{\lambda\ldots\nu}^{\alpha\ldots\gamma}$ may be related to those with respect to $\delta_{\boldsymbol{\alpha}}^\alpha$ by

$$A_{\hat{\boldsymbol{\lambda}}\ldots\hat{\boldsymbol{\nu}}}^{\hat{\boldsymbol{\alpha}}\ldots\hat{\boldsymbol{\gamma}}} = A_{\boldsymbol{\lambda}\ldots\boldsymbol{\nu}}^{\boldsymbol{\alpha}\ldots\boldsymbol{\gamma}}\delta_{\boldsymbol{\alpha}}^{\hat{\boldsymbol{\alpha}}}\ldots\delta_{\boldsymbol{\gamma}}^{\hat{\boldsymbol{\gamma}}}\delta_{\hat{\boldsymbol{\lambda}}}^{\boldsymbol{\lambda}}\ldots\delta_{\hat{\boldsymbol{\nu}}}^{\boldsymbol{\nu}}. \tag{2.3.24}$$

(To obtain this, we simply 'plug' the basis elements $\delta_{\hat{\boldsymbol{\alpha}}}^\alpha, \delta_\alpha^{\hat{\boldsymbol{\alpha}}}$ into (2.3.14).) This is just the tensor component transformation law (2.1.6). Thus, whenever a basis exists, the definition of a tensor given in §2.1 does agree with those used here. (Note that although the interpretation is again different, (2.3.24) formally resembles (2.3.13), (2.3.14) and (2.3.21). In each case, the delta symbol just substitutes one index for another.)

When expressed in terms of components, the four operations of addition, outer multiplication, index substitution and contraction have exactly the same appearance as when expressed directly in terms of the abstract tensors, except that all the indices are bold face. Thus, in the basis $\delta_{\boldsymbol{\alpha}}^{\alpha}$, the components of $A_{\lambda\ldots\nu}^{\alpha\ldots\gamma} + B_{\lambda\ldots\nu}^{\alpha\ldots\gamma}$ are $A_{\boldsymbol{\lambda}\ldots\boldsymbol{\nu}}^{\boldsymbol{\alpha}\ldots\boldsymbol{\gamma}} + B_{\boldsymbol{\lambda}\ldots\boldsymbol{\nu}}^{\boldsymbol{\alpha}\ldots\boldsymbol{\gamma}}$; the components of $A_{\lambda\ldots\nu}^{\alpha\ldots\gamma} D_{\phi\ldots\psi}^{\rho\ldots\tau}$ are $A_{\boldsymbol{\lambda}\ldots\boldsymbol{\nu}}^{\boldsymbol{\alpha}\ldots\boldsymbol{\gamma}} D_{\boldsymbol{\phi}\ldots\boldsymbol{\psi}}^{\boldsymbol{\rho}\ldots\boldsymbol{\tau}}$; the components of $A^{\gamma\ldots\alpha}{}_{\ldots\nu\ldots\lambda}$ are $A^{\boldsymbol{\gamma}\ldots\boldsymbol{\alpha}}{}_{\ldots\boldsymbol{\nu}\ldots\boldsymbol{\lambda}}$; and the components of $A_{\lambda\ldots\mu\zeta}^{\alpha\ldots\beta\zeta}$ are $A_{\boldsymbol{\lambda}\ldots\boldsymbol{\mu}\boldsymbol{\zeta}}^{\boldsymbol{\alpha}\ldots\boldsymbol{\beta}\boldsymbol{\zeta}}$. All these facts are immediate consequences of the definitions and agree with the operations of §2.1. Thus, the entire algebra of abstract tensors is identical with the algebra of arrays of tensor components; the only difference, at this stage, lies in the conceptual interpretation of the quantities involved. It might be felt that a difference of conceptual interpretation alone would hardly justify the use of two alternative alphabets for the indices. However, we shall see in Chapter 4 that when *differentiation* is involved the parallelism breaks down and the two types of indices behave quite differently. What are purely conceptual differences in the case of tensor *algebra* lead naturally to essential formal differences within tensor or spinor *calculus*.

## 2.4 The total reflexivity of $\mathfrak{S}^{\bullet}$ on a manifold

When a finite basis exists for $\mathfrak{S}^{\bullet}$, its total reflexivity has been established in §2.3. But, as we have seen, bases need not exist for vector fields on a manifold (e.g. on $S^2$). Since the total reflexivity of $\mathfrak{S}^{\bullet}$ is an important general property, we shall devote the present section to giving an argument for it that applies to any (Hausdorff, paracompact) manifold $\mathscr{M}$ on which (i) the differentiability conditions on the scalars $\mathfrak{S}$ – assumed to be fields of complex numbers – are sufficiently non-restrictive (say $C^0$, $C^1, \ldots,$ or $C^\infty$, but *not* $C^\omega$) to allow 'partitions of unity' (see (2.4.4) below); and on which (ii) $\mathfrak{S}^{\bullet}$ has bases *locally*. (The reader who is happy to assume on physical grounds that a basis exists, may prefer to pass on to §2.5.)

The arguments in this chapter so far have been *algebraic*, in the sense of being concerned only with the algebra of the tensors generated from the module $\mathfrak{S}^{\bullet}$. No properties of the 'point set' $\mathscr{M}$ on which $\mathfrak{S}$ may be defined, not even the very existence of such a point set, have been assumed. In Chapter 4, where differential operations are discussed, we must examine $\mathscr{M}$ more closely. But here we are concerned only with the two above-mentioned 'manifold' properties of $\mathscr{M}$, which will allow us to establish the total reflexivity of $\mathfrak{S}^{\bullet}$ in the spirit of this chapter, namely algebraically.

To give an algebraic definition of 'tensors defined locally' on $\mathscr{M}$, we need the notion of tensors restricted to some (open) subset of $\mathscr{M}$, a 'neighbourhood' of the point of interest.* Consider any element $f \in \mathfrak{S}$, and the open set $\mathscr{F} \subset \mathscr{M}$ of points of $\mathscr{M}$ for which $f \neq 0$. Define an $f$-equivalence relation between tensors thus:

$$T^{\alpha\ldots\gamma}_{\lambda\ldots\nu} \equiv W^{\alpha\ldots\gamma}_{\lambda\ldots\nu} \tag{2.4.1}$$

iff

$$f T^{\alpha\ldots\gamma}_{\lambda\ldots\nu} = f W^{\alpha\ldots\gamma}_{\lambda\ldots\nu}, \tag{2.4.2}$$

for the equality of tensor fields when 'restricted to $\mathscr{F}$'. Note that without loss of generality we can assume $f$ real non-negative, since (2.4.2) holds if and only if (2.4.2) multiplied by $\bar{f}$ also holds.

Now let the symbol $\mathfrak{S}^{\alpha\ldots\gamma}_{\lambda\ldots\nu}(f)$ denote the set of $f$-equivalence classes (2.4.1). One easily verifies that, with the natural definition of sums and products, $\mathfrak{S}(f)$ is a commutative ring with identity (i.e., the product of the $f$-equivalence class of $a \in \mathfrak{S}$ with that of $b \in \mathfrak{S}$ is that of $ab$; if $a \equiv c$ and $b \equiv d$, we have $fab = fcb = cfb = cfd = fcd$, whence $ab \equiv cd$; etc.) Furthermore, $\mathfrak{S}^{\alpha}(f)$ is a module over $\mathfrak{S}(f)$, as is again easily verified.**

The only new property of the $\mathfrak{S}$-module $\mathfrak{S}^{\bullet}$ that we shall require (and for which we stipulated the properties (i) and (ii) in the first paragraph) is this:

(2.4.3) PROPERTY:
*There exists a finite set of non-negative elements* $\overset{0}{u}, \overset{1}{u}, \ldots, \overset{m}{u} \in \mathfrak{S}$ *such that*

$$\overset{0}{u} + \overset{1}{u} + \cdots + \overset{m}{u} = 1, \tag{2.4.4}$$

*and such that there exists a basis for each module* $\mathfrak{S}^{\alpha}(\overset{\mathrm{i}}{u})$, $(\mathrm{i} = 0, 1, \ldots, m)$.

Let us first see why this would hold for any (Hausdorff, paracompact) manifold $\mathscr{M}$ on which the scalars $\mathfrak{S}$ and the (vector or spinor) fields $\mathfrak{S}^{\bullet}$ are, say, $C^{\infty}$. For each i, the region where $\overset{\mathrm{i}}{u} \neq 0$ is an open set $\mathscr{U}_{\mathrm{i}} \subset \mathscr{M}$, and by (2.4.4) we see that these sets cover $\mathscr{M}$:

$$\mathscr{U}_0 \cup \mathscr{U}_1 \cup \ldots \cup \mathscr{U}_m = \mathscr{M}. \tag{2.4.5}$$

Conversely, if any finite covering of $\mathscr{M}$ by open sets $\mathscr{U}_0, \ldots, \mathscr{U}_m$ exists, such that each $\mathscr{U}_{\mathrm{i}}$ can be defined by the non-vanishing of a non-negative

* The procedure being adopted here is discussed more fully in §4.1. Logically, this section should be presented *after* §4.1, but since total reflexivity is so important to the algebraic theory of tensors, there is good *motivational* reason for presenting our derivation at this stage.

** It should be pointed out, however, that (if $\varnothing \neq \mathscr{F} \neq \mathscr{M}$) the module $\mathfrak{S}(f)$ will not be the same as the module of complex-valued $C^{\infty}$ scalar fields on $\mathscr{F}$, considered as a submanifold of $\mathscr{M}$, but will be a submodule of it. This is because the latter module includes also scalar fields that do not extend smoothly into $\mathscr{M}$. A similar remark applies to $\mathfrak{S}^{\alpha\ldots\gamma}_{\lambda\ldots\nu}(f)$.

real $\overset{i}{v} \in \mathfrak{S}$, then (2.4.4) follows. For defining

$$V = \overset{0}{v} + \overset{1}{v} + \cdots + \overset{m}{v}, \tag{2.4.6}$$

we have $V > 0$ everywhere on $\mathcal{M}$; so $V^{-1}$ exists, and we can satisfy (2.4.4) by setting

$$\overset{i}{u} = V^{-1} \overset{i}{v}. \tag{2.4.7}$$

Such a system of *us* on $\mathcal{M}$ is called a *partition of unity*.

To find a suitable cover (2.4.5) for $\mathcal{M}$, consider the following construction. Choose a triangulation of $\mathcal{M}$ which is sufficiently fine so that the star of each vertex (i.e. the union of all $n$-simplexes through that vertex) has the property that a basis exists for fields restricted to its interior. Now choose a smoothly bounded open neighbourhood of each vertex (e.g. a coordinate ball), small enough so that all these neighbourhoods are disjoint and each lies within the star of the vertex in question. If $\mathfrak{S}$ contains the $C^\infty$ fields, the existence of 'bump functions' (*cf.* (4.1.5)) implies that each chosen neighbourhood can be defined by the non-vanishing of a non-negative function $f$. Obviously we can add these functions, getting, say, $\overset{0}{v}$, and then define the union $\mathcal{U}_0$ of all these neighbourhoods by $\overset{0}{v} \neq 0$, with $\overset{0}{v} \geqslant 0$. A basis will exist for $\mathfrak{S}^\alpha$ restricted to each neighbourhood in turn. Taking all these bases together, we get a basis for $\mathfrak{S}^\alpha(\overset{0}{v})$.

The portions of the edges of the triangulation not lying in $\mathcal{U}_0$ form a disconnected system of closed segments which can be covered by an open set $\mathcal{U}_1$, where again $\mathcal{U}_1$ is a union of disconnected open sets (with smooth boundaries), each covering an edge segment and lying within the star of that edge (i.e. the union of $n$-simplexes through the edge). Again we can arrange that $\mathcal{U}_1$ is defined as $\overset{1}{v} \neq 0$ for some function $\overset{1}{v} \geqslant 0$, and a basis will exist for $\mathfrak{S}^\alpha(\overset{1}{v})$.

The portions of the faces (2-simplexes) not contained in $\mathcal{U}_0 \cup \mathcal{U}_1$ will be disconnected and, as before, we can cover them by a system of disjoint open sets whose union constitutes $\mathcal{U}_2$. As before, we can arrange that $\mathcal{U}_2$ is defined as $\overset{2}{v} \neq 0$, with $\overset{2}{v} \geqslant 0$, and a basis will exist for $\mathfrak{S}^\alpha(\overset{2}{v})$. The portions of the 3-simplexes not in $\mathcal{U}_0 \cup \mathcal{U}_1 \cup \mathcal{U}_2$ will again be disconnected, and the process continues until the $n$-simplexes are covered. By the preceding argument, (2.4.3) is therefore established (with $m = n$, the dimension of $\mathcal{M}$), since clearly $\mathfrak{S}^\alpha(\overset{i}{v}) = \mathfrak{S}^\alpha(\overset{i}{u})$, by (2.4.7).

By use of (2.4.3) we can now prove the total reflexivity of $\mathfrak{S}^\alpha$ essentially along the lines of §2.3, where the existence of a basis for $\mathfrak{S}^\alpha$ was assumed. For this purpose we shall first give a more complete definition of type III tensors (*cf.* (2.3.13)) which is in essence the 'classical' definition. Consider each $\mathfrak{S}^{\alpha}(\overset{\mathrm{i}}{u})$ and denote its basis and dual basis by the equivalence classes of

$$\overset{\mathrm{i}}{\delta}{}^{\alpha}_{\boldsymbol{\alpha}} \quad\text{and}\quad \overset{\mathrm{i}}{\delta}{}^{\boldsymbol{\alpha}}_{\alpha}, \tag{2.4.8}$$

respectively. (In terms of $\mathcal{M}$, these would be the basis and dual basis inside $\mathcal{U}_{\mathrm{i}}$, but arbitrary outside $\mathcal{U}_{\mathrm{i}}$.) We have

$$\overset{\mathrm{i}}{u}\overset{\mathrm{i}}{\delta}{}^{\boldsymbol{\beta}}_{\alpha}\overset{\mathrm{i}}{\delta}{}^{\alpha}_{\boldsymbol{\alpha}} = \overset{\mathrm{i}}{u}\delta^{\boldsymbol{\beta}}_{\boldsymbol{\alpha}}. \tag{2.4.9}$$

(There is no summation over i.) For each valence $\left[{}^{p}_{q}\right]$, consider the arrays

$$\overset{\mathrm{i}}{A}{}^{\boldsymbol{\alpha}\dots\boldsymbol{\gamma}}_{\boldsymbol{\lambda}\dots\boldsymbol{\nu}} \in \mathfrak{S}, \tag{2.4.10}$$

(where $\boldsymbol{\alpha},\dots,\boldsymbol{\gamma}$ are $p$ in number and $\boldsymbol{\lambda},\dots,\boldsymbol{\nu}$ are $q$ in number) whose $\overset{\mathrm{i}}{u}$-equivalence classes define arbitrary sets of $n^{p+q}$ elements of $\mathfrak{S}(\overset{\mathrm{i}}{u})$. Also consider the corresponding type II tensors

$$\overset{\mathrm{i}}{A}{}^{\alpha\dots\gamma}_{\lambda\dots\nu} = \overset{\mathrm{i}}{A}{}^{\boldsymbol{\alpha}\dots\boldsymbol{\gamma}}_{\boldsymbol{\lambda}\dots\boldsymbol{\nu}}\overset{\mathrm{i}}{\delta}{}^{\alpha}_{\boldsymbol{\alpha}}\dots\overset{\mathrm{i}}{\delta}{}^{\gamma}_{\boldsymbol{\gamma}}\overset{\mathrm{i}}{\delta}{}^{\boldsymbol{\lambda}}_{\lambda}\dots\overset{\mathrm{i}}{\delta}{}^{\boldsymbol{\nu}}_{\nu} \tag{2.4.11}$$

(*cf.* (2.3.14)), elements of $\mathfrak{S}^{\alpha\dots\gamma}_{\lambda\dots\nu}(\overset{\mathrm{i}}{u})$. For each j in place of i we can consider arrays like (2.4.10) and tensors like (2.4.11). In order that the two arrays (2.4.10) be *compatible*, we require that the classical tensor transformation law (2.3.24) hold in the 'overlap region':

$$\overset{\mathrm{i}}{u}\overset{\mathrm{j}}{u}\overset{\mathrm{i}}{A}{}^{\boldsymbol{\alpha}\dots\boldsymbol{\gamma}}_{\boldsymbol{\lambda}\dots\boldsymbol{\nu}} = \overset{\mathrm{i}}{u}\overset{\mathrm{j}}{u}\overset{\mathrm{j}}{A}{}^{\boldsymbol{\alpha}_0\dots\boldsymbol{\gamma}_0}_{\boldsymbol{\lambda}_0\dots\boldsymbol{\nu}_0}\overset{\mathrm{ij}}{\delta}{}^{\boldsymbol{\alpha}}_{\boldsymbol{\alpha}_0}\dots\overset{\mathrm{ij}}{\delta}{}^{\boldsymbol{\gamma}}_{\boldsymbol{\gamma}_0}\overset{\mathrm{ji}}{\delta}{}^{\boldsymbol{\lambda}_0}_{\boldsymbol{\lambda}}\dots\overset{\mathrm{ji}}{\delta}{}^{\boldsymbol{\nu}_0}_{\boldsymbol{\nu}}, \tag{2.4.12}$$

where

$$\overset{\mathrm{ij}}{\delta}{}^{\boldsymbol{\alpha}}_{\boldsymbol{\beta}} = \overset{\mathrm{i}}{\delta}{}^{\boldsymbol{\alpha}}_{\alpha}\overset{\mathrm{j}}{\delta}{}^{\alpha}_{\boldsymbol{\beta}} \quad (\mathrm{i},\mathrm{j} = 0, 1, \dots, m). \tag{2.4.13}$$

(Owing to the presence of $n+1$ simultaneous coordinate systems, the normal notational convention for component indices in different systems is temporarily suspended here.) The compatibility condition (2.4.12) ensures that the corresponding tensors (2.4.11) agree on the overlap region:

$$\overset{\mathrm{i}}{u}\overset{\mathrm{j}}{u}\overset{\mathrm{i}}{A}{}^{\alpha\dots\gamma}_{\lambda\dots\nu} = \overset{\mathrm{i}}{u}\overset{\mathrm{j}}{u}\overset{\mathrm{j}}{A}{}^{\alpha\dots\gamma}_{\lambda\dots\nu} \tag{2.4.14}$$

(*cf.* (2.4.11), (2.4.9)). A type III tensor consists of one array (2.4.10) of $n^{p+q}$ elements from *each* of $\mathfrak{S}(\overset{0}{u}), \dots, \mathfrak{S}(\overset{m}{u})$, where the arrays are related to one another according to (2.4.12).

To show the equivalence of the three types of tensor (given the bases), we again find three maps (II$\mapsto$I), (I$\mapsto$III), (III$\mapsto$II), as in §2.3, the first assigning a unique type I tensor to each type II tensor, etc., and then we show that each of the three cyclic compositions of these maps gives the identity. The maps that serve our present purpose are closely related to those of §2.3. Map (II$\mapsto$I) is actually the same, namely that given by

(2.2.17). To define the map (I $\mapsto$ III) we need only specify – for each $\mathfrak{S}(\overset{i}{u})$ – the components of a multilinear map $\overset{i}{A}{}^{\alpha\ldots\gamma}_{\lambda\ldots\nu}$ in the standard way (*cf.* (2.3.13))

$$\overset{i}{A}{}^{\boldsymbol{\alpha}\ldots\boldsymbol{\gamma}}_{\boldsymbol{\lambda}\ldots\boldsymbol{\nu}} = A^{\alpha\ldots\gamma}_{\lambda\ldots\nu}\overset{i}{\delta}{}^{\boldsymbol{\alpha}}_{\alpha}\ldots\overset{i}{\delta}{}^{\boldsymbol{\gamma}}_{\gamma}\overset{i}{\delta}{}^{\lambda}_{\boldsymbol{\lambda}}\ldots\overset{i}{\delta}{}^{\nu}_{\boldsymbol{\nu}}, \tag{2.4.15}$$

the $\overset{i}{u}$-equivalence class being independent (by the multilinearity) of the particular representative (2.4.8) of the basis and dual basis for $\mathfrak{S}^{\alpha}(\overset{i}{u})$. (This can be seen by multiplying (2.4.15) through by $\overset{i}{u}$ and shifting $\overset{i}{u}$ to each basis element in turn.) Clearly the compatibility condition (2.4.12) holds (by (2.4.9)), and so we have a unique type III tensor. Next we define (III $\mapsto$ II). As in (2.4.11), let each array $\overset{i}{A}{}^{\boldsymbol{\alpha}\ldots\boldsymbol{\gamma}}_{\boldsymbol{\lambda}\ldots\boldsymbol{\nu}}$ define a corresponding map $\overset{i}{A}{}^{\alpha\ldots\gamma}_{\lambda\ldots\nu}$, and set

$$A^{\alpha\ldots\gamma}_{\lambda\ldots\nu} = \overset{0}{u}\overset{0}{A}{}^{\alpha\ldots\gamma}_{\lambda\ldots\nu} + \cdots + \overset{m}{u}\overset{m}{A}{}^{\alpha\ldots\gamma}_{\lambda\ldots\nu}. \tag{2.4.16}$$

Since each (2.4.11) is a linear sum of outer products, so is (2.4.16), and a type II tensor is thereby defined. All three required maps have now been specified.

The fact that I $\mapsto$ III $\mapsto$ II $\mapsto$ I gives the identity on the set of type I tensors is obtained essentially as in (2.3.15), except that now a sum $\sum_{\mathrm{i}=0}^{m}\overset{i}{u}(\cdots)$ appears in the initial expression. For each i we expand $Q_{\alpha_0}, \ldots, W^{\nu_0}$ in terms of their components $\overset{i}{Q}_{\boldsymbol{\alpha}}, \ldots, \overset{i}{W}{}^{\boldsymbol{\nu}}$ in the respective bases for $\mathfrak{S}_{\alpha_0}(\overset{i}{u}), \ldots,$ $\mathfrak{S}^{\nu_0}(\overset{i}{u})$, noting that

$$\overset{i}{u}\overset{i}{Q}_{\boldsymbol{\alpha}} = \overset{i}{u}Q_{\alpha}\overset{i}{\delta}{}^{\alpha}_{\boldsymbol{\alpha}} \quad \text{yields} \quad \overset{i}{u}\overset{i}{\delta}{}^{\boldsymbol{\alpha}}_{\alpha}\overset{i}{Q}_{\boldsymbol{\alpha}} = \overset{i}{u}Q_{\alpha}, \tag{2.4.17}$$

and

$$\overset{i}{u}\overset{i}{W}{}^{\boldsymbol{\alpha}} = \overset{i}{u}W^{\alpha}\overset{i}{\delta}{}^{\boldsymbol{\alpha}}_{\alpha} \quad \text{yields} \quad \overset{i}{u}\overset{i}{\delta}{}^{\alpha}_{\boldsymbol{\alpha}}W^{\boldsymbol{\alpha}} = \overset{i}{u}W^{\alpha}. \tag{2.4.18}$$

Using (2.4.4) at the last step, we get the same final expression as in (2.3.15), which was to be shown. Similarly, the chain of maps II $\mapsto$ I $\mapsto$ III $\mapsto$ II leads to the identity on type II tensors by an argument which is essentially that of (2.3.17) except for the incorporation of a sum $\sum_{\mathrm{j}=0}^{m}\overset{j}{u}(\ldots)$. We use (2.4.17), (2.4.18), (2.4.9), and finally (2.4.4), to obtain our result.

The equivalence between type I and type II tensors – and hence the total reflexivity of $\mathfrak{S}^{\bullet}$ – is now established. Nevertheless it is of interest to show that the chain III $\mapsto$ II $\mapsto$ I $\mapsto$ III also gives the identity on III, for that shows that the 'classical' type III definition of tensors is equivalent to the other two. In fact, the 'classical' tensor transformation law (2.4.12) has not even been essentially used as yet. Without it, the map III $\mapsto$ II still gives us a type II tensor, but one which is a weighted sum of the now different type III tensors on each 'overlap region'. The chain

III $\mapsto$ II $\mapsto$ I $\mapsto$ III would lead us back to this weighted sum instead of giving the identity on type III. But if (2.4.12) is assumed, this chain also gives the identity, because, for each i,

$$\overset{\mathrm{i}}{u}\left(\sum_{\mathrm{j}}^{m}\overset{\mathrm{j}}{u}\,\overset{\mathrm{j}}{A}{}^{\boldsymbol{\alpha}\cdots}_{\cdots\boldsymbol{\nu}}\,\overset{\mathrm{j}}{\delta}{}^{\alpha}_{\boldsymbol{\alpha}}\ldots\overset{\mathrm{j}}{\delta}{}^{\boldsymbol{\nu}}_{\nu}\right)\overset{\mathrm{i}}{\delta}{}^{\boldsymbol{\alpha}_0}_{\alpha}\ldots\delta^{\nu}_{\boldsymbol{\nu}_0}=\overset{\mathrm{i}}{u}\overset{\mathrm{i}}{A}{}^{\boldsymbol{\alpha}\cdots}_{\cdots\boldsymbol{\nu}},$$

by (2.4.12), (2.4.13), and (2.4.4). The complete equivalence of the three types of tensors is thus established.

## 2.5 Spinor algebra

To construct spinor algebra we shall employ the theory of the preceding sections, applying it to the case of a basic module $\mathfrak{S}^{\bullet}$ consisting of $C^{\infty}$ spin-vector fields on a space–time manifold. We shall also be interested in the case when $\mathfrak{S}^{\bullet}$ consists of the spin-vectors at one point in the space–time, so that $\mathfrak{S}^{\bullet}$ becomes the spin-space at that point. In this second case, the ring of scalars $\mathfrak{S}$ is the division ring of complex numbers. In the first case it is the ring of $C^{\infty}$ complex scalar fields.

We recall that there are three basic algebraic operations that can be performed on spin-vectors. These are *scalar multiplication* (1.6.1), (1.6.4), *addition* (1.6.2), (1.6.5) and an anti-symmetrical *inner product* (1.6.3), (1.6.6). These operations can be performed between spin-vectors at any one point (so that the spin-vectors refer to the same Minkowski vector space, namely the tangent space at that point), and then the properties (1.6.8)–(1.6.19) hold. (These properties assert, in particular that spin-space is a complex two-dimensional vector space.) We may extend these operations so that they apply to spin-vector *fields* on the space–time, simply by applying them to the spin-vectors at each point separately. The properties (1.6.8)–(1.6.19) will then remain true for spin-vector fields. Now, properties (1.6.8)–(1.6.15) assert that $\mathfrak{S}^{\bullet}$ is a *module* over the complex scalars $\mathfrak{S}$. Thus, introducing a labelling system

$$\mathscr{L}=(A,B,C,\ldots,Z,A_0,B_0,\ldots,A_1,\ldots) \tag{2.5.1}$$

we can apply the theory of §2.2 and obtain canonically isomorphic copies of $\mathfrak{S}^{\bullet}$ denoted by $\mathfrak{S}^{A}$, $\mathfrak{S}^{B},\ldots,\mathfrak{S}^{A_0},\ldots$. Each spin-vector (field) $\boldsymbol{\kappa}\in\mathfrak{S}^{\bullet}$ will have images $\kappa^{A}\in\mathfrak{S}^{A}$, $\kappa^{B}\in\mathfrak{S}^{B},\ldots$. As before, we can define the duals of these $\mathfrak{S}$-modules: $\mathfrak{S}_{A}$, $\mathfrak{S}_{B},\ldots,\mathfrak{S}_{A_0},\ldots$; and consequently, general sets like $\mathfrak{S}^{AB},\ldots,\mathfrak{S}^{A}_{B},\ldots,\mathfrak{S}^{P\ldots R}_{S\ldots U},\ldots$ in terms of multilinear maps (or, equivalently, as equivalence classes of formal sums of outer products.) The module $\mathfrak{S}^{A}$ is totally reflexive, by the arguments of §2.4. The elements of the general sets $\mathfrak{S}^{P\ldots R}_{S\ldots U}$ are called *spinors*. These are not the most general spinors however. We shall define general spinors shortly. But before doing so, it will be worthwhile for us to examine properties of these particular spinors first.

### The ε-spinors

Properties (1.6.16)–(1.6.18) establish the inner product as an (anti-symmetrical) $\mathfrak{S}$-bilinear map from $\mathfrak{S}^{\bullet} \times \mathfrak{S}^{\bullet}$ to $\mathfrak{S}$, so there must be a unique element $\varepsilon_{AB} \in \mathfrak{S}_{AB}$ such that

$$\{\boldsymbol{\kappa}, \boldsymbol{\omega}\} = \varepsilon_{AB}\kappa^A\omega^B = -\{\boldsymbol{\omega}, \boldsymbol{\kappa}\} \tag{2.5.2}$$

for all $\boldsymbol{\kappa}, \boldsymbol{\omega} \in \mathfrak{S}^{\bullet}$, where $\varepsilon_{AB}$ is anti-symmetrical:

$$\varepsilon_{AB} = -\varepsilon_{BA}. \tag{2.5.3}$$

The quantity $\varepsilon_{AB}$ is an essential part of the spinor algebra. It plays a role somewhat analogous to that played by the metric tensor in Cartesian (or Riemannian) tensor theory, but there are important differences arising from its anti-symmetry.

To begin with, we note that $\varepsilon_{AB}$ establishes a canonical mapping (actually an isomorphism) between the modules $\mathfrak{S}^A, \mathfrak{S}^B, \ldots$ and the dual modules $\mathfrak{S}_A, \mathfrak{S}_B, \ldots$:

$$\kappa^B \leftrightarrow \kappa_B = \kappa^A\varepsilon_{AB}. \tag{2.5.4}$$

(To put this another way, the element of the dual of $\mathfrak{S}^{\bullet}$ which corresponds to $\boldsymbol{\kappa}$ is $\{\boldsymbol{\kappa},\ \}$.) The same kernel symbol will be used for an element of $\mathfrak{S}_A$ and for its corresponding element in $\mathfrak{S}^A$. Thus (by analogy with classical Riemannian tensor analysis) we may regard $\varepsilon_{AB}$ as 'lowering the index' of $\kappa^A$ in (2.5.4). The fact that (2.5.4) *is* an isomorphism, and not merely some module homomorphism which is not one-to-one, follows from the component form (1.6.6) of the inner product:

$$\{\boldsymbol{\kappa}, \boldsymbol{\omega}\} = \kappa^0\omega^1 - \kappa^1\omega^0 \tag{2.5.5}$$

(using a coordinate system for spin-vectors at each point, as in Chapter 1); so we have, by (2.5.2) and (2.5.4),

$$\{\boldsymbol{\kappa}, \boldsymbol{\omega}\} = \kappa_B\omega^B = \kappa_{\mathbf{B}}\omega^{\mathbf{B}} = \kappa_0\omega^0 + \kappa_1\omega^1, \tag{2.5.6}$$

where the components* $\kappa_0, \kappa_1$ of $\kappa_B$ are related to those of $\kappa^B$ by

$$\kappa_0 = -\kappa^1, \quad \kappa_1 = \kappa^0, \tag{2.5.7}$$

the one-to-one nature of which is evident. Thus, the inverse map from

* We recall that our spinor component indices range over 0, 1 (or 0′, 1′ as later required) rather than 1, 2. This is visually consistent with our use of $\boldsymbol{o}, \boldsymbol{\iota}$ for a basis.

$\mathfrak{S}_B$ to $\mathfrak{S}^A$ which assigns $\kappa^A$ to $\kappa_B$ must exist (and be $\mathfrak{S}$-linear), and there must be an element $\varepsilon^{AB} \in \mathfrak{S}^{AB}$ (*cf.* (2.2.37)) which effects it:

$$\kappa^A = \varepsilon^{AB}\kappa_B. \tag{2.5.8}$$

The fact that (2.5.4) and (2.5.8) are inverses of each other may be expressed in the equations

$$\varepsilon_{AB}\varepsilon^{CB} = \delta_A^C, \quad \varepsilon^{AB}\varepsilon_{AC} = \delta_C^B \tag{2.5.9}$$

(where we map $\mathfrak{S}^A \to \mathfrak{S}_B \to \mathfrak{S}^C$ and $\mathfrak{S}_B \to \mathfrak{S}^A \to \mathfrak{S}_C$, respectively), the symbols $\delta_A^C$ and $\delta_C^B$ expressing the canonical isomorphisms between $\mathfrak{S}^A$ and $\mathfrak{S}^C$, and between $\mathfrak{S}_B$ and $\mathfrak{S}_C$, respectively (*cf.* (2.2.43), (2.2.44)). However we shall not, henceforth, use the symbol $\delta_A^B$, preferring instead to write this $\varepsilon_A{}^B$(or $-\varepsilon^B{}_A$):

$$\delta_A^B = \varepsilon_A{}^B = -\varepsilon^B{}_A. \tag{2.5.10}$$

In fact, we regard the first term in (2.5.9) as $\varepsilon^{CB}$ acting on $\varepsilon_{AB}$ to 'raise' its second index, and similarly for the third term, in accordance with the raising and lowering conventions (2.5.4) and (2.5.8). We may regard $\varepsilon_A{}^B$ *either* as $\varepsilon_{AB}$ with its second index raised (first equation (2.5.9)) *or* as $\varepsilon^{AB}$ with its first index lowered (second equation (2.5.9)). Combining these two interpretations of $\varepsilon_A{}^B$, we see that $\varepsilon^{AB}$ is $\varepsilon_{AB}$ with *both* indices raised, as the notation suggests. The anti-symmetry

$$\varepsilon^{AB} = -\varepsilon^{BA} \tag{2.5.11}$$

is one consequence of this. The relation $\varepsilon_A{}^B = -\varepsilon^B{}_A$ is also an expression of the anti-symmetry of $\varepsilon_{AB}$ and $\varepsilon^{AB}$. It emphasizes, in addition, the necessity to stagger spinor indices. Each lower index must have a position to which it can be unambiguously raised, and each upper index a position to which it can be unambiguously lowered.

Collecting together our various relations, we have

$$\varepsilon_{AB}\varepsilon^{CB} = -\varepsilon_{AB}\varepsilon^{BC} = \varepsilon_{BA}\varepsilon^{BC} = -\varepsilon_{BA}\varepsilon^{CB} = \varepsilon_A{}^C = -\varepsilon^C{}_A, \tag{2.5.12}$$

and

$$\psi^{\mathscr{C}}{}_A\varepsilon_B{}^A = \psi^{\mathscr{C}}{}_B, \quad \psi^{\mathscr{C}A}\varepsilon_A{}^B = \psi^{\mathscr{C}B}. \tag{2.5.13}$$

Because of the anti-symmetry of the εs, we must exercise care, when raising and lowering indices, to see that the correct index of ε is contracted. Thus

$$\psi^{\mathscr{C}A} = \varepsilon^{AB}\psi^{\mathscr{C}}{}_B = -\psi^{\mathscr{C}}{}_B\varepsilon^{BA} \tag{2.5.14}$$

and

$$\psi^{\mathscr{C}}{}_B = \psi^{\mathscr{C}A}\varepsilon_{AB} = -\varepsilon_{BA}\psi^{\mathscr{C}A}. \tag{2.5.15}$$

We can relate (2.5.14), (2.5.15) to (2.5.13) by means of the spinor 'see-saw':

$$\chi^{\dots}{}_{\dots A}{}^{\dots}{}_{\dots}{}^{A}{}_{\dots} = -\chi^{\dots}{}_{\dots}{}^{A\dots}{}_{\dots A\dots}. \tag{2.5.16}$$

The minus sign is again a consequence of the anti-symmetry of $\varepsilon$. One way of remembering the arrangement of signs involved in (2.5.14) and (2.5.15) is simply to remember the signs in (2.5.10) and use the 'see-saw'.

As in (2.5.6), the inner product can be written:

$$\{\boldsymbol{\kappa}, \boldsymbol{\omega}\} = \kappa_A\omega^A = -\kappa^A\omega_A. \tag{2.5.17}$$

The final relation (1.6.19) is now

$$\kappa_A\omega^A\tau^B + \omega_A\tau^A\kappa^B + \tau_A\kappa^A\omega^B = 0. \tag{2.5.18}$$

Another way of expressing this is

$$(\varepsilon_{AB}\varepsilon_C{}^D + \varepsilon_{BC}\varepsilon_A{}^D + \varepsilon_{CA}\varepsilon_B{}^D)\kappa^A\omega^B\tau^C = 0, \tag{2.5.19}$$

for all $\boldsymbol{\kappa}, \omega, \boldsymbol{\tau}$; hence (*cf.* (2.2.40))

$$\varepsilon_{AB}\varepsilon_C{}^D + \varepsilon_{BC}\varepsilon_A{}^D + \varepsilon_{CA}\varepsilon_B{}^D = 0. \tag{2.5.20}$$

Equivalently, lowering $D$, we have the important identity

$$\varepsilon_{AB}\varepsilon_{CD} + \varepsilon_{BC}\varepsilon_{AD} + \varepsilon_{CA}\varepsilon_{BD} = 0. \tag{2.5.21}$$

Alternatively, we can raise the $C$ in (2.5.20) to obtain

$$\varepsilon_A{}^C\varepsilon_B{}^D - \varepsilon_B{}^C\varepsilon_A{}^D = \varepsilon_{AB}\varepsilon^{CD}. \tag{2.5.22}$$

This implies that

$$\phi_{\mathscr{D}AB} - \phi_{\mathscr{D}BA} = \phi_{\mathscr{D}C}{}^C\varepsilon_{AB} \tag{2.5.23}$$

(transvecting (2.5.22) with $\phi_{\mathscr{D}CD}$). Thus if $\phi_{\mathscr{D}AB}$ is skew in $A, B$ ($\phi_{\mathscr{D}AB} = -\phi_{\mathscr{D}BA}$) then

$$\phi_{\mathscr{D}AB} = \tfrac{1}{2}\phi_{\mathscr{D}C}{}^C\varepsilon_{AB}. \tag{2.5.24}$$

(This relation evidently applies also when $A, B$ are non-adjacent indices; for example, if $\psi_{\mathscr{A}A\mathscr{B}B\mathscr{C}}$ is skew in $A, B$, we can define $\phi_{\mathscr{A}\mathscr{B}\mathscr{C}AB} := \psi_{\mathscr{A}A\mathscr{B}B\mathscr{C}}$ and apply (2.5.24).) Note, as a particular case of (2.5.24), that all anti-symmetrical elements of $\mathfrak{S}_{AB}$ are proportional to $\varepsilon_{AB}$. Note that we can raise the indices $A, B$ in (2.5.23) and (2.5.24) to obtain alternative versions of these results. Note also that $\binom{D}{C}$-contraction of (2.5.20) yields

$$\varepsilon_A{}^A = 2 = -\varepsilon^A{}_A. \tag{2.5.25}$$

(Because of (2.5.10), this is an expression of the two-dimensionality of spin-space.)

*Complex conjugation*

The spinor algebra that we have set up so far is self-contained but it is inadequate for physics. We wish to have the algebra of world-vectors and world-tensors incorporated in our spinor algebra. This is not possible as things stand. The essential reason for this can be seen in the expression (1.2.15) for world-vector components in terms of spin-vector components: the *complex conjugates* of the spin-vector components are necessarily involved. Thus, to incorporate world-vectors, our spinor algebra must include an operation of complex conjugation. We must be able to apply this operation to any element of $\mathfrak{S}^A$. However, the result cannot be just another element of $\mathfrak{S}^A$. For if it were, we should have a property of *reality* for elements of $\mathfrak{S}^A$. Some elements of $\mathfrak{S}^A$ would be real and others purely imaginary (e.g., a spin-vector plus or minus its complex conjugate). The different elements of $\mathfrak{S}^A$ would then cease to be on an equal footing and the Lorentz covariance of the algebraic operations would thereby be lost. (We saw in §1.4 that any two spin-vectors at a point could be transformed one into the other by a Lorentz transformation.) Thus, the complex conjugate of an element $\kappa^A \in \mathfrak{S}^A$ must be an entity of a new type. Let us denote the operation of complex conjugation by a bar and write

$$\overline{\kappa^A} = \bar{\kappa}^{A'} \in \mathfrak{S}^{A'} \tag{2.5.26}$$

for the complex conjugate* of $\kappa^A$. The label $A'$ may be regarded as the complex conjugate of the label $A$. We have, therefore, in addition to the labelling set $\mathscr{L}$ of (2.5.1), another labelling set $\mathscr{L}'$ consisting of the conjugates of the labels belonging to $\mathscr{S}$,

$$\mathscr{L}' = (A', B', C', \ldots, Z', A'_0, B'_0, \ldots, A'_1, \ldots). \tag{2.5.27}$$

The set $\mathfrak{S}^{A'}$ is regarded as the complex conjugate of the set $\mathfrak{S}^A$. The operations of addition and scalar multiplication in $\mathfrak{S}^{A'}$ are defined by the requirement

$$\lambda\kappa^A + \mu\omega^A = \tau^A \Leftrightarrow \bar{\lambda}\bar{\kappa}^{A'} + \bar{\mu}\bar{\omega}^{A'} = \bar{\tau}^{A'}, \tag{2.5.28}$$

where $\lambda, \mu \in \mathfrak{S}$ and $\bar{\lambda}, \bar{\mu}$ are the complex conjugates of $\lambda, \mu$ in the usual sense. It is easy to verify (*cf.* (2.2.3)) that $\mathfrak{S}^{A'}$ is then also an $\mathfrak{S}$-module (since the algebra of complex scalars is sent into itself by the operation of complex conjugation); and $\mathfrak{S}^{A'}$ is *anti-isomorphic* with $\mathfrak{S}^A$, the anti-isomorphism being expressed by (2.5.28). As in (2.5.26), the inverse map from $\mathfrak{S}^{A'}$ to $\mathfrak{S}^A$ is also denoted by a bar over the entire symbol; thus we have

$$\overline{\bar{\tau}^{A'}} = \tau^A, \ \overline{\lambda\kappa^A + \mu\omega^A} = \bar{\lambda}\bar{\kappa}^{A'} + \bar{\mu}\bar{\omega}^{A'}. \tag{2.5.29}$$

* In the original notation of Infeld and van der Waerden (1933), this would be written $\kappa^{\dot{A}}$. The use of a prime rather than a dot has been made for typographical reasons. The use of a bar over the kernel symbol is for notational consistency with what follows. Although it is true that the symbols can tend to get a bit cluttered, the gain in notational consistency and clarity more than compensates for this.

We now use *both* $\mathfrak{S}$-modules $\mathfrak{S}^A$ and $\mathfrak{S}^{A'}$ to generate our spinor system in the manner indicated at the end of §2.2. Thus, in addition to $\mathfrak{S}$-modules $\mathfrak{S}^B, \mathfrak{S}^C, \ldots, \mathfrak{S}^X, \ldots, \mathfrak{S}^{W_3}, \ldots$ each canonically isomorphic to $\mathfrak{S}^A$, we shall have $\mathfrak{S}$-modules $\mathfrak{S}^{B'}, \mathfrak{S}^{C'}, \ldots, \mathfrak{S}^{X'}, \ldots, \mathfrak{S}^{W'_3}, \ldots$ each canonically isomorphic to $\mathfrak{S}^{A'}$ and canonically anti-isomorphic to $\mathfrak{S}^A$. For each $X \in \mathscr{L}$ the complex conjugate of $V^X \in \mathfrak{S}^X$ is $\bar{V}^{X'} \in \mathfrak{S}^{X'}$, that is to say complex conjugation appropriately commutes with index substitution. Each set $\mathfrak{S}^{X'}$ will have a dual $\mathfrak{S}$-module $\mathfrak{S}_{X'}$. The canonical anti-isomorphism between $\mathfrak{S}^X$ and $\mathfrak{S}^{X'}$ induces a canonical anti-isomorphism between $\mathfrak{S}_X$ and $\mathfrak{S}_{X'}$ in which $\tau_X$ corresponds to $\overline{\tau_X} = \bar{\tau}_{X'}$, defined by

$$\bar{\tau}_{X'}\bar{\kappa}^{X'} = \overline{\tau_X\kappa^X}, \tag{2.5.30}$$

the bar on the right denoting ordinary complex conjugation of scalars. We then have

$$\overline{\lambda\alpha_X + \mu\beta_X} = \bar{\lambda}\bar{\alpha}_{X'} + \bar{\mu}\bar{\beta}_{X'}, \overline{\overline{\tau_X}} = \tau_X \tag{2.5.31}$$

(by (2.1.28) and (2.1.29)) as the expression of this anti-isomorphism.

The *general spinor* $\chi_{L\ldots NU'\ldots W'}{}^{A\ldots DP'\ldots R'}$, of valence $\left[\begin{smallmatrix} p & q \\ r & s \end{smallmatrix}\right]$, is defined as an $\mathfrak{S}$-multilinear map from $\mathfrak{S}_A \times \cdots \times \mathfrak{S}_D \times \mathfrak{S}_{P'} \times \cdots \times \mathfrak{S}_{R'} \times \mathfrak{S}^L \times \cdots \times \mathfrak{S}^N \times \mathfrak{S}^{U'} \times \cdots \times \mathfrak{S}^{W'}$ to $\mathfrak{S}$, or, equivalently, in terms of equivalence classes of formal sums of formal products. The arguments of §2.4 can be adapted, with only minor notational complications, to establish that total reflexivity holds not just starting from the module $\mathfrak{S}^A$, but also starting from both modules $\mathfrak{S}^A$, $\mathfrak{S}^{A'}$ together. The subsets $\{A, \ldots, D\}, \{L, \ldots, N\}$ of $\mathscr{L}$ and $\{P', \ldots, R'\}, \{U', \ldots, W'\}$ of $\mathscr{L}'$ (of respective cardinalities $p, r, q, s$) are all disjoint. (The sets $\mathscr{L}$ and $\mathscr{L}'$ are, of course, disjoint from each other in any case.) There is no objection to the same letter appearing in both primed and unprimed versions. For example, $\psi_{ABA'}{}^{B'}$ is an allowable spinor (and no contraction is involved). The set of spinors $\chi_{L\ldots U'\ldots}{}^{A\ldots P'\ldots}$ is denoted by $\mathfrak{S}^{A\ldots P'\ldots}_{L\ldots U'\ldots}$.

As in §2.2, four operations are defined, namely addition, outer multiplication, index substitution and contraction. But now we have a new operation, namely complex conjugation, which is induced by the anti-isomorphism between $\mathfrak{S}^A$ and $\mathfrak{S}^{A'}$. To define the complex conjugate of a spinor $\chi_{L\ldots U'\ldots}{}^{A\ldots P'\ldots}$, using the multilinear map definition, we take the complex conjugate of the result of the map, and replace each of $\mathfrak{S}_A, \ldots, \mathfrak{S}^{U'}, \ldots$ by its complex conjugate $\mathfrak{S}_{A'}, \ldots, \mathfrak{S}^U, \ldots$. This clearly defines an element $\bar{\chi}_{L'\ldots U\ldots}{}^{A'\ldots P\ldots} \in \mathfrak{S}^{P\ldots A'\ldots}_{U\ldots L'\ldots}$ as a multilinear map

$$\bar{\chi}_{L'\ldots U\ldots}{}^{A'\ldots P\ldots}\lambda^{L'}\ldots\pi_P\ldots = \overline{\chi_{L\ldots U'\ldots}{}^{A\ldots P'\ldots}\bar{\lambda}^L\ldots\bar{\pi}_{P'}\ldots}, \tag{2.5.32}$$

the long bar on the right denoting ordinary complex conjugation of scalars.

Let us summarize the various spinor operations and their basic properties. *Addition* assigns an abelian group structure to each spinor set $\mathfrak{S}_{\mathscr{A}}$ (with $\mathscr{A} = A \ldots D' \ldots G^* \ldots J'^* \ldots$, say). *Outer multiplication* is a map from $\mathfrak{S}_{\mathscr{A}} \times \mathfrak{S}_{\mathscr{B}}$ to $\mathfrak{S}_{\mathscr{A}\mathscr{B}}$ for each pair of spinor sets $\mathfrak{S}_{\mathscr{A}}, \mathfrak{S}_{\mathscr{B}}$, where $\mathscr{A}$ and $\mathscr{B}$ involve no common index label (regarding $A, A', B, B', \ldots$ as all distinct but $A$ and $A^*$, or $B'$ and $B'^*$, etc. as dual labels). It is commutative, associative and distributive over addition. A particular case of it, arising when one of the spinor sets is $\mathfrak{S}$, is *scalar multiplication*. This, together with addition, assigns an $\mathfrak{S}$-module structure to each spinor set $\mathfrak{S}_{\mathscr{A}}$. *Index substitution* is induced whenever *separate* permutations are applied to the two labelling sets $\mathscr{L}$ and $\mathscr{L}'$. Thus we can substitute one set of unprimed labels for another set of unprimed labels, and we can substitute one set of primed labels for another set of primed labels, but we cannot substitute primed labels for unprimed ones or unprimed labels for primed ones. The validity of any equation is unaffected if an index substitution is performed throughout the equation. The operation of $\binom{X}{Y}$-*contraction* maps each set $\mathfrak{S}^{X}_{Y\mathscr{A}}$ to $\mathfrak{S}_{\mathscr{A}}$ ($\mathscr{A}$ being any composite index label not involving $X$ or $Y$). The $\binom{X}{Y}$-contraction of, say, $\psi_{\mathscr{A}Y}{}^{X}$ is written $\psi_{\mathscr{A}X}{}^{X}$ or $\psi_{\mathscr{A}Y}{}^{Y}$ or $\psi_{\mathscr{A}Z}{}^{Z}$, where $Z$ is any unprimed label not involved in $\mathscr{A}$. The $\binom{X}{Y}$-contraction of $\psi_{\mathscr{A}Y}{}^{X} + \phi_{\mathscr{A}Y}{}^{X}$ is $\psi_{\mathscr{A}X}{}^{X} + \phi_{\mathscr{A}X}{}^{X}$ and that of $\chi_{\mathscr{B}}\psi_{\mathscr{A}Y}{}^{X}$ is $\chi_{\mathscr{B}}\phi_{\mathscr{A}X}{}^{X}$ (i.e., contraction commutes with addition and, in the appropriate way, with outer multiplication). The $\binom{X}{Y}$-contraction of $\theta_{\mathscr{A}UY}{}^{UX}$ is the same as the $\binom{U}{V}$-contraction of $\theta_{\mathscr{A}VX}{}^{UX}$ and is written $\theta_{\mathscr{A}UX}{}^{UX}$ (contractions commute with other contractions). The $\binom{X}{Y}$-contraction of $\psi_{\mathscr{A}_0 Y}{}^{X}$ is the same as the result of the index substitution $\mathscr{A} \rightarrow \mathscr{A}_0$ applied to $\psi_{\mathscr{A}X}{}^{X}$; the $\binom{X}{Y}$-contraction of $\psi_{\mathscr{A}Y}{}^{X}$ is the same as the $\binom{U}{V}$-contraction of $\psi_{\mathscr{A}V}{}^{U}$ (i.e., contraction appropriately commutes with index substitution). There is also an operation of $\binom{X'}{Y'}$-contraction satisfying corresponding laws. Furthermore the $\binom{X'}{Y'}$-contraction of $\xi_{\mathscr{A}Y'X}{}^{X'X}$ is the same as the $\binom{X}{Y}$-contraction of $\xi_{\mathscr{A}X'Y}{}^{X'X}$, both being written $\xi_{\mathscr{A}X'X}{}^{X'X}$ or $\xi_{\mathscr{A}X'Y}{}^{X'Y}$, etc. Finally, *complex conjugation* is a map from each $\mathfrak{S}_{\mathscr{A}}$ to the corresponding set $\mathfrak{S}_{\mathscr{A}'}$, where if $\mathscr{A} = A \ldots D' \ldots G^* \ldots J'^* \ldots$, then $\mathscr{A}' = A' \ldots D \ldots G'^* \ldots J^* \ldots$. The complex conjugate $\overline{\eta_{\mathscr{A}}}$ of $\eta_{\mathscr{A}}$ is written $\bar{\eta}_{\mathscr{A}'}$, and for scalars this is the standard complex conjugacy relation. When applied twice, the original spinor is recovered: $\overline{\overline{\eta_{\mathscr{A}}}} = \eta_{\mathscr{A}}$ (involutory property). We have $\overline{\eta_{\mathscr{A}} + \zeta_{\mathscr{A}}} = \bar{\eta}_{\mathscr{A}'} + \bar{\zeta}_{\mathscr{A}'}, \overline{\eta_{\mathscr{A}}\chi_{\mathscr{B}}} = \bar{\eta}_{\mathscr{A}'}\bar{\chi}_{\mathscr{B}'}$; also, $\overline{\eta_{\mathscr{A}_0}}$ is the result of the index substitution $\mathscr{A}' \rightarrow \mathscr{A}'_0$ applied to $\bar{\eta}_{\mathscr{A}'}$; finally, $\overline{\psi_{\mathscr{A}X}{}^{X}} = \bar{\psi}_{\mathscr{A}'X'}{}^{X'}$ (i.e., complex conjugation commutes with addition, outer multiplication and, appropriately, with index substitution and contraction).

Since index substitutions which interchange primed and unprimed indices are not permitted, there is no meaning to be attached to the relative order between primed and unprimed indices on a spinor symbol. It is sometimes useful to exploit this fact and allow primed and unprimed indices to be moved across each other, without changing the meaning of the symbol. This applies whether the indices are in upper or lower positions. We can allow indices directly above one another, therefore, *provided* one is primed and the other unprimed. For example,

$$\psi^{AA'}{}_{B'B}{}^{Q} = \psi^{A'A}{}_{B}{}^{Q}{}_{B'} = \psi^{A}{}_{B}{}^{QA'}{}_{B'} = \psi^{AA'Q}{}_{BB'} \neq \psi^{AA'Q}{}_{BB'}. \tag{2.5.33}$$

The element $\varepsilon_{AB} \in \mathfrak{S}_{AB}$ has its complex conjugate $\bar{\varepsilon}_{A'B'} \in \mathfrak{S}_{A'B'}$. It is conventional to omit the bar and write this simply $\varepsilon_{A'B'}$. (We need not regard this as violating our notational conventions. The symbol $\bar{\varepsilon}_{A'B'}$ still correctly stands for $\overline{\varepsilon_{AB}}$. We simply introduce a *new* symbol $\varepsilon_{A'B'}$ – as we are always entitled to do, provided no ambiguity results – which *also* stands for $\overline{\varepsilon_{AB}}$.) The isomorphism between $\mathfrak{S}^{A}$ and $\mathfrak{S}_{B}$, which $\varepsilon_{AB}$ achieves, now induces, via the operation of complex conjugation, an isomorphism between $\mathfrak{S}^{A'}$ and $\mathfrak{S}_{B'}$, achieved by $\varepsilon_{A'B'}$. In effect, this means that $\varepsilon_{A'B'}$, together with its inverse $\varepsilon^{A'B'} (= \bar{\varepsilon}^{A'B'} = \overline{\varepsilon^{AB}})$ can be used for lowering and raising primed indices. The formulae are identical with those of (2.5.8)–(2.5.16), except that the relevant indices are primed. Thus, in particular,

$$\varepsilon_{B'A'} = -\varepsilon_{A'B'}, \quad \varepsilon^{B'A'} = -\varepsilon^{A'B'}, \tag{2.5.34}$$

$$\chi^{\mathscr{C}}{}_{A'}\varepsilon_{B'}{}^{A'} = \chi^{\mathscr{C}}{}_{B'}, \quad \chi^{\mathscr{C}A'}\varepsilon_{A'}{}^{B'} = \chi^{\mathscr{C}B'}, \tag{2.5.35}$$

$$\chi^{\mathscr{C}A'} = \varepsilon^{A'B'}\chi^{\mathscr{C}}{}_{B'} = -\chi^{\mathscr{C}}{}_{B'}\varepsilon^{B'A'}, \tag{2.5.36}$$

$$\chi^{\mathscr{C}}{}_{B'} = \chi^{\mathscr{C}A'}\varepsilon_{A'B'} = -\varepsilon_{B'A'}\chi^{\mathscr{C}A'}, \tag{2.5.37}$$

and (the 'see-saw' property)

$$\chi^{\cdots}{}_{\ldots A'}{}^{\cdots}{}_{\cdots}{}^{A'}{}_{\cdots} = -\chi^{\cdots}{}_{\cdots}{}^{A'\cdots}{}_{\cdots}{}_{\ldots A'\ldots}. \tag{2.5.38}$$

Furthermore, the complex conjugates of (2.5.18)–(2.5.25) all hold, giving corresponding versions with primed indices.

The rule for deciding whether or not a symbol $\chi^{\cdots}_{\cdots}$ with indices is an allowable spinor symbol is the same as when only one initial module is involved, that is, the upper indices must be distinct (where, as we stressed before, $A, A', B, B', \ldots$ are all distinct labels) and the lower indices must be distinct. The symbol $\chi^{\cdots}_{\cdots}$ then represents an element of that spinor set $\mathfrak{S}^{\cdots}_{\cdots}$ whose arrays of upper and lower indices are those of $\chi^{\cdots}_{\cdots}$, but with any duplicated (contracted) upper and lower indices omitted.

## *Spinor Bases*

In (1.6.22) the concept of a *spin-frame* was introduced. This is a pair of spin-vectors $\boldsymbol{o}, \boldsymbol{\iota}$ normalized so that $\{\boldsymbol{o}, \boldsymbol{\iota}\} = 1$. By (2.5.17), we can now write this

$$o_A \iota^A = 1, \tag{2.5.39}$$

or equivalently

$$\iota_A o^A = -1. \tag{2.5.40}$$

By the anti-symmetry of inner product we also have

$$o_A o^A = 0 = \iota_A \iota^A. \tag{2.5.41}$$

Let $\kappa^A \in \mathfrak{S}^A$. We saw in (1.6.24) that as a consequence of the identity (1.6.19) (i.e. of (2.5.18)) the normalization (2.5.39) implies

$$\kappa^A = \kappa^0 o^A + \kappa^1 \iota^A, \tag{2.5.42}$$

where

$$\kappa^0 = -\iota_A \kappa^A, \quad \kappa^1 = o_A \kappa^A. \tag{2.5.43}$$

In fact, *any* $\kappa^0, \kappa^1 \in \mathfrak{S}$ for which (2.5.42) holds, must be given by (2.5.43), as follows from (2.5.39)–(2.5.41) upon transvection of (2.5.42) by $o_A$ and $\iota_A$.

The existence and uniqueness of (2.5.42) thus establishes $o^A$, $\iota^A$ as constituting a basis for $\mathfrak{S}^A$. The normalization condition (2.5.39) is, in itself, sufficient for this, as we have just seen. If we employ the results on parallelizability mentioned in §2.3 above (and *cf.* (1.5.6)) (assuming non-compactness for the space–time) then we may take it that a spin-frame field exists globally, so that $o_A \iota^A = 1$ can indeed be satisfied for some $o^A$, $\iota^A \in \mathfrak{S}^A$. Of course if we are concerned with spinors at just one point, or with spinors in some sufficiently small open subset of the space–time, then it is clear that we may assume that a spin-frame exists.* It is only when we consider the topological structure of the space–time as a whole that the global existence of a spin-frame can come into question.

It is often convenient (as in §2.3) to use a collective symbol $\varepsilon_{\mathbf{A}}{}^A$ for a basis for $\mathfrak{S}^A$. (The use of 'ε' rather than 'δ' is in accordance with (2.5.10).) Then we can set

$$\varepsilon_0{}^A = o^A, \quad \varepsilon_1{}^A = \iota^A. \tag{2.5.44}$$

The components of $\varepsilon_{AB}$ with respect to this basis are

$$\varepsilon_{\mathbf{AB}} = \varepsilon_{AB} \varepsilon_{\mathbf{A}}{}^A \varepsilon_{\mathbf{B}}{}^B = \begin{pmatrix} 0 & \chi \\ -\chi & 0 \end{pmatrix} \tag{2.5.45}$$

where

$$\chi = \varepsilon_{AB} o^A \iota^B = o_A \iota^A. \tag{2.5.46}$$

Thus a condition equivalent to the normalization (2.5.39) for a spin-frame

* However, as we shall see in §§4.14, 4.15, spin-frames with *prescribed geometrical properties* may exist locally, but fail to exist globally.

is that the components $\varepsilon_{\mathbf{AB}}$ of $\varepsilon_{AB}$ should constitute the normal Levi–Civita symbol. It is clear that any basis can be readily converted to a spin-frame if we leave $\varepsilon_0{}^A$ unchanged, but replace $\varepsilon_1{}^A$ by $\chi^{-1}\varepsilon_1{}^A$. The more general spinor basis for which $\chi$ need not be unity is referred to here as a *dyad*.

The dual basis $\varepsilon_A{}^{\mathbf{A}}$ must satisfy

$$\varepsilon_{\mathbf{A}}{}^{A}\varepsilon_{A}{}^{\mathbf{B}} = \varepsilon_{\mathbf{A}}{}^{\mathbf{B}} = \begin{pmatrix} 1 & 0 \\ 0 & 1 \end{pmatrix} \tag{2.5.47}$$

(so $\varepsilon_{\mathbf{A}}{}^{\mathbf{B}}$ is actually a Kronecker delta $\delta_{\mathbf{A}}^{\mathbf{B}}$). The components $\varepsilon^{\mathbf{AB}}$ of $\varepsilon^{AB}$ have to satisfy $\varepsilon^{\mathbf{AB}}\,\varepsilon_{\mathbf{CB}} = \varepsilon_{\mathbf{C}}{}^{\mathbf{A}}$ (*cf.* (2.5.12)); so, for a general dyad,

$$\varepsilon^{\mathbf{AB}} = \begin{pmatrix} 0 & \chi^{-1} \\ -\chi^{-1} & 0 \end{pmatrix}. \tag{2.5.48}$$

Comparing (2.5.47) with (2.5.39)–(2.5.41) we see that *if* the basis is a spin-frame, then

$$\varepsilon_A{}^0 = -\iota_A, \quad \varepsilon_A{}^1 = o_A. \tag{2.5.49}$$

In the general case,

$$\varepsilon_A{}^0 = -\chi^{-1}\iota_A, \quad \varepsilon_A{}^1 = \chi^{-1}o_A. \tag{2.5.50}$$

This agrees with (2.5.43) for the components $\kappa^{\mathbf{A}} = \kappa^A\varepsilon_A{}^{\mathbf{A}}$ of $\kappa^A$ in a spin-frame. The components

$$\kappa_0 = \kappa_A o^A, \quad \kappa_1 = \kappa_A\iota^A \tag{2.5.51}$$

of $\kappa_A$ in the spin-frame are then related to those of $\kappa^A$ by (2.5.7), i.e.

$$\kappa_0 = -\kappa^1, \quad \kappa_1 = \kappa^0. \tag{2.5.52}$$

Note that for a spin-frame

$$\begin{aligned} &\varepsilon_{0A} = o_A = -\varepsilon_{A0}, \quad \varepsilon_{1A} = \iota_A = -\varepsilon_{A1}, \\ &\varepsilon^{0A} = \iota^A = -\varepsilon^{A0}, \quad \varepsilon^{1A} = -o^A = -\varepsilon^{A1}. \end{aligned} \tag{2.5.53}$$

Also, the formulae $\varepsilon^{AB} = \varepsilon^{\mathbf{AB}}\varepsilon_{\mathbf{A}}{}^A\varepsilon_{\mathbf{B}}{}^B$, $\varepsilon_{AB} = \varepsilon_{\mathbf{AB}}\varepsilon_A{}^{\mathbf{A}}\varepsilon_B{}^{\mathbf{B}}$, $\varepsilon_A{}^B = \varepsilon_A{}^{\mathbf{A}}\varepsilon_{\mathbf{A}}{}^B$ can be expressed in the form

$$\varepsilon^{AB} = o^A\iota^B - \iota^A o^B, \quad \varepsilon_{AB} = o_A\iota_B - \iota_A o_B, \quad \varepsilon_A{}^B = o_A\iota^B - \iota_A o^B \tag{2.5.54}$$

in a spin-frame $o^A$, $\iota^A$; and in any basis:

$$\varepsilon^{AB} = \chi^{-1}(o^A\iota^B - \iota^A o^B), \varepsilon_{AB} = \chi^{-1}(o_A\iota_B - \iota_A o_B), \varepsilon_A{}^B = \chi^{-1}(o_A\iota^B - \iota_A o^B). \tag{2.5.55}$$

The only condition on $o^A$, $\iota^A \in \mathfrak{S}^A$ that they constitute a basis for $\mathfrak{S}^A$ (not necessarily as spin-frame) is that they be linearly independent at each point, i.e., that at no point is one a multiple of the other. Another way of putting this is that the $\chi = o_A\iota^A$ of (2.5.46) should vanish nowhere (i.e., $\chi^{-1}$ exists). For, if $o^A$, $\iota^A$ do constitute a basis, then the components of $\varepsilon_{AB}$

cannot all vanish at any point (since the spinor $\varepsilon_{AB}$ vanishes nowhere), so (2.5.45) gives $\chi \neq 0$ at every point. Conversely, it is clear from (2.5.41) that $\chi = o_A \iota^A$ must vanish at any point at which one of $o^A$, $\iota^A$ is a multiple of the other. We may state a closely related result as follows:

(2.5.56) PROPOSITION

*The condition $\alpha_A \beta^A = 0$ at a point is necessary and sufficient for $\alpha_A$, $\beta_A$ to be scalar multiples of each other at that point.**

This tells us that at any point at which $\alpha^A \neq 0 \neq \beta^A$, the scalar $\alpha_A \beta^A$ vanishes iff the flagpole directions of $\alpha^A$ and $\beta^A$ coincide.

Given a basis $\varepsilon_{\mathbf{A}}{}^A$ for $\mathfrak{S}^A$, it is most natural to choose as a basis for $\mathfrak{S}^{A'}$ the complex conjugates of the elements $\varepsilon_{\mathbf{A}}{}^A$ – and, indeed, this is what we shall always do. Let us suppose that $\varepsilon_{\mathbf{A}}{}^A$ is a spin-frame and that $o^A$ and $\iota^A$ are given by (2.5.44). Then we can write

$$o^{A'} := \bar{o}^{A'} = \overline{o^A} = \varepsilon_{0'}{}^{A'}, \quad \iota^{A'} := \bar{\iota}^{A'} = \overline{\iota^A} = \varepsilon_{1'}{}^{A'} \qquad (2.5.57)$$

where, as in the case of $\varepsilon_{A'B'}$ earlier, we have chosen to introduce new symbols $o^{A'}$ and $\iota^{A'}$ so as to avoid the proliferation of bars. (Occasionally it is expedient to omit indices when writing certain expressions. Then such bars must be reinstated.) The dual basis $\varepsilon_{A'}{}^{\mathbf{A}'}$ is related to $\varepsilon_{\mathbf{A}'}{}^{A'}$ by

$$\varepsilon_{\mathbf{A}'}{}^{A'} \varepsilon_{A'}{}^{\mathbf{B}'} = \varepsilon_{\mathbf{A}'}{}^{\mathbf{B}'} \qquad (2.5.58)$$

and we have

$$\varepsilon_{A'}{}^{0'} = -\iota_{A'}, \quad \varepsilon_{A'}{}^{1'} = o_{A'}. \qquad (2.5.59)$$

Note that (since a spin-frame is assumed),

$$\varepsilon_{\mathbf{A}'\mathbf{B}'} = \begin{pmatrix} 0 & 1 \\ -1 & 0 \end{pmatrix} = \varepsilon^{\mathbf{A}'\mathbf{B}'}, \quad \varepsilon_{\mathbf{A}'}{}^{\mathbf{B}'} = \begin{pmatrix} 1 & 0 \\ 0 & 1 \end{pmatrix} \qquad (2.5.60)$$

and

$$o_{A'} \iota^{A'} = 1 = -\iota_{A'} o^{A'}. \qquad (2.5.61)$$

Given any spinor $\chi_{G' \ldots K \ldots}^{A \ldots D' \ldots}$, we obtain its components by transvecting with the basis elements (which need not be normalized):

$$\chi_{\mathbf{G}' \ldots \mathbf{K} \ldots}^{\mathbf{A} \ldots \mathbf{D}' \ldots} = \chi_{G' \ldots K \ldots}^{A \ldots D' \ldots} \varepsilon_A{}^{\mathbf{A}} \ldots \varepsilon_D{}^{\mathbf{D}'} \ldots \varepsilon_{\mathbf{G}'}{}^{G'} \ldots \varepsilon_{\mathbf{K}}{}^{K} \ldots . \qquad (2.5.62)$$

* This is an example of a result which is awkward to state using merely properties of the module $\mathfrak{S}^A$ of spin-vector fields and not mentioning points. For if $\alpha^A$ vanishes in one region and $\beta^A$ vanishes in a separate one, but $\alpha_A \beta^A = 0$ everywhere, then neither $\alpha^A$ nor $\beta^A$ is a multiple of the other by an element of $\mathfrak{S}$. The spinor $\lambda \alpha^A + \mu \beta^A$ vanishes, on the other hand, for some $\lambda, \mu \in \mathfrak{S}$, with $\lambda \neq 0 \neq \mu$; but this is not *sufficient* to imply $\alpha_A \beta^A = 0$, since there may be a region throughout which $\lambda$ and $\mu$ both vanish.

We recover the spinor from its components by using the formula

$$\chi_{G'\dots K\dots}^{A\dots D'\dots} = \chi_{\mathbf{G}'\dots\mathbf{K}\dots}^{\mathbf{A}\dots\mathbf{D}'\dots}\,\varepsilon_{\mathbf{A}}{}^{A}\dots\varepsilon_{\mathbf{D}'}{}^{D'}\dots\varepsilon_{G'}{}^{\mathbf{G}'}\dots\varepsilon_{K}{}^{\mathbf{K}}\dots. \tag{2.5.63}$$

As with tensors in general, addition, outer multiplication, index substitution and contraction of spinors all commute with the operation of taking components. The operation of complex conjugation also commutes with that of taking components, i.e.,

$$\overline{\chi_{\mathbf{G}'\dots\mathbf{K}\dots}^{\mathbf{A}\dots\mathbf{D}'\dots}} = \bar{\chi}_{\mathbf{G}\dots\mathbf{K}'\dots}^{\mathbf{A}'\dots\mathbf{D}\dots} \tag{2.5.64}$$

(recall that the basis for $\mathfrak{S}^{A'}$ is chosen to be the complex conjugate of the basis for $\mathfrak{S}^{A}$), where the bar on the left side means that the complex conjugation is applied to each scalar of the array, while the bar on the right side means that the complex conjugation operation is applied to the spinor itself before the components are evaluated. Inspection of (2.5.64) shows that in spinor component equations involving a conjugation bar over an entire symbol (and *only* in these) it is necessary to make the convention that $\mathbf{A} = \mathbf{A}'$, $\mathbf{B} = \mathbf{B}'$, etc., numerically. It is therefore important to avoid the use of *both* $\mathbf{A}$ and $\mathbf{A}'$, or of *both* $\mathbf{B}$ and $\mathbf{B}'$, etc., under a conjugation bar. *Because* of (2.5.64), however, it is generally possible to avoid symbols with conjugation bars over the indices.*

The components of any spinor which possesses only *lower* indices may be obtained by 'plugging in' $o^A, \iota^A, o^{A'}, \iota^{A'}$ for each numerical index 0, 1, 0′, 1′, respectively, so this can be remembered easily. For example,

$$\psi_{0111'0'} = \psi_{ABCD'E'}\, o^A \iota^B \iota^C \iota^{D'} o^{E'}. \tag{2.5.65}$$

For a spinor possessing some upper indices, we can remember

$$\begin{aligned}
\psi_{\dots}{}^{\dots 0\dots}{}_{\dots}{}^{\dots} &= \psi_{\dots}{}^{\dots}{}_{1\dots}{}^{\dots} \\
\psi_{\dots}{}^{\dots 1\dots}{}_{\dots}{}^{\dots} &= -\psi_{\dots}{}^{\dots}{}_{0\dots}{}^{\dots} \\
\psi_{\dots}{}^{0'\dots}{}_{\dots}{}^{\dots} &= \psi_{\dots 1'}{}^{\dots}{}_{\dots}{}^{\dots} \\
\psi_{\dots}{}^{1'\dots}{}_{\dots}{}^{\dots} &= -\psi_{\dots 0'}{}^{\dots}{}_{\dots}{}^{\dots}
\end{aligned} \tag{2.5.66}$$

in the case of a *spin frame*, and then use (2.5.65).

Finally, let us examine how spinor components transform under change of basis (*cf.* (2.3.22)–(2.3.24).) Let $\varepsilon_{\mathbf{A}}{}^{A}$ and $\varepsilon_{\hat{\mathbf{A}}}{}^{A}$ be two bases for $\mathfrak{S}^{A}$. Let $\varepsilon_{A}{}^{\mathbf{A}}$ and $\varepsilon_{A}{}^{\hat{\mathbf{A}}}$ be the respective dual bases and let $\varepsilon_{\mathbf{A}'}{}^{A'}$, $\varepsilon_{A'}{}^{\mathbf{A}'}$ and

* For example, we have $\overline{u^{AB'}} = \bar{u}^{A'B} = \bar{u}^{BA'}$ and $\overline{u^{AA'}} = \bar{u}^{A'A} = \bar{u}^{AA'}$. But to write $\overline{u^{\mathbf{AA}'}} = \bar{u}^{\mathbf{AA}'}$ is misleading, since, for example, $\overline{u^{01'}} \neq \bar{u}^{01'}$: so we write $\overline{u^{\mathbf{AB}'}} = \bar{u}^{\mathbf{BA}'}$, from which, for example, we correctly obtain $\overline{u^{01'}} = \bar{u}^{10'}$.

$\varepsilon_{\hat{\mathbf{A}}'}{}^{A'}$, $\varepsilon_{A'}{}^{\hat{\mathbf{A}}'}$ be the respective complex conjugates of these. Define matrices

$$\varepsilon_{\mathbf{A}}{}^{\hat{\mathbf{A}}} = \varepsilon_{\mathbf{A}}{}^{A}\varepsilon_{A}{}^{\hat{\mathbf{A}}}, \quad \varepsilon_{\hat{\mathbf{A}}}{}^{\mathbf{A}} = \varepsilon_{\hat{\mathbf{A}}}{}^{A}\varepsilon_{A}{}^{\mathbf{A}} \tag{2.5.67}$$

and

$$\varepsilon_{\mathbf{A}'}{}^{\hat{\mathbf{A}}'} = \varepsilon_{\mathbf{A}'}{}^{A'}\varepsilon_{A'}{}^{\hat{\mathbf{A}}'}, \quad \varepsilon_{\hat{\mathbf{A}}'}{}^{\mathbf{A}'} = \varepsilon_{\hat{\mathbf{A}}'}{}^{A'}\varepsilon_{A'}{}^{\mathbf{A}'}. \tag{2.5.68}$$

Then the matrices $\varepsilon_{\mathbf{A}}{}^{\hat{\mathbf{A}}}$ and $\varepsilon_{\hat{\mathbf{A}}}{}^{\mathbf{A}}$ are inverses of one another while $\varepsilon_{\mathbf{A}'}{}^{\hat{\mathbf{A}}'}$ and $\varepsilon_{\hat{\mathbf{A}}'}{}^{\mathbf{A}'}$ are the respective complex conjugate matrices of these two. Now transvect (2.5.63) with the appropriate basis $\varepsilon_{\hat{\mathbf{A}}}{}^{A}$, dual basis $\varepsilon_{A}{}^{\hat{\mathbf{A}}}$ or complex conjugate of these. We thus obtain

$$\chi^{\hat{\mathbf{A}}\ldots\hat{\mathbf{D}}'\ldots}_{\hat{\mathbf{G}}'\ldots\hat{\mathbf{K}}\ldots} = \chi^{\mathbf{A}\ldots\,\mathbf{D}'\ldots}_{\mathbf{G}'\ldots\mathbf{K}\ldots}\varepsilon_{\mathbf{A}}{}^{\hat{\mathbf{A}}}\ldots\varepsilon_{\mathbf{D}'}{}^{\hat{\mathbf{D}}'}\ldots\varepsilon_{\hat{\mathbf{G}}'}{}^{\mathbf{G}'}\ldots\varepsilon_{\hat{\mathbf{K}}}{}^{\mathbf{K}}\ldots. \tag{2.5.69}$$

Equation (2.5.69) gives the transformation law for spinor components under transformation from one general basis to another. We are normally only interested in the case when the bases are both spin-frames. Then the matrices $\varepsilon_{\mathbf{AB}}$ and $\varepsilon_{\hat{\mathbf{A}}\hat{\mathbf{B}}}$ are the same, being the Levi–Civita symbol in each case (i.e. (2.5.45) with $\chi = 1$), and we have

$$1 = \varepsilon_{\hat{0}\hat{1}} = \varepsilon_{\mathbf{AB}}\varepsilon_{\hat{0}}{}^{\mathbf{A}}\varepsilon_{\hat{1}}{}^{\mathbf{B}} = \det(\varepsilon_{\hat{\mathbf{A}}}{}^{\mathbf{A}}), \tag{2.5.70}$$

showing that the complex matrix $\varepsilon_{\hat{\mathbf{A}}}{}^{\mathbf{A}}$ is *unimodular*. It is thus a *spin-matrix* and, therefore, so also are $\varepsilon_{\mathbf{A}}{}^{\hat{\mathbf{A}}}$, $\varepsilon_{\hat{\mathbf{A}}'}{}^{\mathbf{A}'}$ and $\varepsilon_{\mathbf{A}'}{}^{\hat{\mathbf{A}}'}$. Then (2.5.69) gives what is the familiar form of transformation law for spinor components.

# 3

# Spinors and world-tensors

## 3.1 World-tensors as spinors

In this section we show how world-tensors and, in particular, world-vectors may be regarded as special cases of spinors. The algebra of world-tensors thus emerges as being *embedded* in the spinor algebra of §2.5. This embedding of one kind of tensor algebra in another is a particular example of the procedure described at the end of §2.2. Accordingly, the index labels of the embedded system are composite indices, consisting of certain groups of labels of the original system clumped together. In the particular case of this procedure that will concern us here, the world-tensor labels will be clumped *pairs* of spinor labels, one of which is unprimed and the other primed. The essential reason for this can be seen in the formulae (1.2.15) and (1.2.23) which express world-vector components in terms of spin-vector components. The world-vector components are bilinear in the spin-vector components and in the complex conjugate spin-vector components.

Let us define a world-tensor labelling set

$$\mathscr{K} = \{a, b, c, \ldots, z, a_0, b_0, \ldots, a_1, \ldots\} \tag{3.1.1}$$

from the spinor labelling sets $\mathscr{L}$, $\mathscr{L}'$ (see (2.5.1), (2.5.27)), where

$$a = AA', b = BB', c = CC', \ldots, z = ZZ', a_0 = A_0A'_0, \ldots, a_1 = A_1A'_1, \ldots \tag{3.1.2}$$

Then, for example, we can label the spinor of (2.5.33) variously as:

$$\psi^{AA'}{}_{BB'}{}^{Q} = \psi^{a}{}_{BB'}{}^{Q} = \psi^{a}{}_{b}{}^{Q} = \psi^{A'A}{}_{b}{}^{Q} = \psi^{a}{}_{B}{}^{Q}{}_{B'}. \tag{3.1.3}$$

We do *not* here adopt the convention (which was normal in the general case of composite indices) that composite indices should not implicitly involve single indices occurring elsewhere in an expression. This is because here the clumping scheme (3.1.2) has been made quite definite, so no ambiguity can arise. For example, each of the following equivalent contracted expressions is equally allowable:

$$\psi^{a}{}_{AA'}{}^{Q} = \psi^{AA'}{}_{a}{}^{Q} = \psi^{a}{}_{a}{}^{Q} = \psi^{AA'}{}_{AA'}{}^{Q}, \tag{3.1.4}$$

and again,

$$\psi^{AA'}{}_{B}{}^{B}{}_{B'} = \psi^{a}{}_{b}{}^{B} = \psi^{AA'}{}_{b}{}^{B} = -\psi^{aB}{}_{b} = -\psi^{aB}{}_{BB'}. \tag{3.1.5}$$

Certain spinors can be labelled *entirely* with elements taken from $\mathscr{K}$:

$$\chi^{a\ldots d}{}_{p\ldots r} = \chi^{AA'\ldots DD'}{}_{PP'\ldots RR'} = \chi^{A\ldots DA'\ldots D'}{}_{P\ldots RP'\ldots R'}. \tag{3.1.6}$$

These belong to the spinor sets $\mathfrak{S}, \mathfrak{S}^a, \mathfrak{S}^b, \ldots, \mathfrak{S}_a, \ldots, \mathfrak{S}^{bq_0}_{x_3}, \ldots$, where $\mathfrak{S}^a = \mathfrak{S}^{AA'}, \mathfrak{S}^b = \mathfrak{S}^{BB'}, \ldots, \mathfrak{S}_a = \mathfrak{S}_{AA'}, \ldots, \mathfrak{S}^{bq_0}_{x_3} = \mathfrak{S}^{BQ_0B'Q'_0}_{X_3X'_3}, \ldots$ Spinor sets of this kind play a special role because they are sent to themselves by the operation of complex conjugation. We shall refer to an element $\chi^{a\ldots d}{}_{p\ldots r}$ of such a set as a *complex world-tensor*. Thus the complex conjugate of such a spinor,

$$\begin{aligned}\overline{\chi^{a\ldots d}{}_{p\ldots r}} = \overline{\chi^{AA'\ldots DD'}{}_{PP'\ldots RR'}} &= \bar{\chi}^{A'A\ldots D'D}{}_{P'P\ldots R'R} \\ &= \bar{\chi}^{AA'\ldots DD'}{}_{PP'\ldots RR'} = \bar{\chi}^{a\ldots d}{}_{p\ldots r},\end{aligned} \tag{3.1.7}$$

is another spinor of the same type. Certain complex world-tensors will actually be invariant under complex conjugation and these will be called *real world-tensors* or, simply, *world-tensors*. (This terminology will be justified shortly.) A real world-tensor thus satisfies

$$\chi^{a\ldots d}{}_{p\ldots r} = \bar{\chi}^{a\ldots d}{}_{p\ldots r}. \tag{3.1.8}$$

We denote the subsets of $\mathfrak{S}, \mathfrak{S}^a, \ldots, \mathfrak{S}_x, \ldots, \mathfrak{S}^{a\ldots d}_{p\ldots r}, \ldots$, which consist of real world-tensors, by $\mathfrak{T}, \mathfrak{T}^a, \ldots, \mathfrak{T}_x, \ldots, \mathfrak{T}^{a\ldots d}_{p\ldots r}, \ldots$, respectively.* (The set $\mathfrak{T}$ is the ring of real scalar fields on the space–time – or the division ring of real numbers, in case we are concerned with spinors at a single point.) The system $(\mathfrak{T}, \mathfrak{T}^a, \mathfrak{T}^b, \ldots, \mathfrak{T}^{a\ldots}_{p\ldots}, \ldots)$ is the tensor system generated, in the manner of §2.2, from the $\mathfrak{T}$-module $\mathfrak{T}^a$. Each $\mathfrak{T}^{\cdots}_{\cdots}$ is then a $\mathfrak{T}$-module and the whole system is closed under the tensor operations of addition, outer multiplication, index substitution and contraction. The elements of $\mathfrak{T}^a$ (or $\mathfrak{T}^b$, etc.) are called (real) world-vectors.

If we define the particular real world-tensors

$$g_{ab} = \varepsilon_{AB}\varepsilon_{A'B'}, \tag{3.1.9a}$$

$$g_a{}^b = \varepsilon_A{}^B\varepsilon_{A'}{}^{B'} \tag{3.1.9b}$$

$$g^{ab} = \varepsilon^{AB}\varepsilon^{A'B'} \tag{3.1.9c}$$

then, from the properties (2.5.3), (2.5.9), (2.5.11), (2.5.25), (2.5.34), we have

$$g_{ab} = g_{ba}, g^{ab} = g^{ba}, g_{ab}g^{bc} = g_a{}^c, g_{ab}g^{ab} = 4. \tag{3.1.10}$$

* In conventional terminology (*cf.* footnote on p. 90) the complex world-tensors here arise as the tensors on the $\mathfrak{S}$-module $\mathfrak{S}^{\cdot} \otimes_{\mathfrak{S}} \bar{\mathfrak{S}}^{\cdot}$ and the real world tensors as those on the $\mathfrak{T}$-module of its Hermitian elements. We are allowed to say 'real' here, rather than 'Hermitian', because abstract-index tensor product is commutative.

From (2.5.13) and (2.5.35) we have

$$\chi^{\mathscr{C}}{}_a = \chi^{\mathscr{C}}{}_b g_a{}^b, \chi^{\mathscr{C}a} = \chi^{\mathscr{C}b} g_b{}^a, \tag{3.1.11}$$

and so $g_a{}^b$ plays the role of a Kronecker delta symbol. (We prefer $g_a{}^b$ to $\delta_a^b$ here for notational consistency with $g_{ab}$, and because it avoids a possible confusion when basis frames are introduced.) Furthermore, from (2.5.14), (2.5.15), (2.5.36), (2.5.37) we have

$$\chi^{\mathscr{C}a} = \chi^{\mathscr{C}}{}_b g^{ab}, \chi^{\mathscr{C}}{}_a = \chi^{\mathscr{C}b} g_{ab}, \tag{3.1.12}$$

so that $g_{ab}$ and $g^{ab}$ play roles formally identical to those of the metric tensor and its inverse in lowering and raising world-tensor indices. In fact, we shall presently *identify* the above $g_{ab}$ and $g^{ab}$ with the metric tensor and its inverse.

In the usual approach to the description of space–time $\mathscr{M}$ (as in Chapter 1), the world-vectors and world-tensors are given first. The metric is introduced as a specific world-tensor defining the 'geometry' of $\mathscr{M}$, and only thereafter is the spinor concept defined. Moreover, certain global topological requirements need to hold for $\mathscr{M}$ (*cf*. §1.5) in order that this spinor concept be globally consistent. The spinors can then be interpreted (as in Chapter I) in terms of somewhat complicated space–time geometry, except that there remains an overall sign ambiguity for the interpretation. But we may ask ourselves whether Nature is really so complicated, since spinor fields *are* a part of Nature as described by contemporary physical theory.

The complication seems to be largely due to the tensorial approach. If, as we shall tend to do in this book, one regards the spin-vectors as more basic than the world-vectors – as, perhaps, something more primitive than the space–time structure itself, from which that particular structure can be *deduced* – then these complications largely evaporate. Thus, if we *start* from spinors, we have no sign ambiguity (since the signs are part of the given structure, not something that has to be derived). The resulting space–time is *automatically* time- and space-oriented and has spin-structure (which properties may be regarded as highly desirable in view of various experimental facts; *cf*. remarks at the end of §1.5). Even the dimension and signature of space–time are 'consequences' of our particular spinor formalism. The spinor algebra has *in itself* a certain simplicity. The complications always seem to arise when we try to interpret the spinor operations in space–time terms. We shall see good examples of this in §3.4.

*Spin-frames and their related tetrads*

At this stage we shall not be much concerned with the differential or global properties of the system (spinor derivatives are left to Chapter 4). Our present concern is the local algebraic structure implied by the existence of the spinor system. To see explicitly how the 'world-vectors' arising here are consistent ('isomorphic') with the usual ones arising in relativity theory, we introduce a spin-frame $o^A, \iota^A$ with the standard normalization

$$o_A \iota^A = 1. \tag{3.1.13}$$

Next we define a *null tetrad** of world-vectors $l^a, n^a, m^a, \bar{m}^a$ by

$$l^a = o^A o^{A'}, \qquad n^a = \iota^A \iota^{A'}$$
$$m^a = o^A \iota^{A'}, \quad \bar{m}^a = \iota^A o^{A'}. \tag{3.1.14}$$

These are all *null* vectors with respect to our $g_{ab}$ metric:

$$l^a l_a = n^a n_a = m^a m_a = \bar{m}^a \bar{m}_a = 0. \tag{3.1.15}$$

(For example, $m^a m_a = (o^A \iota^{A'})(o_A \iota_{A'}) = o^A o_A \iota^{A'} \iota_{A'} = 0$, (*cf.* (2.5.41).) Furthermore,

$$l^a n_a = 1, \quad m^a \bar{m}_a = -1 \tag{3.1.16}$$

(for example, $m^a \bar{m}_a = (o^A \iota^{A'})(\iota_A o_{A'}) = (o^A \iota_A)(\iota^{A'} o_{A'}) = (-1) \times (1) = -1$), while the other scalar products vanish:

$$l^a m_a = l^a \bar{m}_a = n^a m_a = n^a \bar{m}_a = 0. \tag{3.1.17}$$

Evidently $l^a$ and $n^a$ are real,

$$l^a = \bar{l}^a, n^a = \bar{n}^a, \tag{3.1.18}$$

and $m^a$ and $\bar{m}^a$ are complex conjugates.

The null tetrad $l^a, n^a, m^a, \bar{m}^a$ constitutes a basis, over $\mathfrak{S}$, for $\mathfrak{S}^a$, the dual basis being $n_a, l_a, -\bar{m}_a, -m_a$. This follows from (3.1.15)–(3.1.17), the relation

$$g_a{}^b = n_a l^b + l_a n^b - \bar{m}_a m^b - m_a \bar{m}^b \tag{3.1.19}$$

(which is a direct consequence of (2.5.55)), and (3.1.9) (*cf.* (2.3.19) *et seq.*). This basis for $\mathfrak{S}^a$ is the one *induced* by the basis $o^A, \iota^A$ for $\mathfrak{S}^A$ (*cf.* (2.3.18)).

To obtain a basis (over $\mathfrak{T}$) for $\mathfrak{T}^a$ from $o^A$ and $\iota^A$, we need *real* world vectors. Thus $m^a$ and $\bar{m}^a$ have to be split into real and imaginary parts. It is

* This is a standard (and very useful) concept in Minkowski geometry. See, for example, Sachs (1961), Newman and Penrose (1962), Trautman (1965), p. 57, Kramer, Stephani, MacCallum and Herlt (1980).

convenient also to form linear combinations of $l^a$ and $n^a$, and write

$$t^a = \frac{1}{\sqrt{2}}(l^a + n^a) = \frac{1}{\sqrt{2}}(o^A o^{A'} + \iota^A \iota^{A'})$$

$$x^a = \frac{1}{\sqrt{2}}(m^a + \bar{m}^a) = \frac{1}{\sqrt{2}}(o^A \iota^{A'} + \iota^A o^{A'}) \tag{3.1.20}$$

$$y^a = \frac{\mathrm{i}}{\sqrt{2}}(m^a - \bar{m}^a) = \frac{\mathrm{i}}{\sqrt{2}}(o^A \iota^{A'} - \iota^A o^{A'})$$

$$z^a = \frac{1}{\sqrt{2}}(l^a - n^a) = \frac{1}{\sqrt{2}}(o^A o^{A'} - \iota^A \iota^{A'}).$$

The inverse relations are

$$l^a = \frac{1}{\sqrt{2}}(t^a + z^a) = o^A o^{A'}$$

$$n^a = \frac{1}{\sqrt{2}}(t^a - z^a) = \iota^A \iota^{A'} \tag{3.1.21}$$

$$m^a = \frac{1}{\sqrt{2}}(x^a - \mathrm{i}y^a) = o^A \iota^{A'}$$

$$\bar{m}^a = \frac{1}{\sqrt{2}}(x^a + \mathrm{i}y^a) = \iota^A o^{A'}$$

From (3.1.15)–(3.1.17) and (3.1.19) we have the orthogonality relations

$$t^a x_a = t^a y_a = t^a z_a = x^a y_a = y^a z_a = z^a x_a = 0 \tag{3.1.22}$$

and the normalizations

$$t^a t_a = 1, x^a x_a = y^a y_a = z^a z_a = -1. \tag{3.1.23}$$

Furthermore, (3.1.19) gives us

$$g_a{}^b = t_a t^b - x_a x^b - y_a y^b - z_a z^b. \tag{3.1.24}$$

These relations imply that $t^a$, $x^a$, $y^a$, $z^a$ do, indeed, constitute a basis, over $\mathfrak{T}$, for $\mathfrak{T}^a$, with dual basis $t_a$, $-x_a$, $-y_a$, $-z_a$. In fact, (3.1.22) and (3.1.23) are identical with the conditions (1.1.7) and (1.1.8) for a *Minkowski tetrad.**

* Note that the Minkowski tetrad (or, equivalently, the null tetrad) defines the spin frame $o^A$, $\iota^A$ locally up to an overall sign. For $l^a$ and $n^a$ define the two flagpoles; knowledge of $m^a$ reduces the freedom to $(o_A, \iota^A) \mapsto e^{i\theta}(o_A, \iota^A)$ ($\theta$ real); the normalization $o_A \iota^A = 1$ then fixes $\theta$ to be a multiple of $\pi$.

Let us set

$$g_0{}^a = t^a, \quad g_1{}^a = x^a, \quad g_2{}^a = y^a, \quad g_3{}^a = z^a \tag{3.1.25}$$

and dually,

$$g_a{}^0 = t_a, \quad g_a{}^1 = -x_a, \quad g_a{}^2 = -y_a, \quad g_a{}^3 = -z_a. \tag{3.1.26}$$

Then the components of $g_{ab}, g^{ab}$ and $g_a{}^b$ in the basis $g_{\mathbf{a}}{}^a$ and dual basis $g_a{}^{\mathbf{a}}$ are*, by (3.1.22), (3.1.23),

$$g_{\mathbf{ab}} = \begin{pmatrix} 1 & & & \\ & -1 & & \\ & & -1 & \\ & & & -1 \end{pmatrix} = g^{\mathbf{ab}}, \quad g_{\mathbf{a}}{}^{\mathbf{b}} = \begin{pmatrix} 1 & & & \\ & 1 & & \\ & & 1 & \\ & & & 1 \end{pmatrix} \tag{3.1.27}$$

At each point we therefore have a standard representation of $\mathfrak{T}^a$ as a Minkowski vector space referred to a Minkowski tetrad.

Note that we can define the concepts of *time-* and *space-orientation* for $\mathfrak{T}^a$ by the specification that the Minkowski tetrad (3.1.20) be deemed to be *restricted* (*cf*. §1.1). (We shall adopt a slightly different and more 'invariant' approach in §3.2.) We saw at the end of §2.5 that changing the spin-frame $o^A, \iota^A$ to another one, at a point, is the result of a spin transformation and is continuous with the identity. This shows that the resulting tetrads (3.1.20) are all continuous with one another – indeed, related at each point by the corresponding restricted Lorentz transformation – so the resulting orientations for $\mathfrak{T}^a$ are *intrinsic* and not dependent upon the choice of $o^A, \iota^A$.

Let $K^a \in \mathfrak{T}^a$. Then in terms of the above basis we have

$$K^a = K^{\mathbf{a}} g_{\mathbf{a}}{}^a = K^0 t^a + K^1 x^a + K^2 y^a + K^3 z^a, \tag{3.1.28}$$

where

$$K^0 = K^a t_a, \; K^1 = -K^a x_a, \; K^2 = -K^a y_a, \; K^3 = -K^a z_a. \tag{3.1.29}$$

Now $K^a = K^{AA'} \in \mathfrak{S}^{AA'}$, so we can also refer $K^a$ to the spinor basis $\varepsilon_0{}^A = o^A, \varepsilon_1{}^A = \iota^A$:

$$\begin{aligned} K^a &= K^{\mathbf{AA'}} \varepsilon_{\mathbf{A}}{}^A \varepsilon_{\mathbf{A'}}{}^{A'} \\ &= K^{00'} o^A o^{A'} + K^{01'} o^A \iota^{A'} + K^{10'} \iota^A o^{A'} + K^{11'} \iota^A \iota^{A'} \\ &= K^{00'} l^a + K^{11'} n^a + K^{01'} m^a + K^{10'} \bar{m}^a. \end{aligned} \tag{3.1.30}$$

* Note that the signature of the metric comes out *automatically* as $(+,-,-,-)$. If it had been desired to obtain the signature $(-,+,+,+)$ for the space–time metric, then the definition $g_{ab} = -\varepsilon_{AB}\varepsilon_{A'B'}$ would have to have been used. This would have lead to difficulties with the spinor index raising and lowering conventions.

So, comparing this with (3.1.28) and using (3.1.20), we get, upon equating coefficients of $l^a, \ldots, \bar{m}^a$:

$$\frac{1}{\sqrt{2}}\begin{pmatrix} K^0 + K^3 & K^1 + \mathrm{i}K^2 \\ K^1 - \mathrm{i}K^2 & K^0 - K^3 \end{pmatrix} = \begin{pmatrix} K^{00'} & K^{01'} \\ K^{10'} & K^{11'} \end{pmatrix}. \tag{3.1.31}$$

The Lorentz group action on $\mathfrak{T}^a$ can be invoked to transform any given future-null vector into any other. Thus, if $K^a$ happens to be future-null, then like $l^a$ (or $n^a$) above (*cf.* (3.1.14)) it is the product of some spinor, say $\kappa^A$, with its complex conjugate, so $K^a$ has the form

$$K^a = \kappa^A \bar{\kappa}^{A'}. \tag{3.1.32}$$

From this, if we set

$$\xi = \kappa^0, \quad \eta = \kappa^1, \tag{3.1.33}$$

we find

$$K^{00'} = \xi\bar{\xi}, K^{01'} = \xi\bar{\eta}, K^{10'} = \eta\bar{\xi}, K^{11'} = \eta\bar{\eta}, \tag{3.1.34}$$

so equation (3.1.31) becomes

$$\frac{1}{\sqrt{2}}\begin{pmatrix} T + Z & X + \mathrm{i}Y \\ X - \mathrm{i}Y & T - Z \end{pmatrix} = \begin{pmatrix} \xi \\ \eta \end{pmatrix}(\bar{\xi} \quad \bar{\eta}) \tag{3.1.35}$$

where we have put

$$T = K^0, X = K^1, Y = K^2, Z = K^3. \tag{3.1.36}$$

Equation (3.1.35) is precisely the same as (1.2.23), which formed the cornerstone of the discussion in Chapter 1. This shows that if we *start* with $\mathcal{M}$ and its metric $g_{ab}$, and if $\mathcal{M}$ satisfies the global conditions that allow us to construct spinors as we did in Chapter 1, then the $g_{ab}$ resulting from the algebra of those spinors via Equations (3.1.9) is the same as the original metric $g_{ab}$.

### *Infeld–van der Waerden symbols*

Note that we have allowed ourselves to use a basis for $\mathfrak{S}^a$ which is *not* the one induced by our basis for $\mathfrak{S}^A$ (*cf.* (2.3.18)). It is often convenient to exploit this freedom. One such occasion arises when we employ an explicit real coordinate system $(x^i)$ for $\mathcal{M}$. Then the coordinate basis $\partial/\partial x^0, \ldots, \partial/\partial x^3$ is often a convenient one to adopt for the module of tangent vector fields to $\mathcal{M}$, i.e. for the module $\mathfrak{T}^a$. In general, such a basis will have no close relation to any basis for $\mathfrak{S}^A$. So it is useful to allow simultaneous consideration of bases for $\mathfrak{T}^a$ (or $\mathfrak{S}^a$) and for $\mathfrak{S}^A$, which are completely unrelated to one another.

Let $\varepsilon_{\mathbf{A}}{}^{A}$, $\varepsilon_{A}{}^{\mathbf{A}}$ be a basis and dual basis for $\mathfrak{S}^{A}$ and let $g_{\mathbf{a}}{}^{a}$, $g_{a}{}^{\mathbf{a}}$ be an unrelated basis and dual basis for $\mathfrak{T}^{a}$. Then any world-tensor $\chi_{a\ldots c}{}^{d\ldots f}$ can be expressed in terms of components with respect to $g_{\mathbf{a}}{}^{a}$ or, reading it as $\chi_{AA'\ldots CC'}{}^{DD'\ldots FF'}$, in terms of components with respect to $\varepsilon_{\mathbf{A}}{}^{A}$. The relationship between the two sets of components is expressed by means of the *Infeld–van der Waerden symbols*, defined by

$$g_{\mathbf{a}}{}^{\mathbf{AA'}} := g_{\mathbf{a}}{}^{a}\varepsilon_{A}{}^{\mathbf{A}}\varepsilon_{A'}{}^{\mathbf{A'}},$$

$$g_{\mathbf{AA'}}{}^{\mathbf{a}} := \varepsilon_{\mathbf{A}}{}^{A}\varepsilon_{\mathbf{A'}}{}^{A'}g_{a}{}^{\mathbf{a}}. \tag{3.1.37}$$

Note that a contraction is taking place between $a$ and $AA'$, whereas there is no contraction between $\mathbf{a}$ and $\mathbf{AA'}$. Thus, each equation (3.1.37) represents 16 scalar equations. (It should be stressed that, according to our conventions, contractions are implied *only* between (i) *identical* indices of *all* kinds (e.g. $B_{a}{}^{a}$, $B_{\mathbf{a}}{}^{\mathbf{a}}$, $D_{\mathscr{C}}{}^{\mathscr{C}}$, $E_{A}{}^{A}$, $E_{\mathbf{A}}{}^{\mathbf{A}}$) and (ii) an *abstract* composite index and one of its implicit constituents (e.g., $F_{a}G^{A}$, $F_{a}G^{A}H^{A'}$).) We may view the Infeld–van der Waerden symbols as being simply the 'Kronecker delta' tensor $g_{a}{}^{b}$ with each index referred to a different kind of basis. From (3.1.37) we obtain the formulae

$$\chi_{\mathbf{a}\ldots\mathbf{c}}{}^{\mathbf{d}\ldots\mathbf{f}} = \chi_{\mathbf{AA'}\ldots\mathbf{CC'}}{}^{\mathbf{DD'}\ldots\mathbf{FF'}}g_{\mathbf{a}}{}^{\mathbf{AA'}}\ldots g_{\mathbf{c}}{}^{\mathbf{CC'}}g_{\mathbf{DD'}}{}^{\mathbf{d}}\ldots g_{\mathbf{FF'}}{}^{\mathbf{f}} \tag{3.1.38}$$

and

$$\chi_{\mathbf{AA'}\ldots\mathbf{CC'}}{}^{\mathbf{DD'}\ldots\mathbf{FF'}} = \chi_{\mathbf{a}\ldots\mathbf{c}}{}^{\mathbf{d}\ldots\mathbf{f}}g_{\mathbf{AA'}}{}^{\mathbf{a}}\ldots g_{\mathbf{CC'}}{}^{\mathbf{c}}g_{\mathbf{d}}{}^{\mathbf{DD'}}\ldots g_{\mathbf{f}}{}^{\mathbf{FF'}}, \tag{3.1.39}$$

which are really just special cases of (2.3.24).

Since we have chosen $g_{\mathbf{a}}{}^{a}\in\mathfrak{T}^{a}$, this basis is real and we have

$$\overline{g_{\mathbf{a}}{}^{\mathbf{AB'}}} = g_{\mathbf{a}}{}^{a}\varepsilon_{A'}{}^{\mathbf{A'}}\varepsilon_{A}{}^{\mathbf{B}} = g_{\mathbf{a}}{}^{\mathbf{BA'}}, \tag{3.1.40}$$

which is to say that each of the matrices $g_{0}{}^{\mathbf{AB'}},\ldots,g_{3}{}^{\mathbf{AB'}}$ is Hermitian. (Note that, in accordance with the convention enunciated after equation (2.5.64), we avoid the use of $\mathbf{AA'}$ together under the conjugation bar in equation (3.1.40) and assume $\mathbf{A}=\mathbf{A'}$, $\mathbf{B}=\mathbf{B'}$, numerically.) Similarly, each of $g_{\mathbf{AB'}}{}^{0},\ldots,g_{\mathbf{AB'}}{}^{3}$ is Hermitian. This implies that if $\chi_{a\ldots c}{}^{d\ldots f}$ is any real world tensor, then the spinor (dyad) components* must form an array which is Hermitian in the sense

$$\overline{\chi_{\mathbf{AP'}\ldots\mathbf{CR'}}{}^{\mathbf{DS'}\ldots\mathbf{FU'}}} = \chi_{\mathbf{PA'}\ldots\mathbf{RC'}}{}^{\mathbf{SD'}\ldots\mathbf{UF'}} \tag{3.1.41}$$

The equation for 'reality' of a *spinor*, $\overline{\chi_{\ldots}{}^{\ldots}} = \chi_{\ldots}{}^{\ldots}$, is in general meaningless, since its two members belong to non-comparable spinor modules

* In the literature a distinction is often made between spinor components and dyad components (Newman & Penrose 1962). Because of the use of abstract labels for actual spinors, no such distinction is necessary here.

(e.g., $\mathfrak{S}^A$ and $\mathfrak{S}^{A'}$). However, for spinors of valence-type $[^p_r\,^p_r]$ this equation becomes meaningful, and indeed important. Recall that, by definition,

$$\overline{\chi_{CD'\ldots}{}^{AB'\ldots}} = \bar{\chi}_{C'D\ldots}{}^{A'B\ldots};$$

so if

$$\overline{\chi_{CD'\ldots}{}^{AB'\ldots}} = \chi_{C'D\ldots}{}^{A'B\ldots}, \tag{3.1.42}$$

then we have

$$\bar{\chi}_{C'D\ldots}{}^{A'B\ldots} = \chi_{C'D\ldots}{}^{A'B\ldots}. \tag{3.1.43}$$

There is a certain conflict between two terminologies which suggest themselves for a spinor satisfying the evidently equivalent conditions (3.1.41) (with real $g_{\mathbf{a}}{}^a$), (3.1.42), and (3.1.43). While (3.1.41) and (3.1.42) suggest that $\chi_{C'D\ldots}{}^{A'B\ldots}$ be called 'Hermitian', (3.1.43) and the already adopted terminology for world tensors suggest 'real'. Here we adopt what seems to be the logical compromise and say that $\chi_{C'D\ldots}{}^{A'B\ldots}$ is indeed *Hermitian*, while the term 'real' is reserved for the corresponding elements $\chi_{CC'\ldots}{}^{AA'\ldots}$ of the self-conjugate sets $\mathfrak{S}^{AA'\ldots}_{CC'\ldots} = \mathfrak{S}^{a\ldots}_{c\ldots}$, which satisfy (3.1.7), i.e.,

$$\bar{\chi}_{CC'\ldots}{}^{AA'\ldots} = \chi_{CC'\ldots}{}^{AA'\ldots}. \tag{3.1.44}$$

Thus 'real' really means 'equal to its complex conjugate'.

The basic equation satisfied by the Infeld–van der Waerden symbols is the component version of the fundamental relation (3.1.9*a*), that is,

$$g_{\mathbf{ab}} = \varepsilon_{\mathbf{AB}}\varepsilon_{\mathbf{A'B'}}g_{\mathbf{a}}{}^{\mathbf{AA'}}g_{\mathbf{b}}{}^{\mathbf{BB'}}, \tag{3.1.45}$$

or, equivalently,

$$g_{\mathbf{ab}}g_{\mathbf{AA'}}{}^{\mathbf{a}}g_{\mathbf{BB'}}{}^{\mathbf{b}} = \varepsilon_{\mathbf{AB}}\varepsilon_{\mathbf{A'B'}}. \tag{3.1.46}$$

The equivalence of (3.1.45) with (3.1.46) is established via the relations

$$g_{\mathbf{a}}{}^{\mathbf{AA'}}g_{\mathbf{AA'}}{}^{\mathbf{b}} = g_{\mathbf{a}}{}^{\mathbf{b}}, \quad g_{\mathbf{AA'}}{}^{\mathbf{a}}g_{\mathbf{a}}{}^{\mathbf{BB'}} = \varepsilon_{\mathbf{A}}{}^{\mathbf{B}}\varepsilon_{\mathbf{A'}}{}^{\mathbf{B'}}. \tag{3.1.47}$$

An alternative equation which is often used in place of (3.1.45), owing to its connection with the anticommutator 'Clifford algebra' equation for Dirac $\gamma$-matrices,* is

$$g_{\mathbf{a}}{}^{\mathbf{A}}{}_{\mathbf{A'}}g_{\mathbf{bB}}{}^{\mathbf{A'}} + g_{\mathbf{b}}{}^{\mathbf{A}}{}_{\mathbf{A'}}g_{\mathbf{aB}}{}^{\mathbf{A'}} = -\varepsilon_{\mathbf{B}}{}^{\mathbf{A}}g_{\mathbf{ab}}. \tag{3.1.48}$$

This follows from (3.1.45), for if we add (3.1.45) to the same equation with **a** and **b** interchanged, we get an expression anti-symmetrical in **A**, **B**, trans-

* We may express (3.1.48) as the $(2\times 2)$ matrix equation $\mathbf{G_aG_b^*} + \mathbf{G_bG_a^*} = -g_{\mathbf{ab}}\mathbf{I}_2$ where the asterisk denotes Hermitian conjugation. If we set $\gamma_{\mathbf{a}} = 2^{\frac{1}{2}}\begin{pmatrix} 0 & \mathbf{G_a} \\ \mathbf{G_a^*} & 0 \end{pmatrix}$, we obtain a solution of the $(4\times 4)$ Dirac matrix equation $\gamma_{\mathbf{a}}\gamma_{\mathbf{b}} + \gamma_{\mathbf{b}}\gamma_{\mathbf{a}} = -2g_{\mathbf{ab}}\mathbf{I}_4$. See footnote on p. 221 and the Appendix to Vol. 2 for more detail.

vected with $\varepsilon_{\mathbf{AB}}$. Multiplying through by $\varepsilon^{\mathbf{CD}}$ and using (2.5.22), we can move the unprimed $\varepsilon$ to the other side of the equation. This gives (3.1.48). Conversely, starting from (3.1.48), if we transvect through by $\frac{1}{2}\varepsilon^{\mathbf{B}}{}_{\mathbf{A}}$, we re-obtain (3.1.45).

If we use the standard Minkowski tetrad $t^a$, $x^a$, $y^a$, $z^a$ related to the spin-frame $o^A$, $\iota^A$ by (3.1.20), then denoting $g_{\mathbf{a}}{}^a$ by (3.1.25) and $g_a{}^{\mathbf{a}}$ by (3.1.26) we get

$$g_0{}^{\mathbf{AB'}} = \frac{1}{\sqrt{2}}\begin{pmatrix} 1 & 0 \\ 0 & 1 \end{pmatrix} = g_{\mathbf{AB'}}{}^0, \ g_1{}^{\mathbf{AB'}} = \frac{1}{\sqrt{2}}\begin{pmatrix} 0 & 1 \\ 1 & 0 \end{pmatrix} = g_{\mathbf{AB'}}{}^1,$$

$$g_2{}^{\mathbf{AB'}} = \frac{1}{\sqrt{2}}\begin{pmatrix} 0 & \mathrm{i} \\ -\mathrm{i} & 0 \end{pmatrix} = -g_{\mathbf{AB'}}{}^2, \ g_3{}^{\mathbf{AB'}} = \frac{1}{\sqrt{2}}\begin{pmatrix} 1 & 0 \\ 0 & -1 \end{pmatrix} = g_{\mathbf{AB'}}{}^3, \qquad (3.1.49)$$

as is readily seen by comparing (3.1.31) with $K^{\mathbf{a}}g_{\mathbf{a}}{}^{\mathbf{AB'}} = K^{\mathbf{AB'}}$. The matrices (3.1.49) are, apart from the factor $2^{-\frac{1}{2}}$, the familiar Pauli spin-matrices and the unit matrix.

It is sometimes convenient to use a basis $g_{\mathbf{a}}{}^a$ for $\mathfrak{S}^a$ *not* all of whose elements are real. This situation could arise in the case of a coordinate basis if complex coordinates for $\mathcal{M}$ are used. Another example arises with a null tetrad $g_0{}^a = l^a$, $g_1{}^a = n^a$, $g_2{}^a = m^a$, $g_3{}^a = \bar{m}^a$; here we find

$$g_0{}^{\mathbf{AB'}} = \begin{pmatrix} 1 & 0 \\ 0 & 0 \end{pmatrix} = g_{\mathbf{AB'}}{}^0, \quad g_1{}^{\mathbf{AB'}} = \begin{pmatrix} 0 & 0 \\ 0 & 1 \end{pmatrix} = g_{\mathbf{AB'}}{}^1,$$

$$g_2{}^{\mathbf{AB'}} = \begin{pmatrix} 0 & 1 \\ 0 & 0 \end{pmatrix} = g_{\mathbf{AB'}}{}^2, \quad g_3{}^{\mathbf{AB'}} = \begin{pmatrix} 0 & 0 \\ 1 & 0 \end{pmatrix} = g_{\mathbf{AB'}}{}^3. \qquad (3.1.50)$$

Since the vectors $g_{\mathbf{a}}{}^a$ are now not all real, the fact that the matrices $g_0{}^{\mathbf{AB'}}, \ldots, g_3{}^{\mathbf{AB'}}$ are not all Hermitian is to be expected. In the general case, with a complex basis $g_{\mathbf{a}}{}^a$ for $\mathfrak{S}^a$, all the equations (3.1.37)–(3.1.48) hold as before, *except* the Hermiticity condition (3.1.40).

## 3.2 Null flags and complex null vectors

Thus far, whenever we have desired to interpret a spinor (or spinor operation) in space–time terms, we have had to rely on the detailed geometrical discussion of Chapter 1. On the other hand, we may prefer just to accept the *abstract* existence of spinors, which form an algebra as described in §2.5, and among which the *real* spinors (in the sense of §3.1) are identified as world-tensors. It is of some interest that the geometric null-flag interpretation of a spin-vector in terms of such world-tensors can then be obtained very rapidly using the spinor algebra. The main procedure is a particular example of the general method, to be given in §3.4, of interpreting spinors in terms of world-tensors.

For simplicity let us assume, in what follows, that the sets $\mathfrak{S}^{\cdots}_{\cdots}$ refer to spinors at a single point $P \in \mathcal{M}$ only, so $\mathfrak{S} \cong \mathbb{C}$ and $\mathfrak{S}^A$ is a two-dimensional vector space. Let

$$\kappa^A \in \mathfrak{S}^A. \tag{3.2.1}$$

Then the most obvious world-tensor we can construct from $\kappa^A$ alone is the world-vector

$$K^a = \kappa^A \bar{\kappa}^{A'}. \tag{3.2.2}$$

The *reality* of $K^a$ is obvious; so also is the fact that $K^a$ is *null*, since

$$K^a K_a = \kappa^A \bar{\kappa}^{A'} \kappa_A \bar{\kappa}_{A'} = |\kappa_A \kappa^A|^2 = 0.$$

In fact, *every* real null world-vector has either the form (3.2.2) or else* the form

$$K^a = -\kappa^A \bar{\kappa}^{A'}. \tag{3.2.4}$$

This follows from the fact that a complex world-vector $\chi^a$ is null, in the sense**

$$\chi^a \chi_a = 0, \tag{3.2.5}$$

if and only if it has the form

$$\chi^a = \kappa^A \xi^{A'}. \tag{3.2.6}$$

For, accepting (3.2.6), we have $\kappa^A \xi^{A'} = \bar{\xi}^A \bar{\kappa}^{A'}$ if $\chi^a$ is real. Transvecting with $\bar{\xi}_A$ gives $(\bar{\xi}_A \kappa^A)\xi^{A'} = 0$, so, by (2.5.56), $\bar{\xi}^A$ must be a multiple of $\kappa^A$ (unless $\kappa^A = 0$). Thus we have $\chi^a = q\kappa^A \bar{\kappa}^{A'}$ where $q$ must be real. If $q > 0$, we absorb $q^{\frac{1}{2}}$ into $\kappa^A$ and get (3.2.2). If $q < 0$ we absorb $(-q)^{\frac{1}{2}}$ and correspondingly get (3.2.4). (If $q = 0$ we get both.)

Conversely, we observe that (3.2.5) has the spinor form

$$\varepsilon_{AB} \varepsilon_{A'B'} \chi^{AA'} \chi^{BB'} = 0, \tag{3.2.7}$$

the left-hand side of which is just twice the determinant of $\chi^{\mathbf{AA'}}$ (in components: $2\chi^{00'}\chi^{11'} - 2\chi^{01'}\chi^{10'}$). The vanishing of this determinant asserts that the rank of $\chi^{\mathbf{AA'}}$ is less than 2, that is, $\chi^{AA'}$ is an outer product (3.2.6) of two spin-vectors. (In any one spin-frame, this is clearly true for the components. Therefore it is true independently of the spin-frame.)

* A complication that arises if we try to apply the discussion to spinor *fields* is that, for a given null vector field $K^a$, in some regions of space–time (3.2.2) might hold and in others (3.2.4), so that neither could be assumed to hold globally. (For this $K^a$ would have to vanish in some regions.)

** There *is* another possible sense, *cf.* (3.2.25) below.

Next, we observe that the existence of the spinor system gives us an absolute distinction between the two null half-cones at the point $P$ of interest. For, we can *define* the *future-pointing* and *past-pointing* null vectors to be those for which, respectively, (3.2.2) and (3.2.4) hold ($\kappa^A \neq 0$). It is clear from this that the scalar product of two future-pointing null vectors or of two past-pointing null vectors must be positive (or zero, if they are proportional) and that the scalar product of two null vectors of different kinds must be negative (or zero, if proportional), as required (*cf.* §1.1). (This choice is consistent with the $t^a$ of (3.1.20) being future-pointing.) Thus when we pass to spinor *fields* on $\mathcal{M}$, the global requirement (1.5.1) (that $\mathcal{M}$ be time-orientable) must hold.

Equation (3.2.2) tells us that any non-zero spin-vector defines a unique future-pointing null vector, which we call its *flagpole.* However, many distinct spin-vectors have the same flagpole since the $K^a$ defined by (3.2.2) allows the freedom

$$\kappa^A \mapsto e^{i\theta}\kappa^A. \tag{3.2.8}$$

To obtain a more complete tensorial realization of $\kappa^A$ than that afforded by (3.2.2), we can form the 'square' $\kappa^A\kappa^B$ of $\kappa^A$. Then, to get as many primed indices as unprimed ones, we multiply by $\varepsilon^{A'B'}$: this gives us a complex world-tensor. To obtain a real world-tensor we can add the complex conjugate:

$$P^{ab} = \kappa^A\kappa^B\varepsilon^{A'B'} + \varepsilon^{AB}\bar{\kappa}^{A'}\bar{\kappa}^{B'}. \tag{3.2.9}$$

Then we have

$$P^{ab} = \bar{P}^{ab} = -P^{ba}. \tag{3.2.10}$$

Furthermore, $P^{ab}$ is 'simple', i.e., of the form

$$P^{ab} = K^aL^b - L^aK^b, \tag{3.2.11}$$

where $L^a$ is any vector* of the form

$$L^a = \kappa^A\bar{\tau}^{A'} + \tau^A\bar{\kappa}^{A'} \tag{3.2.12}$$

for which

$$\kappa_A\tau^A = 1. \tag{3.2.13}$$

To establish (3.2.11), we observe that since $\kappa_A$, $\tau^A$ constitute a spin-frame, we have, from (2.5.54),

$$\varepsilon^{AB} = \kappa^A\tau^B - \tau^A\kappa^B. \tag{3.2.14}$$

Then (3.2.11) follows upon substitution of (3.2.14) into (3.2.9).

* This $L^a$ is not quite the same as the $\boldsymbol{L}$ of §1.4, but it is serves a similar purpose. We can relate the discussion here to that of §1.4 by putting $2^{\frac{1}{2}}(OR/OP)\boldsymbol{L}$ for $L^a$.

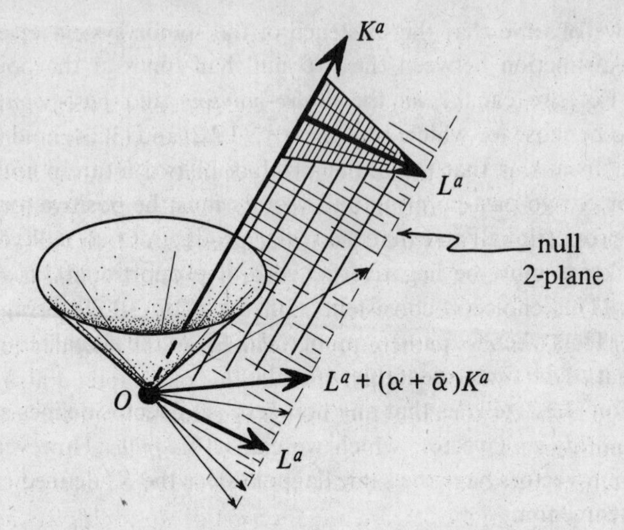


Fig. 3-1. The null flag representing $\kappa^A$, and its relation to $K^a$ and $L^a$.

The vector $L^a$ is real, spacelike of length $\sqrt{2}$, and orthogonal to $K^a$:

$$\bar{L}^a = L^a, \quad L^aL_a = -2, \quad K^aL_a = 0. \tag{3.2.15}$$

It is defined by $P^{ab}$ up to the addition of a real multiple of $K^a$: under

$$\tau^A \to \tau^A + \alpha\kappa^A \tag{3.2.16}$$

(the transformations of $\tau^A$ which leave (3.2.13) invariant) we have

$$L^a \to L^a + (\alpha + \bar{\alpha})K^a. \tag{3.2.17}$$

The positive multiples of these vectors $L^a$ lie on a (two-dimensional) half-plane through the origin in the Minkowski vector space $\mathfrak{T}^a$, which (since $L^a$ is orthogonal to the null vector $K^a$) is tangent to the null cone in $\mathfrak{T}^a$, along the line consisting of multiples of $K^a$ (see Fig. 3-1). This half-plane is the *flag plane* of $\kappa^A$. (Agreement of this construction with that of §1.4 is easy to verify.)

The phase transformation (3.2.8) can be seen to correspond to a rotation through $2\theta$ of the flag plane about the flagpole. For, setting

$$M^a = \mathrm{i}\kappa^A\bar{\tau}^{A'} - \mathrm{i}\tau^A\bar{\kappa}^{A'} \tag{3.2.18}$$

we see that

$$L^a \cos 2\theta + M^a \sin 2\theta = (\mathrm{e}^{\mathrm{i}\theta}\kappa^A)(\mathrm{e}^{\mathrm{i}\theta}\bar{\tau}^{A'}) + (\mathrm{e}^{-\mathrm{i}\theta}\tau^A)(\mathrm{e}^{-\mathrm{i}\theta}\bar{\kappa}^{A'}). \tag{3.2.19}$$

Thus, under (3.2.8) (which must be accompanied by $\tau^A \mapsto \mathrm{e}^{-\mathrm{i}\theta}\tau^A$ to preserve (3.2.13)), the vector $L^a$ is rotated, in the $(L^a, M^a)$-plane, through an angle $2\theta$, to become the vector of (3.2.19). Note also that under $\kappa^A \mapsto k\kappa^A$ (with

$\tau^A \mapsto k^{-1}\tau^A, k > 0$) the flagpole gets multiplied by $k^2$ but the flag plane is unchanged.

It is of some interest to observe that this gives a direct way of obtaining the conformal structure of the space $\mathscr{S}^+$ of future null directions at the origin in $\mathfrak{T}^a$. Any spin-vector $\kappa^A$ defines a point $K$ of $\mathscr{S}^+$ (the direction of the flagpole) and also a tangent direction $\mathbf{L}$ to $\mathscr{S}^+$ at $K$ (the direction of the flag plane). If we have two tangent directions $\mathbf{L}$ and $\mathbf{L}'$ at the same point $K$ of $\mathscr{S}^+$, these will be given, respectively, by $\kappa^A$ and by $e^{i\theta}\kappa^A$ for some $\theta$. We now obtain the angle between $\mathbf{L}$ and $\mathbf{L}'$ as $2\theta$. Having an invariant concept of angle on $\mathscr{S}^+$ we therefore have an invariantly defined conformal structure for $\mathscr{S}^+$. (Of course the constructions of Chapter 1 also achieve this, but here the conformal structure arises directly from the interpretation of spin-vectors.)

In fact, this conformal structure also gives an invariantly defined *orientation* for $\mathscr{S}^+$, and consequently for $\mathscr{M}$. For we can *define* the notion of *right-handedness* on $\mathscr{S}^+$ by simply specifying that 'right-handed' is the sense in which the flag plane of $e^{i\theta}\kappa^A$ rotates as $\theta$ increases (*cf.* remarks before (1.5.2)). (This is consistent with the $x^a$, $y^a$, $z^a$ of (3.1.20) being right-handed.) Again we observe that passage to a spinor *field* implies a global restriction on $\mathscr{M}$, namely that, in addition to being time-orientable, it must be *space-orientable*. As a topological manifold, therefore, $\mathscr{M}$ must be (space–time) orientable (*cf.* (1.5.2)).

We observe that each of (3.2.2) and (3.2.9) is invariant under

$$\kappa^A \mapsto -\kappa^A. \tag{3.2.20}$$

If this is achieved continuously via $e^{i\theta}\kappa^A$ as $\theta$ varies from 0 to $\pi$, the flag plane executes one complete revolution through $2\pi$ and returns to its original state, so $\kappa^A$ is indeed a spinorial object (*cf.* §1.5). Thus the passage to spinor fields requires also that $\mathscr{M}$ have spin-structure (*cf.* (1.5.3)). Our algebraic approach, therefore, when we apply it to spinor fields (as we shall in Chapter 4), requires $\mathscr{M}$ to have spinor structure (*cf.* end of §1.5).

### *Properties of complex null vectors*

Let us now return to the representation (3.2.6) of a general non-zero complex null vector in terms of a pair of spin-vectors. We observe, first, that given $\chi^a$, the decomposition $\chi^a = \kappa^A\xi^{A'}$ is unique up to

$$\kappa^A \mapsto \lambda\kappa^A, \quad \xi^{A'} \mapsto \lambda^{-1}\xi^{A'}. \tag{3.2.21}$$

For, if $\kappa^A\xi^{A'} = \mu^A\nu^{A'}$, then transvection by $\mu_A$ gives $\mu_A\kappa^A = 0$, so by (2.5.56) $\kappa^A$ and $\mu^A$ must be proportional. Similarly, $\xi^{A'}$ and $\nu^{A'}$ must be proportional. Note that this implies that a complex null vector $\chi^a$ defines an ordered pair of real null directions (not necessarily distinct), i.e., two points of $S^+$, namely those given by the flagpoles of $\kappa^A$ and $\xi^{A'}$, respectively. The complex conjugate $\bar{\chi}^a$ of $\chi^a$ defines the same pair of null directions (points of $S^+$) but in the reverse order ($\bar{\chi}^a = \bar{\xi}^A \bar{\kappa}^{A'}$). The two null directions coincide if and only if $\chi^a$ is a complex multiple of a *real* vector. Conversely, this ordered pair of real null directions defines, quite generally, the complex vector $\chi^a$ up to a proportionality.

A number of properties of complex null vectors can be read off easily from (3.2.6). For example, if two complex null vectors $\chi^a$, $\psi^a$ are orthogonal,

$$\chi_a \psi^a = 0, \tag{3.2.22}$$

this means that $\chi^a$ and $\psi^a$ have simultaneously the form $\chi^a = \kappa^A \xi^{A'}$, $\psi^a = \kappa^A \eta^{A'}$ or simultaneously the form $\chi^a = \kappa^A \xi^{A'}$, $\psi^a = \tau^A \xi^{A'}$. It follows that if $\chi^a$, $\psi^a$, $\phi^a$ are three complex null vectors which are orthogonal to one another, then they must be linearly dependent. For, they must simultaneously all have either the forms

$$\chi^a = \kappa^A \xi^{A'}, \quad \psi^a = \kappa^A \eta^{A'}, \quad \phi^a = \kappa^A \zeta^{A'} \tag{3.2.23}$$

or the complex conjugates of these forms. Assume, without loss of generality, that it is (3.2.23) which holds. Owing to the two-dimensionality of spin-space, there must be a linear relation $\lambda \xi^{A'} + \mu \eta^{A'} + \nu \zeta^{A'} = 0$ with not all of $\lambda$, $\mu$, $\nu$ zero. Thus, by (3.2.23), $\lambda \chi^a + \mu \psi^a + \nu \phi^a = 0$.

Suppose, on the other hand, that the complex null vectors $\psi^a$ and $\phi^a$ are each orthogonal to the complex null vector $\chi^a$ but not to one another. Then these vectors must have the forms

$$\chi^a = \kappa^A \xi^{A'}, \quad \psi^a = \kappa^A \eta^{A'}, \quad \phi^a = \omega^A \xi^{A'} \tag{3.2.24}$$

or the complex conjugates of these forms. It follows that there is a unique complex null vector $\theta^a$ which is orthogonal to each of $\psi^a$ and $\phi^a$, and for which $\theta_a \chi^a = \phi_a \psi^a$. For, assuming without loss of generality that it is (3.2.24) which holds, we can set $\theta^a = -\omega^A \eta^{A'}$, the uniqueness of which is easily established.

The *linear set* of complex null vectors of the form $\chi^a = \kappa^A \xi^{A'}$, where $\kappa^A$ is held fixed and $\xi^{A'}$ is allowed to vary, gives a way of representing the spin-vector $\kappa^A$ (up to proportionality) in *complex* terms. In some contexts (for example, in parts of twistor theory, *cf.* §§6.2, 7.3, 7.4, 9.3) it is important to give descriptions in terms of complex quantities not involving the notion of reality or of complex conjugation. In such contexts, the association of a spin-vector with a linear set of complex null vectors becomes much more significant than its association with a single (real) null vector, namely its flagpole, since the latter description requires the notion of complex conjugation. On the other hand, the complex null bivector $\kappa^A\kappa^B\varepsilon^{A'B'}$ could be of significance in these contexts. We do not propose to pursue this matter further here, except for one point: its relation to the structure of the complexified 2-sphere.

We have seen above that each complex null direction can be represented by an ordered pair of points on the sphere $S^+$ – topologically a 2-sphere $S^2$. If we wish to attach no significance to the notion of reality, then (since the coincidence of the two points signifies reality of the null direction) we should really consider each of the two points to lie on a *separate* sphere $S^2$. Thus, the *space of complex null directions* at a point has the structure of a topological product $S^2 \times S^2$. The points of the first $S^2$ are the spin-vectors $\kappa^A$ up to proportionality and the points of the second $S^2$ are the conjugate spin-vectors $\xi^{A'}$ up to proportionality, giving the required null directions as those defined by $\chi^a = \kappa^A\xi^{A'}$. Since the real null directions at a point constitute a conformal sphere $\mathscr{S}^+$, we may regard the complex null directions as constituting a *complexified sphere*. From the above, we see that this complexified sphere has a structure $S^2 \times S^2$. When a point of one of the $S^2$s is held fixed and the other is allowed to vary, we get, up to proportionality, one of the above linear sets (associated with $\kappa^A$ or with $\xi^{A'}$) on the complexified sphere called a *generator*. Thus, the complexified sphere is ruled by two systems of such generators.* This sort of description in fact forms the basis of the general geometrical $n$-dimensional discussion of spinors. But we shall not go into this here.

### *Hermitian-null vectors*

Let us end this section by briefly examining the other type of complex 'null' vector, namely a complex vector $\gamma^a$ satisfying

$$\gamma^a\bar{\gamma}_a = 0. \tag{3.2.25}$$

Choose any real $\theta$ and write

$$\gamma^a = e^{i\theta}(U^a + iV^a), \tag{3.2.26}$$

where $U^a$ and $V^a$ are real. Substituting (3.2.26) into (3.2.25) we get

$$U_aU^a + V_aV^a = 0. \tag{3.2.27}$$

Thus, either both $U^a$ and $V^a$ are null, or one is spacelike and the other time-

* This fact ceases to be surprising when we recall that when complexified, no distinction exists between a sphere and a hyperboloid. A hyperboloid of one sheet is well-known to be generated by two systems of straight lines.

like. Suppose $U^a$ is timelike for $\theta = 0$. Then $U^a$ for $\theta = \pi/2$ (being the same as $V^a$ for $\theta = 0$) must be spacelike. Since $U^a$ varies continuously with $\theta$, it follows that $U^a$ must be null for some $\theta$ between 0 and $\pi/2$. Thus $V^a$ is also null for this same value of $\theta$. By choosing the appropriate one of $\theta, \theta + \pi/2, \theta + \pi$, or $\theta + 3\pi/2$, we can arrange that *both* $U^a$ and $V^a$ are *null and future-pointing* in the representation (3.2.26). This representation of $\gamma^a$ is then *unique* except when $U^a$ or $V^a$ vanishes or when both are null and proportional. In this latter case, we can regain uniqueness (assuming $\gamma^a \neq 0$) by replacing (3.2.26) by

$$\gamma^a = e^{i\theta} U^a, \tag{3.2.28}$$

where $U^a$ is null, real and future-pointing. The spinor representation of a $\gamma^a$, which is null in the sense (3.2.25) (with $\gamma^a \neq 0$) is then

$$\gamma^a = e^{i\theta}(\alpha^A \bar{\alpha}^{A'} + i\beta^A \bar{\beta}^{A'}), \quad \alpha_A \beta^A \neq 0 \tag{3.2.29}$$

or

$$\gamma^a = e^{i\theta} \alpha^A \bar{\alpha}^{A'} \tag{3.2.30}$$

in the above two cases, respectively.

## 3.3 Symmetry operations

Two operations of great value in tensor and spinor algebra are symmetrization and anti-symmetrization. But it turns out that owing to the two-dimensionality of spin-space, the latter operation almost disappears for spinor algebra. It will be useful, however, first to discuss these two operations briefly for more general tensor systems. The only condition on the module $\mathfrak{S}^\bullet$ that we shall need, at first, in addition to the assumed total reflexivity – which latter is not used before (3.3.22) – is that $\mathfrak{S}$ contains a subring isomorphic with the rationals. After (3.3.23) we specialize to the case where a (normally two-dimensional) basis exists.

We adopt the conventional notation that round and square brackets surrounding a collection of indices denote, respectively, symmetrization and anti-symmetrization (sometimes called skew-symmetrization) over the indices enclosed. Thus we have

$$\chi_{\mathscr{P}(\alpha\beta)\mathscr{Q}} = \frac{1}{2!}(\chi_{\mathscr{P}\alpha\beta\mathscr{Q}} + \chi_{\mathscr{P}\beta\alpha\mathscr{Q}}), \tag{3.3.1}$$

$$\psi_{\mathscr{P}(\alpha\beta\gamma)\mathscr{Q}} = \frac{1}{3!}(\psi_{\mathscr{P}\alpha\beta\gamma\mathscr{Q}} + \psi_{\mathscr{P}\beta\alpha\gamma\mathscr{Q}} + \psi_{\mathscr{P}\beta\gamma\alpha\mathscr{Q}} + \psi_{\mathscr{P}\alpha\gamma\beta\mathscr{Q}} + \psi_{\mathscr{P}\gamma\alpha\beta\mathscr{Q}} + \psi_{\mathscr{P}\gamma\beta\alpha\mathscr{Q}}), \tag{3.3.2}$$

etc., and

$$\chi_{\mathscr{P}[\alpha\beta]\mathscr{Q}} = \frac{1}{2!}(\chi_{\mathscr{P}\alpha\beta\mathscr{Q}} - \chi_{\mathscr{P}\beta\alpha\mathscr{Q}}), \tag{3.3.3}$$

$$\psi_{\mathscr{P}[\alpha\beta\gamma]\mathscr{Q}} = \frac{1}{3!}(\psi_{\mathscr{P}\alpha\beta\gamma\mathscr{Q}} - \psi_{\mathscr{P}\beta\alpha\gamma\mathscr{Q}} + \psi_{\mathscr{P}\beta\gamma\alpha\mathscr{Q}} - \psi_{\mathscr{P}\alpha\gamma\beta\mathscr{Q}} + \psi_{\mathscr{P}\gamma\alpha\beta\mathscr{Q}} - \psi_{\mathscr{P}\gamma\beta\alpha\mathscr{Q}}), \tag{3.3.4}$$

etc., and similarly for upper indices.

Occasionally, when it is required to omit certain indices in a symmetrization or anti-symmetrization, a vertical bar will be used at both ends of the group of indices to be omitted from the (anti-)symmetry operation. (Indices in reverse position do not require vertical bars, however.) For example:

$$\phi_{\alpha(\beta}{}^{\gamma}{}_{|\lambda|\mu)} = \tfrac{1}{2}(\phi_{\alpha\beta}{}^{\gamma}{}_{\lambda\mu} + \phi_{\alpha\mu}{}^{\gamma}{}_{\lambda\beta}), \tag{3.3.5}$$

$$\theta^{[\alpha|\beta\gamma|\lambda\mu]} = \tfrac{1}{6}(\theta^{\alpha\beta\gamma\lambda\mu} - \theta^{\lambda\beta\gamma\alpha\mu} + \theta^{\lambda\beta\gamma\mu\alpha} - \theta^{\alpha\beta\gamma\mu\lambda} + \theta^{\mu\beta\gamma\alpha\lambda} - \theta^{\mu\beta\gamma\lambda\alpha}). \tag{3.3.6}$$

We may even write

$$\theta^{(\alpha|\beta[\gamma\lambda]|\mu)} = \tfrac{1}{4}(\theta^{\alpha\beta\gamma\lambda\mu} + \theta^{\mu\beta\gamma\lambda\alpha} - \theta^{\alpha\beta\lambda\gamma\mu} - \theta^{\mu\beta\lambda\gamma\alpha}), \tag{3.3.7}$$

etc.

A number of properties of these two symmetry operations follow immediately from the definitions. Among these are the following:

If symmetrization is applied to a number of indices, and if subsequently another symmetrization is applied to the same (and possibly additional) indices, then the first symmetrization can be ignored. For example if $\xi_{\alpha\lambda\mu\nu} = \eta_{\alpha(\lambda\mu)\nu}$ then $\xi_{\alpha(\lambda\mu\nu)} = \eta_{\alpha(\lambda\mu\nu)}$. This could be written $\eta_{\alpha((\lambda\mu)\nu)} = \eta_{\alpha(\lambda\mu\nu)}$. The same result applies to anti-symmetrizations. Thus

$$\chi_{\mathscr{P}(\alpha\ldots(\gamma\ldots\varepsilon)\ldots\eta)\mathscr{Q}} = \chi_{\mathscr{P}(\alpha\ldots\gamma\ldots\varepsilon\ldots\eta)\mathscr{Q}} \tag{3.3.8}$$

and

$$\chi_{\mathscr{P}[\alpha\ldots[\gamma\ldots\varepsilon]\ldots\eta]\mathscr{Q}} = \chi_{\mathscr{P}[\alpha\ldots\gamma\ldots\varepsilon\ldots\eta]\mathscr{Q}}. \tag{3.3.9}$$

If symmetrization is applied to a number of indices, and if subsequently an anti-symmetrization is applied to two or more of those indices (and possibly additional indices) then the resulting expression vanishes. For example, if $\xi_{\alpha\lambda\mu\nu} = \eta_{\alpha(\lambda\mu)\nu}$ then $\xi_{\alpha[\lambda\mu\nu]} = 0$. The same remark applies with the roles of symmetrization and anti-symmetrization reversed. Thus, assuming that $\gamma, \ldots, \varepsilon$ are more than two in number,

$$\chi_{\mathscr{P}[\alpha\ldots(\gamma\ldots\varepsilon)\ldots\eta]\mathscr{Q}} = 0, \quad \chi_{\mathscr{P}(\alpha\ldots[\gamma\ldots\varepsilon]\ldots\eta)\mathscr{Q}} = 0. \tag{3.3.10}$$

It is *not true* in general that symmetry operations commute with each other. However, two such operations obviously *do* commute if they act on totally different indices. For example, we can form unambiguous expressions like $\psi_{(\alpha\beta)\gamma[\lambda\mu\nu]}$. But expressions like $\psi_{\alpha(\beta\gamma[\lambda)\mu\nu]}$ are not permissible, since it would not be clear whether the symmetry or anti-symmetry operation is to be performed first. The same holds for upper indices, of course. As in (3.3.5), the symmetry operators apply to upper and lower indices independently, for example:

$$\theta^{[\alpha}{}_{(\lambda}{}^{\beta]}{}_{\mu)}{}^{\gamma} = \tfrac{1}{4}(\theta^{\alpha}{}_{\lambda}{}^{\beta}{}_{\mu}{}^{\gamma} - \theta^{\beta}{}_{\lambda}{}^{\alpha}{}_{\mu}{}^{\gamma} + \theta^{\alpha}{}_{\mu}{}^{\beta}{}_{\lambda}{}^{\gamma} - \theta^{\beta}{}_{\mu}{}^{\alpha}{}_{\lambda}{}^{\gamma}). \tag{3.3.11}$$

We shall also have occasion to apply (anti-) symmetry operators to composite indices. For example,

$$\chi_{\mathscr{A}[\mathscr{B}_1\mathscr{B}_2]\mathscr{B}_3} = \tfrac{1}{2}(\chi_{\mathscr{A}\mathscr{B}_1\mathscr{B}_2\mathscr{B}_3} - \chi_{\mathscr{A}\mathscr{B}_2\mathscr{B}_1\mathscr{B}_3}). \tag{3.3.12}$$

This notation will only be allowed, however, when the indices within the bracket are of the same valence. Identity of valence is indicated, symbolically, by using the same kernel letter for the indices. If, for example, $\mathscr{A} = \alpha$ and $\mathscr{B} = \beta\gamma^*$, then (3.3.12) reads:

$$\chi_{\mathscr{A}[\mathscr{B}_1\mathscr{B}_2]\mathscr{B}_3} = \tfrac{1}{2}\chi_{\alpha\beta_1}{}^{\gamma_1}{}_{\beta_2}{}^{\gamma_2}{}_{\beta_3}{}^{\gamma_3} - \tfrac{1}{2}\chi_{\alpha\beta_2}{}^{\gamma_2}{}_{\beta_1}{}^{\gamma_1}{}_{\beta_3}{}^{\gamma_3}. \tag{3.3.13}$$

A tensor (or spinor) is said to be [anti-] symmetric in a collection of (possibly composite) indices if it is unchanged when the operation of [anti-] symmetrization is applied to the relevant indices. Thus, if

$$\chi_{\ldots\lambda\ldots\nu\ldots} = \chi_{\ldots(\lambda\ldots\nu)\ldots} \text{ and } \psi_{\ldots\lambda\ldots\nu\ldots} = \psi_{\ldots[\lambda\ldots\nu]\ldots},$$

then we say that $\chi_{\ldots\lambda\ldots\nu\ldots}$ and $\psi_{\ldots\lambda\ldots\nu\ldots}$ are, respectively, symmetric and anti-symmetric in $\lambda\ldots\nu$. Note that, from the above remarks, the *result* of any [anti-] symmetrization is automatically [anti-] symmetric. For example, $\theta^{\alpha[\beta}{}_{(\lambda}{}^{\gamma]}{}_{\mu\nu)}$ is symmetric in $\lambda, \mu, \nu$ and anti-symmetric in $\beta, \gamma$. A tensor is symmetric in a group of indices if and only if it is unaltered when any pair of indices in the group is interchanged. Similarly, it is anti-symmetric in a group of indices if and only if its sign is reversed whenever any pair of indices in the group is interchanged.

A convenient notation for the subspace of an $\mathfrak{S}^{\cdots}_{\cdots}$, consisting of those elements with some specific symmetry, is to employ the same arrangement of round and square brackets to the symbol $\mathfrak{S}^{\cdots}_{\cdots}$ itself as would yield the desired symmetry when applied to elements of $\mathfrak{S}^{\cdots}_{\cdots}$. In particular, the spaces of symmetric and antisymmetric elements of $\mathfrak{S}_{\alpha\ldots\gamma}$ are denoted, respectively, by $\mathfrak{S}_{(\alpha\ldots\gamma)}$ and $\mathfrak{S}_{[\alpha\ldots\gamma]}$ and, for example, we have

$$\chi_{\mathscr{A}}{}^{\alpha}{}_{\lambda\mu}{}^{\beta}{}_{\nu}{}^{\mathscr{B}_1\mathscr{B}_2} \in \mathfrak{S}^{[\alpha\beta](\mathscr{B}_1\mathscr{B}_2)}_{\mathscr{A}(\lambda\mu\nu)} \quad \textit{iff} \quad \chi_{\mathscr{A}}{}^{\alpha}{}_{\lambda\mu}{}^{\beta}{}_{\nu}{}^{\mathscr{B}_1\mathscr{B}_2} = \chi_{\mathscr{A}}{}^{[\alpha}{}_{(\lambda\mu}{}^{\beta]}{}_{\nu)}{}^{(\mathscr{B}_1\mathscr{B}_2)}. \tag{3.3.14}$$

If a tensor is [anti-] symmetric in two *overlapping* groups of indices, then it is [anti-] symmetric in the combination of the two groups together. Thus

$$\begin{aligned} &\textit{if} \quad \chi_{\ldots\alpha\ldots\gamma\ldots\varepsilon\ldots} = \chi_{\ldots(\alpha\ldots\gamma)\ldots\varepsilon\ldots} = \chi_{\ldots\alpha\ldots(\gamma\ldots\varepsilon)\ldots} \\ &\textit{then}\; \chi_{\ldots\alpha\ldots\gamma\ldots\varepsilon\ldots} = \chi_{\ldots(\alpha\ldots\gamma\ldots\varepsilon)\ldots}; \end{aligned} \tag{3.3.15}$$

and

$$\begin{aligned} &\textit{if} \quad \psi_{\ldots\alpha\ldots\gamma\ldots\varepsilon\ldots} = \psi_{\ldots[\alpha\ldots\gamma]\ldots\varepsilon\ldots} = \psi_{\ldots\alpha\ldots[\gamma\ldots\varepsilon]\ldots} \\ &\textit{then}\, \psi_{\ldots\alpha\ldots\gamma\ldots\varepsilon\ldots} = \psi_{\ldots[\alpha\ldots\gamma\ldots\varepsilon]\ldots}. \end{aligned} \tag{3.3.16}$$

This follows from the fact that any permutation can be represented as a product of transpositions only of adjacent indices. Furthermore, if a tensor is symmetric in a group of indices which overlaps another group of indices in which it is anti-symmetric, then the tensor vanishes. To see this, it is sufficient to consider a tensor $\xi_{\mathscr{A}\lambda\mu\nu}$ which is symmetric in $\lambda$, $\mu$ and skew in $\mu$, $\nu$. We have

$$\xi_{\mathscr{A}\lambda\mu\nu} = \xi_{\mathscr{A}\mu\lambda\nu} = -\xi_{\mathscr{A}\mu\nu\lambda} = -\xi_{\mathscr{A}\nu\mu\lambda} = \xi_{\mathscr{A}\nu\lambda\mu} = \xi_{\mathscr{A}\lambda\nu\mu} = -\xi_{\mathscr{A}\lambda\mu\nu}. \tag{3.3.17}$$

The tensor is thus equal its negative and so must vanish.

If an [anti-] symmetric group of indices on a tensor is contracted with another group of indices, then the second group of indices may be [anti-] symmetrized without changing the result of the contraction:

$$\chi_{\mathscr{A}(\rho\ldots\tau)}{}^{\rho\ldots\tau} = \chi_{\mathscr{A}(\rho\ldots\tau)}{}^{(\rho\ldots\tau)} = \chi_{\mathscr{A}\rho\ldots\tau}{}^{(\rho\ldots\tau)} \tag{3.3.18}$$

and

$$\chi_{\mathscr{A}[\rho\ldots\tau]}{}^{\rho\ldots\tau} = \chi_{\mathscr{A}[\rho\ldots\tau]}{}^{[\rho\ldots\tau]} = \chi_{\mathscr{A}\rho\ldots\tau}{}^{[\rho\ldots\tau]}. \tag{3.3.19}$$

These results are easily proved by expanding one of the [anti-] symmetries in the middle term according to (3.3.1)–(3.3.4) and then relabelling the dummy indices. By invoking the [anti-] symmetry of the remaining indices we can make every term of the sum the same. For a group of $N$ indices, there are $N!$ terms, which is compensated for by the $N!$ in the denominator of (3.3.1)–(3.3.4). One implication of the above result is that if a pair of symmetric indices is contracted with a pair of anti-symmetric indices, then the result must be zero. Thus

$$\psi_{\mathscr{A}(\lambda\mu)}{}^{[\lambda\mu]} = 0 = \psi_{\mathscr{A}[\lambda\mu]}{}^{(\lambda\mu)}. \tag{3.3.20}$$

Note that for any $X^\alpha \in \mathfrak{S}^\alpha$, the tensor $X^\alpha X^\beta \ldots X^\delta$ is symmetric in $\alpha, \beta, \ldots, \delta$. Hence

$$\phi_{\alpha\beta\ldots\delta} X^\alpha X^\beta \ldots X^\delta = \phi_{(\alpha\beta\ldots\delta)} X^\alpha X^\beta \ldots X^\delta. \tag{3.3.21}$$

This shows that the 'if' part of the following result holds:

(3.3.22) PROPOSITION

$\phi_{\mathscr{A}\alpha\beta\ldots\delta} X^\alpha X^\beta \ldots X^\delta = 0$ *for all* $X^\alpha \in \mathfrak{S}^\alpha$ *iff* $\phi_{\mathscr{A}(\alpha\ldots\delta)} = 0$.

To demonstrate the 'only if' part, set

$$X^\alpha = Y^\alpha + \lambda Z^\alpha.$$

Then, assuming $\alpha, \beta, \ldots, \delta$ are $N$ in number, we get

$$\phi_{\mathscr{A}\alpha\beta\ldots\delta}X^\alpha X^\beta \ldots X^\delta = \phi_{\mathscr{A}(\alpha\beta\ldots\delta)}Y^\alpha Y^\beta \ldots Y^\delta + N\lambda\phi_{\mathscr{A}(\alpha\beta\ldots\delta)}Z^\alpha Y^\beta \ldots Y^\delta + \cdots + \lambda^N\phi_{\mathscr{A}(\alpha\beta\ldots\delta)}Z^\alpha Z^\beta \ldots Z^\delta.$$

If the left-hand side is to vanish for all $X^\alpha$, then each coefficient of $\lambda^r$ must vanish on the right. (This requires no more than we have assumed for the ring $\mathfrak{S}$, namely that it contains a subring isomorphic to the rationals. For, if we choose $\lambda = 0, 1, \ldots, N$ in turn, we find that $N+1$ linearly independent combinations of the coefficients on the right vanish. Taking *rational* linear combinations, we see that each coefficient must individually vanish.) In particular, $\phi_{\mathscr{A}(\alpha\beta\ldots\delta)}Z^\alpha Y^\beta \ldots Y^\delta$ must vanish. Since it vanishes for all $Z^\alpha$, $\phi_{\mathscr{A}(\alpha\beta\ldots\delta)}Y^\beta \ldots Y^\delta = 0$. Since this must vanish for all $Y^\alpha$, we can repeat the argument and finally obtain $\phi_{\mathscr{A}(\alpha\beta\ldots\delta)} = 0$, as required. It is clear that (3.3.22) also holds with $\alpha, \beta, \ldots, \delta$ replaced by *composite* indices all of the same valence.

Also, it follows from (3.3.22) that

*the function* $\phi_{\mathscr{A}}(X) \equiv \phi_{\mathscr{A}\alpha\ldots\delta}X^\alpha \ldots X^\delta$ *serves to define the tensor* $\phi_{\mathscr{A}(\alpha\ldots\delta)}$ *uniquely* (3.3.23)

since the difference between any two such functions vanishes identically iff the difference between their corresponding symmetrized tensors vanishes.

## *Results specific to spinors and world-tensors*

All results so far apply no matter what the dimension of $\mathfrak{S}^\alpha$. But because spin-space is only two-dimensional, there are special simplifications which occur in the case of the spinor system that we have introduced. These simplifications arise from the following fact: any spinor which is antisymmetric in three or more of its indices (either primed or unprimed) must vanish. This means that for any $\theta_{\mathscr{A}PQR}$ or $\phi_{\mathscr{A}P'Q'R'}$ we have

$$\theta_{\mathscr{A}[PQR]} = 0, \ \phi_{\mathscr{A}[P'Q'R']} = 0. \qquad (3.3.24)$$

(Clearly there is no loss of generality in considering only three indices, because of (3.3.9).) To see that (3.3.24) must hold, we can consider components in any spin-frame, noting that of three numerical spinor indices two at least must be equal. Alternatively, the result is seen to be a particular case of one given earlier, *cf.* after (2.3.1).

Note that the particular case of (3.3.24)

$$\varepsilon_{[AB}\varepsilon_{C]D} = 0 \qquad (3.3.25)$$

leads us back to the identity (2.5.21) considered earlier. In fact (3.3.24) is a consequence of (2.5.21) and its complex conjugate. To see this we may apply the equivalent relation (2.5.24) to $\theta_{\mathscr{A}[PQR]}$ in two ways, first employing the anti-symmetry in $QR$, and secondly the anti-symmetry in $PQ$. In this way we obtain

$$\theta_{\mathscr{A}[PQR]} = \tfrac{1}{2}\theta_{\mathscr{A}[PST]}\varepsilon^{ST}\varepsilon_{QR} = \tfrac{1}{2}\theta_{\mathscr{A}[RST]}\varepsilon^{ST}\varepsilon_{PQ}. \qquad (3.3.26)$$

Upon transvection with $\varepsilon^{PQ}$, the relation $\theta_{\mathscr{A}[PST]}\varepsilon^{ST} = 0$ follows, whence, by (3.3.26), $\theta_{\mathscr{A}[PQR]} = 0$ as required.

For $n$-dimensional tensors, an $n$-element basis $\delta^{\alpha}_{\mathbf{a}}$ for $\mathfrak{S}^{\alpha}$ being assumed, there is an identity corresponding to (3.3.24), namely

$$\chi_{\mathscr{A}[\alpha_0\alpha_1\ldots\alpha_n]} = 0. \qquad (3.3.27)$$

There is also a relation corresponding to (2.5.24). For this, define tensors $\varepsilon_{\alpha_1\ldots\alpha_n}$ and $\varepsilon^{\alpha_1\ldots\alpha_n}$ whose components in the given basis are

$$\varepsilon_{\mathbf{a}_1\ldots\mathbf{a}_n} = \varepsilon^{\mathbf{a}_1\ldots\mathbf{a}_n} = 1 \text{ or } -1 \text{ or } 0, \qquad (3.3.28)$$

according as $\mathbf{a}_1,\ldots,\mathbf{a}_n$ is an even, an odd, or not a permutation of $1,\ldots,n$. (This agrees with (2.3.4).) The quantities (3.3.28) are called Levi–Civita symbols. We have

$$n!\,\delta^{[\alpha_1}_{\beta_1}\ldots\delta^{\alpha_n]}_{\beta_n} = \varepsilon^{\alpha_1\ldots\alpha_n}\varepsilon_{\beta_1\ldots\beta_n}, \qquad (3.3.29)$$

by comparison of components on the two sides. Hence, if $\psi_{\mathscr{A}\alpha_1\ldots\alpha_n}$ is skew in $\alpha_1,\ldots,\alpha_n$ we have

$$\psi_{\mathscr{A}\beta_1\ldots\beta_n} = \psi_{\mathscr{A}\alpha_1\ldots\alpha_n}\delta^{[\alpha_1}_{\beta_1}\ldots\delta^{\alpha_n]}_{\beta_n} = \left(\frac{1}{n!}\psi_{\mathscr{A}\alpha_1\ldots\alpha_n}\varepsilon^{\alpha_1\ldots\alpha_n}\right)\varepsilon_{\beta_1\ldots\beta_n}. \qquad (3.3.30)$$

This shows that (as in (2.5.24)) any set of $n$ anti-symmetric indices can be 'split off' as an $\varepsilon$-tensor.

This applies in the particular case of $\mathfrak{S}_{\alpha_1\ldots\alpha_n}$ (i.e. when $\mathscr{A}$ is vacuous), showing that all totally anti-symmetric elements of $\mathfrak{S}_{\alpha_1\ldots\alpha_n}$ are proportional to one another. (The same applies to $\mathfrak{S}^{\alpha_1\ldots\alpha_n}$.) In the general case there is no reason to single out any one of these, the selection of $\varepsilon_{\alpha_1\ldots\alpha_n}$ being arbitrarily dependent on the particular choice of basis $\delta^{\alpha}_{\mathbf{a}}$ for $\mathfrak{S}^{\alpha}$. However, for spinors, the existence of an *inner product* serves to single out a particular $\varepsilon_{AB}$ (the corresponding bases for which (3.3.28) holds being the spin-frames). Also, the existence of both a world-tensor metric $g_{ab}$ and an orientation serve to select particular anti-symmetric elements $e_{abcd} \in \mathfrak{T}_{abcd}$ and $e^{abcd} \in \mathfrak{T}^{abcd}$, called *alternating* tensors, for special consideration. However, in keeping with our general mode of procedure, which is to construct the world-tensor concepts using the spinor formalism, we shall prefer, in the first instance, to give a definition of $e_{abcd}$ in terms of $\varepsilon_{AB}$, namely

$$e_{abcd} := \mathrm{i}\varepsilon_{AC}\varepsilon_{BD}\varepsilon_{A'D'}\varepsilon_{B'C'} - \mathrm{i}\varepsilon_{AD}\varepsilon_{BC}\varepsilon_{A'C'}\varepsilon_{B'D'}. \quad (3.3.31)$$

This $e_{abcd}$ is a real tensor since complex conjugation interchanges the two terms on the right and replaces i by $-$i. It is skew in $a, b$ since

$$e_{bacd} = \mathrm{i}\varepsilon_{BC}\varepsilon_{AD}\varepsilon_{B'D'}\varepsilon_{A'C'} - \mathrm{i}\varepsilon_{BD}\varepsilon_{AC}\varepsilon_{B'C'}\varepsilon_{A'D'} = -e_{abcd} \quad (3.3.32)$$

and it is similarly skew in $c, d$. Finally consider an interchange of $b$ and $c$. We have, using the $\varepsilon$-identity (2.5.21) and its complex conjugate,

$$\begin{aligned} e_{abcd} + e_{acbd} &= \mathrm{i}(\varepsilon_{AC}\varepsilon_{BD} - \varepsilon_{AB}\varepsilon_{CD})\varepsilon_{A'D'}\varepsilon_{B'C'} - \mathrm{i}\varepsilon_{AD}\varepsilon_{BC}(\varepsilon_{A'C'}\varepsilon_{B'D'} - \varepsilon_{A'B'}\varepsilon_{C'D'}) \\ &= \mathrm{i}(\varepsilon_{AD}\varepsilon_{BC})\varepsilon_{A'D'}\varepsilon_{B'C'} - \mathrm{i}\varepsilon_{AD}\varepsilon_{BC}(\varepsilon_{A'D'}\varepsilon_{B'C'}) = 0. \end{aligned} \quad (3.3.33)$$

Being thus skew in each of the overlapping pairs $ab$, $bc$ and $cd$, it follows that $e_{abcd}$ is totally skew in $a, b, c, d$:

$$e_{abcd} = e_{[abcd]}. \quad (3.3.34)$$

The world-tensor $e^{abcd}$ is obtained from $e_{abcd}$ in the normal way, i.e., raising its indices with $g^{ab}$. Since this corresponds to raising the spinor indices, we have

$$e^{abcd} = \mathrm{i}\varepsilon^{AC}\varepsilon^{BD}\varepsilon^{A'D'}\varepsilon^{B'C'} - \mathrm{i}\varepsilon^{AD}\varepsilon^{BC}\varepsilon^{A'C'}\varepsilon^{B'D'}. \quad (3.3.35)$$

Using (2.5.12) and (2.5.25) we obtain, from (3.3.31), (3.3.35)

$$e_{abcd}e^{abcd} = -24. \quad (3.3.36)$$

Let us now introduce a restricted Minkowski tetrad $g_0{}^a = t^a, g_1{}^a = x^a, g_2{}^a = y^a, g_3{}^a = z^a$. This will be related (locally at least) to a spin-frame $\varepsilon_0{}^A = o^A, \varepsilon_1{}^A = \iota^A$ according to (3.1.20). We obtain

$$\begin{aligned} e_{0123} &= e_{abcd}t^a x^b y^c z^d \\ &= -\tfrac{1}{4}(\varepsilon_{AC}\varepsilon_{BD}\varepsilon_{A'D'}\varepsilon_{B'C'} - \varepsilon_{AD}\varepsilon_{BC}\varepsilon_{A'C'}\varepsilon_{B'D'})(o^A o^{A'} + \iota^A\iota^{A'}) \\ &\quad (o^B\iota^{B'} + \iota^B o^{B'})(o^C\iota^{C'} - \iota^C o^{C'})(o^D o^{D'} - \iota^D\iota^{D'}) \\ &= -\tfrac{1}{4}(-1-1-1-1) = 1. \end{aligned} \quad (3.3.37)$$

Raising the indices (using (3.1.27)) we get

$$e^{0123} = -1. \quad (3.3.38)$$

Thus

$$e_{abcd} = 24 g_{[a}{}^0 g_b{}^1 g_c{}^2 g_{d]}{}^3 = -24\, t_{[a}x_b y_c z_{d]} \quad (3.3.39)$$

and

$$e^{abcd} = -24\, g_0{}^{[a} g_1{}^b g_2{}^c g_3{}^{d]} = -24\, t^{[a}x^b y^c z^{d]}. \quad (3.3.40)$$

Once we have (3.3.36), the computation (3.3.37) is necessary only for determining the *sign* of $e_{0123}$, since the relation (3.3.36) may be expressed as

$$\begin{aligned} -24 &= e_{abcd}e_{pqrs}g^{ap}g^{bq}g^{cr}g^{ds} \\ &= (e_{0123})^2\, 24 \det(g^{\mathbf{ab}}) = -24\,(e_{0123})^2. \end{aligned} \tag{3.3.41}$$

The positive sign of $e_{0123}$ depends essentially on the fact that $t^a$, $x^a$, $y^a$, $z^a$ is a *proper* tetrad. Had we chosen an improper tetrad we would have had $e_{0123} < 0$. For a proper Minkowski tetrad, it is $e_{\mathbf{abcd}}$ and $-e^{\mathbf{abcd}}$ which are the Levi–Civita symbols, while for an improper tetrad it is $-e_{\mathbf{abcd}}$ and $e^{\mathbf{abcd}}$.

From (3.3.29) we get

$$e_{abcd}e^{pqrs} = -24\, g_a{}^{[p}g_b{}^{q}g_c{}^{r}g_d{}^{s]}. \tag{3.3.42}$$

Successively contracting one upper with a lower index (or verifying directly by means of components) we obtain

$$\begin{aligned} e_{abcd}e^{pqrd} &= -6\, g_a{}^{[p}g_b{}^{q}g_c{}^{r]} \\ e_{abcd}e^{pqcd} &= -4\, g_a{}^{[p}g_b{}^{q]} \\ e_{abcd}e^{pbcd} &= -6\, g_a{}^{p} \end{aligned} \tag{3.3.43}$$

and (3.3.36). The tensor

$$e_{ab}{}^{cd} = \mathrm{i}\varepsilon_A{}^C\varepsilon_B{}^D\varepsilon_{A'}{}^{D'}\varepsilon_{B'}{}^{C'} - \mathrm{i}\varepsilon_A{}^D\varepsilon_B{}^C\varepsilon_{A'}{}^{C'}\varepsilon_{B'}{}^{D'} \tag{3.3.44}$$

will play an important role in the next section. Note the relation

$$e_{ab}{}^{cd}e_{cd}{}^{pq} = -4\, g_a{}^{[p}g_b{}^{q]}, \tag{3.3.45}$$

which follows either from (3.3.43), or directly from (3.3.44). Note also, for an arbitrary tensor $H_{cd}$,

$$e_{ab}{}^{cd}H_{cd} = e_{AA'BB'}{}^{CC'DD'}H_{CC'DD'} = \mathrm{i}(H_{AB'BA'} - H_{BA'AB'}). \tag{3.3.46}$$

*Reduction to symmetric spinors*

To close this section we shall demonstrate the important fact that, in a certain sense, any spinor can be expressed in terms of a collection of spinors each of which is totally symmetric in all unprimed indices and totally symmetric in all primed indices.

Let us start by illustrating the procedure in the case of a general spinor $\phi_{AB}$ of valence $\left[\begin{smallmatrix}0 & 0\\ 2 & 0\end{smallmatrix}\right]$. We have

$$\begin{aligned} \phi_{AB} &= \phi_{(AB)} + \phi_{[AB]} \\ &= \theta_{AB} + \lambda\varepsilon_{AB}, \end{aligned} \tag{3.3.47}$$

where

$$\theta_{AB} = \phi_{(AB)} \text{ and } \lambda = \tfrac{1}{2}\phi_C{}^C, \tag{3.3.48}$$

by (2.5.23). The spinor $\theta_{AB}$ is clearly symmetric and so is $\lambda$ (trivially, since it has no indices). The information contained in $\phi_{AB}$ is shared between these two symmetric spinors. (In terms of components, there are four independent quantities $\phi_{\mathbf{AB}}$. This information splits into the three independent quantities $\theta_{00} = \phi_{00}$, $\theta_{01} = \theta_{10} = \frac{1}{2}(\phi_{01} + \phi_{10})$, $\theta_{11} = \phi_{11}$ and the one scalar $\lambda = \frac{1}{2}(\phi_{01} - \phi_{10})$.)

We turn now to the general case. Let us use the symbol $\sim$ between two spinors if their difference is a sum of terms each of which is an outer product of ε-spinors with spinors of lower valence than the original ones. Clearly $\sim$ is an equivalence relation. We wish, first, to show that

$$\phi_{\mathscr{2}AB...F} \sim \phi_{\mathscr{2}(AB...F)} \tag{3.3.49}$$

holds for each $\phi_{\mathscr{2}AB...F}$. We have

$$\phi_{\mathscr{2}(AB...F)} = \frac{1}{r}(\phi_{\mathscr{2}A(BC...F)} + \phi_{\mathscr{2}B(AC...F)} + \phi_{\mathscr{2}C(ABD...F)} + \cdots + \phi_{\mathscr{2}F(AB...E)}), \tag{3.3.50}$$

where $A, B, \ldots, F$ are $r$ in number, so that there are $r$ terms on the right. Consider the difference between the first and any other one of these terms, e.g.,

$$\phi_{\mathscr{2}A(BCD...F)} - \phi_{\mathscr{2}C(ABD...F)} = -\varepsilon_{AC}\phi_{\mathscr{2}}{}^{X}{}_{(BXD...F)}, \tag{3.5.51}$$

by (2.5.23). Substituting from similar equations for all terms after the first on the right in (3.3.50), we find,

$$\phi_{\mathscr{2}(AB...F)} \sim \phi_{\mathscr{2}A(BC...F)}. \tag{3.3.52}$$

Repeating the argument we get

$$\phi_{\mathscr{2}(AB...F)} \sim \phi_{\mathscr{2}A(B...F)} \sim \phi_{\mathscr{2}AB(C...F)} \sim \cdots \sim \phi_{\mathscr{2}AB...D(EF)} \sim \phi_{\mathscr{2}AB...F} \tag{3.3.53}$$

which establishes (3.3.49).

Clearly the result corresponding to (3.3.49) when $A, B, \ldots, F$ are replaced by primed indices is also true. Thus, applying the argument once for the unprimed indices and again for the primed ones, we see that any spinor $\chi_{A...FP'...S'}$ differs from its symmetric part $\chi_{(A...F)(P'...S')}$ by a sum of outer products of εs with spinors of lower valence. A similar remark applies to these spinors of lower valence, and so on. Thus we have the

(3.3.54) PROPOSITION

*Any spinor $\chi_{A...FP'...S'}$ is the sum of the symmetric spinor $\chi_{(A...F)(P'...S')}$ and of outer products of εs with symmetric spinors of lower valence.*

(We call a spinor *symmetric* if, written with lower or upper indices only, it is symmetric in all its unprimed indices, and also symmetric in all its primed indices.)

Let us illustrate this with two examples. We have

$$\chi_{ABC} = \chi_{(ABC)} - \tfrac{1}{3}\varepsilon_{AB}\chi^{D}{}_{(DC)} - \tfrac{1}{3}\varepsilon_{AC}\chi^{D}{}_{(DB)} - \tfrac{1}{2}\varepsilon_{BC}\chi_{A}{}^{D}{}_{D} \qquad (3.3.55)$$

and

$$\theta_{ABA'B'} = \theta_{(AB)(A'B')} - \tfrac{1}{2}\varepsilon_{AB}\theta^{C}{}_{C(A'B')} - \tfrac{1}{2}\varepsilon_{A'B'}\theta_{(AB)}{}^{C'}{}_{C'} + \tfrac{1}{4}\varepsilon_{AB}\varepsilon_{A'B'}\theta^{C}{}_{C}{}^{C'}{}_{C'}. \qquad (3.3.56)$$

If we have a spinor with upper as well as lower indices, clearly we can simply lower all its indices and then proceed as above. We note in this connection the form that a symmetric spinor takes when some of its indices are raised:

(3.3.57) PROPOSITION

*The spinor* $\psi_{A...DP...SU'...W'F'...H'}$ *is symmetric iff* $\psi^{U'...W'P...S}_{A...D\ \ F'...H'}$ *is symmetric in each of its four index sets,*

$$\psi^{U'...W'P...S}_{A...D\ \ F'...H'} = \psi^{(U'...W')(P...S)}_{(A...D)\ \ (F'...H')},$$

*and every contraction over a pair of indices vanishes, for which it is sufficient that*

$$\psi^{U'...W'X...S}_{XB...D\,F'...H'} = 0, \quad \psi^{X'...W'P...S}_{A...D\ \ X'...H'} = 0.$$

The reason for this is to be found in (2.5.23): the vanishing of a contraction, when the upper index is lowered, asserts the vanishing of a skew part, i.e., symmetry in the two relevant indices.

### *Irreducibility*

Symmetric spinors (at a point) are important in that they are *irreducible* under the spin group $SL(2,\mathbb{C})$. Although we shall not concern ourselves in any great detail with questions of the irreducibility of tensors, spinors, etc., a few general remarks will not be out of place here. Suppose we wish to represent a group $\mathscr{G}$ by linear transformations of a vector space $\mathfrak{V}$, which is then called the representation space. If $\mathfrak{V}$ can be expressed as a direct sum

$$\mathfrak{V} = \mathfrak{V}_1 \oplus \mathfrak{V}_2,$$

where $\mathfrak{V}_1 \neq 0$, $\mathfrak{V}_2 \neq 0$, in such a way that the transformations representing $\mathscr{G}$ send the elements of $\mathfrak{V}_1$ into themselves (no matter where the elements of $\mathfrak{V}_2$ go), then the representation is said to be *reducible*; otherwise it is irreducible. (The elements of the representation space $\mathfrak{V}$ are also said to be [ir]reducible under $\mathscr{G}$ if the representation of $\mathscr{G}$ on $\mathfrak{V}$ has this property.) If $\mathfrak{V}$ can be expressed as a finite (or infinite) sum

$$\mathfrak{V} = \mathfrak{V}_1 \oplus \ldots \oplus \mathfrak{V}_k$$

and the transformations act irreducibly on each $\mathfrak{V}_i$ separately, then the representation is said to be *completely reducible.*

Here we are concerned with the group of restricted Lorentz transformations $O^{\uparrow}_{+}(1, 3)$ and its twofold covering, the spin group $SL(2, \mathbb{C})$. For these groups it can be shown (*cf.* Naimark 1964) that any finite-dimensional representation (i.e., one on a finite-dimensional vector space $\mathfrak{V}$) is completely reducible and that every irreducible representation can be realized as (i.e., is linearly isomorphic to) the transformations acting on a symmetric spinor (at a point*), induced by a spin transformation $\xi_A \mapsto t_A{}^B \xi_B$. Symmetric spinors are therefore irreducible under these groups. If $\phi_{A\ldots CD'\ldots F'}$ is a symmetric spinor, the transformations are of the form

$$\phi_{A\ldots CD'\ldots F'} \mapsto t_A{}^{A_0} \ldots t_C{}^{C_0} \bar{t}_{D'}{}^{D'_0} \ldots \bar{t}_{F'}{}^{F'_0} \phi_{A_0\ldots C_0 D'_0 \ldots F'_0},$$

i.e.,

$$\phi_{\mathscr{A}} \mapsto T_{\mathscr{A}}{}^{\mathscr{A}_0} \phi_{\mathscr{A}_0}$$

where $\mathscr{A} = A \ldots F'$, $\mathscr{A}_0 = A_0 \ldots F'_0$. Now consider the matrices of the components of $T_{\mathscr{A}}{}^{\mathscr{A}_0}$, in effect

$$t_{(\mathbf{A}}{}^{(\mathbf{A}_0} \ldots t_{\mathbf{C})}{}^{\mathbf{C}_0)} \bar{t}_{(\mathbf{D}'}{}^{(\mathbf{D}'_0} \ldots \bar{t}_{\mathbf{F}')}{}^{\mathbf{F}'_0)},$$

but with a suitable elimination of multiplicities: e.g., $t_{(A}{}^{(A_0} t_{B)}{}^{B_0)} \phi_{A_0 B_0}$ can be reduced to

$$\begin{pmatrix} T^{00}_{00} & 2T^{01}_{00} & T^{11}_{00} \\ T^{00}_{01} & 2T^{01}_{01} & T^{11}_{01} \\ T^{00}_{11} & 2T^{01}_{11} & T^{11}_{11} \end{pmatrix} \begin{pmatrix} \phi_{00} \\ \phi_{01} \\ \phi_{11} \end{pmatrix},$$

and, in general, identical rows and columns in the $\phi$- and $T$-matrices can be omitted and the remaining columns in the $T$-matrix multiplied by the number of occurrences of the $\phi$ terms on which they act (1!, 2!, 3!, 2!3!, etc.). Then the $T$-matrices give the irreducible matrix representations of these groups. Thus, any finite-dimensional representation of $O^{\uparrow}_{+}(1, 3)$ or $SL(2, \mathbb{C})$ has a representation space which is a direct sum of spaces of symmetric spinors (at a point). The expressions of spinors in terms of

* By contrast, for the representations of the Poincaré group, symmetry conditions on spinors *at a point* are insufficient. The representation space of the Poincaré group consists of fields satisfying various field equations, e.g., Maxwell's equations.

symmetric spinors that we obtained earlier in this section (*cf.* (3.3.55)) are, in fact, examples of splitting a space up into its irreducible parts (since each of the parts transforms into itself under spin transformations). However, this is not the only, nor the most general, way to form a direct sum. One can also string together, formally, spinors with different index structures, e.g., as is done in the case of Dirac spinors or twistors (*cf.* Appendix vol. 2 and §6.1) where the pair $\psi_A$, $\chi_{A'}$ is considered as a single 4-component object $\Psi_\alpha$, and we can write $\Psi_\alpha = \psi_A \oplus \chi_{A'}$.

Another way of expressing the irreducibility of symmetric spinors at a point is to say that they are, in a certain sense, *saturated* with symmetries: if any further symmetry relations are imposed, then either we lose no information, or we get zero. This is perhaps a little clearer in the case of tensors. Evidently any totally symmetric or totally anti-symmetric tensor is irreducible, and 'saturated' in this sense. But so also is a tensor with Riemann-tensor symmetries (*cf.* (4.3.53)–(4.3.56)):

$$R_{abcd} = R_{[ab][cd]} = R_{cdab},\ R_{a[bcd]} = 0$$

For example, the further symmetrization

$$R_{d(ab)c} =: S_{abcd}$$

'loses no information' since

$$\tfrac{4}{3} S_{a[bc]d} = R_{cbad}.$$

But $R_{[abc]d} = R_{a(bcd)} = 0$ etc. This instances *Young tableau* symmetry.

[In the theory of Young tableaux (Young 1900, see also Weyl 1931)* irreducible tensors are constructed (and classified) by first imposing symmetries on certain groups of indices and then 'saturating' the resulting tensor with anti-symmetries. All further (anti-) symmetry operations either lose no information or yield zero. As an example of Young-tableau symmetry, let us write, for the partition (4, 3, 1) of 8,

$$F_{\alpha\beta\gamma\delta\varepsilon\zeta\eta\theta} = F_{\substack{\varepsilon\gamma\eta\zeta \\ \beta\theta\alpha \\ \delta}}$$

and define

$$S_{\alpha\beta\gamma\delta\varepsilon\zeta\eta\theta} = S_{\substack{\varepsilon\gamma\eta\zeta \\ \beta\theta\alpha \\ \delta}} = F_{\substack{(\overline{\varepsilon}\,\overline{\gamma}\,\overline{\eta}\,\zeta) \\ (\beta\,\underline{\theta}\,\underline{\alpha}) \\ \underline{\delta}}}$$

The meaning of the last symbol is that we symmetrize over the indices $\varepsilon\gamma\eta\zeta$, $\beta\theta\alpha$ *first*, and *then* anti-symmetrize over $\varepsilon\beta\delta$, $\gamma\theta$, $\eta\alpha$. (We write the

* For typographical reasons this footnote on Young tableaux is being put in the text. It ends after five paragraphs, on p. 146, when spinors are taken up again.

symmetrized index groups in order of their lengths.) $S_{\alpha\ldots\theta}$ is now 'completely saturated' with symmetries, and irreducible. For example, the Riemann tensor has precisely the following Young-tableau symmetry (though this is not trivially obvious):

$$\tfrac{3}{4}R_{\alpha\beta\gamma\delta} = R_{\substack{(\bar{\alpha}\ \bar{\gamma})\\(\underline{\beta}\ \underline{\delta})}}.$$

However, tableaux do not supply a unique realization of all irreducible symmetries. For example, the tensor $P_{\alpha\beta\gamma} = Q_{\alpha\beta\gamma} + \omega Q_{\beta\gamma\alpha} + \omega^2 Q_{\gamma\alpha\beta}$ (where $\omega = e^{2i\pi/3}$) has the 'symmetry' $P_{\gamma\alpha\beta} = \omega P_{\alpha\beta\gamma}$, and is irreducible; so also is the tensor $M_{[\alpha\beta][\gamma\delta]} - M_{[\gamma\delta][\alpha\beta]}$, but neither tensor has tableau symmetry; nevertheless $P_{(\alpha\beta)\gamma}$ has tableau symmetry $\alpha\beta, \gamma$ and from it $P_{\alpha\beta\gamma}$ itself can be recovered ($\tfrac{3}{2}P_{\alpha\beta\gamma} = P_{(\alpha\beta)\gamma} + \omega P_{(\beta\gamma)\alpha} + \omega^2 P_{(\gamma\alpha)\beta}$), so we say it is 'equivalent' to a tableau. In the same sense the second tensor is equivalent to a tableau with symmetry $\alpha\beta, \gamma, \delta$, in an obvious notation. Any irreducible tensor is equivalent, in this sense to a tensor with Young tableau symmetry.

We can give elegantly (Frame *et al* 1954, Littlewood 1950) the number of *independent* components of any object with Young-tableau symmetry (or equivalent). Make two tables of squares, in the shape of the symmetrized index groups – for example, for $S_{\alpha\ldots\theta}$, thus:

| $n$ | $n+1$ | $n+2$ | $n+3$ |
|---|---|---|---|
| $n-1$ | $n$ | $n+1$ | |
| $n-2$ | | | |

| 6 | 4 | 3 | 1 |
|---|---|---|---|
| 4 | 2 | 1 | |
| 1 | | | |

In the first table write $n$ down the main diagonal (the number of dimensions of the vector space $\mathfrak{S}_\alpha$), $n-1$, $n-2$, etc., in successive diagonals below, and $n+1$, $n+2$, etc., in successive diagonals above the main diagonal. In the second table write in each square the number of squares to the right, plus the number of squares below, plus one. Form the product of *all* the numbers in the first table, and of *all* the numbers in the second, and divide the former by the latter: this gives the required number of components. In the case of $R_{\alpha\beta\gamma\delta}$ it gives, almost instantaneously, $\tfrac{1}{12}n^2(n^2-1)$.

It may be noted that Young tableaux can be equally well formed by *first* anti-symmetrizing the columns of indices, and *then* symmetrizing the rows. With these two alternative conventions, identical looking tableaux are not equal, but they are 'equivalent'. Thus for applying the abovementioned method of determining the number of independent components, the convention is immaterial.

These two alternative types of irreducible tensor find convenient interpretations as the 'coefficients' of two different kinds of 'form'.

Consider, first, a tensor with Young-tableau symmetry of the latter kind, so that it is symmetric rather than skew in its relevant groups of indices. For example, such a tensor $T_{\alpha\beta\gamma\delta\varepsilon\zeta\eta\theta}$ would be obtained by re-applying the 'horizontal' symmetries in the definition of $S_{\alpha\ldots\theta}$ above:

$$T_{\alpha\beta\gamma\delta\varepsilon\zeta\eta\theta} = S_{\substack{(\varepsilon\gamma\eta\zeta)\\(\beta\theta\alpha)\\\delta}}.$$

(Following this by another application of the vertical antisymmetries would yield a non-zero multiple of $S_{\ldots}$ again.) Because of (3.3.23), the information contained in $T_{\ldots}$ is the same as that contained in the polynomial function (or 'form')

$$T(X^\alpha, Y^\alpha, Z^\alpha) = T_{\alpha\beta\gamma\delta\varepsilon\zeta\eta\theta} X^\varepsilon X^\gamma X^\eta X^\zeta Y^\beta Y^\theta Y^\alpha Z^\delta.$$

Now using (3.3.21), we can rewrite this as

$$\begin{aligned} T(X^\alpha, Y^\alpha, Z^\alpha) &= S_{\alpha\beta\gamma\delta\varepsilon\zeta\eta\theta} X^\varepsilon X^\gamma X^\eta X^\zeta Y^\beta Y^\theta Y^\alpha Z^\delta \\ &= S_{\alpha\beta\gamma\delta\varepsilon\zeta\eta\theta} P^{\varepsilon\beta\delta} Q^{\gamma\theta} Q^{\eta\alpha} R^\zeta = S(P^{\alpha\beta\gamma}, Q^{\alpha\beta}, R^\alpha) \end{aligned}$$

the second form following from the antisymmetry of $S_{\ldots}$, with

$$P^{\alpha\beta\gamma} = X^{[\alpha} Y^\beta Z^{\gamma]}, \; Q^{\alpha\beta} = X^{[\alpha} Y^{\beta]}, \; R^\alpha = X^\alpha.$$

The function $S$ of 'simple' skew tensors (*cf.* (3.5.30) below) of this (hierarchical) type gives us the alternative polynomial form referred to above.

The fact that the function $T$ can be re-expressed in terms of these skew products can be stated as the functional relation

$$T(X^\alpha, Y^\alpha, Z^\alpha) = T(X^\alpha, Y^\alpha + \lambda X^\alpha, Z^\alpha + \mu X^\alpha + \nu Y^\alpha)$$

(for all $\lambda$, $\mu$, $\nu$), or as the differential equations

$$X^\alpha \frac{\partial T}{\partial Y^\alpha} = 0, \; X^\alpha \frac{\partial T}{\partial Z^\alpha} = 0, \; Y^\alpha \frac{\partial T}{\partial Z^\alpha} = 0,$$

where the partial derivative expressions have (if necessary) an obvious abstract-index meaning. Applying (3.3.22), we see that these differential equations are equivalent to the condition that if any of the horizontal symmetries in the right-hand expression

$$T_{\alpha\beta\gamma\delta\varepsilon\zeta\eta\theta} = T_{\substack{\varepsilon\gamma\eta\zeta\\\beta\theta\alpha\\\delta}}$$

is extended to include one more index *lower down* in the tableau, then the resulting tensor vanishes (e.g., $T_{\alpha\beta(\gamma\delta\varepsilon\zeta\eta)\theta} = 0$). This condition is (necessary and) sufficient for a tensor $T_{\alpha\ldots\theta}$ – symmetric in its relevant groups of indices (i.e., the horizontal symmetries being given) – to have Young-tableau symmetry. (Thus it expresses the existence of the 'hidden' vertical anti-symmetries.) A corresponding statement also holds for $S_{\dots}$, with 'symmetry' and 'anti-symmetry' suitably interchanged.]

One of the virtues of 2-spinor theory is that, by the earlier discussion of this section, only *symmetries* need be considered and not anti-symmetries, and so the theory of Young tableaux is not needed.

When one speaks simply of *irreducibility* of a tensor without reference to a specific group, then the group in question is understood to be the group of all linear transformations, $GL(n, \mathbb{C})$. But if the relevant group is some subgroup of $GL(n, \mathbb{C})$, further reductions of the tensor may be possible. For the Lorentz group, the metric tensor $g_{ab}$ and its inverse $g^{ab}$ are invariant objects, and 'generalized symmetries' are possible in which these objects are employed in order to 'reduce' a tensor into smaller parts that cannot be obtained by imposing ordinary symmetries. An important example is afforded by the reduction of the Riemann tensor into its three parts (the Weyl tensor $C_{abcd}$, the Ricci scalar $R$, and the trace-free Ricci tensor $R_{ab} - \frac{1}{4}g_{ab}R$ – *cf.* §§4.6, 4.8), irreducible under the Lorentz group; these parts arise from requiring, in addition to symmetries in the ordinary sense, certain 'trace' conditions, such as $g^{ac}C_{abcd} = 0$. If the Lorentz group is specialized still further to the restricted Lorentz group, then further invariant objects arise, namely $e_{abcd}$ and $e^{abcd}$. These generalized symmetries can get extremely complicated. The translation to spinor form therefore effects a considerable simplification.

The reduction of the Riemann tensor in spinor terms is treated in §4.6. Here we briefly consider a much simpler problem, namely the direct tensor translation of a symmetric spinor with an equal number of unprimed and primed indices:

$$\begin{aligned} \phi_{ab\dots f} &= \phi_{AB\dots FA'B'\dots F'} \\ &= \phi_{(AB\dots F)(A'B'\dots F')}. \end{aligned} \tag{3.3.58}$$

Evidently $\phi_{ab\dots f}$ is symmetric in its tensor indices, and trace-free:

$$\phi_{ab\dots f} = \phi_{(ab\dots f)}, \tag{3.3.59}$$

$$\phi^{a}{}_{ac\dots f} = 0. \tag{3.3.60}$$

Conversely, (3.3.59) implies

$$\begin{aligned} \varepsilon^{AB}\phi_{AA'BB'c\dots f} &= \varepsilon^{AB}\phi_{BB'AA'c\dots f} \\ &= -\varepsilon^{AB}\phi_{AB'BA'c\dots f}, \end{aligned} \tag{3.3.61}$$

and so the left member of (3.3.61) is skew in $A'B'$. But, by (3.3.60), its further transvection with $\varepsilon^{A'B'}$ must vanish; hence it is zero, showing that $\phi_{AB\dots FA'B'\dots F'}$ is symmetric in $AB$. Similarly it is symmetric in $BC$, etc., and also in $A'B'$ etc.; hence it is totally symmetric. This shows that the conditions (3.3.59), (3.3.60) on the tensor are completely equivalent to the symmetry (and thus irreducibility) of the spinor. Note also that this argument makes it easy to count the number of independent complex components of a complex tensor $\phi_{a...f}$, subject to (3.3.59) and (3.3.60).* It is $(r+1)^2$ when $\phi_{a...f}$ has $r$ tensor indices, since $\phi_{A...FA'...F'}$ clearly has that many independent components. (**A**...**F** may contain no zero, one zero, etc., up to $r$ zeros; and so it can take $r+1$ 'values'; similarly for **A**$'$...**F**$'$.) This result is not so easily obtained without the spinor translation. Moreover, we have by the same argument

(3.3.62) PROPOSITION:

*If $\varphi_{A...CP'...R'}$ is symmetric of valence $\left[\begin{smallmatrix}0&0\\p&q\end{smallmatrix}\right]$, then it has $(p+1)(q+1)$ independent (complex) components.*

## 3.4 Tensor representation of spinor operations

It follows from the way we have constructed the world-tensor algebra that every operation carried out with tensors may be reinterpreted as a spinor operation. The only difference in these interpretations is the purely formal one of replacing each tensor index by a pair of spinor indices. Thus, manipulations of tensors are merely special cases of manipulations of spinors, in which only certain types of spinors and only certain spinor operations are considered, namely those in which the indices may be consistently clumped together in pairs throughout the whole procedure, one primed and one unprimed index being involved in each clumping. From this viewpoint, the spinor algebra is considerably richer than the ordinary tensor algebra owing to the presence of numerous spinor operations which do not appear to have a direct tensor analogue, such as, for example, contraction over a single pair of spinor indices or the interchange of a single pair of spinor indices. In this section we shall show that these apparently new operations, introduced with the spinor formalism, do in fact have tensor analogues. *Every algebraic spinor operation and every algebraic spinor equation may be written as an operation on, or equation of, tensors,* though sometimes with sign ambiguities. Thus we may view the advantage of the spinor formalism as not so much that new operations are available, but rather that certain operations are suggested by the

* This number is clearly the same as the number of independent *real* components of a *real* tensor subject to the same linear restrictions.

spinor formalism which, because of their extremely complicated appearance when written in terms of tensors, are not readily suggested by the tensor formalism. It is perhaps significant that some of these tensorially apparently 'unnatural' operations are in fact often useful in physical applications. A further significant point is that the translation of a *linear* spinor equation is often a non-linear tensor equation.

*Trace reversals*

Before entering into the general discussion, it will be useful to study some special cases first. Consider an arbitrary symmetric (possibly complex) world-tensor of valence $\begin{bmatrix}0\\2\end{bmatrix}$:

$$T_{ab} = T_{ba}. \tag{3.4.1}$$

In spinor form, (3.4.1) becomes

$$T_{AA'BB'} = T_{BB'AA'}, \tag{3.4.2}$$

which we can rewrite as

$$T_{ABA'B'} = \tfrac{1}{2}(T_{ABA'B'} + T_{ABB'A'}) + \tfrac{1}{2}(T_{BAB'A'} - T_{ABB'A'}). \tag{3.4.3}$$

The first parenthesis is symmetric in $A'$, $B'$ and, because of (3.4.2), also symmetric in $A$, $B$, while the second is skew in $A$, $B$, and, by (3.4.2), also in $A', B'$. By a double application of (2.5.23) to the second parenthesis we therefore get

$$T_{ab} = T_{AA'BB'} = S_{ABA'B'} + \varepsilon_{AB}\varepsilon_{A'B'}\tau \tag{3.4.4}$$

where

$$S_{ABA'B'} = T_{(AB)(A'B')} = T_{(AB)A'B'} = T_{AB(A'B')} \tag{3.4.5}$$

and

$$\tau = \tfrac{1}{4}T_{CC'}{}^{CC'} = \tfrac{1}{4}T_c{}^c. \tag{3.4.6}$$

(The decomposition (3.4.4) is a particular case of that considered at the end of §3.3, *cf.* (3.3.56).) We may rewrite (3.4.4) as

$$T_{ab} = S_{ab} + g_{ab}\tau. \tag{3.4.7}$$

The tensor $T_{ab}$ is real if and only if both $\tau$ and $S_{ab}$ are real. The tensor $S_{ab}$ is clearly symmetric, and trace-free:

$$S_c{}^c = 0 \tag{3.4.8}$$

(by symmetry in $A$, $B$ or in $A'$, $B'$). We call $S_{ab}$ the *trace-free part* of $T_{ab}$. From (3.4.7) and (3.4.6) we have, in fact,

$$S_{ab} = T_{ab} - \tfrac{1}{4}T_c{}^c g_{ab}. \tag{3.4.9}$$

We can go one step further and apply the operation of *trace reversal* to $T_{ab}$:

$$\hat{T}_{ab} := T_{ab} - \tfrac{1}{2}T_c{}^c g_{ab}. \tag{3.4.10}$$

We then have

$$\hat{T}_c{}^c = -T_c{}^c \tag{3.4.11}$$

and, in spinor form,

$$\hat{T}_{ab} = \hat{T}_{ABA'B'} = S_{ABA'B'} - \varepsilon_{AB}\varepsilon_{A'B'}\tau \tag{3.4.12}$$

By comparison with (3.4.4) we see that

$$T_{ab} - \tfrac{1}{2}T_c{}^c g_{ab} = \hat{T}_{ABA'B'} = T_{BAA'B'} = T_{ABB'A'}, \tag{3.4.13}$$

so the trace reversal operation applied to $T_{ab}$ is effected in the spinor formalism simply by interchanging the constituent spinor indices $A$ and $B$, or alternatively by interchanging $A'$ and $B'$.

## *Dualization*

Let us now consider an arbitrary anti-symmetric (possibly complex) world-tensor of valence $\left[\begin{smallmatrix}0\\2\end{smallmatrix}\right]$ – sometimes called a *bivector*:

$$F_{ab} = -F_{ba}. \tag{3.4.14}$$

In spinor form,

$$F_{AA'BB'} = -F_{BB'AA'}, \tag{3.4.15}$$

which allows us to write

$$F_{ABA'B'} = \tfrac{1}{2}(F_{ABA'B'} - F_{ABB'A'}) + \tfrac{1}{2}(F_{ABB'A'} - F_{BAB'A'}) \tag{3.4.16}$$

By (2.5.23) we then get

$$F_{ab} = F_{AA'BB'} = \phi_{AB}\varepsilon_{A'B'} + \varepsilon_{AB}\psi_{A'B'}, \tag{3.4.17}$$

where*

$$\phi_{AB} = \phi_{(AB)} = \tfrac{1}{2}F_{ABC'}{}^{C'}, \ \psi_{A'B'} = \psi_{(A'B')} = \tfrac{1}{2}F_C{}^C{}_{A'B'}. \tag{3.4.18}$$

We note that $\phi_{AB}$ and $\psi_{A'B'}$ are both symmetric on account of (3.4.15). Notice that if $\phi_{AB}$ and $\psi_{A'B'}$ are chosen arbitrarily (but symmetric) then the resulting $F_{ab}$ according to (3.4.17) will necessarily be skew in $a$, $b$. From (3.4.17) we find

$$\bar{F}_{ab} = \bar{F}_{A'B'AB} = \bar{\phi}_{A'B'}\varepsilon_{AB} + \varepsilon_{A'B'}\bar{\psi}_{AB}, \tag{3.4.19}$$

* In the literature, a quantity $S_{AB}{}^{cd}(= S_{(AB)}{}^{[cd]})$ is often employed, for which $\phi_{AB} = S_{AB}{}^{cd}F_{cd}$ and $\psi_{A'B'} = \bar{S}_{A'B'}{}^{cd}F_{cd}$. With our conventions, this quantity can be written $S_{AB}{}^{cd} = \tfrac{1}{2}\varepsilon_{(A}{}^C\varepsilon_{B)}{}^D\varepsilon^{C'D'}$, as is readily seen.

and thus the operation $F_{ab} \mapsto \bar{F}_{ab}$ corresponds to interchanging $\psi_{A'B'}$ with $\bar{\phi}_{A'B'}$. Hence if $F_{ab}$ is real, we have $\psi_{A'B'} = \bar{\phi}_{A'B'}$ and

$$F_{ab} = \phi_{AB}\varepsilon_{A'B'} + \varepsilon_{AB}\bar{\phi}_{A'B'}. \tag{3.4.20}$$

Conversely this form ensures the reality of $F_{ab}$. Thus we see that there exists a one-to-one correspondence, via (3.4.18) (1) and (3.4.20), between real bivectors $F_{ab}$ and symmetric spinors $\phi_{AB}$. An important example of this is the field tensor of the Maxwell field (*cf.* §5.1). (In terms of components: the information of the six real quantities $F_{01}, F_{02}, F_{03}, F_{12}, F_{13}, F_{23}$ is given by the three complex quantities $\phi_{00}, \phi_{01}, \phi_{11}$. See (5.1.59) and (5.1.62) for the explicit formulae.)

The dual $*F_{ab}$ of a (not necessarily real) bivector $F_{ab}$ is defined by

$$*F_{ab} := \tfrac{1}{2}e_{abcd}F^{cd} = \tfrac{1}{2}e_{ab}{}^{cd}F_{cd}. \tag{3.4.21}$$

Thus, applying (3.3.44) to (3.4.17), we get

$$*F_{ab} = *F_{ABA'B'} = -\mathrm{i}\phi_{AB}\varepsilon_{A'B'} + \mathrm{i}\varepsilon_{AB}\psi_{A'B'} \tag{3.4.22}$$

and

$$*F_{ABA'B'} = \mathrm{i}F_{ABB'A'} = -\mathrm{i}F_{BAA'B'} \tag{3.4.23}$$

We observe that dualization is effected in the spinor formalism simply by an interchange of a pair of spinor indices followed by multiplication by $\pm\mathrm{i}$. From these formulae (and also from (3.3.45)) it is evident that the dual of the dual is minus the original:

$$**F_{ab} = -F_{ab}. \tag{3.4.24}$$

It is, of course, possible to 'dualize' on any two skew indices even if they form only a part of all the indices of a tensor. For example, if $G_{ab\mathscr{A}} = G_{[ab]\mathscr{A}}$, We may define $*G_{ab\mathscr{A}}$ by

$$*G_{ab\mathscr{A}} := \tfrac{1}{2}e_{ab}{}^{cd}G_{cd\mathscr{A}}. \tag{3.4.25}$$

A useful lemma in this connection is the following (note that $\mathscr{A}$ has become $c\mathscr{B}$):

$$*G_{[abc]\mathscr{B}} = 0 \Leftrightarrow *G^{ab}{}_{a\mathscr{B}} = 0. \tag{3.4.26}$$

To prove this, we note first that (*cf.* (3.3.43))

$$\begin{aligned} G_{[abc]\mathscr{B}} &= g_{[a}{}^{p}g_{b}{}^{q}g_{c]}{}^{r}G_{pqr\mathscr{B}} = -\tfrac{1}{6}e_{abcd}e^{pqrd}G_{pqr\mathscr{B}} \\ &= -\tfrac{1}{3}e_{abcd}*G^{pd}{}_{p\mathscr{B}} \end{aligned} \tag{3.4.27}$$

and also

$$*G^{ab}{}_{f\mathscr{B}} = \tfrac{1}{2}e^{abcd}G_{cdf\mathscr{B}} = \tfrac{1}{2}e^{abcd}G_{[fcd]\mathscr{B}}. \tag{3.4.28}$$

It is also possible to define operations of dualizing on one or three indices.

Suppose $J_{a\mathscr{A}}$ is arbitrary and $K_{abc\mathscr{B}}$ is skew in $abc$; then we define

$$^{\dagger}J_{abc\mathscr{A}} = e_{abc}{}^{d}J_{d\mathscr{A}} \tag{3.4.29}$$

$$^{\ddagger}K_{a\mathscr{B}} = \tfrac{1}{6}e_{a}{}^{bcd}K_{bcd\mathscr{B}}. \tag{3.4.30}$$

It is now easy to prove, by the above methods, that

$$^{\ddagger\dagger}J_{a\mathscr{A}} = J_{a\mathscr{A}}, \ ^{\dagger\ddagger}K_{abc\mathscr{B}} = K_{abc\mathscr{B}} \tag{3.4.31}$$

and (with $\mathscr{A}$ becoming $d\mathscr{C}$, and $\mathscr{B}$ becoming $d\mathscr{D}$)

$$^{\dagger}J_{[abcd]\mathscr{C}} = \tfrac{1}{4}e_{abcd}J_{f}{}^{f}{}_{\mathscr{C}}, \ K_{[abcd]\mathscr{D}} = \tfrac{1}{4}e_{abcd}{}^{\ddagger}K_{f}{}^{f}{}_{\mathscr{D}}. \tag{3.4.32}$$

Using (3.3.31) we obtain the following respective spinor forms for (3.4.29) and (3.4.30)

$$^{\dagger}J_{abc\mathscr{A}} = \mathrm{i}\varepsilon_{AC}\varepsilon_{B'C'}J_{BA'\mathscr{A}} - \mathrm{i}\,\varepsilon_{BC}\varepsilon_{A'C'}J_{AB'\mathscr{A}} \tag{3.4.33}$$

and

$$^{\ddagger}K_{a\mathscr{B}} = \frac{\mathrm{i}}{3}K_{AB'BA'}{}^{BB'}{}_{\mathscr{B}}. \tag{3.4.34}$$

Returning to $F_{ab}$, if

$$\text{(i)}\,{}^{*}F_{ab} = -\,\mathrm{i}F_{ab}, \text{ or (ii)}\,{}^{*}F_{ab} = \mathrm{i}F_{ab}, \tag{3.4.35}$$

we say in case (i) that $F_{ab}$ is *anti-self-dual*, and in case (ii) that $F_{ab}$ is *self-dual*. By (3.4.23), these conditions are equivalent to

$$\text{(i)}\ F_{ABA'B'} = F_{BAA'B'} = -\,F_{ABB'A'}\ \text{(ii)}\ F_{ABA'B'} = -\,F_{BAA'B'} = F_{ABB'A'}, \tag{3.4.36}$$

respectively. We may write these succinctly as

$$\text{(i)}\ F_{ABA'B'} = F_{(AB)[A'B']}, \text{(ii)}\ F_{ABA'B'} = F_{[AB](A'B')}. \tag{3.4.37}$$

A non-zero self-dual or anti-self-dual bivector is necessarily complex. Clearly the complex conjugate of a self-dual bivector is anti-self-dual and vice versa. If $F_{ab}$ is an arbitrary complex bivector, then

$$^{-}F_{ab} := \tfrac{1}{2}(F_{ab} + \mathrm{i}\,{}^{*}F_{ab}) = \phi_{AB}\varepsilon_{A'B'} \tag{3.4.38}$$

is *anti*-self-dual, and

$$^{+}F_{ab} := \tfrac{1}{2}(F_{ab} - \mathrm{i}^{*}F_{ab}) = \varepsilon_{AB}\psi_{A'B'} \tag{3.4.39}$$

is self-dual.* Consequently every bivector $F_{ab}$ is (uniquely) the sum of an anti-self-dual and a self-dual bivector:

$$F_{ab} = {}^{-}F_{ab} + {}^{+}F_{ab}; \tag{3.4.40}$$

moreover, if $F_{ab}$ is real, these parts are complex conjugates of each other.

* The reason for choosing the terminology thus and not the other way around is that a right-handed photon is described by a (positive frequency) complex Maxwell field $^{+}F_{ab}$ which is self-dual, and a left-handed one by an anti-self-dual field $^{-}F_{ab}$.

Alternative necessary and sufficient conditions for $F_{ab}$ to be (i) anti-self-dual, (ii) self-dual are (i) $\psi_{A'B'}=0$, (ii) $\phi_{AB}=0$, respectively; that is to say,

$$\text{(i) } F_{ab}=\phi_{AB}\varepsilon_{A'B'}, \text{ (ii) } F_{ab}=\varepsilon_{AB}\psi_{A'B'}. \tag{3.4.41}$$

A *duality rotation* $F_{ab}\mapsto{}^{(\theta)}F_{ab}$ is defined by the equation

$$^{(\theta)}F_{ab}:=F_{ab}\cos\theta+{}^*F_{ab}\sin\theta={}^-F_{ab}\mathrm{e}^{-\mathrm{i}\theta}+{}^+F_{ab}\mathrm{e}^{\mathrm{i}\theta} \tag{3.4.42}$$

By (3.4.38) and (3.4.39) we get

$$^{(\theta)}F_{ab}=\mathrm{e}^{-\mathrm{i}\theta}\phi_{AB}\varepsilon_{A'B'}+\mathrm{e}^{\mathrm{i}\theta}\varepsilon_{AB}\psi_{A'B'}. \tag{3.4.43}$$

Thus, in the general case, the operation $F_{ab}\mapsto{}^{(\theta)}F_{ab}$ corresponds to $\phi_{AB}\mapsto \mathrm{e}^{-\mathrm{i}\theta}\phi_{AB}$ and $\psi_{A'B'}\mapsto\mathrm{e}^{\mathrm{i}\theta}\psi_{A'B'}$ i.e., to ${}^-F_{AB}\mapsto{}^-F_{AB}\mathrm{e}^{-\mathrm{i}\theta}$ and ${}^+F_{AB}\mapsto{}^+F_{AB}\mathrm{e}^{\mathrm{i}\theta}$. If $F_{ab}$ is real, $F_{ab}\mapsto{}^{(\theta)}F_{ab}$ corresponds simply to $\phi_{AB}\to\mathrm{e}^{-\mathrm{i}\theta}\phi_{AB}$ (i.e. to ${}^-F_{ab}\mapsto{}^-F_{ab}\mathrm{e}^{-\mathrm{i}\theta}$). We observe that ${}^*F_{ab}$ is a particular case of ${}^{(\theta)}F_{ab}$, with $\theta=\pi/2$.

There are many properties of (anti-) self-dual bivectors which can be easily read off in the spinor formalism, not all of which can be so readily obtained using tensors. If ${}^-F_{ab}$ is any anti-self-dual bivector and ${}^+G_{ab}$ is any self-dual bivector, then, for example,

$$^-F_{ab}{}^+G^{ab}=0 \tag{3.4.44}$$

and

$$^-F_a{}^{b\,+}G_{bc}={}^+G_a{}^{b\,-}F_{bc}. \tag{3.4.45}$$

Equation (3.4.44) is easy to verify either by tensor or by spinor methods, but (3.4.45) is much easier using spinors. Each side of the equation is obviously simply

$$\phi_{AC}\gamma_{A'C'} \tag{3.4.46}$$

where ${}^-F_{ab}=\phi_{AB}\varepsilon_{A'B'}$ and ${}^+G_{ab}=\varepsilon_{AB}\gamma_{A'B'}$. It is also obvious that the quantity (3.4.46) completely determines the following quantity, and is in turn determined by it:

$$^-F_{ab}{}^+G_{cd}=\phi_{AB}\gamma_{C'D'}\varepsilon_{A'B'}\varepsilon_{CD}. \tag{3.4.47}$$

To pass from (3.4.46) to (3.4.47) we need only multiply by $-\varepsilon_{BD}\varepsilon_{B'D'}$ $(=-g_{bd})$ and then interchange $B$ with $C$ and $A'$ with $D'$. Thus from the contracted product (3.4.45) we can pass to the outer product (3.4.47) and thence to the two tensors ${}^-F_{ab}$, ${}^+G_{ab}$ separately, up to proportionality. In purely tensor terms, however, these spinor index permutations are by no means obvious operations. A general method for performing these spinor operations tensorially will be given shortly. For this particular problem, one can tensorially employ Robinson's 'unscrambler' defined by

$$D_{abcd}{}^{pq} := \left\{ g_{[a}{}^{p} g_{b]r} + \frac{\mathrm{i}}{2} e_{ab}{}^{p}{}_{r} \right\} \left\{ g_{[c}{}^{r} g_{d]}{}^{q} - \frac{\mathrm{i}}{2} e_{cd}{}^{rq} \right\}. \tag{3.4.48}$$

It is straightforward to show, using spinor methods, that

$${}^{-}F_{ab}{}^{+}G_{cd} = D_{abcd}{}^{pq}\,{}^{-}F_{p}{}^{r}\,{}^{+}G_{rq} \tag{3.4.49}$$

An example related to (3.4.45) is the equivalence of the various expressions for the electromagnetic energy – momentum tensor (see (5.2.3)) which again is much easier to establish using spinors than tensors.

*General translation procedure*

We have seen above that for a symmetric tensor $T_{ab}$, interchange of the spinor indices $A$, $B$ effects a trace reversal. On the other hand, for a skew tensor $F_{ab}$, interchange of $A$, $B$ effects the (seemingly unrelated) operation of i times dualization. For a general tensor $H_{ab}$ of valence $\left[\begin{smallmatrix}0\\2\end{smallmatrix}\right]$ these two operations must be combined. For, since

$$H_{AA'BB'} = H_{(ab)} + H_{[ab]}, \tag{3.4.50}$$

we have

$$H_{BA'AB'} = \hat{H}_{(ab)} + \mathrm{i}{}^{*}H_{[ab]} \tag{3.4.51}$$

and

$$H_{AB'BA'} = \hat{H}_{(ab)} - \mathrm{i}\,{}^{*}H_{[ab]}. \tag{3.4.52}$$

Writing these operations out explicitly, we get

$$H_{BAA'B'} = \tfrac{1}{2}(H_{ab} + H_{ba} - H_{c}{}^{c} g_{ab} + \mathrm{i}\, e_{abcd} H^{cd}) \tag{3.4.53}$$

and

$$H_{ABB'A'} = \tfrac{1}{2}(H_{ab} + H_{ba} - H_{c}{}^{c} g_{ab} - \mathrm{i}\, e_{abcd} H^{cd}). \tag{3.4.54}$$

The complexity of these tensor expressions is remarkable, considering that they represent the very simplest of spinor operations, namely the interchange of two index labels.

For future reference we mention also the formulae

$$H_{[ab]} = H_{[AB](A'B')} + H_{(AB)[A'B']} = \tfrac{1}{2}\varepsilon_{AB} H_{C}{}^{C}{}_{(A'B')} + \tfrac{1}{2}\varepsilon_{A'B'} H_{(AB)C'}{}^{C'} \tag{3.4.55}$$

$$H_{(ab)} = H_{(AB)(A'B')} + H_{[AB][A'B']} = H_{(AB)(A'B')} + \tfrac{1}{4}\varepsilon_{AB}\varepsilon_{A'B'} H_{CC'}{}^{CC'} \tag{3.4.56}$$

We can re-express (3.4.53) and (3.4.54) in terms of a certain tensor operator as follows. Set

$$\begin{aligned} U_{ab}{}^{cd} &= \varepsilon_{A}{}^{D} \varepsilon_{B}{}^{C} \varepsilon_{A'}{}^{C'} \varepsilon_{B'}{}^{D'} \\ &= \tfrac{1}{2}(g_{a}{}^{c} g_{b}{}^{d} + g_{a}{}^{d} g_{b}{}^{c} - g_{ab} g^{cd} + \mathrm{i}\, e_{ab}{}^{cd}). \end{aligned} \tag{3.4.57}$$

Then

$$H_{BAA'B'} = U_{ab}{}^{cd}H_{cd} \text{ and } H_{ABB'A'} = \bar{U}_{ab}{}^{cd}H_{cd}. \quad (3.4.58)$$

Note that

$$U^{ab}{}_{cd} = U_{cd}{}^{ab} \quad (3.4.59)$$

and

$$U_{ba}{}^{cd} = \bar{U}_{ab}{}^{cd} = U_{ab}{}^{dc}. \quad (3.4.60)$$

Thus, for example, we can write

$$H^{BAA'B'} = U_{cd}{}^{ab}H^{cd},\ H_{ABB'A'} = U_{ba}{}^{cd}H_{cd}, \quad (3.4.61)$$

etc. Using this operation repeatedly we can express *any* spinor index permutation (applied to some world-tensor) in completely tensorial terms. This is because any permutation can be expressed as a product of transpositions. For example, consider the tensor $Q_{abcd} = Q_{AA'BB'CC'DD'}$. To find a tensor expression for $Q_{CABDD'B'C'A'}$ we may proceed by breaking down the index permutation as follows:

$$Q_{ABCDA'B'C'D'} \mapsto Q_{BACDA'B'C'D'} \mapsto Q_{CABDA'B'C'D'} \mapsto Q_{CABDD'B'C'A'} \quad (3.4.62)$$

Thus

$$Q_{CABDD'B'C'A'} = Q_{a_1b_1c_1d_1}U_{a_2b_2}{}^{a_1b_1}U_{bc}{}^{b_2c_1}\bar{U}_{ad}{}^{a_2d_1} \quad (3.4.63)$$

It is clear that there will often be many ways of achieving a particular spinor index permutation as a product of transpositions, so there will be many equivalent expressions like (3.4.63). In the tensor formalism the equivalence of these expressions may be very far from obvious. To take a simple example, since interchanging a pair of unprimed spinor indices always commutes with interchanging a pair of primed indices, we must have

$$U_{ab}{}^{dx}\bar{U}_{xc}{}^{ef} = \bar{U}_{bc}{}^{xf}U_{ax}{}^{de} \quad (3.4.64)$$

This is because upon transvection with an arbitrary tensor $R_{def}$, each side of the equation yields $R_{BA'AC'CB'}$. It is no simple matter to verify (3.4.64) directly.

If an index permutation involving only unprimed spinor indices is broken down into transpositions in different ways, then again we get a relation satisfied by $U_{ab}{}^{cd}$. For example, the permutation $ABC \to CAB$ may be obtained via $ABC \to BAC \to CAB$ or $ABC \to ACB \to CAB$ or $ABC \to CBA \to CAB$. This gives us

$$U_{ac}{}^{xf}U_{xb}{}^{de} = U_{ab}{}^{dx}U_{xc}{}^{ef} = U_{bc}{}^{ex}U_{ax}{}^{df}. \quad (3.4.65)$$

It is readily seen that, with (3.4.60), either one of these equalities is actually equivalent to (3.4.64). In fact, if we include the relation

$$U_{ab}{}^{cd}U_{cd}{}^{ef} = g_a{}^e g_b{}^f, \tag{3.4.66}$$

which expresses the fact that $AB \mapsto BA \mapsto AB$ results in the identity index permutation, then *all* identities in $U_{ab}{}^{cd}$ which are obtainable in *this* way* are reducible to those that we have already found. We shall not demonstrate this fact in detail here. It is merely a question of showing that every breakdown of a spinor index permutation into a product of transpositions can be converted into any other by means solely of transformations of the type which we have considered above.

We now turn to the question of the representation of a general spinor $\chi_{\mathscr{A}}$ in world-tensor terms. Let us first suppose that, for convenience, all indices have been lowered. (This clearly involves no loss or gain in the information contained in $\chi_{\mathscr{A}}$.) If $\chi_{\mathscr{A}} = \chi_{A\ldots EB'\ldots F'}$ has the same number of primed as unprimed indices, then at most an index substitution is required to obtain a complex world-tensor equivalent to $\chi_{\mathscr{A}}$. If desired, this complex world-tensor may be described in terms of two *real* world tensors, namely its real and imaginary parts. Of course, there will always be many different index substitutions which can be used to yield a complex world-tensor, e.g.,

$$\chi_{ABCD'E'F'} \mapsto \chi_{ABCA'B'C'}, \quad \text{or } \chi_{ABCB'C'A'}, \quad \text{or } \chi_{EBDB'E'D'}, \tag{3.4.67}$$

but, by the above discussion, all such world-tensors are *equivalent* to one another via purely tensorial operations. Thus it is immaterial, from the point of view of the general discussion, which tensor equivalent is selected. In practice, the choice need be governed only by considerations of convenience.

Next, let us suppose that $\chi_{\mathscr{A}} = \chi_{A\ldots EB'\ldots F'}$ has an even *total* number of indices, although the numbers of primed and unprimed indices may be different. We call such a spinor an *even spinor*. In this case we take the outer product of $\chi_{\mathscr{A}}$ with a sufficient number of ε-spinors to make the total number of primed and unprimed indices equal. (This clearly involves no loss or gain of information.) The situation is then reduced to the one considered above, and a complex world-tensor (or pair of real world tensors) results which can be used to describe the spinor $\chi_{\mathscr{A}}$. Again, several different complex world-tensors may arise in this way, but all are tensorially equivalent to one another, so any one may be selected. This applies even if more than the necessary number of ε-spinors have been used. For the

* There are other identities satisfied by $U_{ab}{}^{cd}$, however. For example, $U_{ab}{}^{cd} + U_a{}^{dc}{}_b = g_a{}^c g_b{}^d$ expresses the ε-identity (2.5.22).

extra $\varepsilon$-spinors may, by spinor index substitution, be collected into pairs $\varepsilon_{PQ}\varepsilon_{P'Q'}$. Such a pair is just $g_{pq}$ in tensorial terms, which, appearing in an outer product, neither adds nor subtracts information.

Finally, suppose that $\chi_{\mathscr{A}} = \chi_{A...EB'...F'}$ is an *odd spinor*, i.e., it has an odd total number of indices. There can now be no complete tensor analogue since $\chi_{\mathscr{A}}$ is a spinorial object. (We recall that any outer product of an odd number of spin-vectors or conjugate spin-vectors at a point must change sign under a continuous active rotation through $2\pi$ at the point, since each factor does; the spinor $\chi_{\mathscr{A}}$ is a sum of such outer products and so shares this behaviour.) Thus we can only hope to represent $\chi_{\mathscr{A}}$ tensorially up to an overall sign. Accepting this, however, we can apply the above discussion to the even spinor $\chi_{\mathscr{A}_1}\chi_{\mathscr{A}_2}$, which defines $\chi_{\mathscr{A}}$ up to sign. We may note that the procedure of §3.2 for representing a spin-vector is essentially an application of this method. Starting from the spin-vector $\kappa_A$, we 'square' it to obtain the even spinor $\kappa_A\kappa_B$. Multiplication by $\varepsilon_{A'B'}$ gives us the complex world-tensor $\kappa_A\kappa_B\varepsilon_{A'B'}$, in accordance with the above prescription. For purposes of geometrical realization we take (twice) the real part to obtain $P_{ab} = \kappa_A\kappa_B\varepsilon_{A'B'} + \varepsilon_{AB}\bar{\kappa}_{A'}\bar{\kappa}_{B'}$, which is just (3.2.9) with indices lowered. It is not necessary to consider also the imaginary part of $\kappa_A\kappa_B\varepsilon_{A'B'}$ since this is simply (one-half) the dual $^*P_{ab}$ of $P_{ab}$. The geometrical interpretation of $-^*P_{ab}$ adds nothing to that of $P_{ab}$ since it represents a null flag with the same flagpole as that of $\kappa_A$ but with flag plane rotated (positively) through $\pi/2$ (i.e. giving the flag plane of $e^{i\pi/4}\kappa_A$). The complex bivector $\kappa_A\kappa_B\varepsilon_{A'B'}$ itself is just the anti-self-dual* part $^-P_{ab}$ of $P_{ab}$.

We next consider how the various spinor *operations* can be interpreted in purely tensorial terms. We have already considered spinor index permutations above, and general spinor index substitutions effectively add nothing to these. The operation of complex conjugation presents no problem: the tensor equivalent of the complex conjugate of a spinor is the complex conjugate of the tensor equivalent of the spinor. Also, multiplication of an even [odd] spinor by a scalar corresponds to multiplication of the tensor equivalent by the same scalar [square of the scalar].

The operation of outer multiplication of spinors translates, *in effect*, to outer multiplication of the corresponding tensors. That this is not quite so simple as it sounds can perhaps best be clarified by examples. Suppose,

* Each of $P_{ab}$, $^*P_{ab}$ and $^-P_{ab}$ is *null* in a sense which will be described in §3.5 (*cf.* (3.5.28)), so the complex bivector $^-P_{ab}$ is an *anti-self-dual null bivector* – a property which serves to characterize it as having the form $\kappa_A\kappa_B\varepsilon_{A'B'}$. There is thus a close association between such bivectors and spin-vectors.

first, that we have two even spinors $\psi_{AB}$ and $\phi_{AB'C'D'}$. We represent them by complex world-tensors $P_{ab} = \psi_{AB}\varepsilon_{A'B'}$, $Q_{abc} = \phi_{AA'B'C'}\varepsilon_{BC}$. Now by our earlier prescription, the outer product $\psi_{AB}\phi_{CD'E'F'}$ might be represented by $R_{abc} = \psi_{AB}\phi_{CA'B'C'}$. Clearly $R_{abc}$ is not the outer product of $P_{ab}$ with $Q_{abc}$. However, by means of a spinor index permutation applied to the tensor $P_{ab}Q_{cde}$ (i.e., by transvecting with a succession of $U_{\dots}{}^{\dots}$s) we can reduce it to the form $\psi_{AB}\phi_{CA'B'C'}\varepsilon_{DE}\varepsilon_{D'E'}$ which is just $R_{abc}g_{de}$. In this sense, $R_{abc}$ and $P_{ab}Q_{cde}$ are tensorially equivalent.

As a second example, consider the outer product of the even spinor $\psi_{AB}$ with an odd spinor, say $\xi_A$. We have $P_{ab}$ as above and let $X_{ab} = \xi_A\xi_B\varepsilon_{A'B'}$. A tensor representing the outer product $\psi_{AB}\xi_C$ would be $(\psi_{AB}\xi_C)(\psi_{DE}\xi_F)\varepsilon_{A'B'}\varepsilon_{C'D'}\varepsilon_{E'F'}$. This is quadratic in $\psi_{AB}$ and quadratic in $\xi_A$, whereas the outer product $P_{ab}X_{cd}$ is linear in $\psi_{AB}$ and quadratic in $\xi_A$. Thus these two quantities cannot be regarded as 'tensorially equivalent' in the above sense. On the other hand we can choose the tensor $P_{ab}P_{cd}X_{ef}$ to represent the spinor outer product.

As a third example, let $\xi_A$ and $X_{ab}$ be as above and choose a second odd spinor, say $\eta_{AA'B'}$, with $Y_{abcd} = \eta_{AA'B'}\eta_{BC'D'}\varepsilon_{CD}$. The outer product of tensors $X_{ab}Y_{cdef}$ is quadratic both in $\xi_A$ and $\eta_{CC'D'}$, so it is not 'tensorially equivalent, in the above sense to the tensor representing the spinor outer product $\xi_A\eta_{CC'D'}$ since this is an *even* spinor and should not be squared in the construction of its tensor equivalent. However, if $\xi_A$ and $\eta_{CC'D'}$ are each known only up to sign and nothing is known about their *relative* signs, then the even spinor $\xi_A\eta_{CC'D'}$ will itself be known only up to sign. In this case we could only hope to obtain a tensor equivalent of $\xi_A\eta_{CC'D'}$ up to sign. So a tensor equivalent of its square $(\xi_A\eta_{BA'B'})(\xi_C\eta_{DC'D'})$ – which *is* tensorially equivalent to $X_{ab}Y_{cdef}$ – is the best that can be done.

A slightly different situation arises if we have a number of odd spinors whose relative signs *are* known. Then it is not sufficient that tensor equivalents of each of their squares be known. In addition, one must have tensor equivalents of outer products of different odd spinors. To some extent the problem considered in the preceding paragraph then becomes vacuous. It should however, be pointed out in this context that if $\xi_{\mathscr{A}}\,\eta_{\mathscr{B}}$, $\eta_{\mathscr{B}_1}\,\eta_{\mathscr{B}_2}$ and $\eta_{\mathscr{B}}\zeta_{\mathscr{C}}$ are all known then so is $\xi_{\mathscr{A}}\,\zeta_{\mathscr{C}}$ (since $\eta_{\mathscr{B}_1}\,\eta_{\mathscr{B}_2}\,\xi_{\mathscr{A}}\,\zeta_{\mathscr{C}}$ is known.)

Next we come to the operation of spinor addition. If both spinors to be added are even spinors, then it is clear from linearity (assuming their tensor equivalents are both formed in the same way) that the tensor equivalent of their sum is the sum of their tensor equivalents. On the other hand, things are not nearly so simple in the case of odd spinors. Suppose

$\xi_{\mathscr{A}}$ and $\eta_{\mathscr{A}}$ are odd spinors, and consider how we might express the relation

$$\xi_{\mathscr{A}} + \eta_{\mathscr{A}} = \zeta_{\mathscr{A}} \tag{3.4.68}$$

in terms of the squares of $\xi_{\mathscr{A}}, \eta_{\mathscr{A}}$ and $\zeta_{\mathscr{A}}$. If $\xi_{\mathscr{A}}, \eta_{\mathscr{A}}$ and $\zeta_{\mathscr{A}}$ are known only up to sign, the four relations

$$\xi_{\mathscr{A}} \pm \eta_{\mathscr{A}} \pm \zeta_{\mathscr{A}} = 0 \tag{3.4.69}$$

(where the signs are independent) cannot be distinguished from one another. Now consider the outer product of these four relations, symmetrized over their composite indices:

$$(\xi_{(\mathscr{A}_1} + \eta_{(\mathscr{A}_1} + \zeta_{(\mathscr{A}_1})(\xi_{\mathscr{A}_2} + \eta_{\mathscr{A}_2} - \zeta_{\mathscr{A}_2})(\xi_{\mathscr{A}_3} - \eta_{\mathscr{A}_3} + \zeta_{\mathscr{A}_3})$$
$$\times (\xi_{\mathscr{A}_4)} - \eta_{\mathscr{A}_4)} - \zeta_{\mathscr{A}_4)}) = 0 \tag{3.4.70}$$

By (3.5.15) this can vanish only if one of the factors vanishes at each point. (There is a slight problem since different factors might vanish at different points, but we shall ignore this problem here.) Expanding (3.4.70), we get

$$\xi_{\mathscr{A}_1}\xi_{\mathscr{A}_2}\xi_{\mathscr{A}_3}\xi_{\mathscr{A}_4} + \eta_{\mathscr{A}_1}\eta_{\mathscr{A}_2}\eta_{\mathscr{A}_3}\eta_{\mathscr{A}_4} + \zeta_{\mathscr{A}_1}\zeta_{\mathscr{A}_2}\zeta_{\mathscr{A}_3}\zeta_{\mathscr{A}_4} - 2\xi_{(\mathscr{A}_1}\xi_{\mathscr{A}_2}\eta_{\mathscr{A}_3}\eta_{\mathscr{A}_4)}$$
$$- 2\eta_{(\mathscr{A}_1}\eta_{\mathscr{A}_2}\zeta_{\mathscr{A}_3}\zeta_{\mathscr{A}_4)} - 2\zeta_{(\mathscr{A}_1}\zeta_{\mathscr{A}_2}\xi_{\mathscr{A}_3}\xi_{\mathscr{A}_4)} = 0, \tag{3.4.71}$$

which is clearly expressible in terms of outer squares of $\xi_{\mathscr{A}}, \eta_{\mathscr{A}}$ and $\zeta_{\mathscr{A}}$. The tensor equivalent of (3.4.68), in the case of odd spinors, may thus be thought of as (3.4.71) translated into a tensor form, the tensor translations of $\xi_{\mathscr{A}_1}\xi_{\mathscr{A}_2}, \eta_{\mathscr{A}_1}\eta_{\mathscr{A}_2}$ and $\zeta_{\mathscr{A}_1}\zeta_{\mathscr{A}_2}$ being substituted into the expression.

If, on the other hand, the relative signs of $\xi_{\mathscr{A}}$ and $\eta_{\mathscr{A}}$ are known, tensor translations of $\xi_{\mathscr{A}_1}\xi_{\mathscr{A}_2}, \eta_{\mathscr{A}_1}\eta_{\mathscr{A}_2}$ and $\xi_{\mathscr{A}_1}\eta_{\mathscr{A}_2}$ being assumed known, then the situation is much simpler since (3.4.68) can just be squared and translated directly.

Finally we come to the operation of contraction. By (2.5.23), a spinor contraction is simply an anti-symmetrization with $\frac{1}{2}\varepsilon_{XY}$ split off. Thus, having interpreted spinor index permutation and addition (subtraction) in tensorial terms, spinor contraction is effectively also incorporated into the tensor formalism. However, it is just a little easier to express spinor contraction directly using the tensor $U_{ab}{}^{cd}$ defined in (3.4.57). Since $U_a{}^{dc}{}_b = \varepsilon_{AB}\varepsilon^{CD}\varepsilon_{A'}{}^{C'}\varepsilon_{B'}{}^{D'}$, we have

$$\phi_{\mathscr{Q}XA'}{}^{X}{}_{B'}\varepsilon_{AB} = \phi_{\mathscr{Q}cd}U_a{}^{dc}{}_b. \tag{3.4.72}$$

We have now shown that, apart from the slight difficulties arising because of the sign ambiguity for odd spinors, every spinor and spinor operation has a tensor analogue. However, in the process we have also demonstrated the extreme complication that can sometimes arise in so translating even the simplest of spinor operations. In practice, when searching for a tensor equivalent for some spinor relation, it sometimes turns out that owing to this complication, it is easier to obtain the required tensor relation by inspection rather than by the general theory. But no hard and fast rule can be laid down about this. The tensor translations frequently just *are* very complicated. Extra difficulties arise when derivatives of spinors are to be translated. This case will be discussed at the end of §4.4.

## 3.5 Simple propositions about tensors and spinors at a point

In this section we consider a number of miscellaneous results which are valid for spinors (or tensors) at a single point. That is to say, these results can fail for spinor *fields* (although, in practice, it might be only rather exceptional spinor fields for which they would fail). Our restriction to a single point is expressed in the assumption that we make for the purposes of this section: *$\mathfrak{S}$ is the division ring of complex numbers.* (The results up to and including (3.5.17) will actually apply if $\mathfrak{S}$ is any commutative division ring without characteristic and, when suitably formulated, except for (3.5.15), if $\mathfrak{S}$ has any characteristic other than two.)

(3.5.1) PROPOSITION

*If $\psi_{\mathscr{A}}\phi_{\mathscr{B}}=0$ then either $\psi_{\mathscr{A}}=0$ or $\phi_{\mathscr{B}}=0$.*

*Proof.* Since $\mathfrak{S}$ is now a division ring, $(\xi^{\mathscr{A}}\psi_{\mathscr{A}})(\eta^{\mathscr{B}}\phi_{\mathscr{B}})=0$ implies $\xi^{\mathscr{A}}\psi_{\mathscr{A}}=0$ or $\eta^{\mathscr{B}}\phi_{\mathscr{B}}=0$. This holds for all $\xi^{\mathscr{A}}\in\mathfrak{S}^{\mathscr{A}}$, $\eta^{\mathscr{B}}\in\mathfrak{S}^{\mathscr{B}}$, so the result follows.

(3.5.2) PROPOSITION

*If $\psi_{\mathscr{A}}\phi_{\mathscr{B}}=\chi_{\mathscr{A}}\theta_{\mathscr{B}}\neq 0$, then $\psi_{\mathscr{A}}=\kappa\chi_{\mathscr{A}}$, $\phi_{\mathscr{B}}=\kappa^{-1}\theta_{\mathscr{B}}$ for some non-zero $\kappa\in\mathfrak{S}$.*

*Proof*: Since $\phi_{\mathscr{B}}\neq 0$, we can choose $\xi^{\mathscr{B}}$ so that $\lambda:=\phi_{\mathscr{B}}\xi^{\mathscr{B}}\neq 0$. We have $\psi_{\mathscr{A}}\phi_{\mathscr{B}}\xi^{\mathscr{B}}=\chi_{\mathscr{A}}\theta_{\mathscr{B}}\xi^{\mathscr{B}}$, so $\psi_{\mathscr{A}}=\kappa\chi_{\mathscr{A}}$ where $\kappa=\lambda^{-1}\theta_{\mathscr{B}}\xi^{\mathscr{B}}$. Clearly $\kappa\neq 0$ since $\psi_{\mathscr{A}}\neq 0$. Now $0=\psi_{\mathscr{A}}\phi_{\mathscr{B}}-\chi_{\mathscr{A}}\theta_{\mathscr{B}}=\chi_{\mathscr{A}}(\kappa\phi_{\mathscr{B}}-\theta_{\mathscr{B}})$, but $\chi_{\mathscr{A}}\neq 0$, so $\phi_{\mathscr{B}}=\kappa^{-1}\theta_{\mathscr{B}}$ by (3.5.1).

(3.5.3) PROPOSITION

*If $\psi_{\mathscr{A}\mathscr{B}}\phi_{\mathscr{C}\mathscr{D}}=\chi_{\mathscr{A}\mathscr{D}}\theta_{\mathscr{C}\mathscr{B}}\neq 0$, then $\psi_{\mathscr{A}\mathscr{B}}=\alpha_{\mathscr{A}}\beta_{\mathscr{B}}$, $\phi_{\mathscr{C}\mathscr{D}}=\gamma_{\mathscr{C}}\rho_{\mathscr{D}}$, $\chi_{\mathscr{A}\mathscr{D}}=\alpha_{\mathscr{A}}\rho_{\mathscr{D}}$, $\theta_{\mathscr{C}\mathscr{B}}=\gamma_{\mathscr{C}}\beta_{\mathscr{B}}$ for some $\alpha_{\mathscr{A}}$, $\beta_{\mathscr{B}}$, $\gamma_{\mathscr{C}}$, $\rho_{\mathscr{D}}$.*

*Proof*: Since $\phi_{\mathscr{C}\mathscr{D}}\neq 0$, we can choose $\xi^{\mathscr{C}}$ and $\eta^{\mathscr{D}}$ so that $\lambda:=\xi^{\mathscr{C}}\eta^{\mathscr{D}}\phi_{\mathscr{C}\mathscr{D}}\neq 0$.

We have $\psi_{\mathscr{AB}}\lambda = (\chi_{\mathscr{AD}}\eta^{\mathscr{D}})(\xi^{\mathscr{C}}\theta_{\mathscr{CB}})$, so setting $\alpha_{\mathscr{A}} = \lambda^{-1}\chi_{\mathscr{AD}}\eta^{\mathscr{D}}$ and $\beta_{\mathscr{B}} = \xi^{\mathscr{C}}\theta_{\mathscr{CB}}$, we get $\psi_{\mathscr{AB}} = \alpha_{\mathscr{A}}\beta_{\mathscr{B}}$. Similarly $\phi_{\mathscr{CD}} = \mu_{\mathscr{C}}\nu_{\mathscr{D}}$ for some $\mu_{\mathscr{C}}, \nu_{\mathscr{D}}$. Thus $\chi_{\mathscr{AD}}\theta_{\mathscr{CB}} = \alpha_{\mathscr{A}}\nu_{\mathscr{D}}\mu_{\mathscr{C}}\beta_{\mathscr{B}}$ and so, by (3.5.2), $\chi_{\mathscr{AD}} = \kappa\alpha_{\mathscr{A}}\nu_{\mathscr{D}}$ and $\theta_{\mathscr{CB}} = \kappa^{-1}\mu_{\mathscr{C}}\beta_{\mathscr{B}}$ for some $\kappa \in \mathfrak{S}$. Put $\rho_{\mathscr{D}} = \kappa\nu_{\mathscr{D}}$ and $\gamma_{\mathscr{C}} = \kappa^{-1}\mu_{\mathscr{C}}$ and the result follows.

It is worth pointing out various special cases of (3.5.3). For example, if $\mathscr{D} = \mathscr{B}$, $\chi_{\mathscr{AB}} = \psi_{\mathscr{AB}}$ and $\phi_{\mathscr{CB}} = \theta_{\mathscr{CB}}$, then (3.5.3) gives us

$$\psi_{\mathscr{AB}_1}\theta_{\mathscr{CB}_2} = \psi_{\mathscr{AB}_2}\theta_{\mathscr{CB}_1} \neq 0 \textit{ implies } \psi_{\mathscr{AB}} = \alpha_{\mathscr{A}}\beta_{\mathscr{B}},\ \theta_{\mathscr{CB}} = \gamma_{\mathscr{C}}\beta_{\mathscr{B}} \textit{ for some } \alpha_{\mathscr{A}}, \beta_{\mathscr{B}}, \gamma_{\mathscr{C}}. \tag{3.5.4}$$

More particularly,

$$\psi_{\mathscr{A}_1\mathscr{B}_1}\psi_{\mathscr{A}_2\mathscr{B}_2} = \psi_{\mathscr{A}_1\mathscr{B}_2}\psi_{\mathscr{A}_2\mathscr{B}_1} \neq 0 \textit{ implies } \psi_{\mathscr{AB}} = \alpha_{\mathscr{A}}\beta_{\mathscr{B}} \textit{ for some } \alpha_{\mathscr{A}}, \beta_{\mathscr{B}} \tag{3.5.5}$$

Again, if the composite index $\mathscr{D}$ is vacuous, we get:

$$\psi_{\mathscr{AB}}\phi_{\mathscr{C}} = \chi_{\mathscr{A}}\theta_{\mathscr{BC}} \neq 0 \textit{ implies } \psi_{\mathscr{AB}} = \chi_{\mathscr{A}}\zeta_{\mathscr{B}},\ \theta_{\mathscr{BC}} = \zeta_{\mathscr{B}}\phi_{\mathscr{C}} \textit{ for some } \zeta_{\mathscr{B}}. \tag{3.5.6}$$

For, since $\rho_{\mathscr{D}}$ is now a scalar (and non-zero), we can set $\zeta_{\mathscr{B}} = \rho^{-1}\beta_{\mathscr{B}}$ in (3.5.3). As an even more special case we can allow $\mathscr{B}$ to be vacuous also. This gives (3.5.2) again. If we specialize to $\mathscr{A} = \mathscr{B}$, $\psi_{\mathscr{A}} = \chi_{\mathscr{A}}$ and $\phi_{\mathscr{A}} = \theta_{\mathscr{A}}$ in (3.5.2) we obtain a special case of (3.5.4):

$$\psi_{[\mathscr{A}_1}\phi_{\mathscr{A}_2]} = 0 \textit{ implies } \psi_{\mathscr{A}} = \kappa\phi_{\mathscr{A}} \textit{ for some } \kappa \in \mathfrak{S}, \text{ or } \phi_{\mathscr{A}} = 0. \tag{3.5.7}$$

(3.5.8) PROPOSITION

*The following three conditions on* $\lambda_{\mathscr{AB}}{}^{\mathscr{Q}}$ *are equivalent*:

(i) $\lambda_{\mathscr{AB}}{}^{\mathscr{Q}}\xi_{\mathscr{Q}}$ *has the form* $\rho_{\mathscr{A}}\zeta_{\mathscr{B}}$ *for each* $\xi_{\mathscr{Q}} \in \mathfrak{S}_{\mathscr{Q}}$,
(ii) $\lambda_{\mathscr{A}_1[\mathscr{B}_1}{}^{(\mathscr{Q}_1}\lambda_{|\mathscr{A}_2|\mathscr{B}_2]}{}^{\mathscr{Q}_2)} = 0$,
(iii) $\lambda_{\mathscr{AB}}{}^{\mathscr{Q}}$ *has either the form* $\alpha_{\mathscr{A}}\phi_{\mathscr{B}}{}^{\mathscr{Q}}$ *or the form* $\theta_{\mathscr{B}}{}^{\mathscr{Q}}\beta_{\mathscr{A}}$.

*Proof*: Note that (ii) can be written out as

$$\lambda_{\mathscr{A}_1\mathscr{B}_1}{}^{\mathscr{Q}_1}\lambda_{\mathscr{A}_2\mathscr{B}_2}{}^{\mathscr{Q}_2} + \lambda_{\mathscr{A}_2\mathscr{B}_2}{}^{\mathscr{Q}_1}\lambda_{\mathscr{A}_1\mathscr{B}_1}{}^{\mathscr{Q}_2} - \lambda_{\mathscr{A}_1\mathscr{B}_2}{}^{\mathscr{Q}_1}\lambda_{\mathscr{A}_2\mathscr{B}_1}{}^{\mathscr{Q}_2} - \lambda_{\mathscr{A}_2\mathscr{B}_1}{}^{\mathscr{Q}_1}\lambda_{\mathscr{A}_1\mathscr{B}_2}{}^{\mathscr{Q}_2} = 0. \tag{3.5.9}$$

Now assume that (i) holds. Then we have

$$\lambda_{\mathscr{A}_1\mathscr{B}_1}{}^{\mathscr{Q}_1}\xi_{\mathscr{Q}_1}\lambda_{\mathscr{A}_2\mathscr{B}_2}{}^{\mathscr{Q}_2}\xi_{\mathscr{Q}_2} = \rho_{\mathscr{A}_1}\zeta_{\mathscr{B}_1}\rho_{\mathscr{A}_2}\zeta_{\mathscr{B}_2}, \tag{3.5.10}$$

which is symmetric in $\mathscr{B}_1, \mathscr{B}_2$. Thus

$$\lambda_{\mathscr{A}_1[\mathscr{B}_1}{}^{\mathscr{Q}_1}\lambda_{|\mathscr{A}_2|\mathscr{B}_2]}{}^{\mathscr{Q}_2}\xi_{\mathscr{Q}_1}\xi_{\mathscr{Q}_2} = 0. \tag{3.5.11}$$

This holds for all $\xi_{\mathscr{Q}} \in \mathfrak{S}_{\mathscr{Q}}$, so by (3.3.23) we obtain relation (ii). Conversely suppose (ii) holds. Then (3.5.11) holds for any $\xi_{\mathscr{Q}} \in \mathfrak{S}_{\mathscr{Q}}$. This is a relation of the form given in (3.5.3) where $\psi_{\mathscr{A}\mathscr{B}} = \lambda_{\mathscr{A}\mathscr{B}}{}^{\mathscr{Q}}\xi_{\mathscr{Q}} = \phi_{\mathscr{A}\mathscr{B}} = \chi_{\mathscr{A}\mathscr{B}} = \theta_{\mathscr{A}\mathscr{B}}(\mathscr{C} = \mathscr{A}, \mathscr{D} = \mathscr{B})$, whence (i) follows. Thus (i) and (ii) are equivalent. It is obvious that (iii) implies (i). Now suppose (i) holds with $\lambda_{\mathscr{A}\mathscr{B}}{}^{\mathscr{Q}}\xi_{\mathscr{Q}} = \rho_{\mathscr{A}}\zeta_{\mathscr{B}}, \lambda_{\mathscr{A}\mathscr{B}}{}^{\mathscr{Q}}\eta_{\mathscr{Q}} = \sigma_{\mathscr{A}}\mu_{\mathscr{B}}$. Since (ii) also holds, we may transvect (3.5.9) with $\xi_{\mathscr{Q}_1}\eta_{\mathscr{Q}_2}$ to obtain

$$(\rho_{\mathscr{A}_1}\sigma_{\mathscr{A}_2} - \rho_{\mathscr{A}_2}\sigma_{\mathscr{A}_1})(\zeta_{\mathscr{B}_1}\mu_{\mathscr{B}_2} - \zeta_{\mathscr{B}_2}\mu_{\mathscr{B}_1}) = 0. \tag{3.5.12}$$

By (3.5.1), one or the other of these factors must vanish. If it is the first factor, then the part of

$$\lambda_{\mathscr{A}_1\mathscr{B}_1}{}^{\mathscr{Q}_1}\xi_{\mathscr{Q}_1}\lambda_{\mathscr{A}_2\mathscr{B}_2}{}^{\mathscr{Q}_2}\eta_{\mathscr{Q}_2} \tag{3.5.13}$$

skew in $\mathscr{A}_1, \mathscr{A}_2$ must vanish; if it is the second factor, then the part skew in $\mathscr{B}_1, \mathscr{B}_2$ must vanish. Given $\xi_{\mathscr{Q}}$, the $\eta_{\mathscr{Q}}$s for which each of these holds must form a linear space. Since the union of these linear spaces is the whole of $\mathfrak{S}_{\mathscr{Q}}$, one or other of them must be the whole of $\mathfrak{S}_{\mathscr{Q}}$; so (3.5.13) is skew in $\mathscr{A}_1, \mathscr{A}_2$ for *all* $\eta_{\mathscr{Q}} \in \mathfrak{S}_{\mathscr{Q}}$ or else it is skew in $\mathscr{B}_1, \mathscr{B}_2$ for *all* $\eta_{\mathscr{Q}} \in \mathfrak{S}_{\mathscr{Q}}$. The same holds for $\xi_{\mathscr{Q}}$. Thus, either

$$\lambda_{\mathscr{A}_1\mathscr{B}_1}{}^{\mathscr{Q}_1}\lambda_{\mathscr{A}_2\mathscr{B}_2}{}^{\mathscr{Q}_2} \tag{3.5.14}$$

is symmetric in $\mathscr{A}_1, \mathscr{A}_2$ or else it is symmetric in $\mathscr{B}_1, \mathscr{B}_2$. The required form (iii) then follows from (3.5.4).

(3.5.15) PROPOSITION

$\psi_{(\mathscr{A}_1\ldots\mathscr{A}_r}{}^{\mathscr{B}}\phi_{\mathscr{A}_{r+1}\ldots\mathscr{A}_{r+s})}{}^{\mathscr{C}} = 0$ *implies either* $\psi_{(\mathscr{A}_1\ldots\mathscr{A}_2)}{}^{\mathscr{B}} = 0$ *or* $\phi_{(\mathscr{A}_1\ldots\mathscr{A}_s)}{}^{\mathscr{C}} = 0$.

*Proof*: The result follows from (3.3.22) and (3.5.1) as applied to the expression $\psi_{\mathscr{A}_1\ldots\mathscr{A}_r}{}^{\mathscr{B}}\xi^{\mathscr{A}_1}\ldots\xi^{\mathscr{A}_r}\phi_{\mathscr{A}_{r+1}\ldots\mathscr{A}_{r+s}}{}^{\mathscr{C}}\xi^{\mathscr{A}_{r+1}}\ldots\xi^{\mathscr{A}_{r+s}}$.

The results of this section obtained so far hold for systems of any dimension. There are, however, some special results which depend essentially on the two-dimensionality of spin-space. For example, because by (2.5.23) a spinor contraction is equivalent to an anti-symmetrization, we have by (3.5.4):

*If* $\psi_{\mathscr{A}B}, \theta_{\mathscr{C}B} \neq 0$, *then* $\psi_{\mathscr{A}}{}^{B}\theta_{\mathscr{C}B} = 0$ *implies* $\psi_{\mathscr{A}B} = \alpha_{\mathscr{A}}\beta_B, \theta_{\mathscr{C}B} = \gamma_{\mathscr{C}}\beta_B$ *for some* $\beta_B$. (3.5.16)

More particularly (since if $\mathscr{C}$ is vacuous we can set $\chi_{\mathscr{A}} = \gamma^{-1}\alpha_{\mathscr{A}}$ in (3.5.16)), we have:

*If* $\lambda_B \neq 0$, *then* $\psi_{\mathscr{A}B}\lambda^B = 0$ *implies* $\psi_{\mathscr{A}B} = \chi_{\mathscr{A}}\lambda_B$ *for some* $\chi_{\mathscr{A}}$. (3.5.17)

More particularly still, we obtain (2.5.56) when $\mathscr{A}$, also, is vacuous.

*Principal null directions*

In addition to depending on the two-dimensionality of spin-space, the following result uses the fact that the division ring $\mathfrak{C}$ of complex numbers is algebraically closed.

(3.5.18) PROPOSITION

*If* $\phi_{AB...L} = \phi_{(AB...L)} \neq 0$, *then*

$$\phi_{AB...L} = \alpha_{(A}\beta_B \cdots \lambda_{L)}$$

*for some* $\alpha_A, \beta_A, \ldots, \lambda_A \in \mathfrak{S}_A$. *Furthermore, this decomposition is unique up to proportionality or reordering of the factors.*

*Proof*: Choose a spin-frame $o^A$, $\iota^A$ such that $0 \neq \phi_{11...1} = \phi_{AB...L}\iota^A\iota^B \ldots \iota^L$. (This is clearly possible, by (3.3.22).) Let $\xi^A \in \mathfrak{S}^A$ have components $\xi^0 = 1$, $\xi^1 = z$. Then, if $\phi_{AB...L}$ has $n$ indices,

$$\begin{aligned}\phi_{AB...L}\xi^A\xi^B \ldots \xi^L &= \phi_{00...0} + nz\phi_{10...0} + \cdots + z^n\phi_{11...1} \\ &= (\alpha_0 + z\alpha_1)(\beta_0 + z\beta_1)\ldots(\lambda_0 + z\lambda_1), \end{aligned} \tag{3.5.19}$$

by the 'fundamental theorem of algebra', the $n$ factors being unique up to proportionality or reorderings. Regarding $\alpha_{\mathbf{A}}, \beta_{\mathbf{A}}, \ldots, \lambda_{\mathbf{A}}$ as the components of spinors $\alpha_A, \beta_A, \ldots, \lambda_A$, we get $(\alpha_0 + z\alpha_1) = \alpha_{\mathbf{A}}\xi^{\mathbf{A}} = \alpha_A\xi^A$, etc., so

$$\phi_{AB...L}\xi^A\xi^B \ldots \xi^L = (\alpha_A\xi^A)(\beta_B\xi^B)\ldots(\lambda_L\xi^L). \tag{3.5.20}$$

Thus

$$\{\phi_{AB...L} - \alpha_{(A}\beta_B \ldots \lambda_{L)}\}\xi^A\xi^B \ldots \xi^L = 0, \tag{3.5.21}$$

for any $\xi^A$ which is normalized so that $\xi^0 = 1$. But owing to the homogeneity of (3.5.21) in $\xi^A$, it is clear that this normalization is irrelevant, so (3.5.21) holds for *all* $\xi^A$. The result then follows from (3.3.22).

The expression of a totally symmetric spinor $\phi_{A...L} \neq 0$ as a symmetrized product of one-index spinors $\alpha_A, \ldots, \lambda_L$ is called its *canonical decomposition*. Any spinor $\alpha_A, \ldots, \lambda_L$ arising in this way is called a *principal spinor*. Any non-zero multiple of a principal spinor is again a principal spinor. The flagpole directions corresponding to the various principal spinors are called *principal null directions* (PND), and the corresponding null vectors are called *principal null vectors*. Each PND is thus described by a proportionality class of principal spinors. The symmetric $n$-index spinor $\phi_{A...L}(\neq 0)$ uniquely defines the unordered set of $n$ PNDs where, however, multiplicities may occur among these directions. We say that a principal spinor or PND is *k-fold* if it arises from a term of multiplicity $k$ in the factorization (3.5.19), and thus occurs $k$ times (up to proportionality) in the canonical decomposition. The sum of multiplicities of the PND is always equal to $n$.

Any non-zero symmetric spinor $\phi_{A...L}$ is itself determined up to a complex factor by its PNDs. Such a spinor $\phi_{...}$ will exist for any preassigned set of principal null directions with preassigned multiplicities. Note that symmetric $n$-index spinors have $n+1$ complex (i.e., $2n+2$ real) degrees of freedom, since there are $n+1$ independent components $\phi_{00...0}$, $\phi_{10...0}, \ldots, \phi_{11...1}$. This is consistent with the fact that each PND can be specified by one complex number (e.g., $\xi^1/\xi^0$), giving $n$ complex parameters for all the null directions, and there is one final complex parameter which determines the overall multiplier for $\phi_{A...L}$.

We observe from (3.5.20) that if $\xi^A \neq 0$, then

$$\phi_{AB...L}\xi^A\xi^B \ldots \xi^L = 0 \tag{3.5.22}$$

if and only if $\xi^A$ is a principal spinor (*cf.* (2.5.56)). We can say more in the case of a multiple PND. Suppose $\alpha_A$ is a $k$-fold principal spinor:

$$\phi_{AB...DE...L} = \alpha_{(A}\alpha_B \ldots \alpha_D \eta_E \ldots \lambda_{L)}, \tag{3.5.23}$$

so that $\alpha_A$ occurs $k$ times on the right, none of the spinors $\eta_A, \ldots, \lambda_A$ being proportional to $\alpha_A$. Then expanding the symmetrization in (3.5.23) and transvecting with the product $\alpha^E \ldots \alpha^L$ of $n-k$ $\alpha$s, we get

$$\phi_{AB...DE...L}\alpha^E \ldots \alpha^L = \kappa\alpha_A\alpha_B \ldots \alpha_D, \tag{3.5.24}$$

where

$$\kappa = \frac{k!(n-k)!}{n!}(\eta_E\alpha^E) \ldots (\lambda_L\alpha^L) \neq 0. \tag{3.5.25}$$

If, on the other hand, we transvect (3.5.23) with $n-k+1$ $\alpha$s it is clear that the expression vanishes. Thus:

(3.5.26) PROPOSITION

*A necessary and sufficient condition that $\xi_A \neq 0$ be a $k$-fold principal spinor of the non-vanishing symmetric spinor $\phi_{AB...L}$ is that*

$$\phi_{A...GH...L}\xi^H \ldots \xi^L$$

*should vanish if $n-k+1$ $\xi$s are transvected with $\phi_{A...L}$ but not if only $n-k$ $\xi$s are transvected with $\phi_{A...L}$.*

As a corollary we have:

(3.5.27) PROPOSITION

*If $\xi^A \neq 0$, $\phi^{\mathscr{H}}_{A...G} = \phi^{\mathscr{H}}_{(A...G)}$, and*

$$\xi^A \ldots \xi^C\xi^D\phi^{\mathscr{H}}_{A...CDE...G} = 0,$$

*then there exists a* $\psi^{\mathscr{H}}_{A\ldots C}$ *such that*

$$\phi^{\mathscr{H}}_{A\ldots CDE\ldots G} = \psi^{\mathscr{H}}_{(A\ldots C}\xi_D\xi_E\cdots\xi_{G)}.$$

The proof is immediate from (3.5.26) if we choose a basis and replace $\mathscr{H}$ with numerical indices.

We shall say of a symmetric spinor that it is *null* if all its PNDs coincide. The following is a criterion for nullity.

(3.5.28) PROPOSITION

*The symmetric spinor* $\phi_{AB\ldots L}$ *is null if and only if*

$$\phi_{AB\ldots L}\phi^{AB_0\ldots L_0} = 0.$$

The necessity of this condition is immediate. To establish its sufficiency, choose a spinor $\eta_{B_0\ldots L_0}$ such that

$$\xi^A = \phi^{AB_0\ldots L_0}\eta_{B_0\ldots L_0} \neq 0.$$

Then the condition yields

$$\phi_{AB\ldots L}\xi^A = 0,$$

which, by (3.5.27), implies that there exists a scalar $\psi$ such that

$$\phi_{A\ldots L} = \psi\xi_{(A}\cdots\xi_{L)},$$

as required.

Note that in the case of a two-index symmetric spinor $\varphi_{AB} = \alpha_{(A}\beta_{B)}$ we have, by direct calculation,

$$\varphi_{AB}\varphi^{AB} = -\tfrac{1}{2}(\alpha_A\beta^A)^2, \qquad (3.5.29)$$

so that evidently an alternative criterion for nullity is the vanishing of $\varphi_{AB}\varphi^{AB}$.

In some situations it is useful to talk about PNDs or principal spinors of $\phi_{A\ldots L}$ even when $\phi_{A\ldots L} = 0$. The convention will be that then *every* non-zero $\xi_A$ must be regarded as a principal spinor of $\phi_{A\ldots L}$ and *every* null direction as a PND. (So, strictly speaking, a zero spinor is, in this sense, not null.)

It may be remarked that no simple analogue of (3.5.18) – which allows us to classify spinors $\phi_{\ldots}$ according to the multiplicities of their PNDs – exists for symmetric spinors with both primed and unprimed indices. A classification scheme for such spinors will be given in §8.7.

*Simplicity of skew tensors*

We end this section with a result on anti-symmetric tensors in an arbitrary $n$-dimensional vector space $\mathfrak{S}^\alpha$, which bears a similarity to some of the above results on symmetric objects.

(3.5.30) PROPOSITION

*If $F_{\alpha\beta\ldots\rho}$ is skew in all its p indices, then*

$$F_{[\alpha\beta\ldots\rho}F_{\sigma]\tau\ldots\omega} = 0 \Leftrightarrow F_{\alpha\beta\ldots\rho} = a_{[\alpha}b_\beta \ldots r_{\rho]}$$

*for some $a_\alpha, b_\beta, \ldots, r_\rho$.*

(An $F_{\ldots}$ which is such a skew product of vectors is called *simple*.) The *necessity* of the above condition for simplicity is immediate: expand the second $F_{\ldots}$ in terms of $a_\sigma, b_\tau, \ldots$, and then each term in the sum must vanish because $a_{[\alpha} \ldots r_\rho a_{\sigma]} = 0$, etc. To establish the *sufficiency* of the condition, note first that it can be re-written as

$$F_{\alpha\beta\ldots\rho}F_{\sigma\tau\ldots\omega} = pF_{\sigma[\beta\ldots\rho}F_{\alpha]\tau\ldots\omega}. \tag{3.5.31}$$

For we need only expand the anti-symmetrization in the condition and separate the terms according to which $F_{\ldots}$ possesses the index $\sigma$. Transvecting (3.5.31) with $u^\sigma u^\tau$ annihilates the left-hand side and shows that the $(p-1)$-index tensor

$$u^\sigma F_{\sigma\beta\ldots\rho}$$

satisfies the same condition as $F_{\ldots}$ itself. Now we shall assume that the condition in (3.5.30) implies simplicity for $(p-1)$-index tensors (which is evidently the case when $p-1=1$) and deduce that it then also implies simplicity for $p$-index tensors. Thus, by hypothesis, whenever $u^\sigma \neq 0$,

$$u^\sigma F_{\sigma\beta\ldots\rho} = b_{[\beta} \ldots r_{\rho]},$$

for some $b_\beta, \ldots, r_\rho$. Now choose $u^\sigma$ and $G^{\tau\ldots\omega}$ such that

$$w := u^\sigma G^{\tau\ldots\omega}F_{\sigma\tau\ldots\omega} \neq 0,$$

and transvect (3.5.31) with $u^\sigma G^{\tau\ldots\omega}$. This yields

$$F_{\alpha\beta\ldots\rho} = a_{[\alpha}b_\beta \ldots r_{\rho]},$$

as required, where

$$a_\alpha = \frac{p}{w}F_{\alpha\tau\ldots\omega}G^{\tau\ldots\omega}.$$

So the proposition is established by induction.

One easily sees that the condition in (3.5.30) is equivalent to

$$*F^{\delta\ldots\rho\sigma}F_{\sigma\tau\ldots\omega} = 0, \tag{3.5.32}$$

where $^*F^{\cdots}$ (which has $n-p$ indices if $F_{\ldots}$ has $p$) is defined analogously to (3.4.30), (3.4.21) by

$$^*F^{\delta\ldots\sigma} = \frac{1}{p!}\varepsilon^{\delta\ldots\sigma\alpha\ldots\gamma}F_{\alpha\ldots\gamma}. \tag{3.5.33}$$

(An 'alternating tensor' $\varepsilon^{\cdots}$ in $n$ dimensions has $n$ indices, is non-zero, and skew *cf.* (2.3.4).) Since (3.5.32) is symmetrical in $^*F^{\cdots}$ and $F_{\ldots}$ (apart from the manifestly immaterial positions of the indices), we have the following

(3.5.34) PROPOSITION

*$F_{\alpha\ldots\gamma}$ is simple if and only if its dual $^*F^{\delta\ldots\sigma}$ is simple.*

The above results (3.5.30)–(3.5.34) hold in $n$ dimensions and for skew tensors with any number of indices ($\leqslant n$). But we shall have occasion to be especially interested in *bivectors* $F_{ab}$ in *four* dimensions, and then there are several additional criteria for simplicity. In fact, we have

(3.5.35) PROPOSITION

*In four dimensions the bivector $F_{ab}$ is simple if and only if any of the following conditions holds:*

$$\text{(i)}\ F_{[ab}F_{cd]} = 0, \quad \text{(ii)}\ F_{ab}{}^*F^{ab} = 0, \quad \text{(iii)}\ \det(F_{ab}) = 0.$$

*Proof*: One easily sees that

$$F_{[ab}F_{c]d} = F_{[ab}F_{cd]} = q\eta_{abcd} \tag{3.5.36}$$

for some scalar $q$ and an alternating tensor $\eta_{\ldots}$. The first of these identities, in conjunction with (3.5.30), establishes (i). Condition (ii) results on transvecting the second identity in (3.5.36) with $\varepsilon^{abcd}$, provided $\eta_{abcd}\varepsilon^{abcd} \neq 0$. But that can be verified directly by going to a particular frame. The last condition, (iii), results from the well-known theorem stating that the determinant of a skew matrix is a perfect square, and, in fact, in our specific case,

$$\det(F_{ab}) = \tfrac{1}{16}(F_{ab}{}^*F^{ab})^2. \tag{3.5.37}$$

So (iii) is equivalent to (ii), and the proposition is established.

We may point out that Proposition (3.5.30) would be false if taken to refer to *tensor fields* rather than to *tensors at a point*. A remarkably simple counter-example supporting this assertion is provided by the bivector whose components in ordinary Euclidean space, referred to Cartesian coordinates $x$, $y$, $z$, are given by

$$F_{\mathbf{ab}} = \begin{bmatrix} 0 & z & -y \\ -z & 0 & x \\ y & -x & 0 \end{bmatrix}.$$

This is the dual of the position vector $r^{\mathbf{a}} = (x, y, z)$ and thus pointwise simple, by (3.5.34), since $r^a$ is trivially simple. But by (3.5.32) $r^a F_{ab} = 0$. So $F_{ab} = U_{[a}V_{b]}$ would imply $r^a U_a = r^a V_a = 0$, i.e., $U_a$ and $V_a$ would be normal to $r^a$. Then on each sphere of constant radius $U_a$ and $V_a$ would both constitute nowhere-vanishing tangent vector fields. But for topological reasons we know that such fields do not exist ('fixed point' theorem). This shows that the result would be false even in an *arbitrarily* small neighbourhood of the origin.

It should be borne in mind that all other numbered results of this section would also fail, in some analogous way, for fields rather than for tensors or spinors at one point.

## 3.6 Lorentz transformations

As an application of some of the results of the preceding section, we shall here investigate the structure of Lorentz transformations. This constitutes a somewhat different approach from that of §§1.2 and 1.3. There will be some overlap in the results, but this should be helpful in establishing a link between the two points of view.

We give, among other things, a direct proof of the key result (1.2.27) of §1.2 that *to every restricted Lorentz transformation* $L_a{}^b : V^a \mapsto W^b$ *there correspond exactly two spin transformations* $\pm T_A{}^B : \xi^A \mapsto \pm \eta^B$ *and vice versa.* But our discussion goes further than this, in that we treat improper Lorentz transformations as well as proper ones. (As in §3.5, we are concerned with spinors and tensors at a single point only. Thus, $\mathfrak{S}$ and $\mathfrak{T}$ are the division rings of complex and real numbers, respectively.)

Our notation allows us to express the above active transformations in the form

$$L_a{}^b V^a = W^b, \quad T_A{}^B \xi^A = \eta^B. \tag{3.6.1}$$

Note that we require $L_a{}^b \in \mathfrak{T}_a^b$ and $T_A{}^B \in \mathfrak{S}_A^B$. The required relation between these transformations is that they give the same result when applied to each $V^a \in \mathfrak{T}^a$, the effect of the spin transformation being

$$T_A{}^B \bar{T}_{A'}{}^{B'} V^{AA'} = W^{BB'}. \tag{3.6.2}$$

Thus if the elements $L_a{}^b$ and $T_A{}^B \bar{T}_{A'}{}^{B'}$ are to give the same map from $\mathfrak{T}^a$ to $\mathfrak{T}^b$ we require

$$L_a{}^b = T_A{}^B \bar{T}_{A'}{}^{B'}. \tag{3.6.3}$$

What we must show, therefore, is that if $L_a{}^b$ is a restricted Lorentz transformation then $L_a{}^b$ always 'splits' according to (3.6.3), where $T_A{}^B$ is a spin transformation defined uniquely up to sign; and conversely that if $T_A{}^B$ is a spin transformation then the $L_a{}^b$ defined by (3.6.3) is always a restricted Lorentz transformation. The component version of (3.6.3), relative to the standard choice of frames, is (1.2.26),

The condition for $L_a{}^b$ to be a Lorentz transformation is that it be real and leave the metric invariant:

$$\bar{L}_a{}^b = L_a{}^b, \tag{3.6.4}$$

$$g_{ab} = L_a{}^c L_b{}^d g_{cd}. \tag{3.6.5}$$

It is restricted if and only if it belongs to the same continuous family as does the identity $g_a{}^b$.

[A linear transformation between vector spaces $\mathfrak{S}^\alpha, \mathfrak{S}^\beta$ may always be expressed in the form $P^\beta_\alpha X^\alpha = Y^\beta$, so the map $P^\beta_\alpha : \mathfrak{S}^\alpha \to \mathfrak{S}^\beta$ is given by contracted product (*cf.* (2.2.37)). This map induces a linear transformation $\overset{-1}{P}{}^\alpha_\beta : \mathfrak{S}_\alpha \to \mathfrak{S}_\beta$ where $P^\beta_\alpha \overset{-1}{P}{}^\gamma_\beta = \delta^\gamma_\alpha = \overset{-1}{P}{}^\beta_\alpha P^\gamma_\beta$ and, hence, a linear transformation $P^\lambda_\alpha \ldots P^\nu_\gamma \overset{-1}{P}{}^\rho_\phi \ldots \overset{-1}{P}{}^\tau_\psi : \mathfrak{S}^{\alpha\ldots\gamma}_{\rho\ldots\tau} \to \mathfrak{S}^{\lambda\ldots\nu}_{\phi\ldots\psi}$. If $A^{\alpha\ldots\gamma}_{\rho\ldots\tau} \mapsto B^{\lambda\ldots\nu}_{\phi\ldots\psi}$ we have $P^\lambda_\alpha \ldots P^\nu_\gamma \overset{-1}{P}{}^\rho_\phi \ldots \overset{-1}{P}{}^\tau_\psi A^{\alpha\ldots\gamma}_{\rho\ldots\tau} = B^{\lambda\ldots\nu}_{\phi\ldots\psi}$; equivalently $P^\lambda_\alpha \ldots P^\nu_\gamma A^{\alpha\ldots\gamma}_{\rho\ldots\tau} = P^\phi_\rho \ldots P^\psi_\tau B^{\lambda\ldots\nu}_{\phi\ldots\psi}$. Thus, $A^{\cdots}_{\cdots}$ is *invariant* under $P^\beta_\alpha$ iff $P^\lambda_\alpha \ldots P^\nu_\gamma A^{\alpha\ldots\gamma}_{\rho\ldots\tau} = P^\phi_\rho \ldots P^\psi_\tau A^{\lambda\ldots\nu}_{\phi\ldots\psi}$.]

The condition that $T_A{}^B$ be a spin-transformation is that it should have unit determinant. This can be stated in the form

$$\varepsilon_{AB} = T_A{}^C T_B{}^D \varepsilon_{CD}. \tag{3.6.6}$$

For, the right-hand side is skew in $A$, $B$ and is therefore, by (2.5.23), proportional to $\varepsilon_{AB}$, the factor of proportionality being

$$\tfrac{1}{2} T_A{}^C T_B{}^D \varepsilon_{CD} \varepsilon^{AB} = \det(T_A{}^C). \tag{3.6.7}$$

Alternatively, we may simply examine (3.6.6) in component form (*cf.* (2.5.70)). Condition (3.6.6) states that the ε-spinor is invariant under spin transformations.

Now, suppose we are given a spin transformation $T_A{}^B$, and $L_a{}^b$ is defined according to (3.6.3). Clearly $L_a{}^b$ is then real and

$$g_{ab} = \varepsilon_{AB}\varepsilon_{A'B'} = T_A{}^C T_B{}^D \bar{T}_{A'}{}^{C'} \bar{T}_{B'}{}^{D'} \varepsilon_{CD}\varepsilon_{C'D'} = L_a{}^c L_b{}^d g_{cd}, \tag{3.6.8}$$

so $L_a{}^b$ is a Lorentz transformation. Moreover, $L_a{}^b$ is *restricted* because $T_A{}^B$ is continuous with the identity spin transformation* $\varepsilon_A{}^B$; whence $L_a{}^b$ is continuous with $\varepsilon_A{}^B \varepsilon_{A'}{}^{B'} = g_a{}^b$, the identity Lorentz transformation. Alternatively, the fact that $L_a{}^b$ as given by (3.6.3) is restricted will be a consequence of the discussion to follow.

* See remark (iii) after equation (1.2.26).

Suppose, conversely, that $L_a{}^b$ is a Lorentz transformation. Since it preserves the metric, it must send null vectors into null vectors; indeed, it sends *complex* null vectors into complex null vectors. (For, by (3.6.5), $g_{ab}\chi^a\chi^b = (L_a{}^c\chi^a)(L_b{}^d\chi^b)g_{cd}$ whether $\chi^a$ is real or complex.) By (3.2.6) any complex null vector $\chi^a$ has the form $\kappa^A\xi^{A'}$. Thus

$$L_{AA'}{}^{BB'}\kappa^A\xi^{A'} = \tau^B\eta^{B'}. \tag{3.6.9}$$

With any choice of $\kappa^A$ and $\xi^{A'}$ such an equation holds, so we can apply (3.5.8) to obtain $L_{AA'}{}^{BB'}\kappa^A$ in one or other of the forms

$$L_{AA'}{}^{BB'}\kappa^A = \theta^B_{A'}\psi^{B'} \quad \text{or} \quad \zeta^B\mu_{A'}{}^{B'}. \tag{3.6.10}$$

This must hold for all $\kappa^A$. Indeed, the *same* one of relations (3.6.10) must hold for all $\kappa^A$. Otherwise, by continuity, for some non-zero value of $\kappa^A$ both forms would hold simultaneously, so by (3.5.6) we should have $L_{AA'}{}^{BB'}\kappa^A = \rho_{A'}\zeta^B\psi^{B'}$, whence $L_{AA'}{}^{BB'}(\kappa^A\rho^{A'}) = 0$, violating the non-singularity of the transformation $L_a{}^b$: that Lorentz transformations cannot be singular follows from (3.6.5) alone (*cf.* (3.6.19) below). Applying (3.5.8) again to (3.6.10) we see that $L_{AA'}{}^{BB'}$ must necessarily have one of the following four forms

$$\text{(i) } \omega_{AA'}{}^B\psi^{B'}, \quad \text{(ii) } \theta^B_{A'}\lambda^{B'}_A, \quad \text{(iii) } \phi_A{}^B\mu_{A'}{}^{B'}, \quad \text{(iv) } \zeta^B\nu_{AA'}{}^{B'}. \tag{3.6.11}$$

However, we must reject (i) because $L_{AA'}{}^{BB'}(\bar\psi_B\psi_{B'}) = 0$, and (iv) because $L_{AA'}{}^{BB'}(\zeta_B\bar\zeta_{B'}) = 0$: in each case $L_a{}^b$ would be singular.

This leaves us with (ii) and (iii). The reality (3.6.4) of $L_a{}^b$ gives, respectively,

$$\bar\lambda^B_{A'}\bar\theta^{B'}_A = \theta^B_{A'}\lambda^{B'}_A, \quad \bar\mu_A{}^B\bar\phi_{A'}{}^{B'} = \phi_A{}^B\mu_{A'}{}^{B'}. \tag{3.6.12}$$

Thus, by (3.5.2),

$$\bar\lambda^B_{A'} = \alpha\theta^B_{A'}, \bar\theta^{B'}_A = \alpha^{-1}\lambda^{B'}_A; \bar\mu_A{}^B = \beta\phi_A{}^B, \bar\phi_{A'}{}^{B'} = \beta^{-1}\mu_{A'}{}^{B'}, \tag{3.6.13}$$

whence $\alpha$ and $\beta$ must be real. Absorbing the factor $|\alpha|^{1/2}$ into the definition of $\theta^B_{A'}$ and $|\beta|^{1/2}$ into the definition of $\phi_A{}^B$ we get, according as $\alpha$ is positive or negative in case (ii), or as $\beta$ is positive or negative in case (iii), four different possibilities:

$$L_{AA'}{}^{BB'} = \pm\,\theta^B_{A'}\bar\theta_A{}^{B'}, \tag{3.6.14}$$

$$L_{AA'}{}^{BB'} = \pm\,\phi_A{}^B\bar\phi_{A'}{}^{B'}. \tag{3.6.15}$$

Substituting into (3.6.5). we find that $\det(\theta^B_{A'})(=\tfrac{1}{2}\theta^B_{A'}\theta^D_{C'}\varepsilon^{A'C'}\varepsilon_{BD})$ and $\det(\phi_A{}^B)(=\tfrac{1}{2}\phi_A{}^B\phi_C{}^D\varepsilon^{AC}\varepsilon_{BD})$ both have unit modulus. If we normalize these determinants to unity:

$$\det(\theta^B_{A'}) = 1, \quad \det(\phi_A{}^B) = 1 \tag{3.6.16}$$

(absorbing the phase factor into $\theta_{A'}^{B}$ or $\phi_{A}{}^{B}$), we obtain $\theta_{A'}^{B}$ uniquely up to sign in each case (3.6.14), and $\phi_{A}{}^{B}$ uniquely up to sign in each case (3.6.15).

By transvecting each of (3.6.14), (3.6.15) with a future-pointing null vector $\kappa^{A}\bar{\kappa}^{A'}$ we see that the result is future-pointing if and only if the positive sign is chosen in each of (3.6.14), (3.6.15) (see (3.2.2), (3.2.4)). Thus, the *orthochronous* Lorentz transformations are those of the form (3.6.14) and (3.6.15) in which the *positive* sign holds. Those for which the negative sign holds involve a reversal of time-sense. To see which of (3.6.14), (3.6.15) are *proper*, we may examine the effect of $L_{a}{}^{b}$ on the alternating tensor $e_{abcd}$. (A Minkowski tetrad is proper or improper according as $e_{abcd}t^{a}x^{b}y^{c}z^{d}$ is $+1$ or $-1$, *cf.* (3.3.37); thus $e_{abcd}$ defines the orientation of the Minkowski vector space.) We have

$$e_{abcd} = \pm L_{a}{}^{p}L_{b}{}^{q}L_{c}{}^{r}L_{d}{}^{s}e_{pqrs} = \pm \det(L_{p}{}^{q})e_{abcd}, \tag{3.6.17}$$

the positive sign holding if and only if $L_{p}{}^{q}$ is proper. Substituting (3.6.14) and (3.6.15) into (3.6.17) and using (3.3.31) and the equivalent form

$$\varepsilon_{A'B'} = \theta_{A'}{}^{C}\theta_{B'}{}^{D}\varepsilon_{CD}, \quad \varepsilon_{AB} = \phi_{A}{}^{C}\phi_{B}{}^{D}\varepsilon_{CD} \tag{3.6.18}$$

of (3.6.16), we obtain directly the fact that it is (3.6.15) (with either sign) which is proper and (3.6.14) (with either sign) which is improper. Thus the restricted Lorentz transformations are those of the form (3.6.15) with positive sign. Setting $T_{A}{}^{B} = \phi_{A}{}^{B}$, the required form (3.6.3) is obtained.

Alternatively we may see this from the fact that the transformations (3.6.14) cannot be continuous with the identity Lorentz transformation $\varepsilon_{A}{}^{B}\varepsilon_{A'}{}^{B'}$. For, any continuous path of $L_{AA'}{}^{BB'}$s beginning with (3.6.14) and ending at $\varepsilon_{A}{}^{B}\varepsilon_{A'}{}^{B'}$, would at some point have to have simultaneously the forms (3.6.14) and (3.6.15). By (3.5.3), $L_{AA'}{}^{BB'}$ would then be an outer product of four one-index spinors and hence singular. The restricted $L_{a}{}^{b}$s, being those continuous with the identity, and the negatives of the restricted $L_{a}{}^{b}$s, form the class of proper $L_{a}{}^{b}$s. These must therefore be the $L_{a}{}^{b}$s of the class (3.6.15).

We can examine some of the structure of Lorentz transformations in the light of the spinor representations that we have found. We remark, first, that by raising the index $b$ and lowering $d$ in (3.6.5) we obtain $L_{a}{}^{c}L^{b}{}_{c} = g_{a}{}^{b}$ which tells us that the *inverse* $\overset{-1}{L}{}_{a}{}^{b}$ of $L_{a}{}^{b}$ is given by

$$\overset{-1}{L}{}_{a}{}^{b} = L^{b}{}_{a}. \tag{3.6.19}$$

Applying the same procedure to (3.6.18) we get

$$\theta_{A'}^{C}\theta_{C}^{B'} = -\varepsilon_{A'}{}^{B'}, \quad \phi_{A}{}^{C}\phi^{B}{}_{C} = -\varepsilon_{A}{}^{B}, \tag{3.6.20}$$

which tells us that the inverses of the maps $\theta_{A'}^{B}: \mathfrak{S}^{A'} \to \mathfrak{S}^{B}$ and $\phi_{A}{}^{B}: \mathfrak{S}^{A} \to$

$\mathfrak{S}^B$ are given by

$$\overset{-1}{\theta}{}_B{}^{A'} = -\theta^{A'}_B, \quad \overset{-1}{\phi}{}_B{}^A = -\phi^A{}_B. \tag{3.6.21}$$

*Improper transformations*

Let us examine the *improper* Lorentz transformations first. These are given by (3.6.14), so each one is characterized by a quantity $\theta^{B'}_A$ subject to (3.6.16); that is to say, by a *complex world-vector* $\theta^a = \theta^{AA'}$ of length $\sqrt{2}$:

$$\theta^a \theta_a = 2 \tag{3.6.22}$$

(since (3.6.16) gives $\frac{1}{2}\theta^{BA'}\theta_{BA'} = 1$). The relation between $\theta^a$ and the Lorentz transformation $L_a{}^b$ is given in spinor terms by (3.6.14):

$$L_{ab} = \pm\, \theta_{BA'} \bar{\theta}_{AB'}. \tag{3.6.23}$$

Since the right-hand side is $\pm\,\theta_a \bar{\theta}_b$ with $A$ and $B$ interchanged we can use the theory of §3.4 to obtain a purely tensor form of this relation:

$$L_{ab} = \pm\, \theta_c \bar{\theta}_d U_{ab}{}^{cd}, \tag{3.6.24}$$

where $U_{ab}{}^{cd}$ is defined in (3.4.57). Writing (3.6.24) out in full we therefore get

$$\pm\, L_{ab} = \theta_{(a} \bar{\theta}_{b)} - \tfrac{1}{2} g_{ab} \theta^c \bar{\theta}_c + \frac{\mathrm{i}}{2} e_{abcd} \theta^c \bar{\theta}^d \tag{3.6.25}$$

as the *general expression for an improper Lorentz transformation*, $\theta^a$ being subject only to (3.6.22). The positive sign in (3.6.25) corresponds to $L_a{}^b$ orthochronous.

Of particular interest are those improper Lorentz transformations which are *involutory* (i.e., equal to their own inverses) since these correspond to space–time reflections in lines or hyperplanes. By (3.6.19), the condition for $\overset{-1}{L}{}_a{}^b = L_a{}^b$ is

$$L_{ab} = L_{ba}, \tag{3.6.26}$$

so (3.6.23) tells us that $L_a{}^b$ is involutory if and only if $\bar{\theta}_a$ is proportional to $\theta_a$. Because of the normalization (3.6.22), this implies one of the following:

(i) $\theta_a$ *is real and timelike*
(ii) $\mathrm{i}\theta_a$ *is real and spacelike.* (3.6.27)

Then (3.6.25) becomes

$$\pm\, L_{ab} = 2V_a V_b - g_{ab} V^c V_c, \tag{3.6.28}$$

where the *real unit vector* $V^a$ is $2^{-\frac{1}{2}}\,\theta^a$ in case (i) and $\mathrm{i}2^{-\frac{1}{2}}\,\theta^a$ in case (ii).

Again, the orthochronous transformations take the positive sign in (3.6.28). When $V^a$ is timelike, $L_a{}^b$ is a 'space-reflection in a point' if orthochronous, i.e., more correctly, a reflection in a timelike line. If not orthochronous, it is a reflection in the orthogonal spacelike hyperplane. When $V^a$ is spacelike, $L_a{}^b$ is a 'space-reflection in a plane' if orthochronous, i.e., more correctly, a reflection in a timelike hyperplane. If not orthochronous, it is a reflection in the orthogonal spacelike line. It may be pointed out that the *two* choices $\pm V^a$ both give the same $L_{ab}$. This is because the sign in the spinor transformation $\eta^{B'} = \theta_A^{B'}\xi^A$ is not defined simply by the effect of $L_a{}^b$ on vectors. However, in the timelike case, there is an invariant distinction between $V^a$ and $-V^a$, since one is future-pointing and other past-pointing.

### *Proper transformations*

We now come to the *proper* Lorentz transformations. These are given by (3.6.15):

$$L_{ab} = \pm\,\phi_{AB}\bar{\phi}_{A'B'}, \tag{3.6.29}$$

where $\phi_A{}^B$ is a spin-transformation, i.e., by (3.6.18), subject to

$$\phi_{AB}\phi^{AB} = 2. \tag{3.6.30}$$

Expressing $\phi_{AB}$ in terms of its symmetric and skew parts we get

$$\phi_{AB} = \mu\varepsilon_{AB} + \psi_{AB}, \tag{3.6.31}$$

where

$$\psi_{AB} = \psi_{BA}. \tag{3.6.32}$$

Substituting (3.6.31) into (3.6.30) we get

$$\mu^2 - \nu^2 = 1, \tag{3.6.33}$$

where

$$\nu^2 := -\tfrac{1}{2}\psi_{AB}\psi^{AB}. \tag{3.6.34}$$

Thus, apart from an ambiguity of sign in the definition of $\mu$ (absent only if $\nu = \pm\,\mathrm{i}$), the spin transformation $\phi_A{}^B$ is uniquely defined by an arbitrary symmetric spinor $\psi_{AB}$.

If we substitute (3.6.31) into (3.6.29), we obtain a decomposition of $L_{ab}$ as follows:

$$\pm\,L_{ab} = pg_{ab} + F_{ab} + T_{ab}, \tag{3.6.35}$$

where $p$, $F_{ab}$ and $T_{ab}$ are real, with $T_{ab}$ trace-free symmetric and $F_{ab}$ skew, given by

$$p = \mu\bar{\mu}, F_{ab} = \bar{\mu}\psi_{AB}\varepsilon_{A'B'} + \mu\bar{\psi}_{A'B'}\varepsilon_{AB}, T_{ab} = \psi_{AB}\bar{\psi}_{A'B'}. \qquad (3.6.36)$$

The tensor $F_{ab}$ bears the same relation to $(2\pi)^{-1}pT_{ab}$ as does an electromagnetic field tensor to its energy tensor (*cf.* (5.2.4)). This fact is actually *sufficient* to ensure the form of (3.6.36), which in turn implies that $L_{ab}$, as given by (3.6.35), is *proportional to* a proper Lorentz transformation. If, in addition, $\det(L_a{}^b) = 1$ (equivalent to the normalization (3.6.33)), then $L_a{}^b$ *is* a proper Lorentz transformation, orthochronous if the positive sign is taken in (3.6.35) and $p > 0$.

According to (3.5.18) we can express $\psi_{AB}$ as a symmetrized product of one-index spinors, say

$$\psi_{AB} = \alpha_A\beta_B + \beta_A\alpha_B, \qquad (3.6.37)$$

so by (3.6.34) we can take

$$\nu = \alpha_A\beta^A. \qquad (3.6.38)$$

We have $\alpha_A\beta_B - \alpha_B\beta_A = \nu\varepsilon_{AB}$, so provided $\nu \neq 0$ we get, from (3.6.31),

$$\phi_{AB} = \left(1 + \frac{\mu}{\nu}\right)\alpha_A\beta_B + \left(1 - \frac{\mu}{\nu}\right)\beta_A\alpha_B. \qquad (3.6.39)$$

If $\nu = 0$, $\alpha_A$ and $\beta_A$ are proportional and we can take $\beta_A = \frac{1}{2}\zeta\alpha_A$, whence

$$\phi_{AB} = \varepsilon_{AB} + \zeta\alpha_A\alpha_B. \qquad (3.6.40)$$

Notice that in each case (3.6.39), (3.6.40), $\alpha_A$ is an *eigenspinor* of $\phi_A{}^B$ in the sense that $\phi_A{}^B\alpha_B$ is a multiple of $\alpha_A$:

$$\phi_A{}^B\alpha_B = (\mu + \nu)\alpha_A, \qquad (3.6.41)$$

the corresponding eigenvalue being $\mu + \nu$ ($= 1$ in case (3.6.40)). In case (3.6.39) $\beta_A$ is also an eigenspinor of $\phi_A{}^B$, with eigenvalue $\mu - \nu$. In case (3.6.40), $\alpha_A$ is (up to proportionality) the only eigenspinor of $\phi_A{}^B$ – in other words, the two eigendirections of $\phi_A{}^B$ become coincident. The PNDs of $\psi_{AB}$ are thus seen to be *the same as the eigendirections of* $\phi_A{}^B$.

As an alternative argument, not assuming the canonical decomposition of $\psi_{AB}$, one can see quite rapidly that, given $\alpha_A$, the spinor $\phi_{AB}$ can be expressed in one or other of the forms (3.6.39), (3.6.40). For, $\phi_A{}^B\alpha_B = \xi\alpha_A$ implies $(\phi_A{}^B - \xi\varepsilon_A{}^B)\alpha_B = 0$; so, by (3.5.17), $\phi_A{}^B - \xi\varepsilon_A{}^B = \gamma_A\alpha^B$ for some $\gamma_A$. If $\gamma_A$ is proportional to $\alpha_A$ we are led to case (3.6.40). Otherwise, we expand $\varepsilon_A{}^B$ in terms of $\alpha_A$ and $\gamma_A$ and obtain the form (3.6.39), with $\gamma_A$ proportional to $\beta_A$.

The significance of the eigenspinors of a spin-transformation lies in the fact that their flagpole directions are the invariant null directions of the corresponding Lorentz transformation. We have, from (3.6.29) and (3.6.41)

$$L_a{}^b U_b = |\mu + \nu|^2 U_a, \tag{3.6.42}$$

where the null vector $U_a$ is given by

$$U_a := \alpha_A \bar{\alpha}_{A'}. \tag{3.6.43}$$

When the two invariant null directions are distinct, we can choose a spin-frame whose flagpole directions are the invariant null directions, e.g.,

$$\iota_A := \alpha_A, \quad o_A := -\nu^{-1}\beta_A. \tag{3.6.44}$$

Then the matrix of the spin-transformation (3.6.39) takes the following form (where, by (3.6.33), $\mu - \nu = (\mu + \nu)^{-1}$):

$$\phi_{\mathbf{A}}{}^{\mathbf{B}} = \begin{pmatrix} \mu + \nu & 0 \\ 0 & (\mu + \nu)^{-1} \end{pmatrix}. \tag{3.6.45}$$

Comparison with (1.2.31) and (1.2.37) shows that this is a 'rotation about the $z$-axis through $\psi$' if $\mu + \nu = e^{i\psi/2}$ with $\psi$ real; it is a 'boost in the $z$-direction with velocity $v$' if $\mu + \nu = (1 - v)^{-\frac{1}{4}}(1 + v)^{\frac{1}{4}}$ with $v$ real; and it is a 'four-screw about the $z$-axis' in the general case (*cf.* (1.3.4)).

When the two invariant null directions coincide we get a *null rotation*. We can then choose our spin-frame so that

$$\iota_A := \alpha_A, \tag{3.6.46}$$

with $o_A$ arbitrary. Then the matrix of the spin-transformation (3.6.40) takes the form

$$\phi_{\mathbf{A}}{}^{\mathbf{B}} = \begin{pmatrix} 1 & -\zeta \\ 0 & 1 \end{pmatrix}. \tag{3.6.47}$$

(If desired, we could scale $\iota_A = \alpha_A$ so that $-\zeta = 1$.) Comparison with (1.3.7) confirms that this is a 'null rotation about the $z$-axis'. Note that the eigenvalue of $\phi_{\mathbf{A}}{}^{\mathbf{B}}$ is unity. Thus any null rotation preserves both the flagpole and the flag plane of any spinor whose flagpole direction is that of the invariant null direction.

The involutory proper Lorentz transformations also have some special interest since these represent reflections in space–time 2-planes. The involutory condition $L_{ab} = L_{ba}$ (*cf.* (3.6.26)) when applied to (3.6.35) tells us that $F_{ab} = 0$, whence either $\psi_{AB} = 0$ or $\mu = 0$. The case $\psi_{AB} = 0$ is uninteresting since $\phi_A{}^B$ then reduces to $\pm\varepsilon_A{}^B$. Thus the general involutory proper Lorentz transformation has the form

$$\pm L_a{}^b = T_a{}^b = \psi_A{}^B \bar{\psi}_{A'}{}^{B'}, \tag{3.6.48}$$

where, as remarked earlier (and as we shall see in detail later, *cf.* §5.2), the trace-free symmetric tensor $T_{ab}$ has the form of an electromagnetic energy tensor. The choice of the plus sign in (3.6.48) ensures that $L_a{}^b$ is orthochronous, in which case it is a 'space reflection in a line', i.e., more correctly, a reflection in a space–time timelike 2-plane. If not orthochronous, $L_a{}^b$ is a reflection in a spacelike 2-plane. The timelike 2-plane concerned is the plane spanned by the two invariant null directions (i.e., by the flagpoles of $\alpha^A$ and $\beta^A$) and the other is the orthogonal complement of this.

Since $\mu = 0$, (3.6.33) gives $v^2 = -1$. Thus, by (3.6.34), det $(\psi_{AB}) = 1 =$ det $(\psi_A{}^B)$. Note that, as in (3.6.20), this implies

$$\psi_{AC}\psi^{BC} = \varepsilon_A{}^B. \tag{3.6.49}$$

It is sometimes convenient to use a symmetric spinor such as $\psi_{AB}$ to represent a 2-plane element. If this 2-plane element is not null (i.e., $v \neq 0$ in (3.6.34), so that the flagpole directions of $\alpha^A$ and $\beta^A$ are distinct) then it is often useful to use the normalization (3.6.49), or else

$$\psi_{AC}\psi^{BC} = -\varepsilon_A{}^B. \tag{3.6.50}$$

In a similar way, we may use a Hermitian spinor $\theta_{AB'}(=\bar{\theta}_{AB'})$ to denote a line element or the orthogonal complement hyperplane element, and to normalize according to

$$\theta_{AC'}\theta^{BC'} = \pm\varepsilon_A{}^B. \tag{3.6.51}$$

The local geometry at a point can be conveniently discussed using quantities such as $\psi_A{}^B$, $\theta_A{}^{B'}$ since their 'matrix products' represent geometrically simple operations, namely successions of reflections about lines, planes, etc. The above normalizations imply that the 'matrix square' of each basic quantity is plus or minus the identity $\varepsilon_A{}^B$ or $\varepsilon_{A'}{}^{B'}$.

### *Infinitesimal transformations*

We shall close this section with a brief discussion of *infinitesimal* spin-transformations and their associated Lorentz transformations. The fact that infinitesimal Lorentz transformations are essentially skew two-index tensors is well-known. We may derive this fact as follows. Let $L_a{}^b(\lambda)$ be a one-parameter family of Lorentz transformations depending smoothly on the parameter $\lambda$, such that $\lambda = 0$ gives the identity transformation:

$$L_a{}^b(0) = g_a{}^b. \tag{3.6.52}$$

The infinitesimal Lorentz transformation $S_a{}^b$ corresponding to this family is

$$S_a{}^b := \left[\frac{\mathrm{d}}{\mathrm{d}\lambda}L_a{}^b(\lambda)\right]_{\lambda=0}. \tag{3.6.53}$$

Applying (3.6.52) and (3.6.53) to the $\lambda$-derivative of the relation (3.6.5)

$$g_{ab} = L_a{}^c(\lambda)L_b{}^d(\lambda)g_{cd}, \tag{3.6.54}$$

we obtain

$$0 = S_a{}^c g_b{}^d g_{cd} + g_a{}^c S_b{}^d g_{cd}. \tag{3.6.55}$$

That is to say,

$$S_{ab} = -S_{ba}. \tag{3.6.56}$$

To reconstruct a finite Lorentz transformation out of an infinitesimal one, we use 'exponentiation'. Define, quite generally,

$$\exp(P_\alpha{}^\beta) := \delta_\alpha{}^\beta + P_\alpha{}^\beta + \frac{1}{2!}P_\alpha{}^\gamma P_\gamma{}^\beta + \cdots. \tag{3.6.57}$$

Then if

$$P_\alpha{}^\gamma Q_\gamma{}^\beta = Q_\alpha{}^\gamma P_\gamma{}^\beta, \tag{3.6.58}$$

it follows that

$$\exp(P_\alpha{}^\gamma)\exp(Q_\gamma{}^\beta) = \exp(P_\alpha{}^\beta + Q_\alpha{}^\beta), \tag{3.6.59}$$

formally.* Now, given any skew $S_{ab}$, we can define

$$L_a{}^b := \exp(S_a{}^b), \tag{3.6.60}$$

which gives

$$\begin{aligned} L_a{}^c g_{cd} L_b{}^d &= \exp(S_a{}^c) g_{cd} \exp(S_b{}^d) \\ &= \exp(S_a{}^c)\exp(S^e{}_c) g_{eb} \\ &= \exp(S_a{}^c)\exp(-S_c{}^e) g_{eb} \\ &= \exp(S_a{}^e - S_a{}^e) g_{eb} \\ &= g_a{}^e g_{eb} = g_{ab}. \end{aligned} \tag{3.6.61}$$

Thus, $L_a{}^b$ is a Lorentz transformation.

In a similar way, we can define infinitesimal spin transformations $\sigma_A{}^B$, in terms of a smooth one-parameter family of spin transformations $\phi_A{}^B(\lambda)$ for which

$$\phi_A{}^B(0) = \varepsilon_A{}^B, \tag{3.6.62}$$

by

$$\sigma_A{}^B := \left[\frac{\mathrm{d}}{\mathrm{d}\lambda}\phi_A{}^B(\lambda)\right]_{\lambda=0}. \tag{3.6.63}$$

Differentiating the relation

$$\varepsilon_{AB} = \phi_A{}^C(\lambda)\phi_B{}^D(\lambda)\varepsilon_{CD}, \tag{3.6.64}$$

* It can, in fact, be shown without difficulty that (3.6.57) always converges, and that (3.6.59) is always valid, whenever (3.6.59) holds (*cf.* Hochschild 1965).

(*cf.* (3.6.6)), we obtain

$$0 = \sigma_A{}^C\varepsilon_B{}^D\varepsilon_{CD} + \varepsilon_A{}^C\sigma_B{}^D\varepsilon_{CD}$$
$$= \sigma_{AB} - \sigma_{BA}, \tag{3.6.65}$$

so $\sigma_{AB}$ is symmetric. If, conversely, for any symmetric $\sigma_{AB}$ we set

$$\phi_A{}^B := \exp(\sigma_A{}^B), \tag{3.6.66}$$

we get

$$\phi_A{}^C\varepsilon_{CD}\phi_B{}^D = \exp(\sigma_A{}^C)\varepsilon_{CD}\exp(\sigma_B{}^D)$$
$$= \exp(\sigma_A{}^C)\exp(-\sigma^E{}_C)\varepsilon_{EB}$$
$$= \exp(\sigma_A{}^C)\exp(-\sigma_C{}^E)\varepsilon_{EB}$$
$$= \varepsilon_A{}^E\varepsilon_{EB} = \varepsilon_{AB}. \tag{3.6.67}$$

(The second line uses $\varepsilon_{CD}\sigma_B{}^P\sigma_P{}^Q\sigma_Q{}^D = \sigma_B{}^P\sigma_P{}^Q(-\sigma_{QC}) = \sigma_B{}^P(-\sigma_{PQ})(-\sigma^Q{}_C) = (-\sigma_{BP})(-\sigma^P{}_Q)(-\sigma^Q{}_C) = (-\sigma^E{}_P)(-\sigma^P{}_Q)(-\sigma^Q{}_C)\varepsilon_{EB}$.) Thus, $\phi_A{}^B$ is a spin transformation.

To obtain the relation between infinitesimal spin transformations and infinitesimal Lorentz transformations consider the $\lambda$-derivative of

$$L_{ab}(\lambda) = \phi_{AB}(\lambda)\bar{\phi}_{A'B'}(\lambda). \tag{3.6.68}$$

We get

$$S_{ab} = \sigma_{AB}\varepsilon_{A'B'} + \varepsilon_{AB}\bar{\sigma}_{A'B'}. \tag{3.6.69}$$

(*Cf.* (3.4.20).) Conversely, suppose $S_{ab}$ and $\sigma_{AB}$ are related by (3.6.69), or, equivalently,

$$S_a{}^b = \sigma_A{}^B\varepsilon_{A'}{}^{B'} + \varepsilon_A{}^B\bar{\sigma}_{A'}{}^{B'}. \tag{3.6.70}$$

Since the two terms on the right 'commute', we get, on taking exponentials and using (3.6.59),

$$L_a{}^b = \phi_A{}^B\bar{\phi}_{A'}{}^{B'} \tag{3.6.71}$$

as required, where $L_a{}^b$ is given by (3.6.60) and $\phi_A{}^B$ by (3.6.66).

Let us exhibit $\phi_A{}^B$ explicitly in terms of $\sigma_{AB}$. Put

$$\rho^2 := \tfrac{1}{2}\sigma_{AB}\sigma^{AB}. \tag{3.6.72}$$

Then

$$\sigma_{AB}\sigma_C{}^B = \rho^2\varepsilon_{AC}, \tag{3.6.73}$$

so

$$(\rho^{-1}\sigma_A{}^B)(\rho^{-1}\sigma_B{}^C) = -\varepsilon_A{}^C \tag{3.6.74}$$

(assuming, for the moment, that $\rho \neq 0$). The expression $\rho^{-1}\sigma_A{}^B$ behaves formally like 'i', so we get

$$\exp(\mu\rho^{-1}\sigma_A{}^B) = \varepsilon_A{}^B\cos\mu + \rho^{-1}\sigma_A{}^B\sin\mu. \tag{3.6.75}$$

Setting

$$\lambda := \mu\rho^{-1}, \tag{3.6.76}$$

we obtain

$$\begin{aligned}\exp(\lambda\sigma_A{}^B) &= \varepsilon_A{}^B \cos\rho\lambda + \sigma_A{}^B\rho^{-1}\sin\rho\lambda.\\ &=: \phi_A{}^B(\lambda),\end{aligned} \tag{3.6.77}$$

say. If we replace $\rho^{-1}\sin\rho\lambda$ by its limit as $\rho \to 0$, namely by $\lambda$, then we can use (3.6.77) also in the case $\rho = 0$:

$$\exp(\lambda\sigma_A{}^B) = \varepsilon_A{}^B + \sigma_A{}^B\lambda = \phi_A{}^B(\lambda). \tag{3.6.78}$$

And we have

$$\phi_A{}^B = \phi_A{}^B(1) = \varepsilon_A{}^B\cos\rho + \sigma_A{}^B\rho^{-1}\sin\rho. \tag{3.6.79}$$

(Note that both $\cos\rho$ and $\rho^{-1}\sin\rho$ are even functions and so do not depend on the sign of $\rho$ in (3.6.72).) The corresponding Lorentz transformation $L_a{}^b$ can then be constructed as in (3.6.35), where into (3.6.36) we substitute $\psi_{AB} = \sigma_{AB}\rho^{-1}\sin\rho$ and $\mu = \cos\rho$. The case $\rho = 0$ (i.e., (3.6.78)) corresponds to null rotations (*cf.* (3.6.40), (3.6.47)).

# 4
# Differentiation and curvature

## 4.1 Manifolds

In Chapters 2 and 3 we have not been much concerned with the details of the module $\mathfrak{S}^{\cdot}$ of spin-vector fields. Although in §3.1 it was necessary to tie in the spin-vector module with the concept of a world-vector, the essential property of world-vectors – that they belong to *tangent spaces* of the space–time manifold $\mathscr{M}$ – was never used.

In fact, the discussion so far would also apply to situations which are in essence very different from the one contemplated here. We may illustrate this difference by an example taken from elementary particle theory, namely 'isotopic spin-space'. This is a space which, as its name suggests, bears some (superficial) resemblance to spin-space. The states of actual *spin* of a *nucleon* may be expressed as complex-linear combinations of two states, say 'spin-up' and 'spin-down'. In a similar way, the states of *isotopic spin* of a nucleon are complex-linear combinations of two states, namely 'proton' and 'neutron'. But although formally similar, there is a crucial difference between these two situations. This lies in the fact that the directions in spin-space have to do with actual directions in space (-time) i.e., with the relations between a point and its neighbours, whereas the directions in isotopic spin-space have no such association.

It is also possible to produce mathematical examples where the elements of the basic module *do* have an association with the relation between points and their neighbours, but it is the *wrong* association. Consider a manifold of four real dimensions which is a two-complex-dimensional 'surface'. The tangent space at each point has two complex dimensions and can be given a structure identical to that of spin-space. However this situation is very different from the one that we shall be concerned with here. The association between spin-space and directions in the manifold has to be achieved via the intermediary stage of the world-vector space. As we shall see presently, this has the formal implication that the operation of differentiation has two, rather than one, spinor indices.

But how are we to express this relation between space–time points and their neighbours? To do this, we must make the concept of a tangent vector, or field of tangent vectors, precise. The method we follow is to define vectors as *directional derivatives* (*cf.* (1.4.1)) on the manifold, these derivatives acting on scalar fields. Thus, the vector 'points' in the direction of the manifold in which the derivative is measuring the rate of change of a scalar. These directional derivatives are characterized completely by their formal properties as maps of *the system of scalar fields* – which system actually contains all the information necessary to define the manifold structure of $\mathcal{M}$ and the system of tangent vectors to $\mathcal{M}$.

Since scalar fields play such a basic role in this method of development – and since even coordinate systems may be thought of simply as sets of scalar fields – it will be appropriate to state the axioms defining a manifold entirely in terms of properties of the system of scalar fields. For the sake of generality, we shall give the discussion of this section in a form applicable to any $n$-dimensional (Hausdorff, paracompact, connected) manifold. This will not involve us in any extra complication. Only in §4.4 shall we specialize to the case when $\mathcal{M}$ is a space–time. The discussion of the present section will be primarily carried out in terms of *real* $C^\infty$ scalar fields. We use the letter $\mathfrak{T}$ (as before) to denote the system of such fields. The system $\mathfrak{S}$ of complex $C^\infty$ scalar fields can then be defined in terms of $\mathfrak{T}$ (as $\mathfrak{T} \oplus \mathrm{i}\mathfrak{T}$).

We consider $\mathcal{M}$ as an abstract set of points whose structure is defined by a non-empty set $\mathfrak{T}$, each element $f \in \mathfrak{T}$ (called a *scalar*) being a map

$$f : \mathcal{M} \to \mathbb{R}. \tag{4.1.1}$$

The particular choice of the set $\mathfrak{T}$ which is made will serve to characterize the structure of $\mathcal{M}$ completely as a differentiable manifold, once we have given sufficient axioms for $\mathfrak{T}$. The differentiable structure for $\mathcal{M}$ that results will then be such that each element of $\mathfrak{T}$ is, in fact, a $C^\infty$ scalar field. The axiom system we use is derived from that of Chevalley (1946); *cf.* also Nomizu (1956). It is completely equivalent to the more usual definition of a manifold given, for example, in Lang (1972), Kobayashi and Nomizu (1963), Hawking and Ellis (1973) and Hicks (1965).

For the first axiom we take:

(4.1.2) AXIOM

*If* $f_1, f_2, \ldots, f_r \in \mathfrak{T}$ *and if* $F: \mathbb{R}^r \to \mathbb{R}$ *is any* $C^\infty$ *real-valued function of r real variables, then* $F(f_1, f_2, \ldots, f_r)$ *(considered as a function on* $\mathcal{M}$, i.e., $F(f_1, \ldots, f_r)(P) = F(f_1(P), \ldots, f_r(P))$, $P \in \mathcal{M}$*) is also an element of* $\mathfrak{T}$.

We note, in particular, that since any constant is a $C^\infty$ function, any map (4.1.1) which assigns the same real number $k$ to each point of $\mathcal{M}$, i.e., a *constant map*, will be an element of $\mathfrak{T}$. Without danger of ambiguity, we denote this element of $\mathfrak{T}$ also by the letter $k$. The subset of $\mathfrak{T}$ consisting of all constant maps will be denoted by $\mathfrak{R}$. Clearly $\mathfrak{R}$ is a ring isomorphic to $\mathbb{R}$. Since the operations of addition and multiplication are both $C^\infty$ maps: $\mathbb{R}^2 \to \mathbb{R}$, we have, by axiom (4.1.2), operations of addition and multiplication acting on the set $\mathfrak{T}$, defined by

$$(f+g)(P) := f(P) + g(P) \qquad (4.1.3)$$
$$(fg)(P) := f(P)g(P)$$

for all $P \in \mathcal{M}$. This gives $\mathfrak{T}$ the structure of a commutative ring* with unit (*cf.* (2.1.22)), with $0, 1 \in \mathfrak{R} \subset \mathfrak{T}$. It is evident that $\mathfrak{T}$ is also a vector space over $\mathfrak{R}$. Taken together, these properties define $\mathfrak{T}$ as a *commutative algebra* over $\mathfrak{R}$.

To state the next two axioms we need a concept of a neighbourhood of a point $P \in \mathcal{M}$. We define a $\mathfrak{T}$-neighbourhood of $P$ to be the set of points of $\mathcal{M}$ at which $f \neq 0$, for some $f \in \mathfrak{T}$ with $f(P) \neq 0$. (Recall that this was our procedure in §2.4.) Clearly the intersection of two $\mathfrak{T}$-neighbourhoods is again a $\mathfrak{T}$-neighbourhood since if $\mathcal{U}$ is defined by $f \neq 0$ and $\mathcal{V}$ by $g \neq 0$, then $\mathcal{U} \cap \mathcal{V}$ is defined by $fg \neq 0$.

The topology we assign to $\mathcal{M}$ is the one generated by the $\mathfrak{T}$-neighbourhoods. That is to say, a subset of $\mathcal{M}$ will be called open iff it is a union of $\mathfrak{T}$-neighbourhoods.** We can show that, with respect to this topology,

$$\textit{each element of } \mathfrak{T} \textit{ is a continuous function on } \mathcal{M}, \qquad (4.1.4)$$

i.e., that the inverse image of an open interval of $\mathbb{R}$ under each element of $\mathfrak{T}$ is an open set in $\mathcal{M}$. For proof of (4.1.4), let the open interval be $a < x < b$ (where $a < b$) and define the $C^\infty$ 'bump function'

$$B_{a,b}(x) = \begin{cases} 0 & \text{if } \; x \leqslant a \;\text{ or }\; b \leqslant x \\ \exp\left(\dfrac{1}{(x-a)(x-b)}\right) & \text{if } \; a < x < b \end{cases} \qquad (4.1.5)$$

The inverse image of the interval $a < x < b$ under any map $f \in \mathfrak{T}$ is the open set defined by $0 \neq B_{a,b}(f) \in \mathfrak{T}$, and this establishes our assertion.

The next axiom asserts the 'local' character of the restriction on a real-

* It turns out that, given only the *ring structure* of $\mathfrak{T}$ and not the set $\mathcal{M}$, then $\mathcal{M}$ can in fact be completely and uniquely reconstructed.

** In actual fact, once we have imposed all the axioms and restrictions, it will follow that any open set *is* a $\mathfrak{T}$-neighbourhood. However, we shall not require this fact here.

valued function on $\mathcal{M}$ in order that it should belong to $\mathfrak{T}$. (We expect '$C^\infty$' to be one such local restriction.)

(4.1.6) AXIOM

*If $g:\mathcal{M}\to\mathbb{R}$, and if for each $P\in\mathcal{M}$ there exists a $\mathfrak{T}$-neighbourhood $\mathcal{U}$ of $P$ and an element $f\in\mathfrak{T}$ which agrees with $g$ in $\mathcal{U}$, then $g\in\mathfrak{T}$.*

Note that we can restate this axiom as follows: If $g:\mathcal{M}\to\mathbb{R}$ and if for each $P\in\mathcal{M}$ there exist $h,f\in\mathfrak{T}$ with $h(P)\neq 0$, where $hf=hg$, then $g\in\mathfrak{T}$.

Finally, we need an axiom which asserts the 'locally $n$-dimensional Euclidean' character of $\mathcal{M}$. Here, $n$ is a given fixed integer.

(4.1.7) AXIOM

*For each $P\in\mathcal{M}$ there exists a $\mathfrak{T}$-neighbourhood $\mathcal{U}$ of $P$ and $n$ elements $x^1,\ldots,x^n\in\mathfrak{T}$ such that* (i) *given any two points of $\mathcal{U}$, at least one of the $x^\alpha$ has a different value at the two points, and* (ii) *each element $f\in\mathfrak{T}$ may be expressed, in $\mathcal{U}$, as a $C^\infty$ function of $x^1,\ldots,x^n$.*

The scalars $x^1, x^2, \ldots, x^n$ which occur in (4.1.7) are called *local coordinates* about $P$. The set $\mathcal{U}$ is called a *local coordinate neighbourhood* and the pair $(\mathcal{U}, x^{\mathbf{i}})$ a *local coordinate system.* (This terminology, according to which coordinates are referred to as particular examples of *scalar* fields, is somewhat at variance with the classical usage. However, it is perfectly logical within the framework of the modern development, in which vectors, tensors, scalars, etc. are not defined in the classical way in terms of coordinate changes.) Property (i) in (4.1.7) ensures that the values of the coordinates $x^\alpha$ do in fact serve to label the points in $\mathcal{U}$ – in a continuous way, by (4.1.4) – with distinct points being assigned distinct coordinate labels. Property (ii) of (4.1.7) (together with (4.1.2) with $r=n$) ensures that $x^\alpha$ is a *non-singular* coordinate system* for $\mathcal{U}$ since the elements of $\mathfrak{T}$, when restricted to $\mathcal{U}$, are precisely those which are given as $C^\infty$ functions of the coordinates $x^\alpha$. If a second coordinate system $x^{\hat{\alpha}}$ were introduced which covered another $\mathfrak{T}$-neighbourhood $\hat{\mathcal{U}}$ in accordance with (4.1.7), then in the overlap region $\mathcal{U}\cap\hat{\mathcal{U}}$ each system of coordinates would have

* It should be pointed out, however, that *not all* local coordinates are suitable for use in axiom (4.1.7). Consider, for example, the plane in the usual coordinates $r, \theta$. Here, we have a coordinate patch on the plane for $r>0, 0<\theta<2\pi$. However, the coordinate $\theta$ cannot be extended to a $C^\infty$ (or even $C^0$) function over *all* of $\mathcal{M}$, whence $\theta\notin\mathfrak{T}$ and these particular local coordinates are not among those described in Axiom (4.1.7). Nor, indeed, is even the $(r, \theta)$ patch $r>0<\theta<\pi$. For the function $r=(x^2+y^2)^{1/2}$ is not extendible as a $C^\infty$ scalar at the origin, so $r\notin\mathfrak{T}$.

to be describable as $C^\infty$ functions of the other. This provides the link between the present approach to manifold structure and the more usual one in terms of overlapping coordinate charts.

In fact it is not difficult to prove the equivalence of these definitions. The axioms given here are sufficient to establish that the local coordinate systems $(\mathscr{U}, x^\alpha)$ of (4.1.7) comprise a covering of $\mathscr{M}$ by coordinate patches which, as a consequence of Axioms (4.1.2), (4.1.6) and (4.1.7), satisfy all the conditions normally required. Conversely, given a Hausdorff manifold $\mathscr{M}$ according to the more standard definition, we can define the set $\mathfrak{T}$ to consist of just those real-valued functions on $\mathscr{M}$ which, in each coordinate patch, may be expressed as a $C^\infty$ function of the coordinates. Then $\mathfrak{T}$ satisfies the three axioms (4.1.2), (4.1.6), (4.1.7), so $\mathscr{M}$ is a manifold by the present definition.

Our definition of a topology for $\mathscr{M}$, together with (4.1.6) is sufficient to imply that

*$\mathscr{M}$ is a Hausdorff topological space.*

This means that for any pair of distinct points $P, R \in \mathscr{M}$, a pair of disjoint $\mathfrak{T}$-neighbourhoods can be found, each containing one of the points. To establish this we need only find some function $h \in \mathfrak{T}$ which takes distinct values $p$ and $r$ at $P$ and $R$, respectively. For, choosing $q = \frac{1}{2}|p - r|$, we can define $\mathfrak{T}$-neighbourhoods of $P$, $R$ by $B_{p-q,p+q}(h)$, $B_{r-q,r+q}(h) \in \mathfrak{T}$, respectively, to obtain the required disjoint $\mathfrak{T}$-neighbourhoods. To see that $h$ exists we refer to (4.1.7). Either $R$ belongs to $\mathscr{U}$ or it does not. If $R$ belongs to $\mathscr{U}$, then we can use for $h$ a coordinate $x^\alpha$ which differs at $R$ from its value at $P$. If $R$ does not belong to $\mathscr{U}$, then we can use for $h$ a function which defines $\mathscr{U}$ by being non-zero on $\mathscr{U}$ and zero outside $\mathscr{U}$.

It is normal to assume, for a manifold, that its topology has a countable basis (an assumption equivalent here to paracompactness – Kelley 1955, Engelking 1968). We can state this in the form of

(4.1.8) AXIOM

*There is a countable collection of $\mathfrak{T}$-neighbourhoods such that every $\mathfrak{T}$-neighbourhood can be expressed as a union of members of the collection.*

In fact, once a metric (or a connection) has been introduced into the manifold, this assumption becomes redundant (Engelking 1968). However, for present purposes it will be well to make it. This has the implication that the manifold $\mathscr{M}$ is of the 'ordinary' kind, to which the discussion of §2.4 (leading to the total reflexivity of the system $\mathfrak{S}^\cdot$) can be applied. Another way of stating (4.1.8) is: *$\mathscr{M}$ can be covered by a countable collection of coordinate neighbourhoods.*

Another usual assumption in the case of a space–time manifold is

(4.1.9) AXIOM

*$\mathscr{M}$ is connected*

This means that $\mathscr{M}$ is not the union of two non-empty disjoint open sets. In terms of $\mathfrak{T}$ we may express this condition as follows: if $f, g \in \mathfrak{T}$ and if $fg = 0$, then $f(P) = 0 = g(P)$ for some $P \in \mathscr{M}$. It will be convenient to make this assumption here also. Having made these restrictions on the topology of $\mathscr{M}$, we shall henceforth simply use the term neighbourhood for what we previously called a $\mathfrak{T}$-neighbourhood.

## Vector fields

We are now in a position to introduce the concept of a (contravariant) *vector field* $\boldsymbol{V}$ (or field of tangent vectors) on $\mathscr{M}$. We define $\boldsymbol{V}$ as a map

$$\boldsymbol{V}: \mathfrak{T} \to \mathfrak{T} \tag{4.1.10}$$

with the following three properties

$$\begin{aligned} &\text{(i)} \quad \boldsymbol{V}(k) = 0 \quad \text{if} \quad k \in \mathfrak{R} \\ &\text{(ii)} \quad \boldsymbol{V}(f+g) = \boldsymbol{V}(f) + \boldsymbol{V}(g) \quad \text{if} \quad f, g \in \mathfrak{T} \\ &\text{(iii)} \quad \boldsymbol{V}(fg) = f\boldsymbol{V}(g) + g\boldsymbol{V}(f) \quad \text{if} \quad f, g \in \mathfrak{T}. \end{aligned} \tag{4.1.11}$$

Such a map is called a *derivation* on $\mathfrak{T}$, where $\mathfrak{T}$ is regarded as an algebra over $\mathfrak{R}$. The set of all such derivations will be denoted by $\mathfrak{T}^{\cdot}$. Contact with our previous definition of $\mathfrak{T}^{\cdot}$ will be made presently.

Now suppose we have a map $\boldsymbol{W}: \mathfrak{T} \to \mathfrak{T}$ with the property that in any local coordinate system $(\mathscr{U}, x^{\alpha})$, the effect of $\boldsymbol{W}$ may be expressed in the form*

$$\boldsymbol{W}(f) = W^{\alpha} \frac{\partial f}{\partial x^{\alpha}}, \tag{4.1.12}$$

where

$$W^1, W^2, \ldots, W^n \in \mathfrak{T}. \tag{4.1.13}$$

* The meaning of the symbol $\partial/\partial x^{\alpha}$ here should be clear. Even though $f$ is, strictly speaking, a function of a point $P$ of the manifold $\mathscr{M}$, rather than being explicitly a function of variables $x^1, x^2, \ldots, x^n$, the point $P$ can itself be regarded as a function of $x^1, \ldots, x^n$, by virtue of the coordinatization of $\mathscr{U}$. By axiom (4.1.7) (ii), $f$ may be then regarded as a $C^{\infty}$ function of $x^1, \ldots, x^n$.

Then $\boldsymbol{W}$ clearly satisfies all three relations (4.1.11) and so is a derivation. We can write (4.1.12) as

$$\boldsymbol{W} = W^{\alpha} \frac{\partial}{\partial x^{\alpha}}, \tag{4.1.14}$$

where the operators are understood to act on scalar fields. In another local coordinate system $(\hat{\mathscr{U}}, y^{\hat{\alpha}})$, we shall have

$$\boldsymbol{W} = W^{\hat{\alpha}} \frac{\partial}{\partial y^{\hat{\alpha}}}. \tag{4.1.15}$$

If $\mathscr{U}$ and $\hat{\mathscr{U}}$ overlap, then in the intersection region $\mathscr{U} \cap \hat{\mathscr{U}}$ we must have

$$W^{\hat{\beta}} \frac{\partial}{\partial y^{\hat{\beta}}} = W^{\alpha} \frac{\partial}{\partial x^{\alpha}}. \tag{4.1.16}$$

Thus,

$$W^{\hat{\beta}} = W^{\alpha} \frac{\partial y^{\hat{\beta}}}{\partial x^{\alpha}}, \quad W^{\alpha} = W^{\hat{\beta}} \frac{\partial x^{\alpha}}{\partial y^{\hat{\beta}}}. \tag{4.1.17}$$

Regarding $W^{\alpha}$ and $W^{\hat{\alpha}}$ as the *components* of $\boldsymbol{W}$ in the $x^{\alpha}$ and $y^{\hat{\alpha}}$ coordinate systems, respectively, we arrive at the standard classical definition of a contravariant vector. Thus, any classical contravariant vector corresponds uniquely to a map $\boldsymbol{W}: \mathfrak{T} \to \mathfrak{T}$ which, when referred to any local coordinate system, can be expressed linearly in terms of the partial derivative operators with respect to the coordinates. Furthermore, any such map is an example of a derivation on $\mathfrak{T}$. The following result (4.1.18) establishes the converse. We shall then have complete equivalence between the concepts of a derivation on $\mathfrak{T}$, of a linear differential operator on $\mathfrak{T}$ (or directional derivative), and of a classical contravariant vector.

(4.1.18) PROPOSITION

*If $\boldsymbol{V} \in \mathfrak{T}^{\cdot}$, then in any local coordinate system $(\mathscr{U}, x^{\alpha})$, $\boldsymbol{V}$ has the form $\boldsymbol{V} = V^{\alpha} \partial/\partial x^{\alpha}$ for some $V^1, \ldots, V^n \in \mathfrak{T}$.*

*Proof*: We observe first that if we can establish that at *each* point of $\mathscr{U}$, $\boldsymbol{V}$ has the form $V^{\alpha} \partial/\partial x^{\alpha}$, then it necessarily follows that $V^{\alpha} \in \mathfrak{T}$. For we shall have, in $\mathscr{U}$,

$$V^{\alpha} = V^{\beta} \delta^{\alpha}_{\beta} = V^{\beta} \frac{\partial x^{\alpha}}{\partial x^{\beta}} = \boldsymbol{V}(x^{\alpha}) \in \mathfrak{T}, \tag{4.1.19}$$

since the coordinates $x^{\alpha}$ are themselves $C^{\infty}$ scalars ($x^{\alpha} \in \mathfrak{T}$). Now choose a particular point $X \in \mathscr{U}$, with coordinates $X^1, \ldots, X^n$. Let $\mathscr{D}$ be the coordinate $n$-disc $(x^1 - X^1)^2 + \cdots + (x^n - X^n)^2 < \rho^2$ where $\rho$ is chosen small enough that $\mathscr{D} \subset \mathscr{U}$. Choose any point $P \in \mathscr{D}$, with coordinates $x^1, \ldots, x^n$. Then the points with coordinates $X^\alpha + tx^\alpha - tX^\alpha (0 \leqslant t \leqslant 1)$ are also in $\mathscr{D}$. Let $f$ be any member of $\mathfrak{T}$. We have

$$\begin{aligned} f(x^1, \ldots, x^n) &= f(X^1, \ldots, X^n) + \\ &\quad + \int_0^1 \frac{\mathrm{d}}{\mathrm{d}t} f(X^1 + tx^1 - tX^1, \ldots, X^n + tx^n - tX^n)\,\mathrm{d}t \\ &= f(X^1, \ldots, X^n) + (x^\alpha - X^\alpha) \int_0^1 f_\alpha(X^1 + tx^1 - tX^1, \ldots)\,\mathrm{d}t, \end{aligned} \tag{4.1.20}$$

where

$$f_\alpha(x^1, \ldots, x^n) = \frac{\partial}{\partial x^\alpha} f(x^1, \ldots, x^n). \tag{4.1.21}$$

Thus, we have an expression of the form

$$f(x^1, \ldots, x^n) = f(X^1, \ldots, X^n) + (x^\alpha - X^\alpha) g_\alpha(x^1, \ldots, x^n), \tag{4.1.22}$$

where, at $X$,

$$g_\alpha(X^1, \ldots, X^n) = \left[\frac{\partial f}{\partial x^\alpha}\right]_{x^\alpha = X^\alpha}. \tag{4.1.23}$$

Applying $\boldsymbol{V}$ to (4.1.22), where $X$ is kept fixed but where the coordinates $x^\alpha$ define the variable point $P$ at which $\boldsymbol{V}$ is to be evaluated, we obtain

$$\boldsymbol{V}(f) = 0 + \boldsymbol{V}(x^\alpha) g_\alpha + (x^\alpha - X^\alpha) \boldsymbol{V}(g_\alpha), \tag{4.1.24}$$

using (4.1.11). Now specialize to the point $P = X$. We find

$$\boldsymbol{V}(f) = V^\alpha \frac{\partial f}{\partial x^\alpha} \tag{4.1.25}$$

at the point $X$, using (4.1.23) and defining $V^\alpha = \boldsymbol{V}(X^\alpha)$. This same formula holds for each $f \in \mathfrak{T}$ and for each $X \in \mathscr{U}$, thus establishing the result.

One immediate consequence of (4.1.18) is the following property of a derivation $\boldsymbol{V}$:

*If* $h: \mathbb{R}^n \to \mathbb{R}$ *is* $C^\infty$ *and* $f_1, \ldots, f_r \in \mathfrak{T}$, *then* $\boldsymbol{V}(h(f_1, \ldots, f_r)) = \sum_{\mathbf{i}} \frac{\partial h}{\partial f_{\mathbf{i}}} \boldsymbol{V}(f_{\mathbf{i}})$.

(4.1.26)

The properties (4.1.11) are all special cases of (4.1.26). Thus, (4.1.26) is *equivalent* to (4.1.11). Furthermore, setting $f_{\mathbf{i}} = x^{\mathbf{i}}$ we immediately regain (4.1.18).

The concept of a derivation on $\mathfrak{T}$ provides an elegant algebraic characterization of a tangent vector *field* on $\mathscr{M}$. We shall also need the concept of a tangent vector at a single *point* $P\in\mathscr{M}$. To obtain such a concept, we can set up an equivalence relation between derivations, in which $\boldsymbol{U}$ is equivalent to $\boldsymbol{V}$ if and only if $\boldsymbol{U}(f)$ and $\boldsymbol{V}(f)$, *when evaluated at* $P$, give the same real number, for each $f\in\mathfrak{T}$. This equivalence class, denoted by $\boldsymbol{V}[P]$ – or, often, simply by $\boldsymbol{V}$ *at* $P$ – is called the *tangent vector at* $P$ belonging to the vector field $\boldsymbol{V}$. We have:

$$\boldsymbol{U}[P]=\boldsymbol{V}[P] \quad \textit{iff}\ \{\boldsymbol{U}(f)\}(P)=\{\boldsymbol{V}(f)\}(P)\ \textit{for all}\ f\in\mathfrak{T}. \qquad (4.1.27)$$

In terms of local coordinates about $P$, we have $\boldsymbol{U}[P]=\boldsymbol{V}[P]$ iff $U^\alpha\,\partial f/\partial x^\alpha = V^\alpha \partial f/\partial x^\alpha$ at $P$ for every $f\in\mathfrak{T}$, i.e., $U^\alpha(P)=V^\alpha(P)$. Thus the values of the $n$ components of $\boldsymbol{V}$ at $P$, namely $V^\alpha(P)$, may be regarded as the *components* of $\boldsymbol{V}$ in the coordinates $x^\alpha$. Since these are just $n$ real numbers, the tangent vectors at $P$ form a vector space, over $\mathbb{R}$, of dimension $n$, called the *tangent space* to $\mathscr{M}$ at $P$. This space will be denoted by $\mathfrak{T}^\cdot[P]$. Sometimes the notation $\mathfrak{T}^\cdot$ may be used for this vector space, rather than for the set of tangent vector fields. This will be either when it has been explicitly stated that we are working at one point only – or else when it is immaterial whether vectors at one point or vector fields are being considered. Under the same circumstances, $\mathfrak{T}$ may be used to stand for $\mathfrak{T}[P]=\mathbb{R}$.

An alternative definition of a tangent vector at a point $Q$ which is sometimes used, is as a map $\boldsymbol{W}[P]$: $\mathfrak{T}\to\mathbb{R}$ satisfying (4.1.11), with the additional property that if $f, g\in\mathfrak{T}$ are such that they are identical throughout some neighbourhood of $Q$, then $\boldsymbol{W}[Q](f)=\boldsymbol{W}[Q](g)$. The equivalence of this definition to the one given above can be obtained by repeating the argument for (4.1.18), but where $X$ is now fixed at the point $Q$.

We may also define the concept of a vector field on a (suitably non-pathological) subset $\mathscr{S}$ of $\mathscr{M}$. As with the case of a single point, we can set up an equivalence relation between derivations, $\boldsymbol{U}$ being equivalent to $\boldsymbol{V}$ if and only if $\boldsymbol{U}(f)=\boldsymbol{V}(f)$ at each point of $\mathscr{S}$, for any $f\in\mathfrak{T}$. We denote this equivalence class by $\boldsymbol{U}[\mathscr{S}]$ $(=\boldsymbol{V}[\mathscr{S}])$ and refer to it as the *part of the vector field* $\boldsymbol{U}$ *which lies on* $\mathscr{S}$. We denote by $\mathfrak{T}^\cdot[\mathscr{S}]$ the set of vector fields lying on $\mathscr{S}$, and by $\mathfrak{T}[\mathscr{S}]$ the scalar fields restricted to $\mathscr{S}$ (i.e., 'lying on' $\mathscr{S}$).

Given any two derivations $\boldsymbol{U}, \boldsymbol{V}\in\mathfrak{T}^\cdot$ we can define their sum by

$$(\boldsymbol{U}+\boldsymbol{V})(f):=\boldsymbol{U}(f)+\boldsymbol{V}(f)\ \text{for all}\ f\in\mathfrak{T}. \qquad (4.1.28)$$

Clearly $\boldsymbol{U}+\boldsymbol{V}\in\mathfrak{T}^\cdot$. Also, we can define the multiplication of a derivation $\boldsymbol{U}\in\mathfrak{T}^\cdot$ by a scalar $h\in\mathfrak{T}$ according to

$$(h\boldsymbol{U})(f):=h\boldsymbol{U}(f)\ \textit{for all}\ f\in\mathfrak{T}. \qquad (4.1.29)$$

Again it is clear that $h\boldsymbol{U}\in\mathfrak{T}^\cdot$. It is easy to see that, in terms of some coordinate system, the $\alpha$th component of $\boldsymbol{U}+\boldsymbol{V}$ is $U^\alpha + V^\alpha$ and that of $h\boldsymbol{U}$ is $hU^\alpha$. Furthermore, it is evident that, under the operations (4.1.28) and (4.1.29), $\mathfrak{T}^\cdot$ forms a module over $\mathfrak{T}$. Similar remarks apply to $\mathfrak{T}^\cdot[P]$, $\mathfrak{T}[P]$ and to $\mathfrak{T}^\cdot[\mathscr{S}]$, $\mathfrak{T}[\mathscr{S}]$. The arguments of §2.4 establish that $\mathfrak{T}^\cdot$ is totally reflexive.

*Real and complex tensors*

We are now in a position to apply the theory of §2.2. We introduce a labelling set $\mathscr{S} = \{\alpha, \beta, \ldots, \alpha_0, \ldots\}$ and produce canonically isomorphic copies $\mathfrak{T}^\alpha, \mathfrak{T}^\beta, \ldots$ of $\mathfrak{T}^\cdot$. Then we introduce duals $\mathfrak{T}_\alpha, \mathfrak{T}_\beta, \ldots$ and, finally, the sets $\mathfrak{T}^{\alpha\ldots\gamma}_{\lambda\ldots\nu}$. A similar construction applies if we start with $\mathfrak{T}^\cdot[P]$ leading to $\mathfrak{T}^{\alpha\ldots\gamma}_{\lambda\ldots\nu}[P]$, or with $\mathfrak{T}^\cdot[\mathscr{S}]$ leading to $\mathfrak{T}^{\alpha\ldots\gamma}_{\lambda\ldots\nu}[\mathscr{S}]$.

We can also construct complex tensors as members of sets $\mathfrak{S}^{\alpha\ldots\gamma}_{\lambda\ldots\nu}$ (or $\mathfrak{S}^{\alpha\ldots\gamma}_{\lambda\ldots\nu}[P]$ or $\mathfrak{S}^{\alpha\ldots\gamma}_{\lambda\ldots\nu}[\mathscr{S}]$) where each element $C^{\alpha\ldots\gamma}_{\lambda\ldots\nu} \in \mathfrak{S}^{\alpha\ldots\gamma}_{\lambda\ldots\nu}$ is an expression of the form

$$C^{\alpha\ldots\gamma}_{\lambda\ldots\nu} = A^{\alpha\ldots\gamma}_{\lambda\ldots\nu} + \mathrm{i}B^{\alpha\ldots\gamma}_{\lambda\ldots\nu} \tag{4.1.30}$$

with $A^{\alpha\ldots\gamma}_{\lambda\ldots\nu}, B^{\alpha\ldots\gamma}_{\lambda\ldots\nu} \in \mathfrak{T}^{\alpha\ldots\gamma}_{\lambda\ldots\nu}$, the new quantity i, which is introduced, being a constant scalar subject to

$$\mathrm{i}^2 = -1. \tag{4.1.31}$$

The elements of $\mathfrak{S}$ are maps $h: \mathscr{M} \to \mathbb{C}$ where $h = f + \mathrm{i}g$ $(f, g \in \mathfrak{T})$ gives $h(P) = f(P) + \mathrm{i}g(P)$. The ring structure of $\mathfrak{S}$ is defined by $(f + \mathrm{i}g) + (p + \mathrm{i}q) = (f + p) + \mathrm{i}(g + q)$, $(f + \mathrm{i}g)(p + \mathrm{i}q) = (fp - gq) + \mathrm{i}(fq + gp)(p, q \in \mathfrak{T})$. The elements of $\mathfrak{S}^\cdot$ are maps $\boldsymbol{Z}: \mathfrak{S} \to \mathfrak{S}$ with $\boldsymbol{Z} = \boldsymbol{U} + \mathrm{i}\boldsymbol{V}(\boldsymbol{U}, \boldsymbol{V} \in \mathfrak{T}^\cdot)$ giving $\boldsymbol{Z}(f + \mathrm{i}g) = \boldsymbol{Z}(f) + \mathrm{i}\boldsymbol{Z}(g) = \boldsymbol{U}(f) - \boldsymbol{V}(g) + \mathrm{i}\boldsymbol{U}(g) + \mathrm{i}\boldsymbol{V}(f)$. These satisfy the derivation properties (4.1.11) as applied to complex scalars $f + \mathrm{i}g$. The set $\mathfrak{S}^\cdot$ is a module over $\mathfrak{S}$. It is not hard to see that the general $\mathfrak{S}^{\alpha\ldots\gamma}_{\lambda\ldots\nu}$, defined from these as in §2.2, gives simply the elements (4.1.30). Throughout this section we shall tend to work with the sets $\mathfrak{T}^{\cdots}_{\cdots}$ rather than $\mathfrak{S}^{\cdots}_{\cdots}$. It should be clear, however, that the discussion will apply equally well to complex tensors as to real ones.

The extra structure that we now have which was not present in the general discussion of §2.2 is the interpretation of elements of $\mathfrak{T}^\cdot$ (or of $\mathfrak{S}^\cdot$) as derivations on the algebra of scalars. We may regard this as giving the link between the elements of the basic module and 'the relationship between a point and its neighbours' that was mentioned at the beginning of this section. This extra structure leads us to certain concepts of differentiation, such as the 'gradient of a scalar', 'Lie derivative', 'exterior derivative' and some others. The first of these we discuss in a moment, but for the others we shall wait until §4.2, after the concept of a connection has been introduced. This will enable us to obtain a more comprehensive viewpoint.

*Gradient of a scalar*

Given a scalar $f \in \mathfrak{T}$, we define its *gradient* $\mathrm{d}f$, sometimes called its differential, to be an element $\mathrm{d}f$ of the dual $\mathfrak{T}^{\cdot *}$ of $\mathfrak{T}^{\cdot}$ defined by

$$\mathrm{d}f(\boldsymbol{V}) := \boldsymbol{V}(f) \tag{4.1.32}$$

The fact that this is a linear map from $\mathfrak{T}^{\cdot}$ to $\mathfrak{T}$ is a consequence of (4.1.28) and (4.1.29): we have $\mathrm{d}f(g\boldsymbol{V}) = (g\boldsymbol{V})(f) = g\boldsymbol{V}(f) = g\,\mathrm{d}f(\boldsymbol{V})$, and $\mathrm{d}f(\boldsymbol{U}+\boldsymbol{V}) = (\boldsymbol{U}+\boldsymbol{V})(f) = \boldsymbol{U}(f) + \boldsymbol{V}(f) = \mathrm{d}f(\boldsymbol{U}) + \mathrm{d}f(\boldsymbol{V})$.

The connection between this modern concept of differential and the classical notion of an 'infinitesimally small element' is not particularly intuitive, but it is related to the transformation properties of '$\mathrm{d}x^{\alpha}$' under change of coordinate system. Choose a local coordinate system $(\mathscr{U}, x^{\alpha})$. Then (*cf*. (4.1.12), (4.1.18)) the quantities

$$\frac{\partial}{\partial x^1}, \ldots, \frac{\partial}{\partial x^n} \tag{4.1.33}$$

constitute a basis for $\mathfrak{T}^{\cdot}[\mathscr{U}]$. The dual basis elements are the differentials

$$\mathrm{d}x^1, \ldots, \mathrm{d}x^n \tag{4.1.34}$$

of the coordinates $x^1, \ldots, x^n$, since

$$\mathrm{d}x^{\alpha}\left(\frac{\partial}{\partial x^{\beta}}\right) = \frac{\partial}{\partial x^{\beta}}(x^{\alpha}) = \delta^{\alpha}_{\beta}, \tag{4.1.35}$$

by (4.1.32). Under change to new coordinates $y^{\hat{\alpha}}$ we have

$$\frac{\partial}{\partial x^{\beta}} = \frac{\partial y^{\hat{\gamma}}}{\partial x^{\beta}} \frac{\partial}{\partial y^{\hat{\gamma}}}, \tag{4.1.36}$$

so, to preserve (4.1.35), we require

$$\mathrm{d}y^{\hat{\sigma}} = \frac{\partial y^{\hat{\sigma}}}{\partial x^{\alpha}} \mathrm{d}x^{\alpha}, \tag{4.1.37}$$

which is a formally valid expression for classical differentials.

To justify the terminology 'gradient' for $\mathrm{d}f$, let us find its components in the $x^{\alpha}$ system. Since $\mathrm{d}f$ is an element of $\mathfrak{T}^{\cdot *}[\mathscr{U}]$, these components may be found by taking scalar products with the basis for $\mathfrak{T}^{\cdot}[\mathscr{U}]$; the required components are

$$\mathrm{d}f\left(\frac{\partial}{\partial x^{\alpha}}\right) = \frac{\partial f}{\partial x^{\alpha}}, \tag{4.1.38}$$

by (4.1.32). Thus, the concept of 'differential' as described here does coincide with the classical notion of 'gradient'. The expression for $\mathrm{d}f$ in terms of its components now becomes

$$\mathrm{d}f = \frac{\partial f}{\partial x^\alpha}\mathrm{d}x^\alpha, \tag{4.1.39}$$

which is another version of the formally valid classical expression (4.1.37).

Since we wish to make use of the abstract index notation here, the 'differential' notation for covariant vectors is not entirely suitable. Let us employ the symbol $\nabla_\alpha$ to denote the gradient operation on scalars. Then $\nabla_\alpha f$ is the element of $\mathfrak{T}_\alpha$ which is the canonical image of $\mathrm{d}f$ in $\mathfrak{T}^*$. Since $V^\alpha$ is to be the canonical image of $\boldsymbol{V}$ in $\mathfrak{T}^\alpha$, we can re-express (4.1.32) as $V^\alpha\nabla_\alpha f = \boldsymbol{V}(f)$. That is to say,

$$\boldsymbol{V} = V^\alpha\nabla_\alpha, \tag{4.1.40}$$

the operators acting on scalars. The notation (4.1.40) agrees with one often used for a directional derivative.

We saw in (4.1.38) that the components of $\mathrm{d}f$ in the $x^\alpha$ coordinate system are $\partial f/\partial x^\alpha$. So these are the components $\nabla_\alpha f$ of $\nabla_\alpha f$, and we may write

$$\nabla_\alpha = \frac{\partial}{\partial x^\alpha}, \tag{4.1.41}$$

provided the operators act on scalars.

From (4.1.11) we finally get

$$\nabla_\alpha k = 0, \quad \text{i.e.,} \quad \mathrm{d}k = 0 \tag{4.1.42}$$

if $k \in \mathfrak{K}$ and

$$\nabla_\alpha(f+g) = \nabla_\alpha f + \nabla_\alpha g, \quad \text{i.e.,} \quad \mathrm{d}(f+g) = \mathrm{d}f + \mathrm{d}g, \tag{4.1.43}$$

$$\nabla_\alpha(fg) = f\nabla_\alpha g + g\nabla_\alpha f, \quad \text{i.e.,} \quad \mathrm{d}(fg) = f\,\mathrm{d}g + g\,\mathrm{d}f, \tag{4.1.44}$$

if $f, g \in \mathfrak{S}$.

## 4.2 Covariant derivative

We have seen in (4.1.32) that the concept of a gradient of a scalar can be given a unique invariant meaning dependent only on the differentiable structure of the manifold $\mathcal{M}$ (and, in fact, only on the algebraic structure of $\mathfrak{T}$). On the other hand there is no such unique invariant concept as the gradient of a *vector* $V^\alpha \in \mathfrak{T}^\alpha$ or, indeed, of a tensor of any valence other than $\left[\begin{smallmatrix}0\\0\end{smallmatrix}\right]$. But it is possible to impose an *additional* structure on $\mathcal{M}$, which is a 'gradient' operation on vectors, and which uniquely extends from vectors to all tensor fields on $\mathcal{M}$. That structure is called a *connection* and the operation that defines it is called *covariant derivative*. As mentioned in §4.1, there are certain operations involving differentiation (Lie derivative, exterior derivative, etc.) which do not require this additional structure of a connection on $\mathcal{M}$. Nevertheless it will be useful to discuss these operations, too, in terms of covariant derivative, rather than give an independent treatment of each.

A covariant derivative operator $\nabla_\alpha$ can be defined as a map

$$\nabla_\alpha : \mathfrak{T}^\beta \to \mathfrak{T}^\beta_\alpha \tag{4.2.1}$$

subject to the two* conditions:

$$\nabla_\alpha(U^\beta + V^\beta) = \nabla_\alpha U^\beta + \nabla_\alpha V^\beta \tag{4.2.2}$$

for all $U^\beta, V^\beta \in \mathfrak{T}^\beta$, and

$$\nabla_\alpha(fU^\beta) = f\nabla_\alpha U^\beta + U^\beta \nabla_\alpha f \tag{4.2.3}$$

for all $U^\beta \in \mathfrak{T}^\beta$, $f \in \mathfrak{T}$, where $\nabla_\alpha f$ is the ordinary gradient of $f$ defined in §4.1 (see 4.1.32), (4.1.40)). The elements $\nabla_\alpha U^\gamma \in \mathfrak{T}^\gamma_\alpha$, $\nabla_{\xi_3} U^\alpha \in \mathfrak{T}^\alpha_{\xi_3}$, etc., are defined from $\nabla_\alpha U^\beta$ by index substitution. The possibility of making such index substitutions is always available to us on any formula.

We can extend the definition of $\nabla_\alpha$ uniquely to apply to covariant vectors, giving a map

$$\nabla_\alpha : \mathfrak{T}_\beta \to \mathfrak{T}_{\alpha\beta}, \tag{4.2.4}$$

where $\nabla_\alpha A_\beta$ effects the $\mathfrak{T}$-linear map from $\mathfrak{T}^\beta$ to $\mathfrak{T}_\alpha$ (*cf*. (2.2.37)) defined by

$$(\nabla_\alpha A_\beta)V^\beta = \nabla_\alpha(A_\beta V^\beta) - A_\beta \nabla_\alpha V^\beta. \tag{4.2.5}$$

The fact that this is indeed $\mathfrak{T}$-linear follows from (4.1.43) and (4.2.2) and from (4.1.44) and (4.2.3). Note that this definition of $\nabla_\alpha A_\beta$ is forced upon us if we require the derivative of $A_\beta V^\beta$ to satisfy the Leibniz law.

We have

$$\nabla_\alpha(A_\beta + B_\beta) = \nabla_\alpha A_\beta + \nabla_\alpha B_\beta \tag{4.2.6}$$

and

$$\nabla_\alpha(fA_\beta) = f\nabla_\alpha A_\beta + A_\beta \nabla_\alpha f, \tag{4.2.7}$$

by (4.2.5), (4.1.43), (4.1.44).

* In standard modern treatments (*cf.* Hawking and Ellis 1973) covariant derivative is defined by *three*, rather than two requirements. The use of abstract indices enables us to achieve a somewhat greater economy here.

Next consider a general tensor $T^{\alpha\ldots\gamma}_{\lambda\ldots\nu}$. If we require the Leibniz law to hold for the derivative of $T^{\alpha\ldots\gamma}_{\lambda\ldots\nu}A_\alpha \ldots C_\gamma U^\lambda \ldots W^\nu$ we are led to the relation

$$\begin{aligned}(\nabla_\rho T^{\alpha\ldots\gamma}_{\lambda\ldots\nu})A_\alpha \ldots C_\gamma U^\lambda \ldots W^\nu = \nabla_\rho(T^{\alpha\ldots\gamma}_{\lambda\ldots\nu}A_\alpha \ldots C_\gamma U^\lambda \ldots W^\nu) - \\ - T^{\alpha\ldots\gamma}_{\lambda\ldots\nu}(\nabla_\rho A_\alpha) \ldots C_\gamma U^\lambda \ldots W^\nu - \ldots \\ - T^{\alpha\ldots\gamma}_{\lambda\ldots\nu}A_\alpha \ldots C_\gamma U^\lambda \ldots (\nabla_\rho W^\nu).\end{aligned} \tag{4.2.8}$$

This defines $\nabla_\rho T^{\alpha\ldots\gamma}_{\lambda\ldots\nu}$ as a map (which is easily checked to be $\mathfrak{T}$-multilinear) from $\mathfrak{T}_\alpha \times \cdots \times \mathfrak{T}^\nu$ to $\mathfrak{T}_\rho$ (*cf.* (2.2.38)). Thus any operator $\nabla_\alpha$ satisfying (4.2.1), (4.2.2), (4.2.3) extends *uniquely* to

$$\nabla_\rho : \mathfrak{T}^{\alpha\ldots\gamma}_{\lambda\ldots\nu} \to \mathfrak{T}^{\alpha\ldots\gamma}_{\lambda\ldots\nu\rho} \tag{4.2.9}$$

by the one requirement that $\nabla_\rho$ acting on a contracted product of the type $T^{\alpha\ldots\gamma}_{\lambda\ldots\nu}A_\alpha \ldots C_\gamma U^\lambda \ldots W^\nu$ should satisfy a Leibniz law. (We can also check that in the case of $\nabla_\rho : \mathfrak{T}^\alpha \to \mathfrak{T}^\alpha_\rho$, we get back to the original definition.)

By applying the definition (4.2.8) to the two sides of each of the following two equations, we readily verify that:

$$\nabla_\rho(T_{\mathscr{A}} + S_{\mathscr{A}}) = \nabla_\rho T_{\mathscr{A}} + \nabla_\rho S_{\mathscr{A}} \tag{4.2.10}$$

and

$$\nabla_\rho(T_{\mathscr{A}} R_{\mathscr{B}}) = T_{\mathscr{A}} \nabla_\rho R_{\mathscr{B}} + R_{\mathscr{B}} \nabla_\rho T_{\mathscr{A}}. \tag{4.2.11}$$

It is also clear that

$$\nabla_\rho \textit{ commutes with any index substitution not involving } \rho. \tag{4.2.12}$$

To see that $\nabla_\rho$ commutes also with contraction (not involving $\rho$) we can first build up $T^{\cdots}_{\cdots}$ as a sum of outer products of vectors (see (2.2.14)) and then apply linearity (4.2.10) and the Leibniz law (4.2.11) to each of $\nabla_\rho T^{\cdot\cdot\sigma\cdots\cdot}_{\ldots\tau\ldots}$ and $\nabla_\rho(T^{\cdot\cdot\sigma\cdots\cdot}_{\ldots\sigma\ldots})$. Since, by (4.2.11) and (4.2.5), we get a Leibniz expansion both for $\nabla_\rho(X^\sigma D_\tau)$ and for $\nabla_\rho(X^\sigma D_\sigma)$, it follows that the $\binom{\sigma}{\tau}$-contraction of the former is equal to the latter. The $\binom{\sigma}{\tau}$-contraction of $\nabla_\rho T^{\cdot\cdot\sigma\cdots}_{\ldots\tau\ldots}$ is therefore equal to $\nabla_\rho(T^{\cdot\cdot\sigma\cdots}_{\ldots\sigma\ldots})$ as required:

$$\delta^\tau_\sigma(\nabla_\rho T^{\cdot\cdot\sigma\cdots\cdot}_{\ldots\tau\ldots}) = \nabla_\rho T^{\cdot\cdot\sigma\cdots\cdot}_{\ldots\sigma\ldots}. \tag{4.2.13}$$

The properties (4.1.40), (4.2.10), (4.2.11) and (4.2.13) are often used to axiomatize the covariant derivative.

### *Torsion and curvature*

So far, the rules that we have obtained for $\nabla_\rho$ are all formally identical with the corresponding rules for the 'coordinate gradient operator' $\partial/\partial x^\rho$. However, one essential new feature arises here, namely the fact that the $\nabla_\rho$ operators need not commute with one another. To investigate this, set

$$\Delta_{\alpha\beta} := \nabla_\alpha \nabla_\beta - \nabla_\beta \nabla_\alpha = 2\nabla_{[\alpha}\nabla_{\beta]}. \tag{4.2.14}$$

We first observe that

$$\Delta_{\alpha\beta}(A_{\mathscr{D}} + B_{\mathscr{D}}) = \Delta_{\alpha\beta}A_{\mathscr{D}} + \Delta_{\alpha\beta}B_{\mathscr{D}}, \tag{4.2.15}$$

by (4.2.10), and that

$$\Delta_{\alpha\beta}(A_{\mathscr{D}}C_{\mathscr{E}}) = A_{\mathscr{D}}\Delta_{\alpha\beta}C_{\mathscr{E}} + C_{\mathscr{E}}\Delta_{\alpha\beta}A_{\mathscr{D}}, \tag{4.2.16}$$

by (4.2.11) since the 'cross-terms' $(\nabla_\alpha A_{\mathscr{D}})(\nabla_\beta C_{\mathscr{E}})$ and $(\nabla_\alpha C_{\mathscr{E}})(\nabla_\beta A_{\mathscr{D}})$ each cancel out.

Now consider $\Delta_{\alpha\beta}f$ for any scalar $f$. If $X^{\alpha\beta}$ is any element of $\mathfrak{T}^{\alpha\beta}$ we have, by (4.1.42), (4.2.15), (4.2.16)

$$X^{\alpha\beta}\Delta_{\alpha\beta}k = 0, \tag{4.2.17}$$

$$X^{\alpha\beta}\Delta_{\alpha\beta}(f+g) = X^{\alpha\beta}\Delta_{\alpha\beta}f + X^{\alpha\beta}\Delta_{\alpha\beta}g, \tag{4.2.18}$$

$$X^{\alpha\beta}\Delta_{\alpha\beta}(fg) = fX^{\alpha\beta}\Delta_{\alpha\beta}g + gX^{\alpha\beta}\Delta_{\alpha\beta}f, \tag{4.2.19}$$

for each $k \in \mathfrak{K}$, $f, g \in \mathfrak{T}$. Thus, by (4.1.11), $X^{\alpha\beta}\Delta_{\alpha\beta}$ is a derivation, whence, by (4.1.40),

$$X^{\alpha\beta}\Delta_{\alpha\beta} = Y^\gamma \nabla_\gamma \tag{4.2.20}$$

for some unique $Y^\gamma \in \mathfrak{T}^\gamma$, where the operators act on scalars. The map from $\mathfrak{T}^{\alpha\beta}$ to $\mathfrak{T}^\gamma$ which assigns $Y^\gamma$ to $X^{\alpha\beta}$ in (4.2.20) is obviously $\mathfrak{T}$-linear. (If (4.2.20) holds and $Z^{\alpha\beta}\Delta_{\alpha\beta} = W^\gamma\nabla_\gamma$, then $(X^{\alpha\beta}+Z^{\alpha\beta})\Delta_{\alpha\beta} = (Y^\gamma + W^\gamma)\nabla_\gamma$, $(pX^{\alpha\beta})\Delta_{\alpha\beta} = (pY^\gamma)\nabla_\gamma$.) Thus, by (2.2.37), this map is achieved by a tensor $T_{\alpha\beta}{}^\gamma \in \mathfrak{T}^\gamma_{\alpha\beta}$, called the *torsion tensor*:

$$Y^\gamma = X^{\alpha\beta}T_{\alpha\beta}{}^\gamma. \tag{4.2.21}$$

Substituting into (4.2.20) we get $X^{\alpha\beta}\Delta_{\alpha\beta} = X^{\alpha\beta}T_{\alpha\beta}{}^\gamma\nabla_\gamma$ (on scalars). This holds for all $X^{\alpha\beta} \in \mathfrak{T}^{\alpha\beta}$, so

$$\Delta_{\alpha\beta}f = T_{\alpha\beta}{}^\gamma\nabla_\gamma f \tag{4.2.22}$$

for all $f \in \mathfrak{T}$. Notice that by the anti-symmetry of $\Delta_{\alpha\beta}$ (*cf*. (4.2.14)), the torsion tensor also has this anti-symmetry:

$$T_{\alpha\beta}{}^\gamma = -T_{\beta\alpha}{}^\gamma. \tag{4.2.23}$$

If $T_{\alpha\beta}{}^\gamma = 0$, the operator $\nabla_\rho$ is called *torsion-free*.

Next consider the action of $\Delta_{\alpha\beta}$ on a vector. When torsion is present, it is actually rather simpler to work in terms of the operator (Cyrillic 'D')

$$Д_{\alpha\beta} := \Delta_{\alpha\beta} - T_{\alpha\beta}{}^\gamma\nabla_\gamma, \tag{4.2.24}$$

since

$$Д_{\alpha\beta}f = 0 \tag{4.2.25}$$

for all $f \in \mathfrak{T}$, so the relation (4.2.16)

$$\text{Д}_{\alpha\beta}(A_{\mathscr{D}}C_{\mathscr{C}}) = A_{\mathscr{D}}\text{Д}_{\alpha\beta}C_{\mathscr{C}} + C_{\mathscr{C}}\text{Д}_{\alpha\beta}A_{\mathscr{D}} \tag{4.2.26}$$

(*cf*. (4.2.11)) reduces to

$$\text{Д}_{\alpha\beta}(f A_{\mathscr{D}}) = f\,\text{Д}_{\alpha\beta}A_{\mathscr{D}} \tag{4.2.27}$$

when $C_{\mathscr{C}}$ is the scalar $f$. By (4.2.15) and (4.2.10) we also have

$$\text{Д}_{\alpha\beta}(A_{\mathscr{D}} + B_{\mathscr{D}}) = \text{Д}_{\alpha\beta}A_{\mathscr{D}} + \text{Д}_{\alpha\beta}B_{\mathscr{D}}. \tag{4.2.28}$$

By specializing (4.2.27) and (4.2.28) to elements of $\mathfrak{T}^\gamma$ (i.e., $\mathscr{D} = \gamma^*$), we see that the map $\text{Д}_{\alpha\beta} : \mathfrak{T}^\gamma \to \mathfrak{T}^\delta_{\alpha\beta}$ defined by

$$V^\gamma \mapsto \text{Д}_{\alpha\beta}V^\delta \tag{4.2.29}$$

is $\mathfrak{T}$-linear. It is therefore achieved by a tensor $R_{\alpha\beta\gamma}{}^\delta \in \mathfrak{T}^\delta_{\alpha\beta\gamma}$, called the *curvature tensor**:

$$\text{Д}_{\alpha\beta}V^\delta =: R_{\alpha\beta\gamma}{}^\delta V^\gamma. \tag{4.2.30}$$

Writing this out in full, we have

$$(\nabla_\alpha\nabla_\beta - \nabla_\beta\nabla_\alpha - T_{\alpha\beta}{}^\gamma\nabla_\gamma)V^\delta = R_{\alpha\beta\gamma}{}^\delta V^\gamma. \tag{4.2.31}$$

By (4.2.25) we have $\text{Д}_{\alpha\beta}(A_\delta V^\delta) = 0$. From this and (4.2.26) we get $V^\gamma\text{Д}_{\alpha\beta}A_\gamma = -A_\delta\text{Д}_{\alpha\beta}V^\delta$. This in turn, with (4.2.30) substituted into it, gives

$$\text{Д}_{\alpha\beta}A_\gamma = -R_{\alpha\beta\gamma}{}^\delta A_\delta. \tag{4.2.32}$$

To obtain the effect of $\text{Д}_{\alpha\beta}$ on a general tensor $H^{\rho\ldots\tau}_{\lambda\ldots\nu}$, we may expand $H^{\rho\ldots\tau}_{\lambda\ldots\nu}$ as a sum of outer product of vectors (*cf*. (2.2.14)) and use the 'Leibniz rule' (4.2.26), for $\text{Д}_{\alpha\beta}$, on each term to obtain the (*generalized*) *Ricci identity*:

$$\begin{aligned}(\nabla_\alpha\nabla_\beta - \nabla_\beta\nabla_\alpha - T_{\alpha\beta}{}^\gamma\nabla_\gamma)H^{\rho\ldots\tau}_{\lambda\ldots\nu} &= \text{Д}_{\alpha\beta}H^{\rho\ldots\tau}_{\lambda\ldots\nu} = R_{\alpha\beta\rho_0}{}^\rho H^{\rho_0\ldots\tau}_{\lambda\ldots\nu} + \\ + \cdots + R_{\alpha\beta\tau_0}{}^\tau H^{\rho\ldots\tau_0}_{\lambda\ldots\nu} &- R_{\alpha\beta\lambda}{}^{\lambda_0}H^{\rho\ldots\tau}_{\lambda_0\ldots\nu} - \cdots - R_{\alpha\beta\nu}{}^{\nu_0}H^{\rho\ldots\tau}_{\lambda\ldots\nu_0}.\end{aligned} \tag{4.2.33}$$

Observe that, owing to the anti-symmetry of $\text{Д}_{\alpha\beta}$ (*cf*, (4.2.23), (4.2.24)), we have

$$R_{\alpha\beta\gamma}{}^\delta = -R_{\beta\alpha\gamma}{}^\delta. \tag{4.2.34}$$

We obtain a further (*Bianchi*) 'symmetry' condition on $R_{\alpha\beta\gamma}{}^\delta$ if we apply (4.2.32) to the case $A_\gamma = \nabla_\gamma f$. For simplicity, let us compute this explicitly only in the torsion-free case. We have, in that case,

$$2\nabla_{[\alpha}\nabla_{\beta]}\nabla_\gamma f = -R_{\alpha\beta\gamma}{}^\delta\nabla_\delta f: \tag{4.2.35}$$

anti-symmetrizing in $\alpha, \beta, \gamma$ and using (3.3.9) we get

$$\begin{aligned}-\tfrac{1}{2}R_{[\alpha\beta\gamma]}{}^\delta\nabla_\delta f &= \nabla_{[[\alpha}\nabla_{\beta]}\nabla_{\gamma]}f = \nabla_{[\alpha}\nabla_\beta\nabla_{\gamma]}f \\ &= \nabla_{[\alpha}\nabla_{[\beta}\nabla_{\gamma]]}f = \tfrac{1}{2}\nabla_{[\alpha}T_{\beta\gamma]}{}^\delta\nabla_\delta f = 0,\end{aligned} \tag{4.2.36}$$

* Unfortunately there is no general agreement on the sign and index arrangement of this tensor, and almost all possible variations can be found in the literature.

since we assumed vanishing torsion. But at any one point the $\nabla_\delta f$ are arbitrary, so at *every* point we have

$$R_{[\alpha\beta\gamma]}{}^{\delta} = 0. \tag{4.2.37}$$

Taking account of (4.2.34), this reads

$$R_{\alpha\beta\gamma}{}^{\delta} + R_{\beta\gamma\alpha}{}^{\delta} + R_{\gamma\alpha\beta}{}^{\delta} = 0. \tag{4.2.38}$$

If the torsion is not assumed to vanish, the computation proceeds along the same lines but is more elaborate. The result is:

$$R_{[\alpha\beta\gamma]}{}^{\delta} + \nabla_{[\alpha}T_{\beta\gamma]}{}^{\delta} + T_{[\alpha\beta}{}^{\rho}T_{\gamma]\rho}{}^{\delta} = 0. \tag{4.2.39}$$

Next, expand $\nabla_{[\alpha}\nabla_{\beta}\nabla_{\gamma]}V^{\delta}$ (as in (4.2.36)) in two different ways, using once the anti-symmetry in $\alpha, \beta$, once the anti-symmetry in $\beta, \gamma$. In the first case we apply (4.2.33) to $\nabla_{\gamma}V^{\delta}$ and in the second case we use the derivative of (4.2.31). As above, we shall only do this explicitly in the torsion-free case. We have, in that case,

$$2\nabla_{[[\alpha}\nabla_{\beta]}\nabla_{\gamma]}V^{\delta} = R_{[\alpha\beta|\rho|}{}^{\delta}\nabla_{\gamma]}V^{\rho} - R_{[\alpha\beta\gamma]}{}^{\rho}\nabla_{\rho}V^{\gamma} \tag{4.2.40}$$

and

$$2\nabla_{[\alpha}\nabla_{[\beta}\nabla_{\gamma]]}V^{\delta} = \nabla_{[\alpha}(R_{\beta\gamma]\rho}{}^{\delta}V^{\rho}) = R_{[\beta\gamma|\rho|}{}^{\delta}\nabla_{\alpha]}V^{\rho} + V^{\rho}\nabla_{[\alpha}R_{\beta\gamma]\rho}{}^{\delta}. \tag{4.2.41}$$

Substracting these two expansions and using (4.2.37) we get *Bianchi's identity*:

$$\nabla_{[\alpha}R_{\beta\gamma]\rho}{}^{\sigma} = 0. \tag{4.2.42}$$

If the torsion does not vanish the computation is similar but more complicated. (See Appendix, Fig. A-9.) The result is:

$$\nabla_{[\alpha}R_{\beta\gamma]\rho}{}^{\sigma} + T_{[\alpha\beta}{}^{\delta}R_{\gamma]\delta\rho}{}^{\sigma} = 0. \tag{4.2.43}$$

An alternative method of deriving (4.2.39) and (4.2.43) will be given shortly. (*cf.* (4.2.52) *et seq.*)

### *Change of derivative operator*

Suppose, now, that we have a second covariant derivative operator $\tilde{\nabla}_{\alpha}: \mathfrak{T}^{\beta} \to \mathfrak{T}^{\beta}_{\alpha}$ which also satisfies (4.2.2) and (4.2.3). Consider the difference between this and $\nabla_{\alpha}$. The map

$$(\tilde{\nabla}_{\alpha} - \nabla_{\alpha}): \mathfrak{T}^{\beta} \to \mathfrak{T}^{\beta}_{\alpha} \tag{4.2.44}$$

satisfies $(\tilde{\nabla}_{\alpha} - \nabla_{\alpha})\ (U^{\beta} + V^{\beta}) = (\tilde{\nabla}_{\alpha} - \nabla_{\alpha})U^{\beta} + (\tilde{\nabla}_{\alpha} - \nabla_{\alpha})V^{\beta}$ by (4.2.2); also $(\tilde{\nabla}_{\alpha} - \nabla_{\alpha})\,(fU^{\beta}) = f(\tilde{\nabla}_{\alpha} - \nabla_{\alpha})U^{\beta}$ by (4.2.3) and by the fact that the operators must agree on scalars:

$$\tilde{\nabla}_{\alpha}f = \nabla_{\alpha}f \tag{4.2.45}$$

(see (4.1.40)). The map (4.2.44) is therefore $\mathfrak{T}$-linear and so by (2.2.37) we have

$$(\tilde{\nabla}_\alpha - \nabla_\alpha)U^\beta = Q_{\alpha\gamma}{}^\beta U^\gamma \tag{4.2.46}$$

for some $Q_{\alpha\gamma}{}^\beta \in \mathfrak{T}^\beta_{\alpha\gamma}$. Conversely, given an arbitrary $Q_{\alpha\gamma}{}^\beta$ and a covariant derivative operator $\nabla_\alpha$, *any* $\tilde{\nabla}_\alpha$ defined by (4.2.46) will also be a covariant derivative operator.

Since $(\tilde{\nabla}_\alpha - \nabla_\alpha)(A_\beta U^\beta) = 0$, by (4.2.45), we have, by (4.2.5).

$$(\tilde{\nabla}_\alpha - \nabla_\alpha)A_\beta = -Q_{\alpha\beta}{}^\gamma A_\gamma. \tag{4.2.47}$$

Now any tensor $H^{\alpha\ldots\gamma}_{\lambda\ldots\nu}$ is a sum of outer products of vectors, so it follows from the Leibniz rule and linearity that

$$\begin{aligned}(\tilde{\nabla}_\rho - \nabla_\rho)H^{\alpha\ldots\gamma}_{\lambda\ldots\nu} = {} & Q_{\rho\alpha_0}{}^\alpha H^{\alpha_0\ldots\gamma}_{\lambda\ldots\nu} + \cdots + Q_{\rho\gamma_0}{}^\gamma H^{\alpha\ldots\gamma_0}_{\lambda\ldots\nu} \\ & - Q_{\rho\lambda}{}^{\lambda_0} H^{\alpha\ldots\gamma}_{\lambda_0\ldots\nu} - \cdots - Q_{\rho\nu}{}^{\nu_0} H^{\alpha\ldots\gamma}_{\lambda\ldots\nu_0}.\end{aligned} \tag{4.2.48}$$

Let $\tilde{T}_{\alpha\beta}{}^\gamma$ and $\tilde{R}_{\alpha\beta\gamma}{}^\delta$ be the torsion and curvature tensors, respectively, defined by $\tilde{\nabla}_\rho$. We have

$$\begin{aligned}\tilde{T}_{\alpha\beta}{}^\gamma \nabla_\gamma f = \tilde{T}_{\alpha\beta}{}^\gamma \tilde{\nabla}_\gamma f &= 2\tilde{\nabla}_{[\alpha}\tilde{\nabla}_{\beta]} f = 2\tilde{\nabla}_{[\alpha}\nabla_{\beta]} f \\ &= 2\nabla_{[\alpha}\nabla_{\beta]} f - 2Q_{[\alpha\beta]}{}^\gamma \nabla_\gamma f \\ &= \{T_{\alpha\beta}{}^\gamma - 2Q_{[\alpha\beta]}{}^\gamma\}\nabla_\gamma f\end{aligned} \tag{4.2.49}$$

whence

$$\tilde{T}_{\alpha\beta}{}^\gamma = T_{\alpha\beta}{}^\gamma - Q_{\alpha\beta}{}^\gamma + Q_{\beta\alpha}{}^\gamma, \text{ i.e., } \tilde{T}_{\alpha\beta}{}^\gamma - T_{\alpha\beta}{}^\gamma = -2Q_{[\alpha\beta]}{}^\gamma. \tag{4.2.50}$$

The calculation for $\tilde{R}_{\alpha\beta\gamma}{}^\delta$ proceeds along similar lines by consideration of $2\tilde{\nabla}_{[\alpha}\tilde{\nabla}_{\beta]}V^\delta$, but is somewhat more complicated. The result is

$$\tilde{R}_{\alpha\beta\gamma}{}^\delta = R_{\alpha\beta\gamma}{}^\delta - T_{\alpha\beta}{}^\rho Q_{\rho\gamma}{}^\delta + 2\nabla_{[\alpha}Q_{\beta]\gamma}{}^\delta + 2Q_{[\alpha|\rho|}{}^\delta Q_{\beta]\gamma}{}^\rho. \tag{4.2.51}$$

A particular case of special interest is given by $Q_{\alpha\beta}{}^\gamma = \frac{1}{2}T_{\alpha\beta}{}^\gamma$; for $Q_{\alpha\beta}{}^\gamma$ is now anti-symmetric and it follows from (4.2.50) that $\tilde{\nabla}_\rho$ is torsion-free. Thus we have a canonical prescription* for constructing a torsion-free $\tilde{\nabla}_\rho$ from any covariant derivative operator $\nabla_\rho$. In this case we have

$$\tilde{R}_{\alpha\beta\gamma}{}^\delta = R_{\alpha\beta\gamma}{}^\delta + \nabla_{[\alpha}T_{\beta]\gamma}{}^\delta - \tfrac{1}{2}T_{\rho[\alpha}{}^\delta T_{\beta]\gamma}{}^\rho - \tfrac{1}{2}T_{\alpha\beta}{}^\rho T_{\rho\gamma}{}^\delta. \tag{4.2.52}$$

This formula may be used to obtain an alternative derivation of (4.2.39) and (4.2.43). We just substitute $\tilde{R}_{\alpha\beta\gamma}{}^\delta$ (and $\tilde{\nabla}_\rho$) into the simpler formulae (4.2.37) and (4.2.42) which hold for the torsion-free case. The results are (4.2.39) and (4.2.43), respectively.

Until now we have said nothing about the *existence* of a covariant derivative operator on a given manifold $\mathcal{M}$. It is in fact a theorem

* In the presence of a metric, however, there is a *different* 'canonical' prescription which may be preferred, namely passing to the unique torsion-free operator satisfying (4.3.46) (*cf.* also §4.7)

(*cf.* Kobayashi and Nomizu 1963) that a connection exists globally on *any* manifold which, as here, has a countable basis for its topology. Once one connection has been found, many others may be derived from it by use of arbitrary elements $Q_{\alpha\beta}{}^{\gamma} \in \mathfrak{T}^{\gamma}_{\alpha\beta}$ in the manner described above. We shall not require any deep existence theorems here, however, since the existence of a physical metric will shortly be assumed, and a metric (of any signature) has associated with it a uniquely defined torsion-free connection (*cf.* (4.3.47) below).

### *Coordinate derivative*

It is sometimes convenient to introduce 'arbitrary' connections – which need have no special relation to any preassigned structure on $\mathcal{M}$ – for the purposes of facilitating calculations. As we shall see in a moment, many such connections exist locally. A useful example is a connection arising from the concept of 'coordinate derivative' in some coordinate system. There are two ways of arriving at the concept of coordinate derivative within the present framework. The more straightforward of these is simply to express all tensor quantities in terms of their components with respect to a coordinate system, and then to consider the collection of partial derivatives of these components with respect to the coordinates $x^{\mathbf{a}}$. The result is a set of scalar fields, or, equivalently, a set of functions of the $x^{\mathbf{a}}$. This is what one normally requires when explicit calculations are involved. The alternative point of view is to take this collection of scalar fields and to reconstruct tensor quantities from them, the scalar fields being regarded as the components of the tensors in the given coordinate system. Thus, according to this second point of view, the coordinate derivative provides a means of passing from tensors to tensors; in short, it provides us with a connection on the manifold $\mathcal{M}$. This connection is coordinate-dependent, however, in the sense that a different coordinate system would provide us with a different connection. (We work locally until the end of §4.2.)

Let us examine this in more detail. Consider a local coordinate system $(\mathscr{U}, x^{\mathbf{a}})$. When we take components of a tensor with respect to this coordinate system, we employ the *coordinate basis* for $\mathfrak{T}^{\cdot}$ (sometimes called a 'natural' basis) given in (4.1.33):

$$\boldsymbol{\delta}_1 = \frac{\partial}{\partial x^1}, \ldots, \boldsymbol{\delta}_n = \frac{\partial}{\partial x^n}. \tag{4.2.53}$$

The dual basis (4.1.34) is

$$\boldsymbol{\delta}^1 = \mathrm{d}x^1, \ldots, \boldsymbol{\delta}^n = \mathrm{d}x^n. \tag{4.2.54}$$

The canonical images in $\mathfrak{T}^\alpha$ and in $\mathfrak{T}_\alpha$, of $\delta_{\mathbf{a}} \in \mathfrak{T}^\bullet$ and $\delta^{\mathbf{a}} \in \mathfrak{T}^{\bullet *}$ respectively, are $\delta^\alpha_{\mathbf{a}}$ and $\delta^{\mathbf{a}}_\alpha$. Thus, using the notation $\nabla_\alpha$ for d (*cf.* (4.1.40)) we can re-express (4.2.54) as

$$\delta^{\mathbf{a}}_\alpha = \nabla_\alpha x^{\mathbf{a}}. \tag{4.2.55}$$

Then we may regard $\delta^\alpha_{\mathbf{a}}$ as being defined from $\delta^{\mathbf{a}}_\alpha$ by the equation $\delta^{\mathbf{a}}_\alpha \delta^\alpha_{\mathbf{b}} = \delta^{\mathbf{a}}_{\mathbf{b}}$.

The components of a tensor $H^{\alpha\ldots\gamma}_{\lambda\ldots\mu}$ in this coordinate system are then

$$H^{\mathbf{a}\ldots\mathbf{c}}_{\mathbf{l}\ldots\mathbf{m}} = H^{\alpha\ldots\gamma}_{\lambda\ldots\mu}\delta^{\mathbf{a}}_\alpha \ldots \delta^{\mathbf{c}}_\gamma \delta^\lambda_{\mathbf{l}} \ldots \delta^\mu_{\mathbf{m}}, \tag{4.2.56}$$

by (2.3.13). The partial derivatives with respect to $x^{\mathbf{p}}$ are

$$\frac{\partial}{\partial x^{\mathbf{p}}} H^{\mathbf{a}\ldots\mathbf{c}}_{\mathbf{l}\ldots\mathbf{m}}. \tag{4.2.57}$$

The set of scalars (4.2.57) gives the *first* notion of coordinate derivative discussed above. To obtain the *second* notion we reconstruct a tensor, whose components are (4.2.57), by means of the standard procedure (2.3.14). This defines a tensor, which we write

$$\partial_\rho H^{\alpha\ldots\gamma}_{\lambda\ldots\mu} := \left(\frac{\partial}{\partial x^{\mathbf{p}}} H^{\mathbf{a}\ldots\mathbf{c}}_{\mathbf{l}\ldots\mathbf{m}}\right)\delta^{\mathbf{p}}_\rho \delta^\alpha_{\mathbf{a}} \ldots \delta^\gamma_{\mathbf{c}} \delta^{\mathbf{l}}_\lambda \ldots \delta^{\mathbf{m}}_\mu \tag{4.2.58}$$

The operator $\partial_\rho$ defines a map from $\mathfrak{T}^{\alpha\ldots\gamma}_{\lambda\ldots\mu}$ to $\mathfrak{T}^{\alpha\ldots\gamma}_{\lambda\ldots\mu\rho}$ which clearly satisfies all the properties required of a covariant derivative operator. But it is an operator of no intrinsic interest in general, since the definition of $\partial_\rho$ is tied to the particular* coordinate system $x^{\mathbf{a}}$.

On the other hand, $\partial_\rho$ may sometimes be introduced as a convenience, since it has especially simple properties. These arise from the fact that partial derivatives $\partial/\partial x^{\mathbf{a}}$ *commute*: $\partial^2/\partial x^{\mathbf{a}}\,\partial x^{\mathbf{b}} = \partial^2/\partial x^{\mathbf{b}}\,\partial x^{\mathbf{a}}$. Thus we have

$$\partial_\alpha \partial_\beta = \partial_\beta \partial_\alpha, \tag{4.2.59}$$

so the torsion and curvature defined by $\partial_\alpha$ both *vanish.* (In fact, it can be shown (*cf.* Dodson and Poston 1977) that *any* covariant derivative operator for which the curvature and torsion both vanish must *locally* be of the above form for some coordinates $x^{\mathbf{a}}$.)

The significance of $\partial_\alpha$ lies in its particularly simple expression in terms of components in the $x^{\mathbf{a}}$-system. Let us now examine the somewhat less simple expression of a *given* covariant derivative operator $\nabla_\alpha$ in terms of components. For increased generality, we shall consider a basis $\delta^\alpha_{\mathbf{a}}$ for $\mathfrak{T}^\alpha$, and dual basis $\delta^{\mathbf{a}}_\alpha \in \mathfrak{T}_\alpha$, which need *not* be naturally obtainable from a coordinate system by means of (4.2.53)–(4.2.55). A basis which is so

* It may be observed, however, that any other coordinates $y^{\hat{\mathbf{a}}}$ which are related to $x^{\mathbf{a}}$ by *constant* linear expressions (i.e., so that $\partial y^{\hat{\mathbf{a}}}/\partial x^{\mathbf{b}}$ are constants) give rise to the same operator $\partial_\rho$

obtainable from *some* coordinate system is sometimes called *holonomic*, and in the contrary case, *non-holonomic*.* Thus we shall include non-holonomic bases into our discussion. Many more manifolds admit globally defined non-holonomic bases than admit globally defined holonomic ones. (An example is $S^3$, see footnotes on pp. 92, 93.)

We shall need the components of the covariant derivatives of the basis elements. Define the *connection symbols*** by

$$\Gamma_{\boldsymbol{\alpha\beta}}{}^{\boldsymbol{\gamma}} := \delta^{\boldsymbol{\gamma}}_{\beta} \nabla_{\boldsymbol{\alpha}} \delta^{\beta}_{\boldsymbol{\beta}}, \tag{4.2.60}$$

where $\nabla_{\boldsymbol{\alpha}}$ stands for $\delta^{\alpha}_{\boldsymbol{\alpha}} \nabla_{\alpha}$. Since $\delta^{\boldsymbol{\gamma}}_{\boldsymbol{\beta}}$ are just constants 0 and 1, we have $0 = \nabla_{\boldsymbol{\alpha}} \delta^{\boldsymbol{\gamma}}_{\boldsymbol{\beta}} = \nabla_{\boldsymbol{\alpha}} (\delta^{\boldsymbol{\gamma}}_{\beta} \delta^{\beta}_{\boldsymbol{\beta}}) = \delta^{\boldsymbol{\gamma}}_{\beta} \nabla_{\boldsymbol{\alpha}} \delta^{\beta}_{\boldsymbol{\beta}} + \delta^{\beta}_{\boldsymbol{\beta}} \nabla_{\boldsymbol{\alpha}} \delta^{\boldsymbol{\gamma}}_{\beta}$. Thus we also have

$$\Gamma_{\boldsymbol{\alpha\beta}}{}^{\boldsymbol{\gamma}} = -\delta^{\beta}_{\boldsymbol{\beta}} \nabla_{\boldsymbol{\alpha}} \delta^{\boldsymbol{\gamma}}_{\beta} \tag{4.2.61}$$

Now, consider the components $(\nabla_{\alpha} V^{\beta}) \delta^{\alpha}_{\boldsymbol{\alpha}} \delta^{\boldsymbol{\beta}}_{\beta}$ of the covariant derivative $\nabla_{\alpha} V^{\beta}$ of the vector $V^{\beta}$. We have

$$\begin{aligned}(\nabla_{\alpha} V^{\beta}) \delta^{\alpha}_{\boldsymbol{\alpha}} \delta^{\boldsymbol{\beta}}_{\beta} &= \delta^{\boldsymbol{\beta}}_{\beta} \nabla_{\boldsymbol{\alpha}} (\delta^{\beta}_{\boldsymbol{\gamma}} V^{\boldsymbol{\gamma}}) \\ &= \delta^{\boldsymbol{\beta}}_{\beta} \delta^{\beta}_{\boldsymbol{\gamma}} \nabla_{\boldsymbol{\alpha}} V^{\boldsymbol{\gamma}} + V^{\boldsymbol{\gamma}} \delta^{\boldsymbol{\beta}}_{\beta} \nabla_{\boldsymbol{\alpha}} \delta^{\beta}_{\boldsymbol{\gamma}} \\ &= \nabla_{\boldsymbol{\alpha}} V^{\boldsymbol{\beta}} + V^{\boldsymbol{\gamma}} \Gamma_{\boldsymbol{\alpha\gamma}}{}^{\boldsymbol{\beta}}\end{aligned} \tag{4.2.62}$$

by (4.2.60). Similarly, for the components $(\nabla_{\alpha} A_{\beta}) \delta^{\alpha}_{\boldsymbol{\alpha}} \delta^{\beta}_{\boldsymbol{\beta}}$ of $\nabla_{\alpha} A_{\beta}$ we have

$$\begin{aligned}(\nabla_{\alpha} A_{\beta}) \delta^{\alpha}_{\boldsymbol{\alpha}} \delta^{\beta}_{\boldsymbol{\beta}} &= \delta^{\beta}_{\boldsymbol{\beta}} \nabla_{\boldsymbol{\alpha}} (\delta^{\boldsymbol{\gamma}}_{\beta} A_{\boldsymbol{\gamma}}) = \delta^{\beta}_{\boldsymbol{\beta}} \delta^{\boldsymbol{\gamma}}_{\beta} \nabla_{\boldsymbol{\alpha}} A_{\boldsymbol{\gamma}} + A_{\boldsymbol{\gamma}} \delta^{\beta}_{\boldsymbol{\beta}} \nabla_{\boldsymbol{\alpha}} \delta^{\boldsymbol{\gamma}}_{\beta} \\ &= \nabla_{\boldsymbol{\alpha}} A_{\boldsymbol{\beta}} - A_{\boldsymbol{\gamma}} \Gamma_{\boldsymbol{\alpha\beta}}{}^{\boldsymbol{\gamma}},\end{aligned} \tag{4.2.63}$$

by (4.2.61). For the components of the covariant derivative of a general tensor $H^{\alpha\ldots\gamma}_{\lambda\ldots\nu}$ we have

$$\begin{aligned}&(\nabla_{\rho} H^{\alpha\ldots\gamma}_{\lambda\ldots\nu}) \delta^{\rho}_{\boldsymbol{\rho}} \delta^{\boldsymbol{\alpha}}_{\alpha} \ldots \delta^{\boldsymbol{\gamma}}_{\gamma} \delta^{\lambda}_{\boldsymbol{\lambda}} \ldots \delta^{\nu}_{\boldsymbol{\nu}} \\ &= \delta^{\boldsymbol{\alpha}}_{\alpha} \ldots \delta^{\nu}_{\boldsymbol{\nu}} \nabla_{\boldsymbol{\rho}} (\delta^{\alpha}_{\boldsymbol{\alpha}_0} \ldots \delta^{\boldsymbol{\nu}_0}_{\nu} H^{\boldsymbol{\alpha}_0\ldots\boldsymbol{\gamma}_0}_{\boldsymbol{\lambda}_0\ldots\boldsymbol{\nu}_0}) \\ &= \nabla_{\boldsymbol{\rho}} H^{\boldsymbol{\alpha}\ldots\boldsymbol{\gamma}}_{\boldsymbol{\lambda}\ldots\boldsymbol{\nu}} + H^{\boldsymbol{\alpha}_0\ldots\boldsymbol{\gamma}}_{\boldsymbol{\lambda}\ldots\boldsymbol{\nu}} \Gamma_{\boldsymbol{\rho\alpha}_0}{}^{\boldsymbol{\alpha}} + \cdots + H^{\boldsymbol{\alpha}\ldots\boldsymbol{\gamma}_0}_{\boldsymbol{\lambda}\ldots\boldsymbol{\nu}} \Gamma_{\boldsymbol{\rho}\gamma_0}{}^{\gamma} \\ &\quad - H^{\boldsymbol{\alpha}\ldots\boldsymbol{\gamma}}_{\boldsymbol{\lambda}_0\ldots\boldsymbol{\nu}} \Gamma_{\boldsymbol{\rho\lambda}}{}^{\boldsymbol{\lambda}_0} - \cdots - H^{\boldsymbol{\alpha}\ldots\boldsymbol{\gamma}}_{\boldsymbol{\lambda}\ldots\boldsymbol{\nu}_0} \Gamma_{\boldsymbol{\rho\nu}}{}^{\boldsymbol{\nu}_0}.\end{aligned} \tag{4.2.64}$$

In the particular case when $\delta^{\alpha}_{\boldsymbol{\alpha}}$ is a *coordinate basis* (with $\nabla_{\alpha} x^{\boldsymbol{\alpha}} = \delta^{\boldsymbol{\alpha}}_{\alpha}$), the operator $\nabla_{\boldsymbol{\rho}}$ may be written as $\partial/\partial x^{\boldsymbol{\rho}}$, in accordance with (4.1.41), since it acts on scalars. In this case the part of $\Gamma_{\boldsymbol{\alpha\beta}}{}^{\gamma}$ which is skew in $\boldsymbol{\alpha}, \boldsymbol{\beta}$ defines the torsion components since

$$\begin{aligned}2\Gamma_{[\boldsymbol{\alpha\beta}]}{}^{\boldsymbol{\gamma}} &= 2\delta^{\alpha}_{[\boldsymbol{\alpha}} \nabla_{\boldsymbol{\beta}]} \delta^{\boldsymbol{\gamma}}_{\alpha} = 2\delta^{\alpha}_{[\boldsymbol{\alpha}} \delta^{\beta}_{\boldsymbol{\beta}]} \nabla_{\beta} \nabla_{\alpha} x^{\boldsymbol{\gamma}} \\ &= 2\delta^{\alpha}_{\boldsymbol{\alpha}} \delta^{\beta}_{\boldsymbol{\beta}} \nabla_{[\beta} \nabla_{\alpha]} x^{\boldsymbol{\gamma}} = \delta^{\alpha}_{\boldsymbol{\alpha}} \delta^{\beta}_{\boldsymbol{\beta}} T_{\beta\alpha}{}^{\gamma} \nabla_{\gamma} x^{\boldsymbol{\gamma}} \\ &= T_{\boldsymbol{\beta\alpha}}{}^{\boldsymbol{\gamma}}\end{aligned} \tag{4.2.65}$$

* The (local) condition for a basis $\boldsymbol{\delta}_{\boldsymbol{\alpha}}$ to be holonomic is $[\boldsymbol{\delta}_{\boldsymbol{\alpha}}\ \boldsymbol{\delta}_{\boldsymbol{\iota}}] = 0$, [ ] being the Lie bracket operation of (4.3.2), (4.3.26) below. It may be remarked that a derivative operator $\partial_{\rho}$ can also be defined for a non-holonomic basis using (4.2.58), (4.2.56); this $\partial_{\rho}$ has torsion but no curvature.

** We are using a non-standard ordering of the indices on $\Gamma_{\boldsymbol{\alpha\beta}}{}^{\gamma}$, this being more compatible with our other conventions (notably the use of $\nabla$ rather than a semicolon).

Thus $\Gamma_{\alpha\beta}{}^{\gamma}$ is symmetric in $\alpha, \beta$ if the torsion vanishes. On the other hand, if $\delta^{\alpha}_{\boldsymbol{\alpha}}$ is not assumed to be a coordinate basis – that is to say, in the general non-holonomic case – there is no fixed relation between $\Gamma_{[\alpha\beta]}{}^{\gamma}$ and the torsion. Indeed, in the general case, there is no way of determining the torsion tensor from knowledge of the scalar fields $\Gamma_{\alpha\beta}{}^{\gamma}$ alone. One needs additional information such as the expression of these quantities in terms of some coordinate scalars $x^{\boldsymbol{\alpha}}$.

Let us assume, for the moment, that $\nabla_{\rho}$ is torsion-free but that the $\Gamma_{\alpha\beta}{}^{\gamma}$ are quite general (non-holonomic). We may compute the curvature tensor components from the following formula

$$R_{\rho\sigma\boldsymbol{\alpha}}{}^{\beta} = \delta^{\alpha}_{\boldsymbol{\alpha}}(\nabla_{\rho}\nabla_{\sigma} - \nabla_{\sigma}\nabla_{\rho})\delta^{\beta}_{\boldsymbol{\alpha}} \tag{4.2.66}$$

(*cf.* (4.2.31)). We have

$$\begin{aligned} \tfrac{1}{2}R_{\boldsymbol{\rho\sigma\alpha}}{}^{\boldsymbol{\beta}} &= \delta^{\rho}_{\boldsymbol{\rho}}\delta^{\sigma}_{\boldsymbol{\sigma}}\delta^{\boldsymbol{\beta}}_{\beta}\nabla_{[\rho}\nabla_{\sigma]}\delta^{\beta}_{\boldsymbol{\alpha}} \\ &= \delta^{\rho}_{\boldsymbol{\rho}}\delta^{\sigma}_{\boldsymbol{\sigma}}\delta^{\boldsymbol{\beta}}_{\beta}\nabla_{[\rho}\{\delta^{\boldsymbol{\tau}}_{\sigma]}\delta^{\beta}_{\boldsymbol{\gamma}}\Gamma_{\boldsymbol{\tau\alpha}}{}^{\boldsymbol{\gamma}}\} \\ &= \nabla_{[\boldsymbol{\rho}}\Gamma_{\boldsymbol{\sigma}]\boldsymbol{\alpha}}{}^{\boldsymbol{\beta}} + \Gamma_{[\boldsymbol{\sigma}|\boldsymbol{\alpha}|}{}^{\boldsymbol{\gamma}}\Gamma_{\boldsymbol{\rho}]\boldsymbol{\gamma}}{}^{\boldsymbol{\beta}} - \Gamma_{[\boldsymbol{\rho\sigma}]}{}^{\boldsymbol{\tau}}\Gamma_{\boldsymbol{\tau\alpha}}{}^{\boldsymbol{\beta}}. \end{aligned} \tag{4.2.67}$$

If torsion is present, it is clear from (4.2.31) that the modification of this formula which is required is the inclusion of an additional term

$$-\tfrac{1}{2}T_{\boldsymbol{\rho\sigma}}{}^{\boldsymbol{\gamma}}\Gamma_{\boldsymbol{\gamma\alpha}}{}^{\boldsymbol{\beta}} \tag{4.2.68}$$

on the right. If, on the other hand, torsion is absent and also the basis is specialized to a coordinate basis (4.2.53), then we get the familiar classical formula

$$R_{\boldsymbol{\rho\sigma\alpha}}{}^{\boldsymbol{\beta}} = \frac{\partial\Gamma_{\boldsymbol{\alpha\sigma}}{}^{\boldsymbol{\beta}}}{\partial x^{\boldsymbol{\rho}}} - \frac{\partial\Gamma_{\boldsymbol{\alpha\rho}}{}^{\boldsymbol{\beta}}}{\partial x^{\boldsymbol{\sigma}}} + \Gamma_{\boldsymbol{\alpha\sigma}}{}^{\boldsymbol{\gamma}}\Gamma_{\boldsymbol{\gamma\rho}}{}^{\boldsymbol{\beta}} - \Gamma_{\boldsymbol{\alpha\rho}}{}^{\boldsymbol{\gamma}}\Gamma_{\boldsymbol{\gamma\sigma}}{}^{\boldsymbol{\beta}}. \tag{4.2.69}$$

There is another way of obtaining these formulae in the case of a holonomic (coordinate) basis. This is to employ (4.2.51) and take $\tilde{\nabla}_{\rho} = \partial_{\rho}$, as defined in (4.2.58). Comparison of (4.2.62) with the component version of (4.2.46) yields

$$Q_{\boldsymbol{\alpha\beta}}{}^{\boldsymbol{\gamma}} = -\Gamma_{\boldsymbol{\alpha\beta}}{}^{\boldsymbol{\gamma}}. \tag{4.2.70}$$

We have $\tilde{R}_{\boldsymbol{\alpha\beta\gamma}}{}^{\boldsymbol{\delta}} = 0$, so taking components of (4.2.51) and substituting (4.2.70) we obtain the required formula for $R_{\boldsymbol{\alpha\beta\gamma}}{}^{\boldsymbol{\delta}}$ (whether or not the torsion vanishes). Similarly the expression (4.2.65) for $T_{\boldsymbol{\alpha\beta}}{}^{\boldsymbol{\gamma}}$ in terms of the skew part of $\Gamma_{\boldsymbol{\alpha\beta}}{}^{\boldsymbol{\gamma}}$ in a coordinate basis may be obtained by taking components of (4.2.50), where $\tilde{T}_{\boldsymbol{\alpha\beta}}{}^{\boldsymbol{\gamma}} = 0$.

It is sometimes notationally convenient to express formulae involving coordinate derivatives in their *abstract-index* versions, involving $\partial_{\rho}$ (i.e., to employ the second way of viewing the coordinate derivative that was mentioned above). The reason is that when taking components of a covariant derivative one must display the basis $\delta$s explicitly, whereas these basis $\delta$s may be eliminated in the abstract-index versions of the same formulae. Thus, for example, the fundamental equation (4.2.62) in a coordinate basis,

$$(\nabla_{\boldsymbol{\alpha}} V^{\beta})\delta^{\alpha}_{\boldsymbol{\alpha}}\delta^{\boldsymbol{\beta}}_{\beta} = \frac{\partial V^{\boldsymbol{\beta}}}{\partial x^{\boldsymbol{\alpha}}} + V^{\boldsymbol{\gamma}}\Gamma_{\boldsymbol{\alpha}\boldsymbol{\gamma}}{}^{\boldsymbol{\beta}}, \tag{4.2.71}$$

may be re-expressed in terms of abstract indices as

$$\nabla_{\alpha} V^{\beta} = \partial_{\alpha} V^{\beta} + V^{\gamma}\Gamma_{\alpha\gamma}{}^{\beta}, \tag{4.2.72}$$

where the tensor $\Gamma_{\alpha\gamma}{}^{\beta}$ is defined by

$$\Gamma_{\alpha\gamma}{}^{\beta} := \Gamma_{\boldsymbol{\alpha}\boldsymbol{\gamma}}{}^{\boldsymbol{\beta}}\, \delta^{\boldsymbol{\gamma}}_{\gamma}\delta^{\boldsymbol{\alpha}}_{\alpha}\delta^{\beta}_{\boldsymbol{\beta}} = \delta^{\boldsymbol{\gamma}}_{\gamma}\nabla_{\alpha}\delta^{\beta}_{\boldsymbol{\gamma}}. \tag{4.2.73}$$

Of course, like $\partial_{\rho}$, the tensor $\Gamma_{\alpha\gamma}{}^{\beta}$ is dependent on the particular choice of coordinate system $x^{\boldsymbol{\alpha}}$. Nevertheless it *is* a tensor (in the same sense that $x^3$ is a scalar). In fact, with $\tilde{\nabla}_{\rho} = \partial_{\rho}$, we have $\Gamma_{\alpha\gamma}{}^{\beta} = -Q_{\alpha\gamma}{}^{\beta}$ (*cf.* (4.2.70)), and (4.2.72) becomes simply a case of (4.2.46).

It is perhaps worth stressing that, whereas in our approach coordinates like $x^2$ and tensor components like $A_{331}$ are *scalars* (in the sense of being elements of $\mathfrak{T}$), in the classical approach they are not: classically, scalars are invariants under coordinate transformations. For example, in the classical notation the components of $\nabla_{\alpha} A_{\beta}$ are normally written $A_{\boldsymbol{\beta};\boldsymbol{\alpha}}$, but we cannot treat an individual component $A_3$ of $A_{\beta}$ as a scalar $\phi$ and substitute $\phi$ for $A_{\boldsymbol{\beta}}$ in $A_{\boldsymbol{\beta};\boldsymbol{\alpha}}$ to obtain $\phi_{;\boldsymbol{\alpha}} = \partial\phi/\partial x^{\boldsymbol{\alpha}}$ for $A_{3;\boldsymbol{\alpha}}$. Covariant derivative is *not* an operation on individual components but on the tensor as a whole. To that extent the notation $A_{\boldsymbol{\beta},\boldsymbol{\alpha}}$ is illogical and can lead to computational errors if one is careless. (On the other hand, the notation $A_{\boldsymbol{\beta}.\boldsymbol{\alpha}}$ for *coordinate* derivative is perfectly logical: ',α' acts on individual components.) The point of view and notation adopted in this book avoids these ambiguities. If we write $\nabla_2 A_3$ we do in fact mean $\partial A_3/\partial x^2$ (in a coordinate basis). The component with $\boldsymbol{\alpha} = 2$, $\boldsymbol{\beta} = 3$ in the classical expression $A_{\boldsymbol{\beta},\boldsymbol{\alpha}}$ would here have to be written $\delta^{\beta}_{3}\nabla_2 A_{\beta}$, which differs from $\nabla_2 A_3$ by the expression $-A_{\beta}\nabla_2\delta^{\beta}_{3} = -A_{\boldsymbol{\beta}}\Gamma_{23}{}^{\boldsymbol{\beta}}$. The abstract index $\beta$ cannot, of course, be given a numerical value.

## 4.3 Connection-independent derivatives

One of the advantages of having a connection on a manifold is that it makes differentiation a completely systematic procedure. Those operations which can be specified *independently* of any particular connection, on the other hand, constitute a sort of special 'menagerie'. Nevertheless, some of these operations are particularly important ones. We shall first give a list of several known connection-independent operations and discuss the more important of these in a little detail afterwards.

Each of the following expressions (4.3.1)–(4.3.6) (and also (4.3.45) below) is *independent* of the choice of *torsion-free* covariant derivative operator $\nabla_\rho$:

$$\nabla_{[\rho}A_{\alpha\ldots\gamma]}; \tag{4.3.1}$$

$$U^\alpha\nabla_\alpha V^\beta - V^\alpha\nabla_\alpha U^\beta; \tag{4.3.2}$$

$$\begin{aligned}\underset{V}{\pounds}H_{\lambda\ldots\nu}^{\alpha\ldots\gamma} := V^\rho\nabla_\rho H_{\lambda\ldots\nu}^{\alpha\ldots\gamma} - H_{\lambda\ldots\nu}^{\alpha_0\ldots\gamma}\nabla_{\alpha_0}V^\alpha - \cdots - H_{\lambda\ldots\nu}^{\alpha\ldots\gamma_0}\nabla_{\gamma_0}V^\gamma \\ + H_{\lambda_0\ldots\nu}^{\alpha\ldots\gamma}\nabla_\lambda V^{\lambda_0} + \cdots + H_{\lambda\ldots\nu_0}^{\alpha\ldots\gamma}\nabla_\nu V^{\nu_0};\end{aligned} \tag{4.3.3}$$

$$\begin{aligned}pA^{\rho(\alpha_1\ldots\alpha_{p-1}}\nabla_\rho B^{\alpha_p\ldots\alpha_{p+q-1})} - qB^{\rho(\alpha_1\ldots\alpha_{q-1}}\nabla_\rho A^{\alpha_q\ldots\alpha_{p+q-1})} \\ \textit{if } A^{\alpha_1\ldots\alpha_p} = A^{(\alpha_1\ldots\alpha_p)} \quad \textit{and} \quad B^{\alpha_1\ldots\alpha_q} = B^{(\alpha_1\ldots\alpha_q)};\end{aligned} \tag{4.3.4}$$

$$\begin{aligned}pA^{\rho[\alpha_1\ldots\alpha_{p-1}}\nabla_\rho B^{\alpha_p\ldots\alpha_{p+q-1}]} + (-1)^{pq}qB^{\rho[\alpha_1\ldots\alpha_{q-1}}\nabla_\rho A^{\alpha_q\ldots\alpha_{p+q-1}]} \\ \textit{if } A^{\alpha_1\ldots\alpha_p} = A^{[\alpha_1\ldots\alpha_p]} \quad \textit{and} \quad B^{\alpha_1\ldots\alpha_q} = B^{[\alpha_1\ldots\alpha_q]};\end{aligned} \tag{4.3.5}$$

$$\begin{aligned}(\nabla_\rho A^\gamma_{[\alpha_1\ldots\alpha_p})B^\rho_{\beta_1\ldots\beta_q]} - (\nabla_\rho B^\gamma_{[\beta_1\ldots\beta_q})A^\rho_{\alpha_1\ldots\alpha_p]} \\ + pA^\gamma_{\rho[\alpha_2\ldots\alpha_p}\nabla_{\alpha_1}B^\rho_{\beta_1\ldots\beta_q]} - qB^\gamma_{\rho[\beta_2\ldots\beta_q}\nabla_{\beta_1}A^\rho_{\alpha_1\ldots\alpha_p]} \\ \textit{if } A^\gamma_{\alpha_1\ldots\alpha_p} = A^\gamma_{[\alpha_1\ldots\alpha_p]} \quad \textit{and} \quad B^\gamma_{\beta_1\ldots\beta_q} = B^\gamma_{[\beta_1\ldots\beta_q]}.\end{aligned} \tag{4.3.6}$$

The verification of the invariance of each of these expressions under change of the torsion-free operator $\nabla_\rho$ is a simple and straightforward application of the above discussion. Taking the difference between each expression and the corresponding one involving $\tilde{\nabla}_\rho$ we get, using (4.2.48), a sum of terms involving $Q_{\alpha\beta}{}^\gamma$ which *vanishes* in each case, by virtue of the symmetry

$$Q_{\alpha\beta}{}^\gamma = Q_{\beta\alpha}{}^\gamma, \tag{4.3.7}$$

this symmetry expressing the fact that the torsion tensors of $\nabla_\alpha$ and $\tilde{\nabla}_\alpha$ are identical (both being zero), see (4.2.50).

The expressions (4.3.1)–(4.3.6) have the virtue that they have particularly simple representations in terms of components. We may use the $\partial_\rho$ operator arising from any coordinate system in place of $\nabla_\rho$. Then when we take components with respect to these coordinates, the operators simply translate into $\partial/\partial x^\mathbf{\rho}$.

### *Exterior calculus*

Let us examine (4.3.1) in detail. We have

$$\tilde{\nabla}_{[\rho}A_{\alpha\ldots\gamma]} - \nabla_{[\rho}A_{\alpha\ldots\gamma]} = Q_{[\rho\alpha}{}^\delta A_{|\delta|\ldots\gamma]} + \cdots + Q_{[\rho\gamma}{}^\delta A_{\alpha\ldots]\delta} = 0 \tag{4.3.8}$$

since each term vanishes by (4.3.7). The connection-independent operation (4.3.1) is the basis of a self-contained calculus, namely Cartan's *exterior calculus of differential forms.* This is concerned with anti-symmetrical covariant tensors. The operations are simple enough for indices not to be necessary or (in most instances) even helpful.

We can obtain agreement between Cartan's notation and ours if we employ a device for suppressing indices such as the following. Select a particular infinite subclass of index labels, say

$$\iota_1, \iota_2, \iota_3, \ldots \tag{4.3.9}$$

as the indices which may be suppressed. Here we shall allow ourselves only to suppress *lower* indices and they must occur in their natural order starting with $\iota_1$ and continuing successively. If we are to remain strictly within the exterior calculus, then we should operate entirely with tensors whose *only* index labels are an anti-symmetrical set of lower indices $\iota_1, \iota_2, \ldots, \iota_p$. Such a tensor will be called a *p-form* and we write*

$$\boldsymbol{A} := A_{\iota_1\iota_2\ldots\iota_p}. \tag{4.3.10}$$

The 0-forms are simply scalars and the 1-forms are just covectors. By (3.3.30) every $n$-form is a scalar multiple of $\varepsilon_{\iota_1\ldots\iota_n}$; by (3.3.27) every $p$-form with $p > n$ is zero.

One sometimes considers *tensor-valued p-forms* and then a notation such as

$$\boldsymbol{H}_{\lambda\ldots\nu}{}^{\alpha\ldots\gamma} := H_{\iota_1\iota_2\ldots\iota_p\lambda\ldots\nu}{}^{\alpha\ldots\gamma} \tag{4.3.11}$$

is useful, where $H_{\ldots}{}^{\cdots}$ is anti-symmetrical in $\iota_1, \iota_2, \ldots, \iota_p$. Strictly speaking, however, this takes us outside the exterior calculus.

Differential forms are subject to three operations, namely addition, exterior product and exterior derivative. We allow a $p$-form and a $q$-form to be added only if $p = q$. (In some versions of the exterior calculus, formal sums are also permitted when $p \neq q$. See also Appendix, Vol. 2.) The sum of two $p$-forms $\boldsymbol{A}$ and $\boldsymbol{B}$ is another $p$-form defined by

$$\boldsymbol{A} + \boldsymbol{B} := A_{\iota_1\ldots\iota_p} + B_{\iota_1\ldots\iota_p}. \tag{4.3.12}$$

The exterior product of a $p$-form $\boldsymbol{A}$ with a $q$-form $\boldsymbol{C}$ is a $(p+q)$-form $\boldsymbol{A} \wedge \boldsymbol{C}$ defined by

$$\boldsymbol{A} \wedge \boldsymbol{C} := A_{[\iota_1\ldots\iota_p} C_{\iota_{p+1}\ldots\iota_{p+q}]}. \tag{4.3.13}$$

The exterior derivative of a $p$-form $\boldsymbol{A}$ is a $(p+1)$-form $\mathrm{d}\boldsymbol{A}$ defined, for torsion-free $\nabla_\rho$, by

$$\mathrm{d}\boldsymbol{A} := \nabla_{[\iota_1} A_{\iota_2\ldots\iota_{p+1}]}, \tag{4.3.14}$$

* The possibly more familiar 'd$x$' notation is introduced in (4.3.19), (4.3.22) below.

this being independent of the choice of $\nabla_\rho$, as we have seen. Observe that if $p = 0$, this notation agrees with that of the gradient of a scalar defined in (4.1.32). When the operators act on scalars we may in fact write $\mathrm{d} = \nabla_{\iota_1}$. The following relations hold between differential forms. (Whenever a sum is written it is assumed that the forms involved have the same valence.)

(i) $\boldsymbol{A} + \boldsymbol{B} = \boldsymbol{B} + \boldsymbol{A}$

(ii) $(\boldsymbol{A} + \boldsymbol{B}) + \boldsymbol{C} = \boldsymbol{A} + (\boldsymbol{B} + \boldsymbol{C})$

(iii) $\boldsymbol{A} \wedge (\boldsymbol{B} + \boldsymbol{C}) = \boldsymbol{A} \wedge \boldsymbol{B} + \boldsymbol{A} \wedge \boldsymbol{C}$

(iv) $\boldsymbol{A} \wedge \boldsymbol{B} = (-1)^{pq} \boldsymbol{B} \wedge \boldsymbol{A}$ if $\boldsymbol{A}$ is a $p$-form and $\boldsymbol{B}$ is a $q$-form

(v) $(\boldsymbol{A} \wedge \boldsymbol{B}) \wedge \boldsymbol{C} = \boldsymbol{A} \wedge (\boldsymbol{B} \wedge \boldsymbol{C})$

(vi) $\mathrm{d}(\boldsymbol{A} + \boldsymbol{B}) = \mathrm{d}\boldsymbol{A} + \mathrm{d}\boldsymbol{B}$

(vii) $\mathrm{d}(\boldsymbol{A} \wedge \boldsymbol{B}) = (\mathrm{d}\boldsymbol{A}) \wedge \boldsymbol{B} + (-1)^p \boldsymbol{A} \wedge \mathrm{d}\boldsymbol{B}$ if $\boldsymbol{A}$ is a $p$-form

(viii) $\mathrm{d}(\mathrm{d}\boldsymbol{A}) = 0$ (4.3.15)

The verification of each of these expressions is a simple consequence of the definitions (4.2.12)–(4.2.14). For example, to verify (v), we have

$$(\boldsymbol{A} \wedge \boldsymbol{B}) \wedge \boldsymbol{C} = A_{[[\ldots} B_{\ldots]} C_{\ldots]} = A_{[\ldots} B_{\ldots} C_{\ldots]}$$
$$= A_{[\ldots} B_{[\ldots} C_{\ldots]]} = \boldsymbol{A} \wedge (\boldsymbol{B} \wedge \boldsymbol{C}). \quad (4.3.16)$$

To verify (viii) we choose a particular $\nabla_\rho$, say $\partial_\rho$, whose curvature and torsion both vanish. (This may only be possible locally, but that is sufficient.) We have

$$\mathrm{d}(\mathrm{d}\boldsymbol{A}) = \partial_{[.} \partial_{[.} A_{\ldots]]} = \partial_{[.} \partial_{.} A_{\ldots]} = \partial_{[[.} \partial_{.]} A_{\ldots]} = 0 \quad (4.3.17)$$

since $\partial_{[\alpha} \partial_{\beta]} = 0$. This result is sometimes known as the Poincaré lemma, and sometimes as the converse of the Poincaré lemma (the deeper result being that *locally* $\mathrm{d}\boldsymbol{X} = 0$ implies $\boldsymbol{X} = \mathrm{d}\boldsymbol{A}$ for some $\boldsymbol{A}$, *cf.* (6.5.27)). Exterior derivative generalizes the notion of 'curl'. The well-known formulae of vector analysis div curl $\boldsymbol{V} = 0$ and curl grad $\phi = 0$ may be viewed as particular cases of Poincaré's lemma.

In a coordinate basis we can write

$$\boldsymbol{A} = A_{\iota_1 \ldots \iota_p} = A_{\alpha_1 \ldots \alpha_p} \delta^{\alpha_1}_{\iota_1} \ldots \delta^{\alpha_p}_{\iota_p}$$
$$= A_{\alpha_1 \ldots \alpha_p} \delta^{\alpha_1}_{[\iota_1} \ldots \delta^{\alpha_p}_{\iota_p]}, \quad (4.3.18)$$

the components $A_{\alpha_1 \ldots \alpha_p}$ forming an anti-symmetrical array. Since we can write

$$\delta^{\alpha}_{\iota_1} = \nabla_{\iota_1} x^\alpha = \mathrm{d}x^\alpha, \quad (4.3.19)$$

the expression (4.3.18) can be written in differential form notation as

$$\boldsymbol{A} = A_{\alpha_1 \ldots \alpha_p} \mathrm{d}x^{\alpha_1} \wedge \cdots \wedge \mathrm{d}x^{\alpha_p}. \quad (4.3.20)$$

(Logically there should perhaps be a '$\wedge$' between the $A_{\dots}$ and the first $\mathrm{d}x$, but scalar multiplication, although a special case of exterior product, is normally written without the '$\wedge$'.)

We should remark that, in the literature, a slightly different convention is frequently employed, in that the quantities

$$a_{\alpha_1 \dots \alpha_p} := p! A_{\alpha_1 \dots \alpha_p} \tag{4.3.21}$$

rather than our $A_{\alpha_1 \dots \alpha_p}$ are used to denote the components of a $p$-form. This is simpler only if the summation convention is not employed, the expression (4.3.20) being then written

$$\boldsymbol{A} = \sum_{\alpha_1 < \dots < \alpha_p} a_{\alpha_1 \dots \alpha_p} \mathrm{d}x^{\alpha_1} \wedge \cdots \wedge \mathrm{d}x^{\alpha_p}. \tag{4.3.22}$$

The definition we use here, together with the square bracket notation for anti-symmetrization, serves to avoid some of the awkward factors that appear when the notation (4.3.22) is adopted. Here, exterior product and exterior derivative have the respective component forms

$$A_{[\alpha_1 \dots \alpha_p} B_{\alpha_{p+1} \dots \alpha_{p+q}]} \quad \text{and} \quad \frac{\partial}{\partial x^{[\alpha_1}} A_{\alpha_2 \dots \alpha_{p+1}]} \tag{4.3.23}$$

One of the most important applications of differential forms occurs in the *fundamental theorem of exterior calculus*.* If $\mathscr{P}$ is an oriented $p$-dimensional surface in an oriented $\mathscr{M}$, we define the integral of the $p$-form $\boldsymbol{A}$ over $\mathscr{P}$ to be

$$\begin{aligned} \int_{\mathscr{P}} \boldsymbol{A} &= \int_{\mathscr{P}} A_{\alpha_1 \dots \alpha_p} \mathrm{d}x^{\alpha_1} \wedge \cdots \wedge \mathrm{d}x^{\alpha_p} = \int_{\mathscr{P}} a_{1 \dots p} \mathrm{d}x^1 \wedge \cdots \wedge \mathrm{d}x^p \\ &= p! \int_{\mathscr{P}} A_{1 \dots p} \mathrm{d}x^1 \wedge \cdots \wedge \mathrm{d}x^p = p! \int \left( \dots \left( \int A_{1 \dots p} \mathrm{d}x^1 \right) \dots \right) \mathrm{d}x^p \end{aligned} \tag{4.3.24}$$

whenever $\mathscr{P}$ is such that it can be defined by $x^{p+1} = \cdots = x^n = 0$ in some coordinate system. Otherwise we split $\mathscr{P}$ into pieces where such coordinates exist for each piece, and add the integrals for the separate pieces. The result is independent of the particular choice of coordinates. The fundamental theorem of exterior calculus states that

$$\int_{\mathscr{Q}} \mathrm{d}\boldsymbol{A} = \oint_{\partial \mathscr{Q}} \boldsymbol{A}, \tag{4.3.25}$$

where $\mathscr{Q}$ is a *compact* $(p + 1)$-surface with boundary $\partial \mathscr{Q}$.

* Various versions of this result go under the names of Ostrogradski, Gauss, Green, Kelvin, Stokes, Cartan and probably others. Our adopted terminology was suggested to us by N.M.J. Woodhouse.

*Lie brackets and Lie derivatives*

We now come to (4.3.2), the second connection-independent operation in our list. This is the *Lie bracket* and is the common special case of all the remaining operations (4.3.3)–(4.3.6) in the list. We can arrive at this operation in a more basic way as follows. Let $\boldsymbol{U}$ and $\boldsymbol{V}$ be derivations on the algebra $\mathfrak{T}$. The the map $\boldsymbol{W}:\mathfrak{T}\rightarrow\mathfrak{T}$ defined by the commutator

$$\boldsymbol{W}=[\boldsymbol{U},\boldsymbol{V}]:=\boldsymbol{U}\circ\boldsymbol{V}-\boldsymbol{V}\circ\boldsymbol{U}, \tag{4.3.26}$$

i.e., by

$$\boldsymbol{W}(f)=\boldsymbol{U}(\boldsymbol{V}(f))-\boldsymbol{V}(\boldsymbol{U}(f)), \tag{4.3.27}$$

is also a derivation. (Relations (i) and (ii) of (4.1.11) are obviously satisfied. The verification of (4.1.11) (iii) is a straightforward calculation.) Let $\nabla_\rho$ be any covariant derivative operator. We shall find $W^\alpha$ in terms of $U^\alpha$ and $V^\alpha$. We have

$$\begin{aligned} W^\beta\nabla_\beta f &= U^\alpha\nabla_\alpha(V^\beta\nabla_\beta f)-V^\beta\nabla_\beta(U^\alpha\nabla_\alpha f)\\ &= V^\beta U^\alpha\nabla_\alpha\nabla_\beta f+U^\alpha(\nabla_\alpha V^\beta)\nabla_\beta f\\ &\quad -U^\alpha V^\beta\nabla_\beta\nabla_\alpha f-V^\alpha(\nabla_\alpha U^\beta)\nabla_\beta f\\ &= (U^\alpha\nabla_\alpha V^\beta-V^\alpha\nabla_\alpha U^\beta+U^\alpha V^\gamma T_{\alpha\gamma}{}^\beta)\nabla_\beta f. \end{aligned} \tag{4.3.28}$$

Thus

$$W^\beta=U^\alpha\nabla_\alpha V^\beta-V^\alpha\nabla_\alpha U^\beta+U^\alpha V^\gamma T_{\alpha\gamma}{}^\beta, \tag{4.3.29}$$

which reduces to (4.3.2) when $\nabla_\alpha$ is torsion-free (and, incidentally, shows how (4.3.2) must be modified when it is not).

The Lie bracket satisfies a number of familiar relations common to all commutators, namely

$$\begin{gathered}[\boldsymbol{U},\boldsymbol{V}]=-[\boldsymbol{V},\boldsymbol{U}],\quad [\boldsymbol{U},\boldsymbol{V}+\boldsymbol{X}]=[\boldsymbol{U},\boldsymbol{V}]+[\boldsymbol{U},\boldsymbol{X}],\\ [\boldsymbol{U},[\boldsymbol{V},\boldsymbol{X}]]+[\boldsymbol{V},[\boldsymbol{X},\boldsymbol{U}]]+[\boldsymbol{X},[\boldsymbol{U},\boldsymbol{V}]]=0,\end{gathered} \tag{4.3.30}$$

the last being referred to as the Jacobi identity. The Lie bracket plays a frequent role in modern differential geometry. This often arises from the fact that the commutator of *directional covariant derivatives*

$$\underset{X}{\nabla}:=X^\alpha\nabla_\alpha \tag{4.3.31}$$

(not necessarily acting on scalars) involves a Lie bracket. Essentially repeating the calculation (4.3.28), we have, in fact, (*cf.* (4.2.24))

$$\underset{X}{\nabla}\underset{Y}{\nabla}-\underset{Y}{\nabla}\underset{X}{\nabla}=\underset{[X,Y]}{\nabla}+X^\alpha Y^\beta\sqcap_{\alpha\beta}, \tag{4.3.32}$$

and so, by (4.2.31),

$$(\underset{X}{\nabla}\underset{Y}{\nabla}-\underset{Y}{\nabla}\underset{X}{\nabla}-\underset{[X,Y]}{\nabla})Z^\delta=R_{\alpha\beta\gamma}{}^\delta X^\alpha Y^\beta Z^\gamma. \tag{4.3.33}$$

The third connection-independent operation in our list is (4.3.3), the *Lie derivative* of the tensor $H^{\alpha\ldots\gamma}_{\lambda\ldots\nu}$ with respect to the vector $\boldsymbol{V}$. This operation may be generated as follows. We define

$$\underset{V}{\pounds}\boldsymbol{X} := [\boldsymbol{V}, \boldsymbol{X}] \tag{4.3.34}$$

as the Lie derivative of a contravariant vector $\boldsymbol{X}$ with respect to $\boldsymbol{V}$. The Lie derivative of a scalar $f$ shall simply be its directional derivative defined by $\boldsymbol{V}$:

$$\underset{V}{\pounds}f := V(f) = V^\rho \nabla_\rho f. \tag{4.3.35}$$

The Lie derivative of a covariant vector is then defined from (4.3.34), (4.3.35) by the requirement that the Leibniz rule apply to the Lie derivative of $A_\alpha X^\alpha$:

$$\underset{V}{\pounds}(A_\alpha X^\alpha) = A_\alpha \underset{V}{\pounds} X^\alpha + X^\alpha \underset{V}{\pounds} A_\alpha \tag{4.3.36}$$

This implies

$$\underset{V}{\pounds}A_\alpha = V^\rho \nabla_\rho A_\alpha + A_\beta \nabla_\alpha V^\beta, \tag{4.3.37}$$

where we assume $\nabla_\rho$ to be torsion-free. For a general tensor $H^{\alpha\ldots\gamma}_{\lambda\ldots\nu}$, if we demand that the Leibniz rule shall apply to the Lie derivative of $H^{\alpha\ldots\gamma}_{\lambda\ldots\nu} X^\lambda \ldots Z^\nu A_\alpha \ldots C_\gamma$, or alternatively, if we expand $H^{\alpha\ldots\gamma}_{\lambda\ldots\nu}$ as a sum of outer products of vectors and demand linearity and that the Leibniz rule shall apply to each term, then we are uniquely led to the expression (4.3.3) as the definition of the Lie derivative $\underset{V}{\pounds}H^{\alpha\ldots\gamma}_{\lambda\ldots\nu}$ of the tensor $H^{\alpha\ldots\gamma}_{\lambda\ldots\nu}$, where $\nabla_\rho$ is torsion-free. It is easily verified that the commutator of two Lie derivatives satisfies

$$\underset{UV}{\pounds\pounds} - \underset{VU}{\pounds\pounds} = \underset{[U,V]}{\pounds}. \tag{4.3.38}$$

The geometrical meaning of the Lie derivative of $H^{\alpha\ldots\gamma}_{\lambda\ldots\nu}$ is that it represents the infinitesimal 'dragging' of the tensor $H^{\cdots}_{\cdots}$ along the integral curves of $\boldsymbol{V}$ in $\mathcal{M}$. To represent a finite 'dragging' of $H^{\cdots}_{\cdots}$ along these curves to parameter value $u$, we form the expression $\exp(u \underset{V}{\pounds}) H^{\alpha\ldots\gamma}_{\lambda\ldots\nu}$. For details, see Hawking and Ellis (1973), Choquet–Bruhat, DeWitt–Morette and Dillard–Bleick (1977)

The final three connection-independent operations (4.3.4)–(4.3.6) do not have the same importance that the first three have. We shall say no more about them here except to point out a certain particular case of (4.3.6):

$$A^\rho_\beta \nabla_\rho A^\gamma_\alpha - A^\rho_\alpha \nabla_\rho A^\gamma_\beta + A^\gamma_\rho \nabla_\alpha A^\rho_\beta - A^\gamma_\rho \nabla_\beta A^\rho_\alpha. \tag{4.3.39}$$

This expression has significance in the theory of complex manifolds.

### Riemannian geometry

Let us now consider the implications of introducing a metric into $\mathscr{M}$. A *metric* is a symmetric tensor of valence $\left[\begin{smallmatrix}0\\2\end{smallmatrix}\right]$, denoted by $g_{\alpha\beta}$, which is non-singular in the sense that another tensor $g^{\alpha\beta}$ exists such that

$$g^{\alpha\beta}g_{\beta\gamma} = \delta^{\alpha}_{\gamma}. \tag{4.3.40}$$

We have

$$g^{\alpha\beta} = g^{\beta\alpha}, \quad g_{\alpha\beta} = g_{\beta\alpha} \tag{4.3.41}$$

Let $V^{\alpha}$ be an arbitrary contravariant vector and let $\nabla_{\rho}$ be a *torsion-free* operator. Then the Lie derivative of $g_{\alpha\beta}$ with respect to $V^{\alpha}$ is

$$V^{\alpha}\nabla_{\alpha}\, g_{\beta\gamma} + g_{\alpha\gamma}\nabla_{\beta}V^{\alpha} + g_{\beta\alpha}\nabla_{\gamma}V^{\alpha}. \tag{4.3.42}$$

Furthermore, twice the exterior derivative of $V^{\alpha}g_{\alpha\beta}$ is

$$\begin{aligned}&\nabla_{\gamma}(V^{\alpha}g_{\alpha\beta}) - \nabla_{\beta}(V^{\alpha}g_{\alpha\gamma})\\ &\quad = V^{\alpha}\nabla_{\gamma}g_{\alpha\beta} + g_{\alpha\beta}\nabla_{\gamma}V^{\alpha} - V^{\alpha}\nabla_{\beta}g_{\alpha\gamma} - g_{\alpha\gamma}\nabla_{\beta}V^{\alpha}\end{aligned} \tag{4.3.43}$$

Adding these two connection-independent expressions together,

$$2g_{\alpha\beta}\nabla_{\gamma}V^{\alpha} + V^{\alpha}(\nabla_{\alpha}g_{\beta\gamma} + \nabla_{\gamma}g_{\alpha\beta} - \nabla_{\beta}g_{\alpha\gamma}), \tag{4.3.44}$$

and transvecting with $\frac{1}{2}g^{\beta\sigma}$, we get

$$\nabla_{\gamma}V^{\sigma} + V^{\alpha}\{\tfrac{1}{2}g^{\beta\sigma}(\nabla_{\alpha}g_{\beta\gamma} + \nabla_{\gamma}g_{\alpha\beta} - \nabla_{\beta}g_{\alpha\gamma})\} \tag{4.3.45}$$

which must therefore also be connection-independent. Now suppose $\nabla_{\rho}$, in addition to being torsion-free, satisfies

$$\nabla_{\rho}g_{\alpha\beta} = 0. \tag{4.3.46}$$

Then the connection-independence of (4.3.45) tells us that $\nabla_{\rho}$ is *unique*, since (4.3.45) is equal to the same expression with $\partial_{\rho}$ replacing $\nabla_{\rho}$:

$$\nabla_{\gamma}V^{\sigma} = \partial_{\gamma}V^{\sigma} + V^{\alpha}\{\tfrac{1}{2}g^{\beta\sigma}(\partial_{\alpha}g_{\beta\gamma} + \partial_{\gamma}g_{\alpha\beta} - \partial_{\beta}g_{\alpha\gamma})\}, \tag{4.3.47}$$

where $\partial_{\rho}$ is *any* torsion-free operator – in particular where $\partial_{\rho}$ is the 'co-ordinate derivative' operator associated with some local coordinates $x^{\alpha}$. When written in terms of components, (4.3.47) is the familiar classical expression for covariant derivative in terms of Christoffel symbols.

Conversely, we can show that a torsion-free $\nabla_{\rho}$ *exists* satisfying (4.3.46). We define $\nabla_{\gamma}V^{\sigma}$ (locally) by (4.3.47). Then we have an expression of the form (4.2.46) where $\partial_{\gamma} = \tilde{\nabla}_{\gamma}$ and

$$Q_{\gamma\alpha}{}^{\sigma} = -\tfrac{1}{2}g^{\beta\sigma}(\partial_{\alpha}g_{\beta\gamma} + \partial_{\gamma}g_{\alpha\beta} - \partial_{\beta}g_{\alpha\gamma}). \tag{4.3.48}$$

Thus, by (4.2.48), (4.3.40), (4.3.41),

$$\begin{aligned}\nabla_\rho g_{\alpha\beta} &= \partial_\rho g_{\alpha\beta} + Q_{\rho\alpha}{}^\gamma g_{\gamma\beta} + Q_{\rho\beta}{}^\gamma g_{\alpha\gamma} \\ &= \partial_\rho g_{\alpha\beta} + 2Q_{\rho(\alpha}{}^\gamma g_{\beta)\gamma} \\ &= \partial_\rho g_{\alpha\beta} - (\partial_{(\alpha} g_{\beta)\rho} + \partial_\rho g_{(\alpha\beta)} - \partial_{(\beta} g_{\alpha)\rho}) \\ &= 0.\end{aligned} \quad (4.3.49)$$

This torsion-free operator $\nabla_\rho$, uniquely defined by $g_{\alpha\beta}$, we shall call the *Christoffel* covariant derivative operator. (The existence was proved locally, but local existence, together with uniqueness, implies *global* existence.)

The metric may be used for raising and lowering indices in the standard manner (*cf.* §3.1) (i.e., $g_{\alpha\beta}$ establishes a canonical isomorphism between $\mathfrak{T}^\alpha$ and $\mathfrak{T}_\beta$). Since $g_{\alpha\beta}$ is 'covariantly constant', $\nabla_\rho g_{\alpha\beta} = 0$, the operation of raising or lowering an index will commute with $\nabla_\rho$. That is,

$$\nabla_\rho H_{\mathscr{A}\sigma} = K_{\rho\mathscr{A}\sigma} \quad \textit{iff} \quad \nabla_\rho H_{\mathscr{A}}{}^\sigma = K_{\rho\mathscr{A}}{}^\sigma. \quad (4.3.50)$$

Finally, let us examine the curvature defined by the Christoffel $\nabla_\rho$. We can lower the final index of $R_{\alpha\beta\gamma}{}^\delta$ to obtain the *Riemann* (*–Christoffel*) *tensor*

$$R_{\alpha\beta\gamma\delta} = R_{\alpha\beta\gamma}{}^\lambda g_{\lambda\delta}. \quad (4.3.51)$$

(Owing to the arrangement of indices that we have chosen, we have arrived at a Riemann tensor of a sign which differs from that used in much of the literature, although both signs are commonly adopted. The sign chosen here agrees with the one which has usually been employed in connection with spinor decompositions.)

Applying a commutator of derivatives to $g_{\gamma\delta}$ we obtain, by (4.3.46), (4.2.33) and the vanishing of torsion,

$$\begin{aligned}0 = \Delta_{\alpha\beta} g_{\gamma\delta} &= -R_{\alpha\beta\gamma}{}^\sigma g_{\sigma\delta} - R_{\alpha\beta\delta}{}^\sigma g_{\gamma\sigma} \\ &= -2R_{\alpha\beta(\gamma\delta)}\end{aligned} \quad (4.3.52)$$

showing that $R_{\alpha\beta\gamma\delta}$ is skew in $\gamma, \delta$. By (4.2.34), $R_{\alpha\beta\gamma\delta}$ is also skew in $\alpha, \beta$, so

$$R_{\alpha\beta\gamma\delta} = R_{[\alpha\beta][\gamma\delta]}. \quad (4.3.53)$$

Also, by (4.2.38),

$$R_{\alpha\beta\gamma\delta} + R_{\beta\gamma\alpha\delta} + R_{\gamma\alpha\beta\delta} = 0. \quad (4.3.54)$$

These properties imply

$$\begin{aligned}2R_{\alpha\beta\gamma\delta} &= R_{\alpha\beta\gamma\delta} + R_{\beta\alpha\delta\gamma} \\ &= -R_{\beta\gamma\alpha\delta} - R_{\gamma\alpha\beta\delta} - R_{\alpha\delta\beta\gamma} - R_{\delta\beta\alpha\gamma} \\ &= -R_{\gamma\beta\delta\alpha} - R_{\beta\delta\gamma\alpha} - R_{\alpha\gamma\delta\beta} - R_{\delta\alpha\gamma\beta} \\ &= R_{\delta\gamma\beta\alpha} + R_{\gamma\delta\alpha\beta} = 2R_{\gamma\delta\alpha\beta},\end{aligned} \quad (4.3.55)$$

so $R_{\alpha\beta\gamma\delta}$ also possesses the 'interchange symmetry'

$$R_{\alpha\beta\gamma\delta} = R_{\gamma\delta\alpha\beta}. \tag{4.3.56}$$

(Note that the full relation (4.3.53) is not used in the derivation of (4.3.56), the symmetry $R_{\alpha\beta\gamma\delta} = R_{\beta\alpha\delta\gamma}$ being all that is required.) By virtue of the symmetries (4.3.53) and (4.3.54) (and hence (4.3.56)) the total number of independent components $R_{\alpha\beta\gamma\delta}$ at each point turns out to be $\frac{1}{12}n^2(n^2-1)$. (See p. 144 above.)

The tensor

$$R_{\alpha\beta} := R_{\alpha\gamma\beta}{}^{\gamma} \tag{4.3.57}$$

is called the *Ricci tensor*. By (4.3.56) (and the symmetry of the metric),

$$R_{\alpha\beta} = R_{\beta\alpha}. \tag{4.3.58}$$

The Ricci tensor has $\frac{1}{2}n(n+1)$ independent components $R_{\alpha\beta}$ at each point. The *scalar curvature* is given by

$$R := R_{\alpha}{}^{\alpha} = R_{\alpha\beta}{}^{\alpha\beta}. \tag{4.3.59}$$

Rewriting the Bianchi identity (4.2.42),

$$\nabla_{\alpha}R_{\beta\gamma\rho\sigma} + \nabla_{\beta}R_{\gamma\alpha\rho\sigma} + \nabla_{\gamma}R_{\alpha\beta\rho\sigma} = 0, \tag{4.3.60}$$

and transvecting with $g^{\alpha\rho}g^{\gamma\sigma}$, we obtain the important relation

$$\nabla^{\alpha}(R_{\alpha\beta} - \tfrac{1}{2}Rg_{\alpha\beta}) = 0. \tag{4.3.61}$$

In four-dimensional space–time this forms the mathematical basis for Einstein's field equations.

## 4.4 Differentiation of spinors

We shall now specialize the discussion given in §§4.1, 2, 3 to the case of a four-dimensional space–time* $\mathcal{M}$ and shall extend the concept of covariant derivative so that it applies to spinors. We shall find that in

* It is common 'modern' practice to denote a space–time not by a single symbol, such as $\mathcal{M}$ here, but by a pair, such as $(\mathcal{M}, \boldsymbol{g})$ or $(\mathcal{M}, g_{ab})$, where $\mathcal{M}$ is the manifold and $\boldsymbol{g}$ (or $g_{ab}$) its metric. While this may have some rationale in areas of pure mathematics for which the specific choice of metric is of importance secondary to that of the manifold itself, this is less true for physical space–time, where one hardly ever wishes to consider metrics (or at least conformal metrics) other than the physical one. It should be pointed out that our use of notation such as '$P\in\mathcal{M}$', while actually an abuse of notation – since if $\mathcal{M} = (\mathcal{M}_0, \boldsymbol{g})$, then '$P\in\mathcal{M}$' strictly means '$P = \mathcal{M}_0$ or $P = \boldsymbol{g}$' – it is no more so than the common notation '$P\in\mathcal{M}_0$', where $\mathcal{M}_0$ *is* the manifold. For a manifold is not a point-set either, being itself a collection of sets, mappings, etc. denoting the topology and/or differentiable structure, so '$P\in\mathcal{M}_0$' would mean, strictly, *not* that $P$ is a point of the manifold, but is one of these entire structures instead!

fact the Christoffel derivative of §4.3 extends in a canonical way to spinor fields. This should, perhaps, come as no surprise since spinors can be defined in a clear geometrical way in terms of tensorial objects (up to a sign – but that ambiguity should cause no difficulty for *differential* properties).

Our point of view, however, is ultimately to regard the spinors as more primitive than world-tensors. In §3.1 we laid down the way in which world-tensors are to be regarded as special cases of spinors. To complete that association, we must also make the identification of world-vectors with *tangent vectors* to the space–time manifold $\mathcal{M}$. Thus world-vector fields must, in effect, be identified with derivations on the algebra $\mathfrak{T}$ of $C^\infty$ real scalar fields on $\mathcal{M}$. There is danger of a logical circularity here, since the world-vector fields are to be constructed in two different ways. We shall avoid any actual logical inconsistency by taking the space $\mathfrak{T}^\cdot$ of world vector fields $\boldsymbol{U}$, $\boldsymbol{V}$,... to be the space of derivations on the algebra $\mathfrak{T}$, while each space $\mathfrak{T}^a$, $\mathfrak{T}^b$,... containing elements $U^a$, $V^a$,..., or $U^b$, $V^b$,..., is to be identified as $\mathfrak{T}^{AA'}$, $\mathfrak{T}^{BB'}$,..., respectively, in accordance with §3.1.

We recall from §3.1 that there are two logically distinct ways of approaching (and regarding) the spinor algebra on $\mathcal{M}$, the constructive and the axiomatic. The constructive approach is that which we followed in Chapter I. The four-dimensional space–time manifold $\mathcal{M}$ is taken as *given*, with its $(+ - - -)$ signature, $C^\infty$ metric, and with the three global properties of time-orientability (1.5.1), orientability (1.5.2), and existence of spin-structure (1.5.3). Then spin-vectors can be defined in terms of *geometry* (up to an unimportant overall sign for the whole of $\mathfrak{S}^A$ if $\mathcal{M}$ is simply-connected, while if $\mathcal{M}$ is not simply-connected, the definition of spin-vectors may contain global ambiguities which require a number of discrete choices to be made, *cf.* §1.5). In this approach we need a concept of '$C^\infty$' for a spin-vector field, in order to characterize the elements of the basic module $\mathfrak{S}^A$. This is a local characterization and it can be given in various equivalent ways. For example, in a simply-connected neighbourhood of each given point of $\mathcal{M}$ we can set up a $C^\infty$ system of restricted orthonormal tetrads of tangent vectors and describe a spin-vector $\kappa^A$ with respect to these in terms of standard components $\kappa^0$, $\kappa^1$ as in Chapter 1; then the requirement is that $\kappa^0$, $\kappa^1$ be $C^\infty$ throughout each such neighbourhood, i.e., that they be local complex scalar fields. Equivalently, in each such neighbourhood we can use the spin frame $o^A$, $\iota^A$ defined by the tetrad field (with signs fixed by continuity) and decree these to be $C^\infty$, so that $\kappa^A = \kappa^0 o^A + \kappa^1 \iota^A$ is $C^\infty$ in that neighbourhood whenever $\kappa^0$ and

$\kappa^1$ are $C^\infty$. The $C^\infty$ concept is obviously independent of the particular choice of orthonormal tetrad, so it serves to characterize the geometrical entities that constitute $\mathfrak{S}^A$. Having thus defined $\mathfrak{S}^A$, which is now clearly a module over the ring of $C^\infty$ complex scalar fields $\mathfrak{S}$, and totally reflexive by §2.4, we build up $\mathfrak{S}^{A\ldots C'}_{D\ldots F'}$ as in §2.5. The elements of $\mathfrak{T}^a$, defined to be the real elements of $\mathfrak{S}^{AA'}$, are clearly in 1-1 canonical correspondence with the tangent vectors $\mathfrak{T}^{\cdot}$, because locally we have set up a canonical association of the tangent vector tetrad with the spin-frame.

Alternatively we can choose the axiomatic approach. Here we simply postulate the algebraic requirements for the spinor system. The space–time structure, e.g., metric, signature, topolological requirements, are then thought of as *derived* properties. One may regard the constructive approach as justifying the axioms chosen for the spinor system, since they are satisfied for any space–time in which the constructive approach works. On the other hand, the existence of the spinor algebra might be regarded as providing a 'deeper' reason for the particular space–time structure that arises. It is this axiomatic approach that we shall be essentially following here. Thus, we postulate the existence of a spinor algebra of the type set up in §2.5, and then demand that its systems $\mathfrak{T}$ and $\mathfrak{T}^a = \mathfrak{T}^{AA'}$ be isomorphic, respectively with the scalar fields $\mathfrak{T}$, and their derivations $\mathfrak{T}^{\cdot}$, on a manifold $\mathcal{M}$ defined according to the axioms of §4.1.

The required isomorphism between $\mathfrak{T}^{\cdot}$ and $\mathfrak{T}^a = \mathfrak{T}^{AA'}$ states that every derivation $\boldsymbol{U} \in \mathfrak{T}^{\cdot}$ corresponds to a unique element $U^{AA'} = U^a \in \mathfrak{T}^a$ and vice versa. (Having asserted this canonical isomorphism, we can, without risk of confusion, refer to $U^a$, $V^a, \ldots$ also as world-vector fields on $\mathcal{M}$.) We use the symbol $\nabla_a$ (or, equivalently, $\nabla_{AA'}$) to denote this isomorphism, i.e., $\nabla_a$ effects $U^a \mapsto \boldsymbol{U}$, this being written

$$U^a \nabla_a = \boldsymbol{U}. \tag{4.4.1}$$

These are operators acting on real scalars, so $U^a \nabla_a$ is a map (derivation) from $\mathfrak{T}$ to $\mathfrak{T}$. We can extend the range of this operator to the complex scalars $\mathfrak{S} = \mathfrak{T} \oplus \mathrm{i}\mathfrak{T}$, giving a map from $\mathfrak{S}$ to $\mathfrak{S}$, by

$$U^a \nabla_a (f + \mathrm{i}g) := \boldsymbol{U}(f) + \mathrm{i}\,\boldsymbol{U}(g) =: \boldsymbol{U}(f + \mathrm{i}g);\, f, g \in \mathfrak{T}. \tag{4.4.2}$$

Furthermore, we can define

$$(U^a + \mathrm{i}V^a)\nabla_a h := \boldsymbol{U}(h) + \mathrm{i}\,\boldsymbol{V}(h);\ h \in \mathfrak{S}. \tag{4.4.3}$$

For any *given* $h \in \mathfrak{S}$ we therefore have a map from $\mathfrak{S}^a$ to $\mathfrak{S}$, defined by $W^a \mapsto W^a \nabla_a h$ (with $W^a \in \mathfrak{S}^a$) which is evidently $\mathfrak{S}$-linear; hence, for $h \in \mathfrak{S}$, we have an element

$$\nabla_a h \in \mathfrak{S}_a. \tag{4.4.4}$$

It is clear from (4.4.2) and (4.4.3) that $\overline{W^a\nabla_a h} = \bar{W}^a\nabla_a \bar{h}$, hence

$$\overline{\nabla_a h} = \nabla_a \bar{h}. \tag{4.4.5}$$

In particular, if $h \in \mathfrak{T}$ then $\nabla_a h \in \mathfrak{T}_a$. From the derivation properties we must also have

$$\nabla_a k = 0, \nabla_a(g + h) = \nabla_a g + \nabla_a h, \nabla_a(gh) = g\nabla_a h + h\nabla_a g, \tag{4.4.6}$$

where $k \in \mathfrak{R} \oplus \mathrm{i}\mathfrak{R}$ and $g, h \in \mathfrak{S}$.

We next wish to extend the definition of $\nabla_a$ so that it applies to any spinor. We shall follow closely the development given in §4.2. A *spinor covariant derivative operator* will be defined by a map

$$\nabla_a : \mathfrak{S}^B \to \mathfrak{S}^B_{AA'} \tag{4.4.7}$$

satisfying

$$\nabla_a(\xi^B + \eta^B) = \nabla_a \xi^B + \nabla_a \eta^B, \tag{4.4.8}$$

$$\nabla_a(f\xi^B) = f\nabla_a \xi^B + \xi^B \nabla_a f, \tag{4.4.9}$$

for each $\xi^B, \eta^B \in \mathfrak{S}^B, f \in \mathfrak{S}$; the definition of $\nabla_a f$ being as given above. We can, of course, write $\nabla_{AA'}$ for $\nabla_a$ and apply index substitutions to define, e.g., $\nabla_{X_3X'_3}\xi^{Q_0} = \nabla_{x_3}\xi^{Q_0}$, etc. This possibility will always be assumed. (However, we must be careful to bear in mind that $\nabla_a \xi^A$ is *not* an index substitution of $\nabla_a \xi^B$ but its $\binom{B}{A}$-contraction, namely $\nabla_{AA'}\xi^A$; see remarks after (3.1.37)).

We extend $\nabla_a$ to give a map from $\mathfrak{S}_B$ to $\mathfrak{S}_{ABA'}$, defined by the requirement that the derivative of a contracted product $\alpha_B \xi^B$ should satisfy a Leibniz law:

$$(\nabla_a \alpha_B)\xi^B = \nabla_a(\alpha_B \xi^B) - \alpha_B \nabla_a \xi^B. \tag{4.4.10}$$

The left-hand side of this equation defines $\nabla_a \alpha_B$ as a map (easily seen to be $\mathfrak{S}$-linear, by (4.4.6), (4.4.8), (4.4.9) from $\mathfrak{S}^B$ to $\mathfrak{S}_{AA'}$). We readily verify, from (4.4.6), that

$$\nabla_a(\alpha_B + \beta_B) = \nabla_a \alpha_B + \nabla_a \beta_B \tag{4.4.11}$$

and

$$\nabla_a(f\alpha_B) = f\nabla_a \alpha_B + \alpha_B \nabla_a f \tag{4.4.12}$$

hold for each $\alpha_B, \beta_B \in \mathfrak{S}_B, f \in \mathfrak{S}$. We define the action of $\nabla_a$ on $\mathfrak{S}^{B'}$ and on $\mathfrak{S}_{B'}$ – giving maps from $\mathfrak{S}^{B'}$ to $\mathfrak{S}^{B'}_{AA'}$ and from $\mathfrak{S}_{B'}$ to $\mathfrak{S}_{AA'B'}$, respectively – by means of complex conjugation:

$$\nabla_a \zeta^{B'} = \overline{\nabla_a \bar{\zeta}^B}, \quad \nabla_a \omega_{B'} = \overline{\nabla_a \bar{\omega}_B}. \tag{4.4.13}$$

Then it is clear from (4.4.5) and the complex conjugate of (4.4.10) that a Leibniz law applies also to contracted products $\omega_{B'}\zeta^{B'}$. Clearly, moreover, the linearity and Leibniz properties corresponding to (4.4.8), (4.4.9), (4.4.11), (4.4.12) must also hold for $\nabla_a \zeta^{B'}$ and $\nabla_a \omega_{B'}$, by virtue of (4.4.13).

We are now in a position to define $\nabla_a$ as applied to a general spinor $\chi_{B\ldots F'}{}^{P\ldots S'}$. As in (4.2.8), we simply demand that a Leibniz law shall apply to contracted products of the form $\chi_{B\ldots F'}{}^{P\ldots S'}\beta^B \ldots \phi^{F'}\pi_P \ldots \sigma_{S'}$; then

$$\begin{aligned}(\nabla_a \chi_{B\ldots F'}{}^{P\ldots S'})\beta^B \ldots \phi^{F'}\pi_P \ldots \sigma_{S'} &= \nabla_a(\chi_{B\ldots F'}{}^{P\ldots S'}\beta^B \ldots \sigma_{S'}) \\ &- \chi_{B\ldots F'}{}^{P\ldots S'}(\nabla_a \beta^B) \ldots \sigma_{S'} - \cdots - \chi_{B\ldots F'}{}^{P\ldots S'}\beta^B \ldots (\nabla_a \sigma_{S'})\end{aligned} \tag{4.4.14}$$

defines $\nabla_a \chi_{B\ldots F'}{}^{P\ldots S'}$ as effecting an $\mathfrak{S}$-multilinear map from $\mathfrak{S}^B \times \cdots \times \mathfrak{S}^{F'} \times \mathfrak{S}_P \times \cdots \times \mathfrak{S}_{S'}$ to $\mathfrak{S}_{AA'}$. Thus for each spinor set $\mathfrak{S}^{P\ldots S'}_{B\ldots F'}$, $\nabla_a$ defines a map

$$\nabla_a : \mathfrak{S}^{P\ldots S'}_{B\ldots F'} \to \mathfrak{S}^{P\ldots S'}_{AA'B\ldots F'}. \tag{4.4.15}$$

As in (4.2.10)–(4.2.13), we have the properties

$$\nabla_a(\psi_{\mathscr{D}} + \chi_{\mathscr{D}}) = \nabla_a \psi_{\mathscr{D}} + \nabla_a \chi_{\mathscr{D}}, \tag{4.4.16}$$

$$\nabla_a(\psi_{\mathscr{D}}\phi_{\mathscr{C}}) = \psi_{\mathscr{D}}\nabla_a \phi_{\mathscr{C}} + \phi_{\mathscr{C}}\nabla_a \psi_{\mathscr{D}}. \tag{4.4.17}$$

Also,

$\nabla_a$ *commutes with any index substitution not involving* $A$ *or* $A'$. (4.4.18)

Furthermore, $\nabla_a$ commutes with contraction (not involving $A$ or $A'$):

$$\begin{aligned}\varepsilon_P{}^Q(\nabla_a \psi_{\ldots Q\ldots}{}^{\ldots P\ldots}) &= \nabla_a \psi_{\ldots P\ldots}{}^{\ldots P\ldots}, \\ \varepsilon_{P'}{}^{Q'}(\nabla_a \psi_{\ldots Q'\ldots}{}^{\ldots P'\ldots}) &= \nabla_a \psi_{\ldots P'\ldots}{}^{\ldots P'\ldots}.\end{aligned} \tag{4.4.19}$$

In addition, because of the definitions (4.4.13), it follows that $\nabla_a$ commutes with the operation of complex conjugation:

$$\overline{\nabla_a \psi_{\mathscr{D}}} = \nabla_a(\overline{\psi_{\mathscr{D}}}). \tag{4.4.20}$$

(Formally, this means that $\nabla_a$ is a *real* operator: $\bar{\nabla}_a = \nabla_a$.)

In the particular case of real world-vectors $U^b = U^{BB'} \in \mathfrak{T}^b$, the operator $\nabla_a$ defines a *tensor* covariant derivative $U^b \mapsto \nabla_a U^b$, satisfying $\nabla_a(U^b + V^b) = \nabla_a U^b + \nabla_a V^b$, $\nabla_a(fU^b) = f\nabla_a U^b + U^b \nabla_a f$ in accordance with (4.2.2), (4.2.3). The extension of this $\nabla_a$ to real world-tensors clearly agrees with that given above since the rules (4.2.10)–(4.2.13) (which define it uniquely) are all satisfied by virtue of (4.4.16)–(4.4.20).

### *Uniqueness*

Let us investigate the question of the uniqueness of a spinor covariant derivative operator subject to these rules. Let $\nabla_a$ and $\tilde{\nabla}_a$ be two such operators and consider the map

$$(\tilde{\nabla}_a - \nabla_a) : \mathfrak{S}^B \to \mathfrak{S}^B_{AA'}. \tag{4.4.21}$$

Then (as in (4.2.44)) this map is $\mathfrak{S}$-linear because of (4.4.8), (4.4.9) and the fact that when acting on scalars the operators $\tilde{\nabla}_a$ and $\nabla_a$ must agree:

$$\tilde{\nabla}_a f = \nabla_a f. \tag{4.4.22}$$

Thus there is an element $\Theta_{AA'B}{}^C \in \mathfrak{S}^C_{ABA'}$ such that

$$\tilde{\nabla}_{AA'}\xi^C = \nabla_{AA'}\xi^C + \Theta_{AA'B}{}^C\xi^B. \tag{4.4.23}$$

Since $(\tilde{\nabla}_a - \nabla_a)(\alpha_B\xi^B) = 0$ by (4.4.22), we have, by (4.4.23),

$$\tilde{\nabla}_{AA'}\alpha_B = \nabla_{AA'}\alpha_B - \Theta_{AA'B}{}^C\alpha_C. \tag{4.4.24}$$

Taking complex conjugates of (4.4.23), (4.4.24) we get

$$\tilde{\nabla}_{AA'}\zeta^{C'} = \nabla_{AA'}\zeta^C + \bar{\Theta}_{AA'B'}{}^{C'}\zeta^{B'}. \tag{4.4.25}$$

$$\tilde{\nabla}_{AA'}\omega_{B'} = \nabla_{AA'}\omega_{B'} - \bar{\Theta}_{AA'B'}{}^{C'}\omega_{C'}. \tag{4.4.26}$$

Hence, for a general spinor

$$\begin{aligned}\tilde{\nabla}_{AA'}\chi_{B\ldots F'}{}^{P\ldots S'} = \nabla_{AA'}\chi_{B\ldots F'}{}^{P\ldots S'} &- \Theta_{AA'B}{}^X\chi_{X\ldots F'}{}^{P\ldots S'} - \ldots \\ &- \bar{\Theta}_{AA'F'}{}^{X'}\chi_{B\ldots X'}{}^{P\ldots S'} + \Theta_{AA'X}{}^P\chi_{B\ldots F'}{}^{X\ldots S'} + \ldots \\ &+ \bar{\Theta}_{AA'X'}{}^{S'}\chi_{B\ldots F'}{}^{P\ldots X'}.\end{aligned} \tag{4.4.27}$$

Now consider the special case when $\chi_{\ldots}{}^{\cdots}$ is a world-vector $U^b$. We get

$$\begin{aligned}(\tilde{\nabla}_a - \nabla_a)U^{BB'} &= \Theta_{aC}{}^BU^{CB'} + \bar{\Theta}_{aC'}{}^{B'}U^{BC'} \\ &= (\Theta_{aC}{}^B\varepsilon_{C'}{}^{B'} + \bar{\Theta}_{aC'}{}^{B'}\varepsilon_C{}^B)U^{CC'} \\ &= Q_{ac}{}^bU^c,\end{aligned} \tag{4.4.28}$$

where the quantity

$$Q_{ac}{}^b = \Theta_{AA'C}{}^B\varepsilon_{C'}{}^{B'} + \bar{\Theta}_{AA'C'}{}^{B'}\varepsilon_C{}^B \tag{4.4.29}$$

agrees with that defined in (4.2.46). By (4.2.50), the difference between the torsion tensors $\tilde{T}_{ab}{}^c$, $T_{ab}{}^c$, defined by $\tilde{\nabla}_a$ and $\nabla_a$, respectively (where $(\nabla_a\nabla_b - \nabla_b\nabla_a)f = T_{ab}{}^c\nabla_c f$, and similarly for $\tilde{\nabla}_a$), is given by

$$\tilde{T}_{ab}{}^c - T_{ab}{}^c = Q_{ba}{}^c - Q_{ab}{}^c, \tag{4.4.30}$$

with $Q_{ab}{}^c$ as in (4.4.29).

Consider, next, the derivative of $\varepsilon_{AB}$. We have

$$\begin{aligned}(\tilde{\nabla}_a - \nabla_a)\varepsilon_{BC} &= -\Theta_{AA'B}{}^D\varepsilon_{DC} - \Theta_{AA'C}{}^D\varepsilon_{BD} \\ &= -\Theta_{AA'BC} + \Theta_{AA'CB}.\end{aligned} \tag{4.4.31}$$

If we demand of our operators $\nabla_a$, $\tilde{\nabla}_a$ that the spinor $\varepsilon_{AB}$ *be covariantly constant*,

$$\nabla_a\varepsilon_{BC} = 0 \tag{4.4.32}$$

$$\tilde{\nabla}_a\varepsilon_{BC} = 0, \tag{4.4.33}$$

then $\Theta_{AA'BC}$ must be symmetric in $B, C$:

$$\Theta_{AA'BC} = \Theta_{AA'CB}. \tag{4.4.34}$$

Since we have

$$Q_{abc} = \Theta_{AA'BC}\varepsilon_{B'C'} + \bar{\Theta}_{AA'B'C'}\varepsilon_{BC} \tag{4.4.35}$$

by (4.4.29), the anti-symmetry

$$Q_{abc} = -Q_{acb} \tag{4.4.36}$$

follows from (4.4.34). (This anti-symmetry is more directly a consequence of $0 = (\tilde{\nabla}_a - \nabla_a)g_{bc} = -Q_{ab}{}^d g_{dc} - Q_{ac}{}^d g_{bd} = -Q_{abc} - Q_{acb}$, which follows from (4.4.32), (4.4.33) because $g_{bc} = \varepsilon_{BC}\varepsilon_{B'C'}$.) If $\tilde{\nabla}_a$ is torsion-free and if $\varepsilon_{AB}$ is covariantly constant with respect to both $\nabla_a$ and $\tilde{\nabla}_a$, then the torsion $T_{abc}$ of $\nabla_a$ is given by

$$\begin{aligned} T_{abc} &= Q_{abc} - Q_{bac} \\ &= \Theta_{A'ABC}\varepsilon_{B'C'} + \bar{\Theta}_{AA'B'C'}\varepsilon_{BC} - \Theta_{B'BAC}\varepsilon_{A'C'} - \bar{\Theta}_{BB'A'C'}\varepsilon_{AC}. \end{aligned} \tag{4.4.37}$$

This has significance for the Einstein–Cartan–Sciama–Kibble theory as discussed in §4.7.

We are now in a position to show that *uniqueness* of $\nabla_a$ follows from the two requirements that $\varepsilon_{AB}$ be covariantly constant (4.4.32), and that the torsion $T_{ab}{}^c$ should vanish:

$$\nabla_a \nabla_b f = \nabla_b \nabla_a f \quad \textit{for all } f \in \mathfrak{S}. \tag{4.4.38}$$

For, the latter implies (assuming the same for $\tilde{\nabla}_a$) that $Q_{abc}$ is *symmetric* in $a, b$ (*cf.* (4.4.30)) whereas the former implies that it is *skew* in $b, c$ (*cf.* (4.4.36)). Thus, by (3.3.17), $Q_{abc} = 0$. Taking the part of (4.4.35) which is symmetric in $B, C$ and using (4.4.34), we obtain the uniqueness condition $\Theta_{AA'BC} = 0$, as required.

It is of some interest, in view of our concern in the next chapter with charged fields and with conformal transformations, to examine the nature of the non-uniqueness of $\nabla_a$ when merely the vanishing-torsion condition (4.4.38) is assumed, and not the covariant constancy of $\varepsilon_{AB}$. If $\nabla_a$ and $\tilde{\nabla}_a$ are both torsion-free (or, indeed, if their torsions are equal), we have, by (4.4.29) and the symmetry $Q_{abc} = Q_{bac}$ implied by (4.4.30),

$$\varepsilon_{B'C'}\Theta_{A'ABC} + \varepsilon_{BC}\bar{\Theta}_{AA'B'C'} = \varepsilon_{A'C'}\Theta_{B'BAC} + \varepsilon_{AC}\bar{\Theta}_{BB'A'C'}. \tag{4.4.39}$$

Symmetrizing over $A, B, C$ and transvecting with $\varepsilon^{B'C'}$ we obtain

$$\Theta_{A'(ABC)} = 0. \tag{4.4.40}$$

Applying (3.3.49), we obtain the result that $\Theta_{A'ABC}$ has the form

$$\Theta_{A'ABC} = \lambda_{A'A}\varepsilon_{BC} + \mu_{A'B}\varepsilon_{AC} + \nu_{A'C}\varepsilon_{AB}. \tag{4.4.41}$$

By use of the identity (2.5.20) we can re-express the final term as $\nu_{A'A}\varepsilon_{CB} +$

$v_{A'B}\varepsilon_{AC}$, so we have a relation of the form

$$\Theta_{A'ABC} = \Lambda_{A'A}\varepsilon_{BC} + \Upsilon_{A'B}\varepsilon_{AC}. \tag{4.4.42}$$

Now if we symmetrize (4.4.39) over $A, C$ and over $A', C'$ we get, upon substitution of (4.4.42),

$$-\varepsilon_{B'(C'}\Lambda_{A')(A}\varepsilon_{C)B} - \varepsilon_{B'(C'}\bar{\Lambda}_{A')(A}\varepsilon_{C)B} = 0, \tag{4.4.43}$$

which, by use of (3.5.15), or else by transvection with $\varepsilon^{B'C'}\varepsilon^{BC}$, yields

$$\Lambda_{A'A} + \bar{\Lambda}_{AA'} = 0. \tag{4.4.44}$$

So the vector $\Lambda_a$ is *pure imaginary*, say $\Lambda_a = \mathrm{i}\Pi_a$, where $\Pi_a$ is real. Similarly, symmetrizing (4.4.39) over $A, C$ and over $B', C'$ we get, when (4.4.42) is substituted,

$$-\varepsilon_{B(C}\bar{\Upsilon}_{A)(B'}\varepsilon_{C')A'} = -\varepsilon_{B(C}\Upsilon_{A)(B'}\varepsilon_{C')A'}, \tag{4.4.45}$$

and by an argument similar to the above we find

$$\Upsilon_{AA'} - \bar{\Upsilon}_{AA'} = 0, \tag{4.4.46}$$

i.e., the vector $\Upsilon_a$ is *real*. Collecting these relations together we have

$$\Theta_{AA'B}{}^{C} = \mathrm{i}\Pi_{AA'}\varepsilon_B{}^C + \Upsilon_{A'B}\varepsilon_A{}^C; \quad \Pi_a, \Upsilon_a \in \mathfrak{T}_a. \tag{4.4.47}$$

This is the complete solution to the problem, for substituting (4.4.47) back into (4.4.39) we find that that relation is identically satisfied. We shall see in (5.6.14) that a quantity $\Upsilon_a$ arises in connection with the change in covariant derivative under *conformal rescalings*. The quantity $\Pi_a$ had, in the early literature, been associated with an electromagnetic vector potential (see Infeld and van der Waerden 1933). Our approach in §5.1 will be somewhat different, however.

### *Construction from the Christoffel derivative*

Let us assume, henceforth, that $\nabla_a$ is torsion-free, (4.4.38), *and* that $\varepsilon_{AB}$ is covariantly constant under $\nabla_a$, (4.4.32). Then $\nabla_a$ is unique. Furthermore the operation of raising or lowering an index will commute with $\nabla_a$. That is,

$$\nabla_a \chi_{\mathscr{D}B} = \psi_{a\mathscr{D}B} \quad \textit{iff} \quad \nabla_a \chi_{\mathscr{D}}{}^{B} = \psi_{a\mathscr{D}}{}^{B}. \tag{4.4.48}$$

Of course, we have to show that an operator $\nabla_a$ *exists* with these properties. The existence of such a $\nabla_a$ may be inferred in various ways, (e.g. by means of the somewhat complicated explicit formulae of §4.5). One way is to use the results of §§4.2, 4.3 (*cf.* (4.3.45)) to establish the existence of a (Christoffel) operator $\nabla_a$ whose action on real world-tensors is defined, which satisfies (4.2.10)–(4.2.13), whose torsion vanishes and for which the metric $g_{ab} = \varepsilon_{AB}\varepsilon_{A'B'}$ is covariantly constant: $\nabla_a g_{bc} = 0$.

Then we have to extend the domain of definition to include spinors.

To do this, we must first extend $\nabla_a$ so that it applies to *complex* world vectors. This is easily achieved if we define

$$\nabla_a(H_{\mathscr{D}} + \mathrm{i}G_{\mathscr{D}}) = \nabla_a H_{\mathscr{D}} + \mathrm{i}\nabla_a G_{\mathscr{D}}\,;\, H_{\mathscr{D}}, G_{\mathscr{D}} \in \mathfrak{T}_{\mathscr{D}}. \tag{4.4.49}$$

The composite index $\mathscr{D}$ must (at this stage) involve equal numbers of primed and unprimed indices both in lower and in upper position, but need not be restricted further than this. (For example, $\nabla_a H_{DB'}{}^{BC'}$ may be defined from $\nabla_a H_d{}^c$ by index substitution.)

Now, consider the following expression:

$$\alpha_{B'}\nabla_a(\xi^B\beta^{B'}) - \beta_{B'}\nabla_a(\xi^B\alpha^{B'}) - \xi^B\nabla_a(\alpha_{B'}\beta^{B'}). \tag{4.4.50}$$

This is well-defined since $\nabla_a$ acts only on complex world-vectors or scalars. Furthermore, it is readily verified that (4.4.50) is $\mathfrak{S}$-linear both in $\alpha^{B'}$ and in $\beta^{B'}$. (These verifications just use (4.2.2), (4.2.3).) Thus, for each $\xi^B$, (4.4.50) defines an $\mathfrak{S}$-bilinear map from $\mathfrak{S}^{B'} \times \mathfrak{S}^{C'}$ to $\mathfrak{S}^B_a$, achieved by means of contracted product with an element $\theta_a{}^B{}_{B'C'} \in \mathfrak{S}^B_{aB'C'}$ (where $\theta_a{}^B{}_{B'C'}$ depends on $\xi^B$), (4.4.50) being given by $\theta_a{}^B{}_{B'C'}\alpha^{B'}\beta^{C'}$. Notice that (4.4.50) is anti-symmetrical under interchange of $\alpha^{B'}$ with $\beta^{B'}$. This means that $\theta_a{}^B{}_{B'C'}$ is skew in $B'C'$, so it has the form $\theta_a{}^B{}_{B'C'} = \phi_a{}^B\varepsilon_{B'C'}$ where $\phi_a{}^B$ is a function of $\xi^B$. Write $\tilde{\nabla}_a\xi^B := \frac{1}{2}\phi_a{}^B$; then (4.4.50) is equal to

$$2(\tilde{\nabla}_a\xi^B)\alpha_{B'}\beta^{B'}. \tag{4.4.51}$$

We want to verify that the map $\tilde{\nabla}_a : \mathfrak{S}^B \to \mathfrak{S}^B_a$ defined by $\xi^B \mapsto \tilde{\nabla}_a\xi^B$ satisfies $\tilde{\nabla}_a(\xi^B + \eta^B) = \tilde{\nabla}_a\xi^B + \tilde{\nabla}_a\eta^B$ and $\tilde{\nabla}_a(f\xi^B) = f\tilde{\nabla}_a\xi^B + \xi^B\tilde{\nabla}_a f$. These properties follow at once by substitution of $\xi^B + \eta^B$ and $f\xi^B$, for $\xi^B$, in (4.4.50). Thus $\tilde{\nabla}_a$ defines a spinor covariant derivative operator.

We next check that $\tilde{\nabla}_a\varepsilon_{BC} = 0$. We have

$$(\tilde{\nabla}_a\varepsilon_{BC})\xi^B\eta^C = \nabla_a(\varepsilon_{BC}\xi^B\eta^C) - \varepsilon_{BC}\xi^B\tilde{\nabla}_a\eta^C - \varepsilon_{BC}\eta^C\tilde{\nabla}_a\xi^B. \tag{4.4.52}$$

We can multiply this by $2\varepsilon_{B'C'}\alpha^{B'}\beta^{C'}$ and use

$$2\varepsilon_{B'C'}\alpha^{B'}\beta^{C'}\tilde{\nabla}_a\eta^C = \varepsilon_{B'C'}\alpha^{B'}\nabla_a(\eta^C\beta^{C'}) + \varepsilon_{B'C'}\beta^{C'}\nabla_a(\eta^C\alpha^{B'}) - \eta^C\nabla_a(\varepsilon_{B'C'}\alpha^{B'}\beta^{C'}) \tag{4.4.53}$$

(which is just (4.4.51) equated to (4.4.50)), and the corresponding equation for $2\varepsilon_{B'C'}\alpha^{B'}\beta^{C'}\tilde{\nabla}_a\xi^B$. We must also use the relations

$$\begin{aligned}\eta^C\beta^{C'}\nabla_a(\xi^B\alpha^{B'}) + \xi^B\alpha^{B'}\nabla_a(\eta^C\beta^{C'}) &= \nabla_a(\eta^C\beta^{C'}\xi^B\alpha^{B'})\\ &= \eta^C\alpha^{B'}\nabla_a(\xi^B\beta^{C'}) + \xi^B\beta^{C'}\nabla_a(\eta^C\alpha^{B'})\end{aligned} \tag{4.4.54}$$

and

$$\begin{aligned}\nabla_a(\xi^B\eta^C\varepsilon_{BC}\alpha^{B'}\beta^{C'}\varepsilon_{B'C'}) - \varepsilon_{BC}\varepsilon_{B'C'}\nabla_a(\xi^B\eta^C\alpha^{B'}\beta^{C'})\\ = \xi^B\eta^C\alpha^{B'}\beta^{C'}\nabla_a(\varepsilon_{BC}\varepsilon_{B'C'})\\ = \xi^B\eta^C\alpha^{B'}\beta^{C'}\nabla_a g_{bc} = 0.\end{aligned} \tag{4.4.55}$$

In this way, (4.4.52) yields

$$(\tilde{\nabla}_a \varepsilon_{BC})\xi^B \eta^C = 0 \tag{4.4.56}$$

for all $\xi^B, \eta^C$, whence $\tilde{\nabla}_a \varepsilon_{BC} = 0$ as required. It follows, as in (4.3.50), that the spinor covariant derivative $\tilde{\nabla}_a$ commutes with the operation of raising and lowering spinor indices. Hence, the expression (4.4.50) with $\nabla_a$ replaced by $\tilde{\nabla}_a$ throughout, must also be equal to (4.4.51). Taking the difference between these two versions of (4.4.50) we obtain

$$\alpha_{B'}(\tilde{\nabla}_a - \nabla_a)(\xi^B \beta^{B'}) - \beta_{B'}(\tilde{\nabla}_a - \nabla_a)(\xi^B \alpha^{B'}) = 0.$$

That is, since each of $\tilde{\nabla}_a, \nabla_a$ is a tensor covariant derivative operator when acting on $\mathfrak{S}^b$,

$$\alpha_{B'} Q_{aCC'}{}^{BB'} \xi^C \beta^{C'} = \beta_{B'} Q_{aCC'}{}^{BB'} \xi^C \alpha^{C'},$$

where $Q_{ac}{}^b$ is as in (4.2.46). Hence $Q_{aCC'BB'}$ is symmetric in $C'$, $B'$. Since $Q_{acb}$ is real (each of $\tilde{\nabla}_a, \nabla_a$ maps real world-tensors to real world-tensors), it follows that $Q_{aC'CB'}$ is also symmetric in $C$, $B$. Hence $Q_{acb}$ is symmetric in $c, b$. But $g_{bc}(= \varepsilon_{BC}\varepsilon_{B'C'})$ is covariantly constant with respect to *each* of $\tilde{\nabla}_a, \nabla_a$ (since $\tilde{\nabla}_a \varepsilon_{B'C'} = \tilde{\nabla}_a \varepsilon_{BC} = 0$). Thus, by (4.4.36), $Q_{acb}$ is also *skew* in $c, b$, so $Q_{acb} = 0$. This establishes the identity of $\tilde{\nabla}_a$ and $\nabla_a$ when the operators act on tensors (and, incidentally, shows that the torsion of $\tilde{\nabla}_a$ must vanish). We can therefore write $\nabla_a := \tilde{\nabla}_a$ when the operators act on *any* spinor, and the desired (unique) spinor derivative operator, satisfying the conditions laid down after (4.4.48), is thereby obtained.

*Tensor translation of spinor differential equations*

We complete this section by returning to the discussion given at the end of §3.4 concerning the translation of algebraic spinor operations into tensor form. We are now in a position to extend that discussion to the translation of derivatives of spinors, and so of spinor differential equations. (The converse problem – translating tensor derivatives and differential equations into spinor form – is of course straightforward.) We shall show that, in principle, every spinor differential equation has an equivalent tensor form (which, however, may be quite complicated), apart from certain intrinsic sign ambiguities. Also, since the tensor equations often involve squares of the corresponding spinors, it may happen in non-simply-connected regions of space–time that global solutions exist to the tensor equation while no consistent sign can be assigned to the solution of the spinor equation. So the tensor and spinor equations may be equivalent locally but not globally.

Spinor differential equations arise most naturally in quantum theory, where it is presumed – from the theory of group representations and from the requirements of Lorentz invariance and linear superposition of states – that the basic equations for (free) particles are *linear* spinor laws. On translation, these usually become *non*-linear tensor laws. Thus the attempt to regard the tensor laws, and the tensors occurring in them, as fundamental, would negate the whole standard *linear* character and formalism of quantum theory. With this disclaimer, the possibility of translating spinor into tensor differential equations is nevertheless a significant theoretical result.

Before discussing the general translation question, we shall deal with two specific and important examples – the Dirac–Weyl neutrino* equation and the Dirac electron equation. For this purpose it will be useful to establish two preliminary lemmas for an arbitrary spinor $\phi_A$. The first is the identity

$$\nabla_p(\phi_A\phi_B) + \varepsilon_{AB}\phi_C\nabla_p\phi^C = 2\phi_A\nabla_p\phi_B, \tag{4.4.57}$$

which is established at once by expanding the first term on the left side, and replacing the second** by $-2\nabla_p\phi_{[A}\phi_{B]}$. The second lemma is this: if $F_{ab}$ is the anti-self-dual null bivector corresponding to $\phi_A$,

$$F_{ab} = \phi_A\phi_B\varepsilon_{A'B'}, \tag{4.4.58}$$

and $M_p$ is an auxiliary vector defined by

$$M_p = \phi_A\nabla_p\phi^A, \tag{4.4.59}$$

then

$$F_{ab}\nabla_p F_c{}^b = F_{ac}M_p. \tag{4.4.60}$$

The proof again devolves upon the Leibniz expansion of the derivative:

$$\begin{aligned}\text{LHS} &= \phi_A\phi_B\varepsilon_{A'B'}\nabla_p(\phi_C\phi^B)\varepsilon_{C'}{}^{B'}\\ &= \phi_A\phi_C\varepsilon_{A'C'}\phi_B\nabla_p\phi^B = \text{RHS}.\end{aligned}$$

Now the Dirac–Weyl equation (Dirac 1928, Weyl 1929, Dirac 1982) is

$$\nabla_{AA'}\phi^A = 0. \tag{4.4.61}$$

To translate it to tensor form, we first introduce another auxiliary vector,

$$R_a = \phi_A\nabla_{BA'}\phi^B, \tag{4.4.62}$$

* We use the term 'neutrino' consistently here for a *massless* (uncharged) spin-$\frac{1}{2}$ particle, which is thus taken to satisfy the Dirac–Weyl equation. This is not intended to prejudice the issue of whether or not *physical* neutrinos actually possess mass.

** Here, and at certain other places in these volumes, we make use of a convention that a differential operator (e.g. denoted $\nabla$, $d$, $\eth$, ...) acts only on the symbol (or bracketed expression) which immediately follows it – unless this also is a differential operator. Thus $\nabla AB$ would mean $(\nabla A)B$ and not $\nabla(AB)$; $\nabla\nabla AB$ would mean $(\nabla(\nabla A))B$ etc.

whose vanishing will correspond to the Dirac–Weyl equation, except possibly in regions where $\phi_A = 0$ (but those we can bridge by continuity provided $\phi_A$ is assumed smooth). Next, in (4.4.57) we raise $B$, replace $p$ by $b$ (thus contracting over $B$), and multiply by $\varepsilon_{A'}{}^{B'}$; the resulting equation is purely tensorial:

$$\nabla_b F_a{}^b + M_a = 2R_a. \tag{4.4.63}$$

The Dirac–Weyl equation, being equivalent to $R_a = 0$, is therefore equivalent to the tensor equation

$$\nabla_d F_c{}^d + M_c = 0. \tag{4.4.64}$$

But this still contains the auxiliary vector $M_a$. It is precisely for its elimination that we established our second lemma. Thus, multiplying (4.4.64) by $F_{ab}$ and referring to (4.4.60), we finally obtain

$$F_{ab}\nabla_d F_c{}^d + F_{ad}\nabla_c F_b{}^d = 0. \tag{4.4.65)(a}$$

And, of course, the form of (4.4.58) implies the following additional conditions on $F_{ab}$ (namely, that it be skew, anti-self-dual, and null):

$$F_{ab} = -F_{ba}, \quad F_{ab} = \mathrm{i}\,{}^*F_{ab}, \quad F_{ab}F^{ab} = 0. \tag{4.4.65)(b}$$

The set of equations (4.4.65)(*a*) and (*b*) is the tensor equivalent of the Dirac–Weyl equation. Its structure is evidently much more complicated than that of the original spinor equation; in particular, it is non-linear.

Dirac's equation can be written in the form of two coupled 2-component spinor equations* (Dirac 1928, van der Waerden 1929, Infeld and van

* It is, of course, more usual to write the Dirac equation in terms of *four*-component spinors. The detailed relation between Dirac 4-spinors and the 2-spinors that we use exclusively in this volume will be given as part of the general discussion of spinors in $n$ dimensions that we give in the Appendix to Vol. 2. The reader who has familiarity with 4-spinors may, however, make direct contact with our notation by taking note that it is the *pair* of 2-spinors $(\phi_A, \psi_{A'})$ that constitutes a single Dirac 4-spinor $\mathbf{\Psi}$. The two members of this pair are obtained by operating on $\mathbf{\Psi}$ by $\frac{1}{2}(\mathbf{I} + \mathrm{i}\gamma_5)$ and $\frac{1}{2}(\mathbf{I} - \mathrm{i}\gamma_5)$, respectively, where $\gamma_5 = \gamma_0\gamma_1\gamma_2\gamma_3$, in a standard orthonormal frame, the Dirac matrices $\gamma_0, \ldots, \gamma_3, \gamma_5$ being given by

$$(\gamma_p)_\alpha{}^\beta = \gamma_{p\alpha}{}^\beta = \sqrt{2}\begin{bmatrix} 0 & \varepsilon_{PA}\varepsilon_{P'}{}^{B'} \\ \varepsilon_{P'A'}\varepsilon_P{}^B & 0 \end{bmatrix} \quad (\gamma_5)_\alpha{}^\beta = \begin{bmatrix} -\mathrm{i}\varepsilon_A{}^B & 0 \\ 0 & \mathrm{i}\varepsilon_{A'}{}^{B'} \end{bmatrix},$$

with $\alpha = A \oplus A'$, $\beta = B \oplus B'$. One directly verifies that the Clifford–Dirac equation

$$\gamma_p\gamma_q + \gamma_q\gamma_p = -2g_{pq}\mathbf{I}, \quad \text{i.e.,} \quad \gamma_{p\alpha}{}^\beta\gamma_{q\beta}{}^\gamma + \gamma_{q\alpha}{}^\beta\gamma_{p\beta}{}^\gamma = -2g_{pq}\delta_\alpha^\gamma$$

is satisfied (compare footnote on p. 124). An advantage of the 2-spinor description is that the $\gamma$-matrices disappear completely – and complicated $\gamma$-matrix identities simply evaporate!

der Waerden 1933):

$$\left.\begin{aligned}\nabla^A_{A'}\phi_A &= \mu\chi_{A'}\\ \nabla^{A'}_A\chi_{A'} &= \mu\phi_A,\end{aligned}\right\} \tag{4.4.66}$$

where $\mu$ is a real constant ($\mu = 2^{-\frac{1}{2}}m\hbar^{-1}$). Replacing $A$ by $B$ in the first of equations (4.4.66), and multiplying by $\phi_A$, we see that that equation is equivalent to

$$\phi_A\nabla_{BA'}\phi^B = -\mu\phi_A\chi_{A'} \tag{4.4.67}$$

(except possibly in regions where $\phi_A = 0$). The left member of (4.4.67) we now recognize as the $R_a$ of (4.4.62). So if we define another vector

$$C_a = \phi_A\chi_{A'}, \tag{4.4.68}$$

and refer to (4.4.63), we see that (4.4.67) – and with it (4.4.66)(1) – is equivalent to

$$\nabla_b F_a{}^b + M_a = -2\mu C_a. \tag{4.4.69}$$

In a completely analogous way, (4.4.66)(2) can be shown to be equivalent to

$$\nabla_b G_a{}^b + N_a = -2\mu C_a \tag{4.4.70}$$

where $\bar{G}_{ab}$ and $\bar{N}_a$ are related to $\bar{\chi}_A$ as are $F_{ab}$ and $M_a$ to $\phi_A$. By use of (4.4.60) and its $\bar{\chi}_A$-analogue we can finally eliminate the auxiliary vectors $M_a$, $N_a$ from these equations to obtain the (complicated, coupled, non-linear) tensor differential equations equivalent to Dirac's equation:

$$\begin{aligned}F_{ab}\nabla_d F_c{}^d + F_{ad}\nabla_c F_b{}^d &= -2\mu F_{ab}C_c\\ G_{ab}\nabla_d G_c{}^d + G_{ad}\nabla_c G_b{}^d &= -2\mu G_{ab}C_c\end{aligned} \tag{4.4.71}$$

These, of course, must still be augmented by two sets of algebraic restrictions: (4.4.65)(*b*) and its analogue for $\bar{G}_{ab}$. The bivectors $F_{ab}$ and $G_{ab}$ are algebraically independent of each other, but $C_a$ is a 'secondary' vector determined by $F_{ab}$ and $G_{ab}$ up to sign. In fact,

$$\phi_A\chi_{A'}\phi_B\chi_{B'} = \phi_A\phi^C\varepsilon_{A'}{}^{C'}\chi_{C'}\chi_{B'}\varepsilon_{CB}$$

i.e.,

$$C_aC_b = F_a{}^cG_{cb}. \tag{4.4.72}$$

Our argument is adapted from Whittaker (1937) whose final result, however, is the somewhat unnatural combination of (4.4.69) and the complex conjugate of (4.4.70) into a single equation, their sum.

We next briefly consider the general case. Derivatives of even spinors (even number of indices) present no new difficulties (*cf*. end of §3.4). Contracted derivatives of one-index spinors can be dealt with along the lines of our treatment of the Dirac–Weyl equation. The possibility of translating spinor differential equations involving derivatives of odd spinors generally, rests on the following identity:

$$2\phi^{\mathscr{A}_1}\phi^{\mathscr{A}_2}\nabla_p\phi^{\mathscr{A}_3} = \phi^{\mathscr{A}_1}\nabla_p(\phi^{\mathscr{A}_2}\phi^{\mathscr{A}_3}) + \phi^{\mathscr{A}_2}\nabla_p(\phi^{\mathscr{A}_1}\phi^{\mathscr{A}_3}) - \phi^{\mathscr{A}_3}\nabla_p(\phi^{\mathscr{A}_1}\phi^{\mathscr{A}_2}). \tag{4.4.73}$$

To translate a first-order differential spinor equation, we therefore multiply the equation by all the appropriate odd spinors a sufficient number of times. Then (4.4.73) implies that we can express the differentiated odd spinors in terms of their differentiated tensor equivalents. The translation into tensor form can then be completed by the technique described at the end of §3.4.

Second-order differential equations can be dealt with by generalizing (4.4.73). For example,

$$\begin{aligned}2\phi^{\mathscr{A}_1}\phi^{\mathscr{A}_2}\nabla_p\nabla_q\phi^{\mathscr{A}_3} &= \phi^{\mathscr{A}_1}\nabla_p\nabla_q(\phi^{\mathscr{A}_2}\phi^{\mathscr{A}_3}) + \phi^{\mathscr{A}_2}\nabla_p\nabla_q(\phi^{\mathscr{A}_1}\phi^{\mathscr{A}_3})\\ &\quad - \phi^{\mathscr{A}_3}\nabla_p\nabla_q(\phi^{\mathscr{A}_1}\phi^{\mathscr{A}_2}) - 2\phi^{\mathscr{A}_1}\nabla_{(p}\phi^{\mathscr{A}_2}\nabla_{q)}\phi^{\mathscr{A}_3}\\ &\quad - 2\phi^{\mathscr{A}_2}\nabla_{(p}\phi^{\mathscr{A}_1}\nabla_{q)}\phi^{\mathscr{A}_3} + 2\phi^{\mathscr{A}_3}\nabla_{(p}\phi^{\mathscr{A}_1}\nabla_{q)}\phi^{\mathscr{A}_2},\end{aligned} \tag{4.4.74}$$

and so on for higher orders.

## 4.5 Differentiation of spinor components

In §4.4 we obtained the operation of spinor covariant derivative in an abstract frame-independent fashion. In this section we investigate the effect of that operation on the spinor components. The resulting explicit expressions can, if desired, be used to provide an alternative proof of existence of $\nabla_a$ acting on spinors, with the desired properties (4.4.7)–(4.4.20). Henceforth (except in §4.7) $\nabla_a$ is always *torsion-free.*

Let $\varepsilon_{\mathbf{A}}{}^{A} = (o^A, \iota^A)$ be a spinor dyad (not necessarily normalized), with dual $\varepsilon_A{}^{\mathbf{A}}$, and let $\varepsilon_{\mathbf{A}'}{}^{A'}$ and its dual $\varepsilon_{A'}{}^{\mathbf{A}'}$ be the complex conjugate spinor basis and dual basis. Let $\kappa^A \in \mathfrak{S}^A$ have components $\kappa^{\mathbf{A}} = \kappa^A\varepsilon_A{}^{\mathbf{A}}$ and write

$$\varepsilon_{\mathbf{A}}{}^{A}\varepsilon_{\mathbf{A}'}{}^{A'}\nabla_{AA'} = \nabla_{\mathbf{AA}'}.$$

Then the components of $\nabla_{AA'}\kappa^B$ are

$$\begin{aligned}\varepsilon_{\mathbf{A}}{}^{A}\varepsilon_{\mathbf{A}'}{}^{A'}\varepsilon_B{}^{\mathbf{B}}\nabla_{AA'}\kappa^B &= \varepsilon_B{}^{\mathbf{B}}\nabla_{\mathbf{AA}'}(\kappa^{\mathbf{C}}\varepsilon_{\mathbf{C}}{}^{B})\\ &= \varepsilon_{\mathbf{C}}{}^{B}\varepsilon_B{}^{\mathbf{B}}\nabla_{\mathbf{AA}'}\kappa^{\mathbf{C}} + \kappa^{\mathbf{C}}\varepsilon_B{}^{\mathbf{B}}\nabla_{\mathbf{AA}'}\varepsilon_{\mathbf{C}}{}^{B}\\ &= \nabla_{\mathbf{AA}'}\kappa^{\mathbf{B}} + \kappa^{\mathbf{C}}\gamma_{\mathbf{AA}'\mathbf{C}}{}^{\mathbf{B}},\end{aligned} \tag{4.5.1}$$

where

$$\gamma_{\mathbf{AA}'\mathbf{C}}{}^{\mathbf{B}} := \varepsilon_A{}^{\mathbf{B}}\nabla_{\mathbf{AA}'}\varepsilon_{\mathbf{C}}{}^{A} = -\varepsilon_{\mathbf{C}}{}^{A}\nabla_{\mathbf{AA}'}\varepsilon_A{}^{\mathbf{B}}, \tag{4.5.2}$$

since $\varepsilon_A{}^{\mathbf{B}}\varepsilon_{\mathbf{C}}{}^A = \varepsilon_{\mathbf{C}}{}^{\mathbf{B}}$, the Kronecker delta, so that $\nabla_a \varepsilon_{\mathbf{C}}{}^{\mathbf{B}} = 0$. Note that

$$\bar{\gamma}_{\mathbf{AA'C'}}{}^{\mathbf{B'}} (= \bar{\gamma}_{\mathbf{A'AC'}}{}^{\mathbf{B'}}) = \varepsilon_{A'}{}^{\mathbf{B'}} \nabla_{\mathbf{AA'}} \varepsilon_{\mathbf{C'}}{}^{A'} = -\varepsilon_{\mathbf{C'}}{}^{A'} \nabla_{\mathbf{AA'}} \varepsilon_{A'}{}^{\mathbf{B'}} \tag{4.5.3}$$

If we specialize $\varepsilon_{\mathbf{A}}{}^A$ to a *spin-frame*, i.e., normalize $o_A \iota^A = 1$, so that

$$\varepsilon_{\mathbf{BC}} = \varepsilon_{\mathbf{B}A} \varepsilon_{\mathbf{C}}{}^A = \begin{pmatrix} 0 & 1 \\ -1 & 0 \end{pmatrix}; \tag{4.5.4}$$

then we have the symmetry

$$\gamma_{\mathbf{AA'BC}} = \gamma_{\mathbf{AA'CB}}, \tag{4.5.5}$$

since

$$\begin{aligned} 0 = \nabla_{\mathbf{AA'}} \varepsilon_{\mathbf{BC}} &= \nabla_{\mathbf{AA'}} (\varepsilon_{\mathbf{B}}{}^A \varepsilon_{A\mathbf{C}}) \\ &= \varepsilon_{A\mathbf{C}} \nabla_{\mathbf{AA'}} \varepsilon_{\mathbf{B}}{}^A - \varepsilon_{\mathbf{B}A} \nabla_{\mathbf{AA'}} \varepsilon^A{}_{\mathbf{C}} = \gamma_{\mathbf{AA'BC}} - \gamma_{\mathbf{AA'CB}}. \end{aligned}$$

(Remember that spinor indices may be raised or lowered without regard to preceding $\nabla$-operations.) In this case the quantities $\gamma_{\mathbf{AA'BC}}$ constitute 12 independent complex numbers at each point. They have been called *spin-coefficients* and find many applications in practical calculations. (*cf*. Newman and Penrose 1962, Newman and Unti 1962, Papapetrou 1974, Campbell and Wainwright 1977, Carmeli 1977, Chandrasekhar 1979, Kramer, Stephani, MacCallum and Herlt 1980).

When the normalization (4.5.4) is not maintained, the symmetry (4.5.5) fails and we have 16 independent complex numbers $\gamma_{\mathbf{AA'B}}{}^{\mathbf{C}}$ at each point. There are situations in which it is useful to admit this additional flexibility, and since the resulting formulae are hardly more complicated in this case, we shall, for the most part, present our equations in this more general form. The $\gamma$'s are still called 'spin-coefficients'. Setting

$$\chi = o_A \iota^A, \tag{4.5.6}$$

we have, as in (2.5.45),

$$\varepsilon_{\mathbf{BC}} = \varepsilon_{\mathbf{B}A} \varepsilon_{\mathbf{C}}{}^A = \begin{pmatrix} 0 & \chi \\ -\chi & 0 \end{pmatrix} \tag{4.5.7}$$

in place of (4.5.4), the dual basis to $(o^A, \iota^A)$ being $\varepsilon_A{}^0 = -\chi^{-1}\iota_A$, $\varepsilon_A{}^1 = \chi^{-1} o_A$, as in (2.5.50). Repeating the calculation (4.5.5) with (4.5.7) in place of (4.5.4) we get

$$\gamma_{\mathbf{AA'BC}} - \gamma_{\mathbf{AA'CB}} = \nabla_{\mathbf{AA'}} \varepsilon_{\mathbf{BC}} = \varepsilon_{\mathbf{BC}} \chi^{-1} \nabla_{\mathbf{AA'}} \chi, \tag{4.5.8}$$

i.e.,

$$\gamma_{\mathbf{AA'}01} - \gamma_{\mathbf{AA'}10} = \nabla_{\mathbf{AA'}} \chi. \tag{4.5.9}$$

Equivalently,

$$\gamma_{\mathbf{AA'B}}{}^{\mathbf{B}} = 2\chi^{-1} \nabla_{\mathbf{AA'}} \chi. \tag{4.5.10}$$

It is important to note that whereas with spin-frame indices (i.e., when $\chi = 1$) the operation of raising and lowering commutes with differentiation, this is not so for spinor components in the general case ($\chi \neq$ constant), for we have, for example,

$$\psi_{\mathbf{AA'}\ldots\mathbf{M}}{}^{\cdots} = \nabla_{\mathbf{AA'}}\phi_{\ldots\mathbf{M}}{}^{\cdots} \Leftrightarrow \psi_{\ldots}{}^{\mathbf{M}\ldots} = \nabla_{\mathbf{AA'}}\phi_{\ldots}{}^{\mathbf{M}\ldots} + \phi_{\ldots}{}^{\mathbf{M}\ldots}\chi^{-1}\nabla_{\mathbf{AA'}}\chi,$$
$$\tilde{\psi}_{\mathbf{AA'}\ldots}{}^{\mathbf{M'}\ldots} = \nabla_{\mathbf{AA'}}\phi_{\ldots}{}^{\mathbf{M'}\ldots} \Leftrightarrow \tilde{\psi}_{\mathbf{AA'}\ldots\mathbf{M'}}{}^{\cdots} = \nabla_{\mathbf{AA'}}\phi_{\ldots\mathbf{M'}}{}^{\cdots} - \phi_{\ldots\mathbf{M'}}{}^{\cdots}\bar{\chi}^{-1}\nabla_{\mathbf{AA'}}\bar{\chi}. \tag{4.5.11}$$

Now, starting with a *covariant* spin-vector $\mu_B$ we find the components of $\nabla_{AA'}\mu_B$ to be

$$\varepsilon_{\mathbf{A}}{}^{A}\varepsilon_{\mathbf{A'}}{}^{A'}\varepsilon_{\mathbf{B}}{}^{B}\nabla_{AA'}\mu_B = \nabla_{\mathbf{AA'}}\mu_{\mathbf{B}} - \mu_{\mathbf{C}}\gamma_{\mathbf{AA'B}}{}^{\mathbf{C}} \tag{4.5.12}$$

by a computation analogous to that of (4.5.1). For primed spin-vectors $\phi^{A'}, \zeta_{A'}$ we find the corresponding relations

$$\varepsilon_{\mathbf{A}}{}^{A}\varepsilon_{\mathbf{A'}}{}^{A'}\varepsilon_{B'}{}^{\mathbf{B'}}\nabla_{AA'}\phi^{B'} = \nabla_{\mathbf{AA'}}\phi^{\mathbf{B'}} + \phi^{\mathbf{C'}}\bar{\gamma}_{\mathbf{AA'C'}}{}^{\mathbf{B'}} \tag{4.5.13}$$

$$\varepsilon_{\mathbf{A}}{}^{A}\varepsilon_{\mathbf{A'}}{}^{A'}\varepsilon_{\mathbf{B'}}{}^{B'}\nabla_{AA'}\zeta_{B'} = \nabla_{\mathbf{AA'}}\zeta_{\mathbf{B'}} - \zeta_{\mathbf{C'}}\bar{\gamma}_{\mathbf{AA'B'}}{}^{\mathbf{C'}} \tag{4.5.14}$$

by taking conjugates in (4.5.1) and (4.5.12). And for a general spinor $\psi^{B\ldots E'\ldots}_{H\ldots K'\ldots}$ we have

$$\begin{aligned}\varepsilon_{\mathbf{A}}{}^{A}\ldots\varepsilon_{\mathbf{K'}}{}^{K'}\ldots\nabla_{AA'}\psi^{B\ldots E'\ldots}_{H\ldots K'\ldots} &= \nabla_{\mathbf{AA'}}\psi^{\mathbf{B}\ldots\mathbf{E'}\ldots}_{\mathbf{H}\ldots\mathbf{K'}\ldots} + \psi^{\mathbf{B}_0\ldots\mathbf{E'}\ldots}_{\mathbf{H}\ldots\mathbf{K'}\ldots}\gamma_{\mathbf{AA'B}_0}{}^{\mathbf{B}} \\ &\quad + \cdots + \psi^{\mathbf{B}\ldots\mathbf{E}'_0\ldots}_{\mathbf{H}\ldots\mathbf{K'}\ldots}\bar{\gamma}_{\mathbf{AA'E}'_0}{}^{\mathbf{E'}} + \cdots - \psi^{\mathbf{B}\ldots\mathbf{E'}\ldots}_{\mathbf{H}_0\ldots\mathbf{K'}\ldots}\gamma_{\mathbf{AA'H}}{}^{\mathbf{H}_0} \\ &\quad - \cdots - \psi^{\mathbf{B}\ldots\mathbf{E'}\ldots}_{\mathbf{H}\ldots\mathbf{K}'_0}\bar{\gamma}_{\mathbf{AA'K'}}{}^{\mathbf{K}'_0} - \cdots.\end{aligned} \tag{4.5.15}$$

This can be verified directly by taking components of the expansion of $\nabla_{AA'}(\psi^{\mathbf{B}_0\ldots\mathbf{E}'_0\ldots}_{\mathbf{H}_0\ldots\mathbf{K}'_0\ldots}\varepsilon_{\mathbf{B}_0}{}^{B}\ldots\varepsilon_{H}{}^{\mathbf{H}'_0}\ldots)$.

### *Expressions for individual spin-coefficients*

In explicit calculations it is often convenient to assign single letters to each of the 16 quantities $\gamma_{\mathbf{AA'B}}{}^{\mathbf{C}}$. A standard notation (slightly modified) is given in the following table:

$\gamma_{\mathbf{AA'B}}{}^{\mathbf{C}} =$ (4.5.16)

| AA' \ BC | 00 | 10 | 01 | 11 |
|---|---|---|---|---|
| 00' | $\varepsilon$ | $-\kappa$ | $-\tau'$ | $\gamma'$ |
| 10' | $\alpha$ | $-\rho$ | $-\sigma'$ | $\beta'$ |
| 01' | $\beta$ | $-\sigma$ | $-\rho'$ | $\alpha'$ |
| 11' | $\gamma$ | $-\tau$ | $-\kappa'$ | $\varepsilon'$ |

(The choice of signs for $\kappa, \ldots, \tau$, though perhaps unfortunate, conforms to what has by now become standard notation.) The significance of the use of the 'primed' symbols is that under the replacement

$$o^A \mapsto \mathrm{i}\iota^A,\ \iota^A \mapsto \mathrm{i}o^A,\ o^{A'} \mapsto -\mathrm{i}\iota^{A'},\ \iota^{A'} \mapsto -\mathrm{i}o^{A'} \tag{4.5.17}$$

(which preserves the relation $o_A \iota^A = \chi$) the primed and unprimed quantities get interchanged. With a *prime denoting the above operation*, two primes in succession leave the spin-coefficients unchanged but change the dyad into its negative. Evidently the prime operation commutes with complex conjugation, so that, for example, $\bar{\lambda}'$ can stand for both $(\bar{\lambda})'$ and $\overline{(\lambda')}$. The prime also commutes with addition and multiplication. For future use we note (*cf.* (4.5.19) below)

$$(l^a)' = n^a,\ (m^a)' = \bar{m}^a,\ (\bar{m}^a)' = m^a,\ (n^a)' = l^a. \tag{4.5.18}$$

Explicitly, in terms of the basis spinors $o^A, \iota^A$, we get from (4.5.16) and (4.5.2) the first expressions, (4.5.21), for the spin coefficients in the display below; the second expressions, (4.5.22), in terms of the (unnormalized) null tetrad

$$l^a = o^A o^{A'},\ m^a = o^A \iota^{A'},\ \bar{m}^a = \iota^A o^{A'},\ n^a = \iota^A \iota^{A'}, \tag{4.5.19}$$

as in (3.1.14) but with

$$l^a n_a = \chi\bar{\chi} = -m^a \bar{m}_a, \tag{4.5.20}$$

can be checked directly by substituting for $l^a$ etc., in terms of $o^A$ and $\iota^A$, and using (4.5.24) and (4.5.25). Note that in the spin-coefficient formalism the role of the 'vector' covariant derivative operator is played by the four 'scalar' operators $\nabla_{\mathbf{AA'}}$ (the so-called *intrinsic derivatives* in the tetrad directions), for which special symbols are used as defined in (4.5.23) below. (The symbol $\Delta$ has frequently been used for $D'$.)

$$\begin{array}{|cccc|} \kappa & \varepsilon & \gamma' & \tau' \\ \rho & \alpha & \beta' & \sigma' \\ \sigma & \beta & \alpha' & \rho' \\ \tau & \gamma & \varepsilon' & \kappa' \end{array} = \chi^{-1} \times \begin{array}{|c|c|c|c|} o^A D o_A & \iota^A D o_A & -o^A D \iota_A & -\iota^A D \iota_A \\ o^A \delta' o_A & \iota^A \delta' o_A & -o^A \delta' \iota_A & -\iota^A \delta' \iota_A \\ o^A \delta o_A & \iota^A \delta o_A & -o^A \delta \iota_A & -\iota^A \delta \iota_A \\ o^A D' o_A & \iota^A D' o_A & -o^A D' \iota_A & -\iota^A D' \iota_A \end{array} \tag{4.5.21}$$

$$= \chi^{-1}\bar{\chi}^{-1} \times \begin{array}{|llll|} m^a D l_a & \tfrac{1}{2}(n^a D l_a + m^a D \bar{m}_a + \bar{\chi} D \chi) & \tfrac{1}{2}(l^a D n_a + \bar{m}^a D m_a + \bar{\chi} D \chi) & \bar{m}^a D n_a \\ m^a \delta' l_a & \tfrac{1}{2}(n^a \delta' l_a + m^a \delta' \bar{m}_a + \bar{\chi} \delta' \chi) & \tfrac{1}{2}(l^a \delta' n_a + \bar{m}^a \delta' m_a + \bar{\chi} \delta' \chi) & \bar{m}^a \delta' n_a \\ m^a \delta l_a & \tfrac{1}{2}(n^a \delta l_a + m^a \delta \bar{m}_a + \bar{\chi} \delta \chi) & \tfrac{1}{2}(l^a \delta n_a + \bar{m}^a \delta m_a + \bar{\chi} \delta \chi) & \bar{m}^a \delta n_a \\ m^a D' l_a & \tfrac{1}{2}(n^a D' l_a + m^a D' \bar{m}_a + \bar{\chi} D' \chi) & \tfrac{1}{2}(l^a D' n_a + \bar{m}^a D' m_a + \bar{\chi} D' \chi) & \bar{m}^a D' n_a \end{array} \tag{4.5.22}$$

where *

$$\begin{aligned} D &:= \nabla_{00'} = o^A o^{A'} \nabla_{AA'} = l^a \nabla_a = \bar{D} \\ \delta &:= \nabla_{01'} = o^A \iota^{A'} \nabla_{AA'} = m^a \nabla_a = \bar{\delta}' \\ \delta' &:= \nabla_{10'} = \iota^A o^{A'} \nabla_{AA'} = \bar{m}^a \nabla_a = \bar{\delta} \\ D' &:= \nabla_{11'} = \iota^A \iota^{A'} \nabla_{AA'} = n^a \nabla_a = \bar{D}'. \end{aligned} \tag{4.5.23}$$

The relations which verify (4.5.22) are the following, obtained by applying the Leibniz rule to $\nabla_{BB'} l_a = \nabla_{BB'}(o^A o^{A'})$, etc.:

$$\begin{aligned} & m^a \nabla_{BB'} l_a = \bar{\chi} o^A \nabla_{BB'} o_A (= -l^a \nabla_{BB'} m_a) \\ & \bar{m}^a \nabla_{BB'} n_a = \bar{\chi} \iota^A \nabla_{BB'} \iota_A (= -n^a \nabla_{BB'} \bar{m}_a) \\ & n^a \nabla_{BB'} l_a + m^a \nabla_{BB'} \bar{m}_a + \bar{\chi} \nabla_{BB'} \chi = 2\bar{\chi} \iota^A \nabla_{BB'} o_A \\ & l^a \nabla_{BB'} n_a + \bar{m}^a \nabla_{BB'} m_a + \bar{\chi} \nabla_{BB'} \chi = 2\bar{\chi} o^A \nabla_{BB'} \iota_A \end{aligned} \tag{4.5.24}$$

and

$$\nabla_{BB'} \chi = \iota^A \nabla_{BB'} o_A - o^A \nabla_{BB'} \iota_A. \tag{4.5.25}$$

Note that, since the tetrad $l^a$, $m^a$, $\bar{m}^a$, $n^a$ determines only the modulus of $\chi$ (cf. (4.5.20)), the tetrad by itself does not completely define $\alpha$, $\beta$, $\gamma$ and $\varepsilon$, unless $\chi$ is taken to be real, say.

Some useful formulae, effectively equivalent to (4.5.21) and their complex conjugates, are the following:

$$\begin{aligned} D o^A &= \varepsilon o^A - \kappa \iota^A & D \iota^A &= \gamma' \iota^A - \tau' o^A \\ \delta' o^A &= \alpha o^A - \rho \iota^A & \delta' \iota^A &= \beta' \iota^A - \sigma' o^A \\ \delta o^A &= \beta o^A - \sigma \iota^A & \delta \iota^A &= \alpha' \iota^A - \rho' o^A \\ D' o^A &= \gamma o^A - \tau \iota^A & D' \iota^A &= \varepsilon' \iota^A - \kappa' o^A \end{aligned} \tag{4.5.26}$$

and

$$\begin{aligned} D o^{A'} &= \bar{\varepsilon} o^{A'} - \bar{\kappa} \iota^{A'} & D \iota^{A'} &= \bar{\gamma}' \iota^{A'} - \bar{\tau}' o^{A'} \\ \delta o^{A'} &= \bar{\alpha} o^{A'} - \bar{\rho} \iota^{A'} & \delta \iota^{A'} &= \bar{\beta}' \iota^{A'} - \bar{\sigma}' o^{A'} \\ \delta' o^{A'} &= \bar{\beta} o^{A'} - \bar{\sigma} \iota^{A'} & \delta' \iota^{A'} &= \bar{\alpha}' \iota^{A'} - \bar{\rho}' o^{A'} \\ D' o^{A'} &= \bar{\gamma} o^{A'} - \bar{\tau} \iota^{A'} & D' \iota^{A'} &= \bar{\varepsilon}' \iota^{A'} - \bar{\kappa}' o^{A'}, \end{aligned} \tag{4.5.27}$$

We recall that for the quantities used here (spin-coefficients and directional derivatives) the bar and prime commute. From (3.1.14), (4.5.26) and

* Consistently with our earlier notation (4.1.14), we *could* write ***l***, ***m***, $\bar{\boldsymbol{m}}$, ***n*** for $D$, $\delta$, $\delta'$, $D'$, respectively, when they act on *scalars*. However, we shall reserve this notation for the corresponding 1-forms used in §4.13 below.

(4.5.27) we now obtain

$$\begin{aligned} Dl^a &= (\varepsilon+\bar{\varepsilon})l^a - \bar{\kappa}m^a - \kappa\bar{m}^a, & Dm^a &= (\varepsilon+\bar{\gamma}')m^a - \bar{\tau}'l^a - \kappa n^a \\ \delta l^a &= (\beta+\bar{\alpha})l^a - \bar{\rho}m^a - \sigma\bar{m}^a, & \delta m^a &= (\beta+\bar{\beta}')m^a - \bar{\sigma}'l^a - \sigma n^a \\ \delta' l^a &= (\alpha+\bar{\beta})l^a - \bar{\sigma}m^a - \rho\bar{m}^a, & \delta' m^a &= (\alpha+\bar{\alpha}')m^a - \bar{\rho}'l^a - \rho n^a \\ D'l^a &= (\gamma+\bar{\gamma})l^a - \bar{\tau}m^a - \tau\bar{m}^a, & D'm^a &= (\gamma+\bar{\varepsilon}')m^a - \bar{\kappa}'l^a - \tau n^a \end{aligned}$$

$$\begin{aligned} D\bar{m}^a &= (\gamma'+\bar{\varepsilon})\bar{m}^a - \tau'l^a - \bar{\kappa}n^a, & Dn^a &= (\gamma'+\bar{\gamma}')n^a - \tau'm^a - \bar{\tau}'\bar{m}^a. \\ \delta\bar{m}^a &= (\alpha'+\bar{\alpha})\bar{m}^a - \rho'l^a - \bar{\rho}n^a, & \delta n^a &= (\alpha'+\bar{\beta}')n^a - \rho'm^a - \bar{\sigma}'\bar{m}^a \\ \delta'\bar{m}^a &= (\beta'+\bar{\beta})\bar{m}^a - \sigma'l^a - \bar{\sigma}n^a, & \delta' n^a &= (\beta'+\bar{\alpha}')n^a - \sigma'm^a - \bar{\rho}'\bar{m}^a \\ D'\bar{m}^a &= (\varepsilon'+\bar{\gamma})\bar{m}^a - \kappa'l^a - \bar{\tau}n^a, & D'n^a &= (\varepsilon'+\bar{\varepsilon}')n^a - \kappa'm^a - \bar{\kappa}'\bar{m}^a \end{aligned} \quad (4.5.28)$$

Note that some of these formulae may be derived from others by employing (4.5.18). A short-hand notation for (4.5.26)–(4.5.28) will be given in §4.12 (equations (4.12.28) *et seq.*).

In most practical applications of spin-coefficients the spinor basis is normalized ($\chi = 1$), and then the expressions (4.5.22) simplify slightly, and $\alpha = -\beta'$, $\varepsilon = -\gamma'$, $\beta = -\alpha'$, $\gamma = -\varepsilon'$ by the symmetry (4.5.5). It has been customary, also, to employ the symbols $\pi$, $\lambda$, $\mu$, $\nu$ for $-\tau'$, $-\sigma'$, $-\rho'$, $-\kappa'$, respectively. So we have, in this case,

$\gamma_{\mathbf{AA'BC}} =$

| AA′ \ BC | 00 | 10 or 01 | 11 |
|---|---|---|---|
| 00′ | $\kappa$ | $\varepsilon = -\gamma'$ | $\pi = -\tau'$ |
| 10′ | $\rho$ | $\alpha = -\beta'$ | $\lambda = -\sigma'$ |
| 01′ | $\sigma$ | $\beta$ | $\mu = -\rho'$ |
| 11′ | $\tau$ | $\gamma$ | $\nu = -\kappa'$ |

(4.5.29)

It is possible to obtain a geometrical picture of the meaning of most of the (normalized) spin-coefficients in suitable circumstances, in terms of the congruence of curves to which the flagpoles of $o^A$ and $\iota^A$ are tangent. But this will be left to Chapter 7 (*cf.* §7.1).

*Relations to Infeld–van der Waerden and Christoffel symbols*

We may wish to express quantities in terms of derivatives $\nabla_{\mathbf{a}}$ referred to an arbitrary tensor basis $g_{\mathbf{a}}{}^a$. (If, in particular, that basis is the coordinate basis, we have $\nabla_{\mathbf{a}} = \partial/\partial x^{\mathbf{a}}$ when acting on scalars.) Then we simply translate $\nabla_{\mathbf{AA'}}$ into $\nabla_{\mathbf{a}}$ using the Infeld–van der Waerden symbols $g_{\mathbf{a}}{}^{\mathbf{AA'}}$ (*cf.* (3.1.37)):

$$\nabla_{\mathbf{a}} = g_{\mathbf{a}}{}^{\mathbf{AA'}}\nabla_{\mathbf{AA'}} \quad (4.5.30)$$

If we define quantities $\gamma_{\mathbf{aB}}{}^{\mathbf{C}}$ by

$$\gamma_{\mathbf{aB}}{}^{\mathbf{C}} := \gamma_{\mathbf{AA'B}}{}^{\mathbf{C}} g_{\mathbf{a}}{}^{\mathbf{AA'}}, \tag{4.5.31}$$

then we obtain, from (4.5.1),

$$g_{\mathbf{a}}{}^{a}\varepsilon_{B}{}^{\mathbf{B}}\nabla_{a}\kappa^{B} = \nabla_{\mathbf{a}}\kappa^{\mathbf{B}} + \kappa^{\mathbf{C}}\gamma_{\mathbf{aC}}{}^{\mathbf{B}} \tag{4.5.32}$$

for the components of $\nabla_{a}\kappa^{B}$, where **a** refers to the tensor basis and **B** to the dyad. Similarly, from (4.5.12), we get

$$g_{\mathbf{a}}{}^{a}\varepsilon_{\mathbf{B}}{}^{B}\nabla_{a}\mu_{B} = \nabla_{\mathbf{a}}\mu_{\mathbf{B}} - \mu_{\mathbf{C}}\gamma_{\mathbf{aB}}{}^{\mathbf{C}} \tag{4.5.33}$$

for the corresponding components of $\nabla_{a}\mu_{B}$.

In the case of a general spinor, i.e., for the components of $\nabla_{a}\psi_{H\ldots K'\ldots}^{B\ldots E'\ldots}$ relative to the tensor basis and the dyad, we have, from (4.5.15),

$$\begin{aligned}\nabla_{\mathbf{a}}\psi_{\mathbf{H}\ldots\mathbf{K'}\ldots}^{\mathbf{B}\ldots\mathbf{E'}\ldots} &+ \psi_{\mathbf{H}\ldots\mathbf{K'}\ldots}^{\mathbf{B}_0\ldots\mathbf{E'}\ldots}\gamma_{\mathbf{aB}_0}{}^{\mathbf{B}} + \cdots + \psi_{\mathbf{H}\ldots\mathbf{K'}\ldots}^{\mathbf{B}\ldots\mathbf{E}'_0\ldots}\bar{\gamma}_{\mathbf{aE}'_0}{}^{\mathbf{E'}} + \ldots\\ &- \psi_{\mathbf{H}_0\ldots\mathbf{K'}\ldots}^{\mathbf{B}\ldots\mathbf{E'}\ldots}\gamma_{\mathbf{aH}}{}^{\mathbf{H}_0} - \cdots - \psi_{\mathbf{H}\ldots\mathbf{K}'_0\ldots}^{\mathbf{B}\ldots\mathbf{E'}\ldots}\bar{\gamma}_{\mathbf{aK'}}{}^{\mathbf{K}'_0} - \cdots\end{aligned} \tag{4.5.34}$$

When a quantity possesses both spinor and tensor indices, e.g., $\theta_{B}{}^{c}$, we may wish to obtain the components of its covariant derivative $\nabla_{a}\theta_{B}{}^{c}$, where $a$ and $c$ are to be referred to the tensor basis and $B$ to the dyad. In this case we have (with (4.2.60))

$$g_{\mathbf{a}}{}^{a}g_{c}{}^{\mathbf{c}}\varepsilon_{\mathbf{B}}{}^{B}\nabla_{a}\theta_{B}{}^{c} = \nabla_{\mathbf{a}}\theta_{\mathbf{B}}{}^{\mathbf{c}} - \theta_{\mathbf{D}}{}^{\mathbf{c}}\gamma_{\mathbf{aB}}{}^{\mathbf{D}} + \theta_{\mathbf{B}}{}^{\mathbf{e}}\Gamma_{\mathbf{ae}}{}^{\mathbf{c}}, \tag{4.5.35}$$

as follows from taking components of the expansion of $\nabla_{a}(\theta_{\mathbf{D}}{}^{\mathbf{e}}\varepsilon_{B}{}^{\mathbf{D}}g_{\mathbf{e}}{}^{c})$.

Observe that we could alternatively obtain (4.5.35) by treating $\theta_{B}{}^{c}$ as $\theta_{B}{}^{CC'}$ and using (4.5.34), finally translating back to tensor components by using Infeld–van der Waerden symbols (3.1.37). Instead of one term in $\Gamma$ we would then find two terms in $\gamma$ and $\bar{\gamma}$, and the derivative would be $\nabla_{\mathbf{a}}\theta_{\mathbf{B}}{}^{\mathbf{CC'}}$ which differs from $\nabla_{\mathbf{a}}\theta_{\mathbf{B}}{}^{\mathbf{c}}$ by a term involving derivatives of $g_{\mathbf{c}}{}^{\mathbf{CC'}}$. Hence there must be a relation between $\gamma$, $\Gamma$, and the derivatives of $g_{\mathbf{c}}{}^{\mathbf{CC'}}$. This can also be seen as follows: we have

$$\begin{aligned}\Gamma_{\mathbf{cb}}{}^{\mathbf{a}} &= g_{a}{}^{\mathbf{a}}\nabla_{\mathbf{c}}g_{\mathbf{b}}{}^{a} = g_{a}{}^{\mathbf{a}}\nabla_{\mathbf{c}}g_{\mathbf{b}}{}^{AA'}\\ &= g_{a}{}^{\mathbf{a}}\nabla_{\mathbf{c}}(g_{\mathbf{b}}{}^{\mathbf{AA'}}\varepsilon_{\mathbf{A}}{}^{A}\varepsilon_{\mathbf{A'}}{}^{A'})\\ &= g_{a}{}^{\mathbf{a}}\varepsilon_{\mathbf{A}}{}^{A}\varepsilon_{\mathbf{A'}}{}^{A'}\nabla_{\mathbf{c}}g_{\mathbf{b}}{}^{\mathbf{AA'}} + g_{a}{}^{\mathbf{a}}g_{\mathbf{b}}{}^{\mathbf{AA'}}(\varepsilon_{\mathbf{A'}}{}^{A'}\nabla_{\mathbf{c}}\varepsilon_{\mathbf{A}}{}^{A} + \varepsilon_{\mathbf{A}}{}^{A}\nabla_{\mathbf{c}}\varepsilon_{\mathbf{A'}}{}^{A'})\\ &= g_{\mathbf{AA'}}{}^{\mathbf{a}}\nabla_{\mathbf{c}}g_{\mathbf{b}}{}^{\mathbf{AA'}} + g_{\mathbf{BB'}}{}^{\mathbf{a}}\varepsilon_{A}{}^{\mathbf{B}}\varepsilon_{A'}{}^{\mathbf{B'}}g_{\mathbf{b}}{}^{\mathbf{AA'}}(\varepsilon_{\mathbf{A'}}{}^{A'}\nabla_{\mathbf{c}}\varepsilon_{\mathbf{A}}{}^{A} + \varepsilon_{\mathbf{A}}{}^{A}\nabla_{\mathbf{c}}\varepsilon_{\mathbf{A'}}{}^{A'}).\end{aligned} \tag{4.5.36}$$

Consequently, after changing some dummies on the right, we get

$$g_{\mathbf{a}}{}^{\mathbf{BB'}}g_{\mathbf{AA'}}{}^{\mathbf{b}}\Gamma_{\mathbf{cb}}{}^{\mathbf{a}} = g_{\mathbf{AA'}}{}^{\mathbf{b}}\nabla_{\mathbf{c}}g_{\mathbf{b}}{}^{\mathbf{BB'}} + \varepsilon_{\mathbf{A'}}{}^{\mathbf{B'}}\gamma_{\mathbf{cA}}{}^{\mathbf{B}} + \varepsilon_{\mathbf{A}}{}^{\mathbf{B}}\bar{\gamma}_{\mathbf{cA'}}{}^{\mathbf{B'}}, \tag{4.5.37}$$

and contraction over $\mathbf{A'}, \mathbf{B'}$ and use of (4.5.10) yields the required relation:

$$\gamma_{\mathbf{cA}}{}^{\mathbf{B}} = \tfrac{1}{2}\Gamma_{\mathbf{cAA'}}{}^{\mathbf{BA'}} - \varepsilon_{\mathbf{A}}{}^{\mathbf{B}}\bar{\chi}^{-1}\nabla_{\mathbf{c}}\bar{\chi} - \tfrac{1}{2}g_{\mathbf{AA'}}{}^{\mathbf{b}}\nabla_{\mathbf{c}}g_{\mathbf{b}}{}^{\mathbf{BA'}}. \tag{4.5.38}$$

In the (usual) case of a normalized spin-frame, the term in $\bar{\chi}$ will, of course, vanish. For the rest of this section, we assume this to be the case. If we also assume that a coordinate basis is chosen as our tensor basis, we have, by (4.2.70) and (4.3.48),

$$\Gamma_{\mathbf{ca}}{}^{\mathbf{b}} = \tfrac{1}{2}g^{\mathbf{bd}}(\nabla_{\mathbf{a}}g_{\mathbf{cd}} + \nabla_{\mathbf{c}}g_{\mathbf{ad}} - \nabla_{\mathbf{d}}g_{\mathbf{ac}}), \tag{4.5.39}$$

i.e., the $\Gamma$'s are the usual Christoffel symbols, and

$$\nabla_{\mathbf{a}}g_{\mathbf{cd}} = \nabla_{\mathbf{a}}(g_{\mathbf{c}}{}^{\mathbf{DD}'}g_{\mathbf{dDD}'}) = 2[\nabla_{\mathbf{a}}g_{(\mathbf{c}}{}^{\mathbf{DD}'}]g_{\mathbf{d})\mathbf{DD}'} \tag{4.5.40}$$

(since we may raise and lower dyad indices under the operator $\nabla_{\mathbf{a}}$ whenever $\chi = 1$, *cf.* (4.5.11).) Thus we can now express the spin-coefficients in terms of the Infeld–van der Waerden symbols and their coordinate derivatives. Explicitly, substituting (4.5.40) into (4.5.39) yields

$$\begin{aligned}\Gamma_{\mathbf{ac}}{}^{\mathbf{b}} = \tfrac{1}{2}g^{\mathbf{bd}}(&g_{\mathbf{cDD}'}\nabla_{\mathbf{a}}g_{\mathbf{d}}{}^{\mathbf{DD}'} + g_{\mathbf{dDD}'}\nabla_{\mathbf{a}}g_{\mathbf{c}}{}^{\mathbf{DD}'}\\ &+ g_{\mathbf{aDD}'}\nabla_{\mathbf{c}}g_{\mathbf{d}}{}^{\mathbf{DD}'} + g_{\mathbf{dDD}'}\nabla_{\mathbf{c}}g_{\mathbf{a}}{}^{\mathbf{DD}'}\\ &- g_{\mathbf{aDD}'}\nabla_{\mathbf{d}}g_{\mathbf{c}}{}^{\mathbf{DD}'} - g_{\mathbf{cDD}'}\nabla_{\mathbf{d}}g_{\mathbf{a}}{}^{\mathbf{DD}'}).\end{aligned} \tag{4.5.41}$$

Converting $\mathbf{a}, \mathbf{b}$ into $\mathbf{AA}', \mathbf{BB}'$, contracting over $\mathbf{A}', \mathbf{B}'$, and substituting this into (4.5.38) with $\mathbf{B}$ lowered, gives (when $\chi = 1$)

$$\begin{aligned}\gamma_{\mathbf{cAB}} = \tfrac{1}{4}g_{\mathbf{B}}{}^{\mathbf{A}'\mathbf{d}}(&g_{\mathbf{cDD}'}\nabla_{\mathbf{AA}'}g_{\mathbf{d}}{}^{\mathbf{DD}'} + \nabla_{\mathbf{c}}g_{\mathbf{dAA}'} - \nabla_{\mathbf{d}}g_{\mathbf{cAA}'} - g^{\mathbf{a}}{}_{\mathbf{AA}'}g_{\mathbf{cDD}'}\nabla_{\mathbf{d}}g_{\mathbf{a}}{}^{\mathbf{DD}'})\\ &+ \tfrac{1}{4}\nabla_{\mathbf{AA}'}g_{\mathbf{cB}}{}^{\mathbf{A}'} - \tfrac{1}{4}g^{\mathbf{a}}{}_{\mathbf{AA}'}\nabla_{\mathbf{c}}g_{\mathbf{aB}}{}^{\mathbf{A}'},\end{aligned} \tag{4.5.42}$$

whence finally (for normalized spin-frame and coordinate tensor basis),

$$\begin{aligned}\gamma_{\mathbf{CC}'\mathbf{AB}} = \tfrac{1}{4}g_{\mathbf{B}}{}^{\mathbf{A}'\mathbf{d}}(&\nabla_{\mathbf{AA}'}g_{\mathbf{dCC}'} + \nabla_{\mathbf{CC}'}g_{\mathbf{dAA}'} - g^{\mathbf{c}}{}_{\mathbf{CC}'}\nabla_{\mathbf{d}}g_{\mathbf{cAA}'} - g^{\mathbf{a}}{}_{\mathbf{AA}'}\nabla_{\mathbf{d}}g_{\mathbf{aCC}'})\\ &+ \tfrac{1}{4}g^{\mathbf{c}}{}_{\mathbf{CC}'}\nabla_{\mathbf{AA}'}g_{\mathbf{cB}}{}^{\mathbf{A}'} - \tfrac{1}{4}g^{\mathbf{a}}{}_{\mathbf{AA}'}\nabla_{\mathbf{CC}'}g_{\mathbf{aB}}{}^{\mathbf{A}'}.\end{aligned} \tag{4.5.43}$$

(A similar formula may be obtained involving derivatives of the Infeld–van der Waerden symbols with their tensor index up.) This shows how a knowledge of the $g_{\mathbf{a}}{}^{\mathbf{BB}'}$, as functions of the coordinates $x^{\mathbf{a}}$, may be used to calculate the spin-coefficients explicitly as functions of the $x^{\mathbf{a}}$. An alternative method to this end (as well as other techniques useful in certain practical computations) will be given in §4.13, where differential forms are discussed.

Observe that if the 16 quantities $g_{\mathbf{a}}{}^{\mathbf{BB}'}$ are given arbitrarily as functions of the coordinates (equivalent to the choice of 16 real quantities owing to Hermiticity), then this serves to specify the metric components (via $g_{\mathbf{ab}} = g_{\mathbf{a}}{}^{\mathbf{AA}'}g_{\mathbf{b}}{}^{\mathbf{BB}'}\varepsilon_{\mathbf{AB}}\varepsilon_{\mathbf{A}'\mathbf{B}'}$) as well as the normalized null tetrad (with components $l_{\mathbf{a}} = g_{\mathbf{a}}{}^{00'}, m_{\mathbf{a}} = g_{\mathbf{a}}{}^{01'}, n_{\mathbf{a}} = g_{\mathbf{a}}{}^{11'}$) – i.e., in effect, the spin-frame. The covariant derivative of spinors (or tensors) can now be computed explicitly in terms of components (spin-frame, or tetrad, or some of each). The uniqueness of $\gamma_{\mathbf{AA}'\mathbf{BC}}$ as given by equation (4.5.43) is an illustration of the fact, demonstrated in the last section, that the covariant derivative operator $\nabla$ acting on any spinor is uniquely defined by the requirements that $\varepsilon_{AB}$ be covariantly constant and the torsion vanish. Note that these properties also determined the (Christoffel) form (4.5.39) of the $\Gamma$s and the symmetry (4.5.5) of the $\gamma_{\mathbf{AA'BC}}$ (*cf.* Bergmann 1957).

## 4.6 The curvature spinors

We saw in §4.2 how the concept of covariant differentiation of tensors leads to the definition of the curvature tensor $R_{abc}{}^d$. Thus we expect to obtain a spinor analogue of $R_{abc}{}^d$ which arises from the concept of covariant differentiation within the spinor formalism. While it is possible to develop the theory of spinor curvature entirely within the spinor formalism, this route turns out to be complicated. We shall follow the easier procedure of taking $R_{abc}{}^d$ and its algebraic and differential properties over from tensor theory, and deriving the properties of the spinor curvature from it.

We begin by breaking down the spinor

$$R_{AA'BB'CC'DD'} = R_{abcd}$$

into simpler parts, namely spinors which are totally symmetric in all primed and in all unprimed indices. This will be an illustration of the facts outlined in §3.3 (equations (3.3.47) *et seq*). However, we shall adopt a more direct procedure than the one used there, which has the additional advantage of yielding a useful intermediate stage of the reduction.

The fact that $R_{abcd}$ is skew in $ab$ enables us to employ the decomposition (3.4.17) to obtain

$$R_{abcd} = \tfrac{1}{2}R_{AX'B}{}^{X'}{}_{cd}\varepsilon_{A'B'} + \tfrac{1}{2}R_{XA'}{}^{X}{}_{B'cd}\varepsilon_{AB}.$$

The anti-symmetry in $cd$ then gives

$$\begin{aligned} R_{abcd} = {} & \mathrm{X}_{ABCD}\varepsilon_{A'B'}\varepsilon_{C'D'} + \Phi_{ABC'D'}\varepsilon_{A'B'}\varepsilon_{CD} \\ & + \bar{\Phi}_{A'B'CD}\varepsilon_{AB}\varepsilon_{C'D'} + \bar{\mathrm{X}}_{A'B'C'D'}\varepsilon_{AB}\varepsilon_{CD}, \end{aligned} \tag{4.6.1}$$

where

$$\mathrm{X}_{ABCD} = \tfrac{1}{4}R_{AX'B}{}^{X'}{}_{CY'D}{}^{Y'}, \quad \Phi_{ABC'D'} = \tfrac{1}{4}R_{AX'B}{}^{X'}{}_{YC'}{}^{Y}{}_{D'} \tag{4.6.2}$$

The spinors (4.6.2) are called *curvature spinors*. Their complex conjugates occur in (4.6.1) because of the reality of $R_{abcd}$ – *cf.* (3.4.20). We have the following obvious symmetries (*cf.* (3.4.18)) from the anti-symmetries of $R_{abcd}$:

$$\mathrm{X}_{ABCD} = \mathrm{X}_{(AB)(CD)}, \quad \Phi_{ABC'D'} = \Phi_{(AB)(C'D')}. \tag{4.6.3}$$

The interchange symmetry $R_{abcd} = R_{cdab}$ is evidently equivalent to

$$\mathrm{X}_{ABCD} = \mathrm{X}_{CDAB}, \quad \bar{\Phi}_{ABC'D'} = \Phi_{ABC'D'}. \tag{4.6.4}$$

The second of these equations implies that $\Phi_{AA'BB'}$ corresponds to a *real* tensor $\Phi_{ab}$ while (4.6.3) (2) implies that $\Phi_{ab}$ is symmetric and trace-free:

$$\Phi_{AA'BB'} = \Phi_{ab} = \Phi_{ba} = \bar{\Phi}_{ab}, \quad \Phi_a{}^a = 0. \tag{4.6.5}$$

Note also that the symmetries (4.6.3) (4.6.4) on $\mathrm{X}_{ABCD}$ imply

$$\mathrm{X}_{A(BC)}{}^A = 0. \tag{4.6.6}$$

To translate the cyclic identity satisfied by $R_{abcd}$ into spinor form it is useful first to discuss its various 'duals'. By dualizing on one or both of the skew index pairs of $R_{abcd}$ we can form the following three tensors, whose spinor forms we also exhibit (*cf.* (3.4.23)):

$$\begin{aligned} R^*{}_{abcd} &= \tfrac{1}{2} e_{cd}{}^{pq} R_{abpq} = \mathrm{i} R_{AA'BB'CD'DC'} \\ {}^*R_{abcd} &= \tfrac{1}{2} e_{ab}{}^{pq} R_{pqcd} = \mathrm{i} R_{AB'BA'CC'DD'} \\ {}^*R^*{}_{abcd} &= \tfrac{1}{4} e_{ab}{}^{pq} e_{cd}{}^{rs} R_{pqrs} = -R_{AB'BA'CD'DC'}. \end{aligned} \tag{4.6.7}$$

Clearly all three duals share the anti-symmetries of $R_{abcd}$ ($R_{abcd} = R_{[ab][cd]}$). In addition, we can easily verify that ${}^*R^*_{abcd}$ possesses the interchange symmetry

$${}^*R^*{}_{abcd} = {}^*R^*{}_{cdab} \tag{4.6.8}$$

and satisfies the cyclic identity

$${}^*R^*{}_{a[bcd]} = 0, \tag{4.6.9}$$

while

$${}^*R_{abcd} = R^*{}_{cdab}. \tag{4.6.10}$$

For future reference, we collect (4.6.1) and the corresponding formulae resulting from (4.6.7) into the following scheme, where $\mathrm{X}, \Phi, \bar{\Phi}, \bar{\mathrm{X}}$ represent (temporarily) the terms on the right side of (4.6.1):

$$\begin{aligned} R_{abcd} &= \mathrm{X} + \Phi + \bar{\Phi} + \bar{\mathrm{X}} \\ R^*{}_{abcd} &= -\mathrm{i}\mathrm{X} + \mathrm{i}\Phi - \mathrm{i}\bar{\Phi} + \mathrm{i}\bar{\mathrm{X}} \\ {}^*R_{abcd} &= -\mathrm{i}\mathrm{X} - \mathrm{i}\Phi + \mathrm{i}\bar{\Phi} + \mathrm{i}\bar{\mathrm{X}} \\ {}^*R^*{}_{abcd} &= -\mathrm{X} + \Phi + \bar{\Phi} - \bar{\mathrm{X}} \end{aligned} \tag{4.6.11}$$

We define duality rotations of the Riemann tensor analogously to (3.4.42):

$$\begin{aligned} R^{(\theta)}{}_{abcd} &= R_{abcd} \cos\theta + R^*{}_{abcd} \sin\theta, \\ {}^{(\theta)}R_{abcd} &= R_{abcd} \cos\theta + {}^*R_{abcd} \sin\theta. \end{aligned} \tag{4.6.12}$$

Analogously to (3.4.43) we get, from (4.6.11),

$$R^{(\theta)}{}_{abcd} = e^{-i\theta}X + e^{i\theta}\Phi + e^{-i\theta}\bar{\Phi} + e^{i\theta}\bar{X}$$
$${}^{(\theta)}R_{abcd} = e^{-i\theta}X + e^{-i\theta}\Phi + e^{i\theta}\bar{\Phi} + e^{i\theta}\bar{X}. \tag{4.6.13}$$

We are now ready to translate the last symmetry of the Riemann tensor, $R_{[abc]d} = 0$ or, equivalently, $R_{a[bcd]} = 0$, into spinors. The necessary calculation can be simplified if we observe, from (3.4.26), that the equation $R_{a[bcd]} = 0$ is equivalent to

$$R^*{}_{ab}{}^{bc} = 0 \tag{4.6.14}$$

and that $R_{[abc]d} = 0$ is equivalent to

$${}^*R^{ab}{}_{bc} = 0 \tag{4.6.15}$$

Incidentally, one can now see that ${}^*R_{abcd}$ satisfies the cyclic identity ${}^*R_{[abc]d} = 0$ only under special circumstances: the necessary and sufficient condition is ${}^{**}R^{ab}{}_{bc} = 0$, i.e., $-R^{ab}{}_{bc} = 0$, i.e., the vanishing of the Ricci tensor. The same remark applies to $R^*{}_{abcd}$. On the other hand, ${}^*R^*{}_{abcd}$ always satisfies the cyclic identity since ${}^*R^{**ab}{}_{bc} = -{}^*R^{ab}{}_{bc} = 0$.

To obtain the spinor form of the cyclic identity, we apply (4.6.14) to (4.6.11); if use is made of the already established symmetries of $\Phi_{ABC'D'}$, this shows that (4.6.14) is equivalent to

$$X_{AB}{}^{B}{}_{C}\varepsilon_{A'C'} = \bar{X}_{A'B'}{}^{B'}{}_{C'}\varepsilon_{AC}. \tag{4.6.16}$$

Raising $C$ and $C'$ and contracting with $A$ and $A'$ we obtain

$$\Lambda = \bar{\Lambda}, \tag{4.6.17}$$

where (for later convenience inserting a factor 1/6)

$$\Lambda := \tfrac{1}{6}X_{AB}{}^{AB} \tag{4.6.18}$$

Now (*cf.* (2.5.24))

$$X_{ABC}{}^{B} = 3\Lambda\varepsilon_{AC}, \tag{4.6.19}$$

since the symmetry (4.6.4) implies that $X_{ABC}{}^{B}$ is skew in $AC$. Hence condition (4.6.17) can be seen to imply (4.6.16) and is therefore equivalent to the cyclic identity of the Riemann tensor, once the other symmetries of $X_{ABCD}$ and $\Phi_{ABC'D'}$ are established. (Note that, in terms of components, the cyclic identity is just one algebraic condition, namely $R_{0123} + R_{1203} + R_{2013} = 0$; it is therefore not surprising that it reduces to only one real condition in spinor form.) We have now found the spinor equivalents of all the symmetries of $R_{abcd}$: (4.6.3), (4.6.4) and (4.6.17).

We next compute the Ricci tensor $R_{ac} = R_{abc}{}^{b}$ in spinor form. From (4.6.1) we get

$$R_{ab} = 6\Lambda\varepsilon_{AB}\varepsilon_{A'B'} - 2\Phi_{ABA'B'}, \tag{4.6.20}$$

which may also be written in the form

$$R_{ab} = 6\Lambda g_{ab} - 2\Phi_{ab}. \tag{4.6.21}$$

Hence, for the scalar curvature $R = R_a{}^a$ we find, using (4.6.5),

$$R = 24\Lambda, \tag{4.6.22}$$

and for the trace-free Ricci tensor

$$R_{ab} - \tfrac{1}{4}Rg_{ab} = -2\Phi_{ABA'B'} = -2\Phi_{ab}. \tag{4.6.23}$$

For obvious reasons $\Phi_{ABC'D'}$ is sometimes called the *Ricci spinor*. The Einstein tensor $G_{ab}$ (the trace-reversed Ricci tensor – see (3.4.10)) is given by

$$\begin{aligned} G_{ab} = \hat{R}_{ab} = R_{ab} - \tfrac{1}{2}Rg_{ab} &= R_{AB'BA'} \\ &= -6\Lambda\varepsilon_{AB}\varepsilon_{A'B'} - 2\Phi_{ABA'B'} \end{aligned} \tag{4.6.24}$$

or

$$G_{ab} = -6\Lambda g_{ab} - 2\Phi_{ab}. \tag{4.6.25}$$

We note that

$${}^*R^*{}_{abc}{}^b = G_{ac}, \tag{4.6.26}$$

which follows from (4.6.11) since the passage from $R_{abcd}$ to ${}^*R^*{}_{abcd}$ is equivalent to $\mathrm{X} \mapsto -\mathrm{X}$, which in turn implies $\Lambda \mapsto -\Lambda$, and so, by (4.6.20) and (4.6.24), $R_{ab} \mapsto G_{ab}$.

*Einstein's equations*

We take space–time to be governed by Einstein's field equations, which, for empty space, take the form

$$R_{ab} = 0. \tag{4.6.27}$$

By splitting (4.6.20) into its symmetric and skew parts in $AB$ equation (4.6.27) is seen to be equivalent to

$$\Phi_{ab} = \Phi_{ABA'B'} = 0, \quad \Lambda = 0. \tag{4.6.28}$$

If a cosmological term is included in the field equations, so that

$$R_{ab} = \lambda g_{ab}$$

in empty space, $\lambda$ being called the *cosmological constant*, then we have equivalently, in spinor form,

$$\Phi_{ab} = \Phi_{ABA'B'} = 0, \quad \Lambda = \tfrac{1}{6}\lambda. \tag{4.6.29}$$

In the general case, when sources are present, the field equations with cosmological term are

$$G_{ab} + \lambda g_{ab} = -8\pi\gamma T_{ab}, \tag{4.6.30}$$

where the speed of light is (as always!) unity, $\gamma$ is Newton's constant of gravitation, and $T_{ab}$ is the energy–momentum tensor of the sources. This translates into

$$\Phi_{ab} + (3\Lambda - \tfrac{1}{2}\lambda)g_{ab} = 4\pi\gamma T_{ab}, \tag{4.6.31}$$

i.e.,

$$\Phi_{ab} = 4\pi\gamma(T_{ab} - \tfrac{1}{4}T_q^q g_{ab}), \quad \Lambda = \tfrac{1}{3}\pi\gamma T_q^q + \tfrac{1}{6}\lambda. \tag{4.6.32}$$

The vanishing of $\Lambda$, as in (4.6.28), implies (*cf.* (4.6.19)) that $\mathrm{X}_{ABCD}$ is symmetric in $B$ and $D$; since it is also symmetric in $AB$ and $CD$, it must then be symmetric in all its indices. It is a remarkable fact that with the apparently arbitrary dimension four and signature $+---$ of our space–time, where Einstein's field equations $R_{ab} = 0$ for vacuum are satisfied the curvature can be fully characterized by such a simple and physically natural object as a totally symmetric four-index spinor. We shall see in Chapter 8 how this leads to a very simple algebraic classification scheme for curvature. (If the signature had been $++--$, for example, the classification would have been far more complicated, since *two real* symmetric four-index spinors would be required to describe the curvature; the reality itself would be a complication, since algebra over the complex field is much simpler.)

From (4.6.11) we observe, incidentally, that whenever $\Phi_{ABC'D'} = 0$ (as when $R_{ab} = 0$), *then*

$$^*R^*{}_{abcd} = -R_{abcd}, \quad ^*R_{abcd} = R^*{}_{abcd} \tag{4.6.33}$$

and so, because of (4.6.10), $^*R_{abcd}$ and $R^*{}_{abcd}$ possess the interchange symmetry of $R_{abcd}$. They possess the full symmetries of $R_{abcd}$ (i.e., the cyclic symmetry as well) if, in addition, $\Lambda = 0$; for then $R_{ab} = 0$.

In the general case ($\Lambda \neq 0$) we can isolate the totally symmetric part of $\mathrm{X}_{ABCD}$ as follows (bearing in mind the symmetries (4.6.3), (4.6.4) and the relation (2.5.24)):

$$\begin{aligned}\mathrm{X}_{ABCD} &= \tfrac{1}{3}(\mathrm{X}_{ABCD} + \mathrm{X}_{ACDB} + \mathrm{X}_{ADBC}) + \tfrac{1}{3}(\mathrm{X}_{ABCD} - \mathrm{X}_{ACBD}) + \tfrac{1}{3}(\mathrm{X}_{ABCD} - \mathrm{X}_{ADCB}) \\ &= \mathrm{X}_{(ABCD)} + \tfrac{1}{3}\varepsilon_{BC}\mathrm{X}_{AE}{}^{E}{}_{D} + \tfrac{1}{3}\varepsilon_{BD}\mathrm{X}_{AEC}{}^{E}.\end{aligned}$$

Thus, using (4.6.19),

$$\mathrm{X}_{ABCD} = \Psi_{ABCD} + \Lambda(\varepsilon_{AC}\varepsilon_{BD} + \varepsilon_{AD}\varepsilon_{BC}), \tag{4.6.34}$$

where

$$\Psi_{ABCD} := \mathrm{X}_{(ABCD)} = \mathrm{X}_{A(BCD)}. \tag{4.6.35}$$

The spinor $\Psi_{ABCD}$ plays a very important role in the theory. We call it the *gravitational spinor* since it represents the local degrees of freedom of the free gravitational field: it is the part of $R_{abcd}$ that survives in the absence of matter (if $\lambda = 0$). For reasons to be discussed in §4.8 it is also called the *Weyl conformal spinor*.

The three spinors $\Psi_{ABCD}$, $\Phi_{ABC'D'}$, $\Lambda$ together determine $R_{abcd}$. With $\bar{\Psi}_{A'B'C'D'}$ they constitute the set of totally symmetric spinors into which $R_{abcd}$ may be decomposed according to the scheme of (3.3.47) *et. seq.*

Substitution of (4.6.34) into (4.6.1) immediately shows that $R_{abcd}$ equals an expression similar to (4.6.1) with $\Psi_{ABCD}$ taking the place of $X_{ABCD}$, plus a multiple of $\Lambda$, namely

$$\Lambda(\varepsilon_{AC}\varepsilon_{BD} + \varepsilon_{AD}\varepsilon_{BC})\varepsilon_{A'B'}\varepsilon_{C'D'} + \textit{its complex conjugate}. \quad (4.6.36)$$

Expanding the multiplier of the parenthesis by use of the ε-identity

$$\varepsilon_{A'B'}\varepsilon_{C'D'} + \varepsilon_{A'D'}\varepsilon_{B'C'} - \varepsilon_{A'C'}\varepsilon_{B'D'} = 0 \ (cf.\ (2.5.21))$$

makes this into

$$\Lambda(\varepsilon_{AC}\varepsilon_{BD} + \varepsilon_{AD}\varepsilon_{BC})(\varepsilon_{A'C'}\varepsilon_{B'D'} - \varepsilon_{A'D'}\varepsilon_{B'C'}) + \text{its complex conjugate}$$

i.e.,

$$2\Lambda(\varepsilon_{AC}\varepsilon_{BD}\varepsilon_{A'C'}\varepsilon_{B'D'} - \varepsilon_{AD}\varepsilon_{BC}\varepsilon_{A'D'}\varepsilon_{B'C'}) \quad (4.6.37)$$

Thus we have

$$\begin{aligned} R_{abcd} = {} & \Psi_{ABCD}\varepsilon_{A'B'}\varepsilon_{C'D'} + \bar{\Psi}_{A'B'C'D'}\varepsilon_{AB}\varepsilon_{CD} \\ & + \Phi_{ABC'D'}\varepsilon_{A'B'}\varepsilon_{CD} + \bar{\Phi}_{A'B'CD}\varepsilon_{AB}\varepsilon_{C'D'} \\ & + 2\Lambda(\varepsilon_{AC}\varepsilon_{BD}\varepsilon_{A'C'}\varepsilon_{B'D'} - \varepsilon_{AD}\varepsilon_{BC}\varepsilon_{A'D'}\varepsilon_{B'C'}). \end{aligned} \quad (4.6.38)$$

By a somewhat different manipulation we can convert the term (4.6.36) into the alternative forms

$$2\Lambda(\varepsilon_{AC}\varepsilon_{BD}\varepsilon_{A'B'}\varepsilon_{C'D'} + \varepsilon_{AB}\varepsilon_{CD}\varepsilon_{A'D'}\varepsilon_{B'C'}) \quad (4.6.39)$$

or

$$2\Lambda(\varepsilon_{A'C'}\varepsilon_{B'D'}\varepsilon_{AB}\varepsilon_{CD} + \varepsilon_{A'B'}\varepsilon_{C'D'}\varepsilon_{AD}\varepsilon_{BC}), \quad (4.6.40)$$

which, however, unlike (4.6.37), do not *obviously* exhibit the symmetries of $R_{abcd}$.

Let us now introduce the following tensors (*cf.* (3.4.38), (3.4.39)), of which the first, fourth and fifth are real, and all of which evidently share all the symmetries of $R_{abcd}$:

$$C_{abcd} := \Psi_{ABCD}\varepsilon_{A'B'}\varepsilon_{C'D'} + \bar{\Psi}_{A'B'C'D'}\varepsilon_{AB}\varepsilon_{CD} \quad (4.6.41)$$

$$^{-}C_{abcd} := \Psi_{ABCD}\varepsilon_{A'B'}\varepsilon_{C'D'} \quad (4.6.42)$$

$$^{+}C_{abcd} := \bar{\Psi}_{A'B'C'D'}\varepsilon_{AB}\varepsilon_{CD} \quad (4.6.43)$$

$$E_{abcd} := \Phi_{ABC'D'}\varepsilon_{A'B'}\varepsilon_{CD} + \bar{\Phi}_{A'B'CD}\varepsilon_{AB}\varepsilon_{C'D'} \quad (4.6.44)$$

$$g_{abcd} := \varepsilon_{AC}\varepsilon_{BD}\varepsilon_{A'C'}\varepsilon_{B'D'} - \varepsilon_{AD}\varepsilon_{BC}\varepsilon_{A'D'}\varepsilon_{B'C'} = 2g_{a[c}g_{d]b}. \quad (4.6.45)$$

Then the reduction (4.6.38) of $R_{abcd}$ into irreducible parts (see §3.3) is equivalent to

$$R_{abcd} = C_{abcd} + E_{abcd} + 2\Lambda g_{abcd} \tag{4.6.46}$$

over the real field, or

$$R_{abcd} = {}^{-}C_{abcd} + {}^{+}C_{abcd} + E_{abcd} + 2\Lambda g_{abcd} \tag{4.6.47}$$

over the complex field.

In the language of representation theory, ${}^{-}C\ldots$, ${}^{+}C\ldots$, $E\ldots$, and $g\ldots$ belong to representation spaces for the $D(2,0)$, $D(0,2)$, $D(1,1)$, $D(0,0)$ irreducible representations of the Lorentz group (strictly, $SL(2,\mathbb{C})$); these numbers are half the numbers of symmetric spinor indices (*cf.* §3.3).

## 4.7 Spinor formulation of the Einstein–Cartan–Sciama–Kibble theory

We shall briefly digress at this point to examine a modification of Einstein's theory due, independently, to Einstein–Cartan and, more explicitly to Sciama and Kibble (Cartan 1923, 1924, 1925, Kibble 1961, Sciama 1962, Trautman 1972, 1973; for a general review see Hehl, von der Heyde, Kerlick and Nester 1976), in which the torsion tensor, rather than assumed to be zero, is equated with a certain tensor expression arising from the spin density of matter. We do not enter into the question of the physical reasonableness of this theory, or, indeed, of whether the theory has physical implications different from Einstein's; we merely look at its spinor formulation.

In the ECSK theory – in contrast to the unified field theories – space–time has a real symmetric metric $g_{ab}$ of the usual type, so that our 2-spinor formalism can be employed. The difference from general relativity lies in the nature of the operator $\nabla_a$ that is used. The condition of covariant constancy of the metric,

$$\nabla_a g_{bc} = 0, \tag{4.7.1}$$

is retained, but there is, in general, a non-zero torsion tensor $T_{ab}{}^{c}$ for which (*cf.* (4.2.22))

$$(\nabla_a \nabla_b - \nabla_b \nabla_a)\phi = T_{ab}{}^{c}\nabla_c \phi. \tag{4.7.2}$$

The torsion of space–time is now related to the spin density $S_{abc}$ of its matter content by the equation

$$T_{ab}{}^{c} = -8\pi\gamma(S_{ab}{}^{c} + g_{[a}{}^{c}S_{b]d}{}^{d}), \tag{4.7.3}$$

with $S_{abc}$ satisfying

$$S_{ab}{}^{c} = -S_{ba}{}^{c}, \tag{4.7.4}$$

and $\gamma$, as before, denoting Newton's constant of gravitation. From (4.7.3) we derive

$$T_{ab}{}^{b} = 4\pi\gamma S_{ab}{}^{b}, \tag{4.7.5}$$

so we can reverse (4.7.3):

$$T_{ab}{}^{c} + 2g_{[a}{}^{b}T_{b]d}{}^{d} = -8\pi\gamma S_{ab}{}^{c}. \tag{4.7.6}$$

This equation *supplements* the Einstein field equation

$$R_{ab} - \tfrac{1}{2}g_{ab}R = -8\pi\gamma E_{ab} \tag{4.7.7}$$

(where, to avoid confusion with $T_{ab}{}^{c}$ we now use $E_{ab}$ for the energy–momentum tensor), in which, however, $R_{ab}$ is now a (generally) non-symmetric Ricci tensor* (and $R = R_{a}{}^{a}$), defined as

$$R_{ab} = R_{acb}{}^{c}, \tag{4.7.8}$$

while the curvature tensor $R_{abc}{}^{d}$ is defined by $\nabla_a$ according to (4.2.32). The symmetry relation

$$R_{abcd} = R_{[ab][cd]} \tag{4.7.9}$$

still holds, by (4.7.1) and the arguments given in §4.2, but now

$$R_{[abc]}{}^{d} = -\nabla_{[a}T_{bc]}{}^{d} - T_{[ab}{}^{e}T_{c]e}{}^{d}, \tag{4.7.10}$$

and so the interchange symmetry $R_{abcd} = R_{cdab}$ also fails – which explains the lack of symmetry in $R_{ab}$. Note that this implies that the energy–momentum tensor $E_{ab}$ is also non-symmetric, *cf*. (4.7.7). From (4.7.10) we can deduce the spin 'conservation law':

$$\nabla_c S_{ab}{}^{c} - T_{dc}{}^{d}S_{ab}{}^{c} = E_{ba} - E_{ab}. \tag{4.7.11}$$

To represent the spin density spinorially, we take advantage of (4.7.4) and introduce the spinor

$$\begin{aligned}\sigma_{AB}{}^{CC'} &:= \tfrac{1}{2}S_{AA'B}{}^{A'CC'}\\ &= \sigma_{(AB)}{}^{CC'},\end{aligned} \tag{4.7.12}$$

so that

$$S_{ab}{}^{c} = \sigma_{AB}{}^{CC'}\varepsilon_{A'B'} + \bar{\sigma}_{A'B'}{}^{CC'}\varepsilon_{AB}. \tag{4.7.13}$$

In terms of this spinor, the torsion has the somewhat unremarkable expression

$$\begin{aligned}T_{abc} = -4\pi\gamma\{&\sigma_{ABCC'}\varepsilon_{A'B'} - \sigma_{CABA'}\varepsilon_{B'C'} + \sigma_{BCAB'}\varepsilon_{A'C'}\\ &+ \bar{\sigma}_{A'B'C'C}\varepsilon_{AB} - \bar{\sigma}_{C'A'B'A}\varepsilon_{BC} + \bar{\sigma}_{B'C'A'B}\varepsilon_{AC}\}.\end{aligned} \tag{4.7.14}$$

* Owing to the conventions adopted in this book, the ordering of indices in $R_{ab}$ is the reverse of those in the main references cited above.

However, equivalent information is contained in the tensor $Q_{ab}{}^{c}$ which can be used, according to (4.2.46), to translate back from $\nabla_a$ to the standard Christoffel derivative operator $\tilde{\nabla}_a$:

$$\tilde{\nabla}_a U^b = \nabla_a U^b + Q_{ac}{}^{b} U^c, \tag{4.7.15}$$

where $\tilde{\nabla}_a$ still satisfies (4.7.1) but has vanishing torsion. As in (4.4.37), we have

$$T_{ab}{}^{c} = 2Q_{[ab]}{}^{c} \tag{4.7.16}$$

by (4.2.50) or (4.4.30). Furthermore, as in (4.4.36), equation (4.7.1) and its tilde version give

$$Q_{abc} = -Q_{acb}. \tag{4.7.17}$$

The last two equations determine $Q_{abc}$ uniquely as

$$Q_{abc} = T_{a[bc]} - \tfrac{1}{2}T_{bca}. \tag{4.7.18}$$

Unfortunately the translation of (4.7.11) into $\tilde{\nabla}_a$ terms, using $Q_{abc}$, still does not give simply a divergence of $S_{ab}{}^{c}$ by itself.

In the spinor formalism the role of $Q_{abc}$ is played by the quantity $\Theta_{AA'BC}$ (*cf.* (4.4.23)), whose symmetry $\Theta_{AA'BC} = \Theta_{AA'CB}$ follows from the covariant constancy of $\varepsilon_{AB}, \tilde{\varepsilon}_{AB}$ – as we have seen in (4.4.32)–(4.4.34) – and which is given by

$$Q_{abc} = \Theta_{AA'BC}\varepsilon_{B'C'} + \bar{\Theta}_{AA'B'C'}\varepsilon_{BC}, \tag{4.7.19}$$

as we have seen in (4.4.35). Substituting (4.7.14) and (4.7.19) into (4.7.18), we obtain, after some manipulation, the strikingly simple relation

$$\Theta_{AA'BC} = 8\pi\gamma\sigma_{A(CB)A'} \tag{4.7.20}$$

This may also be obtained more directly from (4.4.37). Having (4.7.20), we can relate the standard Riemann–Christoffel tensor $\tilde{R}_{abcd}$, the standard (symmetric) Ricci tensor $\tilde{R}_{ab}$, etc., to the present $R_{abcd}$, $R_{ab}$, etc., by means of (4.2.51). Note also that (4.7.20) can be inverted to

$$\sigma_{ABCC'} = \frac{1}{8\pi\gamma}(\Theta_{C'(AB)C} - \tfrac{1}{2}\Theta_{C'CAB}) \tag{4.7.21}$$

Spinors may also be used to good effect in the study of the more general type of curvature tensor $R_{abcd}$ that arises here. The skew part $R_{[ab]}$ of the Ricci tensor may be represented as in (3.4.20) by a spinor $\Sigma_{AB} = \Sigma_{(AB)}$, the remaining information in $R_{abcd}$ being expressible in terms of a suitably defined 'Weyl spinor' $\Psi_{ABCD} = \Psi_{(ABCD)}$ and two *complex* quantities $\Phi_{ABA'B'} = \Phi_{(AB)(A'B')}$ and $\Lambda$. When $T_{ab}{}^{c} = 0$, the spinor $\Sigma_{AB}$ vanishes and the others reduce to the standard quantities of §4.6. We do not pursue the matter further here, but *cf.* Penrose (1983).

## 4.8 The Weyl tensor and the Bel–Robinson tensor

The tensor $C_{abcd}$ of (4.6.41) is called the *Weyl conformal tensor* (and so $\Psi_{ABCD}$ is often called the Weyl conformal spinor). It comprises the conformally invariant part of the curvature tensor. (We shall see in (6.8.5) that $C_{abc}{}^{d}$ is invariant under conformal rescalings; in (6.9.23) we shall show that its vanishing is actually necessary and sufficient for space–time to be patchwise conformally flat.) In space–time restricted by the Einstein field equations (without cosmological term) $R_{abcd}$ reduces to $C_{abcd}$ in vacuum. Notice that the 'Ricci tensor' formed from $C_{abcd}$ vanishes:

$$C_{adb}{}^{d} = 0. \tag{4.8.1}$$

From this and the fact that $C_{abcd}$ shares the symmetries of $R_{abcd}$, and differs from $R_{abcd}$ by terms involving the Ricci tensor and the scalar curvature (*cf.* (4.6.41), (4.6.38)), one derives the expression

$$C_{ab}{}^{cd} = R_{ab}{}^{cd} - 2R_{[a}{}^{[c} g_{b]}{}^{d]} + \tfrac{1}{3} R g_{[a}{}^{c} g_{b]}{}^{d}. \tag{4.8.2}$$

(The form of the terms on the right side is determined by the symmetries, while the coefficients $-2, \frac{1}{3}$ follow from $C_{adb}{}^{d} = 0$.*)

Note that, because of (4.6.33),

$$*C_{abcd} = C^{*}{}_{abcd} \tag{4.8.3}$$

Let ${}^{+}C_{abcd}$ and ${}^{-}C_{abcd}$ be the self-dual and anti-self-dual parts of $C_{abcd}$, respectively (*cf.* (3.4.35), (4.6.42), (4.6.43)). By (4.8.3) we need not distinguish between left and right self-duals or anti-self-duals.) Then

$$C_{abcd} = {}^{+}C_{abcd} + {}^{-}C_{abcd}, \tag{4.8.4}$$

and

$$*{}^{+}C_{abcd} = \mathrm{i}\,{}^{+}C_{abcd}, \quad *{}^{-}C_{abcd} = -\mathrm{i}\,{}^{-}C_{abcd}. \tag{4.8.5}$$

One of the great simplifications that the spinor formalism achieves in relativity theory is that it describes the very important but somewhat complicated quantity $C_{abcd}$ by such a simple object as a totally symmetric spinor (namely $\Psi_{ABCD}$). As we mentioned above, this leads, for example, to a very transparent curvature classification scheme. In Chapter 8 we shall analyse the structure of $C_{abcd}$ in considerable detail. For the moment we shall note only a few of its algebraic properties and relate it to a tensor known as the Bel–Robinson tensor. These examples may illustrate the strength of the spinor method.

* In the $n$-dimensional case these coefficients are $-4/(n-2)$ and $2/(n-1)(n-2)$. The corresponding Weyl tensor is also invariant under conformal rescalings. Provided $n \geqslant 4$ its vanishing is necessary and sufficient for conformal flatness.

Consider any two tensors $M_{abcd}$, $N_{abcd}$ having *all* the symmetries of $C_{abcd}$, i.e., being of the form (4.6.41) with $\Psi_{ABCD}$ totally symmetric. Then we have two relations that are analogous to (3.4.44) and (3.4.45):

$$^{-}M_{ab}{}^{pq}\,{}^{+}N_{cdpq} = 0 \tag{4.8.6}$$

and

$$^{-}M_{a\ b}^{\ p\ q}\,{}^{+}N_{cpdq} = {}^{+}N_{a\ b}^{\ p\ q}\,{}^{-}M_{cpdq}. \tag{4.8.7}$$

The proof of (4.8.6) is immediate and identical to that of (3.4.44), while the proof of (4.8.7) is similar to that of (3.4.45): for if we set

$$^{-}M_{abcd} = \mu_{ABCD}\varepsilon_{A'B'}\varepsilon_{C'D'},\quad {}^{+}N_{abcd} = \nu_{A'B'C'D'}\varepsilon_{AB}\varepsilon_{CD},$$

each side of the equation is simply

$$\mu_{ABCD}\nu_{A'B'C'D'}. \tag{4.8.8}$$

(As in the earlier case, a direct tensor proof of (4.8.7) is not trivial.) And again, either of the contracted products in (4.8.7) determines the outer product of these tensors and so determines each tensor separately up to a factor. (These results are due to I. Robinson.)

By specializing (4.8.7) to the Weyl tensor we obtain the so-called Bel-Robinson tensor $T_{abcd}$:

$$T_{abcd} := {}^{-}C_{a\ b}^{\ p\ q}\,{}^{+}C_{cpdq} = {}^{+}C_{a\ b}^{\ p\ q}\,{}^{-}C_{cpdq} = \Psi_{ABCD}\bar{\Psi}_{A'B'C'D'}. \tag{4.8.9}$$

An alternative expression for $T_{abcd}$ in terms of real tensors is the following:

$$T_{abcd} = \tfrac{1}{4}(C_{a\ b}^{\ p\ q}C_{cpdq} + {}^{*}C_{a\ b}^{\ p\ q}{}^{*}C_{cpdq}). \tag{4.8.10}$$

This can easily be verified by using $C_{abcd} = {}^{+}C_{abcd} + {}^{-}C_{abcd}$, ${}^{*}C_{abcd} = \mathrm{i}({}^{+}C_{abcd} - {}^{-}C_{abcd})$.

The symmetry properties of $T_{abcd}$ are by no means apparent from the tensor formula, but they follow directly from the spinor expression in (4.8.9). Thus we see at once that $T_{abcd}$ is totally symmetric and trace-free:

$$T_{abcd} = T_{(abcd)}, \tag{4.8.11}$$

$$T^{a}{}_{abc} = 0. \tag{4.8.12}$$

Indeed, these two relations conversely imply that

$$T_{AA'BB'CC'DD'} = T_{(ABCD)(A'B'C'D')}$$

(see after (3.3.61)). The fact that $T_{AA'BB'CC'DD'}$ 'factorizes' according to (4.8.9) is equivalent to the relation

$$T_{ABCD\,A'B'C'C'}T_{EFGH\,E'F'G'H'} = T_{ABCD\,E'F'G'H'}T_{EFGH\,A'B'C'D'} \tag{4.8.13}$$

(see (3.5.5)). Thus, applying the methods of §3.4, we can obtain a quadratic tensor identity satisfied by $T_{abcd}$. The full tensor expression of this identity is complicated. However, a reduced expression which contains only part of the information of (4.8.13) is the following:

$$T_{abce}T^{abcf} = \tfrac{1}{4}T_{abcd}T^{abcd}g_e{}^f. \tag{4.8.14}$$

To prove this, we first observe that the left side must be proportional to $g_e{}^f$ since, using (2.5.24),

$$\Psi_{ABCE}\Psi^{ABC}{}_F = \tfrac{1}{2}\Psi_{ABCD}\Psi^{ABCD}\varepsilon_{EF},$$

the left side of *this* equation being obviously skew in $EF$ (see-saw $A$, $B$, $C$). The multiplier then follows from contraction over $E$, $F$.

Another property of $T_{abcd}$ is its invariance under duality rotations of $C_{abcd}$ (see (4.6.12)):

$$\begin{aligned} C_{abcd} \mapsto {}^{(\theta)}C_{abcd} &= C_{abcd}\cos\theta + {}^*C_{abcd}\sin\theta \\ &= \mathrm{e}^{-\mathrm{i}\theta}\,{}^-C_{abcd} + \mathrm{e}^{\mathrm{i}\theta}\,{}^+C_{abcd}. \end{aligned} \tag{4.8.15}$$

This is equivalent to the replacement

$$\Psi_{ABCD} \mapsto \mathrm{e}^{-\mathrm{i}\theta}\Psi_{ABCD}, \tag{4.8.16}$$

under which $T_{abcd}$ is evidently invariant.

The spinor formalism also allows one immediately to recognize the uniqueness of $T_{abcd}$ on the basis of suitable criteria. For example, it is (up to proportionality) the only four-index tensor quadratic in $C_{abcd}$ which is invariant under duality rotations of $C_{abcd}$. Again, up to proportionality, it is the only trace-free totally symmetric tensor (of valence greater than zero) which is quadratic in $C_{abcd}$. One merely needs to examine the possible spinor terms that could arise in order to see that (4.8.9) is the only possibility.

The Bel–Robinson tensor has certain positive-definiteness properties which, too, are direct consequences of the spinor form (4.8.9). These will be discussed later (see (5.2.14), (5.2.15)).

## 4.9 Spinor form of commutators

Since the Riemann curvature tensor $R_{abcd}$ appears when a commutator of derivatives $\nabla_a$ is applied to vectors and tensors, we may expect that the spinors which represent $R_{abcd}$ appear when such commutators are applied to spinors. This is indeed the case. Consider the decomposition (3.4.20) applied to the commutator $\Delta_{ab}$ defined in (4.2.14):

$$\Delta_{ab} = 2\nabla_{[a}\nabla_{b]} = \varepsilon_{A'B'}\square_{AB} + \varepsilon_{AB}\square_{A'B'}, \tag{4.9.1}$$

where

$$\square_{AB} = \nabla_{X'(A}\nabla_{B)}{}^{X'}, \quad \square_{A'B'} = \nabla_{X(A'}\nabla_{B')}{}^{X}. \tag{4.9.2}$$

Note that $\square_{A'B'}$ is the complex conjugate of $\square_{AB}$ in the sense that, for any spinor $\chi_{\mathscr{C}}$,

$$\overline{\square_{A'B'}\chi_{\mathscr{C}}} = \square_{AB}(\overline{\chi_{\mathscr{C}}}) = \square_{AB}\bar{\chi}_{\mathscr{C}'} \tag{4.9.3}$$

(*cf.* (4.4.20)). From (4.2.15) and (4.2.16) we have

$$\square_{AB}(\chi_{\mathscr{C}} + \phi_{\mathscr{C}}) = \square_{AB}\chi_{\mathscr{C}} + \square_{AB}\phi_{\mathscr{C}} \tag{4.9.4}$$

$$\square_{AB}(\chi_{\mathscr{C}}\phi_{\mathscr{D}}) = \phi_{\mathscr{D}}\square_{AB}\chi_{\mathscr{C}} + \chi_{\mathscr{C}}\square_{AB}\phi_{\mathscr{D}}, \tag{4.9.5}$$

and similarly for $\square_{A'B'}$. To find the effect of $\square_{AB}$ and $\square_{A'B'}$ on a spinor, say $\kappa^A$, we begin by forming the self-dual null bivector

$$k^{ab} = \kappa^A\kappa^B\varepsilon^{A'B'}. \tag{4.9.6}$$

Then from the (torsion-free) Ricci identity (*cf.* (4.2.33)) we obtain

$$\Delta_{ab}k^{cd} = R_{abe}{}^{c}k^{ed} + R_{abe}{}^{d}k^{ce}.$$

(The non-reality of $k^{ab}$ does not affect this.) Now we substitute (4.9.6) into this relation and use (4.2.16) (which applies to spinors as well as to tensors), and find

$$\kappa^C\varepsilon^{C'D'}\Delta_{ab}\kappa^D + \kappa^D\varepsilon^{C'D'}\Delta_{ab}\kappa^C = R_{abEE'}{}^{CC'}\kappa^E\kappa^D\varepsilon^{E'D'} + R_{abEE'}{}^{DD'}\kappa^C\kappa^E\varepsilon^{C'E'},$$

i.e.,

$$2\varepsilon^{C'D'}\kappa^{(C}\Delta_{ab}\kappa^{D)} = -R_{abE}{}^{D'CC'}\kappa^E\kappa^D + R_{abE}{}^{C'DD'}\kappa^C\kappa^E.$$

Replacing the Riemann tensor by its spinor form (4.6.1) and cancelling $2\varepsilon^{C'D'}$ from each side gives

$$\kappa^{(C}\Delta_{ab}\kappa^{D)} = \{\varepsilon_{A'B'}\mathrm{X}_{ABE}{}^{(C} + \varepsilon_{AB}\Phi_{A'B'E}{}^{(C}\}\kappa^{D)}\kappa^E.$$

Applying (3.5.15) (with $r = s = 1$) to the difference between the left and right sides of this equation, and taking $\kappa^C \neq 0$, we obtain

$$\Delta_{ab}\kappa^C = \{\varepsilon_{A'B'}\mathrm{X}_{ABE}{}^{C} + \varepsilon_{AB}\Phi_{A'B'E}{}^{C}\}\kappa^E, \tag{4.9.7}$$

which, on symmetrizing and skew-symmetrizing over $AB$ yields the equations

$$\square_{AB}\kappa^C = \mathrm{X}_{ABE}{}^{C}\kappa^E, \quad \square_{A'B'}\kappa^C = \Phi_{A'B'E}{}^{C}\kappa^E. \tag{4.9.8}$$

The corresponding formulae for primed spin-vectors are obtained from (4.9.7) and (4.9.8) by taking complex conjugates and using (4.9.3) (and replacing $\bar{\kappa}^{C'}$ by $\tau^{C'}$ for generality):

$$\Delta_{ab}\tau^{C'} = \{\varepsilon_{AB}\bar{\mathrm{X}}_{A'B'E'}{}^{C'} + \varepsilon_{A'B'}\Phi_{ABE'}{}^{C'}\}\tau^{E'}, \tag{4.9.9}$$

$$\square_{AB}\tau^{C'} = \Phi_{ABE'}{}^{C'}\tau^{E'}, \quad \square_{A'B'}\tau^{C'} = \bar{\mathrm{X}}_{A'B'E'}{}^{C'}\tau^{E'}. \tag{4.9.10}$$

Lowering the index $C$ (or $C'$), we also get

$$\square_{AB}\kappa_C = -\mathrm{X}_{ABC}{}^{E}\kappa_E, \quad \square_{A'B'}\kappa_C = -\Phi_{A'B'C}{}^{E}\kappa_E, \tag{4.9.11}$$

$$\square_{AB}\tau_{C'} = -\Phi_{ABC'}{}^{E'}\tau_{E'}, \quad \square_{A'B'}\tau_{C'} = -\bar{\mathrm{X}}_{A'B'C'}{}^{E'}\tau_{E'} \tag{4.9.12}$$

To derive the action of $\square_{AB}$ and $\square_{A'B'}$ on many-index spinors, e.g., $\theta^{C}{}_{D}{}^{E'}{}_{F'}$, we expand $\theta^{C}{}_{D}{}^{E'}{}_{F'}$ as a sum of outer products of spin-vectors and use the properties (4.9.4), (4.9.5). Thus, typically,

$$\square_{AB}\theta^{C}{}_{D}{}^{E'}{}_{F'} = \square_{AB}(\Sigma\alpha^{C}\beta_{D}\gamma^{E'}\delta_{F'}) = \Sigma(\beta_{D}\gamma^{E'}\delta_{F'}\square_{AB}\alpha^{C} + \alpha^{C}\gamma^{E'}\delta_{F'}\square_{AB}\beta_{D} + \alpha^{C}\beta_{D}\delta_{F'}\square_{AB}\gamma^{E'} + \alpha^{C}\beta_{D}\gamma^{E'}\square\delta_{F'}),$$

which, when we substitute from the above formulae for $\square_{AB}\kappa^{C}$ etc, gives

$$\square_{AB}\theta^{C}{}_{D}{}^{E'}{}_{F'} = \mathrm{X}_{ABQ}{}^{C}\theta^{Q}{}_{D}{}^{E'}{}_{F'} - \mathrm{X}_{ABD}{}^{Q}\theta^{C}{}_{Q}{}^{E'}{}_{F'} + \Phi_{ABQ'}{}^{E'}\theta^{C}{}_{D}{}^{Q'}{}_{F'} - \Phi_{ABF'}{}^{Q'}\theta^{C}{}_{D}{}^{E'}{}_{Q'}. \tag{4.9.13}$$

This shows the pattern in the general case. Taking complex conjugates, using (4.9.3) (and replacing $\bar{\theta}^{C'}{}_{D'}{}^{E}{}_{F}$ by $\phi^{C'}{}_{D'}{}^{E}{}_{F}$), we obtain the corresponding formula

$$\square_{A'B'}\phi^{C'}{}_{D'}{}^{E}{}_{F} = \bar{\mathrm{X}}_{A'B'Q'}{}^{C'}\phi^{Q'}{}_{D'}{}^{E}{}_{F} - \bar{\mathrm{X}}_{A'B'D'}{}^{Q'}\phi^{C'}{}_{Q'}{}^{E}{}_{F} + \Phi_{A'B'Q}{}^{E}\phi^{C'}{}_{D'}{}^{Q}{}_{F} - \Phi_{A'B'F}{}^{Q}\phi^{C'}{}_{D'}{}^{E}{}_{Q}. \tag{4.9.14}$$

We may substitute (4.6.34) into the above equations, obtaining, for example, from (4.9.11),

$$\square_{AB}\kappa_{C} = -\Psi_{ABC}{}^{D}\kappa_{D} - \Lambda(\varepsilon_{AC}\kappa_{B} + \varepsilon_{BC}\kappa_{A}), \tag{4.9.15}$$

from which the terms in $\Psi_{ABC}{}^{D}$ and $\Lambda$ can be singled out:

$$\square_{(AB}\kappa_{C)} = -\Psi_{ABC}{}^{D}\kappa_{D}, \tag{4.9.16}$$

$$\square_{AB}\kappa^{B} = -3\Lambda\kappa_{A}. \tag{4.9.17}$$

Further relations may be found from these by taking complex conjugates.

$$\square_{(A'B'}\tau_{C')} = -\bar{\Psi}_{A'B'C'}{}^{D'}\tau_{D'}, \quad \square_{A'B'}\tau^{B'} = -3\Lambda\tau_{A'}. \tag{4.9.18}$$

We can obtain expressions for the curvature spinors analogous to (4.2.66), by substituting $\varepsilon_{C}{}^{\mathbf{C}}$ for $\kappa_{C}$ in (4.9.11):

$$\mathrm{X}_{ABCD} = \varepsilon_{D\mathbf{C}}\square_{AB}\varepsilon_{C}{}^{\mathbf{C}}, \quad \Phi_{A'B'CD} = \varepsilon_{D\mathbf{C}}\square_{A'B'}\varepsilon_{C}{}^{\mathbf{C}}, \tag{4.9.19}$$

the first of which can be decomposed into two parts (*cf.* (4.6.34), (4.6.35)):

$$\Psi_{ABCD} = \varepsilon_{D\mathbf{C}}\square_{(AB}\varepsilon_{C)}{}^{\mathbf{C}}, \quad \Lambda = \tfrac{1}{6}\varepsilon_{A\mathbf{C}}\square^{AB}\varepsilon_{B}{}^{\mathbf{C}}. \tag{4.9.20}$$

The spinor formulae of this section are considerably more involved than the corresponding tensor formulae from which they are derived. Nevertheless one not infrequently encounters the particular combination of $\nabla_{\mathbf{a}}$ operators which occurs in $\square_{AB}$ and $\square_{A'B'}$. In these circumstances the above formulae can be particularly useful. Some applications will be given in §5.11; *cf.* also (5.8.1) and numerous applications in Vol. 2.

## 4.10 Spinor form of the Bianchi identity

Recall that there was a consistency relation to be satisfied between the curvature and the covariant derivative operator, namely the Bianchi identity (4.2.42). Let us determine the spinor form of this. By reference to (3.4.26) it is seen that the Binachi identity

$$\nabla_{[a}R_{bc]de} = 0 \tag{4.10.1}$$

is equivalent to

$$\nabla^a {}^*R_{abcd} = 0. \tag{4.10.2}$$

When we substitute from (4.6.11) into this, it becomes

$$-\mathrm{i}\varepsilon_{C'D'}\nabla^A_{B'}\mathrm{X}_{ABCD} - \mathrm{i}\varepsilon_{CD}\nabla^A_{B'}\Phi_{ABC'D'} + \mathrm{i}\varepsilon_{C'D'}\nabla^{A'}_B\Phi_{CDA'B'} + \mathrm{i}\varepsilon_{CD}\nabla^{A'}_B\bar{\mathrm{X}}_{A'B'C'D'} = 0$$

and by separating this last equation into parts which are, respectively, skew and symmetric in $C'D'$, we find it to be equivalent to

$$\nabla^A_{B'}X_{ABCD} = \nabla^{A'}_B\Phi_{CDA'B'} \tag{4.10.3}$$

and its complex conjugate. Relation (4.10.3) is thus the *spinor form of the Bianchi identity*.

It may be asked whether (4.10.3) can be derived directly as a consistency relation for the commutators of §4.9, analogously to our derivation of (4.2.42). In fact, it can. First observe the identity:

$$\begin{aligned}&\varepsilon^{A(B}\varepsilon_D{}^{C)}\varepsilon_{D'}{}^{A'}\varepsilon^{B'C'} - \varepsilon_D{}^A\varepsilon^{BC}\varepsilon^{A'(B'}\varepsilon_{D'}{}^{C')}\\ &- \varepsilon^{AB}\varepsilon_D{}^C\varepsilon_{D'}{}^{(A'}\varepsilon^{B')C'} + \varepsilon_D{}^{(A}\varepsilon^{B)C}\varepsilon^{A'B'}\varepsilon_{D'}{}^{C'} = 0.\end{aligned} \tag{4.10.4}$$

This may be proved by using the ε-identity (2.5.20) which implies:

$$\varepsilon^{A(B}\varepsilon_D{}^{C)} = \varepsilon^{AB}\varepsilon_D{}^C + \tfrac{1}{2}\varepsilon_D{}^A\varepsilon^{BC}. \tag{4.10.5}$$

Forming the contracted product of (4.10.4) with

$$\nabla_{AA'}\nabla_{BB'}\nabla_{CC'}\kappa^E$$

and using (4.9.2) and (4.9.8), we get

$$\kappa^C(\nabla^B_{D'}\mathrm{X}_{BDC}{}^E - \nabla^{B'}_D\Phi_{B'D'C}{}^E) = 0,$$

from which (4.10.3) again follows, $\kappa^C$ being arbitrary.

We can re-express (4.10.3) in terms of $\Psi_{ABCD}$ and $\Lambda$ by use of (4.6.34):

$$\nabla^A_{B'}\Psi_{ABCD} = \nabla^{A'}_B\Phi_{CDA'B'} - 2\varepsilon_{B(C}\nabla_{D)B'}\Lambda. \tag{4.10.6}$$

It is sometimes useful to split this relation into two irreducible parts, symmetric and skew in $BC$ respectively. Note that it is symmetric in $CD$ already. We get

$$\nabla^A_{B'}\Psi_{ABCD} = \nabla_{(B}{}^{A'}\Phi_{CD)A'B'} \tag{4.10.7}$$

and, contracting the skew part,

$$\nabla^{CA'}\Phi_{CDA'B'} + 3\nabla_{DB'}\Lambda = 0. \tag{4.10.8}$$

This last relation is the spinor form of the important result that the Einstein tensor is divergence-free: $\nabla^a G_{ab} = 0$, as can be seen from (4.6.24).

When the Einstein vacuum field equations hold (with or without the $\Lambda$-term – *cf.* (4.6.29)), we have $\Phi_{ABC'D'} = 0$, $\Lambda = \frac{1}{6}\lambda$ and so the first Bianchi equation (4.10.7) becomes

$$\nabla^{AA'}\Psi_{ABCD} = 0. \tag{4.10.9}$$

This 'field equation', together with Einstein's field equations in the proper sense,

$$\Phi_{ABC'D'} = 0, \quad \Lambda = \tfrac{1}{6}\lambda, \tag{4.10.10}$$

governs the propagation of curvature in vacuum. In particular, (4.10.9), which includes information from (4.10.10), is in a certain sense analogous to an actual field equation. It has significance as being formally identical with the wave equation for a massless (zero rest-mass) spin 2 particle (in our case, the 'graviton') and as such it will be discussed at greater length in §5.7. One simple consequence of it may be noted here, however. It is that the Bel–Robinson tensor (4.8.9) is divergence-free:

$$\nabla^a T_{abcd} = 0. \tag{4.10.11}$$

In the non-vacuum case, when any energy-momentum tensor $T_{ab}$ acts as a source for the gravitational field according to the Einstein field equation (4.6.31), then (4.10.7) becomes

$$\nabla^A_{B'}\Psi_{ABCD} = 4\pi\gamma\nabla^{A'}_{(B}T_{CD)A'B'}, \tag{4.10.12}$$

showing that the *derivative* of $T_{ab}$ may be regarded as a source for the gravitational spinor field $\Psi_{ABCD}$.

## 4.11 Curvature spinors and spin-coefficients

In this section we return to the component description introduced in §4.5 and show how the components of the curvature spinors may be related to the spin-coefficients $\gamma_{\mathbf{BCAA'}}$, and to the intrinsic derivatives $\nabla_{\mathbf{AA'}}$. We first calculate the commutator of intrinsic derivatives. For this we use some of our earlier results on Lie brackets. Recall (*cf.* (4.3.26), (4.3.29), with the torsion equal to zero), that, for a *scalar* $f$,

$$[U, V]f = \{U^p\nabla_p V^q - V^p\nabla_p U^q\}\nabla_q f. \tag{4.11.1}$$

Now consider the expression

$$[\nabla_{\mathbf{AB'}}, \nabla_{\mathbf{CD'}}]f = [\varepsilon_{\mathbf{A}}{}^P\varepsilon_{\mathbf{B'}}{}^{P'}\nabla_p, \varepsilon_{\mathbf{C}}{}^Q\varepsilon_{\mathbf{D'}}{}^{Q'}\nabla_q]f,$$

Replacing $U^p$ and $V^q$ in (4.11.1) by $\varepsilon_{\mathbf{A}}{}^{P}\varepsilon_{\mathbf{B}'}{}^{P'}$ and $\varepsilon_{\mathbf{C}}{}^{Q}\varepsilon_{\mathbf{D}'}{}^{Q'}$, respectively, we obtain

$$[\nabla_{\mathbf{AB}'}, \nabla_{\mathbf{CD}'}]f = \{\nabla_{\mathbf{AB}'}(\varepsilon_{\mathbf{C}}{}^{Q}\varepsilon_{\mathbf{D}'}{}^{Q'}) - \nabla_{\mathbf{CD}'}(\varepsilon_{\mathbf{A}}{}^{Q}\varepsilon_{\mathbf{B}'}{}^{Q'})\}\nabla_q f. \qquad (4.11.2)$$

Expanding the factor { } by use of the Leibniz rule, referring to (4.5.2), and incorporating $\nabla_q$, we derive

$$[\nabla_{\mathbf{AB}'}, \nabla_{\mathbf{BD}'}]f = (\gamma_{\mathbf{AB}'\mathbf{C}}{}^{\mathbf{Q}}\nabla_{\mathbf{QD}'} - \gamma_{\mathbf{CD}'\mathbf{A}}{}^{\mathbf{Q}}\nabla_{\mathbf{QB}'} + \bar{\gamma}_{\mathbf{AB}'\mathbf{D}'}{}^{\mathbf{Q}'}\nabla_{\mathbf{BQ}'} - \bar{\gamma}_{\mathbf{CD}'\mathbf{B}'}{}^{\mathbf{Q}'}\nabla_{\mathbf{AQ}'})f. \qquad (4.11.3)$$

Now, by reference to (4.3.32) and (4.2.24) (in the absence of torsion) it is seen that our previous equation (4.9.7), when transvected with $X^aY^b$, is equivalent to

$$(\underset{X}{\nabla}\underset{Y}{\nabla} - \underset{Y}{\nabla}\underset{X}{\nabla} - \underset{[X,Y]}{\nabla})\kappa^F = X^aY^b\{\varepsilon_{A'B'}X_{ABE}{}^{F} + \varepsilon_{AB}\Phi_{A'B'E}{}^{F}\}\kappa^E. \qquad (4.11.4)$$

In this equation, choose

$$X^a = \varepsilon_{\mathbf{A}}{}^{A}\varepsilon_{\mathbf{B}'}{}^{A'}, \quad Y^b = \varepsilon_{\mathbf{C}}{}^{B}\varepsilon_{\mathbf{D}'}{}^{B'}, \quad \kappa^F = \varepsilon_{\mathbf{E}}{}^{F},$$

and then transvect with $\varepsilon_F{}^{\mathbf{F}}$. The first term on the left becomes

$$\begin{aligned}\varepsilon_F{}^{\mathbf{F}}\nabla_{\mathbf{AB}'}\nabla_{\mathbf{CD}'}\varepsilon_{\mathbf{E}}{}^{F} &= \varepsilon_F{}^{\mathbf{F}}\nabla_{\mathbf{AB}'}(\gamma_{\mathbf{CD}'\mathbf{E}}{}^{\mathbf{Q}}\varepsilon_{\mathbf{Q}}{}^{F}) \\ &= \nabla_{\mathbf{AB}'}\gamma_{\mathbf{CD}'\mathbf{E}}{}^{\mathbf{F}} + \gamma_{\mathbf{CD}'\mathbf{E}}{}^{\mathbf{Q}}\gamma_{\mathbf{AB}'\mathbf{Q}}{}^{\mathbf{F}},\end{aligned}$$

and the second term is the same but with $\mathbf{AB}'$ and $\mathbf{CD}'$ interchanged. Since $\underset{[X,Y]}{\nabla}\kappa^F$ is essentially *defined* as the operator in (4.11.3) acting on $\kappa^F$ instead of $f$, we can use the same calculation as that which give us (4.11.3) for the third term on the left of (4.11.4); but now the operators act on $\varepsilon_{\mathbf{E}}{}^{F}$, and so each intrinsic derivative (with the new factor $\varepsilon_F{}^{\mathbf{F}}$) gives a $\gamma_{..\mathbf{E}}{}^{\mathbf{F}}$, and we get

$$\gamma_{\mathbf{AB}'\mathbf{C}}{}^{\mathbf{Q}}\gamma_{\mathbf{QD}'\mathbf{E}}{}^{\mathbf{F}} - \gamma_{\mathbf{CD}'\mathbf{A}}{}^{\mathbf{Q}}\gamma_{\mathbf{QB}'\mathbf{E}}{}^{\mathbf{F}} + \bar{\gamma}_{\mathbf{AB}'\mathbf{D}'}{}^{\mathbf{Q}'}\gamma_{\mathbf{CQ}'\mathbf{E}}{}^{\mathbf{F}} - \bar{\gamma}_{\mathbf{CD}'\mathbf{B}'}{}^{\mathbf{Q}'}\gamma_{\mathbf{AQ}'\mathbf{E}}{}^{\mathbf{F}}$$

for this term. On the right of (4.11.4) we substitute (4.6.34) and go over to dyad components. Collecting all these terms together and rearranging them, we finally obtain

$$\begin{aligned}\nabla_{\mathbf{AB}'}\gamma_{\mathbf{CD}'\mathbf{E}}{}^{\mathbf{F}} - \nabla_{\mathbf{CD}'}\gamma_{\mathbf{ABE}}{}^{\mathbf{F}} = {}& \gamma_{\mathbf{AB}'\mathbf{E}}{}^{\mathbf{Q}}\gamma_{\mathbf{CD}'\mathbf{Q}}{}^{\mathbf{F}} - \gamma_{\mathbf{CD}'\mathbf{E}}{}^{\mathbf{Q}}\gamma_{\mathbf{AB}'\mathbf{Q}}{}^{\mathbf{F}} + \gamma_{\mathbf{AB}'\mathbf{C}}{}^{\mathbf{Q}}\gamma_{\mathbf{QD}'\mathbf{E}}{}^{\mathbf{F}} \\ & - \gamma_{\mathbf{CD}'\mathbf{A}}{}^{\mathbf{Q}}\gamma_{\mathbf{QB}'\mathbf{E}}{}^{\mathbf{F}} + \bar{\gamma}_{\mathbf{AB}'\mathbf{D}'}{}^{\mathbf{Q}'}\gamma_{\mathbf{CQ}'\mathbf{E}}{}^{\mathbf{F}} - \bar{\gamma}_{\mathbf{CD}'\mathbf{B}'}{}^{\mathbf{Q}'}\gamma_{\mathbf{AQ}'\mathbf{E}}{}^{\mathbf{F}} \\ & + \varepsilon_{\mathbf{B}'\mathbf{D}'}\varepsilon^{\mathbf{FQ}}\Psi_{\mathbf{ACEQ}} + \varepsilon_{\mathbf{B}'\mathbf{D}'}(\varepsilon_{\mathbf{AE}}\varepsilon_{\mathbf{C}}{}^{\mathbf{F}} + \varepsilon_{\mathbf{A}}{}^{\mathbf{F}}\varepsilon_{\mathbf{CE}})\Lambda \\ & + \varepsilon_{\mathbf{AC}}\varepsilon^{\mathbf{FQ}}\Phi_{\mathbf{EQB}'\mathbf{D}'}. \end{aligned} \qquad (4.11.5)$$

This expression can also be obtained *directly* from (4.9.7), by going over to components throughout, and applying (4.5.1) etc. where required. The calculation is essentially similar to the one given above, though slightly longer.

The relations (4.11.3) and (4.11.5) are the fundamental equations of the spin-coefficient formalism. They have a complicated appearance, but look somewhat simpler when specific components are picked out. We shall use the standard notation of (4.5.16) and (4.5.23) for the spin-coefficients and intrinsic derivative operators, and, in addition, special symbols for the dyad components of the Weyl and Ricci spinors, $\Psi_{ABCD}$ and $\Phi_{ABC'D'}$, according to the following scheme:

$$\Psi_0 := \chi^{-1}\bar{\chi}\Psi_{0000}, \quad \Psi_1 := \chi^{-1}\bar{\chi}\Psi_{0001}, \quad \Psi_2 := \chi^{-1}\bar{\chi}\Psi_{0011},$$
$$\Psi_3 := \chi^{-1}\bar{\chi}\Psi_{0111}, \quad \Psi_4 := \chi^{-1}\bar{\chi}\Psi_{1111} \tag{4.11.6}$$
$$\Pi := \chi\bar{\chi}\Lambda \tag{4.11.7}$$
$$\Phi_{00} := \Phi_{000'0'} \quad \Phi_{01} := \Phi_{000'1'} \quad \Phi_{02} := \Phi_{001'1'}$$
$$\Phi_{10} := \Phi_{010'0'} \quad \Phi_{11} := \Phi_{010'1'} \quad \Phi_{12} := \Phi_{011'1'}$$
$$\Phi_{20} := \Phi_{110'0'} \quad \Phi_{21} := \Phi_{110'1'} \quad \Phi_{22} := \Phi_{111'1'}. \tag{4.11.8}$$

These quantities can also be expressed in terms of the null tetrad, by references to (4.6.41), (4.6.20), (4.5.20), and (4.11.7):

$$\Psi_0 = \chi^{-1}\bar{\chi}^{-1}C_{abcd}l^a m^b l^c m^d, \qquad \Psi_1 = \chi^{-1}\bar{\chi}^{-1}C_{abcd}l^a m^b l^c n^d$$
$$\Psi_2 = \chi^{-1}\bar{\chi}^{-1}C_{abcd}l^a m^b \bar{m}^c n^d, \qquad \Psi_3 = \chi^{-1}\bar{\chi}^{-1}C_{abcd}l^a n^b \bar{m}^c n^d$$
$$\Psi_4 = \chi^{-1}\bar{\chi}^{-1}C_{abcd}\bar{m}^a n^b \bar{m}^c n^d, \tag{4.11.9}$$

$$\Phi_{00} = -\tfrac{1}{2}R_{ab}l^a l^b \quad \Phi_{01} = -\tfrac{1}{2}R_{ab}l^a m^b \quad \Phi_{02} = -\tfrac{1}{2}R_{ab}m^a m^b$$
$$\Phi_{10} = -\tfrac{1}{2}R_{ab}l^a \bar{m}^b \quad \Phi_{11} = -\tfrac{1}{2}R_{ab}l^a n^a + 3\Pi \quad \Phi_{12} = -\tfrac{1}{2}R_{ab}m^a n^b$$
$$\Phi_{20} = -\tfrac{1}{2}R_{ab}\bar{m}^a \bar{m}^b \quad \Phi_{21} = -\tfrac{1}{2}R_{ab}\bar{m}^a n^b \quad \Phi_{22} = -\tfrac{1}{2}R_{ab}n^a n^b. \tag{4.11.10}$$

In terms of these quantities, the commutator relations (4.11.3) now become, explicitly,

$$D'D - DD' = (\gamma + \bar{\gamma})D - (\gamma' + \bar{\gamma}')D' - (\tau - \bar{\tau}')\delta' + (\tau' - \bar{\tau})\delta$$
$$\delta D - D\delta = (\beta + \bar{\alpha} + \bar{\tau}')D + \kappa D' - \sigma\delta' - (\varepsilon + \bar{\gamma}' + \bar{\rho})\delta$$
$$\delta D' - D'\delta = \bar{\kappa}'D + (\tau + \bar{\beta}' + \alpha')D' - \bar{\sigma}'\delta' - (\bar{\varepsilon}' + \gamma + \rho')\delta$$
$$\delta'\delta - \delta\delta' = (\rho' - \bar{\rho}')D - (\rho - \bar{\rho})D' - (\alpha' + \bar{\alpha})\delta' + (\alpha + \bar{\alpha}')\delta. \tag{4.11.11}$$

And similarly, the equations (4.11.5) become

$$D\rho - \delta'\kappa = \rho^2 + \sigma\bar{\sigma} - \bar{\kappa}\tau - \kappa(\tau' + 2\alpha + \bar{\beta} - \beta') + \rho(\varepsilon + \bar{\varepsilon}) + \Phi_{00} \tag{a}$$
$$D'\rho' - \delta\kappa' = \rho'^2 + \sigma'\bar{\sigma}' - \bar{\kappa}'\tau' - \kappa'(\tau + 2\alpha' + \bar{\beta}' - \beta) + \rho'(\varepsilon' + \bar{\varepsilon}') + \Phi_{22} \tag{a'}$$
$$D\sigma - \delta\kappa = \sigma(\rho + \bar{\rho} + \bar{\gamma}' - \gamma' + 2\varepsilon) - \kappa(\tau + \bar{\tau}' + \bar{\alpha} - \alpha' + 2\beta) + \Psi_0 \tag{b}$$
$$D'\sigma' - \delta'\kappa' = \sigma'(\rho' + \bar{\rho}' + \bar{\gamma} - \gamma + 2\varepsilon') - \kappa'(\tau' + \bar{\tau} + \bar{\alpha}' - \alpha + 2\beta') + \Psi_4 \tag{b'}$$
$$D\tau - D'\kappa = \rho(\tau - \bar{\tau}') + \sigma(\bar{\tau} - \tau') + \tau(\bar{\gamma}' + \varepsilon) - \kappa(\bar{\gamma} + 2\gamma - \varepsilon') + \Psi_1 + \Phi_{01} \tag{c}$$
$$D'\tau' - D\kappa' = \rho'(\tau' - \bar{\tau}) + \sigma'(\bar{\tau}' - \tau) + \tau'(\bar{\gamma} + \varepsilon') - \kappa'(\bar{\gamma}' + 2\gamma' - \varepsilon) + \Psi_3 + \Phi_{21} \tag{c'}$$

$$\delta\rho - \delta'\sigma = \tau(\rho - \bar{\rho}) + \kappa(\bar{\rho}' - \rho') + \rho(\bar{\alpha} + \beta) - \sigma(\bar{\alpha}' + 2\alpha - \beta') - \Psi_1 + \Phi_{01} \quad (d)$$

$$\delta'\rho' - \delta\sigma' = \tau'(\rho' - \bar{\rho}') + \kappa'(\bar{\rho} - \rho) + \rho'(\bar{\alpha}' + \beta') - \sigma'(\bar{\alpha} + 2\alpha' - \beta) - \Psi_3 + \Phi_{21} \quad (d')$$

$$\delta\tau - D'\sigma = -\rho'\sigma - \bar{\sigma}'\rho + \tau^2 + \kappa\bar{\kappa}' + \tau(\beta + \bar{\beta}') - \sigma(2\gamma - \varepsilon' + \bar{\varepsilon}') + \Phi_{02} \quad (e)$$

$$\delta'\tau' - D\sigma' = -\rho\sigma' - \bar{\sigma}\rho' + \tau'^2 + \kappa'\bar{\kappa} + \tau'(\beta' + \bar{\beta}) - \sigma'(2\gamma' - \varepsilon + \bar{\varepsilon}) + \Phi_{20} \quad (e')$$

$$D'\rho - \delta'\tau = \rho\bar{\rho}' + \sigma\sigma' - \tau\bar{\tau} - \kappa\kappa' + \rho(\gamma + \bar{\gamma}) - \tau(\alpha + \bar{\alpha}') - \Psi_2 - 2\Pi \quad (f)$$

$$D\rho' - \delta\tau' = \rho'\bar{\rho} + \sigma\sigma' - \tau'\bar{\tau}' - \kappa\kappa' + \rho'(\gamma' + \bar{\gamma}') - \tau'(\alpha' + \bar{\alpha}) - \Psi_2 - 2\Pi \quad (f')$$

$$D'\beta - \delta\gamma = \tau\rho' - \kappa'\sigma - \bar{\kappa}'\varepsilon + \alpha\bar{\sigma}' + \beta(\rho' + \bar{\varepsilon}' + \gamma) - \gamma(\bar{\beta}' + \alpha' + \tau) - \Phi_{12} \quad (g)$$

$$D\beta' - \delta'\gamma' = \tau'\rho - \kappa\sigma' - \bar{\kappa}\varepsilon' + \alpha'\bar{\sigma} + \beta'(\rho + \bar{\varepsilon} + \gamma') - \gamma'(\bar{\beta} + \alpha + \tau') - \Phi_{10} \quad (g')$$

$$\delta'\varepsilon - D\alpha = \tau'\rho - \kappa\sigma' + \bar{\kappa}\gamma - \beta\bar{\sigma} - \alpha(\rho + \bar{\varepsilon} + \gamma') + \varepsilon(\bar{\beta} + \alpha + \tau') - \Phi_{10} \quad (h)$$

$$\delta\varepsilon' - D'\alpha' = \tau\rho' - \kappa'\sigma + \bar{\kappa}'\gamma' - \beta'\bar{\sigma}' - \alpha'(\rho' + \bar{\varepsilon}' + \gamma) + \varepsilon'(\bar{\beta}' + \alpha' + \tau) - \Phi_{12} \quad (h')$$

$$D\beta - \delta\varepsilon = \kappa(\rho' - \gamma) - \sigma(\tau' - \alpha) + \beta(\bar{\rho} + \bar{\gamma}') - \varepsilon(\bar{\tau}' + \bar{\alpha}) + \Psi_1 \quad (i)$$

$$D'\beta' - \delta'\varepsilon' = \kappa'(\rho - \gamma') - \sigma'(\tau - \alpha') + \beta'(\bar{\rho}' + \bar{\gamma}) - \varepsilon'(\bar{\tau} + \bar{\alpha}') + \Psi_3 \quad (i')$$

$$\delta'\gamma - D'\alpha = \kappa'(\rho + \varepsilon) - \sigma'(\tau + \beta) - \alpha(\bar{\rho}' + \bar{\gamma}) + \gamma(\bar{\tau} + \bar{\alpha}') + \gamma\beta' - \varepsilon'\alpha + \Psi_3 \quad (j)$$

$$\delta\gamma' - D\alpha' = \kappa(\rho' + \varepsilon') - \sigma(\tau' + \beta') - \alpha'(\bar{\rho} + \bar{\gamma}') + \gamma'(\bar{\tau}' + \bar{\alpha}) + \gamma'\beta - \varepsilon\alpha' + \Psi_1 \quad (j')$$

$$D\gamma - D'\varepsilon = \kappa\kappa' - \tau\tau' - \beta(\tau' - \bar{\tau}) - \alpha(\bar{\tau}' - \tau) - \varepsilon(\gamma + \bar{\gamma}) + \gamma(\gamma' + \bar{\gamma}') + \Psi_2 + \Phi_{11} - \Pi \quad (k)$$

$$D'\gamma' - D\varepsilon' = \kappa\kappa' - \tau\tau' - \beta'(\tau - \bar{\tau}') - \alpha'(\bar{\tau} - \tau') - \varepsilon'(\gamma' + \bar{\gamma}') + \gamma'(\gamma + \bar{\gamma}) + \Psi_2 + \Phi_{11} - \Pi \quad (k')$$

$$\delta'\beta - \delta\alpha = \rho\rho' - \sigma\sigma' - \alpha\bar{\alpha} + \beta\bar{\alpha}' + \alpha(\beta - \alpha') + \gamma(\bar{\rho} - \rho) + \varepsilon(\rho' - \bar{\rho}') + \Psi_2 - \Phi_{11} - \Pi \quad (l)$$

$$\delta\beta' - \delta'\alpha' = \rho\rho' - \sigma\sigma' - \alpha'\bar{\alpha}' + \beta'\bar{\alpha} + \alpha'(\beta' - \alpha) + \gamma'(\bar{\rho}' - \rho') + \varepsilon'(\rho - \bar{\rho}) + \Psi_2 - \Phi_{11} - \Pi \quad (l')$$

(4.11.12)

Note that equations (4.11.12) go in pairs, such that in each pair one is related to the other by the 'priming' operation described after (4.5.16). It is necessary, however, to take into account also the following index correspondences, which result at once from the definitions (4.11.6) and (4.11.8):

$$\Psi_r \mapsto \Psi_s : 0 \leftrightarrow 4,\ 1 \leftrightarrow 3,\ 2 \leftrightarrow 2$$
$$\Phi_{rs} \mapsto \Phi_{tu} : 0 \leftrightarrow 2,\ 1 \leftrightarrow 1. \qquad (4.11.13)$$

Equations (4.11.12) simplify somewhat when the dyad is normalized so that $\chi = 1$, with the consequent identifications $\alpha = -\beta'$, $\varepsilon = -\gamma'$, (*cf.* (4.5.29)); in that case, in particular, the pairs (*h*), (*h'*) and (*j*), (*j'*) become identical with the pairs (*g'*), (*g*) and (*i'*), (*i*) respectively, and (*k*) and (*l*) become identical with their primed versions. So the set reduces from twelve prime-related pairs to eight such pairs plus two self-prime related equations.

The equations (4.11.11) and (4.11.12) are most useful when several of the spin-coefficients vanish (e.g., owing to possible symmetries of a specific problem.) For then they often simplify considerably. However, in §4.12 a somewhat different approach is given which leads to simplifications without specializations having to be made. Expressions for the Bianchi identities and Maxwell equations in spin-coefficient form have also found much useful application in the literature. These we shall defer until the end of §4.12 so that the simplifications which result from the compacted formalism of that section can be taken into account.

## 4.12 Compacted spin-coefficient formalism

In §§4.5 and 4.11 we introduced the formalism of spin-coefficients. The advantages of the use of spin-coefficients are partly those which arise also with any tetrad or component formalism, namely that one operates entirely with scalar quantities, scalars being easily manipulated and able to take on numerical values or the form of explicit functions where necessary. But, in addition, there is the special advantage when spin-coefficients are used that they are all *complex*. Thus each spin-coefficient carries the information of *two* real numbers and a considerable economy of notation is thereby achieved. One needs only 12 such quantities rather than the 24 equivalent real quantities which would be needed if a conventional orthonormal tetrad were used, or the 40 independent coefficients of the Christoffel symbols which play a corresponding role when a conventional coordinate approach is used. Of course, explicit formalisms such as these are at their most advantageous when the basis frames which are introduced can be tied to the geometry or physics in a natural way. If a timelike vector field occurs naturally in a problem (e.g., as tangents to fluid flow lines) then it is often advantageous to choose this as a tetrad vector and employ a tetrad formalism. In the same way, if one or more null vectors occur naturally (as is often the case in radiation problems) then that null vector may, with advantage, be used as a tetrad vector. In such situations one may also go further and use that null vector as the flagpole of one of the dyad spinors of a spin-coefficient formalism.

It will be seen, however, that in all these situations there may be much freedom left in the choice of basis frame. It is not often that the problem defines a complete basis in a suitably natural way. Usually some freedom remains in the choice of basis. This has the effect that many of the quantities involved in the calculation do not have direct geometrical or physical meaning. Instead they are of the nature of 'gauge quantities' whose values transform in certain ways as the basis frame is varied in accordance with the freedom that remains. As a general rule, the presence of too many such 'gauge quantities' may detract considerably from the value of a formalism – especially if the gauge transformation behaviour is complicated. It is one of the virtues of a covariant approach that such complicated gauge behaviour is avoided completely, and so the geometrical or physical content of a formula is likely to be much more immediately apparent. In explicit problems, a fully covariant approach may not always be convenient. But likewise, it may not be convenient to fix the basis system completely, and sometimes a partially covariant formalism can be adopted.

In the case of spin-coefficients, there are two types of 'gauge freedom' which are likely to be encountered. In the first place one may be concerned with a problem in which one null direction only is singled out in a natural way. This is the situation, in particular, when the geometry in relation to a null hypersurface* (or wave front) is being studied. The dyad spinor $o^A$ may be chosen to be in this null direction, but the direction of the remaining dyad spinor $\iota^A$ may be completely free. It is possible to develop a partially covariant formalism for such a situation (Penrose 1972a) but it seems that a high price must be paid in complication if the formalism is to be made generally applicable. We shall not give a complete discussion of this situation here, although certain quantities with the required covariance will be described later (*cf.* §7.1 and (5.12.12)) The second type of gauge freedom frequently encountered with spin-coefficient problems occurs when *two* null directions are singled out but there is no further

* A hypersurface whose normal vector is null, *cf.* §7.1; also §5.12.

information which naturally fixes the flag planes or extents of the flag poles of the corresponding dyad spinors. This situation arises, for example, when the geometry of a spacelike 2-surface is being studied (*cf.* §4.14). There are two null directions orthogonal to the surface at each point of the surface and it is convenient to choose the flagpoles of $o^A$ and $\iota^A$ pointing in these directions. There are, of course, many other situations where two null directions are naturally singled out at each point. The compacted spin-coefficient method that we shall describe is ideally suited to the study of such situations (Geroch, Held and Penrose 1973; *cf.* also Stewart and Walker 1974, Stewart 1979, Held and Voorhees 1974, Fordy 1977, Held 1974, 1975). But in addition, the formalism is sufficiently close to the original spin-coefficient approach that it may, if desired, be used in place of the original formalism, the various formulae encountered being regarded as shorthand expressions for ordinary spin-coefficient formulae. Indeed, the compacted expressions are, for the most part, considerably simpler than their full spin-coefficient counterparts, but sufficiently close to them that a translation back is a simple matter. For this reason we do not always give the spin-coefficient expressions in full but rely on the compacted expression to convey the information required.

*Weighted scalars*

We suppose that two future-pointing null directions are assigned at each point of the space–time $\mathscr{M}$. Let $o^A$ and $\iota^A$ be a pair of spinor fields on $\mathscr{M}$, whose flagpole directions are the given null directions at each point. As in (4.5.6), we set

$$o_A \iota^A = \chi. \tag{4.12.1}$$

Now the most general change of dyad which leaves these two null directions invariant is

$$o^A \mapsto \lambda o^A, \quad \iota^A \mapsto \mu \iota^A, \tag{4.12.2}$$

where $\lambda$ and $\mu$ are arbitrary (nowhere vanishing) complex scalar fields. Under (4.12.2) we have

$$\chi \mapsto \lambda \mu \chi. \tag{4.12.3}$$

If the usual normalization condition $\chi = 1$ is adopted, then to preserve this we require the restriction

$$\mu = \lambda^{-1}. \tag{4.12.4}$$

For greater flexibility in applications we shall not generally adopt (4.12.4), but rather allow $\lambda$ and $\mu$ to be independent quantities. The specialization to (4.12.4) and $\chi = 1$ is easily made if required. In terms of the null tetrad

$$l^a = o^A \bar{o}^{A'}, \quad m^a = o^A \bar{\iota}^{A'}, \quad \bar{m}^a = \iota^A \bar{o}^{A'}, \quad n^a = \iota^A \bar{\iota}^{A'}, \tag{4.12.5}$$

the transformation (4.12.2) effects

$$l^a \mapsto \lambda\bar{\lambda} l^a, \quad m^a \mapsto \lambda\bar{\mu} m^a, \quad \bar{m}^a \to \mu\bar{\lambda}\bar{m}^a, \quad n^a \mapsto \mu\bar{\mu} n^a, \tag{4.12.6}$$

and note that $l^a n_a = \chi\bar{\chi} = -m^a \bar{m}_a$.
Adopting, for the moment, $\chi = 1$ and (4.12.4), we can put

$$\lambda^2 = R\mathrm{e}^{\mathrm{i}\theta} = \mu^{-2}$$

$R$ and $\theta$ being real, and re-express (4.12.6) as the boost

$$l^a \mapsto R l^a, \quad n^a \mapsto R^{-1} n^a \tag{4.12.7}$$

combined with the spatial rotation

$$m^a \mapsto \mathrm{e}^{\mathrm{i}\theta} m^a. \tag{4.12.8}$$

Thus we have a two-dimensional 'gauge' freedom at each point, namely the 2-parameter subgroup of the Lorentz group at each point which preserves the two null directions defined by $l^a$ and $n^a$. This 'gauge group' at each point is seen to be – from (4.12.2) and (4.12.4) – just the multiplicative group of complex numbers $\lambda$. In the more general case, where (4.12.4) is not adopted, the gauge group is seen to be the product of two such multiplicative groups.

Our formalism will deal with scalars (and also sometimes tensors or spinors) $\eta$ associated with the (not necessarily normalized) dyad $o^A, \iota^A$, that undergo transformations

$$\eta \mapsto \lambda^{r'} \bar{\lambda}^{t'} \mu^r \bar{\mu}^t \eta \tag{4.12.9}$$

whenever the dyad $o^A, \iota^A$ is transformed according to (4.12.2). Such a quantity will be called a (weighted) quantity of *type* $\{r', r; t', t\}$. In the special case (4.12.4) only the two numbers

$$p = r' - r, \quad q = t' - t \tag{4.12.10}$$

are defined, and we may say that $\eta$ has type $\{p, q\}$, or equivalently, a *spin-weight* $\frac{1}{2}(p - q)$ and a *boost-weight* $\frac{1}{2}(p + q)$. The terms spin-weight and boost-weight may also be used in the general case when (4.12.4) is dropped. We shall sometimes refer to $\eta$ as a $\{r', r; t', t\}$*-scalar* or a $\{p, q\}$*-scalar*.

More precisely* we should think of a weighted scalar $\eta$ as a function

* Those versed in the language of vector bundles (*cf.* §5.4) will recognize a $\{r', r; t', t\}$-scalar as being a (smooth) cross-section of a certain complex line bundle over $\mathscr{M}$. This bundle can be expressed as $\mathscr{B}_0^{-r'} \otimes \mathscr{B}_1^{-r} \otimes \bar{\mathscr{B}}_0^{-t'} \otimes \bar{\mathscr{B}}_1^{-t}$, where $\mathscr{B}_0$ and $\mathscr{B}_1$ are the bundles of spin-vectors whose flagpoles point along $l^a$ and $n^a$, respectively, and $\bar{\mathscr{B}}_0$ and $\bar{\mathscr{B}}_1$ are their complex conjugates. (For example, if $\xi$ has type $\{-1, 0; 0, 0\}$, then $\xi o^A$ is a cross-section of $\mathscr{B}_0$, being an ordinary 'weightless' spin-vector field.)

which assigns a complex scalar field $\eta(o^A, \iota^A)$ to each pair of spinor fields $o^A, \iota^A$, for which the null directions defined by $o^A$ and by $\iota^A$ are given. To be a weighted quantity, the function $\eta$ must be of a very special type, namely one which satisfies (4.12.9) when the dyad is changed according to (4.12.2), i.e.,

$$\eta(\lambda o^A, \mu\iota^A) = \lambda^{r'}\bar{\lambda}^{t'}\mu^r\bar{\mu}^t\eta(o^A, \iota^A).$$

Note that we may regard $o^A, o^{A'}, \iota^A$ and $\iota^{A'}$ themselves as *spinors* of type $\{1, 0; 0, 0\}$, $\{0, 0; 1, 0\}$, $\{0, 1; 0, 0\}$, and $\{0, 0; 0, 1\}$, respectively, and $l^a, m^a, \bar{m}^a, n^a$ as *vectors* of type $\{1, 0; 1, 0\}$, $\{1, 0; 0, 1\}$, $\{0, 1; 1, 0\}$, and $\{0, 1; 0, 1\}$, respectively.

The spin-coefficients may now be divided into two classes according to whether or not they are weighted quantities. In fact, the spin-coefficients in the first and last columns of (4.5.21), namely $\kappa, \rho, \sigma, \tau, \kappa', \rho', \sigma', \tau'$ are such quantities whereas those in the middle two columns, namely $\varepsilon, \alpha, \beta, \gamma, \varepsilon', \alpha', \beta', \gamma'$, are not. Let us illustrate this with two examples:

$$\sigma \mapsto (\lambda o_A \mu\iota^A)^{-1}(\lambda o^A)(\bar{\mu}\bar{\iota}^{A'})(\lambda o^B)\nabla_{AA'}(\lambda o_B) = \lambda^2\mu^{-1}\bar{\mu}\sigma, \quad (4.12.11)$$

but

$$\beta \mapsto (\lambda o_A \mu\iota^A)^{-1}(\lambda o^A)(\bar{\mu}\bar{\iota}^{A'})(\mu\iota^B)\nabla_{AA'}(\lambda o_B) = \lambda\bar{\mu}\beta + \bar{\mu}\delta\lambda. \quad (4.12.12)$$

The types of the weighted spin-coefficients are as follows:

$$\begin{aligned} &\kappa:\{2, -1; 1, 0\}, \quad \sigma:\{2, -1; 0, 1\}, \quad \rho:\{1, 0; 1, 0\}, \quad \tau:\{1, 0; 0, 1\} \\ &\kappa':\{-1, 2; 0, 1\}, \quad \sigma':\{-1, 2; 1, 0\}, \quad \rho':\{0, 1; 0, 1\}, \quad \tau':\{0, 1; 1, 0\}, \end{aligned} \quad (4.12.13)$$

and we note that

$$\chi \textit{ is a } \{1, 1; 0, 0\}\textit{-scalar} \quad (4.12.14)$$

With any spinor field or tensor field on the space–time there is associated a set of scalars ('components') of various types $\{r', r; t', t\}$ which define the spinor or tensor. These are obtained by transvecting the spinor with various combinations of $o^A, \iota^A, o^{A'}, \iota^{A'}$, or the tensor with various combinations of $l^a, m^a, \bar{m}^a, n^a$. Any tensor field may be interpreted as a spinor field, if desired, but we get precisely the same set of scalars whichever way we do it because of the definition (4.12.5) of the null tetrad in terms of the dyad.

Evidently the product of a $\{r', r; t', t\}$-scalar with a $\{v', v; u', u\}$-scalar is a $\{r' + v', r + v; t' + u', t + u\}$-scalar. On the other hand, sums are allowed only when the summands have the same type; the type of the sum is the same as the type of each summand.

*Weighted derivative operators* ð *and* þ

We next wish to introduce derivative operators into the formalism. Unfortunately the operators (4.5.23) of the spin-coefficient formalism are not suitable for this purpose, for, when applied to a scalar of non-zero type, they do not in general produce a weighted scalar. We therefore modify the derivative operators (4.5.23) by the inclusion of terms involving spin-coefficients. Moreover, it is precisely those spin-coefficients which are *not* weighted ($\varepsilon, \alpha, \beta, \gamma; \varepsilon', \alpha', \beta', \gamma'$) that get included with the derivative operators (which also are not weighted) to form weighted objects, and in this way they get withdrawn from free circulation in our present formalism. For a scalar (or tensor, or spinor) $\eta$ of type $\{r', r; t', t\}$ we define*

$$\begin{aligned} \text{þ}\eta &= (D - r'\varepsilon - r\gamma' - t'\bar{\varepsilon} - t\bar{\gamma}')\eta, \\ \eth\eta &= (\delta - r'\beta - r\alpha' - t'\bar{\alpha} - t\bar{\beta}')\eta, \\ \eth'\eta &= (\delta' - r'\alpha - r\beta' - t'\bar{\beta} - t\bar{\alpha}')\eta, \\ \text{þ}'\eta &= (D' - r'\gamma - r\varepsilon' - t'\bar{\gamma} - t\bar{\varepsilon}')\eta. \end{aligned} \tag{4.12.15}$$

These combinations have been so chosen that, under (4.12.2), the terms involving derivatives of $\lambda$ and $\mu$ cancel exactly:

$$\begin{aligned} \text{þ}\eta &\mapsto \lambda^{r'+1}\bar{\lambda}^{t'+1}\mu^{r}\bar{\mu}^{t}\text{þ}\eta, \\ \eth\eta &\mapsto \lambda^{r'+1}\bar{\lambda}^{t'}\mu^{r}\bar{\mu}^{t+1}\eth\eta, \\ \eth'\eta &\mapsto \lambda^{r'}\bar{\lambda}^{t'+1}\mu^{r+1}\bar{\mu}^{t}\eth'\eta, \\ \text{þ}'\eta &= \lambda^{r'}\bar{\lambda}^{t'}\mu^{r+1}\bar{\mu}^{t+1}\text{þ}'\eta. \end{aligned} \tag{4.12.16}$$

Note from these formulae that the operators have the following types:

$$\text{þ}: \{1, 0; 1, 0\}, \quad \eth: \{1, 0; 0, 1\}, \quad \eth': \{0, 1; 1, 0\}, \quad \text{þ}': \{0, 1; 0, 1\}. \tag{4.12.17}$$

(To say that a differential operator has type $\{v', v; u', u\}$ is to say that when it acts on a scalar – or spinor, or tensor – of type $\{r', r; t', t\}$ it produces a quantity of type $\{r' + v', r + v; t' + u', t + u\}$.) In the special case $\chi = 1$ we have $\varepsilon = -\gamma'$ and $\alpha = -\beta'$, so with $p = r' - r$ and $q = t' - t$, as in (4.12.10), we can then write

$$\begin{aligned} \text{þ}\eta &= (D + p\gamma' + q\bar{\gamma}')\eta, \quad & \text{þ}'\eta &= (D' - p\gamma - q\bar{\gamma})\eta, \\ \eth\eta &= (\delta - p\beta + q\bar{\beta}')\eta, \quad & \eth'\eta &= (\delta' + p\beta' - q\bar{\beta})\eta. \end{aligned} \tag{4.12.18}$$

From the definitions (4.12.15) one can easily verify that *the operators* þ, ð, ð′, þ′ *are additive, and, when applied to products, satisfy the Leibniz rule.*

* The symbol þ is pronounced 'thorn' and ð is pronounced 'eth'; þ and ð are the phonetic symbols for the unvoiced and voiced 'th', respectively.

An alternative way to define these operators is in terms of the zero-weight vector operator (acting on a quantity of type $\{r', r; t', t\}$)

$$\Theta_{AA'} = \nabla_{AA'} + \chi^{-1}(ro^B\nabla_{AA'}\iota_B - r'\iota^B\nabla_{AA'}o_B + to^{B'}\nabla_{AA'}\iota_{B'} - t'\iota^{B'}\nabla_{AA'}o_{B'}) \tag{4.12.19}$$

by

$$þ = \Theta_{00'}, \quad þ' = \Theta_{11'}, \quad ð = \Theta_{01'}, \quad ð' = \Theta_{10'}. \tag{4.12.20}$$

Note that

$$\Theta_a = |\chi|^{-2}(l_a þ' + n_a þ - m_a ð' - \bar{m}_a ð). \tag{4.12.21}$$

It can be verified at once from (4.12.19) that

$$\Theta_{AA'}\chi = 0, \tag{4.12.22}$$

whence

$$þ\chi = 0, \quad ð\chi = 0, \quad ð'\chi = 0, \quad þ'\chi = 0. \tag{4.12.23}$$

The basic weighted objects with which we shall work are the eight spin-coefficients $\kappa, \sigma, \rho, \tau; \kappa', \sigma', \rho', \tau'$, the quantity $\chi$, and the four differential operators þ, ð, þ′, ð′. In addition, there are the operations of complex conjugation and of priming (*cf.* (4.5.17)) which both convert weighted quantities into weighted quantities. The prime operation is involutory up to sign: if $\eta$ has type $\{r', r; t', t\}$, then

$$(\eta')' = (-1)^{r'+t'-r-t}\eta. \tag{4.12.24}$$

(For all scalar quantities explicitly defined in this section, $r' + t' - r - t$ is in fact even, so this sign will play no role for scalars here.) Use of the prime not only halves the number of Greek letters needed for the spin-coefficients, but also effectively halves the number of equations, as we already had occasion to note in (4.11.12).

We can combine the above elements with the tetrad components of various tensor fields (e.g., the electromagnetic field tensor or the Riemann tensor), and with the dyad components of various spinor fields to obtain a self-contained calculus. The types of the several spinor components of the Riemann tensor are (*cf.* (4.11.6), (4.11.7), (4.11.8)).

$$\Psi_r : \{3 - r, r - 1; 1, 1\}, \quad \Pi : \{1, 1; 1, 1\}, \quad \Phi_{rt} : \{2 - r, r; 2 - t, t\}. \tag{4.12.25}$$

Moreover, the various components

$$\xi_{r,t} = \xi_{A\ldots D\ldots G'\ldots K'\ldots}\underbrace{o^A\ldots}_{r'}\underbrace{\iota^D\ldots}_{r}\underbrace{o^{G'}\ldots}_{t'}\underbrace{\iota^{K'}\ldots}_{t} \tag{4.12.26}$$

of a symmetric spinor $\xi_{A\ldots M'}$ of valence $\{{}^{0}_{r'+r}\,{}^{0}_{t'+t}\}$ have respective types $\{r'; r; t', t\}$. The corresponding components of the derivatives of $\xi_{A\ldots M'}$

are then

$$\begin{aligned}
(o^A \ldots \iota^{K'} \ldots)D\xi_{A\ldots K'\ldots} &= \text{þ}\xi_{r,t} + r'\kappa\xi_{r+1,t} + r\tau'\xi_{r-1,t} \\
&\quad + t'\bar{\kappa}\xi_{r,t+1} + t\bar{\tau}'\xi_{r,t-1} \\
(o^A \ldots \iota^{K'} \ldots)\delta\xi_{A\ldots K'\ldots} &= \eth\xi_{r,t} + r'\sigma\xi_{r+1,t} + r\rho'\xi_{r-1,t} \\
&\quad + t'\bar{\rho}\xi_{r,t+1} + t\bar{\sigma}'\xi_{r,t-1} \qquad (4.12.27) \\
(o^A \ldots \iota^{K'} \ldots)\delta'\xi_{A\ldots K'\ldots} &= \eth'\xi_{r,t} + r'\rho\xi_{r+1,t} + r\sigma'\xi_{r-1,t} \\
&\quad + t'\bar{\sigma}\xi_{r,t+1} + t\bar{\rho}'\xi_{r,t-1}\,. \\
(o^A \ldots \iota^{K'} \ldots)D'\xi_{A\ldots K'\ldots} &= \text{þ}'\xi_{r,t} + r'\tau\xi_{r+1,t} + r\kappa'\xi_{r-1,t} \\
&\quad t'\bar{\tau}\xi_{r,t+1} + t\bar{\kappa}'\xi_{r,t-1}.
\end{aligned}$$

These relations may be rapidly derived from the useful formulae

$$\begin{aligned}
&\text{þ}o^A = -\kappa\iota^A, \quad \text{þ}\iota^A = -\tau'o^A, \quad \text{þ}o^{A'} = -\bar{\kappa}\iota^{A'}, \quad \text{þ}\iota^{A'} = -\bar{\tau}'o^{A'} \\
&\eth o^A = -\sigma\iota^A, \quad \eth\iota^A = -\rho'o^A, \quad \eth o^{A'} = -\bar{\rho}\iota^{A'}, \quad \eth\iota^{A'} = -\bar{\sigma}'o^{A'} \qquad (4.12.28) \\
&\eth'o^A = -\rho\iota^A, \quad \eth'\iota^A = -\sigma'o^A, \quad \eth'o^{A'} = -\bar{\sigma}\iota^{A'}, \quad \eth'\iota^{A'} = -\bar{\rho}'o^{A'} \\
&\text{þ}'o^A = -\tau\iota^A, \quad \text{þ}'\iota^A = -\kappa'o^A, \quad \text{þ}'o^{A'} = -\bar{\tau}\iota^{A'}, \quad \text{þ}'\iota^{A'} = -\bar{\kappa}o^{A'}
\end{aligned}$$

which themselves may be directly obtained from (4.5.26), (4.5.27), and (4.12.15). Note that the use of þ, ð, ð′, þ′ instead of $D$, $\delta$, $\delta'$, $D'$, respectively, has the effect of eliminating from the right-hand sides of equations (4.5.26) and (4.5.27) precisely those terms (namely all the first terms) that are not weighted. Similarly, it is easily verified by use of (4.12.28) that the þ, ð, ð′, þ′-form of equations (4.5.28) differs from the $D$, $\delta$, $\delta'$, $D'$-form precisely by the disappearance of all unweighted (bracketed) terms on the right-hand sides.

Complex conjugation changes a quantity or operation of type $\{r', r; t', t\}$ into one of type* $\{\bar{t}', \bar{t}; \bar{r}', \bar{r}\}$. Consistently with this remark, and in order to have the desirable relations

$$\overline{\text{þ}\eta} = \bar{\text{þ}}\bar{\eta}, \quad \overline{\eth\eta} = \bar{\eth}\bar{\eta}, \qquad (4.12.29)$$

we define (*cf*. (4.12.15), (4.12.18))

$$\bar{\text{þ}} = \text{þ}, \quad \bar{\text{þ}}' = \text{þ}', \quad \bar{\eth} = \eth', \quad \bar{\eth}' = \eth. \qquad (4.12.30)$$

The prime operation changes a quantity or operation of type $\{r', r; t', t\}$ into one of type $\{r, r'; t, t'\}$. Note that

$$(\text{þ}\eta)' = \text{þ}'\eta', \quad (\text{þ}'\eta)' = \text{þ}\eta', \quad (\eth\eta)' = \eth'\eta', \quad (\eth'\eta)' = \eth\eta'. \qquad (4.12.31)$$

* For normal applications, $r', r, t', t$ will be integers, so that $\bar{r}' = r'$, etc. However, the formalism still works (taking $\chi = 1$ and $p = r' - r$, $q = t' - t$) if $p$ and $q$ are any pair of complex numbers for which $p - q$ is an integer. See, for example, Naimark (1964).

*Compacted equations*

Let us now rewrite the spin-coefficient equations (4.11.12) in terms of the operators (4.11.15). The effect is to eliminate those spin-coefficients, namely $\varepsilon, \alpha, \beta, \gamma, \varepsilon', \alpha', \beta', \gamma'$, which are not weighted quantities, since terms containing them just collect together to form the new derivative operators. From (4.11.12) (*a*), (*b*), (*c*), (*d*), (*e*), (*f*) we get, respectively,

$$
\begin{aligned}
þ\rho - ð'\kappa &= \rho^2 + \sigma\bar{\sigma} - \bar{\kappa}\tau - \tau'\kappa + \Phi_{00} && (a)\\
þ\sigma - ð\kappa &= (\rho + \bar{\rho})\sigma - (\tau + \bar{\tau}')\kappa + \Psi_0 && (b)\\
þ\tau - þ'\kappa &= (\tau - \bar{\tau}')\rho + (\bar{\tau} - \tau')\sigma + \Psi_1 + \Phi_{01} && (c)\\
ð\rho - ð'\sigma &= (\rho - \bar{\rho})\tau + (\bar{\rho}' - \rho')\kappa - \Psi_1 + \Phi_{01} && (d)\\
ð\tau - þ'\sigma &= -\rho'\sigma - \bar{\sigma}'\rho + \tau^2 + \kappa\bar{\kappa}' + \Phi_{02} && (e)\\
þ'\rho - ð'\tau &= \rho\bar{\rho}' + \sigma\sigma' - \tau\bar{\tau} - \kappa\kappa' - \Psi_2 - 2\Pi && (f) \quad (4.12.32)
\end{aligned}
$$

Applying the prime operation to each of these six equations, we obtain six more equations, equivalent to equations (4.11.12) (*a*′), (*b*′), (*c*′), (*d*′), (*e*′), (*f*′). The remaining equations in (4.11.12) concern derivatives of spin-coefficients which are not weighted quantities. They cannot, therefore, be written explicitly in our present formalism as equations like (4.12.32). Instead, they play their role as part of the commutator equations for the differential operators þ, þ′, ð, and ð′. These commutators, when applied to an $\{r', r; t', t\}$-scalar $\eta$ (with $p = r' - r$, $q = t' - t$) are

$$
\begin{aligned}
þþ' - þ'þ = (\bar{\tau} - \tau')ð + (\tau - \bar{\tau}')ð' - p(\kappa\kappa' - \tau\tau' + \Psi_2 + \Phi_{11} - \Pi)\\
- q(\bar{\kappa}\bar{\kappa}' - \bar{\tau}\bar{\tau}' + \bar{\Psi}_2 + \Phi_{11} - \Pi) \qquad (4.12.33)
\end{aligned}
$$

$$
\begin{aligned}
þð - ðþ = \bar{\rho}ð + \sigma ð' - \bar{\tau}'þ - \kappa þ' - p(\rho'\kappa - \tau'\sigma + \Psi_1)\\
- q(\bar{\sigma}'\bar{\kappa} - \bar{\rho}\tau' + \Phi_{01}) \qquad (4.12.34)
\end{aligned}
$$

$$
\begin{aligned}
ðð' - ð'ð = (\bar{\rho}' - \rho')þ + (\rho - \bar{\rho})þ' + p(\rho\rho' - \sigma\sigma' + \Psi_2 - \Phi_{11} - \Pi)\\
- q(\bar{\rho}\bar{\rho}' - \bar{\sigma}\bar{\sigma}' + \bar{\Psi}_2 - \Phi_{11} - \Pi), \qquad (4.12.35)
\end{aligned}
$$

together with the remaining commutator equations obtained by applying the prime operation, complex conjugation, and both, to (4.12.34). Note that the type of $\eta$ enters explicitly on the right-hand side. We must be careful, when applying primes and bars to these equations, to remember that $\eta', \bar{\eta}$, and $\bar{\eta}'$ have types which are not quite those of $\eta$. Under the prime operation, $\{r', r; t', t\}$ becomes $\{r, r'; t, t'\}$, so $p$ becomes $-p$ and $q$ becomes $-q$; under conjugation, $\{r', r; t', t\}$ becomes $\{t', t; r', r\}$ – assuming $r'$ etc. to be real – so $p$ becomes $q$ and $q$ becomes $p$; under combined bar and prime operations, $\{r', r; t', t\}$ becomes $\{t, t'; r, r'\}$, so $p$ becomes $-q$ and $q$ becomes $-p$.

The commutator equations are the one instance where the compacted formalism yields more complicated formulae than the original spin-coefficient formalism. This seems to be the price paid for the very considerable formal simplification obtained for the other equations. But we must bear in mind that these commutators are actually combining information which comes from two different places in the spin-coefficient formalism. There is, moreover, a gain as regards geometric content of the commutators with the present formalism. The extra terms which arise when $p$ or $q$ is non-zero may sometimes be interpreted as curvature quantities connected with submanifolds in the space–time. We shall see this explicitly, for equation (4.12.35), in (4.14.20) below.

The full Bianchi identity (4.10.6) is somewhat complicated when written out in the original spin-coefficient formalism. It is considerably simpler in the compacted formalism and we give it only in this version. The translation to original spin-coefficients is then straightforward. We can obtain the full set of required equations rapidly by taking components of (4.10.6) and using (4.12.27). This yields the equations

$$\begin{aligned}&þ\Psi_1 - ð'\Psi_0 - þ\Phi_{01} + ð\Phi_{00}\\&\quad = -\tau'\Psi_0 + 4\rho\Psi_1 - 3\kappa\Psi_2\\&\qquad + \bar{\tau}'\Phi_{00} - 2\bar{\rho}\Phi_{01} - 2\sigma\Phi_{10} + 2\kappa\Phi_{11} + \bar{\kappa}\Phi_{02},\end{aligned} \tag{4.12.36}$$

$$\begin{aligned}&þ\Psi_2 - ð'\Psi_1 - ð'\Phi_{01} + þ'\Phi_{00} + 2þ\Pi\\&\quad = \sigma'\Psi_0 - 2\tau'\Psi_1 + 3\rho\Psi_2 - 2\kappa\Psi_3\\&\qquad + \bar{\rho}'\Phi_{00} - 2\bar{\tau}\Phi_{01} - 2\tau\Phi_{10} + 2\rho\Phi_{11} + \bar{\sigma}\Phi_{02},\end{aligned} \tag{4.12.37}$$

$$\begin{aligned}&þ\Psi_3 - ð'\Psi_2 - þ\Phi_{21} + ð\Phi_{20} - 2ð'\Pi\\&\quad = 2\sigma'\Psi_1 - 3\tau'\Psi_2 + 2\rho\Psi_3 - \kappa\Psi_4\\&\qquad - 2\rho'\Phi_{10} + 2\tau'\Phi_{11} + \bar{\tau}'\Phi_{20} - 2\bar{\rho}\Phi_{21} + \bar{\kappa}\Phi_{22},\end{aligned} \tag{4.12.38}$$

$$\begin{aligned}&þ\Psi_4 - ð'\Psi_3 - ð'\Phi_{21} + þ'\Phi_{20}\\&\quad = +3\sigma'\Psi_2 - 4\tau'\Psi_3 + \rho\Psi_4\\&\qquad - 2\kappa'\Phi_{10} + 2\sigma'\Phi_{11} + \bar{\rho}'\Phi_{20} - 2\bar{\tau}\Phi_{21} + \bar{\sigma}\Phi_{22}\end{aligned} \tag{4.12.39}$$

$$\begin{aligned}&þ\Phi_{11} + þ'\Phi_{00} - ð\Phi_{10} - ð'\Phi_{01} + 3þ\Pi\\&\quad = (\rho' + \bar{\rho}')\Phi_{00} + 2(\rho + \bar{\rho})\Phi_{11} - (\tau' + 2\bar{\tau})\Phi_{01}\\&\qquad - (2\tau + \bar{\tau}')\Phi_{10} - \bar{\kappa}\Phi_{12} - \kappa\Phi_{21} + \sigma\Phi_{20} + \bar{\sigma}\Phi_{02}\end{aligned} \tag{4.12.40}$$

$$\begin{aligned}&þ\Phi_{12} + þ'\Phi_{01} - ð\Phi_{11} - ð'\Phi_{02} + 3ð\Pi\\&\quad = (\rho' + 2\bar{\rho}')\Phi_{01} + (2\rho + \bar{\rho})\Phi_{12} - (\tau' + \bar{\tau})\Phi_{02}\\&\qquad - 2(\tau + \bar{\tau}')\Phi_{11} - \bar{\kappa}'\Phi_{00} - \kappa\Phi_{22} + \sigma\Phi_{21} + \bar{\sigma}'\Phi_{10}\end{aligned} \tag{4.12.41}$$

together with their primed versions. The last two equations, (4.12.40) and (4.12.41), are in fact the equivalent of the contracted Bianchi equations (4.10.8). (More generally, equations similar to (4.12.40) and (4.12.41) express the 'conservation' equation on an arbitrary symmetric two-index tensor.)

In space–times governed by Einstein's vacuum field equations, $\Pi$ and all $\Phi$s in equations (4.12.32)–(4.12.35) vanish; conversely, these equations with vanishing $\Pi$ and $\Phi$s characterize vacuum solutions of Einstein's field equations and can be used as a method of finding such solutions. The Bianchi identities (4.12.36)–(4.12.39) simplify considerably in this case, and nothing at all remains of (4.12.40) and (4.12.41). Equations (4.12.36)–(4.12.39) are now (with $\Pi = 0$ and $\Phi_{rs} = 0$) a particular case (4.10.9) of the zero rest-mass field equation (*cf.* (5.7.2) below):

$$\nabla^{AA'}\phi_{AB\ldots L} = 0, \tag{4.12.42}$$

with $\phi_{A\ldots L} = \phi_{(A\ldots L)}$. For, putting

$$\phi_r = \phi_{\underbrace{0\ldots0}_{n-r}\underbrace{1\ldots1}_{r}} = \mathrm{i}^{-n}\phi'_{n-r}: \{n-r, r; 0, 0\}, \quad (r = 0, \ldots, n), \tag{4.12.43}$$

we obtain, by (4.12.27),

$$\begin{aligned} \text{þ}\phi_r - \eth'\phi_{r-1} = (r-1)\sigma'\phi_{n-2} - r\tau'\phi_{r-1} + (n-r+1)\rho\phi_r \\ -(n-r)\kappa\phi_{r+1}, \quad (r = 1, \ldots, n), \end{aligned} \tag{4.12.44}$$

together with their primed versions. (The factor $\chi^{-1}\bar{\chi}$ in the definition (4.11.6) of the $\Psi_r$ makes no difference here, because of (4.12.23).) As we shall see in §5.1, Maxwell's free-space equations are also a special case of (4.12.42) and can therefore be written in the form (4.12.44) with $n = 2$. Of some interest also is the compacted spin-coefficient form of the (conformally invariant) wave equation $(\Box + R/6)\varphi = 0$ (*cf.* §6.8), which is obtained from

$$\Box := \nabla_a\nabla^a = 2(\text{þ}'\text{þ} - \eth'\eth - \bar{\rho}'\text{þ} - \rho\text{þ}' + \bar{\tau}\eth + \tau\eth') \text{ on } \{0, 0; 0, 0\}\text{-scalars} \tag{4.12.45}$$

and of the twistor equation (*cf.* §6.1) $\nabla^{(A}_{A'}\omega^{B)} = 0$ (with $\omega^0 = -\omega^A\iota_A$, $\omega^1 = \omega^A o_A$):

$$\begin{aligned} \kappa\omega^0 = \text{þ}\omega^1, \sigma\omega^0 = \eth\omega^1, \eth'\omega^0 = \sigma'\omega^1, \text{þ}'\omega^0 = \kappa'\omega^1, \\ \text{þ}\omega^0 + \rho\omega^0 = \eth'\omega^1 + \tau'\omega^1, \eth\omega^0 + \tau\omega^0 = \text{þ}'\omega^1 + \rho'\omega^1. \end{aligned} \tag{4.12.46}$$

Finally, we remark on the existence of an additional symmetry possessed by the spin-coefficient formalism which was noticed first by Sachs (1962) Consider the asterisk (*) operation

$$o^A \mapsto o^A, \iota^A \mapsto \iota^A, o^{A'} \mapsto \iota^{A'}, \iota^{A'} \mapsto -o^{A'}, \tag{4.12.47}$$

so that

$$(o^A)^* = o^A, (\iota^A)^* = \iota^A, (o^{A'})^* = \iota^{A'}, (\iota^{A'})^* = -o^{A'} \tag{4.12.48}$$

and

$$(l^a)^* = m^a, (m^a)^* = -l^a, (\bar{m}^a)^* = n^a, (n^a)^* = -\bar{m}^a \tag{4.12.49}$$

This operation preserves $\chi$ and $\bar{\chi}$:

$$\chi^* = \chi, \quad (\bar{\chi})^* = \bar{\chi}, \tag{4.12.50}$$

and, consequently, the tetrad 'orthogonality' relations (4.5.20). Clearly the asterisk operation does not commute with complex conjugation. However, we do have (for real $p$ and $q$)

$$(\eta^*)^* = (-1)^q\eta, \quad (\eta')^* = (-1)^q(\eta^*)', \quad \overline{\eta^*} = (-\mathrm{i})^{p+q}(\bar{\eta}')^*, \tag{4.12.51}$$

where $\eta$ is a scalar of type $\{p, q\}$. For an $\eta$ of type $\{r', r; t', t\}$, $\eta^*$ is of type $\{r', r; t, t'\}$.

From (4.5.21), (4.12.20), (4.11.6), (4.11.7), and (4.11.8), respectively, we obtain

$$\begin{aligned} &\kappa^* = \sigma, \quad \sigma^* = -\kappa, \quad \rho^* = \tau, \quad \tau^* = -\rho, \quad \kappa'^* = -\sigma', \\ &\quad \sigma'^* = \kappa', \quad \rho'^* = -\tau', \quad \tau'^* = \rho', \\ &\bar{\kappa}^* = -\bar{\sigma}', \quad \bar{\sigma}^* = -\bar{\kappa}', \quad \bar{\rho}^* = \bar{\tau}', \quad \bar{\tau}^* = \bar{\rho}', \quad \bar{\kappa}'^* = \bar{\sigma}, \\ &\quad \bar{\sigma}'^* = \bar{\kappa}, \quad \bar{\rho}'^* = -\bar{\tau}, \quad \bar{\tau}'^* = -\bar{\rho}, \end{aligned} \tag{4.12.52}$$

$$\text{þ}^* = \eth, \quad \eth^* = -\text{þ}, \quad \text{þ}'^* = -\eth', \quad \eth'^* = \text{þ}', \tag{4.12.53}$$

$$\begin{aligned} &\Psi_0{}^* = \Psi_0, \quad \Psi_1{}^* = \Psi_1, \quad \Psi_2{}^* = \Psi_2, \\ &\quad \Psi_3{}^* = \Psi_3, \quad \Psi_4{}^* = \Psi_4, \end{aligned}$$

$$\begin{aligned} &\bar{\Psi}_0{}^* = \bar{\Psi}_4, \quad \bar{\Psi}_1{}^* = -\bar{\Psi}_3, \quad \bar{\Psi}_2{}^* = \bar{\Psi}_2, \\ &\quad \bar{\Psi}_3{}^* = -\bar{\Psi}_1, \quad \bar{\Psi}_4{}^* = \bar{\Psi}_0, \end{aligned} \tag{4.12.54}$$

$$\begin{aligned} &\Phi_{00}{}^* = \Phi_{02}, \quad \Phi_{01}{}^* = -\Phi_{01}, \quad \Phi_{02}{}^* = \Phi_{00}, \quad \Phi_{10}{}^* = \Phi_{12}, \\ &\quad \Phi_{11}{}^* = -\Phi_{11}, \quad \Phi_{12}{}^* = \Phi_{10} \quad \Phi_{20}{}^* = \Phi_{22}, \\ &\quad \Phi_{21}{}^* = -\Phi_{21}, \quad \Phi_{22}{}^* = \Phi_{20}, \quad \Pi^* = \Pi \end{aligned} \tag{4.12.55}$$

Under the Sachs asterisk operation, the equations in our list (4.12.32) are permuted among themselves; so are those of the lists (4.12.36)–(4.12.39) and (4.12.40)–(4.12.41); and so also are the commutator equations (4.12.33)–(4.12.35). The Sachs operation, together with the prime operation, can be used to simplify the generation of equations; alternatively, it can provide a useful check on the correctness of equations obtained by other means.

## 4.13 Cartan's method

At the end of §4.5 we gave a method for computing the Christoffel symbols, spin-coefficients, etc., from the Infeld–van der Waerden symbols $g_{\mathbf{a}}{}^{\mathbf{AA'}}$, and in §4.11 and §4.12 we showed how the spin-coefficients could be used to calculate the curvature. But the formula (4.5.43) for calculating the spin-coefficients in terms of the $g_{\mathbf{a}}{}^{\mathbf{AA'}}$, though explicit, is somewhat inelegant and often complicated to use in practice. An alternative method that is sometimes useful (Newman and Penrose 1962) is to employ the commutator equations (4.11.3) or (4.11.11) for the intrinsic derivatives, as applied to the various metric-tensor components. That method relates the spin-coefficients to the metric-tensor component derivatives and it works well when the coordinates are aligned in some helpful way relative to a suitably positioned system of null tetrads. We shall here describe yet another method, which relates spinor techniques to Cartan's powerful calculus of differential forms and moving frames. There are often computational advantages in using such methods, in addition to their conceptual elegance.

We recall that in §4.3 we showed how the differential form calculus could be subsumed within the abstract index approach to tensors, provided that a suitable notational device for suppressing indices was adopted. Thus the abstract labels $i_1, i_2, i_3, \ldots$ (in this specific order and with $i_1 = I_1 I'_1, i_2 = I_2 I'_2$, etc.) may be canonically assigned to differential forms, but suppressed whenever the standard Cartan notation for forms is used. Thus a (complex) $p$-form $\boldsymbol{\phi}$ on $\mathcal{M}$ is a completely anti-symmetric tensor

$$\boldsymbol{\phi} = \phi_{i_1 \ldots i_p} \in \mathfrak{S}_{[i_1 \ldots i_p]} \tag{4.13.1}$$

(Clearly $\boldsymbol{\phi} = 0$ if $p \geqslant 5$.) We also use this notation for *tensor*- or, more generally, *spinor-valued p-forms* (*cf.* (4.3.11)). In fact, from now on, when we speak about a $p$-form we shall mean a spinor-valued $p$-form unless otherwise stated. The suppressed skew indices $i_1, \ldots, i_p$ are always considered as attached *first* to the kernel symbol, all other tensor or spinor indices being written to the right of $i_1, \ldots, i_p$ whenever these are not suppressed. Thus,

$$\boldsymbol{\phi}_{A \ldots F'}{}^{K \ldots L'} = \phi_{i_1 \ldots i_p A \ldots F'}{}^{K \ldots L'} \in \mathfrak{S}^{K \ldots L'}_{[i_1 \ldots i_p] A \ldots F'} \tag{4.13.2}$$

is a typical $p$-form. We shall often lump all indices other than the $i$s into a collective index $\mathscr{A}$ or $\mathscr{B}$ etc.; the first equation in (4.13.2) could be written as $\boldsymbol{\phi}^{\mathscr{A}} = \phi_{i_1 \ldots i_p}{}^{\mathscr{A}}$. The standard 'wedge' (exterior product) notation is used also for spinor-valued forms, with the following significance:

$$\boldsymbol{\phi}^{\mathscr{A}} \wedge \boldsymbol{\theta}^{\mathscr{B}} = \phi_{[i_1 \ldots i_p}{}^{\mathscr{A}} \theta_{i_{p+1} \ldots i_{p+q}]}{}^{\mathscr{B}}, \tag{4.13.3}$$

and we clearly have, if $\boldsymbol{\phi}^{\mathscr{A}}$ is a $p$-form and $\boldsymbol{\theta}^{\mathscr{B}}$ is a $q$-form,

$$\boldsymbol{\phi}^{\mathscr{A}} \wedge \boldsymbol{\theta}^{\mathscr{B}} = (-1)^{pq} \boldsymbol{\theta}^{\mathscr{B}} \wedge \boldsymbol{\phi}^{\mathscr{A}}. \tag{4.13.4}$$

Exterior covariant differentiation ('d-differentiation') is defined as follows:

$$\mathrm{d}\boldsymbol{\phi}^{\mathscr{A}} = \nabla_{[i_1} \phi_{i_2 \ldots i_{p+1}]}{}^{\mathscr{A}}, \tag{4.13.5}$$

and it yields a $(p+1)$-form when applied to a $p$-form. Note that 'd' (like the wedge) has a special relation to the index labels $i_1, \ldots, i_p$ which does not apply to the other labels. Note also that, in contrast to its action on ordinary (scalar-valued) $p$-forms (4.3.14), this operation in general is not independent of the connection, as is obvious, for example, in the case $p = 0$. If $\boldsymbol{\phi}^{\mathscr{A}}$ is a $p$-form, we have, as in (4.3.15) (vii),

$$\mathrm{d}(\boldsymbol{\phi}^{\mathscr{A}} \wedge \boldsymbol{\theta}^{\mathscr{B}}) = (\mathrm{d}\boldsymbol{\phi}^{\mathscr{A}}) \wedge \boldsymbol{\theta}^{\mathscr{B}} + (-1)^p \boldsymbol{\phi}^{\mathscr{A}} \wedge \mathrm{d}\boldsymbol{\theta}^{\mathscr{B}}. \tag{4.13.6}$$

Recalling (4.3.19), (4.3.20), we note that if local coordinates $x^0, \ldots, x^3$ are introduced, the various (dual) basis forms can be written

$$\begin{aligned} &\mathrm{d}x^{\mathbf{a}} = g_{i_1}{}^{\mathbf{a}}, \quad \mathrm{d}x^{\mathbf{a}} \wedge \mathrm{d}x^{\mathbf{b}} = g_{[i_1}{}^{\mathbf{a}} g_{i_2]}{}^{\mathbf{b}}, \\ &\mathrm{d}x^{\mathbf{a}} \wedge \mathrm{d}x^{\mathbf{b}} \wedge \mathrm{d}x^{\mathbf{c}} = g_{[i_1}{}^{\mathbf{a}} g_{i_2}{}^{\mathbf{b}} g_{i_3]}{}^{\mathbf{c}}, \\ &\mathrm{d}x^0 \wedge \mathrm{d}x^1 \wedge \mathrm{d}x^2 \wedge \mathrm{d}x^3 = g_{[i_1}{}^0 g_{i_2}{}^1 g_{i_3}{}^2 g_{i_4]}{}^3, \end{aligned} \tag{4.13.7}$$

so that

$$\boldsymbol{\phi}^{\mathscr{A}} = \phi_{\mathbf{i}_1 \ldots \mathbf{i}_p}{}^{\mathscr{A}} \mathrm{d}x^{\mathbf{i}_1} \wedge \cdots \wedge \mathrm{d}x^{\mathbf{i}_p}, \tag{4.13.8}$$

where the components on the *i*s are taken in the coordinate basis.

Repeated d-differentiation does not generally yield zero (as in the case of scalar-valued forms); we have, for example, for a vector-valued $p$-form $\boldsymbol{V}^a$,

$$\begin{aligned} \mathrm{d}^2 \boldsymbol{V}^a &= \nabla_{[i_1} \nabla_{i_2} V_{i_3 \ldots i_{p+2}]}{}^a = \tfrac{1}{2} R_{[i_1 i_2 | b}{}^a V_{| i_3 \ldots i_{p+2}]}{}^b \\ &= \boldsymbol{\Omega}_b{}^a \wedge \boldsymbol{V}^b = \boldsymbol{V}^b \wedge \boldsymbol{\Omega}_b{}^a, \end{aligned} \tag{4.13.9}$$

by (4.2.33) and (4.2.37), where we have adopted the conventional notation* for the *curvature 2-form*

$$\boldsymbol{\Omega}_a{}^b := \tfrac{1}{2} R_{i_1 i_2 a}{}^b. \tag{4.13.10}$$

From (4.13.10) and the constancy of the εs we derive (as in §4.9), for a spin-vector valued $p$-form $\boldsymbol{\xi}^A$,

$$\mathrm{d}^2 \boldsymbol{\xi}^A = \nabla_{[i_1} \nabla_{i_2} \xi_{i_3 \ldots i_{p+2}]}{}^A = \boldsymbol{\xi}^B \wedge \boldsymbol{\Omega}_B{}^A, \tag{4.13.11}$$

where the 2-form $\boldsymbol{\Omega}_B{}^A$ is defined by

$$\boldsymbol{\Omega}_B{}^A := \tfrac{1}{2} \boldsymbol{\Omega}_{BC'}{}^{AC'}. \tag{4.13.12}$$

* However, we adopt unconventional 'staggering', for compatibility with the rest of our notation.

Clearly we have

$$\boldsymbol{\Omega}_b{}^a = \boldsymbol{\Omega}_B{}^A \varepsilon_{B'}{}^{A'} + \bar{\boldsymbol{\Omega}}_{B'}{}^{A'} \varepsilon_B{}^A \tag{4.13.13}$$

and

$$\boldsymbol{\Omega}_A{}^A = 0, \quad \text{i.e., } \boldsymbol{\Omega}_{AB} = \boldsymbol{\Omega}_{BA}. \tag{4.13.14}$$

The equations (4.13.9) and (4.13.11) can be generalized in the standard way to forms of higher spinor valence, e.g.,

$$\mathrm{d}^2\boldsymbol{\eta}^{A'}{}_{Bc} = \boldsymbol{\eta}^{E'}{}_{Bc} \wedge \bar{\boldsymbol{\Omega}}_{E'}{}^{A'} - \boldsymbol{\eta}^{A'}{}_{Ec} \wedge \boldsymbol{\Omega}_B{}^E - \boldsymbol{\eta}^{A'}{}_{Be} \wedge \boldsymbol{\Omega}_c{}^e. \tag{4.13.15}$$

We recall the dualizing operations (3.4.21), (3.4.29), (3.4.30), and define the following duals for $p$-forms:

$$\begin{aligned}
&\boldsymbol{\chi}^{\mathscr{A}}(\text{0-form}): && {}^{\odot}\boldsymbol{\chi}^{\mathscr{A}} = \mathrm{e}_{i_1i_2i_3i_4}\chi^{\mathscr{A}}(\text{4-form})\\
&\boldsymbol{\psi}^{\mathscr{A}}(\text{1-form}): && {}^{\dagger}\boldsymbol{\psi}^{\mathscr{A}} = \mathrm{e}_{i_1i_2i_3}{}^{j}\psi_j{}^{\mathscr{A}}(\text{3-form})\\
&\boldsymbol{\varphi}^{\mathscr{A}}(\text{2-form}): && {}^{*}\boldsymbol{\varphi}^{\mathscr{A}} = \tfrac{1}{2}\mathrm{e}_{i_1i_2}{}^{jk}\varphi_{jk}{}^{\mathscr{A}}(\text{2-form})\\
&\boldsymbol{\theta}^{\mathscr{A}}(\text{3-form}): && {}^{\ddagger}\boldsymbol{\theta}^{\mathscr{A}} = \tfrac{1}{6}\mathrm{e}_{i_1}{}^{jkl}\theta_{jkl}{}^{\mathscr{A}}(\text{1-form})\\
&\boldsymbol{\mu}^{\mathscr{A}}(\text{4-form}): && {}^{\star}\boldsymbol{\mu}^{\mathscr{A}} = \tfrac{1}{24}\mathrm{e}^{jklm}\mu_{jklm}{}^{\mathscr{A}}(\text{0-form})
\end{aligned} \tag{4.13.16}$$

Note that twice repeated dualization of a $p$-form leads back to the original form times a factor $(-1)^{p+1}$ (*cf.* (3.4.24), (3.4.31), (4.6.11)). Now, by (4.6.2) and (4.6.34),

$$\begin{aligned}
\boldsymbol{\Omega}_{AB} + \mathrm{i}{}^*\boldsymbol{\Omega}_{AB} &= \varepsilon_{I_1'I_2'}\mathrm{X}_{I_1I_2AB}\\
&= \varepsilon_{I_1'I_2'}\{\Psi_{I_1I_2AB} + \Lambda(\varepsilon_{I_1A}\varepsilon_{I_2B} + \varepsilon_{I_1B}\varepsilon_{I_2A})\}
\end{aligned} \tag{4.13.17}$$

and

$$\boldsymbol{\Omega}_{AB} - \mathrm{i}{}^*\boldsymbol{\Omega}_{AB} = \varepsilon_{I_1I_2}\Phi_{I_1'I_2'AB}. \tag{4.13.18}$$

Hence the various spinor curvatures are simply related to the $\boldsymbol{\Omega}_{AB}$.

Suppose, next, that some tensor basis $g_{\mathbf{b}}{}^a$ is set up, which is not necessarily the coordinate basis used in (4.13.7). The dual basis $g_a{}^{\mathbf{b}}$ constitutes a system of 1-forms; to avoid possible confusion we conform to the standard notation

$$\boldsymbol{\theta}^{\mathbf{a}} := g_{i_1}{}^{\mathbf{a}} \tag{4.13.19}$$

for these 1-forms. Their d-derivatives are

$$\mathrm{d}\boldsymbol{\theta}^{\mathbf{a}} = \nabla_{[i_1}g_{i_2]}{}^{\mathbf{a}} = -\Gamma_{\mathbf{bc}}{}^{\mathbf{a}}g_{[i_1}{}^{\mathbf{b}}g_{i_2]}{}^{\mathbf{c}} = -\Gamma_{\mathbf{bc}}{}^{\mathbf{a}}\boldsymbol{\theta}^{\mathbf{b}} \wedge \boldsymbol{\theta}^{\mathbf{c}}, \tag{4.13.20}$$

with $\Gamma_{\mathbf{bc}}{}^{\mathbf{a}}$ given by (4.2.60). The $\boldsymbol{\theta}$s anti-commute, by (4.13.6), and so the equation

$$\mathrm{d}\boldsymbol{\theta}^{\mathbf{a}} + \Gamma_{[\mathbf{bc}]}{}^{\mathbf{a}}\boldsymbol{\theta}^{\mathbf{b}} \wedge \boldsymbol{\theta}^{\mathbf{c}} = 0 \tag{4.13.21}$$

serves to define the quantities $\Gamma_{[\mathbf{bc}]}{}^{\mathbf{a}}$. The following notation for the *connection 1-forms*,

$$\boldsymbol{\omega}_{\mathbf{a}}{}^{\mathbf{b}} := \Gamma_{i_1\mathbf{a}}{}^{\mathbf{b}} = \Gamma_{\mathbf{ca}}{}^{\mathbf{b}}\boldsymbol{\theta}^{\mathbf{c}}, \tag{4.13.22}$$

is standard (again, apart from the staggering – *cf.* footnote on p. 263), and so we may re-express (4.13.20) as

$$\mathrm{d}\boldsymbol{\theta}^{\mathbf{a}} = \boldsymbol{\theta}^{\mathbf{b}} \wedge \omega_{\mathbf{b}}{}^{\mathbf{a}}. \qquad (4.13.23)$$

From $\nabla_a g_{bc} = 0$ we derive

$$\nabla_{\mathbf{a}} g_{\mathbf{bc}} = \Gamma_{\mathbf{ab}}{}^{\mathbf{e}} g_{\mathbf{ec}} + \Gamma_{\mathbf{ac}}{}^{\mathbf{e}} g_{\mathbf{be}} = 2\Gamma_{\mathbf{a(bc)}}. \qquad (4.13.24)$$

Hence if we assume some *fixed* normalization for the tensor basis (4.13.19), i.e., $g_{\mathbf{ab}}$ = constant, so that

$$\mathrm{d}g_{\mathbf{ab}} = 0, \qquad (4.13.25)$$

then we obtain

$$\Gamma_{\mathbf{abc}} = -\Gamma_{\mathbf{acb}}. \qquad (4.13.26)$$

(This is Cartan's 'moving frame' condition, or, equivalently, the condition for the $\Gamma$s to be Ricci rotation coefficients. Compare also (4.4.36) and (4.5.5).) In this case we can use (4.13.21) to solve for the $\Gamma_{\mathbf{abc}}$ since, by virtue of (4.13.26),

$$\Gamma_{\mathbf{abc}} = \Gamma_{[\mathbf{ab}]\mathbf{c}} - \Gamma_{[\mathbf{bc}]\mathbf{a}} - \Gamma_{[\mathbf{ac}]\mathbf{b}}. \qquad (4.13.27)$$

In practical computations it is often simpler not to plug the solutions of (4.13.21) into (4.13.27), but rather to guess or otherwise obtain a set of $\Gamma_{\mathbf{abc}}$ which simultaneously satisfy (4.13.21) and (4.13.26). We shall see the spin-coefficient counterpart of this in a moment. To compute curvature, notice first that

$$\begin{aligned} \mathrm{d}g_{\mathbf{a}}{}^{a} &= \nabla_{i_1} g_{\mathbf{a}}{}^{a} = \Gamma_{i_1\mathbf{a}}{}^{a} \\ &= \omega_{\mathbf{a}}{}^{a}, \end{aligned} \qquad (4.13.28)$$

and that, by (4.13.9) (with $p = 0$) and (4.13.6),

$$\begin{aligned} \boldsymbol{\Omega}_{\mathbf{a}}{}^{a} &= g_{\mathbf{a}}{}^{b}\boldsymbol{\Omega}_{b}{}^{a} = \mathrm{d}^2 g_{\mathbf{a}}{}^{b} = \mathrm{d}\omega_{\mathbf{a}}{}^{a} = \mathrm{d}(\boldsymbol{\omega}_{\mathbf{a}}{}^{\mathbf{b}} g_{\mathbf{b}}{}^{a}) \\ &= g_{\mathbf{b}}{}^{a}\mathrm{d}\omega_{\mathbf{a}}{}^{\mathbf{b}} - \omega_{\mathbf{a}}{}^{\mathbf{b}} \wedge \mathrm{d}g_{\mathbf{b}}{}^{a}. \end{aligned} \qquad (4.13.29)$$

Thus the curvature 2-form components can be calculated from

$$\boldsymbol{\Omega}_{\mathbf{a}}{}^{\mathbf{c}} = \mathrm{d}\omega_{\mathbf{a}}{}^{\mathbf{c}} - \omega_{\mathbf{a}}{}^{\mathbf{b}} \wedge \omega_{\mathbf{b}}{}^{\mathbf{c}} \qquad (4.13.30)$$

(which can be related directly to (4.2.67), the non-standard sign in (4.13.30) arising because of our non-standard staggering.)

*Relation to spin-coefficients*

Next we shall consider the relation of the Cartan calculus to spin-coefficients. The tensor basis $g_a{}^{\mathbf{a}}$ will now be regarded as that arising from a dyad $\varepsilon_A{}^{\mathbf{A}}$, which we assume for simplicity to be normalized ($\chi = 1$). Thus, in the above, from (4.13.19) onwards, we can replace each tetrad index **a**, **b**, etc., by a corresponding pair of dyad indices **AA′**, **BB′**, etc. Accordingly we have basis 1-forms

$$\boldsymbol{\theta}^{\mathbf{AA'}} = g_{i_1}{}^{\mathbf{AA'}}, \tag{4.13.31}$$

which we shall label

$$\boldsymbol{\theta}^{00'} = \boldsymbol{n}, \quad \boldsymbol{\theta}^{01'} = -\bar{\boldsymbol{m}}, \quad \boldsymbol{\theta}^{10'} = -\boldsymbol{m}, \quad \boldsymbol{\theta}^{11'} = \boldsymbol{l} \tag{4.13.32}$$

(for consistency with the standard null tetrad notation). The systematic use of dyad indices throughout now frees the symbols **a**, **b**, ... for use as coordinate indices if desired. Then (4.13.31) can be re-expressed in the form (with coordinate tensor basis)

$$\boldsymbol{\theta}^{\mathbf{AA'}} = g_{\mathbf{a}}{}^{\mathbf{AA'}} \mathrm{d}x^{\mathbf{a}} \tag{4.13.33}$$

which shows that the components of $\boldsymbol{\theta}^{AA'}$ are just the Infeld–van der Waerden symbols (*cf.* (3.1.37)).

Now the metric $\mathrm{d}s^2 = g_{\mathbf{ab}}\mathrm{d}x^{\mathbf{a}}\mathrm{d}x^{\mathbf{b}}$, in 'old-fashioned' coordinate differentials, can be re-expressed, using (3.1.45), in terms of the Infeld-van der Waerden symbols:

$$\begin{aligned} \mathrm{d}s^2 &= \varepsilon_{\mathbf{AB}}\varepsilon_{\mathbf{A'B'}} g_{\mathbf{a}}{}^{\mathbf{AA'}} g_{\mathbf{b}}{}^{\mathbf{BB'}} \mathrm{d}x^{\mathbf{a}}\mathrm{d}x^{\mathbf{b}} \\ &= 2(g_{\mathbf{a}}{}^{00'}\mathrm{d}x^{\mathbf{a}})(g_{\mathbf{b}}{}^{11'}\mathrm{d}x^{\mathbf{b}}) - 2(g_{\mathbf{a}}{}^{01'}\mathrm{d}x^{\mathbf{a}})(g_{\mathbf{b}}{}^{10'}\mathrm{d}x^{\mathbf{b}}) \end{aligned} \tag{4.13.34}$$

which, by reference to (4.13.32), can be written as

$$\mathrm{d}s^2 = 2\boldsymbol{ln} - 2\boldsymbol{m}\bar{\boldsymbol{m}}, \tag{4.13.35}$$

where $\boldsymbol{ln}$ stands for $l_{(i_1}n_{i_2)}$ etc., and $\mathrm{d}s^2$ for $g_{i_1 i_2}$.

To calculate the spin-coefficients $\gamma_{\mathbf{AA'B}}{}^{\mathbf{C}}$, we can look for a (necessarily unique) solution of (4.13.20) in dyad index form,

$$\mathrm{d}\boldsymbol{\theta}^{\mathbf{AA'}} + \Gamma_{\mathbf{BB'CC'}}{}^{\mathbf{AA'}} \boldsymbol{\theta}^{\mathbf{BB'}} \wedge \boldsymbol{\theta}^{\mathbf{CC'}} = 0. \tag{4.13.36}$$

For, from (4.5.37), specialized to the case when the tensor basis arises from the dyad and when consequently $g_{\mathbf{b}}{}^{\mathbf{BB'}} = $ constant, we have

$$\Gamma_{\mathbf{BB'CC'}}{}^{\mathbf{AA'}} = \gamma_{\mathbf{BB'C}}{}^{\mathbf{A}} \varepsilon_{\mathbf{C'}}{}^{\mathbf{A'}} + \bar{\gamma}_{\mathbf{BB'C'}}{}^{\mathbf{A'}} \varepsilon_{\mathbf{C}}{}^{\mathbf{A}}, \tag{4.13.37}$$

and from (4.5.5) we have

$$\gamma_{\mathbf{BB'AC}} = \gamma_{\mathbf{BB'CA}}, \tag{4.13.38}$$

which implies that $\Gamma_{\mathbf{abc}}$ has the required anti-symmetry (4.13.26). We can re-express (4.13.36), (4.13.37) in the form*

$$\frac{\partial}{\partial x^{[\mathbf{a}}} g_{\mathbf{b}]}{}^{\mathbf{CC'}} = -(\gamma_{\mathbf{AA'B}}{}^{\mathbf{C}} \varepsilon_{\mathbf{B'}}{}^{\mathbf{C'}} + \bar{\gamma}_{\mathbf{AA'B'}}{}^{\mathbf{C'}} \varepsilon_{\mathbf{B}}{}^{\mathbf{C}}) g_{[\mathbf{a}}{}^{\mathbf{AA'}} g_{\mathbf{b}]}{}^{\mathbf{BB'}} \tag{4.13.39}$$

* We remind the reader that a *numerical* index pair **AA′** is not to be equated with **a**.

which, with (4.13.38), is the equation whose solution (for the $\gamma$'s) is to be 'guessed', or systematically found by use of (4.13.27) or (4.5.43).

Alternatively, if we write

$$\omega_{\mathbf{B}}{}^{\mathbf{C}} = \gamma_{i_1 \mathbf{B}}{}^{\mathbf{C}} = \gamma_{\mathbf{AA'B}}{}^{\mathbf{C}} \theta^{\mathbf{AA'}}, \tag{4.13.40}$$

so that

$$\omega_{\mathbf{BB'}}{}^{\mathbf{CC'}} = \omega_{\mathbf{B}}{}^{\mathbf{C}} \varepsilon_{\mathbf{B'}}{}^{\mathbf{C'}} + \bar{\omega}_{\mathbf{B'}}{}^{\mathbf{C'}} \varepsilon_{\mathbf{B}}{}^{\mathbf{C}}, \tag{4.13.41}$$

and

$$\omega_{\mathbf{B}}{}^{\mathbf{B}} = 0, \tag{4.13.42}$$

we can re-express (4.13.36) in the form

$$\mathrm{d}\theta^{\mathbf{AA'}} = -(\omega_{\mathbf{B}}{}^{\mathbf{A}} \wedge \theta^{\mathbf{BA'}} + \bar{\omega}_{\mathbf{B'}}{}^{\mathbf{A'}} \wedge \theta^{\mathbf{AB'}}). \tag{4.13.43}$$

Taking components in this equation, we obtain

$$\begin{aligned}
\mathrm{d}\boldsymbol{l} = -\{&\boldsymbol{l} \wedge \boldsymbol{m}(\beta' + \bar{\alpha}' + \bar{\tau}) + \boldsymbol{l} \wedge \bar{\boldsymbol{m}}(\alpha' + \tau + \bar{\beta}') \\
&+ \boldsymbol{l} \wedge \boldsymbol{n}(-\gamma' - \bar{\gamma}') + \boldsymbol{m} \wedge \bar{\boldsymbol{m}}(\bar{\rho} - \rho) - \boldsymbol{m} \wedge \boldsymbol{n}\bar{\kappa} - \bar{\boldsymbol{m}} \wedge \boldsymbol{n}\kappa\} \\
\mathrm{d}\boldsymbol{m} = \{&\boldsymbol{l} \wedge \boldsymbol{m}(-\varepsilon' - \bar{\gamma} - \bar{\rho}') - \boldsymbol{l} \wedge \bar{\boldsymbol{m}}\bar{\sigma}' + \boldsymbol{l} \wedge \boldsymbol{n}(\bar{\tau}' - \tau) \\
&+ \boldsymbol{m} \wedge \bar{\boldsymbol{m}}(-\alpha' - \bar{\alpha}) + \boldsymbol{m} \wedge \boldsymbol{n}(\rho + \gamma' + \bar{\varepsilon}) + \bar{\boldsymbol{m}} \wedge \boldsymbol{n}\sigma\} \\
\mathrm{d}\boldsymbol{n} = -\{&\boldsymbol{l} \wedge \boldsymbol{m}\kappa' + \boldsymbol{l} \wedge \bar{\boldsymbol{m}}\bar{\kappa}' + \boldsymbol{l} \wedge \boldsymbol{n}(\gamma + \bar{\gamma}) + \boldsymbol{m} \wedge \bar{\boldsymbol{m}}(\rho' - \bar{\rho}') \\
&+ \boldsymbol{m} \wedge \boldsymbol{n}(-\alpha - \tau' - \bar{\beta}) + \bar{\boldsymbol{m}} \wedge \boldsymbol{n}(-\bar{\alpha} - \bar{\tau}' - \beta)\}
\end{aligned} \tag{4.13.44}$$

as the equations to be solved (necessarily uniquely) for the required spin-coefficients. Note that, since we have assumed the normalization $\chi = 1$, the spin-coefficients in fact satisfy $\beta' = -\alpha$, $\gamma' = -\varepsilon$, *cf.* (4.5.29). Note also that the first and last of equations (4.13.44) are transforms of each other under the prime operation (*cf.* (4.5.18)), and that the middle equation transforms into itself under the combined bar and prime operation.

To summarize our last method: if we are given a metric $g_{\mathbf{ab}}\mathrm{d}x^{\mathbf{a}}\mathrm{d}x^{\mathbf{b}}$, we first express it in the form (4.13.35) – which, of course, can always be done in a variety of ways, some more, some less convenient; then we express $\mathrm{d}\boldsymbol{l}$, $\mathrm{d}\boldsymbol{m}$, $\mathrm{d}\boldsymbol{n}$ as linear combinations of $\boldsymbol{l} \wedge \boldsymbol{m}$, $\boldsymbol{l} \wedge \bar{\boldsymbol{m}}$, etc., and, by comparing coefficients with (4.13.44), we find the spin-coefficients. Once the spin-coefficients are known, the curvature spinors can be calculated from (4.13.30), (4.13.17), and (4.13.18), if desired, although at this stage it is probably more economical to use (4.11.12) or (4.12.32).

## 4.14 Applications to 2-surfaces

The compacted spin-coefficient formalism introduced in §4.12 is particularly useful in the study of *spacelike 2-surfaces*. One reason for this is that

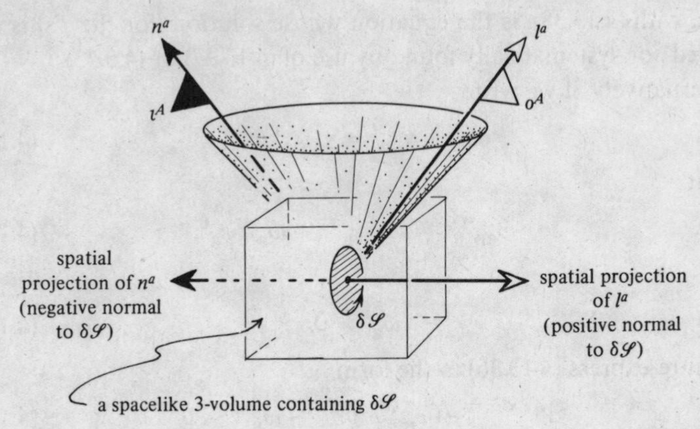


Fig. 4-1. The geometrical relation between the spin-frame and the oriented surface element $\delta\mathscr{S}$.

any spacelike 2-surface element $\delta\mathscr{S}$ at a point uniquely defines two null directions at that point, namely those orthogonal to $\delta\mathscr{S}$, whereas $\delta\mathscr{S}$ does not single out any particular null vectors along these null directions. Also, the assignment of an *orientation* to this element (assuming the usual space and time orientations of $\mathscr{M}$ as given) is equivalent to providing an *ordering* of these two null directions. We take the flagpole directions of $o^A$ and of $\iota^A$, in that order, to correspond to the oriented element $\delta\mathscr{S}$, such that, in the standard correspondence of Chapter 3 ((3.1.20), (3.1.21)), $\delta\mathscr{S}$ corresponds to the local $(X^a, Y^a)$-plane* with the usual orientation. Thus the spatial projection of $l^a$ (with respect to the time-axis direction $l^a + n^a$) provides a *positive* 3-space normal direction to $\delta\mathscr{S}$ (namely that of $Z^a$, i.e., of $l^a - n^a$), while the spatial projection of $n^a$ provides a *negative* 3-space normal direction (namely that of $-Z^a$, i.e., of $n^a - l^a$). (See Fig. 4–1). A normalization of $(o^A, \iota^A)$ to a spin-frame plays no role in this. The 'gauge' transformation $o^A \mapsto \lambda o^A$, $\iota^A \mapsto \mu \iota^A$ clearly leaves the oriented element $\delta\mathscr{S}$ invariant (though it affects the choice of spatial projection). But application of the *prime* operation (*cf.* after (4.5.17)) *reverses* the orientation of $\delta\mathscr{S}$.

In fact, one may also apply the compacted spin-coefficient formalism equally to the study of *timelike* 2-surfaces. Any timelike 2-surface element $\delta\mathscr{S}^*$ defines two null directions in an even more obvious way, namely those two null directions in which it cuts the null cone. Here $\delta\mathscr{S}^*$ is the plane *spanned* by $l^a$ and $n^a$, while in the previous case $\delta\mathscr{S}$ was the ortho-

* For later convenience we here write $X^a, Y^a, Z^a, T^a$ for the $x^a, y^a, z^a, t^a$ of (3.1.20).

gonal complement of this plane, i.e., the (real) plane spanned by $m^a$ and $\bar{m}^a$. We shall not give a detailed discussion of the timelike case, since most useful applications seem to involve only spacelike 2-surfaces. However, it is straightforward to obtain the relevant local formulae for the timelike case from the corresponding spacelike ones, since the Sachs *-operation (*cf.* (4.12.47)–(4.12.55)) can be applied, interchanging $(l^a, n^a)$ with $\pm(m^a, -\bar{m}^a)$. Only certain *global* results for spacelike 2-surfaces (e.g., such as we shall see arising in the study of spherical harmonics) have no analogue for timelike 2-surfaces. As a simple example, a timelike 2-surface cannot have $S^2$-topology. (The existence of a uniquely defined null-direction pair at each point implies that the topology of any *compact* timelike 2-surface must be that of a torus or, if non-oriented, that of a Klein bottle.)

One of the special advantages of the compacted form of the spin-coefficient formalism over the original one is, indeed, that it applies *globally* to an arbitrary spacelike 2-surface $\mathscr{S}$, *whatever its topology*. If we were to use the original formalism of §4.11, then we should have to select a particular tangent direction at each point of $\mathscr{S}$, corresponding to $\mathrm{Re}(m^a) = 2^{-\frac{1}{2}} X^a$ (which, with $\delta\mathscr{S}$, would then also fix $\mathrm{Im}(m^a) = -2^{-\frac{1}{2}} Y^a$). If $\mathscr{S}$ had $S^2$-topology, for example, this could not be done continuously over the whole of $\mathscr{S}$ and there would arise some 'singular' places at which the description breaks down. (This, of course, is a difficulty inherent in any moving-frame or coordinate description and is not specific to the spin-coefficient method.)

In the compacted formalism, no actual choice of $m^a$-vectors need be made and so this difficulty does not arise. The point is, perhaps, a slightly delicate one. For it might be argued that although the compacted formalism is invariant with respect to phase change in the $m^a$-vectors, nevertheless *some* choice has to be made, and since each choice has to break down somewhere on $\mathscr{S}$ the problem has not been avoided. But this objection is inappropriate. One may envisage covering $\mathscr{S}$ with open sets in each of which a smooth choice of $m^a$-vectors is made. Since the formalism is invariant under the transformations which take place in the overlap regions, it will not even 'notice' that such transformations have taken place. This idea can be made mathematically more precise in the language of fibre bundles (*cf.* §5.4 below), but we need not elaborate on it here. Of course, when it comes to making explicit coordinate representations of the $m^a$-vectors, such choices must actually be made, subject to the topological structure of $\mathscr{S}$. But such explicit descriptions are not really part of the 'pure' compacted formalism since they break the stated invariance requirements.* We shall see in the next section (in relation to spherical harmonics) how the explicit descriptions may only hold locally, while the general formalism applies globally.

Of particular significance, in the context of spacelike 2-surfaces, is the commutator (4.12.35):

$$\begin{aligned}\eth\eth' - \eth'\eth = (\bar{\rho}' - \rho')\text{þ} + (\rho - \bar{\rho})\text{þ}' + p(\rho\rho' - \sigma\sigma' + \Psi_2 - \Phi_{11} - \Pi) \\ - q(\bar{\rho}\bar{\rho}' - \bar{\sigma}\bar{\sigma}' + \bar{\Psi}_2 - \Phi_{11} - \Pi)\end{aligned} \tag{4.14.1}$$

acting on a (scalar) quantity of type $\{r', r; t', t\}$, where $p = r' - r$, $q = t' - t$ (*cf.* after (4.12.9)). To apply this formula we need to envisage that the dyad $(o^A, \iota^A)$ is extended to the space–time in some neighbourhood of $\mathscr{S}$ and is not just defined on $\mathscr{S}$ – or rather, in view of the above remarks, that the family of 2-surface elements $\delta\mathscr{S}$ is extended to a field of such surface elements in a neighbourhood of $\mathscr{S}$. But notice that only the operators $\eth$ and $\eth'$ occur on the left side of (4.14.1), these being the derivatives that act entirely *within* $\mathscr{S}$. By virtue of this, (4.14.1) applies to quantities defined only at the points of $\mathscr{S}$ and its effect will actually be independent of the way in which $(o^A, \iota^A)$, or the $\delta\mathscr{S}$-field, is extended outside $\mathscr{S}$. But the operators þ and þ′ appearing on the right side of (4.14.1) do *not* act within $\mathscr{S}$ and their effects *do* depend on the extensions outside $\mathscr{S}$. Indeed, since þ and þ′ act in independent directions (whose span meets $\delta\mathscr{S}$ only in zero), it follows that their coefficients $\bar{\rho}' - \rho'$ and $\rho - \bar{\rho}$ must independently *vanish*. Thus we have:

(4.14.2) PROPOSITION

*If the null vectors $l^a$ and $n^a$ are orthogonal to a spacelike 2-surface $\mathscr{S}$, then $\rho$ and $\rho'$ are both real at $\mathscr{S}$.*

(In Chapter 7 (*cf.* (7.1.48), (7.1.58), (7.1.60)) we shall see the geometrical significance of this result and effectively reobtain it in a more geometrical way.) Applying the Sachs *-operation to the above argument we also incidentally obtain:

(4.14.3) PROPOSITION

*If the null vectors $l^a$ and $n^a$ are tangent to a timelike 2-surface $\mathscr{S}^*$, then $\bar{\tau} = \tau'$ at $\mathscr{S}^*$.*

* In fibre-bundle terminology, the selection of an explicit choice of $m^a$-vectors corresponds to finding a (local) cross-section of a bundle, whereas the global applicability of the formalism as a whole corresponds only to choosing the bundle itself, together with its bundle connection.

*Intrinsic quantities on $\mathscr{S}$*

In order to explain the significance of the other terms on the right side of (4.14.1), we show first that when acting on quantities of zero boost weight, so that

$$p = -q, \tag{4.14.4}$$

the operators ð and ð′ provide, in effect, the two components of *covariant derivative within $\mathscr{S}$*. For simplicity we shall assume in this work, until explicitly stated otherwise, that $(o^A, \iota^A)$ is a spin-frame, i.e., that

$$\chi = 1. \tag{4.14.5}$$

Then the tensor

$$S_a{}^b = -m_a\bar{m}^b - \bar{m}_a m^b, \tag{4.14.6}$$

which satisfies

$$S_a{}^b S_b{}^c = S_a{}^c = \bar{S}_a{}^c = S^c{}_a, \tag{4.14.7}$$

$$S_a{}^b m_b = m_a,\ S_a{}^b \bar{m}_b = \bar{m}_a, \tag{4.14.8}$$

and

$$S_a{}^b l_b = S_a{}^b n_b = 0, \tag{4.14.9}$$

acts as a *projection operator* into the space tangent to $\mathscr{S}$ at each point of $\mathscr{S}$, while $S_{ab}$ acts as the negative definite metric tensor intrinsic to $\mathscr{S}$. If $V^a$ is any vector at a point of $\mathscr{S}$, then

$$V^a S_a{}^b \tag{4.14.10}$$

is its projection into $\mathscr{S}$, being equal to $V^a$ if and only if $V^a$ is tangent to $\mathscr{S}$. If $V^a$ is tangent to $\mathscr{S}$, its covariant derivative in $\mathscr{S}$ is given by the projection of $\nabla_b V^a$ into $\mathscr{S}$, i.e., by

$$S_{a_0}{}^a S_b{}^{b_0} \nabla_{b_0} V^{a_0} =: \Delta_b V^a \tag{4.14.11}$$

(in fact, this may be verified directly by referring to the defining properties for covariant derivative: (4.2.2) and (4.2.3), vanishing torsion and (4.3.46), as applied to tangent vectors to $\mathscr{S}$).

The components of (4.14.11) with respect to $m^a$ and $\bar{m}^a$ are

$$m_a ð V^a,\ \bar{m}_a ð V^a,\ m_a ð' V^a,\ \bar{m}_a ð' V^a, \tag{4.14.12}$$

taking $V^a$ to have type $\{p, q\} = \{0, 0\}$, and noting from (4.12.15) that then $\delta = ð$ and $\delta' = ð'$. In each case we can commute the $m_b$ or $\bar{m}_b$ with the ð or ð′ in (4.14.12), because from (4.12.28) we obtain

$$ð m_a = -\sigma n_a - \bar{\sigma}' l_a,\ ð \bar{m}_a = -\bar{\rho} n_a - \rho' l_a \tag{4.14.13}$$

and their complex conjugates, while

$$n_a V^a = 0 = l_a V^a, \tag{4.14.14}$$

since $V^a$ is tangential to $\mathscr{S}$. The components (4.14.12) of $\Delta_b V^a$ are therefore, respectively,

$$\eth\eta, \eth\tilde{\eta}, \eth'\eta, \eth'\tilde{\eta}, \tag{4.14.15}$$

where the type $\pm\{1, -1\}$ quantities $\eta$ and $\tilde{\eta}$ are defined by

$$\eta = V^a m_a, \quad \tilde{\eta} = V^a \bar{m}_a. \tag{4.14.16}$$

(For real $V^a$, $\tilde{\eta} = \bar{\eta}$.)

In a similar way we can express the components of the $\Delta_u$ derivative of any tensorial quantity in $\mathscr{S}$ of type $\{0, 0\}$,

$$T_{a\ldots c}{}^{d\ldots f},$$

which must be tangential to $\mathscr{S}$:

$$T_{a\ldots c}{}^{d\ldots f} = S_a{}^{a_0} \ldots S_c{}^{c_0} S_{d_0}{}^{d} \ldots S_{f_0}{}^{f} T_{a_0\ldots c_0}{}^{d_0\ldots f_0}, \tag{4.14.17}$$

as

$$\eth\eta_0, \eth'\eta_0, \ldots, \eth\eta_k, \eth'\eta_k, \tag{4.14.18}$$

where $\eta_0, \ldots, \eta_k$ are the various components of $T_{a\ldots c}{}^{d\ldots f}$ with respect to $m^a$ and $\bar{m}^a$. Note that the type of each of $\eta_0, \ldots, \eta_k$ has the form $\{p, -p\}$ for various values of $p$ ranging from minus to plus the total valence of $T_{\cdots}^{\cdots}$. This establishes the statement containing (4.14.4). (*Cf.* also Goldberg *et al.* 1967.)

Taking $\eta$ as in (4.14.16), so that $p = -q = 1$, and applying (4.14.1) to it, we obtain

$$(\eth\eth' - \eth'\eth)\eta = -(K + \bar{K})\eta, \tag{4.14.19}$$

where

$$K = \sigma\sigma' - \Psi_2 - \rho\rho' + \Phi_{11} + \Lambda. \tag{4.14.20}$$

(Since now $\chi = 1$, $\Pi = \Lambda$.) In fact, (4.14.19) tells us that:

(4.14.21) PROPOSITION

*$K + \bar{K}$ is the Gaussian curvature of $\mathscr{S}$.*

This follows from

$$\begin{aligned}(\eth\eth' - \eth'\eth)\eta &= \eth(m_a \bar{m}^c \Delta_c V^a) - \eth'(m_a m^c \Delta_c V^a) \\ &= m_a m^b \bar{m}^c (\Delta_b \Delta_c - \Delta_c \Delta_b) V^a,\end{aligned} \tag{4.14.22}$$

the commutator on the right yielding the 2-space curvature, by (4.2.30):

$$(\Delta_b \Delta_c - \Delta_c \Delta_b) V^a = k(S_{bd} S_c{}^a - S_{cd} S_b{}^a) V^d, \tag{4.14.23}$$

where $k$ is the Gaussian curvature of $\mathscr{S}$. Substituting (4.14.23) into (4.14.22) we obtain $k = K + \bar{K}$, as required for (4.14.21). (It is easily checked in the case of a unit sphere that indeed $K + \bar{K} = 1$, *cf.* (4.15.14) below.)

We see that the operators ð and ð′, as applied to quantities of zero boost-weight (i.e., of type $\{p, -p\}$) on $\mathscr{S}$, provide a neat way of handling the intrinsic geometry of $\mathscr{S}$. Indices can be avoided altogether, the tensorial character of the quantities involved being encoded in the various spin-weights

$$s := \tfrac{1}{2}(p - q) = p \qquad (4.14.24)$$

that arise. In fact, this calculus is effectively the analogue for 2-surfaces of the 2-component spinor calculus for space–times that we have been developing. As described in the Appendix to Vol. 2, the 'reduced spinors' for a $2n$-dimensional space are $2^{n-1}$-component objects. Here $n = 1$, and so we have one-component objects. An 'unprimed' 1-component spinor for $\mathscr{S}$ is thus a spin-weighted scalar of type $\{\frac{1}{2}, -\frac{1}{2}\}$ (i.e., $s = \frac{1}{2}$), while a 'primed' one-component spinor for $\mathscr{S}$ is a scalar of type $\{-\frac{1}{2}, \frac{1}{2}\}$ (i.e., $s = -\frac{1}{2}$). Higher valence tensors (i.e., higher spin-weights) arise when products of these basic 'spinors' are taken. (In Chapter 6 and in §§9.3, 9.4, in connection with twistor theory, we shall see that the case of four-component 'reduced spinors' for a six-dimensional space ($n = 3$) also has importance to us.)

### *Holomorphic coordinates*

The operators ð and ð′ also have significance in relation to complex analysis. Suppose that $\xi$ is a local *holomorphic coordinate* for $\mathscr{S}$; that is, $\xi$ is a complex coordinate defined in some open set $\mathscr{S}'$ in $\mathscr{S}$ such that at any point $Q \in \mathscr{S}'$ a rotation of the 1-form $\mathrm{d}\xi$, at $Q$, through a right angle in the positive sense in the tangent space at $Q$, yields $-\mathrm{i}\mathrm{d}\xi$. Another way of

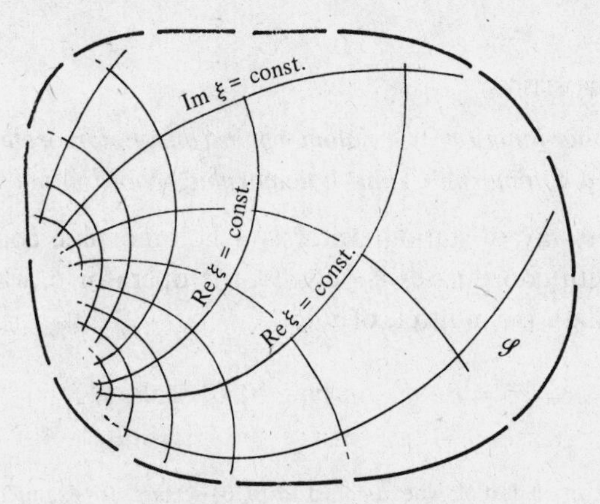


Fig. 4-2. A holomorphic coordinate $\xi$ on $\mathscr{S}$.

putting this (see Fig. 4–2) is that the lines $\mathrm{Re}\xi = \text{constant}$ meet the lines $\mathrm{Im}\xi = \text{constant}$ orthogonally such that at any point the spacing between each of these two families of lines is the same and the $\mathrm{Im}\xi$-increasing direction is a positive rotation of the $\mathrm{Re}\xi$-increasing direction. The most familiar holomorphic coordinate is $\xi = x + \mathrm{i}y$ for the Argand plane, and it is also a familiar result from conformal mapping theory that any holomorphic function of this $\xi$ is itself a holomorphic coordinate for the plane. The complex conjugate of the $\zeta$-coordinate of (1.2.6) is a holomorphic coordinate, in this sense, for the Riemann sphere $S^+$ if we orient the sphere so that its normal points outwards, while $\zeta$ is holomorphic if the normal points inwards (*cf.* Figure 1-3). The reverse holds for the 'celestial sphere' $S^-$ of §1.2.

The definition of a holomorphic coordinate for $\mathscr{S}$ asserts, in effect, that $\mathrm{d}\xi$ is a complex multiple of the differential form $\bar{m}_{i_1}$ (restricted to $\mathscr{S}$) at each point of $\mathscr{S}'$ since, by (3.1.21). $2^{\frac{1}{2}}\bar{m}_a = X_a + \mathrm{i}Y_a$ and a rotation through a right angle in the positive sense then yields $Y_a + \mathrm{i}(-X_a) = -\mathrm{i}2^{\frac{1}{2}}\bar{m}_a$. Now, the index form of $\mathrm{d}\xi$ (in the surface $\mathscr{S}$) is $\Delta_a\xi$, and to assert that this is a multiple of $\bar{m}_a$ amounts to the condition $\bar{m}^a\Delta_a\xi = 0$, so we have,* taking $\xi$ to be a complex coordinate of type $\{0, 0\}$, and using the term *anti-holomorphic coordinate* for the complex conjugate of a holomorphic coordinate:

(4.14.25) PROPOSITION

*$\xi$ is a holomorphic [anti-holomorphic] coordinate for $\mathscr{S}$ iff $\eth'\xi = 0$ $[\eth\xi = 0]$.*

Note that reversing the orientation of $\mathscr{S}$ interchanges holomorphic and anti-holomorphic coordinates. Note also, as an immediate corollary of (4.14.25), that

(4.14.26) PROPOSITION

*Any holomorphic function of a holomorphic [anti-holomorphic] coordinate on $\mathscr{S}$ is again a holomorphic [anti-holomorphic] coordinate on $\mathscr{S}$.*

Yet another way of stating that $\xi$ is a holomorphic coordinate is to assert that, with coordinates $\xi, \bar{\xi}$ for $\mathscr{S}$, the operator $\eth$, when acting on type $\{0, 0\}$ scalars, is a multiple of $\partial/\partial\xi$:

$$\eth = P\frac{\partial}{\partial\xi} \quad \text{(on type } \{0, 0\} \text{ scalars)}, \qquad (4.14.27)$$

* The operators $\eth$ and $\eth'$, when applied to $\{0, 0\}$-scalars, are examples of the $\partial$ and $\bar{\partial}$ operators that occur in complex manifold theory (*cf.* Wells 1973).

where $P$ is of type $\{1, -1\}$. The operator (4.14.27) is, of course, simply $\delta = \nabla_{01'}$. Writing $\boldsymbol{m}$ and $\bar{\boldsymbol{m}}$ for the 1-forms $m_{i_1}$ and $\bar{m}_{i_1}$, respectively, we have (linear maps of vectors)

$$\boldsymbol{m}(\delta) = 0, \quad \bar{\boldsymbol{m}}(\delta) = -1, \quad \mathrm{d}\xi\left(\frac{\partial}{\partial \xi}\right) = 1, \quad \mathrm{d}\bar{\xi}\left(\frac{\partial}{\partial \xi}\right) = 0, \qquad (4.14.28)$$

whence, on $\mathscr{S}$,

$$\bar{\boldsymbol{m}} = -P^{-1}\mathrm{d}\xi, \quad \boldsymbol{m} = -\bar{P}^{-1}\mathrm{d}\bar{\xi}. \qquad (4.14.29)$$

Recall that $S_{ab}$ is the induced metric tensor on $\mathscr{S}$. We can use the standard 'differential' notation $\mathrm{d}s^2$ for this metric; then using juxtaposition of differentials to denote *symmetric* tensor products of forms, (4.14.6) can be rewritten

$$\mathrm{d}s^2 = -2\boldsymbol{m}\bar{\boldsymbol{m}} = -\frac{2\mathrm{d}\xi\,\mathrm{d}\bar{\xi}}{P\bar{P}} = -2S\bar{S}\,\mathrm{d}\xi\,\mathrm{d}\bar{\xi}, \qquad (4.14.30)$$

by (4.14.29), where we have introduced the scalar $S$ defined by (*cf.* (4.14.27))

$$S^{-1} = P = \eth\xi. \qquad (4.14.31)$$

Note that $S$ is of type $\{-1, 1\}$. Also $P$ and $S$ are both 'holomorphic' in the sense that

$$\eth' P = 0, \quad \eth' S = 0. \qquad (4.14.32)$$

These equations are consequences of (4.14.31), which implies

$$\eth' P = \eth'\eth\xi = \eth\eth'\xi = 0,$$

since the commutator (4.14.1) vanishes. Hence also

$$\eth' S = \eth' P^{-1} = -P^{-2}\eth' P = 0.$$

However, we shall see from (4.15.116) below that particular representations of $P$ and $S$ need not 'look' holomorphic in the ordinary sense.

We may use $P$ (or $S$) to convert any $\{s, -s\}$-scalar $\eta$ to a $\{0, 0\}$-scalar, and then use (4.14.32) and the conjugate of (4.14.27) to obtain an expression for $\eth'\eta$:

$$\eth'\eta = \bar{P}P^{s}\frac{\partial}{\partial\bar{\xi}}(P^{-s}\eta). \qquad (4.14.33)$$

Applying this to $\bar{\eta}$ and taking complex conjugates yields

$$\eth\eta = P\bar{P}^{-s}\frac{\partial}{\partial\xi}(\bar{P}^{s}\eta). \qquad (4.14.34)$$

The expressions (4.14.33) and (4.14.34) are useful in explicit representations, as we shall see shortly.*

* See also Newman and Penrose (1966), where the definitions differ with respect to various conventions, *cf.* below after (4.15.107).

We note in passing that (4.14.31) (2) and (4.14.34), when substituted into (4.14.19), provide us, after a short calculation, with an expression for the Gaussian curvature of $\mathscr{S}$:

$$K + \bar{K} = P\bar{P}\frac{\partial^2}{\partial\xi\partial\bar{\xi}}\log(P\bar{P}). \tag{4.14.35}$$

Observe that this formula involves only the modulus of $P$. But this is not surprising since $P$, being spin-weighted (but not boost-weighted) can have its argument (but not its modulus) altered by a rescaling $o^A \mapsto \lambda o^A$, $\iota^A \mapsto \lambda^{-1}\iota^A$. In explicit representations, however, it is usually convenient to fix the choice of $m^a$ to be that which makes $P$ *positive.* Writing $2^{\frac{1}{2}}m^a = X^a - \mathrm{i}Y^a$, we achieve this if

$$\left.\begin{array}{l} X^a \textit{ points along } \mathrm{Im}(\xi) = \textit{constant}, \mathrm{Re}(\xi) \textit{ increasing} \\ Y^a \textit{ points along } \mathrm{Re}(\xi) = \textit{constant}, \mathrm{Im}(\xi) \textit{ increasing}. \end{array}\right\} \tag{4.14.36}$$

This leads to slight simplifications of (4.14.33)–(4.14.35), but it takes us outside the 'strict' compacted spin-coefficient formalism, since $m^a$ is now fixed by the choice of $\xi$.

### *Extrinsic quantities*

So far, we have only interpreted the real part of $K$ in (4.14.20). The imaginary part of $K$ will show up in (4.14.1) only when $p \neq -q$, i.e., when the quantity on which (4.14.1) acts has a non-zero boost-weight. Such quantities are not, in the ordinary sense, entirely intrinsic to $\mathscr{S}$. To investigate this case we can repeat the foregoing discussion, applying (4.14.1) to quantities

$$\eta_1 = V^a l_a, \quad \eta_2 = V^a n_a \tag{4.14.37}$$

instead of (4.14.16). In order that $\eta_1$ and $\eta_2$ should not both vanish, $V^a$ must have components perpendicular to $\mathscr{S}$; and we may as well take $V^a$ in the plane spanned by $l^a$ and $n^a$, so that it is determined by $\eta_1$ (type $\{1, 1\}$) and $\eta_2$ (type $\{-1, -1\}$). In place of (4.14.11) we can consider

$$\tilde{S}_{a_0}{}^{a} S_{b}{}^{b_0} \nabla_{b_0} V^{a_0}, \tag{4.14.38}$$

where

$$\tilde{S}_a{}^b = g_a{}^b - S_a{}^b = l_a n^b + n_a l^b \tag{4.14.39}$$

is the projection operator orthogonal to $S_a{}^b$, i.e., perpendicular to $\mathscr{S}$. Now $V^a \tilde{S}_a{}^b = V^b$ and the (non-vanishing) null tetrad components of (4.14.38) turn out to be

$$ð\eta_1, ð\eta_2, ð'\eta_1, ð'\eta_2 \tag{4.14.40}$$

in place of (4.14.15). The resulting formula analogous to (4.14.22) differs from (4.14.22) only in that $m_a$ is replaced by $l_a$ or $n_a$ and that the $\Delta$ operators now refer to the operation in (4.14.38), according to which vectors perpendicular to $\mathscr{S}$ are transported about $\mathscr{S}$, while previously it had been vectors tangent to $\mathscr{S}$ that were so transported. In place of the $-(K+\bar{K})$ appearing on the right side of (4.14.19) we now have $-(K-\bar{K})$ in the case of $\eta_1$ and $+(K-\bar{K})$ in the case of $\eta_2$.

Thus the *imaginary part* of $K$ is an *extrinsic curvature* quantity concerned with the transport, about $\mathscr{S}$, of vectors perpendicular to $\mathscr{S}$. Recall that the Gaussian curvature can be thought of as a measure of the resultant *rotation* of the *tangent* space as it is carried by parallel transport around a small loop in $\mathscr{S}$. Analogously, we now have a resultant *boost* of the *normal* space as it is carried around the same small loop. While the real part of $2K$ is a measure of the former, the imaginary part is a measure of the latter. We refer to $K$ as the *complex curvature* of $\mathscr{S}$.

We note in passing that $K$ can be expressed as a sum of two parts,

$$\sigma\sigma' - \Psi_2 \quad \text{and} \quad \Phi_{11} + \Lambda - \rho\rho', \tag{4.14.41}$$

the first of which turns out to have simple conformal scaling properties (*cf.* (5.6.28) and (6.8.4), Vol. 2) and the second of which is real. One consequence of this curious fact is that the extrinsic part (i.e., the imaginary part) of the complex curvature is essentially conformally invariant (see §5.6 for definitions). But we shall not pursue this matter further here.

In connection with the above results on curvature, we recall the Gauss–Bonnet theorem which states that if $\mathscr{S}$ is a closed surface of genus* $g$, the integral with respect to surface area of the Gaussian curvature of $\mathscr{S}$ is $4\pi(1-g)$. So in the present case we have

$$\oint_{\mathscr{S}} (K+\bar{K})\mathscr{S} = 4\pi(1-g), \tag{4.14.42}$$

$\mathscr{S}$ being the element of surface area (a 2-form on $\mathscr{S}$). In fact, because of our interpretation for $K-\bar{K}$, it also follows that

$$\oint_{\mathscr{S}} (K-\bar{K})\mathscr{S} = 0. \tag{4.14.43}$$

The reason is that the space of boosts $l^a \mapsto rl^a$, $n^a \mapsto r^{-1}n^a$, $r>0$, is a *topologically trivial* 1-parameter group. The integral $\mathrm{i}\oint(K-\bar{K})\mathscr{S}$ over a

* Recall that the *genus* of a closed (oriented) 2-surface is, roughly speaking, its number of 'handles'. Thus, for a sphere $S^2$ we have $g=0$, while for a torus $g=1$, and for the surface of a standard pretzel $g=3$.

bounded portion of $\mathscr{S}$ provides a measure of the total boost achieved as the boundary of this portion is traversed. When the boundary shrinks to zero this total boost must also shrink to zero – in contrast to what happens in the case of a rotation, when one may end up with a non-trivial total rotation through some multiple of $2\pi$. Combining (4.14.42) with (4.14.43) we obtain

$$\oint_{\mathscr{S}} K\mathscr{S} = 2\pi(1-g). \tag{4.14.44}$$

In view of the remarks made about the quantities (4.14.41), we can also derive the result that

$$\oint_{\mathscr{S}} (\sigma\sigma' - \Psi_2)\mathscr{S} \textit{ is real} \tag{4.14.45}$$

in addition to being (as follows in detail from the discussion of §5.6) a *conformally invariant* number associated with any closed spacelike 2-surface embedded in a space–time. The significance of this result in relation to the Bondi–Sachs mass will be discussed in §9.9.

*Relations to exterior calculus*

We next show how the exterior calculus (*cf.* §4.3) on $\mathscr{S}$ neatly fits in with the present 2-surface formalism. Let

$$\boldsymbol{\alpha} = \alpha_a \mathrm{d}x^a = \alpha_{i_1} \tag{4.14.46}$$

be a 1-form in $\mathscr{M}$, where we are concerned only with its restriction to $\mathscr{S}$. This means we are concerned only with the two components

$$\alpha_{01'} = \alpha_a m^a \quad \text{and} \quad \alpha_{10'} = \alpha_a \bar{m}^a \tag{4.14.47}$$

of respective types $\{1, -1\}$ and $\{-1, 1\}$. If $\boldsymbol{\alpha}$ is real, then all the relevant information is contained in the one type-$\{1, -1\}$ scalar quantity

$$\alpha := \alpha_{01'} \tag{4.14.48}$$

(not to be confused with the spin-coefficient $\alpha$ in (4.5.16)!), since $\alpha_{10'}$ is then its complex conjugate. The condition that $\boldsymbol{\alpha}$ (real or complex) be the exterior derivative

$$\boldsymbol{\alpha} = \mathrm{d}v \tag{4.14.49}$$

of a (type-$\{0, 0\}$) scalar quantity $v$ can be written as

$$\alpha_{01'} = \eth v, \quad \alpha_{10'} = \eth' v, \tag{4.14.50}$$

or, if $v$ and $\boldsymbol{\alpha}$ are real, simply as

$$\alpha = \eth v. \tag{4.14.51}$$

Now suppose $\boldsymbol{\beta}$ is a 2-form

$$\boldsymbol{\beta} = \beta_{ab}\,\mathrm{d}x^a \wedge \mathrm{d}x^b = \beta_{i_1 i_2} \tag{4.14.52}$$

($\beta_{ab} = -\beta_{ba}$). Our concern is only with its restriction to $\mathscr{S}$, and thus with the single (type-$\{0,0\}$) component

$$\tfrac{1}{2}\mathrm{i}\beta := \beta_{01'10'} = -\beta_{10'01'} = \beta_{ab}m^a\bar{m}^b. \tag{4.14.53}$$

Note that $\beta$ is real whenever $\boldsymbol{\beta}$ is real. Indeed, by (3.1.20), we have

$$\beta = 2\beta_{ab}X^aY^b. \tag{4.14.54}$$

The condition that $\boldsymbol{\beta}$ be the exterior derivative

$$\boldsymbol{\beta} = \mathrm{d}\boldsymbol{\alpha} \tag{4.14.55}$$

of some 1-form $\boldsymbol{\alpha}$ is

$$\beta_{ab} = \nabla_{[a}\alpha_{b]}. \tag{4.14.56}$$

To get the restriction of this to $\mathscr{S}$ we take components with respect to $m^a$ and $\bar{m}^a$:

$$\mathrm{i}\beta = 2\beta_{01'10'} = \eth\alpha_{10'} - \eth'\alpha_{01'}, \tag{4.14.57}$$

by (4.14.13), the terms involving $\rho$ and $\rho'$ cancelling because of (4.14.2). For real $\boldsymbol{\alpha}$ this takes the form

$$\beta = 2\mathrm{Im}(\eth'\alpha). \tag{4.14.58}$$

Note that if $\boldsymbol{\alpha} = \mathrm{d}v$ (with $v$ of type $\{0,0\}$), we can substitute (4.14.50) into (4.14.57) which yields $\boldsymbol{\beta} = \mathrm{d}\boldsymbol{\alpha} = 0$, as expected (*cf.* (4.3.15) (viii)), the $\eth$ and $\eth'$ operators commuting by (4.14.1).

The fundamental theorem of exterior calculus (4.3.25) can be applied at two levels on $\mathscr{S}$. First:

$$\int_\Gamma \mathrm{d}\boldsymbol{\alpha} = \oint_{\partial\Gamma} \boldsymbol{\alpha}, \tag{4.14.59}$$

where $\Gamma$ is a compact domain on $\mathscr{S}$, with boundary $\partial\Gamma$; and second:

$$\int_Q^R \mathrm{d}v = v(R) - v(Q), \tag{4.14.60}$$

the integral on the left being taken over any curve $\gamma$ (in the domain of definition of $v$) connecting the points $Q$ and $R$. For the latter case we can introduce a holomorphic coordinate $\xi$ in the neighbourhood of $\gamma$ and rewrite the integral as

$$\int_Q^R \mathrm{d}v = \int_Q^R \left(\frac{\partial v}{\partial \xi}\mathrm{d}\xi + \frac{\partial v}{\partial \bar{\xi}}\mathrm{d}\bar{\xi}\right)$$

$$= \int_Q^R (S\eth v\,\mathrm{d}\xi + \bar{S}\eth' v\,\mathrm{d}\bar{\xi}), \tag{4.14.61}$$

where we have used (4.14.27) and (4.14.31). In particular, if $v$ is *holomorphic* ($\eth' v = 0$), (4.14.60) becomes

$$\int_Q^R S\eth v \, d\xi = v(R) - v(Q). \tag{4.14.62}$$

From (4.14.61) we see, on taking $Q = R$, that

$$\oint_\gamma (S\eth v \, d\xi + \bar{S}\eth' v \, d\bar{\xi}) = 0 \tag{4.14.63}$$

for a *closed* contour $\gamma$, whence, in particular,

$$\oint_\gamma S\eth v \, d\xi = 0 \text{ if } v \textit{ is holomorphic.} \tag{4.14.64}$$

The 2-dimensional integral on the left side of (4.14.59) can be re-expressed in terms of the *surface area element* $\mathscr{S}$. We begin by noting that

$$\begin{aligned} \mathscr{S} &:= (-X_a dx^a) \wedge (-Y_b dx^b) \\ &= \frac{1}{\sqrt{2}}(\boldsymbol{m} + \bar{\boldsymbol{m}}) \wedge \frac{i}{\sqrt{2}}(\boldsymbol{m} - \bar{\boldsymbol{m}}) \\ &= i\bar{\boldsymbol{m}} \wedge \boldsymbol{m}, \end{aligned} \tag{4.14.65}$$

where $\boldsymbol{m} = m_{i_1} = m_a dx^a$, as before. So if $\beta$ is as in (4.14.52), (4.14.53), we have

$$\begin{aligned} \int_\Gamma \boldsymbol{\beta} &= \int_\Gamma \beta_{ab} dx^a \wedge dx^b \\ &= 2\int_\Gamma \beta_{ab} m^a \bar{m}^b \bar{\boldsymbol{m}} \wedge \boldsymbol{m} \\ &= i\int_\Gamma \beta \bar{\boldsymbol{m}} \wedge \boldsymbol{m} \\ &= \int_\Gamma \beta \mathscr{S}, \end{aligned} \tag{4.14.66}$$

where we used the fact that, by (4.14.6),

$$\textit{restricted to } \mathscr{S}: dx^a = S_b{}^a dx^b = -\bar{m}^a \boldsymbol{m} - m^a \bar{\boldsymbol{m}}. \tag{4.14.67}$$

Alternatively, (4.14.66) can be directly obtained by use of (4.14.54). Substituting (4.14.57) in (4.14.59), we get

$$\int_\Gamma (\mathrm{i}\eth'\alpha_{01'} - \mathrm{i}\eth\alpha_{10'})\mathscr{S} = \oint_{\partial\Gamma} \boldsymbol{\alpha}$$

$$= \oint_{\partial\Gamma} (\alpha_{01'} S\,\mathrm{d}\xi + \alpha_{10'}\bar{S}\,\mathrm{d}\bar{\xi}), \tag{4.14.68}$$

where $\xi$ is a holomorphic coordinate in some neighbourhood of $\partial\Gamma$. Since the components $\alpha_{01'}$ and $\alpha_{10'}$ are independent, we have

$$\int_\Gamma \mathrm{i}\eth'\alpha\mathscr{S} = \oint_{\partial\Gamma} \alpha S\,\mathrm{d}\xi \tag{4.14.69}$$

for any type-$\{1, -1\}$ scalar $\alpha$ on $\mathscr{S}$.

As a particular case of (4.14.69) (and its complex conjugate), we note that if $\mathscr{S}$ is a *closed* surface,

$$\oint_{\mathscr{S}} \eth'\alpha\mathscr{S} = 0, \oint_{\mathscr{S}} \eth\tilde{\alpha}\mathscr{S} = 0 \tag{4.14.70}$$

($\alpha$ being of type $\{1, -1\}$ and $\tilde{\alpha}$ of type $\{-1, 1\}$), from which we derive the following useful formulae for integration by parts:

$$\oint_{\mathscr{S}} \chi\eth'\eta\mathscr{S} = -\oint_{\mathscr{S}} \eta\eth'\chi\mathscr{S},$$

$$\oint_{\mathscr{S}} \tilde{\chi}\eth\tilde{\eta}\mathscr{S} = -\oint_{\mathscr{S}} \tilde{\eta}\eth\tilde{\chi}\mathscr{S}, \tag{4.14.71}$$

where the types of $\chi, \eta$ add up to $\{1, -1\}$ and those of $\tilde{\chi}, \tilde{\eta}$ add up to $\{-1, 1\}$.

*On a null hypersurface*

Spacelike 2-surfaces can also play a role in relation to the fundamental theorem of exterior calculus in the next higher dimension, namely

$$\int_\Sigma \mathrm{d}\boldsymbol{\beta} = \oint_{\partial\Sigma} \boldsymbol{\beta}, \tag{4.14.72}$$

where $\boldsymbol{\beta}$ is a 2-form and $\Sigma$ a compact 3-surface with a spacelike boundary $\partial\Sigma$. In some of the most interesting applications of (4.14.72), $\Sigma$ is a portion of a *null hypersurface*, that is, of a 3-surface $\mathscr{N}$ whose normals $n^a$ are *null* vectors. We shall investigate such hypersurfaces in more detail in Vol. 2 (§§7.1, 7.2; *cf.* also §5.11, §5.12 below). Here we merely note that the tangent vectors to $\mathscr{N}$, being the vectors orthogonal to $n^a$, must include

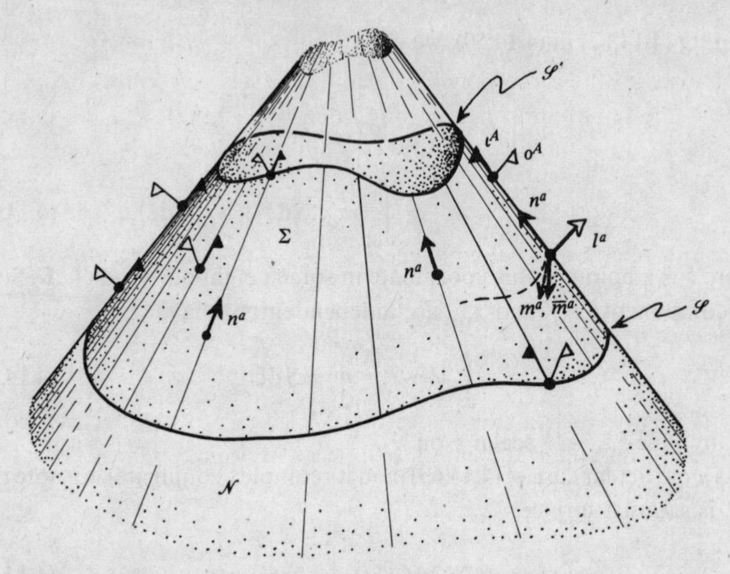


Fig. 4-3. Suitable arrangement of spin-frames on a null hypersurface $\mathscr{N}$, for applications of (4.14.72).

$n^a$ itself, and, furthermore, that any 2-surface element in $\mathscr{N}$, being orthogonal to the null vector $n^a$, must necessarily be *spacelike* unless it contains the direction $n^a$ itself, in which case it is null. The situation we envisage is like that depicted in Fig. 4-3, where $\partial\Sigma$ consists of two closed spacelike 2-surfaces $\mathscr{S}$ and $\mathscr{S}'$:

$$\partial\Sigma = \mathscr{S}' - \mathscr{S}. \tag{4.14.73}$$

We select the spin-frame $(o^A, \iota^A)$ – or, rather, the equivalence class $(o^A, \iota^A) \sim (\lambda o^A, \lambda^{-1}\iota^A)$ – so that the flagpole of $\iota^A$ points along the normal to $\mathscr{N}$ (and so is tangent to $\mathscr{N}$), as has been implicitly assumed in the choice of the letter '$n$' for this normal; also so that $m^a$, $\bar{m}^a$ span the tangent spaces of $\mathscr{S}$ and $\mathscr{S}'$ at $\partial\Sigma$, the choice of the $m^a$, $\bar{m}^a$-planes in the interior of $\Sigma$ being arbitrary except that they must form a smooth family (tangent to $\mathscr{N}$) fitting smoothly on to the given choices at the boundary surfaces $\mathscr{S}$ and $\mathscr{S}'$. The most direct way of achieving this is to require these planes to be tangent to a family of spacelike 2-surfaces on $\mathscr{N}$ which vary smoothly from $\mathscr{S}$ to $\mathscr{S}'$, but a more general choice is also allowed here in which the interior plane elements $\delta\mathscr{S}$ need not be locally 'integrable' to 2-surfaces (i.e., need not constitute a foliation).

Of course we could equally well have chosen $o^A$ instead of $\iota^A$ to have its flagpole normal (i.e., tangent) to $\mathscr{N}$. Our selection is made only for consistency with later notation (*cf.* §5.12 and §9.10). By applying the prime operation to the various formulae was shall develop ((4.14.74)–(4.14.94)) we can obtain the corresponding formulae pertaining to this alternative choice – though we must bear in mind that this will also entail some sign changes because $\mathscr{S}$ and $\mathscr{S}'$ have their natural orientations reversed (*cf.* (4.14.73)).

The relevant commutators are now those involving ð, ð′ and þ′, these being the operations acting tangentially to $\mathscr{N}$. Thus we have (4.14.1) together with (*cf.* (4.12.34))

$$þ'ð - ðþ' = \rho'ð + \bar{\sigma}'ð' - \tau þ' - \bar{\kappa}'þ + p(\sigma\kappa' - \rho'\tau + \Phi_{12}) + q(\bar{\rho}\bar{\kappa}' - \bar{\tau}\bar{\sigma}' + \bar{\Psi}_3) \qquad (4.14.74)$$

and its complex conjugate. Note that we cannot now infer that $\rho = \bar{\rho}$ (since this is the condition that the $\delta\mathscr{S}$ elements be integrable to 2-surfaces) though we do still have

$$\rho' = \bar{\rho}' \qquad (4.14.75)$$

by the vanishing of the coefficient of þ in (4.14.1). The same reasoning applied to (4.14.74) yields

$$\kappa' = 0, \qquad (4.14.76)$$

which (as we shall see in §7.1) is the condition for the integral curves of $n^a$ to be geodesics. These integral curves are referred to as the *generators* of $\mathscr{N}$.

The relevant components of $\boldsymbol{\beta}$ in (4.14.72) are now

$$\beta_{01'11'}, \beta_{10'11'}, \quad \text{and} \quad \frac{\mathrm{i}}{2}\beta = \beta_{01'10'}, \qquad (4.14.77)$$

while any 3-form

$$\gamma = \gamma_{abc}\mathrm{d}x^a \wedge \mathrm{d}x^b \wedge \mathrm{d}x^c = \gamma_{i_1 i_2 i_3} \qquad (4.14.78)$$

($\gamma_{abc} = \gamma_{[abc]}$), when restricted to $\mathscr{N}$, involves only the component

$$\frac{\mathrm{i}}{6}\gamma := \gamma_{01'10'11'}. \qquad (4.14.79)$$

The equation

$$\gamma = \mathrm{d}\boldsymbol{\beta} \qquad (4.14.80)$$

restricted to $\mathscr{N}$ turns out to be, after a short calculation,

$$3\gamma_{01'10'11'} = (ð - \tau)\beta_{10'11'} - (ð' - \bar{\tau})\beta_{01'11'} + (þ' - 2\rho')\beta_{01'10'}. \qquad (4.14.81)$$

We also note that the equation

$$\boldsymbol{\beta} = \mathrm{d}\boldsymbol{\alpha} \qquad (4.14.82)$$

restricted to $\mathscr{N}$ generalizes (4.14.57), the component $\alpha_{11'}$ of the 1-form $\boldsymbol{\alpha}$ being now also involved:

$$2\beta_{01'10'} = \eth\alpha_{10'} - \eth'\alpha_{01'} + (\bar{\rho} - \rho)\alpha_{11'}$$
$$2\beta_{10'11'} = \sigma'\alpha_{01'} - (\text{þ}' - \rho')\alpha_{10'} + (\eth' - \bar{\tau})\alpha_{11'}$$
$$2\beta_{01'11'} = \bar{\sigma}'\alpha_{10'} - (\text{þ}' - \rho')\alpha_{01'} + (\eth - \tau)\alpha_{11'} \qquad (4.14.83)$$

(and it may be directly checked that substituting (4.14.83) into (4.14.81) yields zero, in accordance with $\mathrm{d}^2 = 0$).

A version of the fundamental theorem of exterior calculus appropriate to Fig. 4-3 is obtained by substituting (4.14.81), (4.14.83) into (4.14.72). But a more useful and simple-looking expression can also be obtained by first splitting $\boldsymbol{\beta}$ into its anti-self-dual and self-dual parts (*cf.* (3.4.17)):

$$\boldsymbol{\beta} =: \beta_{I_1 I_2}\varepsilon_{I'_1 I'_2} + \varepsilon_{I_1 I_2}\tilde{\beta}_{I'_1 I'_2},$$
$$\beta_{[AB]} = 0, \quad \tilde{\beta}_{[A'B']} = 0, \qquad (4.14.84)$$

and then setting

$$\mu_A := 2\mathrm{i}\beta_{AB}\iota^B, \quad \tilde{\mu}_{A'} := -2\mathrm{i}\tilde{\beta}_{A'B'}\iota^{B'}. \qquad (4.14.85)$$

Restricting attention to $\mu_A$ and $\tilde{\mu}_{A'}$ does not lose relevant information because the freedom in $\beta_{AB}$ and $\tilde{\beta}_{A'B'}$ (given $\mu_A$ and $\tilde{\mu}_{A'}$) consists merely in the addition of multiples of $\iota_A\iota_B$ and of $\iota_{A'}\iota_{B'}$, respectively, which corresponds to the addition to $\boldsymbol{\beta}$ of components which vanish in $\mathscr{N}$. In fact, we have

$$\beta_{01'10'} = \frac{\mathrm{i}}{2}(\mu_0 + \tilde{\mu}_{0'}), \quad \beta_{10'11'} = -\frac{\mathrm{i}}{2}\mu_1, \quad \beta_{01'11'} = \frac{\mathrm{i}}{2}\tilde{\mu}_{1'}, \qquad (4.14.86)$$

so (4.14.81) gives (with (4.14.79))

$$\gamma = (\text{þ}' - 2\rho')(\mu_0 + \tilde{\mu}_{0'}) - (\eth - \tau)\mu_1 - (\eth' - \bar{\tau})\tilde{\mu}_{1'}. \qquad (4.14.87)$$

In order to apply (4.14.72) we need to interpret 3-surface integrals on $\mathscr{S}$ in terms of the quantities we have been considering. For this, we need also to choose a parameter $u$ (smoothly) on each generator of $\mathscr{N}$, which we take to be scaled in relation to $n^a$ according to

$$n^a\nabla_a u = U \quad (\neq 0). \qquad (4.14.88)$$

In specific representations of the null tetrad (and relaxing the strict compacted formalism) we could choose $U = 1$, but here we simply take $U$ to be a $\{-1, -1\}$ scalar. From (4.14.88) we see that $\mathrm{d}u\ (= \nabla_{i_1} u)$ differs from $U\boldsymbol{l}$ by terms in $\boldsymbol{m}$ and $\bar{\boldsymbol{m}}$ only, so we obtain for the (null) 'volume' element of $\mathscr{N}$ a type-$\{1, 1\}$ 3-form:

$$\mathscr{N} := \mathrm{i}\bar{\boldsymbol{m}} \wedge \boldsymbol{m} \wedge \boldsymbol{l} = U^{-1}\mathscr{S} \wedge \mathrm{d}u. \qquad (4.14.89)$$

Consequently, by (4.14.79),

$$\int_{\Sigma} \gamma = \int_{\Sigma} \gamma U^{-1} \mathscr{S} \wedge \mathrm{d}u = \int_{\Sigma} \gamma \mathscr{N}. \tag{4.14.90}$$

If we now substitute (4.14.80), (4.14.66) in (4.14.90), (4.14.72), and (4.14.73), we obtain

$$\int_{\Sigma} \gamma \mathscr{N} = \oint_{\mathscr{S}'} \beta \mathscr{S} - \oint_{\mathscr{S}} \beta \mathscr{S}, \tag{4.14.91}$$

where $\gamma$ is as in (4.14.87) (and is $\{-1, -1\}$), with the $\mu$s as in (4.14.86). We can separate this expression into one for $\mu_A$ and another for $\tilde{\mu}_{A'}$, which are complex conjugates of one another if $\boldsymbol{\beta}$ is real. Writing the $\mu$-equation out, we finally obtain, in terms of the $\{0,0\}$-scalar $\mu_0$ and the $\{-2, 0\}$-scalar $\mu_1$,

$$\int_{\Sigma} \{(\text{þ}' - 2\rho')\mu_0 - (\eth - \tau)\mu_1\} \mathscr{N} = \oint_{\mathscr{S}'} \mu_0 \mathscr{S} - \oint_{\mathscr{S}} \mu_0 \mathscr{S}, \tag{4.14.92}$$

and the corresponding complex conjugate equation for $\tilde{\mu}_{A'}$. Note that if, in particular,

$$(\text{þ}' - 2\rho')\mu_0 = (\eth - \tau)\mu_1, \tag{4.14.93}$$

then we have the 'conservation law'

$$\oint_{\mathscr{S}} \mu_0 \mathscr{S} = \oint_{\mathscr{S}'} \mu_0 \mathscr{S}. \tag{4.14.94}$$

These relations will have considerable importance for us later (*cf.* §§5.12, 9.9).

## 4.15 Spin-weighted spherical harmonics

As a significant application of the foregoing theory of spacelike 2-surfaces, we examine the case when the surface $\mathscr{S}$ is an ordinary *2-sphere in Minkowski space* $\mathbb{M}$; and we show how the theory of (spin-weighted) spherical harmonics may be developed using these results. We take $(o^A, \iota^A)$ to be normalized to a spin-frame throughout (*cf.* (4.14.5)). Let the point $O \in \mathbb{M}$ be the centre of the sphere $\mathscr{S}$ and let $T^a$ be the future-timelike unit vector (taken constant throughout $\mathbb{M}$) which is orthogonal to the spacelike 3-plane containing $\mathscr{S}$. Let $Q$ be a typical point on $\mathscr{S}$ and $x^a$ its position vector $\overrightarrow{OQ}$. Now $\mathscr{S}$ will be the intersection of a future light cone $\mathscr{L}$, with vertex $L$, and a past light cone $\mathscr{N}$, with vertex $N$. (See Fig. 4-4.) Since $\delta\mathscr{S}$ at $Q$ is orthogonal to the generators of $\mathscr{L}$ and of $\mathscr{N}$,

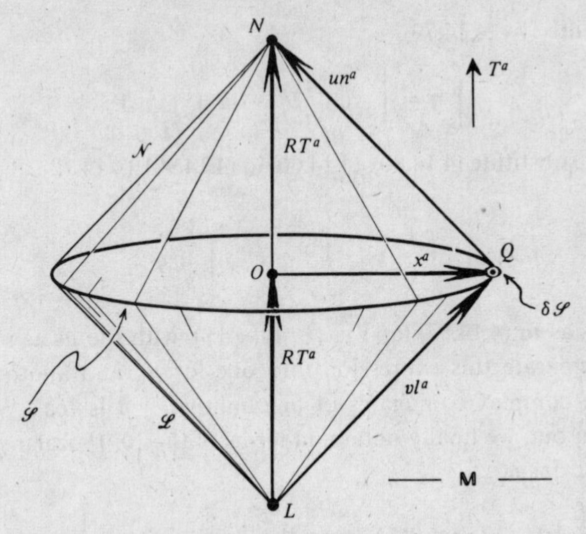


Fig. 4-4. An ordinary spacelike sphere $\mathscr{S}$ in Minkowski space arises as the intersection of two light cones $\mathscr{L}$ and $\mathscr{N}$. (Geometry for (4.15.2).)

our spin-frame $(o^A, \iota^A)$ is determined by

$$(\overrightarrow{LQ})^a = vl^a, \quad (\overrightarrow{QN})^a = un^a, \tag{4.15.1}$$

where $v$ is of type $\{-1, -1\}$ and $u$ of type $\{1, 1\}$. Taking the sphere to have radius $R$, we deduce, from

$$\overrightarrow{OQ} = \overrightarrow{ON} + \overrightarrow{NQ} = \overrightarrow{OL} + \overrightarrow{LQ},$$

the relations

$$\begin{aligned} x^a &= RT^a - un^a = -RT^a + vl^a \\ &= RT^{AA''} - u\iota^A\iota^{A'} = -RT^{AA'} + vo^A o^{A'}, \end{aligned} \tag{4.15.2}$$

since the timelike distances $LO$ and $ON$ must both be equal to the radius $R$. From (4.15.2) we have

$$un^a vl_a = (RT^a - x^a)(RT_a + x_a) = R^2 + R^2,$$

whence

$$uv = 2R^2. \tag{4.15.3}$$

Since $x^a T_a = 0$, transvecting equation (4.15.2) with $T_a$ gives

$$u = \frac{R}{T_{AA'}\iota^A\iota^{A'}}, \quad v = \frac{R}{T_{AA'}o^A o^{A'}}, \tag{4.15.4}$$

while transvecting it with $o_A\iota_{A'}$ gives

$$T^{AA'}o_A\iota_{A'} = 0. \tag{4.15.5}$$

Consequently

$$o^A = -\frac{u}{R} T^{AA'} \iota_{A'}, \tag{4.15.6}$$

and similarly

$$\iota^A = \frac{v}{R} T^{AA'} o_{A'}. \tag{4.15.7}$$

*Using $o^A$ as 'coordinates' for $\mathscr{S}$*

Consider, now, a $\{0,0\}$ quantity $f_{\mathscr{Q}}$ (where $\mathscr{Q}$ is a 'clumped' index) on $\mathscr{S}$. We can adopt a new viewpoint and regard $f_{\mathscr{Q}}$ as a function* of the complex conjugate spinors $o^A, o^{A'}$ (as redundant 'coordinates' on $\mathscr{S}$) and use the chain rule to obtain, from (4.15.2), (4.15.4), (4.15.7), (2.5.54) as applied to $f_{\mathscr{Q}}$:

$$\frac{\partial}{\partial o^A} = \frac{\partial x^b}{\partial o^A}\frac{\partial}{\partial x^b} = \frac{\partial(v o^B o^{B'})}{\partial o^A}\nabla_b = \left(-\frac{v^2}{R} T_{AC'} o^{C'} o^B o^{B'} + v\varepsilon_A{}^B o^{B'}\right)\nabla_b$$

$$= v(\iota_A o^B + \varepsilon_A{}^B)\nabla_{BO'} = v o_A \nabla_{10'}. \tag{4.15.8}$$

(Partial derivatives with respect to abstract-indexed quantities have the obvious meaning here (*cf.* p. 145): one can always consider components in a *constant* spin-frame $(\hat{o}^A, \hat{\iota}^A)$ – such as we shall introduce at the end of this section – and then convert back to abstract indices.)

If, instead, $f_{\mathscr{Q}}$ is a $\{p,q\}$ quantity, with $p$ and $q$ both *non-positive* integers, then

$$f_{\mathscr{Q}} \underbrace{o^D \ldots o^G}_{-p} \underbrace{o^{H'} \ldots o^{K'}}_{-q}$$

is a $\{0,0\}$ quantity and we can apply (4.15.8) to it. Writing the $\nabla_{10'}$ in (4.15.8) as $\eth'$ and using (4.12.28), we thus obtain

$$o^D \ldots o^G o^{H'} \ldots o^{K'} \frac{\partial f_{\mathscr{Q}}}{\partial o^A} - p o^{(D} \ldots o^F \varepsilon_A{}^{G)} o^{H'} \ldots o^{K'} f_{\mathscr{Q}}$$

$$= v o^D \ldots o^G o^{H'} \ldots o^{K'} o_A \eth' f_{\mathscr{Q}} + v\rho p o^{(D} \ldots o^F \iota^{G)} o^{H'} \ldots o^{K'} o_A f_{\mathscr{Q}}$$

$$+ v\bar{\sigma} q o^D \ldots o^G o^{(H'} \ldots o^{J'} \iota^{K')} o_A f_{\mathscr{Q}}.$$

Substituting $\varepsilon_A{}^G = o_A \iota^G - \iota_A o^G$ (*cf.* (2.5.54)) into this equation, we find

* This is consistent with a standard convention that $f(\xi)$ represents a *holomorphic* function of the complex variable $\xi$, whereas $f(\xi, \bar{\xi})$ represents a *general* function of $\xi$, i.e., a function of $\mathrm{Re}(\xi)$ and $\mathrm{Im}(\xi)$.

that it implies

$$\rho = -v^{-1}, \quad \sigma = 0, \tag{4.15.9}$$

and then reduces to

$$\frac{\partial}{\partial o^A} = v o_A \eth' - p \iota_A, \tag{4.15.10}$$

as applied to $f_{\mathscr{Q}}$. For consistency with the Leibniz rule for $\eth'$ we now easily establish that (4.15.10) applies to any $\{p, q\}$ quantity on $\mathscr{S}$ for arbitrary *integral* $p, q$ – and, indeed, fractional $p, q$. (Integral or half-integral $p, q$ will normally be assumed unless otherwise stated.) The complex conjugate of (4.15.10) is

$$\frac{\partial}{\partial o^{A'}} = v o_{A'} \eth - q \iota_{A'}. \tag{4.15.11}$$

Any $\{p, q\}$ quantity on $\mathscr{S}$ can be expressed *either* as a function of $o^A, o^{A'}$ *or* as a function of $\iota^A, \iota^{A'}$, the relation between the two being obtained from (4.15.6), (4.15.7) with (4.15.4). So, repeating the above argument with $\iota^A, \iota^{A'}$ in place of $o^A, o^{A'}$, we can obtain

$$\rho' = u^{-1}, \quad \sigma' = 0. \tag{4.15.12}$$

and

$$\frac{\partial}{\partial \iota^A} = u \iota_A \eth - p o_A, \quad \frac{\partial}{\partial \iota^{A'}} = u \iota_{A'} \eth - q o_A. \tag{4.15.13}$$

We may check these relations against (4.14.20), (4.14.21), noting that the 4-space curvature terms $\Psi_2, \Phi_{11}, \Lambda$ all vanish, and obtain

$$K = -\rho\rho' = u^{-1} v^{-1} = \tfrac{1}{2} R^{-2} \tag{4.15.14}$$

(by (4.15.3)), so that the Gaussian curvature $K + \bar{K}$ of $\mathscr{S}$ is $R^{-2}$, as indeed it should be for an ordinary sphere of radius $R$.

Each of the relations (4.15.10), (4.15.11), (4.15.13) can be split into its two components, yielding

$$\begin{aligned} \eth &= \frac{1}{v} \iota^{A'} \frac{\partial}{\partial o^{A'}} = -\frac{1}{u} o^A \frac{\partial}{\partial \iota^A} \\ &= \frac{1}{R} T^a o_A \frac{\partial}{\partial o^{A'}} = \frac{1}{R} T^a \iota_{A'} \frac{\partial}{\partial \iota^A}, \end{aligned} \tag{4.15.15}$$

$$\begin{aligned} \eth' &= \frac{1}{v} \iota^A \frac{\partial}{\partial o^A} = -\frac{1}{u} o^{A'} \frac{\partial}{\partial \iota^{A'}} \\ &= \frac{1}{R} T^a o_{A'} \frac{\partial}{\partial o^A} = \frac{1}{R} T^a \iota_A \frac{\partial}{\partial \iota^{A'}}, \end{aligned} \tag{4.15.16}$$

and

$$o^A \frac{\partial}{\partial o^A} = p = -\iota^A \frac{\partial}{\partial \iota^A},$$

$$o^{A'} \frac{\partial}{\partial o^{A'}} = q = -\iota^{A'} \frac{\partial}{\partial \iota^{A'}}. \tag{4.15.17}$$

In particular, note that the operators in (4.15.17) are *Euler homogeneity operators*. It follows that:

*If $f_{\mathscr{Q}}$ has type $\{p, q\}$, then when expressed in terms of $o^A, o^{A'}$ it is homogeneous of respective degrees $p, q$, and when expressed in terms of $\iota^A, \iota^{A'}$ it is homogeneous of respective degrees $-p, -q$.* (4.15.18)

## *Conformal motions of $\mathscr{S}$*

There is a significant way of re-interpreting type-$\{p, q\}$ functions on $\mathscr{S}$, when expressed in this way. Since the spinor $o^A$ may be regarded as a (signed) null flag at $L, f_{\mathscr{Q}}$ may be regarded as, in effect, a function of this null flag. The flagpole (namely $l^a$) determines a point on the cone $\mathscr{L}$. When thinking of $f_{\mathscr{Q}}$ as a function of $o^A, o^{A'}$ we need think only in terms of the cone $\mathscr{L}$ and ignore $\mathscr{N}$ completely. The sphere $\mathscr{S}$ itself now becomes an abstract sphere which is the space of generators of $\mathscr{L}$. Recall that in §1.2 we viewed the Riemann sphere in this way. Such a viewpoint is useful when we are interested in conformal transformations of $\mathscr{S}$. For these now arise when an active Lorentz transformation is applied to $\mathbb{M}$, the point $L$ being held fixed. If this is a rotation (with respect to the time-axis $T^a$) then the vertex $N$ of $\mathscr{N}$ is also fixed; but in general $N$ will move. The cone $\mathscr{L}$ (but not $\mathscr{N}$) is mapped to itself, so when $\mathscr{S}$ is viewed as the space of generators of $\mathscr{L}$, it also is mapped to itself. The Lorentz transformations about $L$ thus provide conformal maps of $\mathscr{S}$ to itself.

According to this viewpoint, the significance of the number

$$w = \tfrac{1}{2}(p + q) \tag{4.15.19}$$

is that it provides a *conformal weight* for $f$ (where for simplicity we now assume that $f$ is a weighted *scalar*). Recall that in §1.4 (Fig. 1-11, p. 38) we were able to assign different metrics to the abstract conformal sphere, each compatible with its given conformal structure, simply by taking different cross-sections $\hat{\mathscr{S}}$ of the cone $\mathscr{L}$. (Now see Fig. 4-5.) When $\hat{\mathscr{S}}$ lies in a space-like hyperplane, the metric assigned is that of a metric sphere, but otherwise a more general metric. Consider a particular generator of $\mathscr{L}$ and sup-

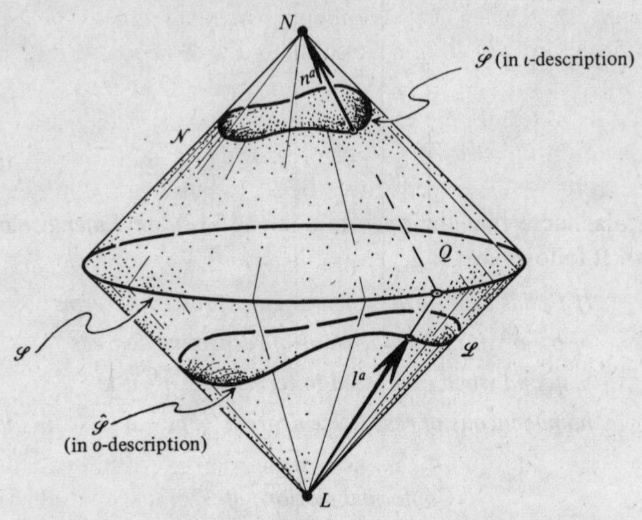


Fig. 4-5. A conformal rescaling of the metric of $\mathscr{S}$ is achieved by moving $\mathscr{S}$ along the generators of $\mathscr{L}$ (or alternatively $\mathscr{N}$) to $\hat{\mathscr{S}}$.

pose $\hat{\mathscr{S}}$ meets it at a point whose position vector from $L$ is $Q^a$. Suppose $\hat{\mathscr{S}}$ is then moved so that the position vector becomes $kQ^a (k > 0)$. The induced metric tensor of the cross-section will correspondingly scale up by the factor $k^2$, on that generator, i.e. the linear scale goes up by $k$. The spinor $o^A$ scales up by $k^{\frac{1}{2}}$, so if $f(o^A, o^{A'})$ has homogeneity degrees $p, q$, then $f$ scales up by $k^{\frac{1}{2}(p+q)}$. This justifies the terminology for (4.15.19) (compare §5.6). Note that the conformal weight $w$ and the *spin-weight* $s = \frac{1}{2}(p - q)$(*cf.* after (4.12.10)) together determine the type $\{p, q\}$ (and vice versa):

$$p = w + s, q = w - s; \quad s = \tfrac{1}{2}(p - q), w = \tfrac{1}{2}(p + q). \qquad (4.15.20)$$

Alternatively, if we view $f$ as a function of $\iota^A, \iota^{A'}$, then $\mathscr{S}$ is interpreted, instead, as the space of generators of $\mathscr{N}$ and the conformal motions of $\mathscr{S}$ arise as a result of applying Lorentz transformations about $N$ (which leave $\mathscr{N}$ invariant). The argument is just as before, and we now find that the conformal weight of $f$ is

$$w' = -\tfrac{1}{2}(p + q). \qquad (4.15.21)$$

Consequently we have

$$p = -w' + s, \quad q = -w' - s. \qquad (4.15.22)$$

To understand this discrepancy with (4.15.19), (4.15.20), we note that the metrics on the abstract conformal sphere are now given by cross sections $\hat{\mathscr{S}}$ of $\mathscr{N}$ and that the correspondence between generators of $\mathscr{L}$ and generators of $\mathcal{N}$ that is given when $Q$ moves about $\mathcal{S}$ involves an *antipodal map*. (See Fig. 4-5; generators of $\mathcal{L}$ may be compared with generators of $\mathcal{N}$ by the translation of $\mathbb{M}$ taking $L$ to $N$.) Thus when a Lorentz transformation is applied to $\mathbb{M}$ (say about $O$, for symmetry's sake) it induces a *different* action on $\mathcal{S}$ regarded as the space of generators of $\mathcal{N}$, from that on $\mathcal{S}$ regarded as the space of generators of $\mathcal{L}$, because of the intercession of this antipodal map. (Though $\mathcal{L}$ and $\mathcal{N}$ are each moved by this action, the abstract spaces of their generator *directions* are each mapped to themselves.)

It should be noted that conformal weight is in general quite a separate concept from the $p$ and $q$ weights here discussed, as will be more fully explained in §5.6. However, in the present work $s$ and $w$ can be used for an alternative type-description and instead of type-$\{p, q\}$ quantities we may speak of type-$[s, w]$ quantities – where we use the $o$-description only, so

$$\{p, q\} = \left[\tfrac{1}{2}(p-q), \tfrac{1}{2}(p+q)\right] = [s, w] = \{w+s, w-s\}. \quad (4.15.23)$$

Let us return to the expressions (4.15.15) and (4.15.16) for ð and ð′. We note that in the lower line of each are expressions involving only $o^A, o^{A'}$ [only $\iota^A, \iota^{A'}$]. If we use these expressions we can stay within the $o^A, o^{A'}$ [*or* $\iota^A, \iota^{A'}$] description and regard the operators as applying within the space of generators of $\mathcal{L}$ [*or* $\mathcal{N}$]. However, owing to the explicit appearance of the vector $T^a$, the operators ð and ð′ are not generally conformally invariant on $\mathcal{S}$.

But it turns out that for each spin-weight $s$ there is a particular conformal weight $w$ for which a given power of ð or of ð′ is effectively conformally invariant (Newman and Penrose 1966, Eastwood and Tod 1982). We use the $o^A, o^{A'}$ description and suppose that $f$ has type $\{p, q\}$ with $p \geqslant 0$. Then

$$\underbrace{\frac{\partial}{\partial o^B} \cdots \frac{\partial}{\partial o^P}}_{p} f \quad (4.15.24)$$

has type $\{0, q\}$, so application of the Euler homogeneity operator $o^A \partial/\partial o^A$ yields zero (*cf.* (4.15.17)):

$$o^A \left\{ \frac{\partial}{\partial o^A} \frac{\partial}{\partial o^B} \cdots \frac{\partial}{\partial o^D} \right\} f = 0.$$

Since the expression $\{\ldots\}$ is symmetric in $AB \ldots D$, it follows from (3.5.27) that

$$\frac{\partial}{\partial o^A} \frac{\partial}{\partial o^B} \cdots \frac{\partial}{\partial o^D} f = \underbrace{o_A o_B \ldots o_D}_{p+1} g \quad (4.15.25)$$

for some scalar $g$, and evidently $g$ is of type $\{-p-2, q\}$. Transvecting this last equation repeatedly with $R^{-1}T^a o_{A'}$ (and noting that this commutes with $\partial/\partial o^E$), we obtain, using also (4.15.16) and (4.15.7),

$$ð'^{p+1}f = v^{-p-1}g. \tag{4.15.26}$$

In the original literature (Newman and Penrose 1966), this formula (or 4.15.30) appears without the '$v$', which is, however, necessary here for the strict applicability of the compacted formalism. (If we scale $o^A$ so that $v = 1$, then $\hat{\mathscr{S}} = \mathscr{S}$.)

At this point it is worth while to list various elementary relations that hold between the quantities we have defined. They are direct consequences of (4.12.28), (4.15.9), (4.15.4), (4.15.6), and (4.15.7):

$$\left.\begin{array}{l} ðo^A = 0,\ ðo^{A'} = v^{-1}\iota^{A'},\ ð\iota^A = -u^{-1}o^A,\ ð\iota^{A'} = 0, \\ ð'o^A = v^{-1}\iota^A,\ ð'o^{A'} = 0,\ ð'\iota^A = 0,\ ð'\iota^{A'} = -u^{-1}o^{A'} \end{array}\right\} \tag{4.15.27}$$

$$ðu = 0,\ ðv = 0,\ ð'u = 0,\ ð'v = 0 \tag{4.15.28}$$

and, from (4.14.1),

$$(ðð' - ð'ð)f = -sR^{-2}f, \tag{4.15.29}$$

where $f$ is any $\{p, q\}$-scalar and $s = \frac{1}{2}(p-q)$.

By (4.15.28) (4) we can rewrite (4.15.26) as

$$g = (vð')^{p+1}f. \tag{4.15.30}$$

Note that (4.15.25) makes no mention of $\iota^A$, $\iota^{A'}$, or $T^a$. The relation between $f$ (type $\{p, q\}$) and $g$ (type $\{-p-q-2, q\}$ expressed by (4.15.30) is therefore Lorentz invariant. As a particular case of (4.15.26) we have that $ð'^{p+1}f = 0$ is a Lorentz invariant equation, as is, similarly, $ð^{q+1}f = 0$ (*cf.* (4.15.32) below).

Since (restricted) Lorentz transformations (centred at $L$) can be identified with the (orientation preserving) conformal motions of $\mathscr{S}$, this Lorentz invariance may be reinterpreted as a *conformal invariance* of the operation in (4.15.30). It is not, however, quite the general local conformal invariance that we shall discuss at length in §5.6. For that, arbitrary rescaling of the metric would be allowed, corresponding, here, to the passage from $\mathscr{S}$ to an arbitrary cross-section $\hat{\mathscr{S}}$ of $\mathscr{L}$, whose induced metric need not be intrinsically that of a sphere. The formula (4.15.30) would not in general hold (except if $p = 0$) for the $ð'$-operator intrinsic to $\hat{\mathscr{S}}$ (i.e. defined with respect to an $n^a$-vector locally orthogonal to $\hat{\mathscr{S}}$). For the particular cases for which $\hat{\mathscr{S}}$ *is* intrinsically metrically a sphere, however, namely those cases where $\hat{\mathscr{S}}$ is the intersection of $\mathscr{L}$ with a spacelike hyperplane (*cf.* Fig. 1-11), then (4.15.30) *does* hold in this sense.

For we need only apply a Lorentz transformation which sends the normal to that hyperplane into the $T^a$-direction and the argument for invariance is as given above.

Assuming, now, that $q \geqslant 0$ (with $p$ unrestricted), we can apply the *complex conjugate* of the above argument to obtain the existence of a $\{p, -q-2\}$-scalar $h$ satisfying

$$\frac{\partial}{\partial o^{A'}} \cdots \frac{\partial}{\partial o^{E'}} f = \underbrace{o_{A'} \ldots o_{E'}}_{q+1} h \tag{4.15.31}$$

and

$$h = (v\eth)^{q+1} f. \tag{4.15.32}$$

If *both* $p \geqslant 0$ and $q \geqslant 0$, then we can apply (4.15.30) to $h$, which yields

$$j = (v\eth')^{p+1} h \tag{4.15.33}$$

for some $\{-p-2, -q-2\}$-scalar $j$. Similarly, we can apply (4.15.32) to $g$ and find, for the *same* $j$,

$$j = (v\eth)^{q+1} g, \tag{4.15.34}$$

as follows at once, if we revert to the forms (4.15.25) and (4.15.31), by the commutativity of $\partial/\partial o^A$ and $\partial/\partial o^{A'}$. This means that

$$\eth^{q+1}\eth'^{p+1} f = \eth'^{p+1}\eth^{q+1} f. \tag{4.15.35}$$

(It is somewhat more involved to obtain this directly from (4.15.29).) Indeed, since we can change the type of $f$ from $\{p+k, q+k\}$ to $\{p, q\}$ by multiplying it by a power of $v$, it is only the value of the difference $p-q$ in (4.15.35) which is relevant. So we can derive the apparently more general form (Newman and Penrose 1966)

$$\eth^a \eth'^b f = \eth'^b \eth^a f, \tag{4.15.36}$$

for any $a, b$ such that $b - a = 2s$, $s$ being the spin-weight of $f$.

We can also parallel the argument leading to (4.15.26) etc. using the $\iota^A, \iota^{A'}$ description. If $F$ is of type $\{p, q\}$ with $p \leqslant 0$ we find

$$\frac{\partial}{\partial \iota^{A}} \cdots \frac{\partial}{\partial \iota^{D}} F = \underbrace{\iota_{A} \ldots \iota_{D}}_{-p+1} G \tag{4.15.37}$$

and

$$G = (u\eth)^{-p+1} F, \tag{4.15.38}$$

with $G$ of type $\{-p-2, q\}$. Similarly, if $q \leqslant 0$ we find

$$\frac{\partial}{\partial \iota^{A'}} \cdots \frac{\partial}{\partial \iota^{E'}} F = \underbrace{\iota_{A'} \ldots \iota_{E'}}_{-q+1} H \tag{4.15.39}$$

and

$$H = (u\eth')^{-q+1} F, \tag{4.15.40}$$

with $H$ of type $\{p, -q-2\}$. That these are exactly the same results as before is readily seen if we re-express $p$ and $q$ in terms of the spin- and conformal weights $s$ and $w$ of (4.15.20). We see that (4.15.32) involves $\eth^{w-s+1}$ acting on a quantity of spin-weight $s$ and conformal weight $w \geqslant -s$, while from (4.15.22) we find that (4.15.38) involves $\eth^{w'-s+1}$ acting on a quantity of spin-weight $s$ and conformal weight $w' \geqslant -s$. The replacement of $v$ in (4.15.32) by $u$ in (4.15.38) is connected to the facts that the conformal scalings act differently under Lorentz transformations, as previously noted, and that $w$ is replaced by $w'$. The correspondence between (4.15.30) and (4.15.40) is precisely similar.

Of particular note are the homogeneous *polynomials*

$$f = f_{A\ldots DE'\ldots H'} \underbrace{o^A \ldots o^D}_{p} \underbrace{o^{E'} \ldots o^{H'}}_{q} \tag{4.15.41}$$

the $o^A, o^{A'}$ description being now reverted to, where $f_{\ldots}$ is *constant* and, without loss of generality, symmetric. We shall sometimes use the notation $\mathbb{S}_{\mathscr{A}}$, in $\mathbb{M}$, for the subsystem (vector space over $\mathbb{C} = \mathbb{S} = \mathfrak{K}$; *cf.* after (4.1.2)) of elements of $\mathfrak{S}_{\mathscr{A}}$ which are *constant* throughout $\mathbb{M}$. Adopting, also, the bracket notation of (3.3.14), we can now write the condition on $f_{\ldots}$ as

$$f_{\underbrace{A\ldots D}_{p}\underbrace{H'\ldots H'}_{q}} \in \mathbb{S}_{(A\ldots D)(E'\ldots H')}, \tag{4.15.42}$$

Clearly $f$ has type $\{p, q\}$, i.e., $[s, w] = [\frac{1}{2}(p-q), \frac{1}{2}(p+q)]$. Under restricted Lorentz transformations about $L$ (i.e., proper under conformal motions of $\mathscr{S}$), these polynomials transform into one another according to a $(p+1)(q+1)$-dimensional complex representation. Such representations, of course, are the symmetric ('irreducible') spinors of §3.5, now playing a new role.*

## *Rotations of $\mathscr{S}$*

Let us now restrict the transformations in question to *rotations* about $T^a$ (leaving $L$, $N$, and $T^a$ invariant). Take $f$ to be given by (4.15.41); we find that the $(p+1)(q+1)$-dimensional space splits up into a direct sum

* In fact the general representation theory of the Lorentz group can be expressed in terms of weighted scalar functions $f$ of type $\{p, q\} = [s, w]$, where $2s = p - q$ is an integer but $w$ is arbitrary complex. The *finite-dimensional* irreducible representations occur when $f$ is a polynomial as in (4.15.41). *Unitary* representations (necessarily infinite-dimensional) occur when $w+1$ is purely imaginary, or when $s = 0$ and $-2 \leqslant w \leqslant 0$. (*cf.* Naimark 1964, Carmeli 1977). (See also footnote on p. 301)

of spaces *each* of which is invariant, and whose dimensions are

$$|p-q|+1, |p-q|+3, |p-q|+5, \ldots, p+q-1, p+q+1. \tag{4.15.43}$$

This can be seen as follows. First we write $o^{A'}$ in terms of $\iota^A$ according to the complex conjugate of (4.15.6),

$$o^{A'} = \frac{u}{R} T_A^{A'} \iota^A, \tag{4.15.44}$$

and substitute it into (4.15.41). Since $T^a$ (and therefore also $u$, *cf.* (4.15.4)) is invariant under rotations, this is an invariant procedure. Keeping the dependence on $u$ explicit, so as to preserve strict invariance under spin-frame rescaling, we arrive at an expression of the form

$$f = u^q t_{A\ldots DE\ldots H} o^A \ldots o^D \iota^E \ldots \iota^H, \tag{4.15.45}$$

with

$$t_{\underbrace{A\ldots D}_{p}\underbrace{E\ldots H}_{q}} \in \mathbb{S}_{(A\ldots D)(E\ldots H)} \tag{4.15.46}$$

Since $t_{A\ldots H}$ is not totally symmetric, it can be reduced, following the procedure of §3.3, into a number of pieces each of which is totally symmetric but with varying numbers of indices:

$$t_{(A\ldots DE\ldots H)}, t_{A(B\ldots D}{}^{A}{}_{F\ldots H)}, t_{AB(C\ldots D}{}^{AB}{}_{G\ldots H)}, \tag{4.15.47}$$

and so on, until one or the other of the two original groups of indices is exhausted. These are symmetric unprimed spinors with, respectively, $p+q, p+q-2, p+q-4, \ldots, |p-q|$ indices, so that they have, respectively, $p+q+1, p+q-1, p+q-3, \ldots, |p-q|+1$ independent components. Denoting the various *totally symmetric* spinors in (4.15.47) by

$$\overset{0}{t}_{A\ldots DE\ldots H}, \overset{1}{t}_{B\ldots DF\ldots H}, \overset{2}{t}_{C\ldots DG\ldots H}, \text{etc.},$$

respectively, we find that the original $t_{A\ldots D}{}^{E\ldots H}$ (writing the second set of indices raised, for notational convenience) can be expressed as a linear combination of

$$\overset{0}{t}_{A\ldots D}{}^{E\ldots H}, \varepsilon_{(A}{}^{(E}\overset{1}{t}_{B\ldots D)}{}^{F\ldots H)}, \varepsilon_{(A}{}^{(E}\varepsilon_{B}{}^{F}\overset{2}{t}_{C\ldots D)}{}^{G\ldots H)}, \text{etc.} \tag{4.15.48}$$

Substituting that into (4.15.45) we obtain $f$ as the corresponding linear combination of terms

$$\begin{aligned}
\overset{0}{f} &= u^q \overset{0}{t}_{A\ldots DE\ldots H} o^A \ldots o^D \iota^E \ldots \iota^H, \\
\overset{1}{f} &= u^q \overset{1}{t}_{B\ldots DF\ldots H} o^B \ldots o^D \iota^F \ldots \iota^H, \\
\overset{2}{f} &= u^q \overset{2}{t}_{C\ldots DG\ldots H} o^C \ldots o^D \iota^G \ldots \iota^H, \text{etc.,}
\end{aligned} \tag{4.15.49}$$

the ε-terms disappearing, since $\varepsilon_{AE}o^A\iota^E = 1$. In fact, it turns out that

$$f = \overset{0}{f} + \frac{pq}{(p+q)}\overset{1}{f} + \frac{p(p-1)q(q-1)}{2(p+q-1)(p+q-2)}\overset{2}{f} + \cdots + \frac{(|p-q|+1)}{(1+\max(p,q))}\overset{\min(p,q)}{f}$$

$$= \sum_{r=0}^{\min(p,q)} \frac{(p+q-2r+1)!\,p!\,q!\,\overset{r}{f}}{(p-r)!(q-r)!\,r!\,(p+q-r+1)!}$$

$$= \sum_{r=0}^{\min(p,q)} \frac{\binom{p}{r}\binom{q}{r}\overset{r}{f}}{\binom{p+q-r+1}{r}}. \tag{4.15.50}$$

The $\overset{i}{f}$s of (4.15.49) can be translated back to the original form (4.15.41) (i.e., in terms of $o^A, o^{A'}$), if desired, by use of (4.15.7). They provide the irreducible pieces of $f$ under rotations that were referred to earlier. And they span invariant subspaces of spin-weighted functions on $\mathscr{S}$ of respective dimensions $p+q+1, p+q-1, p+q-3, \ldots, |p-q|+1$, as was asserted in (4.15.43).

In order to determine whether any given function of this form, say

$$h = u^q h_{A\ldots DE\ldots H} o^A \ldots o^D \iota^E \ldots \iota^H, \tag{4.15.51}$$

actually belongs to one of these subspaces, we need a property that characterizes the possibility of writing (4.15.51) with $h_{A\ldots H}$ totally symmetric:

$$h_{A\ldots DE\ldots H} \in \mathbb{S}_{(A\ldots H)}. \tag{4.15.52}$$

Let the number of indices of $h_{A\ldots H}$ be $2j$, where $j$ is integral or half-integral, so the indices $A\ldots D$ of (4.15.51) are $j+s$ in number and the indices $E\ldots H$ are $j-s$ in number. Note that $j$ is integral if and only if $s$ is, and that

$$-j \leqslant s \leqslant j.$$

Now consider the action of ð on (4.15.51). From (4.15.27), (4.15.28) we obtain

$$ðh = -(j-s)u^{q-1}h_{A\ldots DEF\ldots H}\underbrace{o^A \ldots o^D o^E}_{j+s+1}\underbrace{\iota^F \ldots \iota^H}_{j-s-1}. \tag{4.15.53}$$

If we apply ð′ to this equation, and use (4.15.27), (4.15.28), and (4.15.3), we find

$$ð'ðh = -(j+s+1)(j-s)\tfrac{1}{2}R^{-2}h. \tag{4.15.54}$$

Thus $h$ is an eigenfunction of the operator $\eth'\eth$ with eigenvalue

$$-(j+s+1)(j-s)\tfrac{1}{2}R^{-2}=[s(s+1)-j(j+1)]\tfrac{1}{2}R^{-2}. \quad (4.15.55)$$

From (4.15.29) it follows that $h$ is also an eigenfunction of $\eth\eth'$:

$$\eth\eth' h=-(j-s+1)(j+s)\tfrac{1}{2}R^{-2}h, \quad (4.15.56)$$

now with eigenvalue

$$-(j-s+1)(j+s)\tfrac{1}{2}R^{-2}=[s(s-1)-j(j+1)]\tfrac{1}{2}R^{-2}. \quad (4.15.57)$$

These eigenvalues characterize both $s$ and $j$ since we get $s$ directly from the commutator (4.15.29), whereupon $j(j+1)$ is fixed by (4.15.55), and this determines $j$ since $j\geqslant 0$. For each spin-weight $s$ we refer to the eigenfunctions (4.15.49) of $\eth'\eth$ as *spin-weighted spherical harmonics.** Our discussion here has been confined to polynomial expressions (4.15.41). But it can be shown (though this is beyond the scope of the present work) that *any* (continuous) spin-weighted function on $\mathscr{S}$ can be expressed as an (infinite) sum of such polynomials, so that the spin-weighted spherical harmonics as defined here constitute, in fact, a complete system. (For the case $s=0$, see Courant and Hilbert 1965; completeness for the cases $s\neq 0$ can be readily deduced from that for $s=0$.)

### Linear equations in $\eth$

From (4.15.53) we see that if $h$ is any spin-weighted spherical harmonic with $j=s$, then $\eth h=0$. In fact, the converse is also true:

(4.15.58) PROPOSITION

*If $f$ is any smooth $\{p,q\}$-function on $\mathscr{S}$, then $\eth f=0$* [or $\eth' f=0$] *throughout $\mathscr{S}$ iff $f$ is a spin-weighted spherical harmonic with $j=s$* [or $j=-s$] $(s=\tfrac{1}{2}(p-q))$.

*Proof*: If we *assume* completeness of the polynomial harmonics (4.15.49), the proof is immediate from (4.15.53). But we can also show directly that any solution of $\eth f=0$ must be polynomial in this sense by appealing to results in complex analysis. Consider the function $f_0=u^{-q}f$ which has $q=0$ By (4.15.17), $o^{A'}\partial f_0/\partial o^{A'}=0$. Assume $\eth f=0$. Then by (4.15.15) we also have $\iota^{A'}\partial f_0/\partial o^{A'}=0$ and so $\partial f_0/\partial o^{A'}=0$. Consequently $f_0$ is holomorphic in $o^A$. It is also homogeneous of degree $2s(=p)$ and global

* In the literature (Newman and Penrose 1966, Goldberg *et al.* 1967) this term is usually reserved for the functions that arise after a further reduction with respect to a particular basis in $\mathbb{V}$ has been made, *cf.* (4.15.93).

on $\mathscr{S}$. But the only global homogeneous holomorphic functions on $\mathbb{C}^2$ are polynomials (*cf*. Gunning 1966). From this it follows that $f_0$ is a polynomial and therefore $f$ has the required form

$$f = u^q f_{A\ldots G} o^A \ldots o^G.$$

The argument for the case $\eth' f = 0$ follows from the above by complex conjugation.

As a corollary to (4.15.58) (since there are no spin-weighted spherical harmonics for negative $j$) we deduce

(4.15.59) PROPOSITION

*If $f$, defined on $\mathscr{S}$, has negative [positive] spin-weight then $\eth f = 0$ [or $\eth' f = 0$] implies $f = 0$.*

In the study of spin-weighted spherical harmonics it is useful to contemplate the following array:

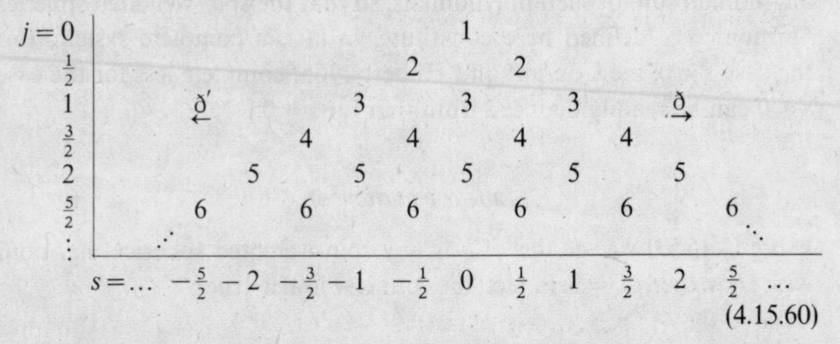

(4.15.60)

The numbers in this triangular array (which extends indefinitely downwards) represent the complex *dimensions* of the various spaces of spin-weighted spherical harmonics, as discussed in (4.15.43) *et seq.* Each of these spaces is characterized by its values of $s$ and $j$, as shown. The dimension *zero* is assigned wherever a blank space appears in the array. The operator $\eth$ carries us a step of one $s$-unit to the right and $\eth'$ one $s$-unit to the left. (From our earlier discussion, the $j$-value is not affected by $\eth$ or $\eth'$.) Whenever such a step carries us off the array, the result of the operator $\eth$ or $\eth'$ is zero. Note that the dimension remains constant whenever it does not drop to, or increase from, zero.

Within the array itself, the operators $\eth$ and $\eth'$ are *invertible*, since by (4.15.54) and (4.15.56), each of $\eth$, $\eth'$ acts as a multiple of the inverse of the other. But at the right-hand sloping edge the effect of $\eth$ is to annihilate one of the spaces, and similarly for $\eth'$ at the left-hand sloping edge.

Just to the left of the left-hand edge, the $\eth$ operator moves us from a zero to a finite dimension. These are the circumstances where the equations

$$\eth f = g \tag{4.15.61}$$

are not soluble. For example, if $g$ has spin-weight $s = -\frac{1}{2}$, and possesses a $j = \frac{1}{2}$ part, then there is no $f$ satisfying (4.15.61). For think of $g$ as occupying points in the $s = -\frac{1}{2}$ column, with a non-zero contribution at 2. The function $f$ would have to lie in the $s = -\frac{3}{2}$ column and could only occupy points 4, 6, ... . The operation $\eth$ cannot manufacture a non-zero $j = \frac{1}{2}$ part so as to produce the '2'. But if $g$ is such that its $j = \frac{1}{2}$ part *vanishes*, then (4.15.61) *is* soluble. In fact it is *uniquely* soluble because, as is clear from (4.15.60), $\eth$ does not annihilate any non-zero $s = -\frac{3}{2}$ quantity.

The situation for (4.15.61) is just the reverse when, say, $s = \frac{3}{2}$. In this case a glance at (4.15.60) shows that (4.15.61) is *always* soluble. For $g$ then lies in the $s = \frac{3}{2}$ column, with dimension numbers 4, 6, ... , and all these numbers are also available in the $s = \frac{1}{2}$ column. But now there is a 2-dimensional space annihilated by $\eth$, so that the solutions of (4.15.61) are non-unique, the non-uniqueness being precisely in the 2-dimensional $j = \frac{1}{2}$ space.

We remark – though we shall not make use of it – that $\eth$ possesses a unique *generalized inverse* $\eth^\dagger$, satisfying

$$\eth\eth^\dagger\eth = \eth, \quad \eth^\dagger\eth\eth^\dagger = \eth^\dagger$$

and

$$\eth\eth^\dagger = \overline{\eth\eth^\dagger}, \quad \eth^\dagger\eth = \overline{\eth^\dagger\eth}$$

(*cf.* Moore 1920, Penrose 1955, Nashed 1976, Exton, Newman and Penrose 1969). The action of $\eth^\dagger$ is that of the (unique) inverse of $\eth$ on those spin-weighted spherical harmonic spaces for which an inverse exists (i.e., for all spaces in (4.15.60) except those represented by the left-hand sloping column); and it is zero otherwise (i.e., *on* that sloping column). Similarly we can define $\eth'^\dagger$, and we find

$$\eth'^\dagger = \overline{\eth^\dagger}$$

(in the usual sense that $\overline{\eth^\dagger} f := \overline{\eth^\dagger \bar{f}}$). By use of $\eth^\dagger$, the general solution of (4.15.61) can be written as

$$f = \eth^\dagger g + (1 - \eth^\dagger\eth)h,$$

where $h$ is arbitrary (but of the appropriate spin-weight), the condition for solubility being $(1 - \eth\eth^\dagger)g = 0$.

### *Conformal behaviour of harmonics*

The array (4.15.60) is useful also in the study of the conformal properties of type-$\{p, q\}$ quantities on $\mathscr{S}$. First, suppose that the conformal weight

$w = \frac{1}{2}(p + q)$ satisfies

$$w \geqslant |s|, \tag{4.15.62}$$

i.e., that $w$ has one of the values

$$|s|, |s| + 1, |s| + 2, \ldots, \tag{4.15.63}$$

where we are adopting the $o^A, o^{A'}$ description, with Lorentz transformations taken about $L$. Then the homogeneities $p, q$ are both non-negative and we have finite-dimensional spaces of polynomials (4.15.41) invariant under conformal motions of $\mathscr{S}$. These spaces reduce, as we have seen, into subspaces (the spin-weighted spherical harmonic subspaces) which are invariant under rotations, but which under non-trivial conformal motions get mixed with one another. In the array (4.15.60), let us fix our attention on the point $(s, j)$, where $j = w$, and consider the set of points of the $s$-column above and including that point. Under conformal motions of $\mathscr{S}$, the spaces represented by these points get mixed with one another, but do not spread to the points below. We note, incidentally, that the unique powers of ð and of ð′ which have conformal invariance properties (namely $ð^{w-s+1}$ and $ð'^{w+s+1}$ – *cf.* (4.15.30), (4.15.33)) annihilate *precisely* the spaces represented by the points under consideration. A space represented by any *other* point of this $s$-column makes contributions along the *entire* column under a general conformal motion of $\mathscr{S}$. The special property of those particular weights $w$ that occur as the allowable $j$-values for a given $s$ is that they provide the *finite-dimentional* representations of the restricted Lorentz group for spin-weight $s$, namely the descriptions in terms of $\mathscr{S}$ of symmetric spinors of valence $\left[\begin{smallmatrix} 0 & 0 \\ p & q \end{smallmatrix}\right]$ with $2s = p - q$.

There is no other Lorentz-invariant subspace of the entire space of $[s, w]$-functions on $\mathscr{S}$ for these choices of $w$. There is, however, a *dual* situation for which $w$ takes one of the values

$$-|s| - 2, -|s| - 3, -|s| - 4, \ldots, \tag{4.15.64}$$

corresponding to (4.15.63) term for term (i.e., with $w$ replaced by $-w-2$). Here we find that the points just considered in (4.15.60) are precisely the ones representing spin-weighted spherical harmonic spaces that do not get contributions *added* to them when a general conformal motion of $\mathscr{S}$ is applied. Thus, in particular, it is conformally invariant to say that *all* the parts of $f$ *vanish* which belong to the spaces represented by these points (with $w$ as in (4.15.64)). These $f$s are those having the form

$$f = (\nu ð)^{s-w-1} g \textit{ for some } [w + 1, s - 1]\textit{-quantity } g, \tag{4.15.65}$$

or, equivalently, those having the form

$$f = (v\eth')^{-s-w-1} g \text{ for some } [-w-1, -s-1]\text{-quantity } g, \tag{4.15.66}$$

as is evident from the table (4.15.60). As we noted in (4.15.30) *et seq*., the operations in (4.15.65) and (4.15.66) are conformally invariant. This establishes our assertion that $f$s of this form will transform among themselves under conformal motions of $\mathscr{S}$.*

The duality that is involved here arises from the fact that there is a conformally invariant *Hermitian scalar product* between $[s, w]$-scalars $f$ and $[s, -w-2]$-scalars $h$ on $\mathscr{S}$ (or, equivalently, between $[s, w]$ and $[-s, -w-2]$-scalars, if we prefer not to incorporate the complex conjugation in the definition and, instead, define a holomorphic rather than, as here, a Hermitian scalar product), namely

$$\langle h, f \rangle := \frac{1}{2\pi} \oint_{\mathscr{S}} \bar{h} f \hat{\mathscr{S}} \tag{4.15.67}$$

where $\hat{\mathscr{S}}$ is the surface-area 2-form for $\hat{\mathscr{S}}$ (*cf.* Fig. 4-5). Taking $\mathscr{S} = \mathrm{i}\bar{\boldsymbol{m}} \wedge \boldsymbol{m}$ to be the surface-area 2-form for $\mathscr{S}$ (as in (4.14.65)) we have

$$\hat{\mathscr{S}} = v^{-2} \mathscr{S} \tag{4.15.68}$$

(using the $o$-description; in the $\iota$-description we would adopt $\hat{\mathscr{S}}' = u^{-2}\mathscr{S}$ in (4.15.67)).

Note that on $\mathscr{S}$ we have

$$\begin{aligned} \mathscr{S} &= \mathrm{i}\bar{\boldsymbol{m}} \wedge \boldsymbol{m} = \mathrm{i}\bar{m}_a dx^a \wedge m_b dx^b \\ &= \mathrm{i}\iota_A o_{A'} d(v o^A o^{A'}) \wedge o_B \iota_{B'} d(v o^B o^{B'}) \\ &= \mathrm{i} v^2 o_{A'} do^{A'} \wedge o_B do^B \end{aligned}$$

whence

$$\hat{\mathscr{S}} = \mathrm{i} o_{A'} do^{A'} \wedge o_B do^B \tag{4.15.69}$$

(showing explicitly that $\hat{\mathscr{S}}$ does not depend on $\iota^A$ or $v$). Since $\hat{\mathscr{S}}$ scales as a $[0, 2]$-quantity, the integrand in (4.15.67), and therefore the integral itself, is conformally invariant. Note that the conformal weights (4.15.63) and (4.15.64) are dual in the sense that when paired they yield a conformally invariant scalar product.

Suppose now that $f$ (with $w$ from the list (4.15.64)) has the form (4.15.65).

* In fact, $[s, w]$-scalars of this form provide an infinite-dimensional *irreducible* representation of the restricted Lorentz group (*cf.* footnote on p. 294). If $w$ is not related to $s$ in either of these ways (indeed, $w$ may be complex), then the *entire* space of $[s, w]$-scalars provides an irreducible infinite-dimensional representation of the restricted Lorentz group (*cf.* Naimark 1964, Gel'fand, Graev and Vilenkin 1966, pp. 141, 156).

Then

$$\langle h, f \rangle = \frac{1}{4\pi}\oint \bar{h}(v\eth)^{s-w-1} g \mathscr{S}$$

$$= \frac{1}{4\pi}\oint g(-v\eth)^{s-w-1}\bar{h}\mathscr{S}$$

$$= \langle k, g \rangle,$$

with

$$k = (-v\eth')^{s-w-1}h,$$

where we have repeatedly used the formula (4.14.71)(2) for integration by parts. Now $k=0$ if and only if $h$ belongs to the subspace of $[s, -w-2]$-scalars annihilated by $(v\eth')^{s-w-1}$, this being one of the finite-dimensional spaces which are spanned by the spin-weighted spherical harmonic spaces discussed earlier, and which transform among themselves under conformal motions of $\mathscr{S}$. Thus any such $h$ is orthogonal to $f$. But since $g$ can be chosen arbitrarily in this argument, we see that the above $f$s are *precisely* the $[s, w]$-scalars orthogonal to all such $[s, -w-2]$-scalars $h$. The conformal invariance of the $f$-space is therefore implied by that of the $h$-space (with $k=0$) and vice versa.

### *Orthogonality of harmonics*

The scalar product (4.15.67) also has importance when we are concerned only with rotationally and not conformally invariant properties of $\mathscr{S}$. Then the 'conformal' weights of $h$ and $f$ are irrelevant and we can revert to our original viewpoint according to which quantities are defined at points of $\mathscr{S}$ (rather than $\hat{\mathscr{S}}$) with respect to local spin-frames $o^A, \iota^A$. The total *boost*-weight of the integrand must be *zero* and the spin-weights of $f$ and $g$ must be equal. Taking $f$ to have type $\{p, q\}$ and $h$ to have type $\{-q, -p\}$ we define

$$\langle h, f \rangle = \frac{1}{4\pi R^2}\oint \bar{h} f \mathscr{S}. \tag{4.15.70}$$

The expressions (4.15.70) and (4.15.67) are consistent with one another, being both special cases of

$$\langle h, f \rangle = \frac{1}{4\pi R^2}\oint \bar{h} f \left(\frac{v}{u}\right)^{c/2} \mathscr{S} \tag{4.15.71}$$

where $h$ and $f$ are such that the product $\bar{h}f$ has type $\{c, c\}$ (*cf.* (4.15.3)).

It is easily seen that the scalar product (4.15.71) (when the weights are such that it is meaningful) has the following standard properties:

$$\langle h, f \rangle = \overline{\langle f, h \rangle} \tag{4.15.72}$$

$$\langle f, f \rangle > 0 \quad \textit{unless} \quad f = 0 \tag{4.15.73}$$

$$\langle h, \lambda f \rangle = \lambda \langle h, f \rangle = \langle \bar{\lambda} h, f \rangle \tag{4.15.74}$$

$$\langle h, f + g \rangle = \langle h, f \rangle + \langle h, g \rangle \tag{4.15.75}$$

$$\langle h + k, f \rangle = \langle h, f \rangle + \langle k, f \rangle \tag{4.15.76}$$

$$\langle h, \eth f \rangle = -\langle \eth' h, f \rangle \tag{4.15.77}$$

$$\langle h, \eth' f \rangle = -\langle \eth h, f \rangle \tag{4.15.78}$$

Moreover, we have the following

(4.15.79) PROPOSITION

*If $f$ and $h$ are spin-weighted spherical harmonics of equal spin-weight $s$ corresponding to different $j$-values, then $\langle h, f \rangle = 0$.*

*Proof*: This is essentially a standard property of eigenfunctions of operators. If we write $j, \hat{j}$ for the respective $j$-values of $f, h$, we have, applying (4.15.54) and (4.15.55) twice, (4.15.77) and (4.15.78),

$$\begin{aligned}\langle h, f \rangle &= \frac{2R^2}{[s(s+1) - j(j+1)]} \langle h, \eth' \eth f \rangle \\ &= \frac{2R^2}{[s(s+1) - j(j+1)]} \langle \eth' \eth h, f \rangle \\ &= \frac{[s(s+1) - \hat{j}(\hat{j}+1)]}{[s(s+1) - j(j+1)]} \langle h, f \rangle,\end{aligned}$$

whence $\langle h, f \rangle = 0$ if $j \neq \hat{j}$.

We next evaluate $\langle h, f \rangle$ explicitly when $h$ and $f$ are, respectively, type-$\{-q, -p\}$ and type-$\{p, q\}$ spin-weighted spherical harmonics with the *same* $j$-values. We can write

$$f = W^q f_{A \ldots DE \ldots K} \underbrace{o^A \ldots o^D}_{j+s} \underbrace{\iota^E \ldots \iota^K}_{j-s} \tag{4.15.80}$$

$$h = W^{-p} h_{A \ldots DE \ldots K} \underbrace{o^A \ldots o^D}_{j+s} \underbrace{\iota^E \ldots \iota^K}_{j-s}, \tag{4.15.81}$$

with $f_{A \ldots K}, h_{A \ldots K} \in \mathfrak{S}_{(A \ldots K)}$, where for the sake of symmetry we have introduced the $\{1, 1\}$-quantity

$$W = \sqrt{\frac{u}{v}} = \frac{u}{R\sqrt{2}} = \left(\frac{v}{R\sqrt{2}}\right)^{-1}. \tag{4.15.82}$$

(Note that $W=1$ would correspond to the standard scaling for which $T^a = 2^{-\frac{1}{2}}(l^a + n^a)$.) Taking the conjugate of (4.15.81), and using (4.15.44) and the conjugate of (4.15.7), we get

$$\bar{h} = W^{-q} H_{A\ldots DE\ldots K} \underbrace{\iota^A \ldots \iota^D}_{j+s} \underbrace{o^E \ldots o^K}_{j-s}, \tag{4.15.83}$$

where

$$H_{A\ldots K} = (-1)^{j-s} 2^j \bar{h}_{A'\ldots K'} T_A^{A'} \ldots T_K^{K'}. \tag{4.15.84}$$

With these expressions we now obtain

$$\langle h, f\rangle = \frac{(-1)^{2j}}{4\pi R^2} \oint H_{A\ldots D}{}^{E_0\ldots K_0} f^{A_0\ldots D_0}{}_{E\ldots K} \iota^A \ldots \iota^D o_{E_0} \ldots o_{K_0} o_{A_0} \ldots o_{D_0} \iota^E \ldots \iota^K \mathscr{S}. \tag{4.15.85}$$

Since $H_{\ldots}{}^{\cdots}$ and $f_{\ldots}{}^{\cdots}$ are both constant, they can be brought outside the integral, which can then be evaluated using the following

(4.15.86) LEMMA

$$\oint \underbrace{o_{A_0} \ldots o_{K_0}}_{r} \underbrace{\iota^A \ldots \iota^K}_{r} \mathscr{S} = \frac{4\pi R^2}{r+1} \varepsilon_{A_0}{}^{(A} \ldots \varepsilon_{K_0}{}^{K)}.$$

We can check this component by component using the formulae (4.15.96) and (4.15.123) to be given below, and simple explicit integration. However, the lemma can also be obtained directly, effectively without calculation, by observing that the LHS is invariant under rotations of the sphere $\mathscr{S}$, whence the RHS must be so also; it must therefore be constructible from $T_{AA'}$ alone by spinor operations (and numerical constants). The elimination of the primed indices on $T_{AA'}$ leads, via

$$T_{AA'} T^{BA'} = \tfrac{1}{2} \varepsilon_A{}^B \tag{4.15.87}$$

($T_a$ being a unit timelike vector), to the elimination of $T_a$ altogether, leaving us with a term proportional to the RHS of (4.15.86) as the only possibility. Finally the numerical coefficient is obtained by taking traces of both sides, observing that

$$\oint \mathscr{S} = 4\pi R^2 \tag{4.15.88}$$

and that the idempotence

$$\varepsilon_{A_1}{}^{(A_0} \ldots \varepsilon_{K_1}{}^{K_0)} \varepsilon_{A_0}{}^{(A} \ldots \varepsilon_{K_0}{}^{K)} = \varepsilon_{A_1}{}^{(A} \ldots \varepsilon_{K_1}{}^{K)}$$

entails that the trace $\varepsilon_A{}^{(A} \ldots \varepsilon_K{}^{K)}$ is equal to the rank of $\varepsilon_{A_0}{}^{(A} \ldots \varepsilon_{K_0}{}^{K)}$, i.e., to the dimension of $\mathfrak{S}^{(A\ldots K)}$, which is $r+1$. (This is using the familiar property of an idempotent matrix that its rank is equal to its trace.)

Substituting (4.15.86) into (4.15.85) yields

$$\langle h, f \rangle = \frac{(-1)^{2j}}{2j+1} H_{A\ldots D}{}^{(E\ldots K} f^{A\ldots D)}{}_{E\ldots K} \tag{4.15.89}$$

If we expand the symmetrization and note that the symmetries of $H_{A\ldots K}$ and of $f_{A\ldots K}$ entail that $H_{\ldots G\ldots}{}^{\ldots G\ldots} = 0 = f_{\ldots G\ldots}{}^{\ldots G\ldots}$, we finally obtain the desired explicit formula for the product:

$$\begin{aligned} \langle h, f \rangle &= \frac{(-1)^{j-s}(j+s)!(j-s)!}{(2j+1)!} H^{A\ldots K} f_{A\ldots K} \\ &= \frac{2^j (j+s)!(j-s)!}{(2j+1)!} \bar{h}_{A'\ldots K'} T^a \ldots T^k f_{A\ldots K}. \end{aligned} \tag{4.15.90}$$

### *Orthonormal basis for spin-weighted functions*

The orthogonality property (4.15.79) can be carried further, and a complete orthonormal basis for spin-weighted scalar functions on $\mathscr{S}$ can be obtained. These are the functions ${}_sY_{j,m}$ defined in the literature (Newman and Penrose 1966, Goldberg *et al.* 1967). We shall here refer to them as the *basic* spin-weighted spherical harmonics. They depend, however, on an (arbitrary) choice of basis for $\mathbb{S}^A$. Let us take this *constant* basis to be a *spin-frame* $\varepsilon_{\hat{\mathbf{A}}}{}^A, = (\hat{o}^A, \hat{\iota}^A)$, for which

$$T^a = \frac{1}{\sqrt{2}}(\hat{o}^A \hat{o}^{A'} + \hat{\iota}^A \hat{\iota}^{A'}), \tag{4.15.91}$$

so that

$$T^{\hat{0}\hat{0}'} = \frac{1}{\sqrt{2}}, \quad T^{\hat{0}\hat{1}'} = 0, \quad T^{\hat{1}\hat{0}'} = 0, \quad T^{\hat{1}\hat{1}'} = \frac{1}{\sqrt{2}}, \tag{4.15.92}$$

and define

$$Z(j,m)_{A\ldots FG\ldots K} = \underbrace{\hat{o}_{(A} \ldots \hat{o}_F}_{j-m} \underbrace{\hat{\iota}_G \ldots \hat{\iota}_{K)}}_{j+m}. \tag{4.15.93}$$

Then for each fixed $j$ and varying $m$ (with $-j \leqslant m \leqslant j$, and $j \pm m$ both being integers) the quantities (4.15.93) clearly span $\mathbb{S}_{(A\ldots K)}$. Moreover, they are orthogonal (but not orthonormal) in the sense that

$$\overline{Z(j,m')}_{A'\ldots K'} T^a \ldots T^k Z(j,m)_{A\ldots K} = \delta_{mm'} \frac{(j-m)!(j+m)!}{(2j)!2^j}. \tag{4.15.94}$$

The components of (4.15.93) in the original (non-constant) spin-frame

$\varepsilon_{\mathbf{A}}{}^{A}$ may be computed by directly expanding the symmetrization, which yields the $\{s, -s\}$-quantity

$$
\begin{aligned}
{}_sZ_{j,m} &= W^{-s}Z(j,m)_{\underbrace{0\ldots0}_{j+s}\underbrace{1\ldots1}_{j-s}} = W^{-s}Z(j,m)_{A\ldots CD\ldots K}\underbrace{o^A\ldots o^C}_{j+s}\underbrace{\iota^D\ldots\iota^K}_{j-s} \\
&= W^{-s}\sum_r \frac{(j+m)!(j-m)!(j+s)!(j-s)!\alpha^r\beta^{j-m-r}\gamma^{j+s-r}\delta^{r+m-s}}{(2j)!r!(j-m-r)!(j+s-r)!(r+m-s)!},
\end{aligned}
\qquad (4.15.95)
$$

where the summation extends over integer values of $r$ in the range $\max(0, s-m) \leqslant r \leqslant \min(j-m, j+s)$, and where

$$
\begin{pmatrix} \alpha & \beta \\ \gamma & \delta \end{pmatrix} = (\varepsilon_{\hat{\mathbf{A}}\mathbf{B}}) = \begin{pmatrix} \hat{o}_A o^A & \hat{o}_A \iota^A \\ \hat{\iota}_A o^A & \hat{\iota}_A \iota^A \end{pmatrix}. \qquad (4.15.96)
$$

The matrix in (4.15.96) must be both *unimodular* and, if $W=1$, also *unitary* because it represents a (passive) spin-transformation which, when $W=1$, is a pure spatial rotation of one spin-frame into another (*cf.* (1.2.29)). The expressions (4.15.95) are, apart from a normalization factor, the basic spin-weighted spherical harmonics ${}_sY_{j,m}$. To obtain the normalization factor, observe that, by (4.15.90) and (4.15.94),

$$
\begin{aligned}
\langle {}_sZ_{j,m}, {}_sZ_{j,m} \rangle &= \frac{2^j(j+s)!(j-s)!}{(2j+1)!}\overline{Z(j,m)_{A'\ldots K'}}T^a\ldots T^k Z(j,m)_{A\ldots K} \\
&= \frac{(j+s)!(j-s)!(j+m)!(j-m)!}{(2j+1)!(2j)!}.
\end{aligned}
\qquad (4.15.97)
$$

Thus we obtain the $\{s, -s\}$-quantity

$$
{}_sY_{j,m} = (-1)^{j+m}{}_sZ_{j,m}\sqrt{\frac{(2j+1)!(2j)!}{4\pi(j+s)!(j-s)!(j+m)!(j-m)!}}, \qquad (4.15.98)
$$

(where the factor $(-1)^{j+m}(4\pi)^{-1/2}$ is inserted to give agreement with the standard literature;* *cf.* Schiff 1955) which obeys the orthonormality conditions (for each $s$):

$$
4\pi\langle {}_sY_{j',m'}, {}_sY_{j,m} \rangle = \delta_{jj'}\delta_{mm'} \qquad (4.15.99)
$$

Although the ${}_sY_{j,m}$ are the standard basic spin-weighted spherical harmonics, for practical purposes the ${}_sZ_{j,m}$ are often easier to use.

From the symmetry of the expressions (4.15.95) and (4.15.97) there

* Although Goldberg *et al.* (1967) differ by a sign for odd $m$.

follows the curious reciprocity relation:*

$$ {}_sZ_{j,m} \leftrightarrow {}_{-m}Z_{j,-s} \text{ and } {}_sY_{j,m} \leftrightarrow (-1)^{m+s}{}_{-m}Y_{j,-s} \text{ under } W\beta \leftrightarrow \gamma, \quad (4.15.100) $$

in addition to the relations of more obvious significance:

$$ {}_sZ_{j,m} \mapsto \mathrm{i}^{2j}{}_sZ_{j,-m} \text{ and } {}_sY_{j,m} \mapsto \mathrm{i}^{-2j}{}_sY_{j,-m} \text{ under } \begin{pmatrix} \alpha & \beta \\ \gamma & \delta \end{pmatrix} \mapsto \mathrm{i}\begin{pmatrix} \gamma & \delta \\ \alpha & \beta \end{pmatrix} \quad (4.15.101) $$

and

$$ {}_sZ_{j,m} \mapsto \mathrm{i}^{2j}{}_{-s}Z_{j,m} \text{ and } {}_sY_{j,m} \mapsto \mathrm{i}^{2j}{}_{-s}Y_{j,m} \text{ under } \begin{pmatrix} \alpha & \beta \\ \gamma & \delta \end{pmatrix} \mapsto \mathrm{i}\begin{pmatrix} W\beta & \alpha/W \\ W\delta & \gamma/W \end{pmatrix}, \quad (4.15.102) $$

these being the result of the prime operation applied to the spin-frames $(\hat{o}^A, \hat{\iota}^A)$ and $(o^A, \iota^A)$, respectively. Moreover, since

$$ \bar{\alpha} = W\delta, \qquad W\bar{\beta} = -\gamma, \quad (4.15.103) $$

which is a consequence of the unitarity (rescaled to incorporate $W$) and unimodularity of (4.15.96), we have from (4.15.95) that

$$ \overline{{}_sZ_{j,m}} = (-1)^{m+s}{}_{-s}Z_{j,-m} \text{ and } \overline{{}_sY_{j,m}} = (-1)^{m+s}{}_{-s}Y_{j,-m}. \quad (4.15.104) $$

We note, also, the effects of ð and ð′ on these quantities, which are readily obtained from (4.15.95) by application of (4.15.27), (4.15.28) (where ð and ð′ annihilate the constant $\hat{o}^A, \hat{\iota}^A$):

$$ ð\,{}_sZ_{j,m} = -\left(\frac{j-s}{R\sqrt{2}}\right){}_{s+1}Z_{j,m}, \quad ð'\,{}_sZ_{j,m} = \left(\frac{j+s}{R\sqrt{2}}\right){}_{s-1}Z_{j,m}, \quad (4.15.105) $$

whence

$$ ð\,{}_sY_{j,m} = -\left(\frac{(j+s+1)(j-s)}{2R^2}\right)^{1/2}{}_{s+1}Y_{j,m}, $$
$$ ð'\,{}_sY_{j,m} = \left(\frac{(j-s+1)(j+s)}{2R^2}\right)^{1/2}{}_{s-1}Y_{j,m}. \quad (4.15.106) $$

There are some discrepancies of convention between the definition of ð as given here and as given originally by Newman and Penrose (1966).

* Spin-weighted spherical harmonics also play a role in the representation theory of $O(4)$, since they can be interpreted as scalar spherical harmonics on $S^3$. (Goldberg *et al.* 1967). In that context the symmetry between $s$ and $-m$ has a clearer geometrical meaning.

To come closest to agreement with the original definition we can take

$$R = \frac{1}{\sqrt{2}}, \tag{4.15.107}$$

the apparently more natural choice of $\mathscr{S}$ as a unit sphere leading to a discrepancy by a factor of $\sqrt{2}$. There is also a *sign difference* which apparently comes about because the metric of $\mathscr{S}$ is here negative-definite (being that induced by the ambient space–time) whereas that of Newman and Penrose (1966) was taken to be positive-definite. Finally, the natural relation between spin-frame and orientation of $\mathscr{S}$ that we have adopted here (*cf.* beginning of §4.14), together with the fact that our $\zeta$-coordinate (*cf.* (1.2.10), (1.2.13)) is *anti*-holomorphic and therefore naturally assigns a negative orientation to $\mathscr{S}$, leads to an apparent interchange of $\eth$ and $\eth'$ with respect to $\zeta$. In effect, the spin-weights arising here are actually (geometrically) the *negatives* of the spin-weights of Newman and Penrose (1966). As it turns out, this is quite fortunate because (*cf.* §§9.7–9.9) the spin-weight concept then agrees with the physical concept of *helicity* of outgoing radiation, rather than with its negative.

### *Explicit coordinate descriptions*

We end this section by giving coordinate descriptions for $\eth$ and ${}_sY_{j,m}$. Two specifications are involved. One is the choice of coordinates for $\mathscr{S}$, and the other is an explicit selection of a spin-frame $(o^A, \iota^A)$ at each point of $\mathscr{S}$ so that the scaling freedom of the strict compacted spin-coefficient formalism is finally eliminated. It is convenient to couple these two choices to one another, and also to take the coordinates on $\mathscr{S}$ to be related in some canonical way to the fixed spin-frame $(\hat{o}^A, \hat{\iota}^A)$. If we assume that $(\hat{o}^A, \hat{\iota}^A)$ is related to the usual Minkowski coordinates $(t, x, y, z)$ for $\mathbb{M}$ in the standard way (*cf.* Chapters 1 and 3), the origin $(0, 0, 0, 0)$ being the centre $O$ of $\mathscr{S}$, then

$$(x^{\hat{\mathbf{A}}\hat{\mathbf{B}}'}) = \frac{1}{\sqrt{2}}\begin{pmatrix} t+z & x+\mathrm{i}y \\ x-\mathrm{i}y & t-z \end{pmatrix}. \tag{4.15.108}$$

We have $N$ at $(R, 0, 0, 0)$ and $L$ at $(-R, 0, 0, 0)$, so $\mathscr{S}$ has equation

$$x^2 + y^2 + z^2 = R^2, \quad t = 0. \tag{4.15.109}$$

We consider two different coordinate systems for $\mathscr{S}$, the *spherical polar* $(\theta, \phi)$ system, for which

$$x = R\sin\theta\cos\phi, \quad y = R\sin\theta\sin\phi, \quad z = R\cos\theta, \tag{4.15.110}$$

and in terms of which the metric of $\mathscr{S}$ is given by

$$ds^2 = -R^2(d\theta^2 + \sin^2\theta\, d\phi^2), \qquad (4.15.111)$$

and the *complex* $(\zeta, \bar{\zeta})$ system, for which (*cf.* (1.2.8))

$$x = \frac{R(\zeta + \bar{\zeta})}{\zeta\bar{\zeta} + 1}, \quad y = \frac{-iR(\zeta - \bar{\zeta})}{\zeta\bar{\zeta} + 1}, \quad z = \frac{R(\zeta\bar{\zeta} - 1)}{\zeta\bar{\zeta} + 1} \qquad (4.15.112)$$

with (*cf.* (1.2.10))

$$\zeta = e^{i\phi} \cot\frac{\theta}{2}, \qquad (4.15.113)$$

in terms of which the metric takes the form

$$ds^2 = -\frac{4R^2\, d\zeta\, d\bar{\zeta}}{(\zeta\bar{\zeta} + 1)^2} \qquad (4.15.114)$$

Let us consider the complex system first. As we remarked above (and see also §1.2), $\zeta$ is an *anti*-holomorphic coordinate for $\mathscr{S}$ (with its standard orientation), and so we can take

$$\xi = \bar{\zeta}, \qquad (4.15.115)$$

with $\xi$ as in §4.14. Then (4.14.31) gives $P = \eth\bar{\zeta}$. The $m^a$-vectors are determined as in (4.14.36) by requiring $P > 0$. Comparing the metric form (4.14.30) with (4.15.114), we obtain

$$P = \frac{\zeta\bar{\zeta} + 1}{R\sqrt{2}}, \qquad (4.15.116)$$

which we can substitute into (4.14.33) to obtain the explicit representations for the actions of $\eth, \eth'$ on an $\eta$ of type $\{s, -s\}$:

$$\eth\eta = \frac{1}{R\sqrt{2}}(\zeta\bar{\zeta} + 1)^{1-s}\frac{\partial}{\partial\bar{\zeta}}((\zeta\bar{\zeta} + 1)^s\eta),$$

$$\eth'\eta = \frac{1}{R\sqrt{2}}(\zeta\bar{\zeta} + 1)^{1+s}\frac{\partial}{\partial\zeta}((\zeta\bar{\zeta} + 1)^{-s}\eta). \qquad (4.15.117)$$

The determination of the $m^a$-vectors is illustrated in Fig. 4-6 which shows the flag plane of $o^A$ (i.e., the direction of $\mathrm{Re}(m^a)$) pointing along $\mathrm{Im}(\zeta) =$ constant, $\mathrm{Re}(\zeta)$ increasing. Reference to Fig. 4-6 and the detailed geometrical constructions of Chapter 1 (for flag planes, as in §1.4, spinor scalar products, etc.) lead us to

$$\begin{pmatrix} \alpha & \beta \\ \gamma & \delta \end{pmatrix} = \frac{i}{\sqrt{\zeta\bar{\zeta} + 1}} \begin{pmatrix} -1 & \bar{\zeta} \\ \zeta & 1 \end{pmatrix} \qquad (4.15.118)$$

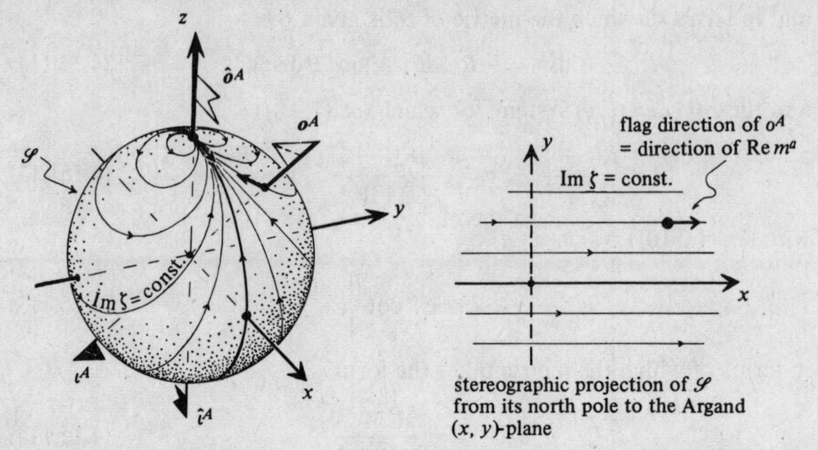


Fig. 4-6. The arrangement of spin-frames and $m$-vectors for the complex stereographic coordinates $(\zeta, \bar{\zeta})$.

(up to an arbitrary choice of overall sign), where we have fixed the real scaling of $o^A$, $\iota^A$ (i.e., the extents of $l^a$, $n^a$) by taking

$$W = 1, \quad \text{i.e.,} \quad u = v = R\sqrt{2}. \tag{4.15.119}$$

Substituting into (4.15.95) and (4.15.98), we can now obtain ${}_sZ_{j,m}$ and ${}_sY_{j,m}$ explicitly.

Finally, we consider the spherical polar system, for which we can take

$$\xi = \log \tan \frac{\theta}{2} + \mathrm{i}\phi \tag{4.15.120}$$

(i.e., $\xi = -\log \bar{\zeta}$). Comparison of (4.14.30) with (4.15.111), where again

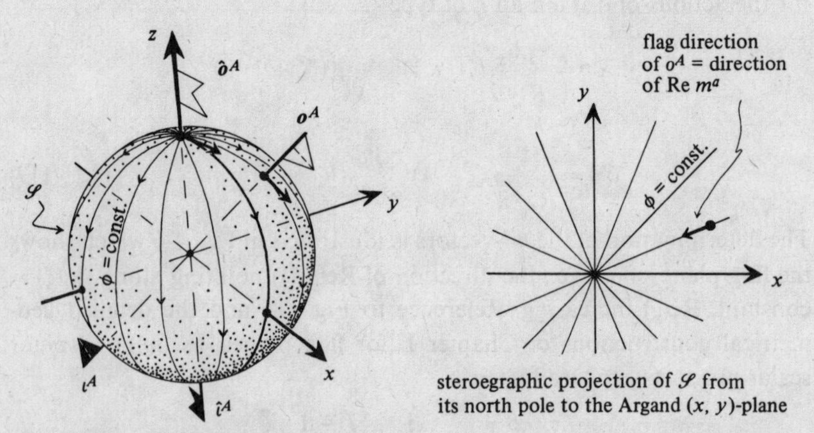


Fig. 4-7. The arrangement of spin-frames and $m$-vectors for the spherical polar coordinates $(\theta, \phi)$.

we adopt (4.14.36) so that $P > 0$, gives, after a short calculation,

$$P = \frac{\sqrt{2}}{R \sin \theta}. \tag{4.15.121}$$

And this, when substituted into (4.14.33), yields, for an $\eta$ of type $\{s, -s\}$,

$$\begin{aligned} ð\eta &= \frac{1}{R\sqrt{2}} \sin^s \theta \left( \frac{\partial}{\partial \theta} - \frac{\mathrm{i}}{\sin \theta} \frac{\partial}{\partial \phi} \right) ((\sin^{-s} \theta) \eta) \\ ð' \eta &= \frac{1}{R\sqrt{2}} \sin^{-s} \theta \left( \frac{\partial}{\partial \theta} + \frac{\mathrm{i}}{\sin \theta} \frac{\partial}{\partial \phi} \right) ((\sin^s \theta) \eta). \end{aligned} \tag{4.15.122}$$

The determination of the $m^a$-vectors is now illustrated by Fig. 4-7, which shows the flag plane of $o^A$ (i.e., the direction of $\mathrm{Re}(m^a)$) pointing downwards along the meridians $\phi = \text{constant}$. For $\alpha, \ldots, \delta$ we now obtain, up to sign,

$$\begin{pmatrix} \alpha & \beta \\ \gamma & \delta \end{pmatrix} = \begin{pmatrix} \mathrm{e}^{-\mathrm{i}\phi/2} \sin \frac{\theta}{2} & \mathrm{e}^{-\mathrm{i}\phi/2} \cos \frac{\theta}{2} \\ -\mathrm{e}^{\mathrm{i}\phi/2} \cos \frac{\theta}{2} & \mathrm{e}^{\mathrm{i}\phi/2} \sin \frac{\theta}{2} \end{pmatrix}. \tag{4.15.123}$$

Again we can fix the extents of $l^a$, $n^a$ by taking $W = 1$ as in (4.15.119). Substituting into (4.15.95) and (4.15.98), we can then obtain ${}_sZ_{j,m}$ and ${}_sY_{j,m}$ explicitly in the $(\theta, \phi)$ system. When $s = 0$ this yields the standard (Legendre) spherical harmonics as a special case.

# 5

# Fields in space–time

## 5.1 The electromagnetic field and its derivative operator

There are many analogies between the gravitational and electromagnetic fields. One of these is brought out particularly strikingly in the spinor formalism, where each of these fields is represented by a symmetric spinor: $\varphi_{AB}$ in the electromagnetic case (*cf.* (3.4.20)) and $\Psi_{ABCD}$ in the gravitational case (*cf.* (4.6.41)). As we shall see shortly, the 'source-free field equations' (where, in the gravitation case, we mean (4.10.9) and not Einstein's vacuum equations (4.10.10)) are also basically identical in the two cases. The analogy goes even further, in that each field quantity can be obtained from a commutator of derivatives. In the gravitational case these are just the covariant derivatives (*cf.* (4.2.30) with (4.2.24), (4.9.15), (4.9.16)). In order to obtain the electromagnetic field in such a way, one must modify the concept of covariant derivative. In flat-space quantum theory this is normally done by adding a multiple of the electromagnetic potential four-vector $\Phi_a$ to the usual flat-space derivative, which we denote temporarily by $\partial_a$, the multiple being $(-\mathrm{i})$ times the charge $e$ of the field on which the derivative acts. Thus the action of the new covariant derivative

$$\nabla_a := \partial_a - \mathrm{i}e\Phi_a \tag{5.1.1}$$

depends on that charge, and so it becomes necessary to specify the charge of each field of the system.

In quantum theory a charged particle is described by a wave function which is such a charged field. (For example, the Dirac field of an electron is a pair of charged spin-vectors each having the same charge; *cf.* (4.4.66) and Corson 1953). The coupling of these charged fields to any Maxwell field present is accomplished by replacing the operator $\partial_a$ in their field equations by the operator (5.1.1).

It is now easily seen how the Maxwell field $F_{ab}$, which is the curl of the potential $\Phi_a$, arises from the commutator $\nabla_{[a}\nabla_{b]}$ acting on a field $\theta^{\mathscr{A}}$ of charge $e$:

$$\mathrm{i}\nabla_{[a}\nabla_{b]}\theta^{\mathscr{A}} = e\theta^{\mathscr{A}}\partial_{[a}\Phi_{b]} = \tfrac{1}{2}e\theta^{\mathscr{A}}F_{ab} \tag{5.1.2}$$

Since we wish to allow electromagnetic and gravitational fields to co-exist, we must make provisions for our operators to act in curved space–time. This is easily achieved by letting the $\partial_a$ of (5.1.1) become the usual ('elementary') covariant derivative operator (whereupon extra curvature terms appear on the right-hand side of (5.1.2), *cf.* (5.1.34) below).

The expression (5.1.1) involves the potential explicitly, however, and that we shall mainly want to avoid, since the value of the potential at any point has no physical significance. This is analogous to the fact that the connection symbols $\Gamma_{\mathbf{ab}}{}^{\mathbf{c}}$ have no direct physical significance in gravitational theory. Similarly, as in the case of the elementary covariant derivative which can be introduced by modifying the coordinate derivative $\partial/\partial x^{\mathbf{a}}$ with $\Gamma$-terms, and then largely discarding $\partial/\partial x^{\mathbf{a}}$ as non-intrinsic; so also in the case of charged fields, the covariant derivative $\partial_a$ upon charged fields can serve as a taking off point but in the development of the theory it is largely ignored as 'gauge'-dependent, and thus physically meaningless. In both cases the physically significant thing that these quantities determine is an *operator*, which we denote universally by $\nabla_a$. Our aim, as in the development of the elementary covariant derivative, is to construct a calculus in which all the objects have physical or geometrical meaning. In this chapter, then, we shall study the properties of the general operator $\nabla_a$ from a *formal algebraic* point of view just as we did in §4.2 for the purely gravitational case. In this way the use of the electromagnetic potentials and of $\partial_a$ can be avoided. (But, like the $\Gamma$s and $\partial/\partial x^{\mathbf{a}}$, they can be brought in, when desired, as a convenience.) The properties of $\nabla_a$ as defined by certain axioms will determine the properties of the tensor $F_{ab}$ in equation (5.1.2) (modified for curved space), just as the properties of the elementary covariant derivative determine the properties of the curvature tensor (*cf.* (4.2.31)). These properties of $F_{ab}$ will allow it to be identified with the Maxwell field tensor.

### *Charged fields*

In order to accomplish the programme outlined above we must make a generalization of the modules $\mathfrak{S}^{\mathscr{A}}$ (of $C^\infty$-smooth spinor fields) defined in Chapter 2, to which the discussion has so far been restricted. It is necessary for each charge value to have a *separate* version of each spinor module $\mathfrak{S}^{\mathscr{A}}$, the derivative $\nabla_a$ acting differently on each version. We thus introduce the *charged* modules

$$\overset{e}{\mathfrak{S}}, \overset{e}{\mathfrak{S}}{}^{A}, \overset{e}{\mathfrak{S}}{}^{B}, \ldots, \overset{e}{\mathfrak{S}}{}^{A\ldots C'\ldots}_{B\ldots D'\ldots}, \ldots \tag{5.1.3}$$

(for each charge $e$) of *charged* $C^\infty$-smooth spinor fields. The original system of §2.5 will correspond to the absence of charge, $e = 0$. The charge $e$ of a field could in principle be taken to be any real (or possibly complex) number, but for various reasons we shall restrict $e$ to be an integer-multiple of some fixed non-zero real number $\varepsilon$, the elementary charge. Thus $e$ takes the values $0, \pm\varepsilon, \pm 2\varepsilon, \pm 3\varepsilon, \ldots$.

An important point in our argument will be that the elements of $\overset{e}{\mathfrak{S}}$, for example, when $e \neq 0$, do not take on *numerical* values at the points $P$ of $\mathscr{M}$. Rather, each $\overset{e}{\mathfrak{S}}[P]$ is an abstract one-dimensional additive complex vector space, with zero element, of course, but no canonically determined unit element. Also there is no canonically determined correspondence between the $\overset{e}{\mathfrak{S}}[P]$ at two different points. (However, a $C^\infty$-smooth correspondence between *neighbouring* $\overset{e}{\mathfrak{S}}{}^{\mathscr{A}}[P]$ is provided by any element of $\overset{e}{\mathfrak{S}}{}^{\mathscr{A}}$.) Nevertheless, just as a 4-vector can be described by four numerical scalars when an arbitrary basis is specified, so the elements of $\overset{e}{\mathfrak{S}}[P]$, for example, can be described by single numerical scalars when an arbitrary 'gauge' is specified (as we shall explain later).

Certain other properties of uncharged fields have no meaningful equivalent for charged fields. For example, the condition that a charged scalar or tensor field be real (at a point) would require the difference between it and its complex conjugate to vanish. But the conjugate has minus the charge of the original field and cannot be subtracted from it according to the rules of the algebra that we shall give in a moment. Somewhat similarly, a charged spin-vector $\kappa^A$ with non-zero charge cannot be represented in the usual way by a null flag. For again the relevant expression (3.2.9) would require two quantities of opposite charge to be added.

As for the algebraic properties of the $\overset{e}{\mathfrak{S}}{}^{\mathscr{A}}$, we require that the operations of addition, outer multiplication, contraction, index substitution and complex conjugation apply to the elements of the systems (5.1.3) as before, subject to the general algebraic rules laid down in §2.5, but with the following additional stipulations:

(5.1.4) *Two charged spinors may be added if and only if their charges are equal; the sum has the same charge as the constituents.*

(5.1.5) *Outer products of charged spinors may be formed whatever their charges are; the charge of the product is the sum of the charges of the constituents.*

(5.1.6) *Contractions may be performed on charged spinors and do not affect the value of the charge.*

(5.1.7) *Index substitutions may be applied to charged spinors and do not affect the value of the charge.*

(5.1.8) *Complex conjugation may be applied to a charged spinor; it reverses the sign of the charge.**

The reasons for these additional axioms become clear when we examine the effect of (5.1.1) (with $\partial_a$ as the elementary covariant derivative) on sums, products, etc., and require that the generalized $\nabla_a$ should satisfy the usual additivity and Leibniz properties (4.4.16), (4.4.17). Additionally we require of $\nabla_a$ to preserve the charge of the spinor on which it acts, i.e., to provide maps

$$\nabla_a : \overset{e}{\mathfrak{S}}{}^{\mathscr{A}} \rightarrow \overset{e}{\mathfrak{S}}{}^{\mathscr{A}}_a \tag{5.1.9}$$

for each charge $e$ and for each composite index $\mathscr{A}$, and to commute with index substitution, contraction, and complex conjugation, as in (4.4.18)–(4.4.20). In fact, such a $\nabla_a$ is uniquely and consistently defined from its action on uncharged fields (which, by the axioms, must be the elementary covariant derivative) as soon as the action

$$\nabla_a : \overset{\varepsilon}{\mathfrak{S}} \rightarrow \overset{\varepsilon}{\mathfrak{S}}_a \tag{5.1.10}$$

has been specified. For if $\alpha \in \overset{\varepsilon}{\mathfrak{S}}$, $\psi^{\mathscr{A}} \in \overset{n\varepsilon}{\mathfrak{S}}{}^{\mathscr{A}}$ (with $\alpha$ nowhere zero, $n$ integral) then, by the Leibniz and additivity properties, we have**

$$\nabla_a \psi^{\mathscr{A}} = \alpha^n \nabla_a(\alpha^{-n}\psi^{\mathscr{A}}) + n\psi^{\mathscr{A}}\alpha^{-1}\nabla_a \alpha, \tag{5.1.11}$$

where $\alpha^{-n}\psi^{\mathscr{A}}$ has charge zero and so, by hypothesis, its derivative is known, as well as that of $\alpha$. The axioms to be satisfied by (5.1.10) are simply

$$\begin{aligned} &\nabla_a(\alpha + \beta) = \nabla_a \alpha + \nabla_a \beta, \\ &\nabla_a(\lambda\alpha) = \lambda\nabla_a \alpha + \alpha\nabla_a \lambda, \quad (\alpha, \beta \in \overset{\varepsilon}{\mathfrak{S}}, \lambda \in \mathfrak{S}). \end{aligned} \tag{5.1.12}$$

We stress that since $\varepsilon_{AB}$, $\varepsilon^{AB}$, $\varepsilon_A{}^B$ are defined as belonging to uncharged modules, $\nabla_a$ acting on them is simply the elementary covariant derivative and so gives zero.***

* If complex charges are considered, the complex conjugated field has minus the complex conjugated charge.

** If $\alpha \in \overset{e}{\mathfrak{S}}$ and is nowhere zero, $\alpha^{-1}$ stands for the unique element of $\overset{-e}{\mathfrak{S}}$ such that the outer product $\alpha\alpha^{-1}$, which belongs to $\mathfrak{S}$ by (5.1.5), is 1, and $\alpha^{-n}$ is defined as $(\alpha^{-1})^n$. It may be remarked that the existence of a *globally* non-vanishing charged scalar is equivalent to the non-existence of 'holes' in the space–time of non-zero effective magnetic charge (Wu and Yang 1976). This is normally a 'physically reasonable' requirement; in any case the discussion of this section can be carried out 'patchwise' using a collection of different $\alpha$s whose non-zero regions together cover $\mathscr{M}$.

*** One could envisage a modified system in which $\varepsilon_{AB}$ possessed a charge $2k$ and $\varepsilon^{AB}$ a charge $-2k$. But this would not make the system more general. For by a redefinition of charge, namely by adding $k$ times [number of: (upper unprimed) – (upper primed) + (lower primed) – (lower unprimed) indices], we can reduce the charge on $\varepsilon_{AB}$ to zero, while still satisfying all the rules.

*Electromagnetic potential*

Let us now see how we can recover a potential $\Phi_a$ and operator $\partial_a$ from a knowledge of $\nabla_a$. Analogously to choosing a coordinate basis, we here choose an arbitrary nowhere vanishing $\alpha \in \overset{\varepsilon}{\mathfrak{S}}$. Then we can *define* a potential

$$\Phi_a := \mathrm{i}(\varepsilon\alpha)^{-1}\nabla_a\alpha, \tag{5.1.13}$$

which is evidently *uncharged*, since $\nabla_a$ preserves charge. We define a corresponding differential operator $\partial_a$ by its action on a spinor $\psi^{\mathscr{A}}$ of charge $e = n\varepsilon$.

$$\partial_a\psi^{\mathscr{A}} := \alpha^n\nabla_a(\alpha^{-n}\psi^{\mathscr{A}}), \tag{5.1.14}$$

so that, by (5.1.11),

$$\nabla_a\psi^{\mathscr{A}} = (\partial_a - \mathrm{i}e\Phi_a)\psi^{\mathscr{A}} \tag{5.1.15}$$

(*cf*. (5.1.1)). We can see from the last equation (by putting $e = 0$) that $\partial_a$ operates on $\psi^{\mathscr{A}}$ as does $\nabla_a$ on an *uncharged* spinor $\psi^{\mathscr{A}}$, namely as the elementary covariant derivative. The operator $\partial_a$ plays a part analogous to that of the coordinate derivative in the standard theory. It satisfies the usual additivity and Leibniz properties. In flat space–time, even in the presence of electromagnetic field, we have

$$\partial_a\partial_b = \partial_b\partial_a. \tag{5.1.16}$$

This follows at once from the remark after (5.1.15); or from (5.1.14) and the fact that the $\nabla_a$ in that expression acts on uncharged fields, so that the $\nabla$s commute. (*Cf*. (4.2.59).) In curved space–time, $\partial_{[a}\partial_{b]}$ involves curvature but no electromagnetic part.

If $\alpha$ is specialized so that

$$\alpha\bar{\alpha} = 1, \quad \alpha \in \overset{\varepsilon}{\mathfrak{S}}, \tag{5.1.17}$$

then we refer to $\alpha$ as a *gauge*. (If (5.1.17) does not initially hold, then it can be achieved by the replacement $\alpha \mapsto \alpha(\alpha\bar{\alpha})^{-\frac{1}{2}}$: $\alpha\bar{\alpha}$ is always a positive uncharged scalar field provided $\alpha$ is nowhere vanishing, so $(\alpha\bar{\alpha})^{-\frac{1}{2}}$ is defined.) For any gauge $\alpha$ we have

$$0 = \nabla_a(\alpha\bar{\alpha}) = \alpha\nabla_a\bar{\alpha} + \bar{\alpha}\nabla_a\alpha, \quad \text{i.e.,}\ \bar{\alpha}^{-1}\nabla_a\bar{\alpha} = -\alpha^{-1}\nabla_a\alpha,$$

whence $\Phi_a$ is real:

$$\bar{\Phi}_a = \Phi_a. \tag{5.1.18}$$

Also, for any gauge $\alpha$, the operator $\partial_a$ is real, in the sense

$$\overline{\partial_a\psi^{\mathscr{A}}} = \partial_a\overline{\psi^{\mathscr{A}}}, \tag{5.1.19}$$

as follows from the definition (5.1.14), the property (5.1.8), the relation $\bar{\alpha}=\alpha^{-1}$, and the reality of $\nabla_a$:

$$\overline{\partial_a \psi^{\mathscr{A}}} = \bar{\alpha}^n \nabla_a (\overline{\alpha^{-n}} \overline{\psi^{\mathscr{A}}}) = \alpha^{-n} \nabla_a (\alpha^n \overline{\psi^{\mathscr{A}}}) = \partial_a \overline{\psi^{\mathscr{A}}}.$$

The gauge $\alpha$ serves to map any charged field (of charge $e = n\varepsilon$) into an uncharged one according to*

$$\psi^{\mathscr{A}} \mapsto \alpha^{-n} \psi^{\mathscr{A}}, \tag{5.1.20}$$

so that, for example, numerical components with respect to some basis frame may then be taken. (Recall that *charged* scalars do not have canonical numerical values.) Thus to specify components for charged fields, both the basis *and* the gauge are required. It is easily verified that the action of $\partial_a$ goes over into the action of $\nabla_a$ under the map (5.1.20).

If the gauge $\alpha$ is replaced by a new one $\alpha'$, then the corresponding uncharged field that $\psi^{\mathscr{A}}$ is mapped to, in (5.1.20), undergoes a *gauge transformation*:

$$\alpha^{-n} \psi^{\mathscr{A}} \mapsto \alpha'^{-n} \psi^{\mathscr{A}} = \mathrm{e}^{in\theta} (\alpha^{-n} \psi^{\mathscr{A}}), \tag{5.1.21}$$

where the real uncharged scalar $\theta$ is defined (possibly only locally) by

$$\mathrm{e}^{\mathrm{i}\theta} = \alpha/\alpha', \tag{5.1.22}$$

and correspondingly we have

$$\Phi_a \mapsto \Phi'_a = \frac{\mathrm{i}}{\varepsilon\alpha'} \nabla_a \alpha' = \frac{\mathrm{i}}{\varepsilon\alpha} \mathrm{e}^{\mathrm{i}\theta} \nabla_a (\alpha \mathrm{e}^{-\mathrm{i}\theta})$$

$$= \Phi_a + \frac{1}{\varepsilon} \nabla_a \theta, \tag{5.1.23}$$

and

$$\partial_a \psi^{\mathscr{A}} \mapsto (\partial_a - \mathrm{i} e \nabla_a \theta) \psi^{\mathscr{A}}. \tag{5.1.24}$$

Observe that (5.1.23) has the usual form of a gauge transformation in electromagnetic theory.

It may be remarked that in both the electromagnetic and gravitational cases we have 'gauge transformations of the second kind'. These are the transformations of the gauge which do not change the associated operator $\partial_a$. In the electromagnetic case they are clearly given by $\alpha \mapsto \alpha' = \mathrm{e}^{-\mathrm{i}\theta}\alpha$ where $\theta$ is real and *constant* (although dropping the reality requirement would affect only the normalization (5.1.17) of $\alpha$ and none of the succeeding equations). In the gravitational case they are given by the linear inhomogeneous transformations $x^{\mathbf{a}} \mapsto A_{\mathbf{b}}{}^{\mathbf{a}} x^{\mathbf{b}} + B^{\mathbf{a}}$, where $A_{\mathbf{b}}{}^{\mathbf{a}}$ and $B^{\mathbf{a}}$ are real

* Such an isomorphism $\alpha : \overset{e}{\mathfrak{S}}{}^{\mathscr{A}} \mapsto \mathfrak{S}^{\mathscr{A}}$ is referred to as a *trivialization*.

and constant with det $(A_{\mathbf{b}}{}^{\mathbf{a}}) \neq 0$. For some purposes one might require $A_{\mathbf{b}}{}^{\mathbf{a}}$ to be a (restricted) Lorentz matrix. Then a choice of 'gauge' $x^{\mathbf{a}}$ (i.e., of coordinates) would correspond to a mapping of the space–time $\mathscr{M}$ to Minkowski space, whose symmetries would be respected by the gauge freedom. This, in effect, is the procedure involved in the 'Lorentz covariant' formulations of general relativity.

### *The Maxwell field tensor*

Next we examine the commutator

$$\Delta_{ab} = \nabla_a \nabla_b - \nabla_b \nabla_a = 2\nabla_{[a}\nabla_{b]}. \tag{5.1.25}$$

Assume that there is no torsion, i.e., for an *uncharged* scalar $\gamma \in \mathfrak{S}$,

$$\Delta_{ab}\gamma = 0 \qquad (\gamma \in \mathfrak{S}). \tag{5.1.26}$$

Suppose $0 \neq \alpha \in \overset{\varepsilon}{\mathfrak{S}}$, $\psi \in \overset{e}{\mathfrak{S}}$, where $e = n\varepsilon$, and where $\alpha$ vanishes nowhere. Then there is a $\gamma \in \mathfrak{S}$ such that

$$\gamma \alpha^n = \psi. \tag{5.1.27}$$

From the axioms, as in (4.2.15), (4.2.16), it follows that $\Delta_{ab}$ satisfies additivity and the Leibniz law, and so $\Delta_{ab}\alpha^n = n\alpha^{n-1}\Delta_{ab}\alpha$, for any integer $n$. Thus, by (5.1.26) and (5.1.27),

$$n\gamma\alpha^{n-1}\Delta_{ab}\alpha = \Delta_{ab}\psi,$$

i.e.,

$$n\psi\alpha^{-1}\Delta_{ab}\alpha = \Delta_{ab}\psi. \tag{5.1.28}$$

If we set

$$F_{ab} := \frac{\mathrm{i}}{\varepsilon\alpha}\Delta_{ab}\alpha, \tag{5.1.29}$$

then we have

$$\mathrm{i}\Delta_{ab}\psi = eF_{ab}\psi. \tag{5.1.30}$$

The equation (5.1.30) shows that, unlike (5.1.13), (5.1.29) is independent of the particular $\alpha$ that is chosen. We call $F_{ab}$ the *Maxwell* or *electromagnetic field tensor** (associated with $\nabla_a$).

If $\psi^{\mathscr{A}} \in \overset{e}{\mathfrak{S}}{}^{\mathscr{A}}$ is an arbitrary charged spinor, we have

* In a non-simply-connected space–time region it is possible to have $F_{ab} = 0$ everywhere, yet $\nabla_a$ to be non-trivial in the sense that no charged scalar $\alpha$ exists with $\nabla_a\alpha = 0$; equivalently, every choice of potential may necessarily be non-zero somewhere (*cf.* Aharonov and Bohm 1959).

(5.1.31) PROPOSITION

*$\Delta_{ab}\psi^{\mathscr{A}}$ differs from the result of a commutator acting on an uncharged $\psi^{\mathscr{A}}$ simply by the additional term $-\mathrm{i}eF_{ab}\psi^{\mathscr{A}}$.*

This fact can also be expressed by the equation

$$\Delta_{ab}\psi^{\mathscr{A}} = 2\partial_{[a}\partial_{b]}\psi^{\mathscr{A}} - \mathrm{i}eF_{ab}\psi^{\mathscr{A}}, \tag{5.1.32}$$

in which the first term on the right is actually covariant though $\partial_a$ by itself is gauge-dependent. For proof, note that if $\psi^{\mathscr{A}} \in \overset{n\varepsilon}{\mathfrak{S}}{}^{\mathscr{A}}$, then by the Leibniz rule,

$$\Delta_{ab}(\alpha^{-n}\psi^{\mathscr{A}}) = -n\alpha^{-n-1}\psi^{\mathscr{A}}\Delta_{ab}\alpha + \alpha^{-n}\Delta_{ab}\psi^{\mathscr{A}},$$

whence

$$\Delta_{ab}\psi^{\mathscr{A}} = \alpha^{n}\Delta_{ab}(\alpha^{-n}\psi^{\mathscr{A}}) - \mathrm{i}eF_{ab}\psi^{\mathscr{A}}. \tag{5.1.33}$$

Since $\alpha^{-n}\psi^{\mathscr{A}}$ is uncharged, $\Delta_{ab}(\alpha^{-n}\psi^{\mathscr{A}})$ is just the elementary covariant derivative commutator. Reference to (4.2.33) or (4.9.1) and (4.9.13) now bears out our assertion. For example (*cf.* (4.2.32)),

$$\Delta_{ab}\psi_c = -R_{abc}{}^{d}\psi_d - \mathrm{i}eF_{ab}\psi_c. \tag{5.1.34}$$

Next we examine some properties of $F_{ab}$. In the first place, from (5.1.29), it is evidently *skew*:

$$F_{ab} = -F_{ba}, \tag{5.1.35}$$

and, since $\nabla_a$ preserves charge, $F_{ab}$ is *uncharged*. Furthermore, since $\bar{\psi} \in \overset{-e}{\mathfrak{S}}$ whenever $\psi \in \overset{e}{\mathfrak{S}}$ ($e$ real), we have, by (5.1.30),

$$\Delta_{ab}\bar{\psi} = \mathrm{i}eF_{ab}\bar{\psi},$$

the complex conjugate of which (using $\bar{\nabla}_a = \nabla_a$), together with (5.1.30), yields

$$\bar{F}_{ab} = F_{ab},$$

i.e., $F_{ab}$ is *real*. Finally, by a process analogous to that (*cf.* (4.2.40)) leading to the Bianchi identity, we have, for $\psi \in \overset{e}{\mathfrak{S}}$,

$$\begin{aligned}\nabla_{[a}\nabla_b\nabla_{c]}\psi &= \nabla_{[a}\nabla_{[b}\nabla_{c]]}\psi = -\tfrac{1}{2}\mathrm{i}e\nabla_{[a}(F_{bc]}\psi) \\ &= -\tfrac{1}{2}\mathrm{i}e\psi\nabla_{[a}F_{bc]} - \tfrac{1}{2}\mathrm{i}eF_{[bc}\nabla_{a]}\psi.\end{aligned}$$

But also (*cf.* (5.1.34))

$$\nabla_{[a}\nabla_b\nabla_{c]}\psi = \nabla_{[[a}\nabla_{b]}\nabla_{c]}\psi = -\tfrac{1}{2}R_{[abc]}{}^{d}\nabla_d\psi - \tfrac{1}{2}\mathrm{i}eF_{[ab}\nabla_{c]}\psi.$$

Subtracting these two expressions and using (4.2.37), we get, after division by $\tfrac{1}{2}\mathrm{i}e\psi$,

$$\nabla_{[a}F_{bc]} = 0. \tag{5.1.36}$$

For an alternative derivation of this equation, let us for the moment consider a particular gauge $\alpha$ and examine the relation between $\Phi_a$ and $F_{ab}$. Forming the curl of $\Phi_a$,

$$\nabla_{[a}\Phi_{b]} = -(\mathrm{i}/\varepsilon\alpha^2)(\nabla_{[a}\alpha)(\nabla_{b]}\alpha) + (\mathrm{i}/\varepsilon\alpha)\nabla_{[a}\nabla_{b]}\alpha = \tfrac{1}{2}F_{ab}$$

(*cf*. (5.1.13), (5.1.29)), we see that

$$F_{ab} = \nabla_a\Phi_b - \nabla_b\Phi_a. \tag{5.1.37}$$

Equation (5.1.36) now follows at once from (5.1.37), being an example of the relation $d^2 = 0$ of the exterior calculus (*cf*. (4.3.15) (viii)).

Equation (5.1.37) is, of course, the same as the usual relation between field tensor and potential vector in Maxwell's theory. It implies that if $F_{ab} = 0$ then (locally at least) $\Phi_a$ can be expressed as a gradient: $\Phi_a = \nabla_a\chi$, where $\chi$ is real and uncharged. (This is not so in the general case, even though, at first sight, it may seem that (5.1.13) would give $\Phi_a$ in this form; but there is no allowed scalar 'log $\alpha$' for charged $\alpha$, whose $\nabla_a$ could yield $\alpha^{-1}\nabla_a\alpha$.) If two distinct potentials $\Phi_a$ and $\Phi'_a$ satisfy (5.1.37), then $\nabla_{[a}(\Phi'_{b]} - \Phi_{b]}) = 0$ and so $\Phi'_a - \Phi_a$ is (locally at least) a gradient: $\Phi'_a - \Phi_a = \varepsilon^{-1}\nabla_a\theta$ with $\theta$ real and uncharged. Thus, by (5.1.23) and (5.1.22), there exists (locally) a gauge $\alpha'$ yielding any given potential satisfying (5.1.37).

Equation (5.1.36) is the first 'half' of Maxwell's equations. If we *define* the charge-current vector $J^a$ (in Gaussian units) by

$$\nabla_a F^{ab} = 4\pi J^b, \tag{5.1.38}$$

this gives the other half. (These two equations are the basis of classical Maxwell electromagnetic theory.) Note that $J^a$ has charge zero. (This is not as paradoxical as it may perhaps seem, since $J^a$ is a 'current' and involves charged fields and their complex conjugates bilinearly, *cf*. equation (5.10.16) below.)

### *The electromagnetic spinor*

Since $F_{ab}$ is real and skew, it has a decomposition of the form (*cf*. (3.4.20))

$$F_{ab} = \varphi_{AB}\varepsilon_{A'B'} + \varepsilon_{AB}\bar{\varphi}_{A'B'}, \tag{5.1.39}$$

where $\varphi_{AB}$ is the *electromagnetic spinor* and

$$\varphi_{AB} = \varphi_{(AB)} = \tfrac{1}{2}F_{ABC'}{}^{C'} \tag{5.1.40}$$

As previously described in §3.4, the tensors ${}^-F_{ab} = \varphi_{AB}\varepsilon_{A'B'}$ and ${}^+F_{ab} = \varepsilon_{AB}\bar{\varphi}_{A'B'}$ are, respectively, the anti-self-dual and self-dual parts of the Maxwell field. Analogously to the way in which the curvature spinors occur in the expansion of the operators $\square_{AB}$, $\square_{A'B'}$ (*cf*. (4.9.13)), the electromagnetic spinor occurs additionally when charged fields are involved (even scalar charged fields). For example, if $\psi \in \overset{e}{\mathfrak{S}}$ (*cf.* (4.9.1),

$$\square_{AB}\psi = \tfrac{1}{2}\varepsilon^{A'B'}\Delta_{ab}\psi = -\tfrac{1}{2}\mathrm{i}e\varepsilon^{A'B'}F_{ab}\psi = -\mathrm{i}e\varphi_{AB}\psi, \qquad (5.1.41)$$

and, similarly,

$$\square_{A'B'}\psi = -\mathrm{i}e\bar{\varphi}_{A'B'}\psi \qquad (5.1.42)$$

When $\psi^{\mathscr{A}} \in \overset{e}{\mathfrak{S}}{}^{\mathscr{A}}$, then $-\mathrm{i}e\varphi_{AB}\psi^{\mathscr{A}}$ [*or* $-\mathrm{i}e\varphi_{A'B'}\psi^{\mathscr{A}}$] gets added to the appropriate uncharged expansion of $\square_{AB}\psi^{\mathscr{A}}$ [*or* $\square_{A'B'}\psi^{\mathscr{A}}$]. For, by transvecting (5.1.33) with $\frac{1}{2}\varepsilon^{A'B'}$, we immediately obtain

$$\square_{AB}\psi^{\mathscr{A}} = \alpha^n \square_{AB}(\alpha^{-n}\psi^{\mathscr{A}}) - \mathrm{i}e\varphi_{AB}\psi^{\mathscr{A}}, \qquad (5.1.43)$$

and this bears out our assertion. (To get the corresponding result for $\square_{A'B'}\psi^{\mathscr{A}}$, we simply transvect (5.1.33) with $\frac{1}{2}\varepsilon^{AB}$.) For example, we have

$$\square_{AB}\psi^D = X_{ABC}{}^D\psi^C - \mathrm{i}e\varphi_{AB}\psi^D \qquad (5.1.44)$$

$$\square_{A'B'}\psi^D = \Phi_{A'B'C}{}^D\psi^C - \mathrm{i}e\bar{\varphi}_{A'B'}\psi^D. \qquad (5.1.45)$$

The spinor form of (5.1.37) is

$$\varphi_{AB} = \nabla_{A'(A}\Phi^{A'}_{B)}, \qquad (5.1.46)$$

as is readily seen from (5.1.39). Alternatively this can be obtained directly from (5.1.13) using (5.1.41).

A gauge condition often imposed on $\Phi_a$ (normally in flat space–time) is the Lorenz* gauge condition:

$$\nabla^a\Phi_a = 0. \qquad (5.1.47)$$

This is equivalent, via (5.1.13), to the following condition on the gauge $\alpha$:

$$\alpha\nabla^a\nabla_a\alpha = (\nabla^a\alpha)(\nabla_a\alpha). \qquad (5.1.48)$$

With this condition, (5.1.46) simplifies to

$$\varphi_{AB} = \nabla_{A'A}\Phi^{A'}_B, \qquad (5.1.49)$$

since the part skew in $AB$ on the right now vanishes.

The charge-current vector (5.1.38) in spinor form is given by

$$\nabla^{A'B}\varphi^A{}_B + \nabla^{AB'}\bar{\varphi}^{A'}{}_{B'} = 4\pi J^{AA'}. \qquad (5.1.50)$$

However, (5.1.36) is equivalent to $\nabla_a{}^*F^{ab} = 0$ (*cf.* (3.4.26)), and so, since ${}^*F^{ab} = -\mathrm{i}\varphi^{AB}\varepsilon^{A'B'} + \mathrm{i}\varepsilon^{AB}\bar{\varphi}^{A'B'}$ (*cf.* (3.4.22)), to

$$\nabla^{A'B}\varphi^A{}_B = \nabla^{AB'}\bar{\varphi}^{A'}{}_{B'}. \qquad (5.1.51)$$

The complete Maxwell equations (5.1.50) and (5.1.51) can now be com-

* This is L. Lorenz (1867) – not H.A. Lorentz! (See Whittaker 1910).

bined into the single equation

$$\nabla^{A'B}\varphi^{A}{}_{B} = 2\pi J^{AA'}, \tag{5.1.52}$$

together with the fact that $J^a$ is real:

$$J^{AA'} = \bar{J}^{AA'}. \tag{5.1.53}$$

The divergence equation

$$\nabla_a J^a = 0 \tag{5.1.54}$$

is a consequence of (5.1.52) or (5.1.38), the curvature terms cancelling out. For example, from (5.1.52) we get (*cf.* (4.9.2), (4.9.13), (4.6.19))

$$\begin{aligned} -2\pi\nabla_a J^a &= \nabla_{AA'}\nabla_B^{A'}\varphi^{AB} = \square_{AB}\varphi^{AB} \\ &= X_{ABQ}{}^{A}\varphi^{QB} + X_{ABQ}{}^{B}\varphi^{AQ} \\ &= 3\Lambda(\varepsilon_{BQ}\varphi^{QB} + \varepsilon_{AQ}\varphi^{AQ}) = 0. \end{aligned}$$

(Alternatively, this result can be simply derived by use of differential forms. For if $\boldsymbol{F} := F_{i_1 i_2}$, $\boldsymbol{J} := J_{i_1}$, then (5.1.38) and (5.1.54) take the respective forms $\mathrm{d}\,{}^*\boldsymbol{F} = \frac{4}{3}\pi{}^\dagger\boldsymbol{J}$ and $\mathrm{d}\,{}^\dagger\boldsymbol{J} = 0$, the latter being a consequence of the former because $\mathrm{d}^2 = 0$. See (5.9.5)–(5.9.13) and (4.3.17) for details.) Combining (5.1.46) with (5.1.52) gives

$$\nabla_B^{A'}\nabla^{C'(A}\Phi^{B)}_{C'} = 2\pi J^{AA'}. \tag{5.1.55}$$

If the Lorenz gauge condition is adopted in the form (5.1.49), then we can express (5.1.55) as follows:

$$\begin{aligned} 2\pi J^{AA'} &= \nabla_B^{A'}\nabla^{C'B}\Phi^{A}_{C'} \\ &= \nabla_B^{(A'}\nabla^{C')B}\Phi^{A}_{C'} + \nabla_B^{[A'}\nabla^{C']B}\Phi^{A}_{C'} \\ &= \square^{A'C'}\Phi^{A}_{C'} + \tfrac{1}{2}\varepsilon^{A'C'}\nabla_{BB'}\nabla^{BB'}\Phi^{A}_{C'} \\ &= -\Phi^{AA'}{}_{CC'}\Phi^{CC'} + 3\Lambda\Phi^{AA'} + \tfrac{1}{2}\nabla_b\nabla^b\Phi^a \end{aligned}$$

(*cf.* (4.9.14) and (4.6.19)) and thus (*cf.* (4.6.21))

$$4\pi J^a = \nabla_b\nabla^b\Phi^a + R^a{}_c\Phi^c. \tag{5.1.56}$$

(It so happens that this particular relation could be somewhat more easily derived in a tensorial way.)

Note that when $J^a = 0$, (5.1.52) can be written

$$\nabla^{AA'}\varphi_{AB} = 0 \tag{5.1.57}$$

which is therefore the spinor version of the complete source-free Maxwell equations $\nabla_{[a}F_{bc]} = 0$, $\nabla_a F^{ab} = 0$. The similarity between (5.1.57) and the spinor form (4.10.3) of the Bianchi identities (when specialized to vacuum) is striking. In fact, (5.1.57) is the spin-1 version of the massless free field equations (5.7.2) which will be considered in §5.7.

*Relation to electric and magnetic 3-vectors*

We shall end this section with some elementary formulae for Maxwell's theory in spinor form. When we refer the field tensor $F_{ab}$ to a standard Minkowski tetrad $t^a, x^a, y^a, z^a$, its components are related *by definition* to the components of the electric and magnetic 3-vector fields $\boldsymbol{E}$ and $\boldsymbol{B}$ as follows:

$$F_{\mathbf{ab}} = \begin{bmatrix} 0 & E_1 & E_2 & E_3 \\ -E_1 & 0 & -B_3 & B_2 \\ -E_2 & B_3 & 0 & -B_1 \\ -E_3 & -B_2 & B_1 & 0 \end{bmatrix} \tag{5.1.58}$$

In terms of the standard spin-frame associated with the Minkowski tetrad under consideration, and using the tensor–spinor translation scheme (3.1.38), (3.1.39), (3.1.49), we then find, form (5.1.40),

$$\begin{aligned} \varphi_{00} &= \tfrac{1}{2}(F_{31} + F_{01} - \mathrm{i}F_{32} - \mathrm{i}F_{02}) = \tfrac{1}{2}(C_1 - \mathrm{i}C_2) \\ \varphi_{01} &= \tfrac{1}{2}(-F_{03} - \mathrm{i}F_{12}) = -\tfrac{1}{2}C_3 \\ \varphi_{11} &= \tfrac{1}{2}(F_{31} - F_{01} + \mathrm{i}F_{32} - \mathrm{i}F_{02}) = -\tfrac{1}{2}(C_1 + \mathrm{i}C_2), \end{aligned} \tag{5.1.59}$$

where

$$\boldsymbol{C} = \boldsymbol{E} - \mathrm{i}\boldsymbol{B}. \tag{5.1.60}$$

Conversely, if we write (*cf.* (4.12.43))

$$\begin{aligned} &\varphi_0 = \varphi_{00}, \varphi_1 = \varphi_{01}, \varphi_2 = \varphi_{11}, \\ &\bar{\varphi}_0 = \bar{\varphi}_{0'0'}, \bar{\varphi}_1 = \bar{\varphi}_{0'1'}, \bar{\varphi}_2 = \bar{\varphi}_{1'1'}, \end{aligned} \tag{5.1.61}$$

we find, from (5.1.39),

$$F_{\mathbf{ab}} = \tfrac{1}{2}\begin{bmatrix} 0 & (\varphi_0 - \varphi_2 + \bar{\varphi}_0 - \bar{\varphi}_2) & (\mathrm{i}\varphi_0 + \mathrm{i}\varphi_2 - \mathrm{i}\bar{\varphi}_0 - \mathrm{i}\bar{\varphi}_2) & (-2\varphi_1 - 2\bar{\varphi}_1) \\ (-\varphi_0 + \varphi_2 - \bar{\varphi}_0 + \bar{\varphi}_2) & 0 & (2\mathrm{i}\varphi_1 - 2\mathrm{i}\bar{\varphi}_1) & (-\varphi_0 - \varphi_2 - \bar{\varphi}_0 - \bar{\varphi}_2) \\ (-\mathrm{i}\varphi_0 - \mathrm{i}\varphi_2 + \mathrm{i}\bar{\varphi}_0 + \mathrm{i}\bar{\varphi}_1) & (-2\mathrm{i}\varphi_1 + 2\mathrm{i}\bar{\varphi}_1) & 0 & (\mathrm{i}\varphi_2 - \mathrm{i}\varphi_0 - \mathrm{i}\bar{\varphi}_2 + \mathrm{i}\bar{\varphi}_0) \\ (2\varphi_1 + 2\bar{\varphi}_1) & (\varphi_0 + \varphi_2 + \bar{\varphi}_0 + \bar{\varphi}_2) & (-\mathrm{i}\varphi_2 + \mathrm{i}\varphi_0 + \mathrm{i}\bar{\varphi}_2 - \mathrm{i}\bar{\varphi}_0) & 0 \end{bmatrix} \tag{5.1.62}$$

*Complex* fields $F_{ab}$ satisfying Maxwell's source-free equations play a role as wave functions of single photons. For this and other reasons it is of interest briefly to include this more general case in our discussion. Instead of (5.1.39) and (5.1.40) we then have

$$F_{ab} = \varphi_{AB}\varepsilon_{A'B'} + \varepsilon_{AB}\tilde{\varphi}_{A'B'}, \tag{5.1.63}$$

and

$$\varphi_{AB} = \tfrac{1}{2}F_{ABC'}{}^{C'}, \quad \tilde{\varphi}_{A'B'} = \tfrac{1}{2}F_C{}^C{}_{A'B'}, \tag{5.1.64}$$

where $\varphi_{AB}, \tilde{\varphi}_{A'B'}$ are now *independent* spinor fields. Formula (5.1.62) with the definitions (5.1.61) still holds, but now $\tilde{\varphi}$ replaces $\bar{\varphi}$ throughout; (5.1.59) holds without change; and a corresponding formula for $\tilde{\varphi}$ is obtained by replacing i by $-$i in (5.1.59) *and* in (5.1.60).

The dual (*cf.* (3.4.21)) $^*F_{\mathbf{ab}}$ of (5.1.58) is easily seen to be

$$^*F_{\mathbf{ab}} = \begin{bmatrix} 0 & -B_1 & -B_2 & -B_3 \\ B_1 & 0 & -E_3 & E_2 \\ B_2 & E_3 & 0 & -E_1 \\ B_3 & -E_2 & E_1 & 0 \end{bmatrix} \tag{5.1.65}$$

Now there are two important scalar invariants associated with any electromagnetic field $F_{ab}$, namely

$$P = \tfrac{1}{2}F_{ab}F^{ab} = -\tfrac{1}{2}{}^*F_{ab}{}^*F^{ab}, \quad Q = \tfrac{1}{2}F_{ab}{}^*F^{ab}, \tag{5.1.66}$$

which are perhaps better known in the forms

$$P = B^2 - E^2, \quad Q = 2\boldsymbol{E}\cdot\boldsymbol{B}, \tag{5.1.67}$$

obtainable at once from the definitions (5.1.66) and (5.1.58), (5.1.65). In spinor form we have, by use of (3.4.38) and (2.5.9),

$$K := \varphi_{AB}\varphi^{AB} = \tfrac{1}{2}{}^-F_{ab}{}^-F^{ab} = \tfrac{1}{2}F_{ab}{}^-F^{ab} = P + \mathrm{i}Q. \tag{5.1.68}$$

In the case of a *real* field $F_{ab}$, $P$ and $Q$ are manifestly real, and so they constitute the real and imaginary parts of the one spinor invariant, $K$. If, on the other hand, $F_{ab}$ is complex, we define, in analogy with (5.1.68),

$$\tilde{K} := \tilde{\varphi}_{A'B'}\tilde{\varphi}^{A'B'} = \tfrac{1}{2}F_{ab}{}^+F^{ab} = P - \mathrm{i}Q \tag{5.1.69}$$

(note $\tilde{K} = \bar{K}$ if the field is real), so that

$$P = (K + \tilde{K})/2, \quad Q = (K - \tilde{K})/2\mathrm{i}. \tag{5.1.70}$$

If $P = Q = 0$ the field is *null*, i.e., the PNDs of $\varphi_{AB}$ (and also those of $\tilde{\varphi}_{A'B'}$) are coincident. This follows at once from (3.5.29).

If $Q = 0$ and $P \neq 0$, we say – in the case of a real field – that the field is either *purely electric* or *purely magnetic* according as $P < 0$ or $P > 0$, respectively. The reason for this terminology is that in these cases one can find a Lorentz transformation (in fact, infinitely many) which 'transforms away' the magnetic or the electric field, as the case may be. It is merely necessary, for example, to apply a boost with velocity $E^{-2}(\boldsymbol{E} \times \boldsymbol{B})$ in the first case, and one with velocity $B^{-2}(\boldsymbol{E} \times \boldsymbol{B})$ in the second case. We shall see later (in §8.5) that in these cases, when one of the fields $\boldsymbol{E}$ or $\boldsymbol{B}$ has been transformed away, the two PNDs of $\varphi_{AB}$ point in opposite directions on the Riemann sphere.

We may also note that $Q = 0$ is the necessary and sufficient condition for the tensor $F_{ab}$ to be *simple* (*cf.* (3.5.30) and (3.5.35)).

Further remarks on the structure of the electromagnetic field will be found at the end of §8.5.

## 5.2 Einstein–Maxwell equations in spinor form

We next consider the spinor form of the combined Einstein–Maxwell 'electrovac' equations, i.e., the field equations of general relativity with the energy tensor of the electromagnetic field as the only source term. First we must find the spinor equivalent of the energy tensor of the electromagnetic field. This is a real symmetric tensor $T_{ab}$ which is quadratic in the electromagnetic field tensor $F_{ab}$ and satisfies

$$\nabla^a T_{ab} = 0 \tag{5.2.1}$$

when the source-free Maxwell equations hold. There is one obvious expression in the spinor formalism which has this property, namely

$$T_{ab} = k\varphi_{AB}\bar{\varphi}_{A'B'}, \tag{5.2.2}$$

where $k$ is a real and, in fact, necessarily positive constant (because of the positive-definiteness requirements that we shall examine shortly). The tensor defined in (5.2.2) is real, symmetric, quadratic in $\varphi_{AB}$ and therefore in $F_{ab}$, and it satisfies (5.2.1) by virtue of (5.1.57). Recall that the Bel–Robinson tensor (4.8.9) satisfies a similar equation (*viz* (4.10.11)) for a similar reason (*viz* (4.10.9)).

The standard tensor expressions for the Maxwell energy tensor are

$$\begin{aligned} T_{ab} &= \frac{1}{4\pi}(\tfrac{1}{4}g_{ab}F_{cd}F^{cd} - F_{ac}F_b{}^c) \\ &= -\frac{1}{8\pi}(F_{ac}F_b{}^c + {}^*F_{ac}{}^*F_b{}^c). \end{aligned} \tag{5.2.3}$$

Substitution of (5.1.39) into either of these expressions does indeed yield (5.2.2) with $k = (2\pi)^{-1}$:

$$T_{ab} = \frac{1}{2\pi}\varphi_{AB}\bar{\varphi}_{A'B'}. \tag{5.2.4}$$

Note that the second expression in (5.2.3) is very similar to the tensor expression (4.8.10) of the Bel–Robinson tensor in terms of the Weyl tensor. Again, $T_{ab}$ is invariant under duality rotations of the electromagnetic field (3.4.42), for these correspond to $\varphi_{AB} \mapsto e^{-i\theta}\varphi_{AB}$. This invariance is rather less immediate from the tensor expressions (5.2.3). Also $T_{ab}$ is trace-free:

$$T_a{}^a = 0, \tag{5.2.5}$$

as follows at once from (5.2.4). Both duality rotation invariance and tracefreeness are also properties of the Bel–Robinson tensor, *cf.* (4.8.16) and (4.8.12). Of course, the electromagnetic energy tensor $T_{ab}$ was discovered long before the Bel–Robinson tensor $T_{abcd}$. By tensor methods $T_{abcd}$ was hard to find, as an analogue of $T_{ab}$. Only in terms of spinors is the one as simple as the other.

Substituting (5.2.4) into Einstein's equations in the form (4.6.32), and bearing in mind that now $T_a{}^a = 0$, we get

$$\Phi_{ABA'B'} = 2\gamma\varphi_{AB}\bar{\varphi}_{A'B'}, \quad \Lambda = \tfrac{1}{6}\lambda, \tag{5.2.6}$$

which, together with the Maxwell source-free equation (5.1.57), provide the spinor form of the *Einstein–Maxwell equations* (and usually we take $\lambda = 0$). On the other hand, substituting (5.2.4) into (4.10.12) gives

$$\begin{aligned}\nabla^A_{B'}\Psi_{ABCD} &= 2\gamma\nabla^{A'}_{(B}[\varphi_{CD)}\bar{\varphi}_{A'B'}] \\ &= 2\gamma[\bar{\varphi}_{A'B'}\nabla^{A'}_{(B}\varphi_{CD)} + \varphi_{(CD}\nabla^{A'}_{B)}\bar{\varphi}_{A'B'}].\end{aligned}$$

Now the second term on the right vanishes because of (5.1.57); and, by an easy argument (*cf.* (5.7.16) below), one shows the symmetrization in the first term on the right to be superfluous. So the Bianchi identity becomes

$$\nabla^A_{B'}\Psi_{ABCD} = 2\gamma\bar{\varphi}_{A'B'}\nabla^{A'}_B\varphi_{CD}. \tag{5.2.7}$$

In passing, we remark that the following modification of the Bel–Robinson tensor for the Einstein–Maxwell equations, though not totally symmetric, has zero divergence:

$$\begin{aligned}T_{abcd} = \Psi_{ABCD}\bar{\Psi}_{A'B'C'D'} &- 2\gamma\nabla_{CD'}\varphi_{AB}\nabla_{DC'}\bar{\varphi}_{A'B'} \\ &+ 6\gamma\nabla_{D(A'|}\varphi_{(AB}\nabla_{C)D'}\bar{\varphi}_{|B'C')} - \lambda\varphi_{(AB}\varepsilon_{C)D}\bar{\varphi}_{(A'B'}\varepsilon_{C')D'};\\ T_{abcd} = T_{(abc)d}, \; T^a{}_{acd} &= 0, \nabla^d T_{abcd} = 0.\end{aligned}$$

## *Positivity properties of Maxwell's energy tensor*

The tensor $T_{ab}$ possesses an important positive-definiteness property. Observe that for any pair of spinors $\mu^A$, $\nu^A$ with corresponding null vectors

$$M^a = \mu^A\bar{\mu}^{A'}, \quad N^a = \nu^A\bar{\nu}^{A'}$$

we have

$$T_{ab}M^aN^b = \frac{1}{2\pi}|\varphi_{AB}\mu^A\nu^B|^2 \geqslant 0, \tag{5.2.8}$$

i.e., this inequality applies to any two future-null vectors. Since any future-causal vector is a sum of future-null vectors, we derive by such expansion and use of (5.2.8) the following:

(5.2.9) PROPOSITION

*For every pair of future-causal vectors* $U^a$, $V^a$,

$$T_{ab}U^aV^a \geqslant 0.$$

This can be expressed slightly differently, in the form:

(5.2.10) PROPOSITION

*For each future-causal vector* $V^a$,

$$V^aT_a{}^bT_{bc}V^c \geqslant 0, \quad V^aT_{ab}V^b \geqslant 0.$$

For (5.2.9) says that the vector $V^aT_a{}^b$ has non-negative scalar product with all future-causal vectors; it is therefore itself causal (*cf.* (5.2.10) (1)), and, indeed, future-causal when $V^a$ is (*cf.* (5.2.10) (2)). Condition (5.2.9) or (5.2.10) is sometimes referred to as the *dominant energy condition.* If $V^a$ is the 4-velocity vector of an observer, then $T^a{}_bV^b$ is his Poynting '4-vector', having component form: (energy, Poynting 3-vector). Thus (5.2.9) states that the velocity of energy flow, as described by the Poynting vector, does not exceed the velocity of light.

Note that the following weakened form of the above energy condition:

$$T_{ab}V^aV^b \geqslant 0 \tag{5.2.11}$$

(sometimes: *weak energy condition*) states that the energy density measured by an observer ($T_{00}$) must be a non-negative-definite function of his 4-velocity $V^a$ (which he measures as $g_0{}^a$). It is of some interest to examine the locus

$$T_{ab}V^aV^b = 0. \tag{5.2.12}$$

First take $V^a$ to be the null vector $N^a = \nu^A\bar{\nu}^{A'} \neq 0$. Then (5.2.8) tells us that $T_{ab}N^aN^b$ vanishes if and only if

$$\varphi_{AB}\nu^A\nu^B = 0,$$

and this occurs (*cf.* (3.5.22)) when the flagpole of $\nu^A$ points in one of the two (possibly coincident) principal null directions of the field $\varphi_{AB}$ (taking $\varphi_{AB} \neq 0$). In fact these are the *only* causal vectors $V^a$ for which (5.2.12) holds. For if $V^a$ is timelike, say future-timelike, we can select any future-null direction $M^a$ which is *not* a principal null direction of $\varphi_{AB}$, and express $V^a$ as a sum of a multiple of $M^a$ and another null vector. Substituting into (5.2.12) and expanding, we get a sum of non-negative terms (by (5.2.9)), at least one of which (namely $T_{ab}M^aM^b$) is strictly positive. Thus we conclude:

(5.2.13) PROPOSITION

$T_{ab}V^aV^b = 0, V^aV_a \geqslant 0, V^a \neq 0$ *iff* $V^a$ *is a principal null vector of* $\varphi_{AB}$.

Similar results hold for the Bel–Robinson tensor. We have, in fact,

(5.2.14) PROPOSITION

$T_{abcd}S^aU^bV^cW^d \geqslant 0$ *for all future-causal vectors* $S^a, U^a, V^a, W^a$,

and also

(5.2.15) PROPOSITION

$T_{abcd}V^aV^bV^cV^d = 0,\ V^aV_a \geqslant 0,\ V^a \neq 0$ *iff* $V^a$ *is a principal null vector of* $\Psi_{ABCD}$.

The reasoning is just the same as for the electromagnetic case above. The principal null directions of $\Psi_{ABCD}$ will later be seen to play a key role in the classification scheme for Weyl tensors (see Chapter 8).

In §4.8 (*cf.* (4.8.13)) we remarked that the Bel–Robinson tensor satisfies a quadratic identity (in addition to being symmetric and trace-free) although the complete tensor expression for this was not found explicitly. In the electromagnetic case we have

$$T_{ABA'B'}T_{CDC'D'} = T_{ABC'D'}T_{CDA'B'} \tag{5.2.16}$$

as an immediate consequence of (5.2.4). The tensor form of this equation is well known in Maxwell's theory. We can obtain it from (5.2.16) by repeated use of (2.5.23):

$$T_{ac}T_b{}^c = \tfrac{1}{4}(T_{cd}T^{cd})g_{ab}. \tag{5.2.17}$$

## 5.3 The Rainich conditions

As we have seen in the preceding section, the Maxwell energy tensor $T_{ab}$ is real, symmetric, and possesses the following

(5.3.1) PROPERTIES

(i) $T_a{}^a = 0$
(ii) $T_{ab}T_c{}^b \propto g_{ac}$
(iii) $T_{ab}U^aV^b \geqslant 0$ *for every pair of future-causal vectors* $U^a, V^a$.

They are all automatic consequences of the spinor form $T_{ab} = k\varphi_{AB}\bar{\varphi}_{A'B'}$ (*cf.* (5.2.2)) with $\varphi_{AB} = \varphi_{(AB)}$ and $k$ real and positive. Now, conversely, for each real symmetric tensor $T_{ab}$ satisfying (5.3.1) at any one point, there exists such a spinor form at that point, and there exist real skew solutions $F_{ab}$ of equation (5.2.3); moreover, all of these solutions are obtained from each other by duality rotations (*cf*. (3.4.42)). This result was first established by Rainich (1925) and equations (5.3.1) are therefore called the *Rainich conditions*. Later the result was rediscovered by Misner and Wheeler (1957) and made the basis of their 'geometrodynamics'. Of course, in order to qualify as an electromagnetic field tensor, $F_{ab}$ must satisfy Maxwell's equations, and thus certain further (differential) restrictions must be imposed on $T_{ab}$ in order for it to be the energy tensor of a Maxwell field. These conditions (for non-null fields) were also first worked out by Rainich and rediscovered by Misner and Wheeler. There seems little doubt that this theory is most simply discussed by means of the spinor calculus, as was effectively done by Witten (1962); our development below follows somewhat different lines from his.

Suppose a real symmetric tensor $T_{ab}$ is given which satisfies the conditions (5.3.1). Referring back to (3.4.4)–(3.4.6) we see that, because of (5.3.1)(i),

$$T_{ABA'B'} = T_{(AB)(A'B')}. \tag{5.3.2}$$

Equation (5.3.1) (ii) in spinor form is

$$T_{AA'BB'}T^{BB'}{}_{CC'} \propto \varepsilon_{AC}\varepsilon_{A'C'}.$$

This, by use of (2.5.23), yields

$$T_{AA'[B}{}^{[B'}T^{D']}{}_{D]CC'} \propto \varepsilon_{AC}\varepsilon_{A'C'}\varepsilon_{BD}\varepsilon^{B'D'},$$

whence

$$T^{(A'|[B'}{}_{(A|[B}T^{D']|C')}{}_{D]|C)} = 0. \tag{5.3.3}$$

Using the symmetry (5.3.2) to interchange $A, B$ positions and $C, D$ positions, and relabelling the indices: $A \mapsto B$, $C \mapsto D$, we have

$$T^{(A'|[B'}{}_{[A|(B}T^{D']|C')}{}_{D)|C]} = 0. \tag{5.3.4}$$

Adding this equation to (5.3.3) and expanding the lower symmetry operations, we find

$$T^{(A'|[B'}{}_{[\mathscr{A}}T^{D']|C')}{}_{\mathscr{C}]} = 0, \tag{5.3.5}$$

where we have written $\mathscr{A} = AB$ and $\mathscr{C} = CD$. Once again using the symmetry (5.3.2), and relabelling indices, we find from (5.3.5) that

$$T^{[A'|(B'}{}_{[\mathscr{A}}T^{D')|C']}{}_{\mathscr{C}]} = 0. \tag{5.3.6}$$

Adding this to (5.3.5) we find, by complete analogy with the process that lead to (5.3.5) itself,

$$T^{[\mathscr{A}'}_{[\mathscr{A}}T^{\mathscr{C}']}_{\mathscr{C}]}=0, \tag{5.3.7}$$

where $\mathscr{A}'=A'B'$, $\mathscr{C}'=C'D'$. This is equivalent to

$$T_{\mathscr{A}\mathscr{A}'}T_{\mathscr{C}\mathscr{C}'}=T_{\mathscr{A}\mathscr{C}'}T_{\mathscr{C}\mathscr{A}'}. \tag{5.3.8}$$

Now we choose an arbitrary non-zero spinor $X^{\mathscr{C}}$ and multiply (5.3.8) by $X^{\mathscr{C}}\bar{X}^{\mathscr{C}'}$, thus obtaining (in regions where $T_{\mathscr{C}\mathscr{C}'}X^{\mathscr{C}}\bar{X}^{\mathscr{C}'}\neq 0$)

$$T_{\mathscr{A}\mathscr{A}'}=(T_{\mathscr{C}\mathscr{C}'}X^{\mathscr{C}}\bar{X}^{\mathscr{C}'})^{-1}T_{\mathscr{A}\mathscr{C}'}\bar{X}^{\mathscr{C}'}T_{\mathscr{A}'\mathscr{C}}X^{\mathscr{C}}, \tag{5.3.9}$$

which, because of the reality of $T_{ab}$, is of the required form

$$T_{ABA'B'}=k\varphi_{AB}\bar{\varphi}_{A'B'} \tag{5.3.10}$$

at each point,* with $k$ a real scalar (*cf.* (3.5.5).) The last of the Rainich conditions, (5.3.1)(iii), implies that any such $k$, as defined by (5.3.10) and (5.3.9), is in fact positive. We can therefore normalize $\varphi_{AB}$ so that $k=1/2\pi$, as in (5.2.4). The $F_{ab}$ as defined in (5.1.39) automatically satisfies (5.2.3), and the existence of *a* solution of (5.2.3) at each point is therefore established. Evidently this solution is not unique, since $\varphi_{AB}\mapsto e^{-i\theta}\varphi_{AB}$ ($\theta$ real) leaves (5.2.4) unchanged and corresponds to a duality rotation $F_{ab}\mapsto{}^{(\theta)}F_{ab}$ (*cf.* (3.4.42), (3.4.43)). On the other hand, this is clearly *all* the freedom allowed by (5.3.10) at each point (*cf.* (3.5.2). So the algebraic part of the Rainich theory is established.

Before proceeding to the differential part of the theory we shall discuss what Misner and Wheeler call the *complexion* of the field. As we have seen, all field tensors $F_{ab}$ in a class with common energy tensor $T_{ab}$ differ from each other by a duality rotation at each point. In each such class there are, except where $K=0$ (null field), exactly two fields, differing only in sign, which are 'purely electric', i.e., which have invariants $P<0, Q=0$ (*cf.* after (5.1.70)). For let $F_{ab}$ be any field in the class, with corresponding spinor $\varphi_{AB}$ and invariant $K=:-e^{-2i\theta}K_0$ where $K_0$ is real and positive. Then $P_{ab}=\pm{}^{(-\theta)}F_{ab}$ have spinors $e^{i\theta}\varphi_{AB}$, and each has invariant $e^{2i\theta}K=-K_0$ and is consequently purely electric, as reference to (5.1.68) shows. Evidently $F_{ab}=\pm{}^{(\theta)}P_{ab}$. The angle $\theta$ defined up to the addition of an integer multiple of $\pi$, is said to be the *complexion of* $F_{ab}$; these are the angles through which $\pm P_{ab}$ have to be duality-rotated in order to coincide with $F_{ab}$. The complexion becomes indeterminate only where the field is null. Now, a 'generic' field $F_{ab}$ becomes null on some 2-surface (since $K=0$ corresponds to two real equations in a four-dimensional space).

* There is no guarantee, at this stage, that $\varphi_{AB}$ can be chosen to be everywhere continuous.

Thus a 'generic' $T_{ab}$, subject to (5.3.1), will also satisfy $T_{cd}T^{cd}=0$ on a 2-surface, on which the complexion therefore becomes indeterminate. It may then happen that the region of space–time from which this surface has been removed is not simply-connected: a 2-surface has just the right dimension to be linked by a curve in four dimensions. By choosing a suitable closed path around this 2-surface it might be possible to take $P_{ab}$ continuously into $-P_{ab}$, so that a constant global choice of sign for the field $P_{ab}$ might be impossible. This difficulty arises even before we consider the particular Rainich field equations. So we now assume, in order to proceed further, that we are concerned only with a region of space–time in which the sign of $P_{ab}$ can be chosen continuously. Moreover, we assume that $P_{ab}$ can be chosen to be smooth – which again does not quite follow from the assumption that $T_{ab}$ is smooth, if there are regions where $T_{ab}$ vanishes.

### *Differential Rainich condition*

Now suppose that, in accordance with these assumptions, we are given an energy tensor $T_{ab}$ satisfying the Rainich conditions (5.3.1). Then at each point of our region of interest we have a smooth purely electric tensor $P_{ab}$ having $T_{ab}$ for its energy tensor. We then ask the question: is it possible to find a tensor $F_{ab}={}^{\theta}P_{ab}$, for some variable real $\theta$, which satisfies Maxwell's equations, and which therefore represents a Maxwell field having energy tensor $T_{ab}$? If $\chi_{AB}$ is the symmetric spinor corresponding to $P_{ab}$, with $\chi=\frac{1}{2}\chi_{AB}\chi^{AB}<0$, then $e^{-i\theta}\chi_{AB}$ will correspond to $F_{ab}$. Applying Maxwell's equations (5.1.57) to this spinor, we get

$$\nabla_{AA'}(e^{-i\theta}\chi^{AB})=e^{-i\theta}\nabla_{AA'}\chi^{AB}-ie^{-i\theta}\chi^{AB}\nabla_{AA'}\theta=0.$$

Cancelling the exponential factor, multiplying by $\chi_{BC}$, and using (2.5.23), we next get

$$\chi_{BC}\nabla_{AA'}\chi^{AB}-i\chi\nabla_{CA'}\theta=0, \tag{5.3.11}$$

which yields, after some relabelling of indices,

$$\nabla_{AA'}\theta=\frac{1}{i\chi}\chi_{AB}\nabla_{CA'}\chi^{BC}=:S_{AA'}, \tag{5.3.12}$$

$S_{AA'}$ being a spinor defined by this equation. Since $\theta$ is real, $S_{AA'}$ is a real vector. Our problem now reduces to finding the tensor equivalent of $S_{AA'}$ and solving (5.3.12) for $\theta$.

We shall proceed synthetically. Differentiating (5.2.4) with $\chi_{AB}$ in place of $\varphi_{AB}$, contracting once and writing $k = 1/2\pi$, we get one term on the right-hand side essentially like $S_{AA'}$:

$$\nabla_{AC'}T^{AA'BB'} = k\bar{\chi}^{A'B'}\nabla_{AC'}\chi^{AB} + k\chi^{AB}\nabla_{AC'}\bar{\chi}^{A'B'}. \tag{5.3.13}$$

Pursuing this synthesis, we multiply (5.3.13) by (5.2.4) with $\chi$ for $\varphi$ and the index $A$ replaced by $D$, and find

$$\begin{aligned} T_{DA'BB'}\nabla_{AC'}T^{AA'BB'} &= 2k^2\bar{\chi}\chi_{DB}\nabla_{AC'}\chi^{AB} + k^2\chi\varepsilon_D{}^A\bar{\chi}_{A'B'}\nabla_{AC'}\bar{\chi}^{A'B'} \\ &= 2\mathrm{i}k^2\chi^2 S_{DC'} + k^2\chi\nabla_{DC'}\chi, \end{aligned} \tag{5.3.14}$$

where we have used the reality of $\chi$. We can eliminate the last term in (5.3.14) by taking complex conjugates, relabelling some indices, and subtracting the resulting equation from (5.3.14); in this way we get

$$\begin{aligned} 4\mathrm{i}k^2\chi^2 S_{DC'} &= T_{DA'BB'}\nabla_{AC'}T^{AA'BB'} - T_{AC'BB'}\nabla_{DA'}T^{AA'BB'} \\ &= \mathrm{i}e_{AA'DC'}{}^{PP'QQ'}T_{BB'PP'}\nabla_{QQ'}T^{AA'BB'}, \end{aligned} \tag{5.3.15}$$

where for the last line we refer to (3.3.46): the effect of the dualizer is precisely to form a difference of two terms which differ from each other by an interchange of the index pairs $AC'$ and $DA'$ as in the line above. Using the value

$$4k^2\chi^2 = T_{ab}T^{ab}$$

obtained from (5.1.68) and (5.2.2), we can now translate (5.3.15) into tensor form (making obvious changes in the indices):

$$S_c = \frac{e_{ac}{}^{pq}T_{bp}\nabla_q T^{ab}}{T_{ab}T^{ab}}. \tag{5.3.16}$$

A *differential condition* on $T_{ab}$ is now obtained by the substitution of (5.3.16) into the integrability condition for equation (5.3.12):

$$\nabla_b S_a - \nabla_a S_b = 0, \tag{5.3.17}$$

which holds since $\theta$ is a scalar. If the condition is satisfied, $\theta$ is determined by (5.3.12) and (5.3.16) to within an additive constant.

A discussion of the implications of Maxwell's equations in regions containing loci on which the field is null (or zero), or which are not simply-connected, is beyond our present scope. Note, however, that the algebraic part of the theory applies equally to non-null and null fields.

## 5.4 Vector Bundles

A viewpoint we have tended to emphasize in this book is one which regards the kind of algebra satisfied by various types of field as basic (abstract index algebra and formal rules for $\nabla$) and the geometric inter-

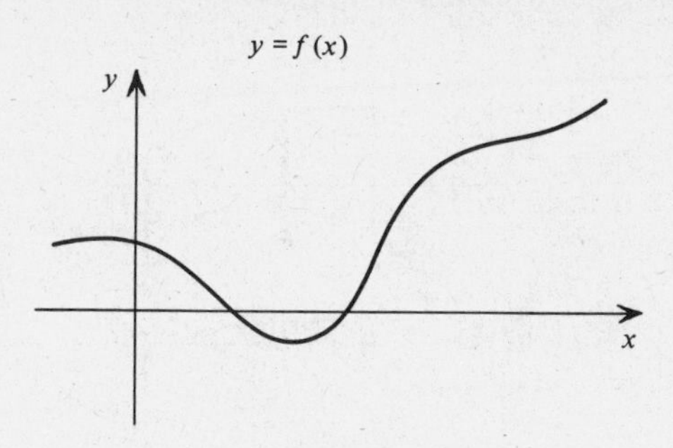


Fig. 5-1. The graph of a function.

pretation of these objects and operations as secondary. It is often useful, on the other hand, to picture things also in a geometrical way and, in fact, for the basic spinors themselves we took such a geometrical viewpoint in Chapter 1. There we considered, but did not concentrate much on, the concept of *vector-bundles* (*cf*. § 1.5), which is a useful one when passing from a local to a global description. Since a rather complete local geometrical picture of the basic spin-vectors can be given, the bundle description can, to some extent, be there avoided. But in the case of the charged fields of electromagnetic theory – and even more in the case of the 'multi-charged' fields of Yang–Mills theory to be discussed presently – the geometric content of the theory is hard to grasp except in the context of vector bundles. Accordingly we give a brief introduction of this concept here.

Let us begin with a very simple idea, that of the *graph* of a function. Consider a real-valued function of a single real variable $f: \mathbb{R} \mapsto \mathbb{R}$. Usually the graph of $f$ is plotted, as in Fig. 5-1, by drawing a horizontal $x$-axis and a vertical $y$-axis, and marking the locus $y = f(x)$. Suppose, however, that we are concerned with 'functions' of a different kind, whose 'values' are not simply numbers, and are not necessarily comparable for different values of the argument $x$. A familiar example of such a function is provided by a tangent-vector-valued function on a manifold, i.e., by a vector field. Thus, in place of the $x$-axis in Fig. 5-1, which was a copy of $\mathbb{R}$, we envisage some manifold $\mathcal{M}$, called the *base space* (which, for definiteness, we could picture to be, for example, a sphere $S^2$). In place of the $y$-axis we need something to represent the tangent space at a typical point of $\mathcal{M}$. But since the tangent spaces at any two different points of $\mathcal{M}$ are not generally in canonical correspondence with one another, they must be thought of not as *identical* but only as *isomorphic* spaces, called *fibres*, one for each point of $\mathcal{M}$. The fibre corresponding to a point $P \in \mathcal{M}$ is called the fibre *above P*. Instead of the simple product space $\mathbb{R} \times \mathbb{R}$ carrying the graph

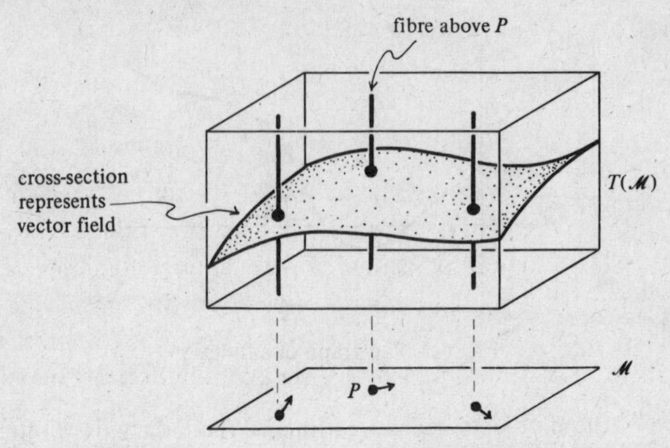


Fig. 5-2. The tangent bundle of $\mathcal{M}$, and one of its cross-sections representing a vector field on $\mathcal{M}$.

of the function $f : \mathbb{R} \mapsto \mathbb{R}$, we now have a more complicated space, called, in this example, the *tangent bundle* $T(\mathcal{M})$ of the manifold $\mathcal{M}$. A vector field on $\mathcal{M}$ can be represented as a kind of graph known as a (smooth) *cross-section* of the bundle $T(\mathcal{M})$. (See Fig. 5-2.)

The tangent bundle is only one very important example of a vector bundle, in which the fibres happen to be the tangent spaces at the various points of $\mathcal{M}$ (and, of course, the bundle concept refers also to base spaces other than the space–times we consider). Tensor or spinor bundles are other examples; here the cross-sections are the elements of $\mathfrak{T}^{\mathscr{A}}$, or of $\mathfrak{S}^{\mathscr{A}}$, for some fixed $\mathscr{A}$, i.e., all the tensor and spinor fields we have been discussing so far: their properties could, in fact, have been developed in bundle terms. Other types of vector bundles over $\mathcal{M}$ can be constructed by choosing, for fibres, copies of *any* real or complex finite-dimensional vector spaces (which may be quite independent of the tangent space to $\mathcal{M}$ or its associated spin-spaces – for example, the isotopic spin space discussed at the beginning of Chapter 4, or the "colour spaces" that are frequently considered in contemporary particle physics). The spaces at different points are to be all isomorphic to one another, and with an important stipulation: loosely speaking, these vector spaces must 'join smoothly' together (so that it makes sense to speak of 'smooth' cross-sections). Thus, although we don't need to know which elements in different fibres correspond, we must know which elements in neighbouring fibres 'differ by little'. In fact, each point of $\mathcal{M}$ must belong to an open neighbourhood $\mathscr{U} \subset \mathcal{M}$ such that the portion of the bundle above $\mathscr{U}$ is smoothly equivalent to a product space, although the entire bundle may not be equivalent to a product space. For an example of such a non-trivial bundle, see Fig. 5-3 below.

*Definition of a vector bundle*

The aim of our procedure will be to provide an abstract tensor algebra suitable for working within a bundle. This will entail the introduction of *bundle indices* (denoted by capital Greek letters) to supplement the ordinary space–time (and spinor) indices. For this, we shall need to construct a system $\mathfrak{T}^{\Phi}$ [*or* $\mathfrak{S}^{\Phi}$] whose elements describe ($C^{\infty}$) cross-sections of the bundle. Our definition will be given initially in terms of coordinate descriptions which hold locally on $\mathcal{M}$ and for this, bold upright capital Greek letters will be used according to our conventions of Chapter 2. The procedure will then lead directly to the global and coordinate-free abstract-index system. (A closely related approach has recently been proposed by Ashtekar, Horowitz and Magnon-Ashtekar 1982.)

We need a more formal definition of a vector bundle (*cf.* Bott and Mather 1968). Given a (Hausdorff, paracompact, $C^{\infty}$) manifold $\mathcal{M}$, a real [complex] $k$-vector bundle over $\mathcal{M}$ is a manifold $\mathcal{B}$, together with a $C^{\infty}$ map

$$\Pi : \mathcal{B} \to \mathcal{M} \tag{5.4.1}$$

(to be thought of as the projection which collapses each fibre to the point of $\mathcal{M}$ 'below' it), such that

$$\text{$\Pi^{-1}(P)$ \textit{is a real [or complex] vector space of dimension $k$, for each} $P \in \mathcal{M}$.} \tag{5.4.2}$$

Furthermore, defining a *cross-section* $\boldsymbol{\lambda}$ of $\Pi^{-1}(\mathcal{U})$, for any open $\mathcal{U} \subset \mathcal{M}$, to be a $C^{\infty}$ map

$$\boldsymbol{\lambda} : \mathcal{U} \to \Pi^{-1}(\mathcal{U}), \tag{5.4.3}$$

such that $\Pi \circ \boldsymbol{\lambda}$ is the identity on $\mathcal{U}$, we require that there is a covering of $\mathcal{M}$ by a family of open sets $\{\mathcal{U}_{\mathrm{i}}\}$, $\mathcal{M} = \bigcup_{\mathrm{i}} \mathcal{U}_{\mathrm{i}}$, such that

$$\text{\textit{for each} $\mathcal{U}_{\mathrm{i}}$ \textit{there is a basis} $\overset{\mathrm{i}}{\boldsymbol{\delta}}_{\boldsymbol{\Phi}} = \overset{\mathrm{i}}{\boldsymbol{\delta}}_1, \ldots, \overset{\mathrm{i}}{\boldsymbol{\delta}}_k$ \textit{of cross-sections of} $\Pi^{-1}(\mathcal{U}_{\mathrm{i}})$} \tag{5.4.4}$$

in terms of which the general cross-section $\overset{\mathrm{i}}{\boldsymbol{\lambda}}$ of $\Pi^{-1}(\mathcal{U}_{\mathrm{i}})$ is uniquely expressible as (with no sum over i)

$$\overset{\mathrm{i}}{\boldsymbol{\lambda}} = \overset{\mathrm{i}}{\lambda}{}^{\boldsymbol{\Phi}} \overset{\mathrm{i}}{\boldsymbol{\delta}}_{\boldsymbol{\Phi}}, \tag{5.4.5}$$

where $\overset{\mathrm{i}}{\lambda}{}^{\boldsymbol{\Phi}} = \overset{\mathrm{i}}{\lambda}{}^1, \ldots, \overset{\mathrm{i}}{\lambda}{}^k \in \mathfrak{T}(\mathcal{U}_{\mathrm{i}})$ [ *or* $\mathfrak{S}(\mathcal{U}_{\mathrm{i}})$]. Addition of two cross-sections and multiplication of cross-sections by scalar fields are defined in the obvious way with respect to the linear structure on the fibres. The condition (5.4.4) says that, in the appropriate sense, $\mathscr{B}$ is locally a product space (locally in $\mathscr{M}$, that is).

Any (global) cross-section $\boldsymbol{\lambda}$ of $\mathscr{B}$ will have the property that its restriction to any open $\mathscr{U} \subset \mathscr{M}$ is a cross-section of $\Pi^{-1}(\mathscr{U})$. However, if $\varnothing \neq \mathscr{U} \neq \mathscr{M}$, there may be cross-sections of $\Pi^{-1}(\mathscr{U})$ that are *not* restrictions of cross-sections of $\mathscr{B}$, namely those which, because of bad differentiability properties at the boundary of $\Pi^{-1}(\mathscr{U})$, are not extendable beyond this boundary. We require the module of *extendable* cross-sections above $\mathscr{U}$. Assume that $\mathscr{U}$ is given by $f \neq 0$ for some $f \in \mathfrak{S}$. Then, as in §4.1, the required extendable cross-sections can be represented as equivalence classes of cross-sections of $\mathscr{B}$ such that $\boldsymbol{\lambda} \sim \boldsymbol{\mu}$ whenever $f\boldsymbol{\lambda} = f\boldsymbol{\mu}$. If $\mathscr{U}$ is an open set in $\mathscr{M}$ and $\mathscr{U}'$ is a 'slightly smaller' open set for which $\bar{\mathscr{U}}'$ (the closure of $\mathscr{U}'$ in $\mathscr{M}$) $\subset \mathscr{U}$, then the restriction to $\mathscr{U}'$ of any cross-section above $\mathscr{U}$ is also the restriction to $\mathscr{U}'$ of some global cross-section (since such a global cross-section can fall smoothly to zero between $\mathscr{U}'$ and $\mathscr{U}$ and then remain zero outside $\mathscr{U}$). In particular, the basis $\overset{\mathrm{i}}{\boldsymbol{\delta}}_{\boldsymbol{\Phi}}$ above $\mathscr{U}_{\mathrm{i}}$ in (5.4.4), restricts to a basis $\overset{\mathrm{i}}{\boldsymbol{\delta}}{}'_{\boldsymbol{\Phi}}$ over a 'slightly smaller' $\mathscr{U}'_{\mathrm{i}}$, where each $\overset{\mathrm{i}}{\boldsymbol{\delta}}{}'_1, \ldots, \overset{\mathrm{i}}{\boldsymbol{\delta}}{}'_k$ extends to a global cross-section. Any covering $\{\mathscr{U}_{\mathrm{i}}\}$ of $\mathscr{M}$ can, in this way, be made 'slighly smaller' to give a covering $\{\mathscr{U}'_{\mathrm{i}}\}$ of $\mathscr{M}$ with $\bar{\mathscr{U}}'_{\mathrm{i}} \subset \mathscr{U}_{\mathrm{i}}$.

By the argument given in §2.4, we can show that a finite covering $\{\mathscr{U}_{\mathrm{i}}\}$ of $\mathscr{M}$ exists and a partition of unity $\overset{\mathrm{i}}{u} \in \mathfrak{T}$, with $\overset{\mathrm{i}}{u} \geqslant 0$ and $\sum_{\mathrm{i}} \overset{\mathrm{i}}{u} = 1$, where $\overset{\mathrm{i}}{u} \neq 0$ defines $\mathscr{U}_{\mathrm{i}}$, and where the basis $\overset{\mathrm{i}}{\boldsymbol{\delta}}_{\boldsymbol{\Phi}}$ may now be assumed to be extendable to global cross-sections. Thus if $\boldsymbol{\lambda}$ is any (global) cross-section of $\mathscr{B}$, we have elements $\overset{\mathrm{i}}{\lambda}{}^{\boldsymbol{\Phi}} \in \mathfrak{T}$ [*or* $\mathfrak{S}$] (components of $\boldsymbol{\lambda}$ in the basis $\overset{\mathrm{i}}{\boldsymbol{\delta}}_{\boldsymbol{\Phi}}$) for which

$$\overset{\mathrm{i}}{u}\boldsymbol{\lambda} = \overset{\mathrm{i}}{u}\overset{\mathrm{i}}{\lambda}{}^{\boldsymbol{\Phi}}\overset{\mathrm{i}}{\boldsymbol{\delta}}_{\boldsymbol{\Phi}}.$$

Hence, summing,

$$\boldsymbol{\lambda} = \sum_{\mathrm{i}} \overset{\mathrm{i}}{u}\overset{\mathrm{i}}{\lambda}{}^{\boldsymbol{\Phi}}\overset{\mathrm{i}}{\boldsymbol{\delta}}_{\boldsymbol{\Phi}}. \tag{5.4.6}$$

The argument of §2.4 shows that the cross-sections of $\mathscr{B}$ form a totally reflexive module over $\mathfrak{T}$ [*or* $\mathfrak{S}$]. We now use capital Greek letters $\Phi, \Psi, \ldots$, as abstract labels for these cross-sections, and denote the module and its isomorphic copies by $\mathfrak{T}^{\Phi}, \mathfrak{T}^{\Psi}, \ldots$ [*or* $\mathfrak{S}^{\Phi}, \mathfrak{S}^{\Psi}, \ldots$]. Then the relation (5.4.5) can be re-expressed as

$$\lambda^{\Phi} = \lambda^{\boldsymbol{\Phi}}\delta_{\boldsymbol{\Phi}}^{\Phi} \in \mathfrak{T}^{\Phi} \quad [or\ \mathfrak{S}^{\Phi}] \tag{5.4.7}$$

in cases where there is a global basis $\delta^{\Phi}_{\mathbf{\Phi}}$ for the cross-sections. Otherwise, by (5.4.6),

$$\lambda^{\Phi} = \sum_{\mathrm{i}} \overset{\mathrm{i}}{u} \overset{\mathrm{i}}{\lambda}{}^{\mathbf{\Phi}} \overset{\mathrm{i}}{\delta}{}^{\Phi}_{\mathbf{\Phi}}. \tag{5.4.8}$$

The module $\mathfrak{T}^{\Phi}$ [*or* $\mathfrak{S}^{\Phi}$] of cross-sections of $\mathscr{B}$ serves to characterize $\mathscr{B}$ as a bundle over $\mathscr{M}$ completely up to isomorphism: for, *by definition*, two bundles over $\mathscr{M}$ are considered isomorphic if and only if their modules of cross-sections are isomorphic, as modules over $\mathfrak{T}$ [*or* $\mathfrak{S}$].

We note that $\lambda^{\Phi}$ (with an abstract index) simply denotes an entire cross-section and does not need to be expressed in terms of patches. This illustrates again one of the advantages of the abstract-index notation: calculations with abstract indices automatically have global significance.

## *Explicit construction of bundles*

To construct a vector bundle $\mathscr{B}$ over $\mathscr{M}$, explicitly, consider the covering of $\mathscr{M}$ by open sets $\mathscr{U}_1, \mathscr{U}_2, \ldots$, over each of which the bundle is a simple product $\mathscr{U}_{\mathrm{i}} \times \mathbb{R}^k$ [*or* $\mathscr{U}_{\mathrm{i}} \times \mathbb{C}^k$], the various basis cross-sections (5.4.4) being

$$\mathscr{U}_{\mathrm{i}} \times (1, 0, \ldots, 0),\ \mathscr{U}_{\mathrm{i}} \times (0, 1, 0, \ldots, 0), \ldots, \mathscr{U}_{\mathrm{i}} \times (0, \ldots, 0, 1).$$

Since every point of $\mathscr{M}$ lies in at least one of the $\mathscr{U}_{\mathrm{i}}$, every point of $\mathscr{B}$ lies in at least one $\mathscr{U}_{\mathrm{i}} \times \mathbb{R}^k$ [*or* $\mathscr{U}_{\mathrm{i}} \times \mathbb{C}^k$]. But some points of $\mathscr{M}$ may lie in two or more of the $\mathscr{U}_{\mathrm{i}}$ and then explicit transformations are needed to specify the patching. Thus, if $P \in \mathscr{U}_{\mathrm{i}} \cap \mathscr{U}_{\mathrm{j}}$, the pairs $(P, \overset{\mathrm{i}}{y}) \in \mathscr{U}_{\mathrm{i}} \times \mathbb{R}^k$ [*or* $\mathscr{U}_{\mathrm{i}} \times \mathbb{C}^k$], $(P, \overset{\mathrm{j}}{y}) \in \mathscr{U}_{\mathrm{j}} \times \mathbb{R}^k$ [*or* $\mathscr{U}_{\mathrm{j}} \times \mathbb{C}^k$] will represent the same point of $\mathscr{B}$ if and only if

$$\overset{\mathrm{i}}{y} = \overset{\mathrm{ij}}{\mathbf{L}}(P)\overset{\mathrm{j}}{y} \quad \text{(no sum)} \tag{5.4.9}$$

where, for each i, j and for each $P$,

$$\overset{\mathrm{ij}}{\mathbf{L}}(P) \in GL(k, \mathbb{R}) \quad [\textit{or } GL(k, \mathbb{C})], \tag{5.4.10}$$

the matrix $\overset{\mathrm{ij}}{\mathbf{L}}(P)$ varying $C^{\infty}$-smoothly with $P$, and $GL$ denoting the group of non-singular $k \times k$ matrices. For consistency, this entails that

$$\overset{\mathrm{ji}}{\mathbf{L}}(P) = (\overset{\mathrm{ij}}{\mathbf{L}}(P))^{-1}, \tag{5.4.11}$$

and that throughout each $\mathscr{U}_{\mathrm{i}} \cap \mathscr{U}_{\mathrm{j}} \cap \mathscr{U}_{\mathrm{k}}$ we have

$$\overset{\mathrm{ij}}{\mathbf{L}}(P)\overset{\mathrm{jk}}{\mathbf{L}}(P) = \overset{\mathrm{ik}}{\mathbf{L}}(P) \quad \text{(no sum)}. \tag{5.4.12}$$

A family of matrices (5.4.10) satisfying (5.4.11) and (5.4.12) serves to define the bundle.

In practice, it may be convenient to combine the piecing together of the fibres with the piecing together of the manifold itself. Then the various $\mathscr{U}_{\mathrm{i}}$ are given as coordinate patches for $\mathscr{M}$, the coordinates for $x \in \mathscr{U}_{\mathrm{i}}$ being

$$\overset{\mathrm{i}}{x}{}^{\alpha} = \overset{\mathrm{i}}{x}{}^{1}, \ldots, \overset{\mathrm{i}}{x}{}^{n} \in \mathbb{R}^{n}, \tag{5.4.13}$$

and coordinate transformations are specified on each $\mathscr{U}_{\mathrm{i}} \cap \mathscr{U}_{\mathrm{j}}$. The $\mathbf{L}$s are $C^{\infty}$ functions of these coordinates, and (5.4.9), (5.4.11), (5.4.12) are replaced, respectively, by

$$\overset{\mathrm{i}}{y} = \overset{\mathrm{ij}}{\mathbf{L}}(\overset{\mathrm{j}}{x}{}^{\alpha})\overset{\mathrm{j}}{y}, \tag{5.4.14}$$

$$\overset{\mathrm{ji}}{\mathbf{L}}(\overset{\mathrm{i}}{x}{}^{\alpha}) = (\overset{\mathrm{ij}}{\mathbf{L}}(\overset{\mathrm{j}}{x}{}^{\alpha}))^{-1} \tag{5.4.15}$$

$$\overset{\mathrm{ij}}{\mathbf{L}}(\overset{\mathrm{j}}{x}{}^{\alpha})\overset{\mathrm{jk}}{\mathbf{L}}(\overset{\mathrm{k}}{x}{}^{\alpha}) = \overset{\mathrm{ik}}{\mathbf{L}}(\overset{\mathrm{k}}{x}{}^{\alpha}) \quad \text{(no sum)}. \tag{5.4.16}$$

The possible non-triviality of the vector bundle concept (as opposed to the simple concept of a global product space) arises because the fibres possess *symmetries* (i.e., non-trivial automorphisms). This is clear for the tangent spaces to points of $S^2$, for example, since these spaces can be rotated into themselves. But even when the fibres are one-dimensional (and real) such symmetries can arise. For example, take each fibre to be a one-dimensional real vector space $\mathscr{V}^{\bullet}$, with no additional structure. Then the only *canonical* element of $\mathscr{V}^{\bullet}$ is the zero element; all other elements of $\mathscr{V}^{\bullet}$ are on an equal footing. The automorphisms of $\mathscr{V}^{\bullet}$ are given by selecting any non-zero element $r \in \mathbb{R}$ and mapping $\mathscr{V}^{\bullet}$ into itself according to $\boldsymbol{y} \mapsto r\boldsymbol{y}$, where $\boldsymbol{y} \in \mathscr{V}^{\bullet}$. As a simple example of how this can lead to a non-trivial vector bundle, we shall consider the Möbius band (see Fig. 5-3). Here the base space $\mathscr{M}$ is the circle $S^1$ and the fibres $\mathscr{V}^{\bullet}$ are one-dimensional real vector spaces. We can take two coordinate patches $\mathscr{U}_1$, $\mathscr{U}_2$ for

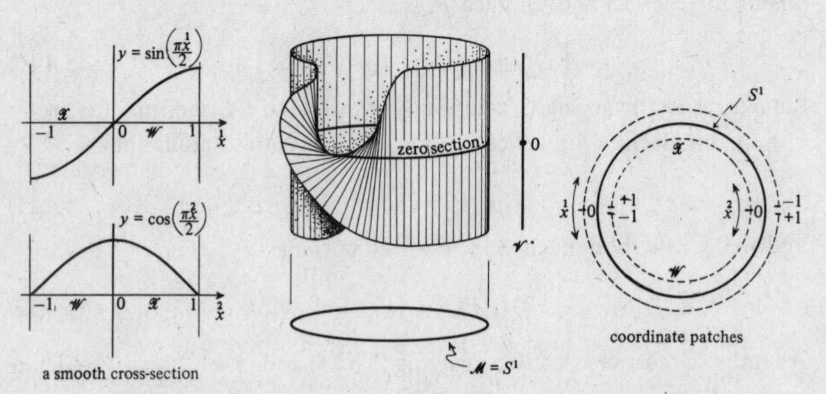


Fig. 5-3. The Möbius band as a one-dimensional vector bundle over $S^1$.

$S^1$ with coordinate $\overset{1}{x} \in (-1, 1)$ for $\mathscr{U}_1$ and coordinate $\overset{2}{x} \in (-1, 1)$ for $\mathscr{U}_2$; in the overlap $\mathscr{U}_1 \cap \mathscr{U}_2 (= \mathscr{X} \cup \mathscr{W})$ we have $\overset{1}{x} = \overset{2}{x} - 1$ if $-1 < \overset{1}{x} < 0$, $0 < \overset{2}{x} < 1$ (region $\mathscr{X}$) and $\overset{1}{x} = \overset{2}{x} + 1$ if $0 < \overset{1}{x} < 1$, $-1 < \overset{2}{x} < 0$ (region $\mathscr{W}$). As fibre coordinate we take $y \in \mathbb{R}$ and specify

$$\overset{12}{\mathbf{L}} = \begin{cases} -1 & \text{in } \mathscr{X} \\ 1 & \text{in } \mathscr{W} \end{cases}$$

Since $\mathscr{X}$ and $\mathscr{W}$ are disjoint, $\overset{12}{\mathbf{L}}$ is clearly $C^\infty$. Observe that the Möbius vector bundle is *topologically* distinct from the product space $S^1 \times \mathbb{R}$: the orientation of the fibre reverses its direction as we pass once around the circle $S^1$. It is necessary to invoke an automorphism of $\mathscr{V}^{\bullet}$ that involves multiplication by a negative number, and thus it is essential that the fibres *not* be canonical copies of $\mathbb{R}$, which does not permit such an automorphism. Another way to illustrate the difference of the Möbius vector bundle from the cylinder $S^1 \times \mathbb{R}$ as a bundle over $S^1$ is to consider that for topological reasons every cross-section of the Möbius bundle must vanish somewhere (i.e., must intersect the zero cross-section), whereas this is clearly not the case for the cylinder bundle.

We could envisage another way of deforming the bundle $S^1 \times \mathbb{R}$. To construct the Möbius bundle we invoked the symmetry $y \mapsto -y$ of the vector space $\mathscr{V}^{\bullet}$. Let us see how we could invoke a symmetry such as $y \mapsto 2y$. Suppose we construct a bundle over $S^1$ and coordinatize it exactly as in the Möbius case, but now we put

$$\overset{12}{\mathbf{L}} = \begin{cases} 1 & \text{in } \mathscr{X} \\ 2 & \text{in } \mathscr{W}. \end{cases}$$

Then, as we pass once around $S^1$, there is a resultant stretching of $\mathscr{V}^{\bullet}$ by a factor 2 (or shrinking, if we go the other way). However, if we apply the criteria for equivalence of bundles that we have adopted, we find that our 'stretch band' does not differ from $S^1 \times \mathbb{R}$. For we can find a family of non-zero $C^\infty$ cross-sections,

$$\overset{1}{y} = a2^{\frac{1}{2}\overset{1}{x}}, \quad \overset{2}{y} = a2^{\frac{1}{2}\overset{2}{x} - \frac{1}{2}}$$

(that gradually take up the factor 2 as $S^1$ is traversed), and this family can be mapped to the constant cross-sections $y = a$ of the cylinder bundle $S^1 \times \mathbb{R}$.

Sometimes, however, it is natural and significant to impose a further structure on a vector bundle $\mathscr{B}$, according to which the above stretch band would, in fact, *differ* from the cylinder $S^1 \times \mathbb{R}$. This is a structure that supplies a means of characterizing certain cross-sections as *locally constant* (or 'horizontal'). For the case of the cylinder $S^1 \times \mathbb{R}$, many locally constant cross-sections exist globally. But in the case of the stretch band, if the cross-section maintains its local constancy, we find an incompatibility by a factor of 2 when we go once around $S^1$; only the zero cross-section is locally constant everywhere.

## *Bundle connections*

To formalize the concept of local constancy of a cross-section, we must define a 'connection' on the bundle $\mathscr{B}$. Now, a gradient operator $\nabla_a$ acting on scalar functions on $\mathscr{M}$ always exists (by the definition of a manifold). In the same way that this operator can be extended to apply to tangent vectors (and tensors and spinors) yielding a 'manifold connection', so also can it be extended to apply to other types of bundle cross-sections (and *their* tensors), yielding a 'bundle connection'. If both these extensions have been defined, the operator can also act on objects with mixed tensor-, spinor-, and cross-section abstract indices. Now consider a curve $\gamma$ on $\mathscr{M}$ with tangent-vector field $\boldsymbol{X}$. If a connection $\nabla_a$ exists on $\mathscr{M}$, a tangent-vector field $\boldsymbol{Z}$ is constant (parallelly propagated) along $\gamma$ if it is annihilated by the operator $\underset{X}{\nabla} = X^a\nabla_a$. And, in the same way, a bundle cross-section $\boldsymbol{\lambda}$ is locally constant if it is annihilated by $\underset{X}{\nabla}$. Normally $\nabla_a$, operating on the cross-sections of $\mathscr{B}$, will be non-commutative (unless the base space $\mathscr{M}$ is one-dimensional) and thus it will have *curvature*. Then the non-integrability that is illustrated by the stretch band on a global level can also occur at the infinitesimal level when a small loop in the base space is traversed. In detail, given the gradient operator $\nabla_a$, a *bundle connection* extends its domain to cross-sections by the requirements

$$\nabla_a : \mathfrak{T}^{\Phi} \to \mathfrak{T}^{\Phi}_a \quad [or\ \mathfrak{S}^{\Phi} \to \mathfrak{S}^{\Phi}_a], \tag{5.4.17}$$

$$\nabla_a(\lambda^{\Phi} + \mu^{\Phi}) = \nabla_a\lambda^{\Phi} + \nabla_a\mu^{\Phi}, \tag{5.4.18}$$

$$\nabla_a(f\lambda^{\Phi}) = \lambda^{\Phi}\nabla_a f + f\nabla_a\lambda^{\Phi}, \quad f \in \mathfrak{T} \quad [or\ \mathfrak{S}]. \tag{5.4.19}$$

And this can be extended, in the usual way, to cross-section tensors, i.e., objects with several capital Greek indices. As before, $\underset{X}{\nabla}$ is defined as $X^a\nabla_a$. Then we can define $\sqcap_{ab}$ by

$$\underset{X}{\nabla}\underset{Y}{\nabla} - \underset{Y}{\nabla}\underset{X}{\nabla} - \underset{[X,Y]}{\nabla} = X^a Y^b \sqcap_{ab} \tag{5.4.20}$$

(*cf.* (4.3.32)), since the LHS is bilinear in $X^a$ and $Y^a$. This gives the *bundle curvature* $K_{ab\Omega}{}^{\Phi} \in \mathfrak{T}^{\Phi}_{[ab]\Omega}$[*or* $\mathfrak{S}^{\Phi}_{[ab]\Omega}$] via

$$\sqcap_{ab}\lambda^{\Phi} =: K_{ab\Omega}{}^{\Phi}\lambda^{\Omega} \tag{5.4.21}$$

(*cf.* (4.2.30)). If the torsion vanishes (as in the cases of most interest to us), we also have

$$\text{Д}_{ab} = \Delta_{ab} := \nabla_a \nabla_b - \nabla_b \nabla_a, \tag{5.4.22}$$

so that

$$K_{ab\Omega}{}^{\Phi}\lambda^{\Omega} = (\nabla_a \nabla_b - \nabla_b \nabla_a)\lambda^{\Phi}. \tag{5.4.23}$$

Further properties of the bundle curvature, including its spinor description will be given at the end of §5.5. Note that $K_{ab\Omega}{}^{\Phi}$ reduces to Riemann's $R_{abc}{}^{d}$ if $\mathscr{B}$ is the tangent bundle and we use the Christoffel connection on it.

There is much more that can be said concerning the use of the abstract index formalism in the context of vector bundles. We end this section by making just a few relevant remarks. We note first that, as in the discussion in §4.2, the change from one bundle connection $\nabla_a$ to another one $\tilde{\nabla}_a$ is described by an element $Q_{a\Phi}{}^{\Omega} \in \mathfrak{T}^{\Omega}_{a\Phi}$ [*or* $\mathfrak{S}^{\Omega}_{a\Phi}$] where

$$(\tilde{\nabla}_a - \nabla_a)\lambda^{\Omega} = Q_{a\Phi}{}^{\Omega}\lambda^{\Phi} \tag{5.4.24}$$

and a formula similar to (4.2.51) for the change in bundle curvature holds. The dependence of general expressions on the choice of bundle connection may be investigated by use of (5.4.24) and its generalization analogous to (4.2.48).

The bundle connection is also of relevance when we consider fields *on $\mathscr{B}$ itself* (which it is sometimes important to do) even in the case of scalar fields on $\mathscr{B}$. For suppose $F$ is such a scalar field. Then $F$ is a function not only of $P \in \mathscr{M}$ but of the 'fibre coordinate' $y^{\Phi} \in \mathfrak{T}^{\Phi}[P]$ [*or* $\mathfrak{S}^{\Phi}[P]$]. The exterior derivative (gradient) d$F$ of $F$ then involves two parts, namely

$$\frac{\partial F}{\partial y^{\Theta}} \text{ and } \nabla_a F. \tag{5.4.25}$$

The first involves holding $P$ fixed and varying $y^{\Phi}$, this being an unambiguous derivative within each vector-space fibre, while the second involves 'holding $y^{\Phi}$ fixed'. The latter concept has an invariant meaning only when a bundle connection is defined. If not, then there is no invariant splitting of d$F$ into two parts like (5.4.25) (although the first by itself is always invariant). Similar, but more complicated, remarks apply to higher derivatives. Note that if $F$ is analytic about the zero section of $\mathscr{B}$, then we can express it as

$$F = f + f_{\Phi}y^{\Phi} + \frac{1}{2!}f_{\Phi\Omega}y^{\Phi}y^{\Omega} + \ldots,$$

where

$$f \in \mathfrak{T}, f_{\Phi} \in \mathfrak{T}_{\Phi}, f_{\Phi\Omega} \in \mathfrak{T}_{(\Phi\Omega)}, \ldots$$

[*or* $\mathfrak{S}, \mathfrak{S}_\Phi, \ldots$], so $F$ can be represented in terms of the infinite collection of bundle tensors $(f, f_\Phi, f_{\Phi\Omega}, \ldots)$. Then we find that the two terms (5.4.25) are represented by $(f_\Theta, f_{\Theta\Phi}, \ldots)$ and $(\nabla_a f, \nabla_a f_\Phi, \ldots)$ respectively.

## 5.5 Yang–Mills Fields

Having discussed the basic properties of vector bundles, we are now in a position to make a certain generalization of the theory of the electromagnetic field as developed in §5.1. The electromagnetic field is the simplest type of a *gauge field*, namely the one corresponding to the group $U(1)$, since the gauge transformations (5.1.21) are achieved by multiplication by complex scalar fields of unit modulus, i.e., by fields of elements of the Lie group $U(1)$. It is possible to construct analogous theories for other Lie groups: the resulting analogues of the Maxwell field are referred to as *Yang–Mills fields* (Yang and Mills 1954). They are thought to have relevance to elementary particle interactions.

We shall show in this section how the abstract-index formalism adapts naturally to the treatment of Yang–Mills fields. While our expressions will sometimes have a more cumbersome appearance than is usual in conventional approaches, our purposes is not to replace these, but merely to show how Yang–Mills fields fall into the general abstract-index scheme. This has a conceptual value, and also a computational one in certain contexts (*cf.* also Ashtekar, Horowitz and Magnon-Ashtekar 1983).

Mathematically, the theory of Yang–Mills fields is intimately bound up with the concept of a vector bundle and of a connection in such a bundle. The charged scalar fields (elements of $\overset{e}{\mathfrak{S}}$) of electromagnetic theory can be regarded as cross-sections of a vector bundle whose fibres are complex one-dimensional vector spaces (a complex line bundle), and then the connection $\nabla_a$ of (5.1.9) is the corresponding bundle connection. The generalization to Yang–Mills fields consists in allowing the fibres to become general abstract vector spaces $\mathscr{V}^\bullet$ (again with no special relation to the tangent spaces of $\mathscr{M}$ or their associated spin-spaces); the cross-sections of the resulting bundle $\mathscr{B}$ are the Yang–Mills-charged space–time-scalar (i.e. space–time-index-free) fields; and the elements of a specified continuous symmetry group of $\mathscr{V}^\bullet$ provide the gauge transformations.

We use capital Greek abstract-index labels for the elements of the vector spaces $\mathscr{V}^\bullet \cong \mathscr{V}^\Phi \cong \mathscr{V}^\Psi \cong \ldots$ and also for the modules $\mathfrak{S}^\Phi \cong \mathfrak{S}^\Psi \cong \ldots$ of cross-sections of $\mathscr{B}$. If $\mathscr{V}^\bullet$ is $n$-dimensional, then an element $\lambda^\Phi$ of $\mathfrak{S}^\Phi$ can be described locally in terms of $n$ scalar component fields

$\lambda^1, \ldots, \lambda^n \in \mathfrak{S}$, but not in any canonical way. Thus, to assign components $\lambda^{\mathbf{\Phi}}$ (locally) to a YM (Yang–Mills)-charged field we require some arbitrarily but smoothly chosen YM basis for $\mathscr{V}^{\bullet}$ at each point of $\mathscr{M}$, giving, though perhaps only locally, a basis of YM-charged fields, $\delta_{\mathbf{\Phi}}^{\Phi} = (\delta_1^{\Phi}, \ldots, \delta_n^{\Phi})$, so that $\lambda^{\Phi} = \lambda^{\mathbf{\Phi}} \delta_{\mathbf{\Phi}}^{\Phi}$. The modules $\mathfrak{S}_{\Phi}, \mathfrak{S}_{\Psi}, \ldots, \mathfrak{S}_{\Theta \ldots \Lambda}^{\Phi \ldots \Omega}, \ldots$ are defined from $\mathfrak{S}^{\Phi}$ in the standard way, and spinor (and tensor) indices can also be included to form modules $\mathfrak{S}^{\mathscr{A}} = \mathfrak{S}_{\Theta \ldots \Lambda G \ldots N'}^{\Phi \ldots \Omega A \ldots F'}$, where now the script letters $\mathscr{A}, \mathscr{B}, \ldots$ may include all these types of indices.

## Structure of $\mathscr{V}^{\bullet}$; Yang–Mills connection

The vector space $\mathscr{V}^{\bullet}$ may be either real, in which case the component fields $\lambda^{\mathbf{\Phi}}$ of $\lambda^{\Phi}$ would normally be chosen real, or else complex. In the real case, the appropriate notation $\mathfrak{T}^{\Phi}$ rather than $\mathfrak{S}^{\Phi}$ should be used for the module of YM-charged fields $\lambda^{\Phi}$, because it is a $\mathfrak{T}$-module (having real scalar fields as coefficients) and not an $\mathfrak{S}$-module (which has complex scalar fields as coefficients). However, even in the real case the corresponding complexification $\mathfrak{S}^{\Phi} = \mathfrak{T}^{\Phi} \oplus \mathrm{i}\mathfrak{T}^{\Phi}$ can be defined, which is an $\mathfrak{S}$-module, and the components $\lambda^{\mathbf{\Phi}}$ of $\lambda^{\Phi} \in \mathfrak{S}^{\Phi}$ are elements of $\mathfrak{S}$. For convenience, one might also sometimes introduce a complex basis $\delta_{\mathbf{\Phi}}^{\Phi} \in \mathfrak{S}^{\Phi}$ even if $\mathscr{V}^{\bullet}$ is real, in which case the components $\lambda^{\mathbf{\Phi}}$ would be complex even for $\lambda^{\Phi} \in \mathfrak{T}^{\Phi}$. This is similar to the situation that arises when a null tetrad $l^a, m^a, \bar{m}^a, n^a \in \mathfrak{S}^a$ is used for describing elements of $\mathfrak{T}^a$, the space of real tangent vectors.

The tensor algebra $(\ldots, \mathfrak{T}^{\mathscr{A}}, \ldots)$ or $(\ldots, \mathfrak{S}^{\mathscr{A}}, \ldots)$ satisfies the rules of §5.4 and Chapter 2. Thus sums, products, contractions, and index permutations may be formed in the usual manner, and, in the case of $\mathfrak{S}^{\mathscr{A}}$, an operation of complex conjugation may be introduced which sends $\mathfrak{S}^{\Phi}$ into an anti-isomorphic system $\mathfrak{S}^{\Phi'}$ with a new index label $\Phi'$. For the case of a real space $\mathscr{V}^{\bullet}$ complex conjugation applies to the complexification $(\ldots, \mathfrak{S}^{\mathscr{A}}, \ldots)$ of the real tensor system, but here we have $\Phi' = \Phi$, the real tensors being those invariant under complex conjugation. (This is analogous to the Latin indices of space–time tensors being unchanged under complex conjugation, whereas spinor indices get primed – or unprimed.)

The space $\mathscr{V}^{\bullet}$ in general has some additional structure imposed on it, characterized by a certain Lie group $\mathscr{G}$ which acts as the linear transformation group on $\mathscr{V}^{\bullet}$ preserving that structure. Thus, for example, $\mathscr{V}^{\bullet}$ might be a three-dimensional real vector space and $\mathscr{G}$ the orthogonal group $O(3)$ acting on $\mathscr{V}^{\bullet}$ in the standard way. In this case there will be an element $g_{\Phi\Psi} \in \mathfrak{T}_{\Phi\Psi}$ which is positive definite ($g_{\Phi\Psi} V^{\Phi} V^{\Psi} > 0$ if $V^{\Phi} \neq 0$) and symmetric ($g_{[\Phi\Psi]}=0$) and which is invariant under $\mathcal{G}$. Conversely, $\mathcal{G}$ is characterized by its leaving $g_{\Phi\Psi}$ invariant and being the largest linear group on $\mathcal{V}^{\bullet}$ with this property. Similarly $\mathcal{G}=SO(3)$ would be characterized by the invariance of the pair of elements $g_{\Phi\Psi}\in\mathfrak{T}_{\Phi\Psi}, e_{\Phi\Psi\Omega}\in\mathfrak{T}_{\Phi\Psi\Omega}$, where $g_{\Phi\Psi}$ is as before and $0\neq e_{\Phi\Psi\Omega}=e_{[\Phi\Psi\Omega]}$. As another example, the scale transformations can be incorporated into $\mathcal{G}$ together with all the elements of $O(3)$ if we specify that it is merely the product $g_{\Phi\Psi}g^{\Lambda\Omega}\in\mathfrak{T}^{\Lambda\Omega}_{\Phi\Psi}$ that is invariant, with $g^{\Lambda\Omega}$ the inverse of $g_{\Phi\Psi}$ (i.e., $g_{\Phi\Psi}g^{\Psi\Omega}=\delta^{\Omega}_{\Phi}$). One may also impose Hermitian-type structures on $\mathcal{V}^{\bullet}$ in the case when $\mathcal{V}^{\bullet}$ is an complex vector space. Thus, for example, the group $\mathcal{G}=U(n)$ arises if a positive definite Hermitian bilinear form $h_{\Phi\Phi'}V^{\Phi}U^{\Phi'}$($>0$ if $U^{\Phi'}=\overline{V^{\Phi}}\neq 0$, $h_{\Phi\Phi'}=\overline{h_{\Phi'\Phi}}$) on $\mathcal{V}^{\Phi}\times\mathcal{V}^{\Phi'}$ (where $\mathcal{V}^{\Phi'}$ is the complex conjugate of $\mathcal{V}^{\Phi}$) is specified as invariant or, equivalently, an invariant isomorphism $U^{\Phi'}\mapsto h_{\Phi\Phi'}U^{\Phi'}$ is specified between $\mathcal{V}^{\Phi'}$ and the dual $\mathcal{V}_{\Phi}$ of $\mathcal{V}^{\Phi}$. If the latter view is adopted, it may be convenient to identify $\mathcal{V}^{\Phi'}$ with $\mathcal{V}_{\Phi}$ and hence dispense with the primed indices altogether (as we shall do in Vol. 2, in a somewhat different context, with twistors, *cf.* Chapter 6, especially §6.9).

In the particular case of electromagnetism, $\mathcal{G}=U(1)$ and $\mathcal{V}^{\bullet}$ is one-dimensional complex-Hermitian. Here the abstract index notation is not worthwhile to adopt. Every $\mathfrak{S}^{\Phi\ldots\Omega}_{\Theta\ldots\Lambda}$ is *one-dimensional*, its elements being all *symmetric* so that index permutation yields nothing new. Contraction loses no information, so $\mathfrak{S}^{\Phi\ldots\Omega\Delta}_{\Theta\ldots\Lambda\Upsilon}$ is canonically equivalent to $\mathfrak{S}^{\Phi\ldots\Omega}_{\Theta\ldots\Lambda}$, etc. (since any element of the former can be contracted over $\Delta$ and $\Upsilon$ without loss of information), and so any such module is canonically equivalent to one of the systems $\mathfrak{S}$ (charge zero) $\mathfrak{S}^{\Psi_1\ldots\Psi_n}$ (charge $n\varepsilon$), or $\mathfrak{S}_{\Psi_1\ldots\Psi_n}$ (charge $-n\varepsilon$). Finally, because of the Hermitian structure on $\mathcal{V}^{\bullet}$, $\mathfrak{S}^{\Psi'}$ may be identified with $\mathfrak{S}_{\Psi}$, so that complex conjugation merely reverses the sign of the charge and yields nothing new. The various $\mathfrak{S}^{\Phi\ldots\Upsilon'}_{\Theta\ldots\Delta'}$ are thus all canonically equivalent to one another for each given charge value $n\varepsilon$ (with $n=$ number of upper unprimed minus lower unprimed minus upper primed plus lower primed indices), and inequivalent for different charge values. The general $\mathfrak{S}^{\mathscr{A}}$, possessing spinor indices as well, is now obtained by taking products with the above 'charged scalars' yielding a charged tensor algebra of the type considered in §5.1.

Thus far, for a general $\mathcal{V}^{\bullet}$ and $\mathcal{G}$, we have merely set up the appropriate abstract tensor algebra for YM-charged fields. The Yang–Mills field itself can be expressed in terms of (or 'as') a bundle connection on $\mathscr{B}$, defined as in (5.4.17)–(5.4.19). That connection can then be extended to the general YM-charged module $\mathfrak{S}^{\mathscr{A}}$ following the same procedure as that given in

§§4.2, 4.4, 5.1, with, in the case of complex $\mathscr{V}^{\bullet}$,

$$\overline{\nabla_a \lambda^{\Psi}} = \nabla_a \bar{\lambda}^{\Psi'}. \qquad (5.5.1)$$

In addition to satisfying (5.4.17)–(5.4.19), however, we require $\nabla_a$ to preserve – under the parallel transport of bundle vectors $\lambda^{\Phi}$ that it defines – the structure of the fibres $\mathscr{V}^{\bullet}$ that is characterized by the group $\mathscr{G}$. One way of doing this, when $\mathscr{G}$ is specified as the largest group that leaves invariant a set of 'canonical' elements of the modules $\mathfrak{S}^{\Phi\ldots\Psi'}_{\Theta\ldots\Omega'}$ (e.g., the $g_{\Phi\Psi}$ and $e_{\Phi\Psi\Omega}$, or $h_{\Psi\Psi'}$, considered earlier), is simply to demand that these elements be *annihilated* by $\nabla_a$. Alternatively, we may state this additional condition on $\nabla_a$ directly in terms of the group $\mathscr{G}$, and this may seem more natural than referring to (elements of) higher valence modules $\mathfrak{S}^{\Phi\ldots\Psi'}_{\Theta\ldots\Omega'}$. For this purpose we use the concept of a *gauge* and of a *gauge transformation* on $\mathfrak{T}^{\Psi}$ [*or* $\mathfrak{S}^{\Psi}$].

Suppose $\mathscr{G}$ is given as an explicit group of matrices $\mathbf{q}$ acting on $\mathbb{R}^n$ [*or* $\mathbb{C}^n$], where $n$ is the real [complex] dimension of $\mathscr{V}^{\bullet}$. The structure of $\mathscr{V}^{\bullet}$ can be expressed as a family of linear maps from $\mathscr{V}^{\bullet}$ to $\mathbb{R}^n$ [*or* $\mathbb{C}^n$], each pair of which is related precisely by an element $\mathbf{q}$ of $\mathscr{G}$. Each such linear map may be thought of as an allowable coordinate system for $\mathscr{V}^{\bullet}$, and is defined by a particular choice of standard basis for $\mathscr{V}^{\bullet}$. Using the abstract index notation, we denote such a standard basis by

$$\alpha_{\mathbf{\Psi}}{}^{\Psi} = (\alpha_1{}^{\Psi}, \ldots, \alpha_n{}^{\Psi}) \in \mathscr{V}^{\Psi}, \qquad (5.5.2)$$

and the transformation from this basis to another, $\alpha_{\hat{\mathbf{\Psi}}}{}^{\Psi}$, is given by

$$\alpha_{\hat{\mathbf{\Psi}}}{}^{\Psi} = q_{\hat{\mathbf{\Psi}}}{}^{\mathbf{\Psi}} \alpha_{\mathbf{\Psi}}{}^{\Psi}, \quad \text{matrix } (q_{\hat{\mathbf{\Psi}}}{}^{\mathbf{\Psi}}) \in \mathscr{G}. \qquad (5.5.3)$$

(We use $\alpha_{\mathbf{\Psi}}{}^{\Psi}$ rather than the $\delta_{\mathbf{\Psi}}{}^{\Psi}$ used earlier, to emphasize that a *standard* basis is chosen now, and to bring out the fact that this procedure generalizes the introduction of the charged scalar $\alpha$ for the electromagnetic field.) The collection of standard bases (5.5.2), related to one another by (5.5.3), provides another way of characterizing the structure imposed on $\mathscr{V}^{\bullet}$ by $\mathscr{G}$.

We now consider *fields* of these standard bases, i.e., sets of $n$ linearly independent cross-sections of $\mathscr{B}$. The statements (5.5.2) and (5.5.3) still hold at each point, but now

$$\alpha_{\mathbf{\Psi}}{}^{\Psi} \in \mathfrak{T}^{\Psi} \text{ [or } \mathfrak{S}^{\Psi}], \quad q_{\hat{\mathbf{\Psi}}}{}^{\mathbf{\Psi}} \in \mathfrak{T} \text{ [or } \mathfrak{S}] \qquad (5.5.4)$$

for each $\mathbf{\Psi}, \hat{\mathbf{\Psi}} = 1, 2, \ldots, n$. Such a set of fields $\alpha_{\mathbf{\Psi}}{}^{\Psi}$ is called a *gauge* for $\mathfrak{T}^{\Psi}$ [*or* $\mathfrak{S}^{\Psi}$] and the matrix of fields $q_{\hat{\mathbf{\Psi}}}{}^{\mathbf{\Psi}}$ provides a *gauge transformation*. A gauge always exists locally but for topological reasons may fail to exist globally. A global gauge provides what in mathematical language is called a *trivialization* of the bundle $\mathscr{B}$.

Now $\nabla_a$ will preserve the structure of $\mathscr{V}^{\bullet}$ if under parallel transport an allowable basis is carried into an allowable basis. So let us consider some aspects of parallel transport that we shall need here and later. Let $\gamma$ be a smooth curve in $\mathscr{M}$ and have tangent vector $t^a$ corresponding to a smooth parameter $u$ on $\gamma$ (which is to say that $t^a$ is scaled so that $t^a\nabla_a u = 1$). A (tensor, spinor YM-charged) field $\lambda^{\mathscr{A}}$ is said to be parallelly transported along $\gamma$ if it is annihilated by the operator $t^a\nabla_a$. We denote by $\exp(vt^a\nabla_a)$ the operation which, when applied to a field $\lambda^{\mathscr{A}}$ defined along $\gamma$, yields a new field

$$\tilde{\lambda}^{\mathscr{A}} = \exp(vt^a\nabla_a)\lambda^{\mathscr{A}} \tag{5.5.5}$$

also defined along $\gamma$, such that $\tilde{\lambda}^{\mathscr{A}}$ at the point $P$ (say with parameter $u_0$) is obtained from $\lambda^{\mathscr{A}}$ at the point $Q$ (with parameter $u_0 + v$) by parallelly transporting it back along $\gamma$ from $Q$ to $P$. This operation is well defined at all points of $\gamma$ for which points still exist on the curve when the parameter is increased by $v$. It also applies to fields on $\mathscr{M}$ if $\gamma$ belongs to a smooth congruence of curves with smoothly varying parametrization. As we shall see later, when $\mathscr{M}, \gamma, t^a$, and $\lambda^{\mathscr{A}}$ are analytic, and $|v|$ is sufficiently small, (5.5.5) can be written as the notation suggests (*cf.* (5.11.6)):

$$\tilde{\lambda}^{\mathscr{A}} = \lambda^{\mathscr{A}} + vt^a\nabla_a\lambda^{\mathscr{A}} + \frac{v^2}{2!}t^a\nabla_a(t^b\nabla_b\lambda^{\mathscr{A}}) + \cdots, \tag{5.5.6}$$

where '$=$' is to be interpreted in terms of parallel transport along $\gamma$. But in the present context only the first two terms are needed, in effect, and analyticity need not be assumed.

In light of the above discussion, $\nabla_a$ preserves the structure of $\mathscr{V}^{\bullet}$ if, by reference to (5.5.3) and (5.5.5),

$$\exp(vt^a\nabla_a)\alpha_{\Psi}{}^{\Psi} = q_{\Psi}{}^{\Theta}(v)\alpha_{\Theta}{}^{\Psi} \tag{5.5.7}$$

for some matrix $(q_{\Psi}{}^{\Theta})$ in $\mathscr{G}$ which tends smoothly to the identity matrix as $v \to 0$. Dividing (5.5.6) by $v$ and going to the limit $v \to 0$ (and therefore using only the first-order terms in $v$), we obtain

$$t^a\nabla_a\alpha_{\Psi}{}^{\Psi} = p_{\Psi}{}^{\Theta}\alpha_{\Theta}{}^{\Psi}, \tag{5.5.8}$$

where

$$p_{\Psi}{}^{\Theta} = \left[\frac{\mathrm{d}}{\mathrm{d}v}q_{\Psi}{}^{\Theta}(v)\right]_{v=0}. \tag{5.5.9}$$

The matrix $(p_{\Psi}{}^{\Theta})$ belongs not to the group $\mathscr{G}$ but to its *Lie algebra* $\mathscr{A}$, from which the elements of $\mathscr{G}$ (close enough to the identity) can be reconstructed by exponentiation Defining $\alpha_{\Psi}{}^{\Psi} \in \mathfrak{S}_{\Psi}$ $(\Psi = 1, \ldots, n)$ as the dual of $\alpha_{\boldsymbol{\Psi}}{}^{\Psi}$,

$$\alpha_{\Psi}{}^{\boldsymbol{\Psi}}\alpha_{\boldsymbol{\Psi}}{}^{\Phi} = \delta_{\Psi}^{\Phi}, \quad \alpha_{\boldsymbol{\Psi}}{}^{\Psi}\alpha_{\Psi}{}^{\boldsymbol{\Phi}} = \delta_{\boldsymbol{\Psi}}^{\boldsymbol{\Phi}}, \tag{5.5.10}$$

we obtain the required condition on $\nabla_a$ in the form

$$\text{matrix}(t^a \alpha_{\Psi}{}^{\boldsymbol{\Theta}} \nabla_a \alpha_{\boldsymbol{\Psi}}{}^{\Psi}) \in \mathscr{A} \tag{5.5.11}$$

for each $t^a$ and $\alpha_{\boldsymbol{\Psi}}{}^{\Psi}$.

## *Yang–Mills potential and metric*

The *Yang–Mills potentials* can now be introduced (in close analogy with (5.1.13)) as

$$\Phi_{a\boldsymbol{\Psi}}{}^{\boldsymbol{\Theta}} = \mathrm{i}\alpha_{\Psi}{}^{\boldsymbol{\Theta}} \nabla_a \alpha_{\boldsymbol{\Psi}}{}^{\Psi}. \tag{5.5.12}$$

The factor i is incorporated here as a convenience when dealing (as one frequently does) with a group $\mathscr{G}$ of *unitary* (or pseudo-unitary) matrices. For then $\Phi_{a\boldsymbol{\Psi}}{}^{\boldsymbol{\Theta}}$ turns out to be *Hermitian* in the sense

$$\overline{\Phi_{a\boldsymbol{\Psi}}{}^{\boldsymbol{\Theta}}} = \Phi_{a\boldsymbol{\Theta}}{}^{\boldsymbol{\Psi}}. \tag{5.5.13}$$

Here we are adopting the convention that when complex conjugation is applied to a lower numerical index $\boldsymbol{\Psi}$, it is moved to the upper position, and vice versa, e.g.,

$$\overline{\alpha_{\boldsymbol{\Psi}}{}^{\Psi}} = \bar{\alpha}^{\boldsymbol{\Psi}\Psi'}, \qquad \overline{\alpha_{\Psi}{}^{\boldsymbol{\Psi}}} = \bar{\alpha}_{\Psi'\boldsymbol{\Psi}}, \qquad \overline{q_{\hat{\boldsymbol{\Theta}}}{}^{\boldsymbol{\Psi}}} = \bar{q}^{\hat{\boldsymbol{\Theta}}}{}_{\boldsymbol{\Psi}}, \tag{5.5.14}$$

so that the unitary condition on $q_{\hat{\boldsymbol{\Psi}}}{}^{\boldsymbol{\Psi}}$ becomes

$$\bar{q}^{\hat{\boldsymbol{\Theta}}}{}_{\boldsymbol{\Lambda}} q_{\hat{\boldsymbol{\Theta}}}{}^{\boldsymbol{\Psi}} = \delta_{\boldsymbol{\Lambda}}^{\boldsymbol{\Psi}}. \tag{5.5.15}$$

Multiplying (5.5.3) by its complex conjugate and contracting over $\hat{\boldsymbol{\Psi}}$, we obtain the result that the quantity

$$h^{\Psi\Psi'} := \alpha_{\boldsymbol{\Psi}}{}^{\Psi}\bar{\alpha}^{\boldsymbol{\Psi}\Psi'} = \overline{h^{\Psi'\Psi}} \tag{5.5.16}$$

is independent of the choice of standard basis. Essentially the same argument applies to (5.5.7) shows that $h^{\Psi\Psi'}$ goes to itself under finite parallel transport. Hence

$$\nabla_a h^{\Psi\Psi'} = 0, \tag{5.5.17}$$

which, when applied to the defining relation (5.5.16), yields

$$\alpha_{\boldsymbol{\Lambda}}{}^{\Psi}\nabla_a \bar{\alpha}^{\boldsymbol{\Lambda}\Psi'} + \bar{\alpha}^{\boldsymbol{\Lambda}\Psi'}\nabla_a \alpha_{\boldsymbol{\Lambda}}{}^{\Psi} = 0. \tag{5.5.18}$$

Transvecting with $\alpha_{\Psi'\boldsymbol{\Psi}}\alpha_{\Psi}{}^{\boldsymbol{\Theta}}$ and using (5.5.12), we obtain the relation (5.5.13) as required.

We can, in fact, use $h^{\Psi\Psi'}$ the *Yang–Mills Hermitian metric* – and its inverse $h_{\Psi\Psi'} = \alpha_{\Psi}{}^{\boldsymbol{\Psi}}\bar{\alpha}_{\Psi'\boldsymbol{\Psi}}$ to eliminate all occurrences of primed YM indices, in the case of a group $\mathscr{G}$ of unitary matrices. For example, if we substitute

$$\lambda_{\Theta'\Psi}{}^{\Lambda'} \mapsto \lambda^{\Theta}{}_{\Psi\Lambda} = \lambda_{\Theta'\Psi}{}^{\Lambda'}h^{\Theta\Theta'}h_{\Lambda\Lambda'}, \tag{5.5.19}$$

we obtain an essentially equivalent YM-charged field. In adopting this convention, we must remember that unitary YM index positions may be reversed rather than primed under complex conjugation (compare also the notation of twistor theory in Vol. 2).

The gauge quantities $\alpha_{\boldsymbol{\Psi}}{}^{\Psi}$ and $\alpha_{\Psi}{}^{\boldsymbol{\Psi}}$ (together with their complex conjugates, if needed) provide a means of assigning components to any YM-charged field; e.g., the components of $\lambda_{\Theta}{}^{\Psi}$ are

$$\lambda_{\boldsymbol{\Theta}}{}^{\boldsymbol{\Psi}} = \lambda_{\Theta}{}^{\Psi}\alpha_{\boldsymbol{\Theta}}{}^{\Theta}\alpha_{\Psi}{}^{\boldsymbol{\Psi}}. \tag{5.5.20}$$

The same procedure can also be applied to any YM-charged *spinor* field, giving its set of component spinor fields, in exact analogy to the procedure (5.1.11) in the electromagnetic case. When a spinor (or tensor) basis is defined as well (quite independently of the $\alpha$s), the components of these spinor fields can then also be taken, so that finally everything can be expressed in terms of scalars.

If component fields are taken with respect to two different gauges, then these will be related by a YM gauge transformation; for example,

$$\lambda_{\boldsymbol{\Theta}}{}^{\boldsymbol{\Psi}} \mapsto \lambda_{\hat{\boldsymbol{\Theta}}}{}^{\hat{\boldsymbol{\Psi}}} = \lambda_{\boldsymbol{\Theta}}{}^{\boldsymbol{\Psi}}q_{\hat{\boldsymbol{\Theta}}}{}^{\boldsymbol{\Theta}}r^{\hat{\boldsymbol{\Psi}}}{}_{\boldsymbol{\Psi}}, \tag{5.5.21}$$

as follows from (5.5.3) and the two corresponding versions of (5.5.20), where the matrix of $r$s ($\in\mathscr{G}$) is the inverse of the matrix of $q$s.

Given the gauge $\alpha_{\boldsymbol{\Psi}}{}^{\Psi}$, we can define a differential operator $\partial_a$ which commutes with itself in flat space–time (or in curved space–time when acting on YM-charged scalars) by analogy with (5.1.14). For example,

$$\partial_a\lambda_{\Theta}{}^{\Psi\mathscr{A}} = \alpha_{\Theta}{}^{\boldsymbol{\Theta}}\alpha_{\boldsymbol{\Psi}}{}^{\Psi}\nabla_a\lambda_{\boldsymbol{\Theta}}{}^{\boldsymbol{\Psi}\mathscr{A}} \tag{5.5.22}$$

(in the unitary case), where $\mathscr{A}$ contains no YM indices. An expression analogous to (5.1.15) can also be written down, but this involves the perhaps unnatural combination of a (gauge-dependent) potential with (gauge-independent) abstract YM indices. So we prefer to write the corresponding fully gauge-dependent expression, possessing numerical YM indices only. We have, for example (with $\mathscr{A}$ free of YM indices),

$$\nabla_a\lambda_{\Theta}{}^{\Psi\mathscr{A}} = \nabla_a(\lambda_{\boldsymbol{\Theta}}{}^{\boldsymbol{\Psi}\mathscr{A}}\alpha_{\Theta}{}^{\boldsymbol{\Theta}}\alpha_{\boldsymbol{\Psi}}{}^{\Psi}), \tag{5.5.23}$$

and expanding the right side, and transvecting the whole equation with $\alpha_{\boldsymbol{\Theta}}{}^{\Theta}\alpha_{\Psi}{}^{\boldsymbol{\Psi}}$ in order to take component fields. we find from (5.5.12)

$$\alpha_{\boldsymbol{\Theta}}{}^{\Theta}\alpha_{\Psi}{}^{\boldsymbol{\Psi}}(\nabla_a\lambda_{\Theta}{}^{\Psi\mathscr{A}}) = \nabla_a\lambda_{\boldsymbol{\Theta}}{}^{\boldsymbol{\Psi}\mathscr{A}} + \mathrm{i}\Phi_{a\boldsymbol{\Theta}}{}^{\boldsymbol{\Delta}}\lambda_{\boldsymbol{\Delta}}{}^{\boldsymbol{\Psi}\mathscr{A}} - \mathrm{i}\Phi_{a\boldsymbol{\Delta}}{}^{\boldsymbol{\Psi}}\lambda_{\boldsymbol{\Theta}}{}^{\boldsymbol{\Delta}\mathscr{A}}. \tag{5.5.24}$$

Note that under the gauge transformation (5.5.21), the potential (5.5.12) undergoes the transformation

$$\Phi_{a\Psi}{}^{\Theta} \mapsto \hat{\Phi}_{a\hat{\Psi}}{}^{\hat{\Theta}} = \Phi_{a\Psi}{}^{\Theta} r_{\Theta}{}^{\hat{\Theta}} q_{\hat{\Psi}}{}^{\Psi} + i r_{\Psi}{}^{\hat{\Theta}} \nabla_a q_{\hat{\Psi}}{}^{\Psi}, \tag{5.5.25}$$

which, together with (5.5.21), preserves the form of (5.5.24).

### *Yang–Mills field tensor*

Let us now assume the torsion vanishes (or else we can use $\sqcap_{ab}$ for $\Delta_{ab}$) and consider the commutator $\Delta_{ab}$. We have

$$\begin{aligned}\Delta_{ab}\mu^{\Psi} &= \Delta_{ab}(\mu^{\Theta}\alpha_{\Theta}{}^{\Psi}) = 2\mu^{\Theta}\nabla_{[a}\nabla_{b]}\alpha_{\Theta}{}^{\Psi} \\ &= -2\mathrm{i}\mu^{\Theta}\nabla_{[a}(\Phi_{b]\Theta}{}^{\Psi}\alpha_{\Psi}{}^{\Psi}) \\ &= -\mathrm{i}\mu^{\Theta}F_{ab\Theta}{}^{\Psi},\end{aligned} \tag{5.5.26}$$

where

$$F_{ab\Theta}{}^{\Psi} := 2\alpha_{\Theta}{}^{\Theta}\alpha_{\Psi}{}^{\Psi}(\nabla_{[a}\Phi_{b]\Theta}{}^{\Psi} - \mathrm{i}\,\Phi_{\Lambda}{}^{\Psi}{}_{[a}\Phi_{b]\Theta}{}^{\Lambda}) \tag{5.5.27}$$

is the *Yang–Mills field tensor* (we have adopted the obviously allowable convention that YM indices and space–time indices can be moved across one another). Its component fields are given by

$$\tfrac{1}{2}F_{ab\Theta}{}^{\Psi} = \nabla_{[a}\Phi_{b]\Theta}{}^{\Psi} - \mathrm{i}\,\Phi_{\Lambda}{}^{\Psi}{}_{[a}\Phi_{b]\Theta}{}^{\Lambda}. \tag{5.5.28}$$

If we transvect this equation with an arbitrary $u^a v^b$, each term on the right becomes a matrix belonging to the Lie algebra $\mathscr{A}$. This follows from (5.5.11), (5.5.12), and the fact that the quadratic term is just a commutator of $\mathscr{A}$ elements. Thus, the same holds true for the Yang–Mills field tensor components on the left.

It follows from the form of (5.5.26) that $F_{ab\Theta}{}^{\Psi}$ is *independent of the gauge* $\alpha_{\Psi}{}^{\Psi}$. Its component fields, therefore, are subject to the standard gauge transformations

$$F_{ab\Theta}{}^{\Psi} \mapsto F_{ab\hat{\Theta}}{}^{\hat{\Psi}} = F_{ab\Theta}{}^{\Psi} r_{\Psi}{}^{\hat{\Psi}} q_{\hat{\Theta}}{}^{\Theta}. \tag{5.5.29}$$

Note that, in contrast to the electromagnetic case, the Yang–Mills field tensor is YM-charged. Equation (5.5.26) yields also

$$\Delta_{ab}\beta_{\Psi} = \mathrm{i}\beta_{\Theta}F_{ab\Psi}{}^{\Theta}, \tag{5.5.30}$$

and, for example,

$$\Delta_{ab}\gamma_{\Psi}{}^{\Theta d} = R_{abc}{}^{d}\gamma_{\Psi}{}^{\Theta c} + \mathrm{i}F_{ab\Psi}{}^{\Lambda}\gamma_{\Lambda}{}^{\Theta d} - \mathrm{i}F_{ab\Lambda}{}^{\Theta}\gamma_{\Psi}{}^{\Lambda d}. \tag{5.5.31}$$

We have

$$F_{ab\Theta}{}^{\Psi} = -F_{ba\Theta}{}^{\Psi}, \tag{5.5.32}$$

and, when $\mathscr{G}$ consists of unitary matrices,

$$F_{ab\Theta\Psi'} := F_{ab\Theta}{}^{\Psi} h_{\Psi\Psi'}$$
$$= \bar{F}_{ab\Psi'\Theta}. \tag{5.5.33}$$

Furthermore, the relation

$$\nabla_{[a} F_{bc]\Theta}{}^{\Psi} = 0 \tag{5.5.34}$$

follows from (5.5.26) (similarly to (5.1.36)).

As in the Maxwell case, therefore, the first 'half' of the field equations is an automatic consequence of the formalism. The second 'half' of the Yang–Mills ('source-free') equations is

$$\nabla^a F_{ab\Theta}{}^{\Psi} = 0, \tag{5.5.35}$$

and, as in the Maxwell case, this has to be *imposed*. We can also consider a RHS to (5.5.35), constituting a Yang–Mills *current*.

*Spinor treatment*

The spinor expressions for the Yang–Mills field follow directly. We have

$$F_{ab\Theta}{}^{\Psi} = \varphi_{AB\Theta}{}^{\Psi} \varepsilon_{A'B'} + \varepsilon_{AB} \chi_{A'B'\Theta}{}^{\Psi}, \tag{5.5.36}$$

where

$$\varphi_{AB\Theta}{}^{\Psi} = \varphi_{(AB)\Theta}{}^{\Psi} = \tfrac{1}{2} F_{ABC'}{}^{C'}{}_{\Theta}{}^{\Psi}$$
$$\chi_{A'B'\Theta}{}^{\Psi} = \chi_{(A'B')\Theta}{}^{\Psi} = \tfrac{1}{2} F_{C}{}^{C}{}_{A'B'\Theta}{}^{\Psi}, \tag{5.5.37}$$

In the unitary case (5.5.33) we have

$$F_{ab\Theta\Psi'} = \varphi_{AB\Theta\Psi'} \varepsilon_{A'B'} + \varepsilon_{AB} \bar{\varphi}_{A'B'\Psi'\Theta}, \tag{5.5.38}$$

where

$$\varphi_{AB\Theta\Psi'} = \varphi_{AB\Theta}{}^{\Psi} h_{\Psi\Psi'}. \tag{5.5.39}$$

Generally, with $\square_{AB}, \square_{A'B'}$ as in (4.9.13), we have

$$\square_{AB} \mu^{\Psi} = -\mathrm{i} \mu^{\Theta} \varphi_{AB\Theta}{}^{\Psi}, \quad \square_{A'B'} \mu^{\Psi} = -\mathrm{i} \mu^{\Theta} \chi_{A'B'\Theta}{}^{\Psi},$$
$$\square_{AB} \lambda_{\Psi} = \mathrm{i} \lambda_{\Theta} \varphi_{AB\Psi}{}^{\Theta}, \quad \square_{A'B'} \lambda_{\Psi} = \mathrm{i} \lambda_{\Theta} \chi_{A'B'\Psi}{}^{\Theta}, \tag{5.5.40}$$

where, in the usual way, the effect of each of these operators on a multi-indexed object is the sum of the effects on each index separately.

The spinor form of (5.5.28) is the pair of equations

$$\varphi_{AB\Theta}{}^{\Psi} = \nabla_{A'(A} \Phi_{B)\Theta}^{A'}{}^{\Psi} - \mathrm{i}\, \Phi_{\Lambda}{}^{\Psi}{}_{A'(A} \Phi_{B)\Theta}^{A'}{}^{\Lambda},$$
$$\chi_{A'B'\Theta}{}^{\Psi} = \nabla_{A(A'} \Phi_{B')\Theta}^{A}{}^{\Psi} - \mathrm{i}\, \Phi_{\Lambda}{}^{\Psi}{}_{A(A'} \Phi_{B')\Theta}^{A}{}^{\Lambda}, \tag{5.5.41}$$

which are complex conjugates of each other in the unitary case (*cf.*

(5.1.46)). The spinor form of (5.5.35) is (*cf.* (5.1.51))

$$\nabla^B_{A'}\varphi_{AB\Theta}{}^{\Psi} = \nabla^{B'}_{A}\chi_{A'B'\Theta}{}^{\Psi}, \tag{5.5.42}$$

which is a consequence of (5.5.41). The independent Yang–Mills field equation (5.5.35) is (*cf.* (5.1.50))

$$\nabla^B_{A'}\varphi_{AB\Theta}{}^{\Psi} + \nabla^{B'}_{A}\chi_{A'B'\Theta}{}^{\Psi} = 0, \tag{5.5.43}$$

and (5.5.42) and (5.5.43) are together equivalent to

$$\nabla^B_{A'}\varphi_{AB\Theta}{}^{\Psi} = 0 = \nabla^{B'}_{A}\chi_{A'B'\Theta}{}^{\Psi}, \tag{5.5.44}$$

these equations being complex conjugates of one another in the unitary case (5.5.33).

In the unitary case we can define a Yang–Mills *energy tensor* in analogy to the electromagnetic energy tensor (5.2.4),

$$T_{ab} = \frac{1}{2\pi}\varphi_{AB\Theta}{}^{\Psi}\bar{\varphi}_{A'B'}{}^{\Theta}{}_{\Psi}, \tag{5.5.45}$$

which possesses the usual properties required of a source term in Einstein's field equations,

$$T_{ab} = \bar{T}_{ab}, \quad T_{[ab]} = 0, \tag{5.5.46}$$

and, as follows from (5.5.44) and $\overline{\chi_{A'B'\Theta}{}^{\Psi}} = \varphi_{AB}{}^{\Psi'}{}_{\Theta'}$

$$\nabla^a T_{ab} = 0 \tag{5.5.47}$$

It also satisfies the trace-free condition characteristic of a massless field:

$$T_a{}^a = 0. \tag{5.5.48}$$

A class of Yang–Mills fields of some special interest, particularly in view of certain interrelations with twistor theory (see Vol. 2, end of §6.10), is that of *self-dual* and *anti-self-dual* fields. The self-dual part of the Yang–Mills field is

$${}^{+}F_{ab\Theta}{}^{\Psi} = \varepsilon_{AB}\chi_{A'B'\Theta}{}^{\Psi} \tag{5.5.49}$$

and the anti-self-dual part is

$${}^{-}F_{ab\Theta}{}^{\Psi} = \varphi_{AB\Theta}{}^{\Psi}\varepsilon_{A'B'}. \tag{5.5.50}$$

(These fields are self-dual or anti-self-dual in the usual sense, i.e., on the space–time indices only.) A self-dual Yang–Mills field is one for which $\varphi_{AB\Theta}{}^{\Psi} = 0$, while for an anti-self-dual one, $\chi_{A'B'\Theta}{}^{\Psi} = 0$. Note that the Yang–Mills field equation (5.5.43), or equivalently (5.5.35), is an automatic consequence of (5.5.42), or equivalently of (5.5.34), in the case of a self-dual or anti-self-dual field. This fact has significance for Ward's construction of such fields, as will be described in §6.10.

## 5.6 Conformal rescalings

In the geometric description of spin-vectors in Chapter 1 much use was made of the null cone structure of the space–time manifold $\mathscr{M}$. The role of the metric itself – which gives rise to that structure – was not quite so fundamental. In fact, it is possible to have spinors when only a *conformal* structure is assumed for $\mathscr{M}$, i.e., when significance is ascribed only to the equivalence class of metrics which can be obtained from a given metric $g_{ab}$ by a *conformal rescaling*

$$g_{ab} \mapsto \hat{g}_{ab} = \Omega^2 g_{ab}. \tag{5.6.1}$$

Here $\Omega$ is any scalar field ($\Omega \in \mathfrak{T}$) which is everywhere positive ($\Omega > 0$). Note that no transformation of *points* is involved. The information contained in the conformal structure is precisely the null cone structure. (Evidently two conformally equivalent metrics share their null directions; conversely, two metrics of Minkowskian signature which share their *real* null directions must be conformal. See, for example, Rindler 1982, equations (6.4)–(6.8).) From a basic physical point of view the null cone structure may be regarded as more primitive than the metric scaling. For example, for the discussion of the basic concept of *causality* between points, it is fully sufficient. In the present section we examine the conformal structure in detail.

In Chapter 1 we gave the geometrical interpretation of a spin-vector $\kappa^A$ at a point $P \in \mathscr{M}$ (up to sign) as a null flag. Its construction involves the geometry of the null cone (in the tangent space to $\mathscr{M}$) at $P$. A conformal metric is, in fact, *necessary* in order for spinors to be defined. But the entire construction is independent of the actual scaling afforded by the particular metric $g_{ab}$. This scaling enters, instead, into the canonical relation between $\mathfrak{S}^A$ and its dual $\mathfrak{S}_A$, i.e., into the (skew) inner product structure on $\mathfrak{S}^A$ given by $\varepsilon_{AB}$. We recall that in §1.6 (*cf.* (1.6.25) *et seq.*) the inner product $\{\boldsymbol{\kappa}, \boldsymbol{\tau}\} = \kappa^A \tau^B \varepsilon_{AB}$ between two spin-vectors was defined purely geometrically, in terms of the geometry of the null cone. The *argument* of this inner product was defined purely in terms of conformal geometry (angles, stereographic projections, etc.), whereas the *modulus* of the inner product required the concept of length. Thus, $\arg\{\boldsymbol{\kappa}, \boldsymbol{\tau}\}$ should be invariant under conformal rescalings (5.6.1), whereas $|\{\boldsymbol{\kappa}, \boldsymbol{\tau}\}|$ could be expected to change. Thus, if we wish to retain our geometric interpretations, $\varepsilon_{AB}$ could be altered under a conformal rescaling, but only to the extent of being multiplied by a *real* number.* In order to preserve (3.1.9): $g_{AB} = \varepsilon_{AB}^{\prime}\varepsilon_{A'B'}$, we therefore *choose* to accompany (5.6.1) by

$$\varepsilon_{AB} \mapsto \hat{\varepsilon}_{AB} = \Omega\varepsilon_{AB} \tag{5.6.2}$$

The only alternative to this choice, *viz* $\hat{\varepsilon}_{AB} = -\Omega\varepsilon_{AB}$ is not continuous with the identity scaling and is therefore rejected.

In the usual *component* (i.e. spin-frame) descriptions one has $\varepsilon_{\mathbf{AB}} = 0, 1, -1, 0$, and this cannot be scaled as in (5.6.2), whereas (5.6.2) is natural in the *abstract index* approach. It has the added advantage of making many conformal transformation formulae simpler than they would otherwise be. Thus abstract indices lead one naturally in a direction which is not the one suggested by the component approach, and some definite advantages are thereby gained. The choice (5.6.2) still leaves us freedom when it comes to introducing spinor components. We consider three particular possibilities. Suppose we have a dyad $o^A, \iota^A$ normalized with respect to $\varepsilon_{AB}$, i.e., $o^A\iota^B\varepsilon_{AB} = 1$ (*cf.* (2.5.39)), i.e., we have a spin-frame. Then the components of $\varepsilon_{AB}$ in this dyad are the standard $\varepsilon_{\mathbf{AB}} = 0, 1, -1, 0$. If we apply a conformal rescaling, and take $\hat{o}^A = o^A, \hat{\iota}^A = \iota^A$, these vectors will cease to be normalized with respect to the new $\hat{\varepsilon}_{AB}$. In fact we shall have $o^A\iota^B\hat{\varepsilon}_{AB} = \Omega$ so that $\hat{\varepsilon}_{\mathbf{AB}} = 0, \Omega, -\Omega, 0$. (It will always be understood that hatted quantities have their components taken with respect to the hatted basis.) In this case the dyad $\hat{o}^A, \hat{\iota}^A$ is no longer a spin-frame. We have $\hat{o}_A = -\hat{\varepsilon}_{AB}\hat{o}^B = -\Omega\varepsilon_{AB}o^B = \Omega o_A$, and similarly $\hat{\iota}_A = \Omega\iota_A$, so for this basis $\hat{\varepsilon}^{\mathbf{AB}} = 0, \Omega, -\Omega, 0$. A second possibility is to define a new dyad $\hat{o}^A = \Omega^{-\frac{1}{2}}o^A, \hat{\iota}^A = \Omega^{-\frac{1}{2}}\iota^A$. This is normalized and $\varepsilon_{01} = 1$, so $\hat{\varepsilon}_{\mathbf{AB}} = \varepsilon_{\mathbf{AB}}$; it also gives $\hat{o}_A = \Omega^{\frac{1}{2}}o_A$, $\hat{\iota}_A = \Omega^{\frac{1}{2}}\iota_A$, whence $\hat{\varepsilon}^{01} = 1$ also, and $\hat{\varepsilon}^{\mathbf{AB}} = \varepsilon^{\mathbf{AB}}$. However, it is often more convenient to make a third and asymmetrical choice: $\hat{o}^A = \Omega^{-1}o^A$, $\hat{\iota}^A = \iota^A$. This implies $\hat{o}_A = o_A$ and $\hat{\iota}_A = \Omega\iota_A$. Again we get a normalized dyad, so that $\hat{\varepsilon}_{\mathbf{AB}} = \varepsilon_{\mathbf{AB}}$ and $\hat{\varepsilon}^{\mathbf{AB}} = \varepsilon^{\mathbf{AB}}$. This choice often turns out to be useful when there is a preferred field $o_A$ (or $\iota^A$), as, for example, in the discussion of conformal infinity given in Vol. 2, particularly §9.7 (and Penrose 1968).

The 'Kronecker delta' quantities $g_a{}^b, \varepsilon_A{}^B, \varepsilon_{A'}{}^{B'}$ must remain unchanged:

$$\hat{g}_a{}^b = g_a{}^b, \quad \hat{\varepsilon}_A{}^B = \varepsilon_A{}^B, \quad \hat{\varepsilon}_{A'}{}^{B'} = \varepsilon_{A'}{}^{B'}, \tag{5.6.3}$$

since they effect index substitutions between the various sets $\mathfrak{S}^{\cdots}_{\cdots}$, or,

* However, with a slight shift in our interpretations we can be led to consider a modification of (5.6.2) in which $\Omega$ is complex and the $\Omega^2$ of (5.6.1) is replaced by $\Omega\bar{\Omega}$. This naturally gives rise to a torsion in $\mathscr{M}$, as discussed in Penrose (1983); *cf.* also footnote on p. 356.

alternatively, since they satisfy such relations as $g_a{}^b g_b{}^c = g_a{}^c$ (*cf*. (3.1.11), (2.5.13)). It follows that we must also have

$$\hat{\varepsilon}_{A'B'} = \Omega\varepsilon_{A'B'}, \quad \hat{\varepsilon}^{AB} = \Omega^{-1}\varepsilon^{AB}, \quad \hat{\varepsilon}^{A'B'} = \Omega^{-1}\varepsilon^{A'B'} \tag{5.6.4}$$

because of the complex conjugate and inverse relations these quantities have with $\hat{\varepsilon}_{AB}$. Similarly we need

$$\hat{g}^{ab} = \Omega^{-2}g^{ab}. \tag{5.6.5}$$

One consequence of the above set of formulae is that the important operations of raising or lowering a tensor or spinor index do *not* commute with conformal rescaling. It must therefore be made clear, when an index is raised or lowered, which $g$ or $\varepsilon$ is being employed. Our convention will be that any hatted kernel symbol must have its indices shifted with $\hat{g}$ or $\hat{\varepsilon}$, while unhatted kernel symbols have their indices shifted with $g$ or $\varepsilon$.

### *Conformal densities*

As we have observed above, a spin-vector $\kappa^A$ has a definite geometric interpretation (flag and flagpole) which is quite independent of any rescaling. A spin-covector $\omega_A$ also has a define geometric interpretation, which, however, is less direct. (See, e.g., the second footnote on p. 72 for the geometric interpretation of the flagpole $W_a = \omega_A\bar{\omega}_{A'}$ of $\omega_A$.) Given only a conformal structure, $\omega_A$ cannot be interpreted via its associated spin-vector $\omega^A$, since that is determined only up to a factor. The basic intrinsic way to regard $\omega_A$ is simply as the mapping $\kappa^A \mapsto \omega_A\kappa^A$ for spin-vectors $\kappa^A$.

Suppose we have a spin-vector $\kappa^A$ which we regard as geometrically determined, and therefore unaffected by rescaling:

$$\hat{\kappa}^A = \kappa^A. \tag{5.6.6}$$

Then for its associated spin-covector we have

$$\hat{\kappa}_A = \hat{\varepsilon}_{BA}\hat{\kappa}^B = \Omega\varepsilon_{BA}\kappa^B = \Omega\kappa_A. \tag{5.6.7}$$

Hence $\kappa_A$ is a *conformal density* of weight 1, i.e., a quantity that gets multiplied by $\Omega^1$ under a rescaling (5.6.1).

Conversely, suppose we have an intrinsically fixed spin-covector $\omega_A$, so that

$$\hat{\omega}_A = \omega_A. \tag{5.6.8}$$

Then

$$\hat{\omega}^A = \hat{\varepsilon}^{AB}\hat{\omega}_B = \Omega^{-1}\varepsilon^{AB}\omega_B = \Omega^{-1}\omega^A, \tag{5.6.9}$$

and this is a conformal density of weight $-1$.

More generally, it is convenient to work with conformal densities of arbitrary weight. We say that $\theta^{\mathscr{A}}$ is a *conformal density of weight* $k$ if it is to change under a rescaling (5.6.1) to

$$\hat{\theta}^{\mathscr{A}} = \Omega^k \theta^{\mathscr{A}}. \tag{5.6.10}$$

We may think of a conformal density as a function not only of a point on the manifold $\mathscr{M}$ but of the particular $g_{ab}$ chosen.* Normally $k$ is an integer, or possibly a half-integer. Observe that $g_{ab}$, $\varepsilon_{AB}$, $\varepsilon_{A'B'}$, $\varepsilon^{AB}$, $\varepsilon^{A'B'}$, $g^{ab}$ have respective conformal weights $2, 1, 1, -1, -1, -2$. Consequently, whenever a spinor index is raised on $\theta^{\cdots}_{\cdots}$ its conformal weight is reduced by unity, and whenever a spinor index is lowered its weight is increased by unity (*cf.* $\varepsilon_{AB}$ and $\varepsilon^{AB}$). Similarly, when a tensor index is raised, the weight is reduced by 2, and when a tensor index is lowered, the weight is increased by 2.

## The associated change in $\nabla_a$

We shall be concerned to a considerable extent in this section with questions of conformal invariance. A system of fields and field equations, etc., will be said to be *conformally invariant* if it is possible to attach conformal weights to all field quantities occurring in the system, in such a way that the field equations remain true after conformal rescaling.** For this, we must first examine the conformal behaviour of the covariant derivative operator $\nabla_a$. Since $g_{ab}$ and $\varepsilon_{AB}$ are altered under conformal rescaling, their covariant constancy before rescaling imposes a different condition on $\nabla_a$ after rescaling. Thus we need two different operators $\nabla_a$ and $\hat{\nabla}_a$ where

$$\nabla_a \varepsilon_{BC} = 0, \quad \hat{\nabla}_a \hat{\varepsilon}_{BC} = 0. \tag{5.6.11}$$

We assume that the torsion vanishes in each case. By the results of §4.4, we find (*cf.* (4.4.22), (4.4.23))

$$\hat{\nabla}_a f = \nabla_a f, \quad \hat{\nabla}_a \xi^C = \nabla_a \xi^C + \Theta_{aB}{}^C \xi^B, \tag{5.6.12}$$

* Thus we may, if we choose to, think of a conformal density as a field defined not on $\mathscr{M}$ itself but on the 5-dimensional manifold which is a bundle over $\mathscr{M}$, the fibres being the one-dimensional spaces of possible choices of conformal scale at each point. A conformal density is a field defined on this bundle which varies up each fibre according to (5.6.10).

** A flat-space theory which is Poincaré invariant and also conformally invariant in this sense, will be invariant under the 15-parameter conformal group. This is because the Poincaré motions of Minkowski space become conformal motions according to any other conformally rescaled flat metric. Conformal motions obtainable in this way are sufficient to generate the full conformal group. This will be discussed fully in Vol. 2 (*cf.* §9.2). But the type of conformal invariance described above is really more general than this, since it applies to curved space–times also.

where (*cf.* (4.4.47))

$$\Theta_{aB}{}^{C} = \mathrm{i}\Pi_{a}\varepsilon_{B}{}^{C} + \Upsilon_{A'B}\varepsilon_{A}{}^{C}, \quad \Pi_{a}, \Upsilon_{a} \in \mathfrak{T}_{a}. \tag{5.6.13}$$

Now, from (5.6.11) and (4.4.27), we have

$$\begin{aligned} 0 = \hat{\nabla}_{a}\hat{\varepsilon}_{BC} &= \hat{\nabla}_{a}(\Omega\varepsilon_{BC}) \\ &= \nabla_{a}(\Omega\varepsilon_{BC}) - \Theta_{aB}{}^{D}\Omega\varepsilon_{DC} - \Theta_{aC}{}^{D}\Omega\varepsilon_{BD} \\ &= \varepsilon_{BC}(\nabla_{a}\Omega - \Omega\Theta_{aD}{}^{D}) \\ &= \varepsilon_{BC}(\nabla_{a}\Omega - 2\mathrm{i}\Omega\Pi_{a} - \Omega\Upsilon_{a}), \end{aligned}$$

whence

$$\Omega^{-1}\nabla_{a}\Omega = \Upsilon_{a} + 2\mathrm{i}\Pi_{a}.$$

But since $\Omega$ is real $\Pi_a = 0$ and so

$$\Upsilon_{a} = \Omega^{-1}\nabla_{a}\Omega = \nabla_{a}\log\Omega. \tag{5.6.14}$$

With these values of $\Pi_a$, $\Upsilon_a$ substituted first into (5.6.13) and then into (4.4.27), we have,* for a generic spinor $\chi_{B\ldots F'\ldots}{}^{P\ldots S'\ldots}$,

$$\begin{aligned} \hat{\nabla}_{AA'}\chi_{B\ldots F'\ldots}^{P\ldots S'\ldots} = \nabla_{AA'}\chi_{B\ldots F'\ldots}^{P\ldots S'\ldots} - \Upsilon_{BA'}\chi_{A\ldots F'\ldots}^{P\ldots S'\ldots} - \cdots - \Upsilon_{AF'}\chi_{B\ldots A'\ldots}^{P\ldots S'\ldots} - \cdots \\ + \varepsilon_{A}{}^{P}\Upsilon_{XA'}\chi_{B\ldots F'\ldots}^{X\ldots S'\ldots} + \cdots + \varepsilon_{A'}{}^{S'}\Upsilon_{AX'}\chi_{B\ldots F'\ldots}^{P\ldots X'\ldots} + \cdots. \end{aligned} \tag{5.6.15}$$

We observe the important fact that if the spinor $\chi_{B\ldots F'\ldots}^{P\ldots S'\ldots}$ is charged, the entire above argument goes through unchanged, and the formulae hold without modification. Moreover, if $\chi_{B\ldots F'\ldots}^{P\ldots S'\ldots}$ has additional Yang–Mills indices this does not affect the validity of (5.6.15), no extra terms arising from the presence of these indices.

It may be remarked that if it is desired merely to verify the above formulae rather than derive them, then the theory used to obtain (5.6.12) in (4.4.47) may be partially circumvented. One merely needs to verify that $\Theta_{AA'B}{}^{D} = \Upsilon_{BA'}\varepsilon_{A}{}^{D}$ with (5.6.14) leads to (5.6.11) and that the torsion of $\hat{\nabla}_a$ is zero with this definition.

We note in passing that (5.6.15) holds also in *Weyl geometry* (Weyl 1923) but with $\Upsilon_a$ merely restricted to be real, not necessarily a gradient. In Weyl geometry there is a well-defined conformal structure (hence spinors), but no preferred metric. There is a covariant derivative operator $\nabla_a$, but it need annihilate no metric. The operator $\nabla_a$ defines parallel transport in the usual way, and so it allows comparisons of length to be made

* If we admit the possibility of a complex $\Omega$, as mentioned in the footnote on p. 353, we find that rather than allowing $\Pi_a \neq 0$ it is more natural to introduce a torsion $\mathrm{i}(\Upsilon_d - \bar{\Upsilon}_d)e_{ab}{}^{cd}$, where $\Upsilon_a$ is given by (5.6.14) – or to add this to a pre-existing torsion – and modify (5.6.15) by using $\bar{\Upsilon}$ in place of $\Upsilon$ in each term where $\Upsilon$ possesses an index $A$ (rather than $A'$). (See Penrose 1983.)

at different points. But this comparison is path-dependent, i.e., non-integrable. On introducing an arbitrary metric $\hat{g}_{ab}$ consistent with the conformal structure, one finds that the associated Christoffel derivative operator $\hat{\nabla}_a$ is related to the Weyl derivative $\nabla_a$ by (5.6.15). Such a $\Upsilon_a$ is 'arbitrary', in the sense that for given $g_{ab}$, *any* choice of $\Upsilon_a \in \mathfrak{T}_a$ yields a corresponding unique Weyl connection $\nabla_a$.

As a particular case of (5.6.15) we can derive the conformal behaviour of the covariant derivative of tensors. Consider first the case of a covector $V_b$:

$$\begin{aligned} \hat{\nabla}_a V_b = \hat{\nabla}_{AA'} V_{BB'} &= \nabla_{AA'} V_{BB'} - \Upsilon_{BA'} V_{AB'} - \Upsilon_{AB'} V_{BA'} \\ &= \nabla_a V_b - \Upsilon_a V_b - \Upsilon_b V_a + g_{ab} \Upsilon_c V^c, \end{aligned} \tag{5.6.16}$$

where we simply applied (3.4.13) to the last two terms in the first row. To discuss the general case, we define the tensor

$$Q_{ab}{}^c = 2\Upsilon_{(a} g_{b)}{}^c - g_{ab} \Upsilon^c. \tag{5.6.17}$$

Then (5.6.16) can be stated

$$\hat{\nabla}_a V_b = \nabla_a V_b - Q_{ab}{}^c V_c, \tag{5.6.18}$$

from which we get (*cf*. (4.2.46), (4.2.47))

$$\hat{\nabla}_a U^b = \nabla_a U^b + Q_{ac}{}^b U^c, \tag{5.6.19}$$

and, generally (*cf*. (4.2.48)),

$$\begin{aligned} \hat{\nabla}_a H^{b\ldots d}_{f\ldots h} = \nabla_a H^{b\ldots d}_{f\ldots h} &+ Q_{ab_0}{}^b H^{b_0\ldots d}_{f\ldots h} + \cdots + Q_{ad_0}{}^d H^{b\ldots d_0}_{f\ldots h} + \cdots \\ &- Q_{af}{}^{f_0} H^{b\ldots d}_{f_0\ldots h} - \cdots - Q_{ah}{}^{h_0} H^{b\ldots d}_{f\ldots h_0}. \end{aligned} \tag{5.6.20}$$

Again, these formulae hold for charged fields as well as for uncharged fields. We observe that the tensorial form (5.6.20) of the $\nabla_a$ transformation is rather more complicated than the spinorial form (5.6.15), in that each $Q$-term really stands for three terms, via (5.6.17). This contributes to the fact that proofs of conformal invariance tend to be easier in spinor than in tensor formalism.

Our simple derivation of the change in *curvature* under conformal rescaling finds a natural place in §6.8, Vol. 2, so we delay our detailed derivations until then and give, here, only the basic formulae

$$\begin{aligned} \hat{\Phi}_{ABA'B'} &= \Phi_{ABA'B'} - \nabla_{A(A'} \Upsilon_{B')B} + \Upsilon_{A(A'} \Upsilon_{B')B} \\ \Omega^2 \hat{\Lambda} &= \Lambda + \tfrac{1}{4} \nabla^a \Upsilon_a + \tfrac{1}{4} \Upsilon^a \Upsilon_a \\ \hat{\Psi}_{ABCD} &= \Psi_{ABCD} \end{aligned}$$

(*cf*. (6.8.24), (6.8.25), (6.8.4)). Note, from the last relation, that $\Psi_{ABCD}$ is *conformally invariant*. Moreover, we shall see in §6.9 that $\Psi_{ABCD}=0$ is necessary and sufficient for $\mathcal{M}$ to be (patchwise) *conformally flat*.

### *Behaviour of spin-coefficients under rescaling*

We next give some formulae that relate conformal transformations to spin-coefficients. First we give the transformations of the spin-coefficients (4.5.21) under the general rescaling of the dyad:

$$\hat{o}_A=\Omega^{w_0+1}o_A,\quad \hat{\iota}_A=\Omega^{w_1+1}\iota_A, \tag{5.6.21}$$

which implies

$$\hat{o}^A=\Omega^{w_0}o^A,\quad \hat{\iota}^A=\Omega^{w_1}\iota^A,\quad \hat{\chi}=\Omega^{w_0+w_1+1}\chi. \tag{5.6.22}$$

It will be convenient to write

$$\omega=\log\Omega \tag{5.6.23}$$

so that, by (5.6.14),

$$\mathrm{D}\omega=\Upsilon_{00'},\quad \delta\omega=\Upsilon_{01'},\quad \delta'\omega=\Upsilon_{10'},\quad \mathrm{D}'\omega=\Upsilon_{11'}. \tag{5.6.24}$$

Then we find directly, by applying (5.6.15) to the definitions (4.5.21), and writing $\Omega^{w_0-w_1}=\Sigma$,

$$\begin{bmatrix}\hat{\kappa} & \hat{\varepsilon} & \hat{\gamma}' & \hat{\tau}'\\ \hat{\rho} & \hat{\alpha} & \hat{\beta}' & \hat{\sigma}'\\ \hat{\sigma} & \hat{\beta} & \hat{\alpha}' & \hat{\rho}'\\ \hat{\tau} & \hat{\gamma} & \hat{\varepsilon}' & \hat{\kappa}'\end{bmatrix} = \tag{5.6.25}$$

$$\Omega^{w_0+w_1}\times\begin{bmatrix}\kappa\Sigma^2 & [\varepsilon+(w_0+1)\mathrm{D}\omega]\Sigma & (\gamma'+w_1\mathrm{D}\omega)\Sigma & \tau'-\delta'\omega\\ (\rho-\mathrm{D}\omega)\Sigma & \alpha+w_0\delta'\omega & \beta'+(w_1+1)\delta'\omega & \sigma'\Sigma^{-1}\\ \sigma\Sigma & \beta+(w_0+1)\delta\omega & \alpha'+w_1\delta\omega & (\rho'-\mathrm{D}'\omega)\Sigma^{-1}\\ \tau-\delta\omega & (\gamma+w_0\mathrm{D}'\omega)\Sigma^{-1} & [\varepsilon'+(w_1+1)\mathrm{D}'\omega]\Sigma^{-1} & \kappa'\Sigma^{-2}\end{bmatrix}$$

The following four particular cases of (5.6.21) are of special interest:

(i) $\hat{o}_A=o_A, \hat{\iota}_A=\iota_A; \hat{o}^A=\Omega^{-1}o^A, \hat{\iota}^A=\Omega^{-1}\iota^A; \hat{\chi}=\Omega^{-1}\chi,$

(ii) $\hat{o}_A=\Omega o_A, \hat{\iota}_A=\Omega\iota_A; \hat{o}^A=o^A, \hat{\iota}^A=\iota^A; \hat{\chi}=\Omega\chi,$

(iii) $\hat{o}_A=\Omega^{\frac{1}{2}}o_A, \hat{\iota}_A=\Omega^{\frac{1}{2}}\iota_A; \hat{o}^A=\Omega^{-\frac{1}{2}}o^A, \hat{\iota}^A=\Omega^{-\frac{1}{2}}\iota^A; \hat{\chi}=\chi,$

(iv) $\hat{o}_A=o_A, \hat{\iota}_A=\Omega\iota_A; \hat{o}^A=\Omega^{-1}o^A, \hat{\iota}^A=\iota^A; \hat{\chi}=\chi,$ (5.6.26)

and it will be worth while to exhibit (5.6.25) as it applies to each of these cases in turn:

$$\begin{array}{|cccc|}\hline \hat\kappa & \hat\varepsilon & \hat\gamma' & \hat\tau' \\ \hat\rho & \hat\alpha & \hat\beta' & \hat\sigma' \\ \hat\sigma & \hat\beta & \hat\alpha' & \hat\rho' \\ \hat\tau & \hat\gamma & \hat\varepsilon' & \hat\kappa' \\ \hline \end{array} = \begin{cases} \text{(i)}: \Omega^{-2}\times \begin{array}{|c|c|c|c|}\hline \kappa & \varepsilon & \gamma' - \mathrm{D}\omega & \tau' - \delta'\omega \\ \rho - \mathrm{D}\omega & \alpha - \delta'\omega & \beta' & \sigma' \\ \sigma & \beta & \alpha' - \delta\omega & \rho' - \mathrm{D}'\omega \\ \tau - \delta\omega & \gamma - \mathrm{D}'\omega & \varepsilon' & \kappa' \\ \hline \end{array} \\[2ex] \text{(ii)}: \begin{array}{|c|c|c|c|}\hline \kappa & \varepsilon + \mathrm{D}\omega & \gamma' & \tau' - \delta'\omega \\ \rho - \mathrm{D}\omega & \alpha & \beta' + \delta'\omega & \sigma' \\ \sigma & \beta + \delta\omega & \alpha' & \rho' - \mathrm{D}'\omega \\ \tau - \delta\omega & \gamma & \varepsilon' + \mathrm{D}'\omega & \kappa' \\ \hline \end{array} \\[2ex] \text{(iii)}: \Omega^{-1}\times \begin{array}{|c|c|c|c|}\hline \kappa & \varepsilon + \frac{1}{2}\mathrm{D}\omega & \gamma' - \frac{1}{2}\mathrm{D}\omega & \tau' - \delta'\omega \\ \rho - \mathrm{D}\omega & \alpha - \frac{1}{2}\delta'\omega & \beta' + \frac{1}{2}\delta'\omega & \sigma' \\ \sigma & \beta + \frac{1}{2}\delta\omega & \alpha' - \frac{1}{2}\delta\omega & \rho' - \mathrm{D}'\omega \\ \tau - \delta\omega & \gamma - \frac{1}{2}\mathrm{D}'\omega & \varepsilon' + \frac{1}{2}\mathrm{D}'\omega & \kappa' \\ \hline \end{array} \\[2ex] \text{(iv)}: \begin{array}{|c|c|c|c|}\hline \Omega^{-3}\kappa & \Omega^{-2}\varepsilon & \Omega^{-2}\gamma' & \Omega^{-1}(\tau' - \delta'\omega) \\ \Omega^{-2}(\rho - \mathrm{D}\omega) & \Omega^{-1}(\alpha - \delta'\omega) & \Omega^{-1}(\beta' + \delta'\omega) & \sigma' \\ \Omega^{-2}\sigma & \Omega^{-1}\beta & \Omega^{-1}\alpha' & \rho' - \mathrm{D}'\omega \\ \Omega^{-1}(\tau - \delta\omega) & \gamma - \mathrm{D}'\omega & \varepsilon' + \mathrm{D}'\omega & \Omega\kappa' \\ \hline \end{array} \end{cases} \qquad (5.6.27)$$

The simplicity of cases (i) and (ii) is somewhat deceptive, since the normalization $\chi = 1$ cannot be preserved (i.e., the rescalings cannot be applied to spin-frames). In cases (iii) and (iv) we *can* set $\chi = \hat{\chi} = 1$ and then the two middle columns of spin-coefficients become negatives of each other (*cf.* (4.5.29)). Note that the 'obvious' choice (iii) preserving this normalization leads to somewhat more complicated formulae than the asymmetric choice (iv). The latter is the more interesting choice, being useful, for example, in the asymptotic analysis of the gravitational and other massless fields (*cf.* §9.7). We may also remark that the reverse-scaled case (iv)': $\hat{\iota}_A = \iota_A$, $\hat{o}^A = o^A$, can easily be read off from (5.6.27) (iv) by simply priming all unprimed quantities and removing the prime from all primed ones (taking $\chi' = \chi$, $\Omega' = \Omega$, $\omega' = \omega$).

Note from (5.6.25) that *whichever* scaling is taken,

*$\kappa$, $\sigma$, $\kappa'$, $\sigma'$ are all conformal densities,*

(of respective weights $3w_0 - w_1$, $2w_0$, $3w_1 - w_0$, $2w_1$) (5.6.28)

and also

*$\tau - \bar{\tau}'$ and the imaginary parts of $\rho$, $\rho'$, $\varepsilon$, $\varepsilon'$, $\gamma$, $\gamma'$*
*are all conformal densities* (of respective weights
$w_0 + w_1, 2w_0, 2w_1, 2w_0, 2w_1, 2w_0, 2w_1$), (5.6.29)

while for *certain* scalings, some selections of $\varepsilon, \alpha, \beta, \gamma, \varepsilon', \alpha', \beta', \gamma'$ can be made to be such.

*Conformally invariant 'eth' and 'thorn'*

We end this section by showing how the compacted spin-coefficient formalism of §4.12 can be further developed so that its operations become conformally invariant. Recall that under a ('gauge') change of dyad (4.12.2):

$$o_A \mapsto \lambda o_A, \; o^A \mapsto \lambda o^A; \; \iota_A \mapsto \mu \iota_A, \; \iota^A \mapsto \mu \iota^A, \tag{5.6.30}$$

a scalar quantity $\eta$ of *type* $\{r', r; t', t\}$ changes as follows (by definition, *cf.* (4.12.9)):

$$\eta \mapsto \lambda^{r'} \bar{\lambda}^{t'} \mu^r \bar{\mu}^t \eta. \tag{5.6.31}$$

Now suppose that $\eta$ also has a *conformal weight* $w$ so that, under a conformal dyad rescaling (5.6.21), $\eta$ changes as follows:

$$\hat{\eta} = \Omega^w \eta. \tag{5.6.32}$$

(Note that this commutes with (5.6.31).) Then we shall define new compacted þ and ð operators, to act on such doubly weighted scalars, by

$$\begin{aligned}
\text{þ}_{\mathscr{C}} &= \text{þ} + [w - r'(w_0 + 1) - rw_1 - t'(w_0 + 1) - tw_1]\rho \\
\text{þ}'_{\mathscr{C}} &= \text{þ}' + [w - r'w_0 - r(w_1 + 1) - t'w_0 - t(w_1 + 1)]\rho' \\
\eth_{\mathscr{C}} &= \eth + [w - r'(w_0 + 1) - rw_1 - t'w_0 - t(w_1 + 1)]\tau \\
\eth'_{\mathscr{C}} &= \eth' + [w - r'w_0 - r(w_1 + 1) - t'(w_0 + 1) - tw_1]\tau'
\end{aligned} \tag{5.6.33}$$

It may be directly verified that the result of these operations is as follows:

$$\begin{aligned}
\widehat{\text{þ}_{\mathscr{C}}\eta} &= \hat{\text{þ}}_{\mathscr{C}}\hat{\eta} = \Omega^{w+2w_0}\text{þ}_{\mathscr{C}}\eta \\
\widehat{\text{þ}'_{\mathscr{C}}\eta} &= \hat{\text{þ}}'_{\mathscr{C}}\hat{\eta} = \Omega^{w+2w_1}\text{þ}'_{\mathscr{C}}\eta \\
\widehat{\eth_{\mathscr{C}}\eta} &= \hat{\eth}_{\mathscr{C}}\hat{\eta} = \Omega^{w+w_0+w_1}\eth_{\mathscr{C}}\eta \\
\widehat{\eth'_{\mathscr{C}}\eta} &= \hat{\eth}'_{\mathscr{C}}\hat{\eta} = \Omega^{w+w_0+w_1}\eth'_{\mathscr{C}}\eta,
\end{aligned} \tag{5.6.34}$$

so that $\text{þ}_{\mathscr{C}}, \text{þ}'_{\mathscr{C}}, \eth_{\mathscr{C}}, \eth'_{\mathscr{C}}$ are conformally weighted operators of respective weights $2w_0, 2w_1, w_0 + w_1, w_0 + w_1$ (in the sense defined, *mutatis mutandis*, after (4.12.17)). But they are also 'gauge' weighted operators of the same types (4.12.17) as þ, þ′, ð, ð′, respectively.

Note that as defined, $\text{þ}_{\mathscr{C}}$ and $\text{þ}'_{\mathscr{C}}$ are not real, in general, nor are $\eth_{\mathscr{C}}$ and $\eth'_{\mathscr{C}}$ complex conjugates of one another*, in contradistinction to (4.12.30). Had we wished, we could have worked with the real operators $\frac{1}{2}(\text{þ}_{\mathscr{C}} + \bar{\text{þ}}_{\mathscr{C}})$, $\frac{1}{2}(\text{þ}'_{\mathscr{C}} + \bar{\text{þ}}'_{\mathscr{C}})$ and with the complex conjugate pair $\frac{1}{2}(\eth_{\mathscr{C}} + \bar{\eth}'_{\mathscr{C}}), \frac{1}{2}(\bar{\eth}_{\mathscr{C}} + \eth'_{\mathscr{C}})$. This would have been slightly more complicated, but essentially equiva-

* A case can be made, however, for defining operators $\text{þ}_{\mathscr{C}} = \bar{\text{þ}}_{\mathscr{C}}, \eth_{\mathscr{C}} = \bar{\eth}'_{\mathscr{C}}$ which act on quantities which have generalized conformal weights described by *two* numbers, where a *complex* $\Omega$ is adopted and these numbers provide the powers to which each of $\Omega$ and $\bar{\Omega}$ is raised under rescaling (see footnotes on pp. 353, 356).

lent, since the differences $\text{þ}_{\mathscr{C}} - \text{þ}'_{\mathscr{C}}$, $\eth_{\mathscr{C}} - \bar{\eth}'_{\mathscr{C}}$, etc., are simply expressible in terms of the conformally weighted quantities $\rho - \bar{\rho}$ and $\tau - \bar{\tau}'$ (*cf.* (5.6.29)), which are 'allowable' elements of the present calculus. It is the non-conformally-weighted quantities such as $\rho + \bar{\rho}$ and $\tau + \bar{\tau}'$, and indeed $\rho$ and $\tau$ individually, that must be 'withdrawn from circulation' (*cf.* just before (4.12.15)).

As we saw in (5.6.22), $\chi$ has conformal weight $w_0 + w_1 + 1$, so that, by (5.6.33) and (4.12.23),

$$\text{þ}_{\mathscr{C}}\chi = \text{þ}'_{\mathscr{C}}\chi = \eth_{\mathscr{C}}\chi = \eth'_{\mathscr{C}}\chi = 0. \tag{5.6.35}$$

When the normalization $\chi$ is unchanged (e.g. when $o^A, \iota^A$ is a spin-frame before and after rescaling), we have $w_0 + w_1 + 1 = 0$. Then, with $p = r' - r$ and $q = t' - t$, as in (4.12.10), the expressions (5.6.33) simplify to

$$\begin{aligned} &\text{þ}_{\mathscr{C}} = \text{þ} + [w + (p+q)w_1]\rho, \quad \text{þ}'_{\mathscr{C}} = \text{þ}' + [w - (p+q)w_0]\rho', \\ &\eth_{\mathscr{C}} = \eth + [w + pw_1 - qw_0]\tau, \quad \eth'_{\mathscr{C}} = \eth' + [w - pw_0 + qw_1]\tau'. \end{aligned} \tag{5.6.36}$$

Using these operators we can simplify the appearance of various conformally invariant equations written in (compacted) spin-coefficient form. We note, for future reference, that the massless free-field equations, which have been given in compacted spin-coefficient form in (4.12.44) – and which we discuss in more detail in the following section – can be written (with $r = 1, \ldots, n$)

$$\begin{aligned} \text{þ}_{\mathscr{C}}\phi_r - \eth'_{\mathscr{C}}\phi_{r-1} &= (r-1)\sigma'\phi_{r-2} - (n-r)\kappa\phi_{r+1}, \\ \eth_{\mathscr{C}}\phi_r - \text{þ}'_{\mathscr{C}}\phi_{r-1} &= (r-1)\kappa'\phi_{r-2} - (n-r)\sigma\phi_{r+1}, \end{aligned} \tag{5.6.37}$$

and that the twistor equation (*cf.* (4.12.46)) becomes

$$\begin{aligned} &\eth'_{\mathscr{C}}\omega^0 = \sigma'\omega^1, \quad \eth_{\mathscr{C}}\omega^1 = \sigma\omega^0, \\ &\text{þ}'_{\mathscr{C}}\omega^0 = \kappa'\omega^1, \quad \text{þ}_{\mathscr{C}}\omega^1 = \kappa\omega^0, \\ &\eth_{\mathscr{C}}\omega^0 = \text{þ}'_{\mathscr{C}}\omega^1, \quad \text{þ}_{\mathscr{C}}\omega^0 = \eth'_{\mathscr{C}}\omega^1. \end{aligned} \tag{5.6.38}$$

For these equations we take $\phi_{A\ldots L}$ and $\omega^A$ to have conformal weights $-1$ and 0, respectively. The resulting values of $w$ possessed by the various components $\phi_r$, $\omega^{\mathbf{A}}$ depend upon the choice of $w^0$ and $w^1$ in (5.6.21), but this makes no difference in the definitions (5.6.33) since the coefficients in the correction terms exactly compensate for changes in $w^0$ and $w^1$.

We note, furthermore, that the exterior derivative equation $\boldsymbol{\gamma} = \mathrm{d}\boldsymbol{\beta}$ of (4.14.80), taking the 3-form $\boldsymbol{\gamma}$ and the 2-form $\boldsymbol{\beta}$ both to have conformal weight zero, as stated in (4.14.81), can be written in the form

$$3\gamma_{01'10'11'} = \eth_{\mathscr{C}}\beta_{10'11'} - \eth'_{\mathscr{C}}\beta_{01'11'} + \text{þ}'_{\mathscr{C}}\beta_{01'10'}, \tag{5.6.39}$$

and that (4.14.92) becomes

$$\oint_{\Sigma}\{\text{þ}'_{\mathscr{C}}\mu_0 - \eth_{\mathscr{C}}\mu_1\}\mathscr{N} = \oint_{\mathscr{S}'}\mu_0\mathscr{S} - \oint_{\mathscr{S}}\mu_0\mathscr{S}. \tag{5.6.40}$$

Further uses of these operators will be found in §§5.12, 9.8, 9.9.

## 5.7 Massless fields

We now examine an important class of spinor equations which turn out to be conformally invariant: the *massless free-field equations for* arbitrary spin $\frac{1}{2}n$, where $n$ is a positive integer. Let $\phi_{AB\ldots L}$ have $n$ indices and be symmetric:

$$\phi_{AB\ldots L} = \phi_{(AB\ldots L)}. \tag{5.7.1}$$

The massless free-field equation for spin $\frac{1}{2}n$ is then taken to be

$$\nabla^{AA'}\phi_{AB\ldots L} = 0. \tag{5.7.2}$$

The complex conjugate form of this equation ($n$ indices)

$$\nabla^{AA'}\theta_{A'B'\ldots L'} = 0, \quad \theta_{A'B'\ldots L'} = \theta_{(A'B'\ldots L')} \tag{5.7.3}$$

also describes a massless free field of spin $\frac{1}{2}n$. When these fields are to represent wave functions in $\mathbb{M}$, it is usual to impose a positive-frequency requirement to the effect that in their Fourier decomposition, only terms in $e^{-ip_a x^a}$ occur for which $p_a$ is future-pointing, $x^a$ being the position vector (*cf.* also §6.10). Then the solutions of (5.7.2) represent *left-handed* massless particles (helicity $-\frac{1}{2}n\hbar$) and the solutions of (5.7.3), *right-handed* massless particles (helicity $+\frac{1}{2}n\hbar$). (See Dirac 1936a, Fierz & Pauli 1939, Fierz 1940, Penrose 1965, Penrose & MacCallum 1972.)

Recall that the Bianchi identity has this form in empty space, with $\Psi_{ABCD}$ taking the place of $\phi_{\ldots}$ (*cf.* (4.10.9)). It is thus a 'curved-space spin-2 field equation', and its close relation to Einstein's field equations has already been noted (see remark after (4.10.10)). Similarly, the source-free Maxwell equations (5.1.57) have this form with $\varphi_{AB}$ taking the place of $\phi_{\ldots}$ (spin 1). The Dirac–Weyl equation for the neutrino (*cf.* (4.4.61)) also falls into this category with $\phi_{\ldots} = \nu_A$ (spin $\frac{1}{2}$), namely

$$\nabla^{AA'}\nu_A = 0.$$

*Spin 2: gravitational perturbations*

The equation (5.7.2) in the case of spin 2 also has interest in $\mathbb{M}$ (*cf.* Fierz & Pauli 1939), as the spinor version of the 'gauge invariant' form of the weak field limit of Einstein's vacuum equations (i.e. linearized Einstein theory, sometimes called the 'fast approximation'). We envisage a smooth 1-parameter family of space–times, satisfying Einstein's vacuum equations, such that the member with parameter $u = 0$ is $\mathbb{M}$. For each fixed value of $u$ we have a spinor field $\Psi_{ABCD}$ on the manifold, satisfying $\nabla^{AA'}\Psi_{ABCD} = 0$. Since this field tends smoothly to zero as $u \to 0$, we would expect that $u^{-1}\Psi_{ABCD}$ has a well-defined limit $\phi_{ABCD}$ as $u \to 0$, i.e., in the Minkowski space $u = 0$, and that it there satisfies the *flat*-space version of (5.7.2). Indeed, this procedure can be carried through, although it is more usual to describe linearized Einstein theory terms of a real symmetric tensor ('potential') field $h_{ab} \in \mathfrak{T}_{(ab)}$, in $\mathbb{M}$, which represents the first-order deviation from flatness of the metric ($g_{ab}(u) = g_{ab} + uh_{ab} + O(u^2)$, where $g_{ab} = g_{ab}(0)$ is, by supposition, the flat-space metric). The computation of the curvature (to first order in $u$) yields the following result:

$$K_{abcd} := \lim_{u \to 0}(u^{-1}R_{abcd}(u)) = 2\nabla_{[a}\nabla_{|[c}h_{d]|b]}, \tag{5.7.4}$$

where $\nabla_a$ is the flat-space derivative operator and so possesses the commutative property.

Obviously $K_{abcd}$ has the Riemann tensor symmetries

$$K_{abcd} = K_{[cd][ab]}, \; K_{[abc]d} = 0, \tag{5.7.5}$$

and the Einstein equations (4.6.30) become

$$K_{abc}{}^{b} - \tfrac{1}{2}g_{ac}K_{bd}{}^{bd} = -8\pi\gamma E_{ac} \tag{5.7.6}$$

where $E_{ab}$ is the linearized theory's version of the energy–momentum tensor $T_{ab}$. In the absence of sources $K_{abcd}$ satisfies

$$K_{abc}{}^{b} = 0, \tag{5.7.7}$$

so it coincides with the first-order Weyl tensor $\lim(u^{-1}C_{abcd}(u))$, and can be expressed in the form (*cf.* (4.6.41))

$$K_{abcd} = \phi_{ABCD}\varepsilon_{A'B'}\varepsilon_{C'D'} + \bar{\phi}_{A'B'C'D'}\varepsilon_{AB}\varepsilon_{CD}, \tag{5.7.8}$$

where $\phi_{ABCD} = \lim(u^{-1}\Psi_{ABCD}(u))$ and is totally symmetric. Evidently $K_{abcd}$ satisfies the Bianchi identity

$$\nabla_{[a}K_{bc]de} = 0 \tag{5.7.9}$$

which, in the case (5.7.7), is equivalent (*cf.* (4.10.9)) to

$$\nabla^{AA'}\phi_{ABCD} = 0. \tag{5.7.10}$$

Thus if $\phi_{ABCD}$ is regarded as a massless field, its field equation (5.7.10) corresponds to the Bianchi identity of $K_{abcd}$, while its symmetry is expressed by the symmetries (5.7.5), (5.7.7) – which involve the Einstein field equations. Physically, $\phi_{ABCD}$ is more significant than $h_{ab}$. For $h_{ab}$

is subject to 'gauge transformations' which leave the physical situation unchanged. These are induced by the 'infinitesimal coordinate transformations' and have the form

$$h_{ab} \mapsto h_{ab} - 2\nabla_{(a}\xi_{b)} \quad \textit{for some } \xi_b. \tag{5.7.11}$$

However, $K_{abcd}$ is invariant, and so, consequently, is $\phi_{ABCD}$. We may think of (5.7.10) as the gauge invariant equation for the weak vacuum gravitational field. The tensor version of this, together with the symmetry of $\phi_{ABCD}$, is all of (5.7.5), (5.7.7) and (5.7.9).

In fact (5.7.9), or in the absence of sources (5.7.10), is *sufficient* for the $K_{abcd}$ of (5.7.8) to be derivable locally from some symmetric $h_{ab}$ as in (5.7.4). Moreover, for the empty regions outside sources, the sufficiency of (5.7.10) holds *globally* if and only if a certain set of 10 integrals vanishes. (See Sachs and Bergmann 1958, Trautman 1962 and §6.4.)

Whether sources are present or not, we always have

$$\phi_{ABCD} = \tfrac{1}{4}K_{(ABCD)A'B'C'D'}\varepsilon^{A'B'}\varepsilon^{C'D'},$$

which becomes, via (5.7.4),

$$\phi_{ABCD} = \tfrac{1}{2}\nabla^{A'}_{(A}\nabla^{B'}_{B}h_{CD)A'B'}, \tag{5.7.12}$$

giving the relation between $h_{ab}$ and $\phi_{ABCD}$. When sources are present, with weak-field energy–momentum tensor $E_{ab}$, the generalization of (5.7.10) is (*cf.* (4.10.12))

$$\nabla^{AA'}\phi_{ABCD} = 4\pi\gamma\nabla^{B'}_{(B}E_{CD)B'}{}^{A'}. \tag{5.7.13}$$

The field equation satisfied by $h_{ab}$ can be written

$$\begin{aligned} -16\pi\gamma E_{ab} &= \Box\hat{h}_{ab} - 2\nabla_{(a}\nabla^c\hat{h}_{b)c} + g_{ab}\nabla^c\nabla^d\hat{h}_{cd} \\ &= \Box h_{AB'BA'} - \nabla_{AB'}\nabla^{CD'}h_{CA'BD'} - \nabla_{BA'}\nabla^{CD'}h_{CB'AD'} \end{aligned}$$

where $\Box = \nabla_a\nabla^a$ and

$$\hat{h}_{ab} = h_{ab} - \tfrac{1}{2}g_{ab}h_c{}^c = h_{AB'BA'} = h_{BA'AB'}$$

(*cf.* (3.4.13); and it reduces to

$$\Box\hat{h}_{ab} = -16\pi\gamma E_{ab}$$

when the 'de Donder gauge condition'

$$\nabla^a\hat{h}_{ab} = 0, \quad \text{i.e.} \quad \nabla^{AB'}h_{ab} = 0$$

holds. With this gauge condition we can, in the absence of sources, drop the symmetry brackets around the indices in (5.7.12) (because symmetry in $AD$ and in $BC$ follow from $\nabla^{B'}_{[B}h_{C]A'DB'} = 0$, while symmetry in $AB$ follows, in vacuum, from $\Box h_{ab} = 0$) and, in the presence of sources, we can write

$$\phi_{ABCD} = \tfrac{1}{2}\nabla^{C'}_{A}\nabla^{D'}_{B}h_{cd} - \tfrac{4}{3}\pi\gamma\varepsilon_{A(B}\varepsilon_{C)D}E_p{}^p.$$

As we shall see in a moment the conformal invariance of this weak field gravitational theory in the absence of sources (namely (5.7.10)) is quite transparent in terms of $\phi_{ABCD}$. But it is by no means easy to see in terms of $h_{ab}$.

An important generalization of the flat-space formula (5.7.4) is obtained if we consider perturbations away from some given *non-flat* space–time $\mathscr{M}$, where we suppose, for simplicity, that both $\mathscr{M}$ and the perturbation satisfy Einstein's *vacuum* equations. Then we have a fixed non-zero $\Psi_{ABCD}$ for the background and some variable $\phi_{ABCD}$ representing the perturbation. However, $\phi_{ABCD}$ is not now 'gauge invariant', in the sense that if the $h_{ab}$ from which it is obtained undergoes (5.7.11), then $\phi_{ABCD}$ changes, in general. Roughly speaking, the reason for this is that there is an uncertainty about which point of $\mathscr{M}$ corresponds to which point of the perturbed space. Since $\phi_{ABCD}$ represents a *difference* between the perturbed curvature and $\Psi_{ABCD}$, this uncertainty will affect the resulting value of $\phi_{ABCD}$ whenever $\Psi_{ABCD} \neq 0$. Furthermore, the massless free field equation (5.7.10) does not in general hold. To describe the perturbation we need to involve the potential quantity $h_{ab}$ explicitly. The (vacuum) field equations are now

$$\nabla_a \nabla^a h_{bc} - \nabla_a \nabla_b h_c{}^a - \nabla_a \nabla_c h_b{}^a + \nabla_b \nabla_c h_a{}^a = 0 \qquad (5.7.14)$$

with $h_{ab}$ subject to the gauge freedom (5.7.11), and in place of (5.7.12) and (5.7.10), respectively, we have

$$\phi_{ABCD} = \tfrac{1}{2} \nabla^{A'}_{(A} \nabla^{B'}_{B} h_{CD)A'B'} + \tfrac{1}{4} h_p{}^p \Psi_{ABCD} \qquad (5.7.15)$$

and

$$\nabla^{AA'} \phi_{ABCD} = \tfrac{1}{2} h^{RSA'B'} \nabla_{BB'} \Psi_{RSCD} - \Psi_{RS(BC} \nabla^{B'}_{D)} h^{RSA'}{}_{B'} - \tfrac{1}{2} \Psi_{RS(BC} \nabla^{RB'} h_{D)}{}^{SA'}{}_{B'}.$$

Under (5.7.11), $\phi_{ABCD}$ transforms as

$$\phi_{ABCD} \mapsto \phi_{ABCD} - \xi^{EE'} \nabla_{E'(A} \Psi_{BCD)E} - 2\Psi_{E(ABC} \nabla_{D)E'} \xi^{EE'}.$$

(These relations are adapted from Curtis 1975.)

*Conformal invariance*

To establish the conformal invariance of (5.7.2), it is convenient first to re-express that equation in a form that will also later be found useful. The equation is equivalent to (*cf.* (2.5.24))

$$\nabla_{M'M} \phi_{AB\ldots L} = \nabla_{M'(M} \phi_{A)B\ldots L}$$

and, consequently (*cf.* (3.3.15)), to

$$\nabla_{M'M} \phi_{AB\ldots L} = \nabla_{M'(M} \phi_{AB\ldots L)}. \qquad (5.7.16)$$

Now *choose* $\phi_{AB\ldots L}$ *to be a conformal density of weight* $-1$:

$$\hat{\phi}_{AB\ldots L} = \Omega^{-1}\phi_{AB\ldots L}. \tag{5.7.17}$$

Then, by (5.6.15),

$$\begin{aligned}\Omega\hat{\nabla}_{M'M}\hat{\phi}_{AB\ldots L} &= \Omega\hat{\nabla}_{M'M}(\Omega^{-1}\phi_{AB\ldots L})\\ &= \nabla_{M'M}\phi_{AB\ldots L} - \Upsilon_{M'M}\phi_{AB\ldots L} - \Upsilon_{M'A}\phi_{MB\ldots L} - \cdots - \Upsilon_{M'L}\phi_{AB\ldots M},\end{aligned} \tag{5.7.18}$$

where we have used the particular case $r = -1$ of the useful relation

$$\Omega^{-r}\nabla_a\Omega^r = r\Upsilon_a, \tag{5.7.19}$$

which follows at once from (5.6.14). Now the RHS of (5.7.18) beyond the first term is automatically symmetric in $MAB\ldots L$. Consequently the LHS is symmetric in $MAB\ldots L$ if and only if (5.7.16) holds. But this means that equation (5.7.16) is conformally invariant. For future reference we note another form of this statement which now also follows at once (on transvecting (5.7.18) with $\varepsilon^{AM}$):

$$\hat{\nabla}^{AA'}\hat{\phi}_{AB\ldots L} = \Omega^{-3}\nabla^{AA'}\phi_{AB\ldots L}. \tag{5.7.20}$$

## 5.8 Consistency conditions

There is an algebraic consistency condition for equation (5.7.2) in curved space (Buchdahl 1958, 1962, Plebanski 1965) if $n > 2$, and another for charged fields $\phi_{\ldots}$ in the presence of electromagnetism if $n > 1$ (Fierz and Pauli 1939). To obtain these relations we apply the operator $\nabla^B_{A'}$ to (5.7.2), assuming that $\phi_{AB\ldots L}$ has charge $e$; then, by use of (5.1.44),

$$\begin{aligned}0 &= \nabla^B_{A'}\nabla^{AA'}\phi_{ABC\ldots L} = \nabla^{(B}_{A'}\nabla^{A)A'}\phi_{ABC\ldots L}\\ &= \Box^{AB}\phi_{ABC\ldots L}\\ &= -\mathrm{i}e\varphi^{AB}\phi_{ABC\ldots L} - \mathrm{X}^{ABM}{}_A\phi_{MBC\ldots L} - \mathrm{X}^{ABM}{}_B\phi_{AMC\ldots L}\\ &\quad - \mathrm{X}^{ABM}{}_C\phi_{ABM\ldots L} - \cdots - \mathrm{X}^{ABM}{}_L\phi_{ABC\ldots M}\end{aligned} \tag{5.8.1}$$

The first two terms involving X vanish since $\mathrm{X}^{A(BM)}{}_A = 0$ (*cf.* (4.6.6)). Since also, by (4.6.35), $\mathrm{X}^{(ABM)}{}_C = \Psi^{ABM}{}_C$, the above calculation yields, for $n \geqslant 2$,

$$(n-2)\phi_{ABM(C\ldots K}\Psi_{L)}{}^{ABM} = -\mathrm{i}e\varphi^{AB}\phi_{ABC\ldots L}. \tag{5.8.2}$$

This constitutes an *algebraic* condition which relates the field to the conformal curvature $\Psi_{ABCD}$ if $n > 2$ and to the electromagnetic field $\varphi_{AB}$ if $e \neq 0$ and $n > 1$.

These algebraic conditions render the field equation (5.7.2) unsatisfactory for those situations where the conditions are non-vacuous. Let us enumerate some possibilities. First, if space–time is Minkowskian, and the electromagnetic field or the charge of $\phi_{A\ldots L}$ vanishes, then the field equation (5.7.2) *is* satisfactory, in the sense that there is as much freedom in its solutions as there is freedom in finding complex solutions of the wave equation (or real solutions of Maxwell's source-free equations). This follows from the work of §§5.10, 5.11. Secondly, suppose that the space–time is curved but (locally) *conformally* Minkowskian, i.e., that a conformal rescaling can be found (locally) which reduces the metric to that of $\mathbb{M}$. Then again (still assuming $e\varphi_{AB} = 0$), equation (5.7.2) is satisfactory since, because of its conformal invariance, the solution procedure can be reduced to finding the solution in $\mathbb{M}$. (In fact, $\Psi_{ABCD}$ vanishes in conformally Minkowskian space, so that (5.8.2) becomes vacuous if its RHS vanishes.) Thirdly, suppose that the space–time is not conformally flat, but still $e\varphi_{AB} = 0$. The Weyl conformal spinor now turns out to be non-zero (*cf.* (6.9.23)) and so the consistency condition (5.8.2) must be contended with. In the cases $n = 1, 2$ (neutrino and Maxwell fields) there is again no restriction and the fields turn out to have the same freedom (apart from possible global problems) as they have in $\mathbb{M}$. However, for $n > 2$ the condition (5.8.2) is very restrictive. For example, Bell and Szekeres (1972) show, among other things, that in a vacuum space–time which is 'algebraically general' (i.e. with distinct gravitational PND, *cf.* after (3.5.21), and §§7.3, 8.1) there can be at most *two* linearly independent solutions of (5.7.2) for $n = 4$, and in general the *only* solutions are multiples of $\Psi_{ABCD}$.

This is the situation when we are looking for solutions of (5.7.2) on a *given* space–time $\mathcal{M}$. Of course, the situation is quite different for Einstein's (full) vacuum equations. If we take $\phi_{\ldots}$ in (5.7.2) to be the Weyl spinor $\Psi_{ABCD}$, then the restriction (5.8.2) becomes

$$\Psi_{ABM(C}\Psi_{D)}{}^{ABM} = 0,$$

and this is vacuous, being automatically satisfied by any totally symmetric $\Psi_{ABCD}$; for if see-saws are applied to the three contractions, the expression on the left is seen to equal minus itself.

The consistency relations (5.8.2) in the presence of charge $e$ and electromagnetic field $\varphi_{AB}$ are less interesting, simply because charged massless fields do not occur in nature. Difficulty would be encountered with the electromagnetic interaction when the spin of the field is greater than $\frac{1}{2}$ $(n > 1)$. However, the same situation (existence of algebraic restrictions) also arises with *massive* charged fields (Fierz and Pauli 1939). Moreover similar difficulties occur, in the presence of gravitation (curvature), when the spin is greater than $1 (n > 2)$ (Buchdahl 1958).

*Energy–momentum tensors*

In the massless case, there appears to be a relation between the above-mentioned difficulties and problems with the construction of a meaningful (symmetric, divergence-free) energy–momentum tensor such as $T_{ab}$ in the gravitational case, or of a charge-current vector such as $J_a$ in the electromagnetic case (these being needed for the RHS of the relevant field equation). We have already seen in (5.2.4) how to construct $T_{ab}$ when the source is a spin-1 zero rest-mass field. For the Dirac–Weyl case of spin $\frac{1}{2}$ one has

$$T_{ab} = k(\mathrm{i}\nu_{(A}\nabla_{B)A'}\bar{\nu}_{B'} - \mathrm{i}\bar{\nu}_{(A'}\nabla_{B')A}\nu_B), \tag{5.8.3}$$

while for the *massive* (Dirac) spin $\frac{1}{2}$ field (4.4.66), the energy–momentum tensor is

$$\begin{aligned} T_{ab} = \frac{\mathrm{i}k}{2}(&\phi_A\nabla_b\bar{\phi}_{A'} - \bar{\phi}_{A'}\nabla_b\phi_A + \phi_B\nabla_a\bar{\phi}_{B'} - \bar{\phi}_{B'}\nabla_a\phi_B \\ &- \bar{\chi}_A\nabla_b\chi_{A'} + \chi_{A'}\nabla_b\bar{\chi}_A - \bar{\chi}_B\nabla_a\chi_{B'} + \chi_{B'}\nabla_a\bar{\chi}_B) \end{aligned}$$

$k$ being a real constant. (It should be borne in mind that these fields are not classical fields. This is true of all half-odd-integer-spin fields, since the exclusion principle applies to such fields *cf.* Bjorken and Drell 1964. Thus the expression (5.8.3) should really be applied in the context of quantum field theory. The lack of positive definiteness for $T_{ab}V^aV^b$, with $V^a$ timelike, is related to this.)

These tensors are obviously symmetric and, in the case of (5.8.3), trace-free, $T_a{}^a = 0$, because (5.7.16) implies its symmetry in $AB$ and $A'B'$. The vanishing divergence condition $\nabla^a T_{ab} = 0$ also holds, although the verification of this fact in curved space–time is not quite immediate. It depends on the cancellation of the curvature terms arising from the commutation of derivatives. The result for (5.8.3) may be established from the following identity, which will also be needed in Vol. 2:

$$\begin{aligned} \nabla^A_{A'}\nabla_{B'(A}\xi_{B)} = \nabla_{B'B}\nabla^A_{A'}\xi_A - \tfrac{1}{2}\nabla_{A'B}\nabla^A_{B'}\xi_A \\ + 3\Lambda\varepsilon_{A'B'}\xi_B - \Phi_{ABA'B'}\xi^A. \end{aligned} \tag{5.8.4}$$

(This follows from (4.9.7), by using the identity (2.5.23) in the form $S^A{}_{AB} = S^A{}_{BA} - S_B{}^A{}_A$.) When $\xi_A = \nu_A$ the two differentiated terms on the right vanish and the symmetrization around $AB$ on the left may be omitted. When $\nabla^a$ is applied to (5.8.3) a term involving (5.8.4) appears, which cancels with the conjugate term. The result then follows easily.

The case of spin 0 deserves certain special considerations which are best delayed until Vol. 2, §6.8. Here we simply state two alternatives for the energy–momentum tensor, namely

$$T_{ab} = \tfrac{1}{2}k\nabla_{AB'}\phi\nabla_{BA'}\phi$$

for the equation $\Box\phi = 0$ ($\phi$ real) and

$$T_{ab} = \tfrac{1}{6}k\{2\nabla_{A(A'}\phi\nabla_{B')B}\phi - \phi\nabla_{A(A'}\nabla_{B')B}\phi + \phi^2\Phi_{ABA'B'}\}$$

for the real conformally invariant equation $(\Box + \frac{1}{6}R)\phi = 0$ (*cf.* (6.8.30)–(6.8.37)). (See Newman and Penrose 1968.)

However, for spins 3/2, 2, ..., there is *no* expression for $T_{ab}$ which has the required properties of symmetry and vanishing divergence, and which depends quadratically on the local field quantity $\phi_{A...L}$. This is not hard to see by examining the various possible terms quadratic or bilinear in $\phi_{A...L}$ and $\bar{\phi}_{A'...L'}$ and their derivatives. In effect, $\phi_{A...L}$ has an excess of indices which, as it turns out, cannot be removed by contraction. Taking derivatives of $\phi_{A...L}$ does not help this difficulty. What one would need to do, essentially, is to *integrate* $\phi_{A...L}$ in order to construct $T_{ab}$. Indeed, expressions for $T_{ab}$ constructed from potentials for $\phi_{A...L}$ do exist. But these are not satisfactory for general relativity because the *local* values of $T_{ab}$ – not merely the integrated total energy – are needed in an essential way in Einstein's field equations. These local values would be 'gauge-dependent' quantities if potentials are used, and thus not physically meaningful. In the case of gravity itself, no local energy–momentum tensor occurs. But it is not needed, since gravity does not contribute to the right-hand side of Einstein's equations. Gravitational energy emerges, instead, as a non-local quantity (*cf.* §§9.9, 9.10).

Although of limited physical interest, in the case of zero rest-mass fields, it is worth noting that for such fields it becomes impossible to define a locally meaningful charge-current vector (on the pattern of (5.10.16), (5.10.21)) at just the same spin value at which difficulties with the consistency relations (5.8.2) are encountered. If the neutrino field were charged, its charge-current vector would be proportional to $\nu_A\bar{\nu}_{A'}$. But for higher spin, local ('gauge invariant') expressions are not possible.

### *Consistent higher spin systems*

Under certain circumstances, consistent massless field equations *can* be given for higher-spin fields in interaction with gravitational or electromagnetic fields. However, these higher-spin fields can no longer be described simply by a gauge invariant spinor subject to some field equations like (5.7.2). For example, we have seen how to construct spin 2 fields on a background space–time $\mathcal{M}$ which satisfies Einstein's vacuum equations, by considering perturbations of the space–time metric which still satisfy the vacuum equations. Thus, in place of the gauge invariant field quantity $\phi_{ABCD}$, we describe the field by $h_{ab} \in \mathfrak{T}_{(ab)}$, subject to (5.7.14) as field equation, where two such quantities $h_{ab}$ are considered to be *equivalent* if and only if they are related by a transformation of the form (5.7.11); then we *define* $\phi_{ABCD}$ by (5.7.15). But if we consider this to be the description of a spin-2 massless field on a given vacuum space–time background, we still have the difficulty that a gauge invariant local energy–momentum tensor does not appear to exist.

The situation is perhaps a little more satisfactory for the case of a spin-3/2 massless field. One of the byproducts of supersymmetry theory (*cf.* Freedman, van Nieuwenhuizen and Ferrara 1976, Deser and Zumino 1976) is a coupled system of equations for a spin-3/2 massless field and the gravitational field. The spin-3/2 field can be given by a potential

$$\chi_{ABC'} \in \mathfrak{S}_{(AB)C'}$$

(which, however, is normally described, in effect, by the 'Majorana 4-spinor-tensor' $(\chi_{Ab}, \bar{\chi}_{A'b})$, the symmetry condition on $\chi_{ABC'}$ being more complicatedly expressed than here) subject to

$$\nabla^{AA'}\chi_{ABC'} = 0.$$

In $\mathbb{M}$ this would imply that the 'field'

$$\phi_{ABC} = \nabla_A^{C'}\chi_{BCC'}$$

satisfies the massless field equation (5.7.2), but in curved space–time there are correction terms involving $\chi_{ABC'}$ and the curvature. The energy–momentum tensor is, modulo a divergence, proportional to

$$\phi_{ABC}\bar{\chi}_{A'B'}{}^{C} - \chi_{AB}{}^{C'}\bar{\phi}_{A'B'C'}$$

plus quartic terms in $\chi$... due to a *torsion* (*cf.* §4.2), proportional to

$$\chi^{C}{}_{AA'}\bar{\chi}^{C'}{}_{B'B} - \chi^{C}{}_{BB'}\bar{\chi}^{C'}{}_{A'A}.$$

(Full supergravity has gauge transformations involving the gravitational and spin-3/2 fields: the metric is altered by a term proportional to $\chi_{AB(A'}\bar{\xi}_{B')} + \xi_{(A}\bar{\chi}_{B)A'B'}$ and the spin-3/2 field by $\nabla_{AC'}\xi_B$, where the 'spin-$\frac{1}{2}$' gauge field satisfies $\nabla^A_{A'}\xi_A = 0$. The spin-3/2 field also possesses certain anti-commutativity properties, and these are needed for the consistency of the above equations. The form of the equation on $\chi_{ABC'}$ that we have given arises when the gauge transformations are restricted so that the symmetry in $AB$ is maintained. See also Aichelburg and Urbantke 1981.)

Other particular systems of consistent equations seem also to be possible (*cf.* Dowker and Dowker 1966, Buchdahl 1962, 1982).

## 5.9 Conformal invariance of various field quantities

We recall that the energy tensors of the Maxwell and Dirac–Weyl fields are trace-free. That, as might be expected, is a property closely related to conformal invariance. Consider, generally, a trace-free symmetric tensor

$$T_{ab} = T_{ABA'B'} = T_{(AB)(A'B')}$$

whose divergence vanishes:

$$\nabla^a T_{ab} = 0. \tag{5.9.1}$$

Because of the quadratic nature of $T_{ab}$ in the field quantities, we might expect $T_{ab}$ to be a conformal density of weight $-2$:

$$\hat{T}_{ab} = \Omega^{-2} T_{ab}. \tag{5.9.2}$$

With this hypothesis, (5.9.1) is indeed a conformally invariant equation. For, by (5.6.15),

$$\begin{aligned} \Omega^2 \hat{\nabla}^a \hat{T}_{ab} &= \Omega^2 \hat{\nabla}^a (\Omega^{-2} T_{ab}) \\ &= \nabla^a T_{ab} - 2\Upsilon^a T_{ab} - \Upsilon_A{}^{A'} T^A{}_{BA'B'} \\ &\quad - \Upsilon_B{}^{A'} T_A{}^A{}_{A'B'} - \Upsilon^A{}_{A'} T_{AB}{}^{A'}{}_{B'} - \Upsilon^A{}_{B'} T_{ABA'}{}^{A'} \\ &= 0 - 2\Upsilon^a T_{ab} + \Upsilon^a T_{ab} - 0 + \Upsilon^a T_{ab} - 0 = 0. \end{aligned}$$

A similar (but slightly shorter) calculation shows that the vanishing divergence condition on a charge-current vector, $\nabla^a J_a = 0$, is also conformally invariant if

$$\hat{J}_a = \Omega^{-2} J_a. \tag{5.9.3}$$

This may be inferred alternatively from other considerations. For example, in a coordinate basis there is the classical expression (*cf.* Schrödinger 1950)

$$\nabla_a J^a = (-g)^{-\frac{1}{2}} \frac{\partial}{\partial x^{\mathbf{a}}} \{(-g)^{\frac{1}{2}} J^{\mathbf{a}}\}, \tag{5.9.4}$$

where $g = \det(g_{\mathbf{ab}})$. Keeping the coordinates fixed when applying a rescaling (5.6.1), we have $\hat{g} = \Omega^8 g$, so that $\{\ \dots\ \}$ has conformal weight zero if $J^{\mathbf{a}}$ has conformal weight $-4$ (which agrees with (5.9.3)). The whole expression is therefore a conformal density (of weight $-4$), showing that its vanishing is a conformally invariant property.

There is also a coordinate-independent way of proving the same thing. Observe that $J^a$ has a dual 3-form

$$^\dagger\boldsymbol{J} = {}^\dagger J_{i_1 i_2 i_3} = e_{i_1 i_2 i_3 a} J^a, \tag{5.9.5}$$

using the notation of (3.4.29) and (4.3.10), specialized to 4 dimensions as in §4.13 (Latin indices!). Thus,

$$\mathrm{d}^{\dagger}\boldsymbol{J} = \nabla_{[i_1}{}^{\dagger}J_{i_2i_3i_4]} = -\tfrac{1}{4}(\nabla_a J^a)e_{i_1i_2i_3i_4}, \tag{5.9.6}$$

using (4.3.14) and (3.4.32). This shows that the exterior derivative of $^{\dagger}\boldsymbol{J}$ is, in effect, simply the divergence of $J^a$. Since, by (3.3.31),

$$\hat{e}_{abcd} = \Omega^4 e_{abcd} \tag{5.9.7}$$

we have, by reference to (5.9.3) and (5.9.5), $^{\dagger}\hat{\boldsymbol{J}} = {}^{\dagger}\boldsymbol{J}$. Hence, $\mathrm{d}^{\dagger}\hat{\boldsymbol{J}} = \mathrm{d}^{\dagger}\boldsymbol{J}$, since exterior derivative does not depend on the choice of covariant derivative. Thus

$$\hat{\nabla}_a \hat{J}^a = \Omega^{-4}\nabla_a J^a,$$

as before.

The expressions $\nabla^a T_{ab}$ and $\nabla^a J_a$ are part of a more general system of conformally invariant expressions that will be considered in Vol. 2 (see (6.7.33)).

Next we establish the conformal invariance of Maxwell's equations. This can be done in many ways. The source-free equations are (*cf.* (5.1.52)) $\nabla^{AA'}\varphi_{AB} = 0$, and we have already seen (*cf.* (5.7.17)) that these are conformally invariant if

$$\hat{\varphi}_{AB} = \Omega^{-1}\varphi_{AB}, \quad \text{i.e., } \hat{\varphi}^{AB} = \Omega^{-3}\varphi^{AB}, \tag{5.9.8}$$

whence, via (5.1.39),

$$\hat{F}_{ab} = F_{ab}, \quad \hat{F}^{ab} = \Omega^{-4}F^{ab}. \tag{5.9.9}$$

When a source term is included, the equations are (*cf.* (5.1.52))

$$\nabla^{AA'}\varphi_{AB} = 2\pi J_B^{A'} \tag{5.9.10}$$

By (5.7.20) and (5.9.3) we see that each side is a conformal density of weight $-3$, and invariance is established.

We may also infer the conformal invariance of Maxwell's equations by using the formalism of §5.1. The definition of $F_{ab}$ via (5.1.13) and (5.1.37) is unaffected by conformal rescaling, and this is consistent with (5.9.9). Thus the first half of Maxwell's equations, (5.1.36), is evidently unaffected by conformal rescaling. In the proof of the second half of Maxwell's equations, (5.1.38), the formalism has nothing to add and we are thrown back essentially on our previous argument. We note that the choice of conformal weight for $F_{ab}$ which naturally arises in the formalism of charged fields is the same as that required for conformal invariance of the Maxwell equations. But this is not a foregone conclusion, since the invariance of Maxwell's equations does not necessarily imply that the link of $F_{ab}$ to the charged fields must be invariant. In the case of (linearized) gravitation the corresponding uniformity does not hold.

We can also use differential forms to re-express Maxwell's equations, and so to re-establish their conformal invariance. If we set

$$\boldsymbol{\Phi} := \Phi_{i_1}, \boldsymbol{F} := F_{i_1 i_2}, {}^*\boldsymbol{F} := {}^*F_{i_1 i_2}, \tag{5.9.11}$$

then, by (4.3.14) and (5.1.37),

$$\mathrm{d}\boldsymbol{\Phi} = \nabla_{[i_1}\Phi_{i_2]} = \tfrac{1}{2}F_{i_1 i_2} = \tfrac{1}{2}\boldsymbol{F},$$

whence

$$\boldsymbol{F} = 2\mathrm{d}\boldsymbol{\Phi}. \tag{5.9.12}$$

By successive use of (4.3.14), (3.4.27), (5.1.38), (5.9.5), we find

$$\mathrm{d}\,{}^*\boldsymbol{F} = \nabla_{[i_1}{}^*F_{i_2 i_3]} = \tfrac{1}{3}e_{i_1 i_2 i_3 a}\nabla_b \boldsymbol{F}^{ba} = \frac{4\pi}{3}e_{i_1 i_2 i_3 a}J^a = \frac{4\pi}{3}{}^\dagger\boldsymbol{J}$$

The Maxwell equation (5.1.38) is thus seen to be equivalent to the second of the following formulae:

$$\mathrm{d}\boldsymbol{F} = 0, \quad \mathrm{d}\,{}^*F = \frac{4\pi}{3}{}^\dagger\boldsymbol{J}, \tag{5.9.13}$$

while the first is directly equivalent to the Maxwell equation (5.1.36). As we have seen, (5.1.13) suggests $\hat{\boldsymbol{\Phi}} = \boldsymbol{\Phi}$, whence, from (5.9.12), $\hat{\boldsymbol{F}} = \boldsymbol{F}$ and consequently ${}^*\hat{\boldsymbol{F}} = {}^*\boldsymbol{F}$; then, taking ${}^\dagger\hat{\boldsymbol{J}} = {}^\dagger\boldsymbol{J}$ (which corresponds to (5.9.3)), the above equations (5.9.13) (and also (5.9.12)) are all equations between terms of zero weight, and are thus conformally invariant. (Note that, despite the appearance of (5.9.13), Maxwell's equations are not completely metric-independent, since the relation between $\boldsymbol{F}$ and ${}^*\boldsymbol{F}$ requires a conformal metric.)

It is interesting to observe that the Lorenz gauge condition,

$$\nabla^a \Phi_a = 0$$

(*cf.* (5.1.47), having the same form as (5.1.54) (which is a consequence of Maxwell's equations), is conformally invariant if $\Phi_a$ (like $J_a$) is assigned a weight $-2$. But this is *not* the weight which makes (5.1.37) conformally invariant. Hence we must regard Maxwell's theory *with* the Lorenz gauge as *not* being a conformally invariant theory.

## 5.10 Exact sets of fields

In this section we shall show how to set up a general framework for the discussion of sets of interacting fields in flat or curved background space–times, or in general relativity itself. The gravitational field is then described by the spinor $\Psi_{ABCD}$ which, for the purposes of the general discussion, can be treated on a similar footing to the other fields under consideration. The key concept will be that of an exact set of interacting fields (Penrose 1963, 1966b). Once one has an exact set, one is ensured that the fields will propagate correctly through space–time; in the case of general relativity, they propagate correctly and simultaneously generate the structure of the space–time. The appropriate form of initial value problem, as discussed in various aspects in §§5.11, 5.12, is based on characteristic (i.e., null) initial hypersurfaces. For an exact set, the initial data will be complete and irredundant (without constraints) so that the counting of degrees of freedom becomes a simple matter. The simplifications and unifications that are obtained are a direct result of the consistent use of two-component spinors. A corresponding tensor treatment would, on the other hand, be exceedingly complicated.

Let us consider a system of fields

$$\psi_{AB\ldots G}^{P'Q'\ldots U'},\ \phi_{AB\ldots H}^{P'Q'\ldots S'},\ldots,\ \chi_{AB\ldots E}^{P'Q'\ldots T'}, \tag{5.10.1}$$

where each spinor is symmetric in all its unprimed indices and symmetric in all its primed indices. Either, or both, sets of indices may be vacuous. (The reason for writing all the primed indices here in contravariant form and the unprimed ones in covariant form is simply notational convenience for what follows.) We have seen in §3.3. that any spinor can be represented in terms of $\varepsilon$s and spinors which, like (5.10.1), are symmetric. Any finite set of interacting locally Lorentz covariant (finite component) fields can therefore be represented as a set (5.10.1).

Suppose the fields (5.10.1) are subject to a set of covariant differential equations involving the operator $\nabla_a$. Then (5.10.1) will be called an *exact set* of fields if, at each point $P$, the following two conditions are satisfied:

($a$) all the symmetrized derivatives

$$\ldots,\nabla_{(J}^{(V'}\ldots\nabla_{L}^{X'}\psi_{A\ldots H)}^{P'\ldots U')},\ldots,\nabla_{(J}^{(V'}\ldots\nabla_{M}^{Y'}\chi_{A\ldots E)}^{P'\ldots T')},\ldots \tag{5.10.2}$$

(including the 'zero times' differentiated fields (5.10.1)) are *independent* (i.e. they can take independently arbitrary* values at $P$) and

($b$) all the unsymmetrized derivatives

$$\ldots,\nabla_{K}^{W'}\ldots\nabla_{N}^{Z'}\psi_{A\ldots H}^{P'\ldots U'},\ldots,\nabla_{J}^{V'}\ldots\nabla_{N}^{Z'}\chi_{A\ldots E}^{P'\ldots T'},\ldots \tag{5.10.3}$$

* We are ignoring such questions as limits on the growth rates of the quantities in the list (5.10.2) (or (5.10.3)). In effect, 'arbitrary' is to mean that members of any finite subset of (5.10.2) can be chosen arbitrarily. Our considerations are here essentially algebraic, and a more complete discussion would require the appropriate notion of Sobelov space.

are *determined*, at $P$, by the values of the symmetrized derivatives (5.10.2) at $P$, by virtue of the differential relations satisfied by the fields.

In effect, when we refer to the spinors (5.10.2) as being independent, we mean that there are no *algebraic* (spinorial) relations connecting them and their complex conjugates. Likewise, the spinors (5.10.3) are to be determined from the spinors (5.10.2) and their complex conjugates by *algebraic* (spinorial) relations.

An exact set (5.10.1) will be called *invariant* if the expressions for (5.10.3) in terms of (5.10.2) are the same whichever point $P$ is chosen and are locally Lorentz covariant (i.e. (5.10.3) are expressed as definite spinorial combinations of $\varepsilon$s and the spinors (5.10.2) involving no extraneous quantities other than scalar constants).

### *Free massless fields*

As a simple example of an invariant exact set, consider a massless free field of spin $\frac{1}{2}n > 0$ in $\mathbb{M}$. Such a field is represented by a single symmetric spinor $\phi_{AB\ldots L}$, and (*cf*. (5.7.2)) is governed by the field equation

$$\nabla^{AA'}\phi_{AB\ldots L} = 0. \tag{5.10.4}$$

We saw in (5.7.16) that this equation is equivalent to the symmetry condition $\nabla^{M'}_{M}\phi_{A\ldots L} = \nabla^{M'}_{(M}\phi_{A\ldots L)}$. Next consider $\nabla^{N'}_{N}\nabla^{M'}_{M}\phi_{A\ldots L}$. The operators $\nabla^{N'}_{N}$, $\nabla^{M'}_{M}$ here comute, so that we have symmetry in $NA\ldots L$ as well as in $MA\ldots L$. Thus $\nabla^{N'}_{N}\nabla^{M'}_{M}\phi_{A\ldots L} = \nabla^{N'}_{(N}\nabla^{M'}_{M}\phi_{A\ldots L)}$. Symmetry in $N'M'$ then also follows from the commuting of the operators. Repeating the argument with higher derivatives, we get, generally,

$$\nabla^{M'}_{M}\nabla^{N'}_{N}\ldots\nabla^{Q'}_{Q}\phi_{A\ldots L} = \nabla^{(M'}_{(M}\nabla^{N'}_{N}\ldots\nabla^{Q')}_{Q}\phi_{A\ldots L)}. \tag{5.10.5}$$

Thus, condition (*b*) for exactness is (trivially) satisfied, as is the condition for invariance. Since (5.10.4) is linear and derivatives commute, *all* algebraic relations satisfied by the derivatives (5.10.5) at $P$ must be linear. Such relations would necessarily emerge as linear operations on the *indices* of (5.10.5) because of the invariance of (5.10.4). But (5.10.5) expresses *complete* symmetry, so no further relations can in fact emerge. Thus condition (*a*) for exactness is also satisfied, so that $\phi_{AB\ldots L}$ forms, by itself, an invariant exact set.

This example covers the Maxwell field ($n = 2$), the Dirac–Weyl neutrino field ($n = 1$) and the linearized Einstein gravitational field ($n = 4$) in $\mathbb{M}$ (*cf*. §5.7). The case $n = 0$ is also essentially the same, but in place of (5.10.4) we must have the second-order wave (D'Alembert) equation in $\mathbb{M}$

$$\Box\phi := \nabla^{AA'}\nabla_{AA'}\phi = 0. \tag{5.10.6}$$

We can restate this equation as

$$\nabla^{A'}_{[A}\nabla^{B'}_{B]}\phi = 0, \tag{5.10.7}$$

the symmetry in $A'B'$ being a consequence (because the derivatives commute) of that imposed on $AB$. Thus $\nabla^{A'}_{A}\nabla^{B'}_{B}\phi$ is symmetric in $AB$ and $A'B'$: $\nabla^{A'}_{A}\nabla^{B'}_{B}\phi = \nabla^{(A'}_{(A}\nabla^{B')}_{B)}\phi$. Then, by the same argument as above,

$$\nabla^{A'}_{A}\nabla^{B'}_{B'}\ldots\nabla^{H'}_{H}\phi = \nabla^{(A'}_{(A}\nabla^{B'}_{B}\ldots\nabla^{H')}_{H)}\phi \tag{5.10.8}$$

and $\phi$ forms an invariant exact set.

A further (rather trivial) generalization of the massless field equation is the following. Let $\theta^{P'\ldots S'}_{A\ldots E}$ be symmetric and subject to the simultaneous conditions in $\mathbb{M}$

$$\nabla^{AA'}\theta^{P'\ldots S'}_{AB\ldots E} = 0, \quad \nabla_{PP'}\theta^{P'Q'\ldots S'}_{A\ldots E} = 0. \tag{5.10.9}$$

Then arguments similar to the above show that

$$\nabla^{T'}_{F}\ldots\nabla^{V'}_{H}\theta^{P'\ldots S'}_{A\ldots E} = \nabla^{(T'}_{(F}\ldots\nabla^{V'}_{H}\theta^{P'\ldots S')}_{A\ldots E)},$$

and $\theta^{\cdots}_{\cdots}$ forms an invariant exact set. This gives us nothing essentially new, however, because we have $\nabla^{T'}_{F}\theta^{P'Q'\ldots S'}_{AB\ldots E} = \nabla^{P'}_{A}\theta^{T'Q'\ldots S'}_{FB\ldots E}$ which expresses a vanishing curl, implying (at least locally) that $\theta^{\cdots}_{\cdots}$ has the form

$$\theta^{P'Q'\ldots S'}_{AB\ldots E} = \nabla^{P'}_{A}\chi^{Q'\ldots S'}_{B\ldots E}.$$

By the symmetry of $\theta^{\cdots}_{\cdots}$, $\chi^{\cdots}_{\cdots}$ also satisfies (5.10.9). Repeating the argument, until one or the other set of indices is exhausted, we see that

(5.10.10) PROPOSITION

*If* (5.10.9) *holds, then* $\theta^{P'\ldots S'}_{A\ldots E} \in \mathfrak{S}^{(P'\ldots S')}_{(A\ldots E)}$ *is an* $r$th *derivative of some massless free field.*

### *Electromagnetic sources*

Consider now a Maxwell field with sources. In place of (5.10.4) (with $n = 2$) we have (*cf.* (5.9.10))

$$\nabla^{AA'}\varphi_{AB} = 2\pi J^{A'}_{B}, \tag{5.10.11}$$

where $J_{AA'} = \bar{J}_{AA'}$ represents the given charge–current vector subject to the divergence condition

$$\nabla^{AA'}J_{AA'} = 0. \tag{5.10.12}$$

Instead of (5.10.5) we now get

$$\nabla^{C'}_{C}\varphi_{AB} = \nabla^{C'}_{(C}\varphi_{AB)} - \frac{4\pi}{3}\varepsilon_{C(A}J^{C'}_{B)}, \tag{5.10.13}$$

$$\nabla_C^{C'}\nabla_D^{D'}\varphi_{AB} = \nabla_{(C}^{(C'}\nabla_D^{D')}\varphi_{AB)} - \frac{4\pi}{3}\varepsilon_{CD}\varepsilon^{C'D'}\nabla_{(A}^{E'}J_{B)E'}$$
$$+\frac{5\pi}{3}\nabla_{(C}^{(C'}\{\varepsilon_{D)(A}J_{B)}^{D')}\} - \frac{\pi}{3}\nabla_{(A}^{(C'}\{\varepsilon_{B)(C}J_{D)}^{D')}\}, \tag{5.10.14}$$

and so on for higher derivatives: the unsymmetrized $r$th derivative of $\varphi_{AB}$ differs from the symmetrized $r$th derivative by an expression linear in the $(r-1)$th derivative of $J_A^{A'}$. This, and the consequent fact that $\varphi_{AB}$ forms an exact set, follows by the same argument as in the source-free case, except for the essential modification entailed when we replace (5.10.4) by (5.10.11); i.e., when a pair of indices is interchanged, a term involving $J_A^{A'}$ may be introduced.

However, it is clear that $\varphi_{AB}$ by itself does not form an *invariant* exact set here, since the extraneous quantity $J_A^{A'}$ appears in (5.10.13), (5.10.14). On the other hand, we may admit $J_A^{A'}$ as a variable field; but then $\varphi_{AB}$ and $J_A^{A'}$ together do not, as things stand, form an exact set. For (5.10.12) restricts only the part of $\nabla_A^{A'}J_B^{B'}$ which is skew both in $A'B'$ and $AB$. The part which is symmetric in $AB$ and skew in $A'B'$, for example, will remain undetermined. Further conditions would have to be imposed on $J_A^{A'}$ to give an exact set. This could be done in many ways, but of most interest are the cases when $J_A^{A'}$ is given by the charge–current vector of a physical field (or fields), say a Dirac field or a Schrödinger–Klein–Gordon field.

Consider, first, the Dirac case. In the two-component spinor form (4.4.66), the Dirac field is represented as a pair of spinors $\psi_A$, $\chi^{A'}$ subject to

$$\nabla^{AA'}\psi_A = \mu\chi^{A'}, \quad \nabla_{AA'}\chi^{A'} = -\mu\psi_A, \tag{5.10.15}$$

where $\mu = m/\hbar\sqrt{2}$, $m$ being the mass and $\hbar$ Planck's constant divided by $2\pi$. In the absence of electromagnetism we can assume that the $\nabla$s commute. It is then not hard to verify that $\psi_A$ and $\chi^{A'}$ together form an exact set. Moreover, if we introduce electromagnetism via (5.1.1), each of $\psi_A$, $\chi^{A'}$ having the same charge $e$, and define the Dirac *charge current vector* by

$$J_A^{A'} = \frac{q}{2\pi}(\psi_A\bar{\psi}^{A'} + \bar{\chi}_A\chi^{A'}), \tag{5.10.16}$$

where $q$ is a simple positive numerical multiple of $e$, whose exact value depends on one's conventions for normalizing the Dirac wave function (e.g. one natural choice would yield $q = 2\pi e$), then

$$\nabla^{AA'}\varphi_{AB} = q(\psi_B\bar{\psi}^{A'} + \bar{\chi}_B\chi^{A'}), \tag{5.10.17}$$

and $\psi_A$, $\chi^{A'}$ and $\varphi_{AB}$ together form an invariant exact set. This may be demonstrated in essentially the same way as for the first case, except that the interchanging of derivatives may give rise to extra terms involving $\varphi_{AB}$, $\bar{\varphi}^{A'B'}$ (*cf*. (5.1.43)). For example,

$$\nabla_B^{A'}\psi_A = \nabla_{(B}^{A'}\psi_{A)} - \tfrac{1}{2}\mu\varepsilon_{BA}\chi^{P'}, \tag{5.10.18}$$

$$\begin{aligned}\nabla_B^{B'}\nabla_C^{C'}\psi_A = {} & \nabla_{(B}^{(B'}\nabla_C^{C')}\psi_{A)} - \tfrac{2}{3}\mathrm{i}e\bar{\varphi}^{B'C'}\varepsilon_{B(C}\psi_{A)} + \\ & + \mathrm{i}e\varepsilon^{B'C'}\{\tfrac{1}{2}\psi_B\varphi_{CA} - \varphi_{B(C}\psi_{A)}\} + \\ & + \tfrac{1}{2}\mu^2\varepsilon^{B'C'}\varepsilon_{BC}\psi_A + \tfrac{2}{3}\mu\varepsilon_{A(C}\nabla_{B)}{}^{(B'}\chi^{C')}. \end{aligned} \tag{5.10.19}$$

The case of a Schrödinger–Klein–Gordon field $\theta$ is similar, but now in place of (5.10.15), (5.10.17) we have (*cf.* (5.10.6))

$$(\square + 2\mu^2)\theta = 0, \tag{5.10.20}$$

the charge current vector being proportional to $\mathrm{i}\theta\nabla_a\bar{\theta} - \mathrm{i}\bar{\theta}\nabla_a\theta$, so

$$\nabla^{AA'}\varphi_{AB} = q(\mathrm{i}\theta\nabla_B^{A'}\bar{\theta} - \mathrm{i}\bar{\theta}\nabla_B^{A'}\theta). \tag{5.10.21}$$

Again, $\theta$, $\varphi_{AB}$ form an invariant exact set.

There is an alternative method of dealing with charged fields which involves introducing the electromagnetic potential $\Phi_A^{A'}$ explicitly and treating it as a new field. In this approach, $\nabla_A^{A'}$ stands for the operator denoted by $\partial_A^{A'}$ in (5.1.1) and in (5.1.14), i.e., it acts on charged fields as though they were uncharged, and commutes with itself. The electromagnetic interaction is expressed through $\nabla_A^{A'} - \mathrm{i}e\Phi_A^{A'}$. Thus $\Phi_A^{A'}$ occurs explicitly in (5.10.15) and in (5.10.20) as a new field, with $\nabla_A^{A'}$ replaced by $\nabla_A^{A'} - \mathrm{i}e\Phi_A^{A'}$. In order to get an exact set we must impose a restriction on $\Phi_A^{A'}$, such as the Lorenz condition: $\nabla_{A'}^{A}\Phi_A^{A'} = 0$. Then (by (5.1.49)) we have $\varphi_{AB} = \nabla_{AA'}\Phi_B^{A'}$. We also have $\bar{\Phi}_{AA'} = \Phi_{AA'}$ (*cf.* (5.1.18)), whence $\bar{\varphi}^{A'B'} = -\nabla^{AA'}\Phi_A^{B'}$. Using these equations, it is not hard to show that $\psi_A$, $\chi^{A'}$, $\varphi_{AB}$, $\Phi_A^{A'}$, and $\theta$, $\varphi_{AB}$, $\Phi_A^{A'}$ each form an invariant exact set.

This alternative approach is perhaps conceptually a little simpler than the one using (5.10.15) and the consequent non-commutativity of the $\nabla$s, and it is useful in some contexts. However, it is more in keeping with the philosophy being adopted here *not* to introduce gauge dependent quantities, such as $\Phi_a$, explicitly into the formalism. In the case of electromagnetism the gauge *independent* approach in fact involves somewhat simpler formulae (e.g., (5.10.19)) than does the approach explicitly involving $\Phi_a$. Moreover the theories which operate in *curved* space–time can apparently be treated according to the present formalism *only* by virtue of the existence of a gauge-independent (i.e., coordinate-independent) method.

Consider, first, the case of a set of fields in a *given* Riemannian background space–time. The curvature quantities $\Psi_{ABCD}$, $\Phi_{ABC'D'}$ and $\Lambda$

(see §4.6) and their derivatives are then given at each event $O$. These quantities enter into the commutation relations for the $\nabla$s (*cf.* (4.9.7), (4.6.34)). Certain flat space–time exact sets can be transcribed for curved space–time by simply adding the appropriate terms (*cf.* (4.9.13), (4.9.14)) to deal with interchanges of indices of $\nabla$s, this resulting in an exact set. For example, in the case of a Maxwell field, in place of (5.10.5) we have

$$\nabla_C^{C'}\nabla_D^{D'}\varphi_{AB} = \nabla_{(C}^{(C'}\nabla_D^{D')}\varphi_{AB)} + \varepsilon^{C'D'}(\tfrac{1}{2}\Psi_{DAB}{}^X\varphi_{CX} - \tfrac{3}{2}\Psi_{C(AB}{}^X\varphi_{D)X} + 3\Lambda\varepsilon_{C(D}\varphi_{AB)}) - \tfrac{3}{2}\varepsilon_{C(D}\Phi_A{}^{XC'D'}\varphi_{B)X}, \tag{5.10.22}$$

etc. and $\varphi_{AB}$ gives an exact set. It is clearly not an invariant exact set, however. Only in the case of de Sitter (or Minkowski) space ($\Lambda =$ given const., $\Psi_{ABCD} = 0$, $\Phi_{AB}^{A'B'} = 0$) does an invariant exact set in fact arise. The Maxwell–Dirac equations can also be transcribed in the same way for curved space–time and an exact set is obtained. On the other hand, the zero rest-mass equation (5.7.2) for $n \geqslant 3$ does *not*, as it stands, lead to an exact set for $\phi_{AB\ldots L}$ for a given space–time which is *not conformally flat*, because of the consistency relation (5.8.2), which shows that the $\phi_{AB\ldots L}$ are not independent unless $\Psi_{ABCD} = 0$.

*Gravitation*

For general relativity proper, the situation is somewhat different. Here, curvature quantities are to be considered as field variables. The degrees of freedom of the gravitational field are to be described by the conformal spinor $\Psi_{ABCD}$. The quantities $\Phi_{AB}^{A'B'}$ and $\Lambda$ are defined directly in terms of the remaining fields – and possibly a cosmological constant $\lambda$ – through their energy–momentum tensor $T_{ab}$ (*cf.* (4.6.30)):

$$4\pi\gamma T_{AA'BB'} = \Phi_{ABA'B'} + (3\Lambda - \tfrac{1}{2}\lambda)\varepsilon_{AB}\varepsilon_{A'B'}. \tag{5.10.23}$$

In particular, the *Einstein–Maxwell equations* (with cosmological constant $\lambda$) are given by (*cf.* (5.2.6)):

$$\Phi_{AB}^{A'B'} = 2\gamma\varphi_{AB}\bar{\varphi}^{A'B'}, \quad \Lambda = \tfrac{1}{6}\lambda, \tag{5.10.24}$$

and the Bianchi identity becomes (*cf.* (5.2.7)):

$$\nabla^{AA'}\Psi_{ABCD} = -2\gamma\bar{\varphi}^{A'B'}\nabla_{BB'}\varphi_{CD}. \tag{5.10.25}$$

We shall see shortly that, together with Maxwell's source-free equations (5.1.57) and the commutator equations ((4.9.13), (4.9.14)), the relations (5.10.24) and (5.10.25) lead to the invariant-exact-set conditions holding for $\Psi_{ABCD}$, $\varphi_{AB}$. There are also many other sources for the gravitational field which, with $\Psi_{ABCD}$, constitute an invariant exact set.

Let us first examine the *Einstein empty-space equations* (with or without cosmological term). We shall show that they imply that $\Psi_{ABCD}$ alone constitutes an invariant exact set. The relevant equations are the vacuum Bianchi identity (*cf.* (4.10.9)):

$$\nabla^{AA'}\Psi_{ABCD}=0 \quad \text{or} \quad \varepsilon^{EA}\nabla^{P'}_{E}\Psi_{ABCD}=0, \tag{5.10.26}$$

and the Ricci identities generated by

$$\Delta_{ab}\kappa^{C}=\varepsilon_{A'B'}\Psi_{ABE}{}^{C}\kappa^{E}-\tfrac{1}{3}\lambda\varepsilon_{A'B'}\kappa_{(A}\varepsilon_{B)}{}^{C} \tag{5.10.27}$$

(*cf.* (4.9.7) with (4.6.29) and (4.6.34)), and its conjugate,

$$\Delta_{ab}\tau^{C'}=\varepsilon_{AB}\bar{\Psi}_{A'B'E'}{}^{C'}\tau^{E'}-\tfrac{1}{3}\lambda\varepsilon_{AB}\tau_{(A'}\varepsilon_{B')}{}^{C'}. \tag{5.10.28}$$

These we can write in form

$$\begin{aligned}
&\varepsilon_{R'S'}\{\nabla^{R'}_{G}\nabla^{S'}_{H}+\nabla^{R'}_{H}\nabla^{S'}_{G}\}\kappa_{A}=2\Psi_{GHAB}\kappa^{B}-\frac{\lambda}{3}\{\kappa_{G}\varepsilon_{HA}+\kappa_{H}\varepsilon_{GA}\}\\
&\varepsilon_{R'S'}\{\nabla^{R'}_{G}\nabla^{S'}_{H}+\nabla^{R'}_{H}\nabla^{S'}_{G}\}\tau^{P'}=0
\end{aligned} \tag{5.10.29}$$

$$\begin{aligned}
&\varepsilon^{GH}\{\nabla^{R'}_{G}\nabla^{S'}_{H}+\nabla^{S'}_{G}\nabla^{R'}_{H}\}\kappa_{A}=0\\
&\varepsilon^{GH}\{\nabla^{R'}_{G}\nabla^{S'}_{H}+\nabla^{S'}_{G}\nabla^{R'}_{H}\}\tau^{P'}=2\bar{\Psi}^{R'S'P'}{}_{Q'}\tau^{Q'}-\frac{\lambda}{3}\{\tau^{R'}\varepsilon^{S'P'}+\tau^{S'}\varepsilon^{R'P'}\}.
\end{aligned} \tag{5.10.30}$$

Now consider the spinors

$$\Psi_{ABCD},\ \nabla^{P'}_{E}\Psi_{ABCD},\ \nabla^{P'}_{E}\nabla^{Q'}_{F}\Psi_{ABCD},\ \ldots. \tag{5.10.31}$$

The various derivatives of (5.10.26) must all hold identically also. Hence the algebraic relations on the spinors (5.10.31) arising from (5.10.26) are

$$\varepsilon^{HA}(\nabla^{P'}_{E}\ldots\nabla^{R'}_{G}\nabla^{S'}_{H}\Psi_{ABCD})=0 \tag{5.10.32}$$

This expresses a condition on (namely, the vanishing of) the part of $\nabla^{P'}_{E}\ldots\nabla^{S'}_{H}\Psi_{ABCD}$ which is skew in $H$, $A$ and says nothing about the part symmetric in $H$, $A$. Moreover the relations (5.10.29) connect

$$\varepsilon_{R'S'}(\nabla^{P'}_{E}\ldots\nabla^{R'}_{G}\nabla^{S'}_{H}\ldots\nabla^{V'}_{K}\Psi_{ABCD})+\varepsilon_{R'S'}(\nabla^{P'}_{E}\ldots\nabla^{R'}_{H}\nabla^{S'}_{G}\ldots\nabla^{V'}_{K}\Psi_{ABCD})$$

with lower derivatives of $\Psi_{ABCD}$, while (5.10.30) connect

$$\varepsilon^{GH}(\nabla^{P'}_{E}\ldots\nabla^{R'}_{G}\nabla^{S'}_{H}\ldots\nabla^{V'}_{K}\Psi_{ABCD})+\varepsilon^{GH}(\nabla^{P'}_{E}\ldots\nabla^{S'}_{G}\nabla^{R'}_{H}\ldots\nabla^{V'}_{K}\Psi_{ABCD})$$

with lower derivatives of $\Psi_{ABCD}$. These express conditions only on parts of $\nabla^{P'}_{E}\ldots\nabla^{V'}_{K}\Psi_{ABCD}$ which are skew in a pair of primed indices or in a pair of unprimed indices. Thus the algebraic relations arising from (5.10.26), (5.10.29) and (5.10.30) connecting the spinors (5.10.31) and their complex conjugates are all concerned with parts of $\nabla^{P'}_{E}\ldots\nabla^{V'}_{K}\Psi_{ABCD}$ which are skew in at least one pair of indices. They imply no conditions on the parts totally symmetric in all primed indices and in all unprimed indices. (It might, perhaps, be thought that other relations could be obtained by expanding skew parts of $\nabla_E^{P'} \ldots \nabla_K^{V'} \Psi_{ABCD}$ in two different ways. However, these all lead back to (5.10.32) which is the only consistency condition implied.) Hence the spinors

$$\Psi_{ABCD}, \Psi_{ABCDE}{}^{P'} = \nabla_{(E}^{P'} \Psi_{ABCD)}, \Psi_{ABCDEF}{}^{P'Q'} = \nabla_{(E}^{(P'} \nabla_F^{Q')} \Psi_{ABCD)}, \ldots \tag{5.10.33}$$

and their complex conjugates are all algebraically independent and can therefore be specified arbitrarily (ignoring convergence questions; *cf.* footnote on p. 374) at any one point $P$.

It remains to be shown, conversely, that all the spinors (5.10.31) can be obtained algebraically from the spinors (5.10.33) and their complex conjugates. An induction argument will be used. We wish to express $\nabla_E^{P'} \ldots \nabla_K^{V'} \Psi_{ABCD}$ in terms of $\Psi_{ABCDE\ldots K}{}^{P'\ldots V'}$ and lower order derivatives of $\Psi_{ABCD}$ since it may be supposed as the inductive hypothesis that all these lower derivatives have already been expressed algebraically in terms of symmetrized derivatives $\Psi_{AB\ldots G}{}^{P'\ldots R'}$ and their complex conjugates. Now, if we add together all the spinors obtained from $\nabla_E^{P'} \ldots \nabla_K^{V'} \Psi_{ABCD}$ by permuting $P', \ldots, V'$ in all possible ways and $A, B, C, D, E, \ldots, K$ in all possible ways, we get a multiple of $\Psi_{AB\ldots K}{}^{P'\ldots V'}$. Thus, if it can be shown that each of the spinors obtained by such permutations differs from $\nabla_E^{P'} \ldots \nabla_K^{V'} \Psi_{ABCD}$ by expressions involving only lower derivatives of $\Psi_{ABCD}$ the result will be proved. The spinor $\nabla_E^{P'} \ldots \nabla_K^{V'} \Psi_{ABCD}$ will then be seen to differ from $\Psi_{AB\ldots K}{}^{P'\ldots V'}$ by a spinor built up from lower derivatives of $\Psi_{ABCD}$.

Any two spinors obtained by such a permutation of indices from

$$\nabla_E^{P'} \ldots \nabla_K^{V'} \Psi_{ABCD}$$

will be called *equivalent* (denoted by $\sim$) if they differ from each other by expressions built up from lower order derivatives of $\Psi_{ABCD}$. This is clearly an equivalence relation. It is required to show that all such spinors are, in fact, equivalent to one another. Now since

$$\nabla_W^{X'} \nabla_Y^{Z'} - \nabla_Y^{Z'} \nabla_W^{X'} \equiv \tfrac{1}{2} \varepsilon^{X'Z'} \varepsilon_{M'N'} \{\nabla_W^{M'} \nabla_Y^{N'} + \nabla_Y^{M'} \nabla_W^{N'}\} + \tfrac{1}{2} \varepsilon_{WY} \varepsilon^{ST} \{\nabla_S^{X'} \nabla_T^{Z'} + \nabla_S^{Z'} \nabla_T^{X'}\}$$

(see (4.9.1)), we have, applying (5.10.29) and (5.10.30),

$$\ldots \nabla_W^{X'} \nabla_Y^{Z'} \ldots \Psi_{ABCD} \sim \ldots \nabla_Y^{Z'} \nabla_W^{X'} \ldots \Psi_{ABCD}.$$

Hence any permutation of the $\nabla_M^{N'}$ symbols gives rise to an equivalent spinor. (Any permutation can be expressed as a product of transpositions of adjacent elements.) That is, any permutation of $P', \ldots, V'$ can be applied to $\nabla_E^{P'} \ldots \nabla_K^{V'} \Psi_{ABCD}$ provided that the same permutation is applied to $E, \ldots, K$ and an equivalent spinor is obtained. It remains to show that $E, \ldots, K, A, B, C, D$ can be permuted independently and an equivalent spinor is still obtained. The symmetry of $\Psi_{ABCD}$ implies that $A, B, C, D$ can be permuted without change. Furthermore, from (5.10.32), $K$ and $A$ can be interchanged in $\nabla_E^{P'} \ldots \nabla_K^{V'} \Psi_{ABCD}$. Also,

$$\begin{aligned} \ldots \nabla_Y^{Z'} \ldots \nabla_K^{V'} \Psi_{ABCD} &\sim \ldots \nabla_K^{V'} \ldots \nabla_Y^{Z'} \Psi_{ABCD} \\ \sim \ldots \nabla_K^{V'} \ldots \nabla_A^{Z'} \Psi_{YBCD} &\sim \ldots \nabla_A^{Z'} \ldots \nabla_K^{V'} \Psi_{YBCD}, \end{aligned}$$

so that $A$ can be interchanged with any other unprimed index and an equivalent spinor is obtained. It follows that any pair of unprimed indices can be interchanged since

$$\begin{aligned} \ldots \nabla_W^{X'} \ldots \nabla_Y^{Z'} \ldots \Psi_{ABCD} &\sim \ldots \nabla_W^{X'} \ldots \nabla_A^{Z'} \ldots \Psi_{YBCD} \\ \sim \ldots \nabla_Y^{X'} \ldots \nabla_A^{Z'} \ldots \Psi_{WBCD} &\sim \ldots \nabla_Y^{X'} \ldots \nabla_W^{Z'} \ldots \Psi_{ABCD}. \end{aligned}$$

Hence all the spinors are equivalent and the result is proved.

*Einstein–Maxwell case*

The case when an electromagnetic field is present in the space can be treated by an extension of the method for empty space described above. The spinors

$$\Psi_{ABCDE\ldots G}{}^{P'\ldots R'} = \nabla_{(E}^{(P'} \ldots \nabla_G^{R')} \Psi_{ABCD)}$$

are defined as before and spinors $\varphi_{AB}, \varphi_{ABC}{}^{P'}, \varphi_{ABCD}{}^{P'Q'}, \ldots$ are introduced, defined similarly by

$$\varphi_{ABC\ldots E}{}^{P'\ldots R'} = \nabla_{(C}^{(P'} \ldots \nabla_E^{R')} \varphi_{AB)}.$$

By the same kind of argument as before, it follows that $\varphi_{AB}, \varphi_{ABC}{}^{P'}, \ldots, \Psi_{ABCD}, \Psi_{ABCDE}{}^{P'}, \ldots$ and their complex conjugates are all algebraically independent. Instead of (5.10.26) we have

$$\varepsilon^{CA} \nabla_C^{P'} \varphi_{AB} = 0 \quad \text{and} \quad -\varepsilon^{EA} \nabla_E^{P'} \Psi_{ABCD} = 2\gamma \bar{\varphi}^{P'}{}_{Q'} \nabla_D^{Q'} \varphi_{BC}$$

from (5.1.57) and (5.2.7) (in suitable units). The first of these states the symmetry of

$$\nabla_C^{P'} \ldots \nabla_E^{R'} \varphi_{AB}$$

in $E, A$, while the second expresses the part of

$$\nabla_E^{P'} \ldots \nabla_G^{R'} \Psi_{ABCD}$$

skew in $G, A$ in terms of derivatives of $\varphi_{AB}$ of at most the same order. They imply no condition on the symmetrized derivatives of $\varphi_{AB}$ or

$\Psi_{ABCD}$. Nor do the equivalents of (5.10.29) and (5.10.30), which differ from them only in that the second relation (5.10.29) is replaced by

$$\varepsilon_{R'S'}\{\nabla_G^{R'}\nabla_H^{S'} + \nabla_H^{R'}\nabla_G^{S'}\}\tau^{P'} = 4\gamma\varphi_{GH}\bar{\varphi}^{P'}{}_{Q'}\tau^{Q'}$$

(see (5.2.6)) and the first relation (5.10.30) by

$$\varepsilon^{GH}\{\nabla_G^{R'}\nabla_H^{S'} + \nabla_G^{S'}\nabla_H^{R'}\}\kappa_A = 4\gamma\bar{\varphi}^{R'S'}\varphi_{AB}\kappa^B.$$

The argument to show that the unsymmetrized derivatives can be expressed algebraically in terms of the symmetrized derivatives and their complex conjugates is exactly analogous to that for the pure gravitational case. The derivative $\nabla_C^{P'}\ldots\nabla_E^{R'}\varphi_{AB}$ differs from $\varphi_{ABC\ldots E}{}^{P'\ldots R'}$ by expressions constructed from lower order derivatives of $\varphi_{AB}$ and $\Psi_{ABCD}$, while $\nabla_E^{P'}\ldots\nabla_G^{R'}\Psi_{ABCD}$ differs from $\Psi_{ABCDE\ldots G}{}^{P'\ldots R'}$ by expressions constructed from derivatives of $\varphi_{AB}$ of the same order or lower, and from lower order derivatives of $\Psi_{ABCD}$. Thus, we can construct $\nabla_C^{P'}\varphi_{AB}$, $\nabla_E^{P'}\Psi_{ABCD}$, $\nabla_C^{P'}\nabla_D^{Q'}\varphi_{AB}$, $\nabla_E^{P'}\nabla_F^{Q'}\Psi_{ABCD}, \ldots$, in that order, from the symmetrized derivatives. This completes our proof.

### *Further examples*

Other examples of exact sets are the Yang–Mills 'free' fields subject to (5.5.40) and (5.5.44) (for various different groups), and Yang–Mills fields with suitable sources. These fall essentially under the heading of exact sets as defined at the beginning of this section provided that the Yang–Mills indices are treated as *abstract* and do not partake of any of the symmetrizations in (5.10.2). If a Yang–Mills basis and Yang–Mills potentials are introduced, then further 'gauge conditions' are needed in order that an exact set may be obtained. The situation is analogous to that of the electromagnetic field, where an extra gauge condition such as the Lorenz gauge is needed in order that the potentials should propagate, if a non-invariant approach is adopted.

Finally, some sets of fields which, by themselves, do not constitute exact sets, can be completed to exact sets by including certain contracted derivatives of the fields as additional fields. A rather trivial example is the Fock–Feynman–Gell-Mann equation for a spin-$\frac{1}{2}$ particle in $\mathbb{M}$ (Fock 1937, Feynman and Gell-Mann 1958). The field is described by a single two-component spinor $\psi_A$, and governed, in the presence of an electromagnetic field, by the field equation

$$(\square + 2\mu^2)\psi_A = -2ie\varphi_{AB}\psi^B. \qquad (5.10.34)$$

By itself (if $\varphi_{AB} = 0$), or with $\varphi_{AB}$, $\psi_A$ does not form an exact set. But, if we include $\chi^{A'} = \mu^{-1}\nabla^{AA'}\psi_A(\mu = m/\hbar\sqrt{2})$ as an additional field, we are back with the Dirac equation (5.10.15) since by (5.1.43)

$$\mu^2\psi_A = -\mu\nabla_{AA'}\chi^{A'} = \nabla_{AA'}\nabla^{A'}_B\psi^B = \square_{AB}\psi^B + \tfrac{1}{2}\varepsilon_{AB}\nabla_{CA'}\nabla^{CA'}\psi^B$$
$$= -\mathrm{i}e\varphi_{AB}\psi^B - \tfrac{1}{2}\square\psi_A,$$

so $\psi_A, \chi^{A'}$ (and $\varphi_{AB}$) provide an exact set.

Somewhat similar is the case of a Dirac (–Fierz) free particle (Dirac 1936*a*, Fierz 1938) of spin $\frac{1}{2}n \geqslant 1$, in $\mathbb{M}$ – which is essentially equivalent to the Rarita–Schwinger (1941) equation and generalizes that of Duffin–Kemmer, etc. (see Corson 1953). Suppose electromagnetism and other interactions to be absent. The field is described by a pair of symmetric spinors $\psi^{Q'\dots T'}_{A\dots D}, \chi^{P'\dots T'}_{B\dots D}$ of respective valences $\left[\begin{smallmatrix}0 & p\\ q+1 & 0\end{smallmatrix}\right], \left[\begin{smallmatrix}0 & p+1\\ q & 0\end{smallmatrix}\right]$ subject to field equations

$$\nabla^{AP'}\psi^{Q'\dots T'}_{AB\dots D} = \mu\chi^{P'Q'\dots T'}_{B\dots D}, \; \nabla_{AP'}\chi^{P'Q'\dots T'}_{B\dots D} = -\mu\psi^{Q'\dots T'}_{AB\dots D}. \tag{5.10.35}$$

Here, as before, $\hbar\mu\sqrt{2}$ is the mass, the spin of the field being $\frac{1}{2}n = \frac{1}{2}(p + q + 1)$. Note that the symmetry of each of $\psi^{\dots}_{\dots}, \chi^{\dots}_{\dots}$, together with (5.10.35) implies that the 'subsidiary conditions'

$$\nabla^A_{Q'}\psi^{Q'R'\dots T'}_{A\,B\,\dots D} = 0, \quad \nabla^B_{P'}\chi^{P'Q'\dots T'}_{B\dots D} = 0 \tag{5.10.36}$$

both hold. Also we have, as above,

$$(\square + 2\mu^2)\psi^{Q'\dots T'}_{A\dots D} = 0, \quad (\square + 2\mu^2)\chi^{P'\dots T'}_{B\dots D} = 0, \tag{5.10.37}$$

(So $m = \hbar\mu\sqrt{2}$ is indeed the mass). In fact, as with the Fock–Feynman–Gell-Mann equation above, we may consider just $\psi^{\dots}_{\dots}$, say, and use (5.10.35) (1) to define $\chi^{\dots}_{\dots}$. This will, provided (5.10.36) (1) holds, give $\chi^{\dots}_{\dots}$ the correct symmetries, (5.10.35) (2) now being a consequence of (5.10.37) (1).

The two fields $\psi^{Q'\dots T'}_{A\,\dots D}, \chi^{P'\dots T'}_{B\,\dots D}$ do not form an exact set (nor does $\psi^{\dots}_{\dots}$ alone). However, we can easily complete the system to an exact set by introducing $n - 1$ new spinors, each being symmetric with $n$ indices but with differing numbers of primed and unprimed indices: in their natural order these $n + 1$ spinors form a linear sequence of spinors each of which is obtainable from its immediate neighbours by a differentiation contracted on one index. Together, these spinors form an invariant exact set. For example, if $n = 2$ we have the invariant exact set $\psi_{AB}, \chi^{A'}_B, \xi^{A'B'}$ with

$$\nabla^{AA'}\psi_{AB} = \mu\chi^{A'}_B, \; \nabla^{AA'}\chi^{B'}_A = \mu\xi^{A'B'},$$
$$\nabla_{AA'}\xi^{A'B'} = -\mu\chi^{B'}_A, \; \nabla_{AA'}\chi^{A'}_B = -\mu\psi_{AB}. \tag{5.10.38}$$

However, if electromagnetism (with $n \geqslant 2$) or gravity (with $n \geqslant 3$) are to be incorporated, the situation becomes more complicated because of the

Fierz–Pauli (1939) and Buchdahl (1958) consistency relations which we encountered in the zero-rest-mass case (*cf*. (5.8.2)). As it stands, (5.10.35) is inconsistent with (5.1.43) (when $n \geqslant 2$) and with (4.9.13) (when $n \geqslant 3$). The modifications that have been considered by various authors to remove these inconsistencies are beyond the scope of this book, and the interested reader is referred to the literature cited here and in §5.8.

## 5.11 Initial data on a light cone

In the preceding section the most important of the explicitly known physical fields, gravitation included, were exhibited in the form of exact sets. Interactions between fields found expression either through 'field equations' such as (5.10.15), (5.10.17), (5.10.25), or through 'commutator equations' for the $\nabla$s (*cf.* (5.1.44), (5.1.45)). (Note that in this formalism, the Bianchi identity counts as a 'field equation'!) The entire formalism is best expressed in terms of *field* quantities rather than gauge dependent potentials. Thus, the electromagnetic potential $\Phi_a$ need never appear explicitly, nor is there any place for the explicit appearance of gravitational potentials (i.e. expressions for $g_{\mathbf{ab}}$ in terms of a coordinate basis). In the present section this "geometrical attitude" is exploited further. Assuming analyticity of all the quantities involved, if all the derivatives (5.10.3) are known at one event $O$, then, by means of power series (i.e., 'Taylor's theorem') we can 'step' from event to event, thus calculating the fields (5.10.1) and their derivatives (5.10.3) at every other event in the space–time. This gives a method of exploring the space–time in a way which is, in principle, completely coordinate-free. However, it is not generally a very practical method, although it is possible that developments in formal technique might lead to a more manageable procedure.

On the other hand, the method does lead to one important deduction concerning the nature of the exact-set condition, and it is this which we shall present here. Roughly speaking, it turns out that in the same way that the derivatives (5.10.3) enable us to 'walk' about the space–time, it is the *symmetrized* derivatives (5.10.2) which enable us to 'walk' up the *light cone* of the event $O$ with that component of each field which is associated with the null direction in the cone (Penrose 1963). In consequence, condition (*a*) for exactness will tell us that, for each field, this component can be specified as an essentially *arbitrary* function of the light cone of $O$, while condition (*b*) tells us that knowledge of this function defines the fields *everywhere** throughout space–time. We thus have a form of initial

* Strictly speaking, this need apply only in some open neighbourhood of $O$, as it is possible for ambiguities and consistency problems to arise globally.

value problem for which the data can be specified in an arbitrary (i.e. 'constraint-free') way on a light cone. The argument is presented here only in the simplest situation, where it is assumed that everything is analytic. The results should have a much wider domain of applicability than this, but proofs become much more difficult when analyticity is dropped. There are important questions concerning domains of dependence, coherence relations, stability, etc., which cannot be treated by arguments of the kind presented here. A full discussion would take us much too far afield from our immediate purposes, but the comparatively simple (analytic) argument presented here should be adequate to illustrate the importance of the exact-set concept.

### The 'Taylor series' on $\mathcal{M}$

We assume, then, that the space–time $\mathcal{M}$ is an analytic manifold with analytic metric $g_{ab}$. The operator $\nabla_a$ must also be analytic (i.e., yield an analytic field whenever applied to an analytic field). Of course, if $\nabla_a$ is just the ordinary Christoffel derivative, this fact follows automatically from the analyticity of $g_{ab}$. Here we envisage that an electromagnetic field may also be present and incorporated into the definition of $\nabla_a$ as applied to charged fields. In effect, the analyticity requirement on $\nabla_a$ states that the electromagnetic (as well as gravitational) field must be analytic. We do not envisage any interaction other than gravitational or electromagnetic incorporated into the definition of the $\nabla_a$ operator (although the inclusion of Yang–Mills fields would not greatly complicate the discussion). For simplicity we shall also exclude the possibility of torsion. The presence of torsion would indirectly affect the definitions of the geodesics in $\mathcal{M}$ and we prefer not to consider this complication here.

Let $\gamma$ be a smooth curve in $\mathcal{M}$ and let $t^a$ be a tangent vector defined at each point of $\gamma$. Suppose for some $\theta^{\cdots}_{\cdots}$,

$$t^a\nabla_a\theta^{\cdots}_{\cdots} = 0 \tag{5.11.1}$$

at each point of $\gamma$. Then we say that $\theta^{\cdots}_{\cdots}$ is *constant* along $\gamma$ (with respect to $\nabla_a$) and write

$$[\theta^{\cdots}_{\cdots}]_P \overset{\gamma}{=} [\theta^{\cdots}_{\cdots}]_Q \tag{5.11.2}$$

for $\theta^{\cdots}_{\cdots}$ evaluated at any two points $P$, $Q$ along $\gamma$. Unless the commutators of the $\nabla$s vanish, the notion of equality defined by (5.11.2) generally depends on the choice of curve $\gamma$ which connects the points in question. (If $\gamma$ is a closed curve originating and terminating at the same point $P(=Q)$, then (5.11.2) would not generally imply that the original and final fields at $P$ are equal in the ordinary sense. For an infinitesimal loop, the discrepancy comes directly from the $\nabla$ commutator term.)

If $t^a$ itself can be chosen constant along $\gamma$, i.e.,

$$t^a \nabla_a t^b = 0, \tag{5.11.3}$$

then $\gamma$ is a *geodesic* in $\mathscr{M}$. Let $v$ be a real parameter on such a geodesic $\gamma$, scaled so that

$$t^a \nabla_a v = 1 \tag{5.11.4}$$

(Then $v$ will be an affine parameter on $\gamma$; $t^a$ and $v$ are 'uncharged'.) Fix a point $O \in \mathscr{M}$ and let $\gamma$ be some geodesic through $O$, with parameter value $v=0$ at $O$. To represent a point $X$ on $\mathscr{M}$ relative to $O$, define a vector $x^a$ at $O$ by

$$x^a = v t^a, \tag{5.11.5}$$

where $v$ is the parameter value at $X$. (It follows at once from (5.11.3) and (5.11.4) that (5.11.5) is independent of the scaling of $t^a$. If $t^a \mapsto k t^a$, then $k$ must be constant along $\gamma$ to preserve (5.11.3), whence $v \mapsto k^{-1} v$.) Thus we may say that $x^a$ gives the *position vector** of $X$ relative to $O$. As $\gamma$ varies among all geodesics through $O$, the components $x^{\mathbf{a}}$ of $x^a$ with respect to some fixed basis at $O$ give, in fact, a system of *normal coordinates* with origin $O$. That is to say, the direction of $x^a$ is that of the geodesic $\gamma$ from $O$ to $X$ and the length (or 'extent' if $x^a$ is null) of $x^a$ is the same as that of $\gamma$ from $O$ to $X$. (This follows directly from (5.11.3), (5.11.4), (5.11.5).)

Let $\psi^{\mathscr{A}}$ be analytic at $O$; then for $X$ not too far from $O$ on the geodesic $\gamma$ we have

$$\begin{aligned}[\psi^{\mathscr{A}}]_X &\stackrel{\gamma}{=} [\psi^{\mathscr{A}}]_O + \frac{x^a}{1!}[\nabla_a \psi^{\mathscr{A}}]_O + \frac{x^a x^b}{2!}[\nabla_a \nabla_b \psi^{\mathscr{A}}]_O + \cdots \\ &= [\exp(x^a \nabla_a)\psi^{\mathscr{A}}]_O \end{aligned} \tag{5.11.6}$$

To see this (compare Synge 1960), we note that if $v'$ is the parameter assigned to a variable point $X'$ between $O$ and $X$ on $\gamma$, the quantity

$$\begin{aligned}\exp((v-v')t^a\nabla_a)\psi^{\mathscr{A}} = \psi^{\mathscr{A}} &+ \frac{(v-v')t^a}{1!}\nabla_a \psi^{\mathscr{A}} + \\ &+ \frac{(v-v')^2 t^a t^b}{2!}\nabla_a \nabla_b \psi^{\mathscr{A}} + \cdots \end{aligned} \tag{5.11.7}$$

defined at $X'$ is constant along $\gamma$ by (5.11.1), (5.11.3), (5.11.4) ($v$ being

* This applies unambiguously provided that $X$ is sufficiently 'close' to $O$. Otherwise, because geodesics can refocus and cross over, there may be more than one 'position vector' at $O$ for a point $X$ (or, sometimes, none at all).

constant and $v'$ satisfying $t^a \nabla_a v' = 1$). The convergence and appropriate uniformity of (5.11.7), for $X$ near enough to $O$, is implicit in the analyticity of the quantities involved but we shall not discuss such matters in detail here. When $v' = 0$ and $v' = v$, we get, respectively, the right- and left-hand sides of (5.11.6).

*The 'Taylor series' on $\mathcal{N}$*

Relation (5.11.6) tells us how knowledge of all the *unsymmetrized derivatives* (5.10.3) of a set of fields at $O$ can be employed in the determination of the fields at all other points (not too far from $O$ in the first instance, but 'distant' points can be reached using several steps). Now suppose that

$$\psi^{\mathscr{A}} = \psi_{A\ldots E}^{P'\ldots S'} = \psi_{(A\ldots E)}^{(P'\ldots S')} \tag{5.11.8}$$

and let us examine the role of the *symmetrized derivatives* (5.10.2). Consider the points $X$ which lie on the *light cone* $\mathcal{N}$ of $O$. Then $\gamma$ will be a null geodesic and $t^a$ a null vector, which is thus of the form

$$t^a = \xi^A \bar{\xi}^{A'} \tag{5.11.9}$$

(choosing $t^a$ as future-pointing). For a specific null geodesic $\gamma$ we select $\xi^A$ to be constant (and uncharged) along $\gamma$. Now multiply (5.11.6) by $\xi^A \xi^B \ldots \xi^E \bar{\xi}_{P'} \bar{\xi}_{Q'} \cdots \bar{\xi}_{S'}$ (where, because of the constancy of $\xi^A$ along $\gamma$, we can unambiguously write the $\xi$s outside the brackets):

$$\begin{aligned}
\xi^A \ldots \xi^E \bar{\xi}_{P'} \ldots \bar{\xi}_{S'} [\psi_{A\ldots E}^{P'\ldots S'}]_X &\overset{\gamma}{=} \xi^A \ldots \xi^E \bar{\xi}_{P'} \ldots \bar{\xi}_{S'} [\psi_{A\ldots E}^{P'\ldots S'}]_O \\
&- \frac{v}{1!} \xi^A \ldots \xi^E \xi^F \bar{\xi}_{P'} \ldots \bar{\xi}_{S'} \bar{\xi}_{T'} [\nabla_F^{T'} \psi_{A\ldots E}^{P'\ldots S'}]_O \\
&+ \frac{v^2}{2!} \xi^A \ldots \xi^E \xi^F \xi^G \bar{\xi}_{P'} \ldots \bar{\xi}_{S'} \bar{\xi}_{T'} \bar{\xi}_{U'} [\nabla_F^{T'} \nabla_G^{U'} \psi_{A\ldots E}^{P'\ldots S'}]_O - \ldots .
\end{aligned} \tag{5.11.10}$$

Because of the symmetry of $\xi^A \ldots \xi^H \bar{\xi}_{P'} \ldots \bar{\xi}_{V'}$, it is really the symmetrized derivatives $[\nabla_{(F}^{(T'} \ldots \nabla_H^{V'} \psi_{A\ldots E)}^{P'\ldots S')}]_O$ that are relevant here (*cf.* (3.3.23)). Knowledge of all these symmetrized derivatives at $O$ will determine the quantity

$$\psi = \xi^A \ldots \xi^E \bar{\xi}^{P'} \ldots \bar{\xi}^{S'} [\psi_{A\ldots EP'\ldots S'}]_X \tag{5.11.11}$$

along any null geodesic through $O$ and, therefore, at each point of $\mathcal{N}$.

The complex number $\psi$ will be called the *null-datum* for the field $\psi^{\mathscr{A}}$ at the point $X$ of $\mathcal{N}$. If we make a definite choice of $\xi^A$ for each null direction at $O$, (say by choosing $\xi^0 = 1$ at $O$ – which excludes only the generator $\xi^0 = 0$), we may regard $\psi$ as a scalar function defined on the light cone of $O$, this function being determined by the symmetrized derivatives of $\psi^{\mathscr{A}}$ at $O$ though, more correctly, $\psi$ is a weighted function, as in §4.12, as we shall see shortly. Conversely, the values of $\psi$ on $\mathcal{N}$ *determine* these symmetrized derivatives. For the light cone $\mathcal{N}$ may be parametrized by the real number $v$ ($v=0$ giving the vertex) and by the complex ratio $\xi=\xi^1/\xi^0$ at $O$, i.e., by the complex number $\xi=\xi^1$, if we choose $\xi^0=1$. The coefficient of $(-v)^r$ in (5.11.10) is

$$\xi^A \ldots \xi^E \xi^F \ldots \xi^H \bar{\xi}_{P'} \ldots \bar{\xi}_{S'} \bar{\xi}_{T'} \ldots \bar{\xi}_{V'} [\nabla_F^{T'} \ldots \nabla_H^{V'} \psi_{A\ldots E}^{P'\ldots S'}]_O, \qquad (5.11.12)$$

which, for each value of $\xi$, will therefore be determined by $\psi$. But the behaviour of (5.11.12) under $\xi^A \mapsto \lambda\xi^A$ is also known, namely it gets multiplied by $\lambda^{p+r}\bar{\lambda}^{q+r}$. Hence (5.11.12) will be determined by $\psi$ as a function of $\xi^A$, $\bar{\xi}_{P'}$. But $\xi^A$ and $\bar{\xi}_{P'}$ are *algebraically independent*. Therefore (5.11.12), being a polynomial in $\xi^0, \xi^1, \bar{\xi}_{0'}, \bar{\xi}_{1'}$, will determine its coefficients $[\nabla_{(F}^{(T'} \ldots \nabla_H^{V'} \psi_{A\ldots E)}^{P'\ldots S')}]_O$ uniquely.

The significance of the concept of an exact set should now be apparent. Condition (*b*) for exactness for a set of fields $\psi^{\mathscr{A}}, \ldots, \chi^{\mathscr{C}}$ ensures that the *values of their null-data* $\psi, \ldots, \chi$, *on the light cone* $\mathcal{N}$, *determine the fields throughout the space–time* $\mathcal{M}$. Condition (*a*) for exactness ensures that the various null data can be chosen freely on $\mathcal{N}$, that is, there are no *constraints*. For any exact set, therefore, the null-data are a *complete irredundant* set of initial data on any light cone.

## *Counting degrees of freedom*

We are now in a position to count the number of degrees of freedom for the fields. For this we invoke what appears to be a general principle in the initial value problem, namely that on a *characteristic* (i.e., a null) initial hypersurface just one-half as many real numbers per point of the hypersurface are required as would be required in the case of a spacelike hypersurface (*cf.* d'Adhémar 1905, Riesz 1949, Hadamard 1952, Duff 1956). In the present case we are concerned with initial data on $\mathcal{N}$ which (not too far from $O$, at least, and excluding $O$ itself) will be a smooth null hypersurface. Thus, we expect the number of real numbers per point that we require for data on $\mathcal{N}$ shall be just one-half the number required for an ordinary spacelike hypersurface. We shall have just one complex number (i.e., two real numbers) per point of $\mathcal{N}$, for each of the fields (5.10.1) belonging to the exact set. This will hold *unless* the field is specified as being *Hermitian* (in which case it must have an equal number of primed and unprimed indices). For a Hermitian field the null-datum must be real. Thus, the number of 'degrees of freedom' for the fields, being defined as one-half the number of real variables required for an initial spacelike hypersurface, is *twice the number of non-Hermitian fields occurring in* (5.10.1) *plus the number of Hermitian fields.*

Let us just briefly check with some known results. For the case of Maxwell's free-space equations, in a given background space–time, we have just one field $\varphi_{AB}$, so the number of degrees of freedom is *two*. (In the usual 'spacelike' formalism, this is one-half the number obtained from: six degrees of freedom for $\boldsymbol{E}$ and $\boldsymbol{B}$ minus two for the constraints div $\boldsymbol{E}=0$, div $\boldsymbol{B}=0$.) In the case of a Dirac field we have two spinors $\psi_A, \chi^{A'}$ in (5.10.1). Thus we have *four* degrees of freedom. (In the usual formalism one thinks of these as given by the four complex components of a single Dirac 4-spinor, that is eight real numbers per point of a spacelike hypersurface.) For a real scalar (Schrödinger–Klein–Gordon or D'Alembert) field we have one 'Hermitian' field, and thus just *one* degree of freedom. (In the usual formalism, the initial data are the field together with its time derivative: two numbers per point.) For the Dirac–Fierz higher spin fields (5.10.35), we require a total of $n+1$ non-Hermitian spinors (*cf.* (5.10.38)) in order to get an exact set, so there are $n+1$ degrees of freedom. (In the usual formalism this comes about in a different way (*cf.* Corson 1953, p. 121). Finally, in the case of general relativity, we see at once that there are just *two* degrees of freedom for the gravitational field, since this is defined by the single non-Hermitian spinor $\Psi_{ABCD}$. (In the usual formalism the result is the same, but is not nearly so directly obtained, there being many redundant variables and constraints to be taken into account (*cf.* Bruhat 1962, Arnowitt, Deser and Misner1962).

### *Regularity at O*

We do not enter into any of the difficult questions concerned, say, with removing the condition of analyticity, or of replacing the light cone $\mathscr{N}$ by a more general characteristic (i.e. null) hypersurface. We should expect, for example, the null-data on the future half ($\mathscr{N}^+$) of $\mathscr{N}$ to define the fields *inside* $\mathscr{N}^+$ (at least, not too far from $O$). But such questions cannot be directly answered using only the sorts of methods we have been considering here. In §5.12 we shall provide some specific formulae which go some way towards answering such technical questions for certain fields. The only general question of detail we examine *here* concerns the nature of the null-data at the vertex $O$ of $\mathscr{N}$. Since $O$ is a singular point of $\mathscr{N}$, it is not immediately clear what kind of smoothness conditions the null-data should satisfy there.

In this connection, the dependence of $\psi$ on $\xi^A$ at a general point of $\mathscr{N}$

should be kept in mind. For under

$$\xi^A \mapsto \lambda \xi^A \tag{5.11.13}$$

we have

$$\psi \mapsto \lambda^p \bar{\lambda}^q \psi, \tag{5.11.14}$$

i.e. in the terminology of §4.12, $\psi$ is a *$\{p, q\}$-scalar*, where we choose a spin-frame at each point of $\mathcal{N}$ with

$$o^A = \xi^A. \tag{5.11.15}$$

(At any point $X$ of $\mathcal{N}$, $\xi^A$ is, of course, defined up to proportionality as it corresponds to the tangent to the generator of $\mathcal{N}$ through $X$.) Recall that the integer or half-odd-integer

$$s = \frac{p-q}{2} \tag{5.11.16}$$

is the *spin-weight of* $\psi$.

Now, let us consider a $\{p, q\}$-scalar $\psi$ defined on $\mathcal{N}$, but without reference to the whole space–time $\mathcal{M}$. We may take for the definition of analyticity for $\psi$, simply that a series of the form

$$[\psi]_X \overset{\gamma}{=} \sum_{n=0}^{\infty} \xi^A \ldots \xi^E \bar{\xi}_{P'} \ldots \bar{\xi}_{S'} x^{a_1} x^{a_2} \ldots x^{a_n} \psi_{(n)A \ldots E a_1 a_2 \ldots a_n}^{P' \ldots S'} \tag{5.11.17}$$

should exist giving $\psi$ on $\mathcal{N}$, convergent in some neighbourhood of $O$ – for $x^a$ a *null* vector defined at $O: x^a = v\xi^A \bar{\xi}^{A'}$ – the $\psi_{(\underline{n})\ldots}^{\cdots}$ being constants defined at $O$. The $\xi^A, \ldots, \xi^E$ are $p$ in number and the $\bar{\xi}_{P'}, \ldots, \bar{\xi}_{S'}$ are $q$ in number, making $\psi$ a $\{p, q\}$-scalar as required. Clearly the $\psi$ defined by (5.11.11), (5.11.10) will, in this sense, be analytic. (Since $\psi$ is a scalar, the relevance of the '$\gamma$' in (5.11.17) is only to charged fields. But we can ignore it in any case if we choose the electromagnetic potential to be analytic – again in the sense of (5.11.17).)

Let us consider the nature of the regularity of $\psi$ at the vertex $O$, implied by (5.11.17). For this, we identify the tangent space at $O$ with the $\mathbb{M}$ of §4.15 and think of the origin $O$ as the point $L$. Setting $o^A = \xi^A$, as in (5.11.15), so $x^a = v o^A o^{A'}$, $t^a = l^a$, we see that the $n$th term in the sum (5.11.17) is precisely of the form (4.15.41). So from the discussion concerning (4.15.43)–(4.15.57) we see that (with respect to an arbitrarily chosen time-axis $T^a$ at $O$) this $n$th term (coefficient of $v^n$) is a linear combination of spin-weighted spherical harmonics for spin-weight $s$ and

$$j = |s|, |s|+1, |s|+2, \ldots, n + \tfrac{1}{2}(p+q). \tag{5.11.18}$$

Let us examine the significance of this in the case $p = q = 0$. Then the function $\psi$ has a power series expansion for which the coefficient of $v^n$ involves ordinary spherical harmonics only of orders $0, 1, \ldots, n$. This may be compared with the behaviour of an analytic function of $x, y, z$ in an ordinary Euclidean 3-space, re-expressed in terms of, say, ordinary spherical polar coordinates $r, \theta, \phi$. In this case, the coefficient of $r^n$ would involve spherical harmonics only of orders $n, n-2, n-4, \ldots, 1$ ($n$ odd) or $n, n-2, \ldots, 0$ ($n$ even). Thus there is, roughly speaking, 'twice as much information' in the analytic $\{0, 0\}$-scalar function $\psi$ on $\mathcal{N}$ as there would be for an ordinary analytic scalar function on Euclidean 3-space. This may be regarded as one 'explanation' of the halving of the number of initial data functions on the null hypersurface $\mathcal{N}$ as compared with what would be needed for a spacelike hypersurface, since each function on $\mathcal{N}$ 'counts for twice as much'. This is, however, very far from a complete explanation.

It is worth while to point out a curious property of the zeros of the null datum $\psi$, on $\mathcal{N}$, in the case $s \neq 0$. In the generic case there will be $m$ lines of such zeros entering the vertex $O$, where

$$|p-q| \leqslant m \leqslant p+q. \tag{5.11.19}$$

This follows partly from topological considerations and partly from the discussion to be given in §8.8.

*Geometry of $\mathcal{N}$*

In the case of gravitation, the null-datum $\Psi$ has a special significance. We shall only indicate this here. A more complete discussion would depend on the geometry that will be introduced in Chapter 7. Choose a complex null vector $m^a$, orthogonal to $l^a$, whose real part (spacelike and of length $2^{-\frac{1}{2}}$) spans, with $l^a$, the null-flag plane of $\xi^A = o^A$ at each point of $\mathcal{N}$. With the usual null-tetrad notation (3.1.14), we have $l^a = t^a$. The tetrad vector $l^a$ thus points in the null direction in $\mathcal{N}$ (and is normal to $\mathcal{N}$) and the real and imaginary parts of $m^a$ are tangent to $\mathcal{N}$ in spacelike directions, these three vectors spanning the tangent spaces to $\mathcal{N}$. The remaining tetrad vector $n^a$ points out of $\mathcal{N}$. Then we can write (*cf.* (4.11.6), (4.11.9)):

$$\Psi = \xi^A \xi^B \xi^C \xi^D \Psi_{ABCD} = \Psi_0 = l^a m^b l^c m^d C_{abcd}. \tag{5.11.20}$$

The discussion of §7.2 tells us that $\Psi$ measures the 'purely astigmatic' part of the geodetic deviation (*cf.* Pirani & Schild 1961, Sachs 1961, Penrose 1966*a*) of the null geodesics in $\mathcal{N}$. If a matter tensor $T_{ab}$ is present, we can put

$$\Phi = \xi^A \xi^B \bar{\xi}^{A'} \bar{\xi}^{B'} \Phi_{ABA'B'} = -\tfrac{1}{2} R_{ab} l^a l^b$$
$$= 4\pi\gamma T_{ab} l^a l^b, \qquad (5.11.21)$$

by (5.10.23) and (4.6.25). Then $\Phi$ similarly measures the 'anastigmatic' part of this geodetic deviation. For the matter fields we have been considering here, it turns out that $\Phi$ is determined directly by the null-data of these matter fields (possibly involving a derivative in the $l^a$ direction). In particular, for the Einstein–Maxwell equations, we have $\Phi = 2\gamma\varphi\bar{\varphi}$, by (5.10.24), where $\varphi$ is the Maxwell null-datum. Thus, we see that, in general relativity, the geometry of $\mathcal{N}$ is itself directly determined by the null-data of all the various fields involved.

In the special case of the Einstein vacuum equations we can go further. The intrinsic geometry of $\mathcal{N}$ is then essentially *equivalent* to the null-datum $\Psi$. (There is, however, a certain subtlety involved here in that the intrinsic metric of $\mathcal{N}$ has vanishing determinant, but this is not serious; *cf.* Penrose 1972a.) Thus, we can say that for a vacuum space–time $\mathcal{M}$ (which is analytic) the *geometry of* $\mathcal{M}$ *is (locally) determined by the intrinsic geometry of the light cone of any one event in* $\mathcal{M}$.

The characteristic initial value problem for general relativity has also been studied for the case of a pair of intersecting null hypersurfaces $\mathcal{N}_1$, $\mathcal{N}_2$ (Sachs 1962*b*, *cf.* also Darmois 1927). In line with our present approach, where the gravitational data on $\mathcal{N}$ is given by the null-datum $\Psi$ (rather than the shear $\sigma$, *cf.* §§7.1, 7.2, or the inner metric) we would need to specify $\Psi$ on each of $\mathcal{N}_1$ and $\mathcal{N}_2$. But also certain data on $\mathcal{N}_1 \cap \mathcal{N}_2$ would be needed, which can take the form of $\rho, \rho', \sigma, \sigma'$ (where the flagpoles of $o^A, \iota^A$ point along the generators of $\mathcal{N}_1, \mathcal{N}_2$, respectively) and, in addition to just the inner metric of $\mathcal{N}_1 \cap \mathcal{N}_2$, the complex curvature quantity $K$ that was introduced in (4.14.20). The matter will not be pursued further here.

## 5.12 Explicit field integrals

There are certain integral expressions that may be used for determining fields explicitly in terms of null-data, the (null) initial data hypersurface $\mathcal{N}$ being not necessarily a null cone. The prototype of these expressions is the Kirchhoff–d'Adhemar integral formula for massless scalar (i.e., d'Alembert) fields in Minkowski space–time $\mathbb{M}$. This formula has a natural generalization to massless fields of arbitrary spin. In the present section we derive this general formula, first for $\mathbb{M}$, and then show how it applies to conformally flat space–time. Finally we indicate how repeated application of the formula leads to a corresponding expression for certain exact sets of interacting fields in $\mathbb{M}$, notably the (classical) Maxwell–Dirac system.

### *The generalized Kirchhoff–d'Adhémar formula*

Let $P$ be a point of $\mathbb{M}$ and suppose that the light cone $\mathscr{C}$ of $P$ is such that its intersection with $\mathscr{N}$ provides a smooth cross-section $\mathscr{S}$ of $\mathscr{C}$ (see Fig. 5-4) – either in the past or future of $P$. Let $Q$ be a typical point of $\mathscr{S}$, and consider spinors $\xi^A$ and $\eta^A$ at $Q$ having flagpoles pointing along generators of $\mathscr{N}$ and $\mathscr{C}$, respectively, so they are orthogonal to $\mathscr{S}$ at $Q$. Let $\mathscr{S}$ be the (2-form) element of surface area of $\mathscr{S}$ at $Q$ (*cf.* (4.14.65)). Also let these spinors be normalized:

$$\eta_A \xi^A = 1. \tag{5.12.1}$$

If we write

$$(\overrightarrow{QP})^a = r\eta^A \bar{\eta}^{A'}, \tag{5.12.2}$$

then the real number $r$ is a measure of the extent of $QP$, and it is positive or negative according as $Q$ lies on the past or future cone of $P$.

Now suppose $\phi_{A\ldots L}$ is a massless free field of spin $\frac{1}{2}n > 0$:

$$\phi_{\underbrace{A\ldots L}_{n}} = \phi_{(A\ldots L)}, \quad \nabla^{AA'}\phi_{AB\ldots L} = 0, \tag{5.12.3}$$

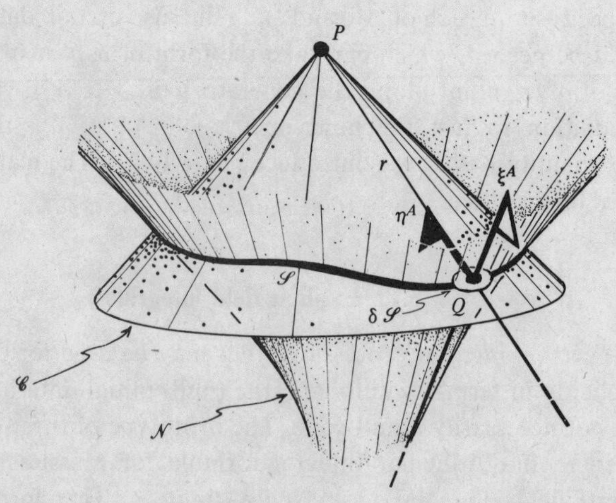


Fig. 5-4. The generalized Kirchhoff–d'Adhémar formula (5.12.6) expresses the massless field at $P$ as an integral over the intersection $\mathscr{S}$, of $\mathscr{C}$ with $\mathscr{N}$, in terms of the null-datum on $\mathscr{N}$.

or of spin zero:

$$(\Box + \tfrac{1}{6}R)\phi = 0 \tag{5.12.4}$$

(the factor $\frac{1}{6}R = 4\Lambda$ being of relevance when we consider curved space–time), in some neighborhood $\mathscr{K}$ of the join of $\mathscr{S}$ to $P$ (i.e. of the region swept out by the null segments $QP$). The null-datum for $\phi_{A\ldots L}$ on $\mathscr{N}$, at $Q$, is

$$\phi := \phi_{A\ldots L}\xi^A \ldots \xi^L. \tag{5.12.5}$$

We shall show that this, together with its derivative in the null direction in $\mathscr{N}$ at $\mathscr{S}$, determines the field at $P$ by the explicit formula (Penrose 1963)

$$\phi_{AB\ldots L}(P) = \frac{1}{2\pi}\oint_{\mathscr{S}} \eta_A \eta_B \ldots \eta_L \frac{\text{þ}_{\mathscr{C}}\phi}{r}\mathscr{S}, \tag{5.12.6}$$

where $\text{þ}_{\mathscr{C}}$ is the conformally invariant modified form (5.6.33) of the compacted spin-coefficient operator þ of (4.12.15), and the spin frame $(o^A, \iota^A)$ is defined by

$$o^A = \xi^A, \quad \iota^A = -\eta^A. \tag{5.12.7}$$

We take $o^A$, $\iota^A$ to have conformal weights $-1, 0$, respectively (which is the case (5.6.26)(iv) – although alternative scalings would do just as well). Since $\phi$ is a $\{n, 0\}$-scalar of conformal weight $-n-1$ (*cf.* (5.7.17)), we have

$$\text{þ}_{\mathscr{C}} = \text{þ} - (n+1)\rho = D - n\varepsilon - (n+1)\rho \tag{5.12.8}$$

when acting on $\phi$, where

$$D = \xi^A \bar{\xi}^{A'} \nabla_a. \tag{5.12.9}$$

The condition for the generators of $\mathscr{N}$ to be *geodesics* can be stated as

$$D\xi_A = \varepsilon\xi_A \tag{5.12.10}$$

(see (5.11.3); and also (7.1.8)). We shall see in §7.1 that null hypersurfaces are *always* generated by null geodesics. Here we simply adopt this fact as part of our assumptions about the nature of the hypersurface $\mathscr{N}$. Thus,

$$\kappa := \xi^A D\xi_A = 0 \tag{5.12.11}$$

whence (compare (7.1.16), (7.1.17))

$$\xi^A \bar{\xi}^{B'} \nabla_b \xi_A = \rho\xi_B, \quad \xi^A \xi^B \nabla_b \xi_A = \sigma\xi_B. \tag{5.12.12}$$

Though $\sigma$ happens not to be directly involved in (5.12.6), it enters into the generalizations (5.12.50), (5.12.54) below. We shall also see later, from (7.1.58)(7.1.61), that null hypersurfaces also satisfy

$$\rho = \bar{\rho}, \tag{5.12.13}$$

though this fact has already been established, in effect, in (4.14.2).

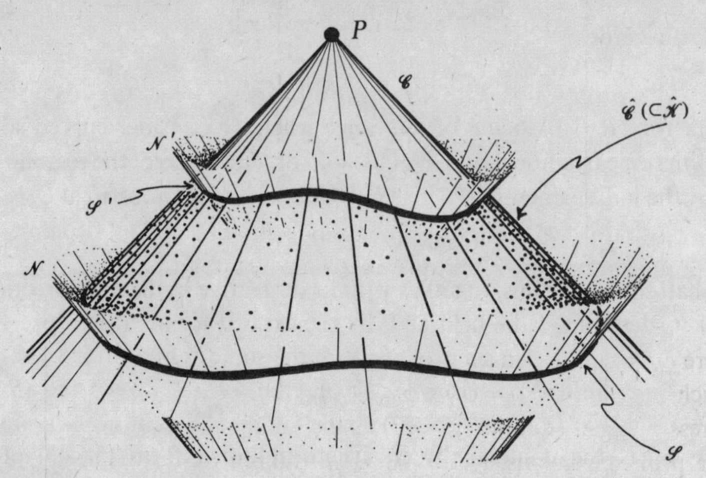


Fig. 5-5. The fundamental theorem of exterior calculus can be invoked to show that the integral over $\mathscr{S}$ is equal to that over $\mathscr{S}'$.

## *Proof of the formula*

With these preliminaries we are ready to prove (5.12.6). Our proof proceeds in two stages. First we show that for any two smooth cross-sections $\mathscr{S}$ and $\mathscr{S}'$ of $\mathscr{C}$ (see Fig. 5-5), for which (5.12.3) or (5.12.4) holds in some neighbourhood $\hat{\mathscr{K}}$, in $\mathbb{M}$, of the portion $\hat{\mathscr{C}}$ of $\mathscr{C}$ bounded by $\mathscr{S}$ and $\mathscr{S}'$, the integral (5.12.6) over $\mathscr{S}$ is equal to that over $\mathscr{S}'$. (We remark that *any* smooth cross-section $\mathscr{S}'$ of $\mathscr{C}$ arises as the intersection of $\mathscr{C}$ with a null hypersurface $\mathscr{N}$. For $\mathscr{N}$ is simply swept out by the null geodesics ('rays'), other than generators of $\mathscr{C}$, meeting $\mathscr{S}'$ orthogonally (*cf.* §§4.14, 7.1).) The second stage consists in showing that (5.12.6) holds in the limit when $\mathscr{S}'$ shrinks down to an infinitesimal sphere at $P$. That then establishes (5.12.6) generally (*cf.* Newman and Penrose, 1968).

To prove the first part, we use the version (4.14.92) – (4.14.94) of the fundamental theorem of exterior calculus described at the end of §4.14, which uses spin- (and boost-) weighted scalars. For this we must first put (5.12.6) into a weighted *scalar* form, and so we choose $\omega^A \in \mathfrak{S}^A$ arbitrarily (*cf.* (4.15.42)) and define the $\{-1,0\}$-scalar.

$$\omega = \omega^A \eta_A = -\omega^A \iota_A = \omega^0, \tag{5.12.14}$$

in terms of which (5.12.6) can be written as

$$\omega^A \omega^B \ldots \omega^L \phi_{AB\ldots L} = \frac{1}{2\pi} \oint_{\varphi} \omega^n r^{-1} þ_{\mathscr{C}} \phi \mathscr{S}. \tag{5.12.15}$$

To apply (4.14.93), (4.14.94), which we actually use in *complex conjugate form*, we also need a $\{0, -2\}$-scalar $\mu_{1'}$, which together with the $\{0,0\}$-scalar

$$\mu_{0'} = \omega^n r^{-1} \text{þ}_{\mathscr{C}} \phi \qquad (5.12.16)$$

satisfies (4.14.93) (in conjugate form):

$$A := (\text{þ}' - 2\bar{\rho}')\mu_{0'} - (\eth' - \bar{\tau})\mu_{1'} = 0 \qquad (5.12.17)$$

Here $\mu_{0'}$ and $\mu_{1'}$ provide the two components of a $\{0, -1\}$-spinor $\mu_{A'}$.

It turns out that (5.12.17) can be satisfied by taking

$$\mu_{1'} = \omega^n r^{-1}(\eth\phi - (n+1)\tau\phi + n\sigma\phi_1), \qquad (5.12.18)$$

where $\phi_1 = \phi_{100\ldots0}$, in accordance with the standard notation of (4.12.43) (which also allows $\phi = \phi_0 = \phi_{00\ldots0}$ and $\phi_2 = \phi_{110\ldots0}$). To see this, we express the field equation (5.12.3) as the family of equations (4.12.44) and their primed versions, (4.12.44)′. Here (taking $n > 1$) we need only the first *two* of equations (4.12.44) and only the last *one* of equations (4.12.44)′, these being, respectively,

$$\begin{aligned} B &:= \text{þ}\phi_1 - \eth'\phi_0 + \tau'\phi_0 - n\rho\phi_1 = 0, \\ C &:= \text{þ}\phi_2 - \eth'\phi_1 + 2\tau'\phi_1 - (n-1)\rho\phi_2 = 0, \\ E &:= \eth\phi_1 - \text{þ}'\phi_0 + \rho'\phi_0 + (n-1)\sigma\phi_2 - n\tau\phi_1 = 0, \end{aligned} \qquad (5.12.19)$$

where we have used $\kappa = 0$ from (5.12.11) and also

$$\sigma' = \kappa' = 0, \qquad (5.12.20)$$

which holds because $\mathscr{C}$ is a light cone in $\mathbb{M}$ (*cf.* (4.14.76), (4.15.12)). Analogously to (5.12.13) we have

$$\rho' = \bar{\rho}' \qquad (5.12.21)$$

and, indeed, from (4.15.12) we get

$$\rho' = \frac{1}{r}. \qquad (5.12.22)$$

(Comparison of (4.15.2) with (5.12.2) shows that $u = r$.) From (5.12.22), (4.12.32)($a'$) and (4.12.32)($d'$) (with $\kappa' = \sigma' = \Phi_{22} = \Phi_{21} = \Psi_3 = 0$) it follows at once that $\text{þ}'r^{-1} = \rho' r^{-1}$, i.e.,

$$\text{þ}'r = -\rho' r \qquad (5.12.23)$$

and

$$\eth' r = 0, \qquad (5.12.24)$$

$r$ being a $\{-1, -1\}$-scalar. Also (4.12.28) gives $\text{þ}'\iota^A = -\kappa' o^A = 0$ and $\eth'\iota^A = -\sigma' o^A = 0$, whence, by (5.12.14),

$$\text{þ}'\omega = 0 = \eth'\omega. \qquad (5.12.25)$$

We can now verify (5.12.17) by considering the combination

$$\omega^{-n} rA - (\eth - n\tau - \bar{\tau}')B - (n-1)\sigma C + (\text{þ} - n\rho - \bar{\rho})E, \qquad (5.12.26)$$

and showing that it vanishes identically in the given circumstances. This is established by a somewhat lengthy but straightforward calculation, which uses the compacted spin-coefficient equations (4.12.32) $(a')$, $(b)$, $(c)$, $(d)$, $(f')$, together with the commutator equations (4.12.35) and (4.12.33) as applied to $\phi_0$ (type $\{n, 0\}$), and (4.12.34) as applied to $\phi_1$ (type $\{n-2, 0\}$), and takes into account the equations (5.12.20), (5.12.21), (5.12.23), (5.12.24), (5.12.25), and

$$\Psi_0 = \Psi_1 = \Psi_2 = 0. \qquad (5.12.27)$$

From the vanishing of (5.12.26), and from $B = C = E = 0$, we deduce $A = 0$, as required. (The cases $n = 0$, 1 are similar – and simpler.) Thus, the integral in (5.12.15) is independent of the particular cross-section $\mathscr{S}$ of $\mathscr{C}$, as required.

Now consider a cross-section $\mathscr{S}_\delta$ which is obtained as the intersection of $\mathscr{C}$ with a spacelike hyperplane that passes almost, but not quite, through the point $P$ (either to the past or future of $P$). Let $\mathscr{S}_\delta$ have radius $|\delta|$ (and take $\delta > 0$ or $< 0$ according as $\mathscr{S}_\delta$ lies just to the past or just to the future of $P$). Then (*cf.* §4.15; in particular (4.15.3), (4.15.9), (4.15.12)) we find

$$\rho r^{-1} = \rho\rho' = -\tfrac{1}{2}\delta^{-2}, \qquad (5.12.28)$$

so $|\rho|$ and $|\rho'|$ each diverge as $O(|\delta|^{-1})$ when $\delta \to 0$. Since the field $\phi_{AB\ldots L}$ is smooth at $P$, and since the area of $\mathscr{S}_\delta$ is $O(\delta^2)$, the contributions from the terms $D\phi$ and $-n\varepsilon\phi$ to $\text{þ}_{\mathscr{C}}\phi$ (*cf.* (5.12.8)) both disappear in the limit, and we are left with the term $-(n+1)\rho\phi$. So the integral (5.12.6) becomes

$$\omega^A \ldots \omega^L \frac{1}{2\pi} \oint \frac{(n+1)}{2\delta^2} \eta_A \ldots \eta_L \xi^{A_0} \ldots \xi^{L_0} \phi_{A_0 \ldots L_0} \mathscr{S}_\delta. \qquad (5.12.29)$$

In the limit, $\phi_{A_0\ldots L_0}$ can come outside the integral, taking for its value that at $P$. We then apply Lemma (4.15.86) with $R = \delta$ (and with (5.12.7)), to obtain the required agreement with (5.12.15). The factor $\omega^A \ldots \omega^L$ may be removed from both sides, $\omega^A$ being arbitrary (*cf.* (3.3.23)). Thus formula (5.12.6) is established.

### *Conformally flat $\mathscr{M}$*

The foregoing discussion applies almost without change when $\mathbb{M}$ is replaced by a *conformally flat* space–time $\mathscr{M}$. We may suppose that by a suitable choice of conformal factor $\Omega$ the metric in the neighbourhood of $\mathscr{C}$ can be rescaled to that in a corresponding neighbourhood of a light cone in $\mathbb{M}$. We must, however, be careful about the interpretation of the parameter $r$. The critical equations are (5.12.23) and (5.12.24). The interpretation of (5.12.23) is that $r$ is a so-called *luminosity parameter* on $\mathscr{C}$ (Bondi, van der Burg and Metzner 1962, Sachs 1962a). We cannot in general maintain (5.12.22), but *that* equation was, in fact, not required for showing that (5.12.26) vanishes (except as a convenience in deriving (5.12.23) in $\mathbb{M}$). The conformal rescaling behaviour of $r$ is given by*

$$\hat{r} = \Omega(P)\Omega^{-1}r, \tag{5.12.30}$$

where $\Omega(P)$ is the conformal factor at $P$ (and where $\Omega = \Omega(Q)$, according to the standard notation of §5.6). The reason that the conformal scale at $P$ enters into the scaling behaviour of $r$ is that the relation (5.12.23) does not, by itself, determine the actual value of $r$, but fixes $r$ only up to an overall scale factor for each separate generator of $\mathscr{C}$. The scale factor is determined by the requirement that, for $Q$ near $P$, the position vector of $Q$ relative to $P$ is $-r\iota^A\iota^{A'}$. This scales as a distance, and since $\iota^A\iota^{A'}$ has been chosen invariant under conformal rescaling, $r$ must also scale as a distance at $P$, i.e., as $\Omega(P)$. The second factor $\Omega^{-1}$ gives compatibility with (5.12.23) and (5.6.27)(iv), where we have adopted (5.6.26)(iv) (*cf.* after (5.12.7)). The scaling for $\omega$ is

$$\hat{\omega} = \Omega\omega, \tag{5.12.31}$$

and this is compatible with the preservation of (5.12.25). In fact, we can replace $\omega^n$ in (5.12.15) by a more general $\{-n, 0\}$-scalar $\Gamma$ (of conformal weight $n$), subject to

$$\text{þ}'_{\mathscr{C}}\Gamma = 0 = \eth'_{\mathscr{C}}\Gamma; \quad \text{i.e., } \text{þ}'\Gamma = 0, \quad \eth'\Gamma = 0. \tag{5.12.32}$$

This would be the result of taking linear combinations of expressions (5.12.15) with different $\omega^A$s. By the discussion of §4.15, in the case when $\mathscr{S}$ is a metric sphere, such $\Gamma$s will be spin-weighted spherical harmonics with $j = -s = \frac{1}{2}n$ (see Proposition (4.15.58)). In the general case, such $\Gamma$s provide a suitable generalization of spin-weighted spherical harmonics for $\mathscr{S}$. In all cases, for the limitingly small spheres $\mathscr{S}_\delta$, these $\Gamma$s will approach the standard spin-weighted spherical harmonics for $\mathscr{S}_\delta$.

This provides one approach to assigning a meaning to the integrals (5.12.6), namely through (5.12.15), or (5.12.15) generalized by the replacement of $\omega^n$ by $\Gamma$ subject to (5.12.32). The difficulty in using (5.12.6) directly

* The apparent discrepancy between (5.12.30) and the more symmetrical-looking form $r = \Omega(P)\Omega r$ given in Newman and Penrose (1968) is due to the difference in scaling conventions for $\iota^A$.

is that it involves integrating a spinor quantity over $\mathscr{S}$ when no natural parallelism over $\mathscr{S}$ has been provided. An alternative procedure for circumventing this problem is to refer the entire integration to the point $P$. This is unambiguous because of the invariance properties of the integrand in (5.12.6), both under the spin- and boost-weight rescaling

$$\eta_A \mapsto \lambda^{-1}\eta_A, \quad \xi_{A'} \mapsto \lambda \xi_{A'}, \tag{5.12.33}$$

and under conformal rescaling, with

$$\eta_A \mapsto \Omega \eta_A, \quad \xi_{A'} \mapsto \xi_{A'}, \tag{5.12.34}$$

since, in the latter case, (5.12.30) holds, $\mathscr{S} \mapsto \Omega^2 \mathscr{S}$ and $\text{þ}_{\mathscr{C}}\phi \mapsto \Omega^{-n-3}\text{þ}_{\mathscr{C}}\phi$ by (5.6.34). (If desired, we can scale $\eta_A$ to be parallelly propagated along the generators of $\mathscr{C}$ and refer quantities at $Q$ to $P$ by parallel propagation.) The integration can then be carried out in the tangent space at $P$, and (by (5.12.30)) the result is a conformal density of weight $-1$.

We remark that although $\Phi_{22} = \Phi_{21} = 0$ was used in the derivation of (5.12.23) and (5.12.24) from (5.12.22), the validity of (5.12.6) does *not* depend on the vanishing of these or any other Ricci spinor components. We merely need assume that (5.12.23), (5.12.24) hold, which does not require (5.12.22). On the other hand, the vanishing of the Weyl spinor components (5.12.27) is necessary for the argument. Moreover, as a consequence of (5.12.23) and (5.12.24), the remaining Weyl spinor components $\Psi_3$ and $\Psi_4$ must also vanish, so in fact we need

$$\Psi_{ABCD} = 0 \tag{5.12.35}$$

on $\mathscr{C}$. Curiously, however, the condition (5.12.13) was not needed for establishing (5.12.17) in the region of $\mathscr{C}$ between $\mathscr{S}$ and $\mathscr{S}'$ (though it is implicitly involved *at* $\mathscr{S}$ and $\mathscr{S}'$).

The conformal invariance of $\Psi_{ABCD}$ has been referred to in §5.6 (and will be established in §6.8). It is of some interest to note that not only is the conformally invariant operator $\text{þ}_{\mathscr{C}}$ (of (5.6.33)) used in (5.12.6) (and (5.12.16)), but the related conformally invariant operator $\eth_{\mathscr{C}}$ is effectively used in (5.12.18), which can be written

$$\mu_{1'} = \omega^n r^{-1}(\eth_{\mathscr{C}}\phi + n\sigma\phi_1). \tag{5.12.36}$$

This shows that $\mu_{1'}$ is a conformal density of weight $-1$ (since $\omega, r, \phi, \sigma$ and $\phi_1$ are conformal densities of respective weights $1, -1, -n-1, -2$, and $-n$) in addition to $\mu_{0'}$ being a conformal density of weight $-2$. Thus we can write (5.12.17) in the conformally invariant form

$$A := \text{þ}'_{\mathscr{C}}\mu_{0'} - \eth'_{\mathscr{C}}\mu_{1'} = 0, \tag{5.12.37}$$

which shows $A$ to be a conformal density of weight $-2$ (*cf.* also the remark at the end of Section 5.6). The conformal invariance of $\mu_{0'}$ and $\mu_{1'}$ can also be seen if we write

$$\mu_{M'} = -\omega^n r^{-1}\eta^M \alpha_{MM'}, \tag{5.12.38}$$

where (with $A, \ldots, L$ being $n$ in number)

$$\alpha_u = \xi^A \ldots \xi^L \xi^M (\xi_U \nabla_{MU'} \phi_{A\ldots L} - (n+1)\phi_{A\ldots L} \nabla_u \xi_M), \tag{5.12.39}$$

since the particular combination (5.12.39) of conformal densities $\phi_{A\ldots L}$ and $\xi_A$ of respective weights $-1$ and 0 is easily seen, from (5.6.15), to be a conformal density of weight $-n-2$ (in addition to scaling as $\alpha_u \mapsto \lambda^{n+2}\alpha_u$ under $\xi_A \mapsto \lambda\xi_A$).*

### *Relevance to a spacelike data hypersurface*

It will have been noticed that in the preceding discussion the hypersurface $\mathscr{N}$ has played a rather small role. In fact, it has served just *one* essential purpose: namely, to define the particular cross-section $\mathscr{S}$ of $\mathscr{C}$. Once this cross-section is chosen, the flagpole direction of the spinor $\xi^A$ is fixed as the unique null normal direction to $\mathscr{S}$ not lying in $\mathscr{C}$. This direction serves not only to single out the particular component of the field $\phi_{A\ldots L}$, on $\mathscr{S}$, which is used (in the case of non-zero spin) as the null datum $\phi$, but it also specifies the direction in which the derivative þ is to act, and the convergence of these null directions provides the spin-coefficient $\rho$. Thus, the formula (5.12.6) can also be applied when suitable data are given on a *spacelike* (or even a *timelike*) hypersurface $\mathscr{H}$. Indeed, the original Kirchhoff (1882) expression for the scalar (D'Alembert) field $\phi$ was given for a *spacelike* $\mathscr{H}$ on which the field together with its normal derivative are specified. The integral expression requires both this normal derivative and a tangential derivative of $\phi$ within the surface, these two combining to give, in effect, $\mathrm{D}\phi$ in the direction of the $\xi^A$-flagpole. In terms of the normal to $\mathscr{H}$, this flagpole direction is such that, when taken together with that of $\eta^A$, it spans a 2-plane containing the normal.

This applies also in D'Adhemar's (1905) expression using a *null* hypersurface $\mathscr{N}$, but now the $\xi^A$-flagpole points *in* the direction of the normal – which is also *tangential* to $\mathscr{N}$, so no additional 'normal derivative' is needed. Moreover, when the point $P$ is moved, keeping $Q$ fixed and $\overrightarrow{QP}$ null (see Fig. 5-6), the $\xi^A$-flagpole direction changes in the case of $\mathscr{H}$, but

* The 1-form $\alpha_u \mathrm{d}x^u$ has some significance as the space–time translation of a certain twistor 1-form arising in relation to an inversion formula for the twistor integrals (6.10.1), due to Bramson, Penrose and Sparling (*cf*. Penrose 1975, p. 314).

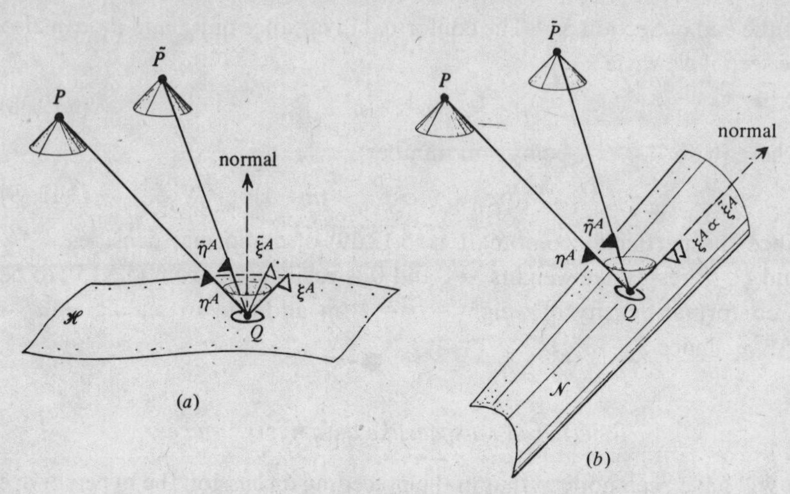


Fig. 5-6. For a spacelike initial hypersurface the relevant data component in (5.12.6) varies as $P$ moves, whereas this is not so for a null initial hypersurface.

not in the case of $\mathscr{N}$. Thus the D'Adhémar form of the integral (essentially (5.12.6) with $n=0$) has a considerably greater economy than that of Kirchhoff. For higher spin (Penrose 1963) the economy is even more pronounced, since in the case of $\mathscr{N}$ we need only *one* (complex) component of the field, and *avoid all constraints* (spin $> \frac{1}{2}$).

## *Background fields*

The generalized Kirchoff–d'Adhémar formula (5.12.6) can also be used in circumstances slightly different from the ones that we have been considering. For example, we may suppose that $\mathscr{M}$ is flat or conformally flat and that, as suffices in the first part of the proof given above, the field equation (5.12.3) or (5.12.4) holds only in some neighbourhood $\hat{\mathscr{K}}$ of the portion $\hat{\mathscr{C}}$ of $\mathscr{C}$ which connects two cross-sections $\mathscr{S}$ and $\mathscr{S}'$ of $\mathscr{C}$ (as in Fig. 5-5). We may, for example, envisage a situation in which a world-tube of sources for $\phi_{A...L}$ threads through the region $\hat{\mathscr{C}}$ (*cf.* Fig. 5-7). The integral (5.12.6) can still be evaluated meaningfully, yielding, say, a spinor $\psi_{A...L}$ at the point $P$. In general $\psi_{A...L}$ will not be equal to $\phi_{A...L}$ at $P$. Indeed, if $P$ lay on the world-line of a point source, $\phi_{A...L}$ would not even be defined. But $\psi_{A...L}$ will be independent of the section $\mathscr{S}$ of $\mathscr{C}$ to the extent that $\mathscr{S}$ can be moved continuously over any region of $\mathscr{C}$ which does not intersect the sources (or other places where $\phi_{A...L}$ is not defined). We regard $\psi_{A...L}$ as the *background field* at $P$ with the contribution from the sources surround-

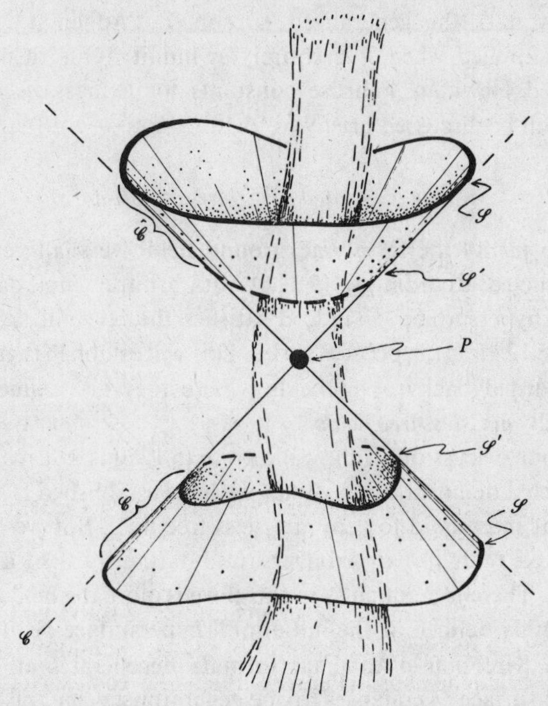


Fig. 5-7. For surface surrounding a world-tube of sources, the generalized Kirchhoff–d'Adhemar formula yields a background field at $P$.

ed by $\mathscr{S}$ subtracted out. Depending on whether $\mathscr{S}$ lies on the *past* or *future* light cone of $P$ we say that $\psi_{A...L}$ is, respectively, the *retarded* or *advanced* background field at $P$.

This gives a very useful procedure, for example in classical electrodynamics when applying the Lorentz force law to a point charge. Some concept of background field is required for this, since the full field diverges to infinity at the charge itself. In the normal Dirac (1939) procedure one employs a background field which amounts to taking one-half the advanced plus one-half the retarded background fields at the point charge. However, in the standard literature this is computed by a 'renormalization' procedure whereby an infinite field quantity is subtracted from the full (divergent) field at the point in question. The method we have just described achieves the same result (Unruh 1976) but in a much more direct way, no infinite quantities appearing at any stage of the calculation.

There is another somewhat more bizarre application of these ideas in which the point $P$ itself does not exist (or may become, in a sense, singular for the space–time). When $\mathscr{M}$ is asymptotically flat (but *not* normally conformally flat), the generalized Kirchhoff–d'Adhémar formula may actually be applied when $\mathscr{C}$ is entirely at infinity. One then obtains the the so-called Newman–Penrose constants for a massless field of spin $\frac{1}{2}n$. These will be discussed briefly in §9.10.

*Fields generated by arbitrary null-data*

In order to justify the term 'background field' we shall verify that the field constructed according to (5.12.6), with arbitrary null-datum $\phi$ on a null initial hypersurface $\mathscr{N}$ indeed satisfies the relevant field equations (5.12.3) or (5.12.4), as the point $P$ varies. This will establish that the advanced and retarded background fields (where they are defined) are both automatically *massless free fields*.

In fact, our discussion of the validity of (5.12.6) is not really complete without such a demonstration. What we have established is a *consistency relation* that must hold for any massless free field. But we have not yet demonstrated that, for an arbitrary null-datum, (5.12.6) always yields such a field. There is a difficulty in this if we require the field we construct to be smoothly defined at the initial null hypersurface $\mathscr{N}$ itself. For the generalized Kirchhoff–d'Adhémar formula degenerates at such points, where the 2-surface $\mathscr{S}$ collapses to one generator segment of $\mathscr{N}$, terminated by $P$ at one end and by a singularity of $\mathscr{N}$ at the other (see Fig. 5-8). Unless we are careful about the consistency conditions which hold at certain of these singular points of $\mathscr{N}$ (and, moreover, for $n \geqslant 2$ we shall generally require extra field components at singular 'crossover' points*

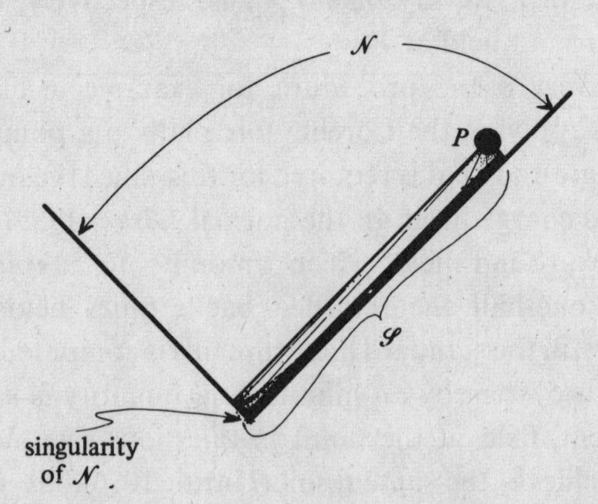


Fig. 5-8. When $P$ approaches $\mathscr{N}$, $\mathscr{S}$ degenerates to a segment of a generator of $\mathscr{N}$.

* The generalization of (5.12.6) that is required to handle such crossover points (e.g., when $\mathscr{N}$ degenerates into two intersecting null hypersurfaces) is discussed in Penrose (1963)

of $\mathcal{N}$), we may obtain a field $\phi_{A\ldots L}$ which is not smooth along $\mathcal{N}$. Recall, from the discussion at the end of §5.11, that in the particular case when $\mathcal{N}$ is a light cone the null-datum has to have a certain characteristic behaviour at the vertex in order that the field be analytic. If the null-datum does not have this general type of behaviour, then the field computed according to (5.12.6) will have singularities along $\mathcal{N}$, though not elsewhere in the region inside $\mathcal{N}$.

The question of the regularity of the field at $\mathcal{N}$ is beyond the scope of the present book. Here we merely show that the field suitably away from (i.e., 'inside') $\mathcal{N}$ indeed satisfies (5.12.3) or (5.12.4). We carry out the argument explicitly for $\mathbb{M}$. The result for conformally flat $\mathcal{M}$ may then be derived by conformal rescaling. Let $O$ be an arbitrary fixed origin in $\mathbb{M}$ and let $x^a$ be the position vector of $P$ relative to $O$. For convenience, choose $\xi^A$ to be constant along each generator $\nu$ of $\mathcal{N}$, and associate with it an affine (i.e., linear) parameter $v$. (See Fig. 5-9.) Thus if $H$ is the point on $\nu$ at which $v = 0$ and $Q$ a typical point on $\nu$ with parameter $v$, then

$$(\overrightarrow{HQ})^a = v\xi^A\bar{\xi}^{A'}. \tag{5.12.40}$$

Taking $h^a$ to be the position vector of $H$ we have (*cf.* Fig. 5.9)

$$x^a = h^a + v\xi^A\bar{\xi}^{A'} + r\eta^A\bar{\eta}^{A'}, \tag{5.12.41}$$

(where $\eta^A$ may be considered to be constant along $QP$). If we keep the generator $\nu$ fixed and vary $P$, then $v, r$, and $\eta^A$ are functions of $x^a$ ($\xi^A$ being constant). So differentiation of (5.12.41) with respect to $x^b$ yields

$$\varepsilon_B{}^A\varepsilon_{B'}{}^{A'} = \xi^A\bar{\xi}^{A'}\frac{\partial v}{\partial x^b} + \eta^A\bar{\eta}^{A'}\frac{\partial r}{\partial x^b} + r\eta^A\frac{\partial\bar{\eta}^{A'}}{\partial x^b} + r\bar{\eta}^{A'}\frac{\partial\eta^A}{\partial x^b}. \tag{5.12.42}$$

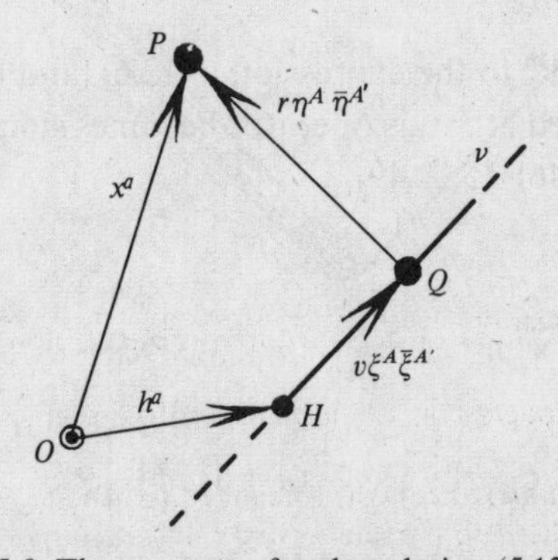


Fig. 5-9. The geometry for the relation (5.12.41).

We transvect this with $\eta_A\bar{\eta}_{A'}$ to obtain

$$\frac{\partial v}{\partial x^b} = \eta_B\bar{\eta}_{B'}, \tag{5.12.43}$$

and with $\xi_A\bar{\xi}_{A'}$ to obtain

$$\frac{\partial r}{\partial x^b} = \xi_B\bar{\xi}_{B'}, \tag{5.12.44}$$

since $\xi_A\eta^A = -1$ implies

$$\xi_A\frac{\partial \eta^A}{\partial x^b} = 0. \tag{5.12.45}$$

Further, if we transvect (5.12.42) with $\eta_A\bar{\xi}_{A'}$ and use (5.12.45), we find

$$\frac{\partial \eta^A}{\partial x^b} = -\frac{1}{r}\eta_B\bar{\xi}_{B'}\xi^A. \tag{5.12.46}$$

Note, from (5.12.43), that any quantity $\psi$ defined along $v$ satisfies

$$\frac{\partial \psi}{\partial x^b} = \eta_B\bar{\eta}_{B'}\mathrm{D}\psi \tag{5.12.47}$$

(*cf.* (5.12.9), (5.12.10); we now have $\varepsilon = 0$ since $\xi^A$ has been chosen constant along $v$). Hence, using (4.11.12) (a), we obtain

$$\begin{aligned}\frac{\partial}{\partial x^b}\{\text{þ}_{\mathscr{C}}\phi\} &= \frac{\partial}{\partial x^b}\{\mathrm{D}\phi - (n+1)\rho\phi\}\\ &= \eta_B\bar{\eta}_{B'}\{\mathrm{D}^2\phi - (n+1)\rho\mathrm{D}\phi - (n+1)(\rho^2 + \sigma\bar{\sigma})\phi\}\end{aligned} \tag{5.12.48}$$

Also (as most easily follows if we use the interpretation of $\rho$ as 'convergence' given in §7.1), we have

$$\frac{\partial}{\partial x^b}\mathscr{S} = \eta_B\bar{\eta}_{B'}\mathrm{D}\mathscr{S} = -2\rho\eta_B\bar{\eta}_{B'}\mathscr{S}. \tag{5.12.49}$$

Thus, applying $\partial/\partial x^m$ to the expression (5.12.6) (and bearing in mind that for a quantity defined at $P$ this operator becomes simply $\nabla_m$), we get, using (5.12.44) and (5.12.46)–(5.12.49),

$$\begin{aligned}&\nabla_M^{M'}\phi_{A\ldots L}(P)\\ &= \frac{1}{2\pi}\oint_{\mathscr{S}}\Big[\frac{1}{r}\eta_A\ldots\eta_L\eta_M\bar{\eta}^{M'}\{\mathrm{D}^2\phi - (n+3)\rho\,\mathrm{D}\phi + (n+1)(\rho^2 - \sigma\bar{\sigma})\phi\}\\ &\quad - \frac{(n+1)}{r^2}\eta_{(A}\ldots\eta_L\xi_{M)}\bar{\xi}^{M'}\{\mathrm{D}\phi - (n+1)\rho\phi\}\Big]\mathscr{S}\end{aligned} \tag{5.12.50}$$

(in $\mathbb{M}$). The right-hand side of this equation is manifestly symmetric in $A, \ldots, L, M$, showing that (5.12.3) is indeed satisfied when $n > 0$. For the case $n = 0$, (5.12.50) must be differentiated one more time, and then we find that (5.12.4) is satisfied.

This completes the argument showing that (5.12.6) always yields a massless free field in $\mathbb{M}$, the null-datum being chosen arbitrarily on $\mathcal{N}$, provided that $P$ lies in a region for which $\mathscr{C}$ meets $\mathcal{N}$ transversely in a smooth closed surface (necessarily topologically $S^2$, in fact). This applies, specifically, to the advanced and retarded background fields just discussed. Note that because of the appearance of a derivative term in (5.12.6) and (5.12.8), there may be a loss of one degree of differentiability in passing from the null-datum to the field. But if $\phi$ and $\mathcal{N}$ are $C^\infty$ then so will the resulting field be (in the interior region referred to). Even when the datum (or $\mathcal{N}$) is insufficiently smooth to produce a smooth field, the equation (5.12.3) or (5.12.4) will still be satisfied in the appropriate *distributional* sense (*cf.* Friedlander 1975).

The above argument was actually given in a little more detail than absolutely necessary, since the precise form of (5.12.48) was not needed in order to obtain the required symmetry of (5.12.50). However, the expression (5.12.50) is of interest in its own right, as it gives a direct integral expression for the derivative of a massless free field. In fact, the field

$$\theta^{M'}_{A\ldots LM} = \nabla^{M'}_{M} \phi_{A\ldots L} \tag{5.12.51}$$

forms by itself an exact set, the field equations being

$$\nabla_{NM'} \theta^{M'}_{A\ldots LM} = 0, \quad \nabla^{MN'} \theta^{M'}_{A\ldots LM} = 0 \tag{5.12.52}$$

(*cf.* (5.10.10)), and we may regard (5.12.50) as providing the analogue of (5.12.6) for the field $\theta^{M'}_{A\ldots M}$. However, we note that the null-datum is now

$$\theta = \mathrm{D}\phi, \tag{5.12.53}$$

whereas $\phi$ appears undifferentiated in (5.12.50). Thus to obtain $\theta^{M'}_{A\ldots M}$ entirely from its null-datum we need to integrate $\theta$ along generators of $\mathcal{N}$ first, before we can apply (5.12.50). This means, in effect, that the field $\theta^{M'}_{A\ldots M}$ *does not satisfy Huygens' principle*, at least not in so strong a sense as does $\phi_{A\ldots L}$. The manifestation of Huygens' principle exhibited by (5.12.6) is that $\phi_{A\ldots L}$ is determined entirely by the null-datum (and its derivative) at points on the light cone of the field point, whereas $\theta^{M'}_{A\ldots M}$ depends, to some extent, upon its null-datum at points lying within this cone.

For completeness, we also give the generalization of (5.12.50) for the

$k$th derivative of (5.12.6) in $\mathbb{M}$:

$$\underbrace{\nabla_M^{M'} \ldots \nabla_U^{U'}}_{k} \phi_{A \ldots L}(P) = \frac{1}{2\pi} \oint_{\mathscr{S}} \sum_{j=0}^{k} \frac{(-1)^{k-j}(n+k)!k!}{(n+j)!(k-j)!j!r^{k-j+1}} \underbrace{\eta_{(A} \ldots \eta_R}_{n+j} \underbrace{\xi_S \ldots \xi_{U)}}_{k-j}$$

$$\underbrace{\bar{\eta}^{(M'} \ldots \bar{\eta}^{R'}}_{j} \underbrace{\bar{\xi}^{S'} \ldots \bar{\xi}^{U')}}_{k-j} \{\mathrm{D}^{j+1}\phi - (n+2j+1)\rho \mathrm{D}^j \phi + j(n+j)(\rho^2 - \sigma\bar{\sigma})\mathrm{D}^{j-1}\phi\} \mathscr{S}$$

*Distributional fields*

An alternative procedure for establishing that (5.12.6) always yields a massless free field can also be given, providing somewhat different insights. We may regard the function $þ_{\mathscr{C}}\phi$ on $\mathscr{N}$ as built up linearly from Dirac $\delta$-function contributions. Each such $\delta$-function in $þ_{\mathscr{C}}\phi$, having support at a single point $R$ of $\mathscr{N}$, yields, by (5.12.6), a distributional field with support on the light cone of $R$. (Clearly (5.12.6) yields a field which is non-zero only when $RP$ is null.) If we verify that this distributional field satisfies the massless free-field equation, then, by linearity, we have an argument showing that (5.12.6) in general satisfies this equation. These distributional solutions of (5.12.3), or (5.12.4), can also be employed as part of a technique which supplies analogues of (5.12.6) for various coupled systems of interacting fields. To end the present section we outline these various ideas, though we make no attempt whatever at completeness or rigour. (See Friedlander 1975 for distributions on a manifold.)

We first need a few properties of some distributional fields. Define the 'step-function' scalar $\Delta_0$ by

$$\Delta_0(x^a) = \begin{cases} +1 & \text{if } x^a \text{ is future-timelike} \\ -1 & \text{if } x^a \text{ is past-timelike} \\ 0 & \text{if } x^a \text{ is spacelike.} \end{cases} \tag{5.12.55}$$

We can also define $\Delta_0 = \frac{1}{2}$ [*or* $-\frac{1}{2}$] on the future [past] light cone of $O$ (with $\Delta_0 = 0$ at $O$), if desired, but this plays no role here. Observe that $\Delta_0$ is invariant under orthochronous Lorentz transformations. It follows that the gradient of $\Delta_0$ must point in the direction of $x^a$, i.e., that

$$\nabla_a \Delta_0 = x_a \Delta_1$$

(with $\nabla_a = \partial/\partial x^a$), for some distribution $\Delta_1$. In fact the support of $\Delta_1$ is entirely the light cone of the origin, $\Delta_1$ being a $\delta$-function on this cone whose strength varies inversely as the extent of $x^a$. From the standard $\delta$-

function relation $u\delta(u) = 0$ we then derive

$$0 = x^a x_a \Delta_1 = x^a \nabla_a \Delta_0, \tag{5.12.57}$$

a fact that also follows from the observation that the 'step' 0 to 1, or $-1$ to 0, is constant along any generator of the light cone, showing that the operator $x^a \nabla_a$ indeed annihilates $\Delta_0$ along this cone.

From the Lorentz invariance argument we similarly derive

$$\nabla_a \Delta_1 = x_a \Delta_2, \tag{5.12.58}$$

for some $\Delta_2$ whose support is also the light cone, but which involves a derivative $\delta$-function along the cone. Continuing, we obtain successive distributions $\Delta_3, \Delta_4, \dots$ defined by

$$\nabla_a \Delta_j = x_a \Delta_{j+1} \quad (j = 0, 1, 2, \dots). \tag{5.12.59}$$

It is sometimes useful also to define the $C^{k-1}$ scalar

$$\Delta_{-k} = \frac{1}{k!}(\tfrac{1}{2} x_a x^a)^k \Delta_0 \quad (k = 1, 2, \dots), \tag{5.12.60}$$

and then (5.12.59) is satisfied for *all* integral $j$.

We can now establish the following:

(5.12.61) PROPOSITION

$$x_a x^a \Delta_{j+1} = -2j \Delta_j \quad (j = \cdots, -2, -1, 0, 1, 2, \dots).$$

*Proof*: For $j < 0$ the proof is immediate from (5.12.60). For $j \geqslant 0$ we operate on the relation in (5.12.61) with $\nabla_b$, substitute (5.12.59), and then divide out by $x_b$. The result is the same relation as before, but with $j$ replaced by $j+1$. Since the case $j = 0$ has been obtained earlier in (5.12.57), (5.12.61) follows by induction.

As a particular consequence of (5.12.61) we can derive

$$\Box \Delta_1 = 0, \tag{5.12.62}$$

for

$$\Box \Delta_1 = \nabla^a \nabla_a \Delta_1 = \nabla^a (x_a \Delta_2) = 4\Delta_2 + x^a x_a \Delta_3 = (4 - 4)\Delta_2 = 0.$$

The following result generalizes (5.12.61) in the cases $j = 1, 2$:

(5.12.63) PROPOSITION

*Let $\alpha_{\mathscr{A}}, \beta_{\mathscr{A}}, \gamma_{\mathscr{A}}$ be continuous; then*

(i) $\alpha_{\mathscr{A}} \Delta_1 + \beta_{\mathscr{A}} \Delta_2 = 0$ *iff* $\beta_{\mathscr{A}} = 0$ *and* $\nabla_b \beta_{\mathscr{A}} = x_b \alpha_{\mathscr{A}}$ *along* $x^a x_a = 0$;

(ii) $$\alpha_{\mathscr{A}}\Delta_1 + \beta_{\mathscr{A}}\Delta_2 + \gamma_{\mathscr{A}}\Delta_3 = 0 \textit{ iff } \gamma_{\mathscr{A}} = 0, \nabla_b\gamma_{\mathscr{A}} = \tfrac{1}{2}x_b\beta_{\mathscr{A}} \textit{ and}$$
$$\nabla_b\nabla_c\gamma_{\mathscr{A}} = x_{(b}\nabla_{c)}\beta_{\mathscr{A}} + \tfrac{1}{2}g_{bc}\beta_{\mathscr{A}} - x_bx_c\alpha_{\mathscr{A}} \textit{ along } x^ax_a = 0.$$

*Proof:* We can drop the clumped index $\mathscr{A}$, which is merely a 'passenger'. From $\alpha\Delta_1 + \beta\Delta_2 = 0$ we obtain, on multiplication by $x_ax^a$ and use of (5.12.61), $-2\beta\Delta_1 = 0$, so $\beta = 0$ on the cone. Differentiating $\beta\Delta_1 = 0$ we get $\Delta_1\nabla_b\beta + x_b\beta\Delta_2 = 0$, so $(\nabla_b\beta - x_b\alpha)\Delta_1 = 0$, whence $\nabla_b\beta = x_b\alpha$ on the cone. Similarly, from $\alpha\Delta_1 + \beta\Delta_2 + \gamma\Delta_3 = 0$ we obtain, on multiplication by $x_ax^a$ and use of (5.12.61), $\beta\Delta_1 + 2\gamma\Delta_2 = 0$, so, by (5.12.63) (i), $\gamma = 0$ and $\nabla_b\gamma = \tfrac{1}{2}x_b\beta$ on the cone. Differentiating $(\nabla_b\gamma - \tfrac{1}{2}x_b\beta)\Delta_1 = 0$ we get

$$(\nabla_b\nabla_c\gamma - \tfrac{1}{2}g_{bc}\beta - \tfrac{1}{2}x_b\nabla_c\beta)\Delta_1 + (x_c\nabla_b\gamma - \tfrac{1}{2}x_bx_c\beta)\Delta_2 = 0.$$

Differentiating $\beta\Delta_1 + 2\gamma\Delta_2 = 0$ and using $\alpha\Delta_1 + \beta\Delta_2 + \gamma\Delta_3 = 0$ we get

$$(x_c\nabla_b\gamma - \tfrac{1}{2}x_bx_c\beta)\Delta_2 = (x_bx_c\alpha - \tfrac{1}{2}x_c\nabla_b\beta)\Delta_1,$$

whence

$$(\nabla_b\nabla_c\gamma - \tfrac{1}{2}g_{bc}\beta - x_{(b}\nabla_{c)}\beta + x_bx_c\alpha)\Delta_1 = 0,$$

and the required final relation follows. It is not hard to reverse these arguments to obtain the converse relations.

We have seen in (5.12.62) that $\Delta_1$ satisfies the (d'Alembert) wave equation. We can also produce a corresponding solution of the massless free-field equations for each spin $\tfrac{1}{2}n$, namely

$$\underbrace{\eta_A \cdots \eta_L}_{n}\Delta_1, \tag{5.12.64}$$

where, for some straight null line $v$ through the origin, $\eta^A$ has flagpole direction along light cones with vertices $Q$ on $v$ (see Fig. 5-10); $\eta^A$ is normalized against $\xi^A$ by $\eta_A\xi^A = 1$ (as in (5.12.1)), where $\xi^A$ is a *constant* spinor with flagpole along $v$; $\eta^A$ is undefined on $v$ and on the null hyperplane $\mathscr{P}$ through $v$. The position vector $x^a$ of a general point $P$ has the form $\overrightarrow{OQ} + \overrightarrow{QP}$, i.e.,

$$x^a = v\xi^A\bar{\xi}^{A'} + r\eta^A\bar{\eta}^{A'}. \tag{5.12.65}$$

This is (5.12.41) with $h^a = 0$, and all the relations (5.12.43)–(5.12.47) hold just as before. Also we have $v = 0$ on the light cone of $O$, so

$$v\Delta_1 = 0. \tag{5.12.66}$$

Moreover,

$$x_ax^a = 2rv, \tag{5.12.67}$$

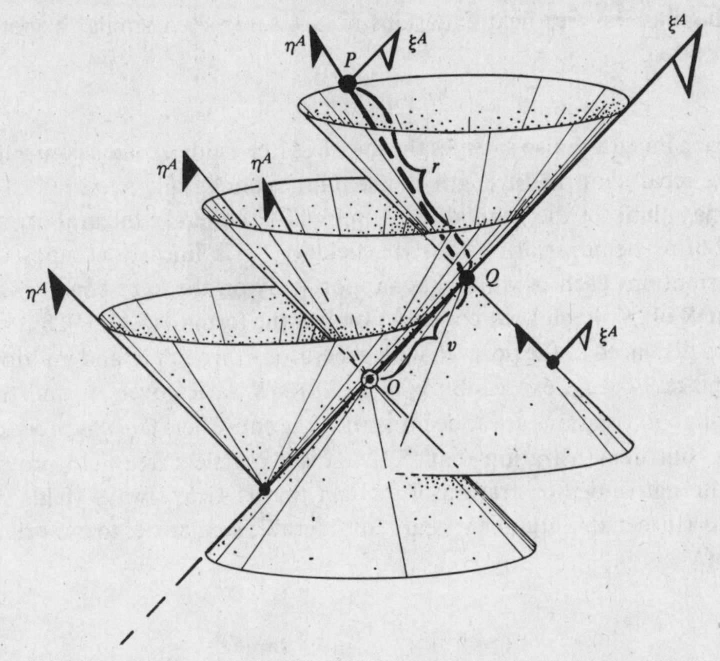


Fig. 5-10. The geometry for (5.12.64) and (5.12.65).

which, when substituted into (5.12.61), gives us

$$v\Delta_{j+1} = -\frac{j}{r}\Delta_j \quad (j = \dots, -1, 0, 1, 2, \dots), \tag{5.12.68}$$

whence, by (5.12.59) and (5.12.65), for all integral $j$,

$$\nabla_a \Delta_j = r\eta_A \bar{\eta}_{A'} \Delta_{j+1} - \frac{j}{r}\xi_A \bar{\xi}_{A'} \Delta_j. \tag{5.12.69}$$

Note that we are here ignoring the region $r = 0$, i.e., the line $v$ – or, more correctly, the entire null hyperplane $\mathscr{P}$.* Differentiating (5.12.64), and using (5.12.46) and (5.12.69), we now find

$$\nabla_{MM'}(\eta_A \dots \eta_L \Delta_1) = r\eta_A \dots \eta_L \eta_M \bar{\eta}_{M'} \Delta_2 - \frac{(n+1)}{r}\eta_{(A} \dots \eta_L \xi_{M)} \bar{\xi}_{M'} \Delta_1, \tag{5.12.70}$$

* Some subtleties arise if we wish the equations also to hold *on* $\mathscr{P}$. In the first place (in connection with the Grgin phenomenon of §9.4) it turns out that $\eta^A$ must be defined to jump by a factor of $i$ as $\mathscr{P}$ is crossed. Secondly, even so, there will be a source along the line $v$ itself, unless this is cancelled, e.g., by a further suitable collection of fields like (5.12.64), but based on other points of $v$. These questions become a little clearer in the context of §§9.2 and 9.4, but take us beyond the scope of the present volumes.

so the massless free-field equations hold *outside* $\mathscr{P}$. A similar argument shows that

$$r^{j-1}\eta_A \ldots \eta_L \Delta_j, \tag{5.12.71}$$

for each integral $j$, also satisfies the massless free-field equations outside $\mathscr{P}$.

We recall that the 'strength' of the $\delta$-function $\Delta_1$ falls off as $r^{-1}$. Thus we may think of the generalized Kirchhoff–d'Adhémar integral formula (5.12.6) as demonstrating that the field $\phi_{A\ldots L}$ is linearly composed of contributions each of which has support only on the light cone of some point $R$ of $\mathscr{N}$. Each such contribution has the form of (5.12.64), but with origin displaced to the point $R$, with generator $v$ through $R$, and multiplied by the null-datum expression $\text{þ}_{\mathscr{C}}\phi$ at $R$. As $R$ varies over $\mathscr{N}$ and these various contributions are added together, the entire field $\phi_{A\ldots L}$ is produced. Thus, our demonstration that (5.12.64) is a massless free field provides an alternative demonstration of the fact that (5.12.6) always yields such a field. (In fact, this argument bears considerable similarities to our original proof.)

## *The Dirac field*

This point of view is useful for an analogous treatment of certain exact sets of *interacting* fields. We shall find that, for these fields, we can express each field linearly in terms of contributions like (5.12.64), but not necessarily centered on points of $\mathscr{N}$. Roughly speaking, the various individual contributions can scatter off one another to produce new such contributions originating in the region interior to $\mathscr{N}$. We may think of the total interacting field as composed of pieces in which fields propagate for a while along null straight lines as massless free fields, but scatter repeatedly at points in this interior region. The novel feature that arises in our approach is the propagation entirely along null lines between scatterings. This is achieved by breaking down the system of fields into pieces for which the integral (5.12.6) can be used, which is then applied to situations where $\mathscr{N}$ is replaced by pieces of light cones that are the supports of various contributions like (5.12.64) from which the field has scattered.

To understand how these ideas arise, let us begin with a very simple example. We consider the free Dirac equation in the form (*cf.* (5.10.15))

$$\nabla^A_{A'}\psi_A = \mu\chi_{A'}, \quad \nabla^{A'}_A\chi_{A'} = \mu\psi_A, \tag{5.12.72}$$

where $\mu = 2^{-1/2}\hbar^{-1}m$, and treat it as describing *a pair of interacting fields*

$\psi_A$ and $\chi_{A'}$. First, we replace the system (5.12.72) by the infinite collection of fields $\overset{0}{\psi}_A, \overset{0}{\chi}_{A'}, \overset{1}{\psi}_{A'}, \overset{1}{\chi}_{A'}, \overset{2}{\psi}_A, \overset{2}{\chi}_{A'}, \ldots$, subject to

$$\nabla^A_{A'}\overset{0}{\psi}_A = 0, \quad \nabla^{A'}_A\overset{0}{\chi}_{A'} = 0, \tag{5.12.73}$$

and

$$\nabla^A_{A'}\overset{i}{\psi}_A = \mu \overset{i-1}{\chi}_{A'}, \quad \nabla^{A'}_A\overset{i}{\chi}_{A'} = \mu \overset{i-1}{\psi}_A \quad (i = 1, 2, \ldots). \tag{5.12.74}$$

This infinite collection constitutes an exact set, and (assuming convergence) a solution to the original Dirac system (5.12.72) is recovered whenever we put

$$\psi_A = \sum_{i=0}^{\infty} \overset{i}{\psi}_A, \quad \chi_{A'} = \sum_{i=0}^{\infty} \overset{i}{\chi}_{A'}. \tag{5.12.75}$$

In fact, if we have a suitable initial null hypersurface $\mathcal{N}$ – say the future light cone of a point $O$, for convenience – then we can take for the respective null-data of our collection of fields the special situation

$$0 = \overset{1}{\psi} = \overset{1}{\chi} = \overset{2}{\psi} = \overset{2}{\chi} = \overset{3}{\psi} = \ldots; \quad \overset{0}{\psi} = \psi, \overset{0}{\chi} = \chi, \tag{5.12.76}$$

where $\psi$ and $\chi$ are given null-data for the Dirac pair $\psi_A, \chi_{A'}$, and then (5.12.75) gives the required solution of (5.12.72) for these given null-data.

Let us apply our proposed technique to solving (5.12.73), (5.12.74) for such data. Since $\overset{0}{\psi}_A$ and $\overset{0}{\chi}_{A'}$ are massless free fields, we can use the methods adopted earlier: $\overset{0}{\psi}_A$ and $\overset{0}{\chi}_{A'}$ at some point $P$, to the future of $O$, can be obtained from their null-data $\psi, \chi$ directly by use of (5.12.6) (and its complex conjugate). The values of the null-data quantities $\text{þ}_{\mathscr{C}}\psi, \text{þ}_{\mathscr{C}}\chi$ at a particular point $R_1$ of $\mathcal{N}$ enter through contributions to $\overset{0}{\psi}_A, \overset{0}{\chi}_{A'}$ with support on the future light cone of $R_1$. Each of these individual contributions may be used to generate contributions to $\overset{1}{\psi}_A, \overset{1}{\chi}_{A'}$, and so on.

Consider the case where $\text{þ}_{\mathscr{C}}\psi$ is a *δ-function* on $\mathcal{N}$ at the point $R_1$, with $\chi = 0$. We get, by our previous discussion,

$$\overset{0}{\psi}_A = \eta_A \Delta_1(1), \quad \overset{0}{\chi}_{A'} = 0, \tag{5.12.77}$$

where here and from now on we adopt the notation $\Delta_j(k)$ for $\Delta_j$ with the origin displaced to the point $R_k$. The next step is to solve the equation (5.12.74) for $\overset{1}{\chi}_{A'}$, namely

$$\nabla^{A'}_A \overset{1}{\chi}_{A'} = \mu\eta_A\Delta_1(1). \tag{5.12.78}$$

Since the null-datum for $\overset{1}{\chi}_{A'}$ is zero, we require a solution of (5.12.78) which vanishes on $\mathscr{N}$ (except at $R_1$ or to the future of $R_1$ on the generator through $R_1$). By (5.12.44) and (5.12.69), such a solution is

$$\overset{1}{\chi}_{A'} = -\mu r^{-1}\bar{\xi}_{A'}\Delta_0(1). \tag{5.12.79}$$

In fact, it is not hard to see that the solution of (5.12.74) for all the remaining fields is given by

$$\overset{2j}{\psi}_A = \frac{(-1)^j\mu^{2j}}{j!}\eta_A\Delta_{1-j}(1), \quad \overset{2j+1}{\psi}_A = 0$$

$$\overset{2j}{\chi}_{A'} = 0, \quad \overset{2j+1}{\chi}_{A'} = \frac{(-1)^{j+1}\mu^{2j+1}}{j!}r^{-1}\bar{\xi}_{A'}\Delta_{-j}(1) \quad (j = 0, 1, 2, \ldots) \tag{5.12.80}$$

(*cf.* (5.12.44), (5.12.46), (5.12.69)). Substituting (5.12.80) into (5.12.75) and using (5.12.60), with (5.12.41), we obtain an explicit expression for the Dirac field resulting from a $\delta$-function in $\text{þ}_{\mathscr{C}}\psi$ at the point $R_1$ in terms of Bessel functions. Similarly we can obtain the explicit Bessel-function form of the Dirac field resulting from a $\delta$-function in $\text{þ}_{\mathscr{C}}\chi$ at the point $R_1$. These two expressions may be convoluted with $\text{þ}_{\mathscr{C}}\psi$ and $\text{þ}_{\mathscr{C}}\chi$, over $\mathscr{N}$, to yield the entire Dirac field with the given null-data.

However, that is not really the purpose of the present discussion. The idea here is to use the future light cone of $R_1$ as an *initial data hypersurface* for $\overset{1}{\chi}_{A'}$ (and then to repeat the process for $\overset{2}{\psi}_A, \overset{3}{\chi}_{A'} \ldots$). The null-datum for $\overset{1}{\chi}_{A'}$ on this light cone is

$$\overset{1}{\chi}_{A'}\eta^{A'} = \mu r^{-1}, \tag{5.12.81}$$

by (5.12.79). (Since we are concerned with the field $\overset{1}{\chi}_{A'}$ to the future of this cone, we can take $\Delta_0(1) = 1$.) The '$\text{þ}_{\mathscr{C}}$' operator that is relevant here is

$$\frac{\partial}{\partial r} + \frac{2}{r}, \tag{5.12.82}$$

since, as in (4.15.9), we have $-r^{-1}$ for the corresponding '$\rho$'. Applying (5.12.82) to (5.12.81), we get

$$\mu r^{-2} \tag{5.12.83}$$

for the resulting '$\text{þ}_{\mathscr{C}}\overset{1}{\chi}$'. Thus, by use of (5.12.6), we can express $\overset{1}{\chi}_{A'}$, at some point $P$ to the future of $R_1$, as an integral of (5.12.83) over the intersection of the past light cone of $P$ with the future light cone of $R_1$. Let $R_2$ be a typical point on this intersection. Then $\overrightarrow{R_1R_2}$ and $\overrightarrow{R_2P}$ as well as

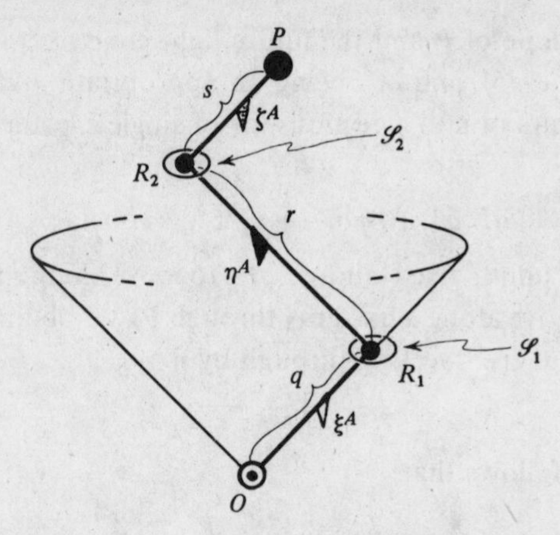


Fig. 5-11. The geometry for (5.12.84) and (5.12.86). (First 'mass scattering' for the Dirac field.)

$\overrightarrow{OR}_1$ are future-null vectors (see Fig. 5-11.), so we can write

$$(\overrightarrow{OR}_1)^a = q\xi^A\bar{\xi}^{A'}, \quad (\overrightarrow{R_1R_2})^a = r\eta^A\bar{\eta}^{A'}, \quad (\overrightarrow{R_2P}) = s\zeta^A\bar{\zeta}^{A'}. \tag{5.12.84}$$

For later convenience we here drop the normalization (5.12.1) and write

$$\eta_A\xi^A = z_1, \quad \zeta_A\eta^A = z_2. \tag{5.12.85}$$

On fitting these various facts together, and denoting the 2-surface element at $R_i$ by $\mathscr{S}_i$, we now have

$$\overset{1}{\chi}_{A'}(P) = \frac{1}{4\pi^2}\int\left(\frac{1}{s\bar{z}_2^2 z_2}\bar{\zeta}_{A'}\frac{\mu}{r^2 z_1}\mathscr{S}_2 \text{þ}_{\mathscr{C}}\psi(R_1)\right)\wedge \mathrm{d}q \wedge \mathscr{S}_1, \tag{5.12.86}$$

where the $z$-factors have been appropriately inserted so that the integrand is invariant under rescalings of $\xi^A$, $\eta^A$ and $\zeta^A$, and where the integration is taken over the 5-dimensional space $\mathscr{K}_{12}$ of all null zig-zags, of the kind we have been considering, that join $O$ and $P$ ($O$, $P$ fixed, and $\overrightarrow{OP}$ future-timelike).

It is convenient to rewrite (5.12.86) in terms of the differential forms

$$\mathscr{K}_1 = \frac{\mathscr{S}_1}{qrz_1\bar{z}_1} \quad \text{and} \quad \mathscr{C}_2 = -\frac{\mathrm{d}s}{s}\wedge\mathscr{S}_2, \tag{5.12.87}$$

or

$$\mathscr{N}_1 = \frac{\mathrm{d}q}{q}\wedge\mathscr{S}_1 \quad \text{and} \quad \mathscr{K}_2 = \frac{\mathscr{S}_2}{rsz_2\bar{z}_2}, \tag{5.12.88}$$

the 3-forms $\mathscr{C}_2$ and $\mathscr{N}_1$ being, respectively, invariant volume forms on the past light cone of $P$ and the future light cone of $O$ (Synge 1957, p. 7 *cf.* also Synge 1955), and $\mathscr{K}_i$ being an appropriate invariant 2-form on the space of pairs of null directions (or of single 2-plane elements) at $R_i$. Note that

$$q\xi^A\bar{\xi}^{A'} + r\eta^A\bar{\eta}^{A'} + s\zeta^A\bar{\zeta}^{A'} = \text{constant}, \tag{5.12.89}$$

so if we hold $\xi^A$ and $\zeta^A$ fixed (allowing $R_1$ to move along a fixed ray through $O$, and $R_2$ to move along a fixed ray through $P$), we obtain, on differentiating (5.12.89) and transvecting through by $\eta_A\bar{\eta}_{A'}$,

$$z_1\bar{z}_1\,\mathrm{d}q = -z_2\bar{z}_2\,\mathrm{d}s \tag{5.12.90}$$

from which it follows that

$$\mathscr{K}_1 \wedge \mathscr{C}_2 = \mathscr{N}_1 \wedge \mathscr{K}_2 =: \mathscr{K}_{12}. \tag{5.12.91}$$

(It is sufficient to keep $\xi^A$ and $\zeta^A$ fixed in the derivation of (5.12.91) because the allowed motions of $R_1$ and $R_2$ are those generated by fixed $\xi^A$, $\zeta^A$, together with others whose differentials have vanishing wedge products with $\mathscr{K}_1$ and $\mathscr{K}_2$.) We can now rewrite (5.12.86) as

$$\overset{1}{\chi}_{A'}(P) = \frac{\mu}{4\pi^2}\int_{\mathscr{K}_{12}} \frac{\psi(1)\bar{\zeta}_{A'}\mathscr{K}_{12}}{z_1 r\bar{z}_2}, \tag{5.12.92}$$

where

$$\psi(1) := q\text{þ}_{\mathscr{C}}\psi = \frac{1}{q}\frac{\partial}{\partial q}(q^2\psi) \quad (\text{at } R_1). \tag{5.12.93}$$

The whole process can now be repeated: we represent $\overset{1}{\chi}_{A'}$ linearly in terms of $\delta$-function contributions involving $\Delta_1(2)$, with $P$ now to the future of $R_2$, and we use the future cone of $R_2$ as an initial data hypersurface for $\overset{2}{\psi}_A$. The result is

$$\overset{2}{\psi}_A(P) = \frac{\mu^2}{8\pi^3}\int_{\mathscr{K}_{123}} \frac{\psi(1)\overset{3}{\xi}_A\mathscr{K}_{123}}{z_1 r_{12}\bar{z}_2 r_{23} z_3}, \tag{5.12.94}$$

where the 8-form

$$\mathscr{K}_{123} := \mathscr{K}_1 \wedge \mathscr{C}_2 \wedge \mathscr{C}_3 = \mathscr{N}_1 \wedge \mathscr{K}_2 \wedge \mathscr{C}_3 = \mathscr{N}_1 \wedge \mathscr{N}_2 \wedge \mathscr{K}_3 \tag{5.12.95}$$

is a natural volume element for the space of null zig-zags

$$R_0 := O, R_1, R_2, R_3, R_4 := P,$$

$\mathscr{K}_i$ in general being the 2-form (corresponding to $\mathscr{K}_1$ and $\mathscr{K}_2$ defined earlier) which gives an invariant volume element on the space of pairs of null directions at $P_i$, and $\mathscr{N}_{i+1}$ and $\mathscr{C}_{i-1}$ being the respective invariant

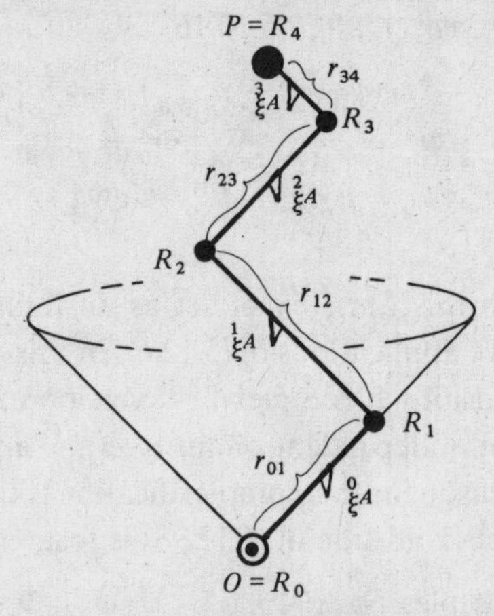


Fig. 5-12. The geometry for (5.12.96) and (5.12.94). (Second 'mass scattering' for the Dirac field.)

volume element 3-forms for the future and past light cones of $P_i$. Also we write

$$(\overrightarrow{R_i R_{i+1}})^a =: r_{i,i+1} \overset{i}{\xi}{}^A \overset{\bar{i}}{\bar{\xi}}{}^{A'} \tag{5.12.96}$$

(see Fig. 5-12), and

$$z_i = \overset{i}{\xi}_A \overset{i-1}{\xi}{}^A, \tag{5.12.97}$$

The general pattern should be clear. Substituting into (5.12.75), we get, for the final field $\psi_A$, the sum of two infinite series, the first involving the null-datum $\psi$ and being over the spaces of null zig-zags with an even number of segments, and the second involving the null-datum $\chi$ and being over the spaces of null zig-zags with an odd number of segments. The result for $\chi_{A'}$ is similar, but the other way about. All this is clearly not intended as a practical procedure, but the result is perhaps suggestive (particularly with respect to the twistor formalism that will be introduced in Vol. 2, Chapter 6, since there descriptions in terms of null lines will have a particular role to play). (Compare Feynman and Hibbs 1965.)

### *Maxwell–Dirac equations*

We end this discussion by briefly indicating a corresponding treatment of the Maxwell–Dirac system. The equations to be satisfied (using the Lorenz gauge) are (with $\varphi_{AB} = \varphi_{BA}$, and constant real $q$, $e$, where we can normalize so that $q = 2\pi e$; *cf*. after (5.10.16)):

$$\nabla^{B}_{A'}\varphi_{AB} = 2\pi e(\psi_A\bar{\psi}_{A'} + \chi_{A'}\bar{\chi}_A) \quad (= \nabla^{B'}_{A}\bar{\varphi}_{A'B'}) \tag{5.12.98}$$

$$\nabla_{AA'}\Phi^{A'}_{B} = \varphi_{AB}, \quad \nabla_{AA'}\Phi^{A}_{B'} = \bar{\varphi}_{A'B'} \tag{5.12.99}$$

$$(\nabla^{A}_{A'} - \mathrm{i}e\Phi^{A}_{A'})\psi_A = \mu\chi_{A'}, \quad (\nabla^{A'}_{A} - \mathrm{i}e\Phi^{A'}_{A})\chi_{A'} = \mu\psi_A. \tag{5.12.100}$$

As before, we rewrite this finite exact set as an infinite one which can be solved by successive applications of (5.12.6). In this, we do not separate $\Phi_a$ from $\varphi_{AB}$, and consider a 'complexified' version of (5.12.99), in which $\bar{\varphi}_{A'B'}$ is replaced by an independent quantity $\tilde{\varphi}_{A'B'}$, and which is to hold for *each* order. (The reason for this complexification is that, at higher order, the reality of the right-hand side of (5.12.98) is lost, *cf*. (5.12.112) below.) Thus we shall have triples $(\overset{0}{\varphi}_{AB}, \overset{0}{\Phi}_a, \overset{0}{\tilde{\varphi}}_{A'B'})$, $(\overset{1}{\varphi}_{AB}, \overset{1}{\Phi}_a, \overset{1}{\tilde{\varphi}}_{A'B'})$, ..., related at each stage by the complexified version of (5.12.99). The various potentials $\overset{0}{\Phi}_a, \overset{1}{\Phi}_a, \ldots$ are fixed in terms of the fields by the requirement that *all* their null-data $\overset{0}{\Phi}, \overset{1}{\Phi}, \ldots$ vanish.

To begin with, we have the 'zeroth order' fields $\overset{0}{\varphi}_{AB}$ (with $\overset{0}{\tilde{\varphi}}_{A'B'} = \overset{0}{\bar{\varphi}}_{A'B'}$), $\overset{0}{\psi}_A, \overset{0}{\chi}_{A'}$ satisfying the massless free-field equations

$$\nabla^{A}_{A'}\overset{0}{\varphi}_{AB} = 0, \quad \nabla^{A}_{A'}\overset{0}{\psi}_A = 0, \quad \nabla^{A'}_{A}\overset{0}{\chi}_{A'} = 0, \tag{5.12.101}$$

together with a real potential $\overset{0}{\Phi}_a\ (= \overset{0}{\bar{\Phi}}_a)$ satisfying

$$\nabla_{AA'}\overset{0}{\Phi}{}^{A'}_{B} = \overset{0}{\varphi}_{AB}. \tag{5.12.102}$$

(Although the 'zeroth order' potential is real, we shall see that this cannot be maintained in our procedure for the 'higher order' potentials.) The solutions of (5.12.101) from null-data are given by (5.12.6), as before, and to proceed to a second stage it is again useful to think of the fields given by (5.12.6) as linearly composed of $\delta$-function contributions centred on the points of the future light cone $\mathscr{N}$ of $O$. In the case of $\overset{0}{\varphi}_{AB}$ we have the contribution

$$z_1^{-2}\eta_A\eta_B\Delta_1(1)\overset{0}{\varphi}(1) \tag{5.12.103}$$

(with $\eta_A\xi^A = z_1$, and $\overset{0}{\varphi}(1) = q^{-2}\partial(q^3\varphi)/\partial q$ at $R_1$, in analogy with (5.12.93)), centered at the point $R_1 \in \mathscr{N}$. And we obtain the solution

$$z_1^{-1}\bar{z}_1^{-1}r^{-1}(z_1^{-1}\eta_B\bar{\xi}_{B'}\varphi(1) + \bar{z}_1^{-1}\xi_B\bar{\eta}_{B'}\bar{\varphi}(1))\Delta_0(1) \tag{5.12.104}$$

of (5.12.102), for the corresponding contribution to the potential $\overset{0}{\Phi}_{BB'}$ (with vanishing null-datum on $\mathscr{N}$).

In contrast to what happens in the case of the free Dirac field discussed earlier, the 'higher order' fields in the Maxwell–Dirac case do not emerge in any very natural linear order, and would more appropriately be labelled by 'trees' rather than by integers. But since we aim only to give an indication here, and not a fully detailed treatment, we shall not so label them. The numbers appearing above the field symbols will therefore have no significance and refer only to the order in which we choose to consider these fields.

At the next stage, in place of (5.12.100), we have the equations

$$\nabla^A_{A'} \overset{1}{\psi}_A = \mu \overset{0}{\chi}_{A'}, \quad \nabla^{A'}_A \overset{1}{\chi}_{A'} = \mu \overset{0}{\psi}_A \tag{5.12.105}$$

and

$$\nabla^A_{A'} \overset{2}{\psi}_A = \mathrm{i}e \overset{0}{\Phi}{}^A_{A'} \overset{0}{\psi}_{A}, \quad \nabla^{A'}_A \overset{2}{\chi}_{A'} = \mathrm{i}e \overset{0}{\Phi}{}^{A'}_A \overset{0}{\chi}_{A'}. \tag{5.12.106}$$

We have already seen how to deal with (5.12.105). To solve the first of equations (5.12.106), we express $\overset{0}{\varphi}_{AB}$ and $\overset{0}{\psi}_A$ in terms of $\delta$-function contributions like (5.12.103) and like

$$z_2^{-1} \zeta_A \Delta_1(2) \overset{0}{\psi}(2), \tag{5.12.107}$$

respectively, where for the notation see Fig. 5-13, and where $z_2 = \zeta_A \theta^A$. (We do not require $R_2$ to lie on $\mathscr{N}$; so the solutions will still apply when $\overset{2}{\psi}_A$ is replaced by a higher order field.) Appropriately transvecting (5.12.104) with (5.12.107), we obtain the required right-hand side of (5.12.106)(1). A solution of that equation can then be found with $\overset{2}{\psi}_A$ in the form

$$U \zeta_A \Delta_1(2) \Delta_0(1) + V_A \Delta_0(2) \Delta_0(1). \tag{5.12.108}$$

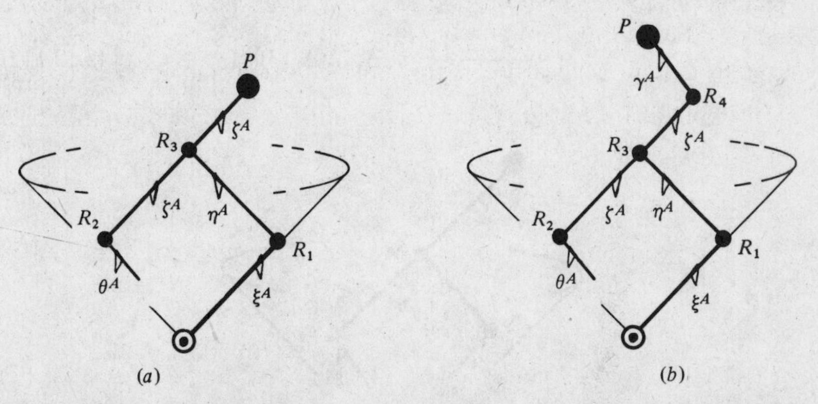


Fig. 5-13. The geometry for (5.12.107) and (5.12.108). (Scattering of the Dirac field by the Maxwell field: (a) phase shift, (b) scattering.)

The first term of (5.12.108) gives, in effect, a 'phase shift' in the field propagating from $R_2$ as it crosses into the region of potential generated from $R_1$, and corresponds to Fig. 5-13*a*; the second term gives the true scattering, and corresponds to Fig. 5-13*b*. We obtain the null-datum for the $V_A$ of this scattering term on an initial hypersurface that consists of the portions of the light cones of $R_1$ and $R_2$ which lie to the future of the intersection of these cones. The null-datum on this portion of $R_1$'s cone vanishes, so the null-datum on this portion of $R_2$'s cone yields the field at $P$ via (5.12.6). The configurations to be integrated over are those shown in Fig. 5-13*b*. The calculations are simple enough, but will be omitted here.

We must also examine the equations arising from (5.12.98). These have the form

$$\nabla^B_{A'}\overset{4}{\varphi}_{AB} = 2\pi e\overset{2}{\psi}_A\overset{3}{\bar{\psi}}_{A'} = \nabla^{B'}_A\overset{4}{\tilde{\varphi}}_{A'B'} \quad\text{and}\quad \nabla^B_{A'}\overset{5}{\varphi}_{AB} = 2\pi e\overset{2}{\chi}_{A'}\overset{3}{\bar{\chi}}_A = \nabla^{B'}_A\overset{5}{\tilde{\varphi}}_{A'B'}. \tag{5.12.109}$$

Let us consider the first of these, and let us regard $\overset{2}{\psi}_A$ and $\overset{3}{\psi}_A$ as composed of $\delta$-function contributions based on points $R_2$ and $R_3$, respectively, not necessarily on $\mathscr{N}$. Then we obtain a source term in (5.12.109)(1) of the form

$$(z_2^{-1}\eta_A\Delta_1(2))(\bar{z}_3^{-1}\bar{\zeta}_{A'}\Delta_1(3)), \tag{5.12.110}$$

multiplied by the appropriate null-datum quantities $\overset{2}{\psi}(2), \overset{3}{\bar{\psi}}(3)$. Taking

$$z_2 = \eta_A\xi^A, \quad z_3 = \zeta_A\theta^A, \quad z_4 = \eta_A\zeta^A, \tag{5.12.111}$$

with $\theta^A$, $\zeta^A$, $\xi^A$, $\eta^A$ as in Fig. 5-14, we find, for a source term (5.12.110), a solution $\overset{4}{\varphi}_{AB}$ of the form

$$\frac{2\pi e}{r_{34}z_4z_2\bar{z}_3}\eta_A\eta_B\Delta_1(2)\Delta_0(3). \tag{5.12.112}$$

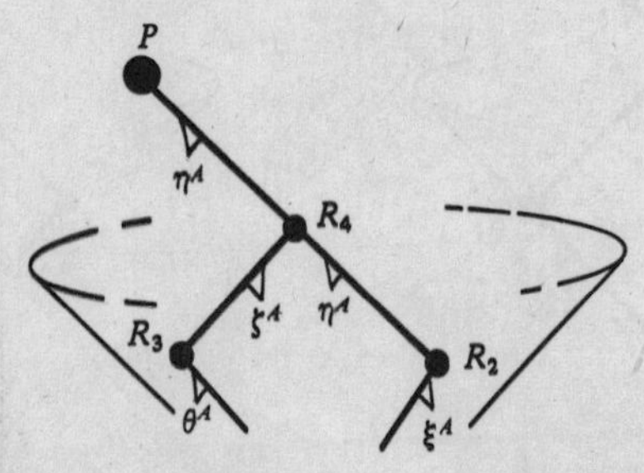


**Fig. 5-14. The geometry for (5.12.110) and (5.12.112). (Production of Maxwell field from Dirac current.)**

This form shows that the field $\overset{4}{\varphi}_{AB}$ at $P$ arises from contributions associated with the configurations indicated in Fig. 5-14.

Since (5.12.110) is not real, $\overset{4}{\tilde{\varphi}}_{A'B'}(\neq \overset{4}{\bar{\varphi}}_{A'B'})$ is given, correspondingly, by

$$\frac{-2\pi e}{r_{24}\bar{z}_4 z_2 \bar{z}_3}\bar{\zeta}_{A'}\bar{\zeta}_{B'}\Delta_0(2)\Delta_1(3). \tag{5.12.113}$$

In place of (5.12.99) we have

$$\nabla_{AA'}\overset{4}{\Phi}{}^{A'}_{B} = \overset{4}{\varphi}_{AB}, \quad \nabla_{AA'}\overset{4}{\Phi}{}^{A}_{B'} = \overset{4}{\tilde{\varphi}}_{A'B'}, \tag{5.12.114}$$

with right-hand sides given by (5.12.112) and (5.12.113), respectively, from which we find, for $\overset{4}{\Phi}_a$,

$$\frac{2\pi e}{(r_{45}r_{46} - r_{35}r_{26})z_4\bar{z}_4 z_2\bar{z}_3}\eta_A\bar{\zeta}_{A'}\Delta_0(2)\Delta_0(3). \tag{5.12.115}$$

Our notation here is similar to that used before, and we have

$$(\overrightarrow{R_4R_5})^a = r_{45}\zeta^A\bar{\zeta}^{A'} = (\overrightarrow{R_6P})^a, \quad (\overrightarrow{R_4R_6})^a = r_{46}\eta^A\bar{\eta}^{A'} = (\overrightarrow{R_5P})^a, \tag{5.12.116}$$

$$(\overrightarrow{R_3R_5})^a = r_{35}\zeta^A\bar{\zeta}^{A'}, \quad (\overrightarrow{R_2R_6})^a = r_{26}\eta^A\bar{\eta}^{A'} \tag{5.12.117}$$

(see Fig. 5-15), the six points $R_2, \ldots, R_6$, and $P$ (at which the potential is being evaluated, lying in one 2-plane.

In verifying that (5.12.115), (5.12.112), and (5.12.113) together satisfy (5.12.114) (and also that (5.12.112), (5.12.110) and (5.12.113) together satisfy (5.12.109)(1)), it is useful to note that, as $P$ varies and $R_2$ and $R_3$ (and $\xi^A, \theta^A$) remain constant, the vector $\overrightarrow{R_3R_4} - \overrightarrow{R_2R_4}$ is constant, while $\overrightarrow{R_3R_5} + \overrightarrow{R_5P}$ and $\overrightarrow{R_2R_6} + \overrightarrow{R_6P}$ differ from the position vector (of $P$)

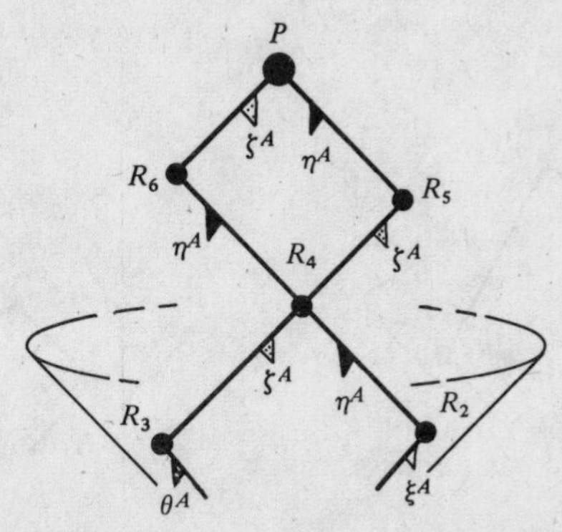


Fig. 5-15. The geometry for (5.12.115)–(5.12.117). (Production of Maxwell potential from Dirac current.)

by a constant. Substituting (5.12.116) and (5.12.117), and differentiating, we obtain the various required relations. In particular, we find that

$$r_{24}r_{34}z_4\bar{z}_4 \quad \text{and} \quad K := r_{34}\bar{z}_3\xi_A\zeta^A - r_{24}z_2\bar{\theta}_{A'}\bar{\eta}^{A'} \qquad (5.12.118)$$

remain constant. The potential (5.12.115), in the field-free region consisting of the common future of the cones of $R_2$ and $R_3$, turns out to be the gradient of the complex scalar

$$\frac{2\pi e}{K}\log\left(\frac{r_{34}\bar{z}_3 z_4}{z_2}\right) = \text{constant} - \frac{2\pi e}{K}\log\left(\frac{r_{24}z_2\bar{z}_4}{z_3}\right). \qquad (5.12.119)$$

Finally, we need to see how the potential (5.12.115) effects the scattering of the two parts of the Dirac field. The procedure is the same as that used for the potential (5.12.104). For $\overset{7}{\psi}_A$ we substitute into (5.12.106)(1) an expression like (5.12.108), but with $R_2$ replaced by $R_7$ (and $R_1$ by, say, $R_3$), the two terms now involving the *triple* products $\Delta_1(7)\Delta_0(2)\Delta_0(3)$ and $\Delta_0(7)\Delta_0(2)\Delta_0(3)$, respectively. The right-hand side is given by (5.12.115) (with $R_2$ and $R_3$ as given) transvected with an expression like (5.12.107) (with $R_2$ replaced by $R_7$), and represents the previous-order $\psi_A$ field emanating from $R_7$. As before, the first term gives merely a 'phase shift' and is represented in terms of a diagram like Fig. 5-16(*a*) (or like Fig. 5-16(*a*) with $R_2$ and $R_3$ interchanged). The scattering is given by the second term and is represented by Fig. 5-16(*b*) (or Fig. 5-16(*b*) with $R_2$ and $R_3$ interchanged).

To evaluate the complete Maxwell–Dirac field, we sum an infinite number of terms, each of which is an integral over a finite-dimensional

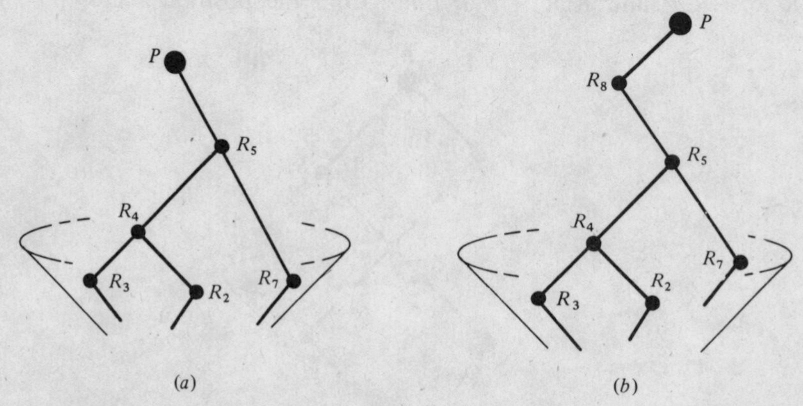


Fig. 5-16. Geometry for scattering of Dirac field by Dirac current, with intermediary Maxwell potential: (*a*) phase shift, (*b*) scattering.

compact space representing forked null zig-zags starting at null-data points on $\mathcal{N}$ and terminating at $P$, which are obtained by appropriately combining the configurations that we have been considering. Each integral is necessarily finite, but the terms increase in complexity as the series extends. We do not discuss the full details of these expressions, nor do we touch upon the question of convergence of the series. These are matters that would repay further investigation.

One of the topics introduced in Vol. 2 is the machinery of twistor theory. We shall see, in particular, that twistors provide a direct and elegant representation of null straight lines in $\mathbb{M}$, and the transcription of the formulae of this section into twistor terms would seem to present an interesting and perhaps significant exercise. It would be interesting also to develop our procedure (with or without twistors) into a full description of quantum electrodynamics (*cf.* Bjorken and Drell 1964).

The diagrams arising here are in many ways analogous to Feynman diagrams. But there is the unusual feature that here we are concerned only with *null* space–time separations, even for massive fields. The view that null separations are more fundamental than spacelike or timelike ones goes hand-in-hand with a philosophy that we have tended to promote in this book, either directly or indirectly, that 2-spinors are to be regarded as more *fundamental* than vectors or tensors in the description of space–time structure.

# Appendix
# Diagrammatic notation

One problem confronting anyone who works extensively with tensors, spinors or similar structures is that of notation. Formulae tend to become encumbered with numerous small indices. Furthermore, each actual letter used in an expression has no importance in itself, only the correspondence between letters in different places having significance. Of course this feature of notation is also present for many other types of expression in mathematics, for example in the integral

$$L(x) = \int_1^x y^{-1} \log y \, \mathrm{d}y$$

'$y$' is a dummy letter and also, with regard to the equation as a whole, the letter '$x$' has no particular significance. However, in complicated tensor expressions, a large number of such 'meaningless' letters may have to be used. This is particularly true of many of the rough calculations which quite properly never find their way into print. The indices tend to be small and consequently may be only barely distinguishable from one another. The all-important associations between the different positions of a letter in an expression may be hard to discern without a careful search.

One kind of way around this notational problem has been often suggested. This involves inventing special index-free notations for particular operations (e.g., certain contracted or anti-symmetrized products) and then attempting to express every other operation of interest in terms of these. The well-known three-dimensional vector algebra involving scalar and vector products is one example. The Grassmann (or Cartan) calculus of skew forms is another (*cf.* §§4.3, 4.13). The scope of such notations for tensors (or spinors) generally, however, appears to be somewhat limited and the transparency of the basic rules of tensor (spinor) algebra may well be lost. Another notation (Penrose 1971, Cvitanović 1976, Cvitanović and Kennedy 1982) which uses explicit *diagrams*, is described here. This avoids the use of indices as such, but retains this transparency of the basic rules of operation. The notation has been found very useful in practice as it greatly simplifies the appearance of complicated tensor or

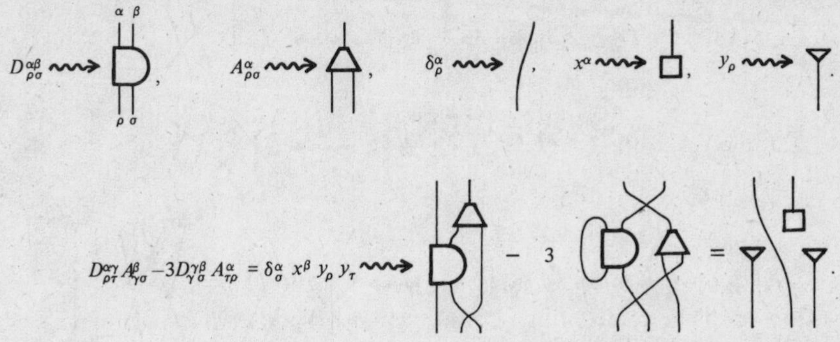

$$D^{\alpha\gamma}_{\rho\tau} A^{\beta}_{\gamma\sigma} - 3 D^{\gamma\beta}_{\gamma\sigma} A^{\alpha}_{\tau\rho} = \delta^{\alpha}_{\sigma} x^{\beta} y_{\rho} y_{\tau}$$

Fig. A-1. Diagrammatic representation of a tensor equation.

spinor equations, the various interrelations expressed being discernable at a glance. Unfortunately the notation seems to be of value mainly for private calculations because it cannot be printed in the normal way.

We first describe the general scheme for representing tensor formulae according to this notation. Some specific suggestions for spinors are made afterwards, though many variations can be made to suit specific needs. The basic idea lies in the representation of a contraction not by a pair of identical letters in different parts of an expression, but by a line connecting the relevant points in the tensor symbols. Each term now becomes a diagram.

In order to keep the appearance of these diagrams as simple as possible, products of tensors or spinors may be represented not merely by the horizontal juxtaposition of symbols but also by vertical or oblique juxtapositions. (The commutative and associative laws of multiplication ensures the consistency of this.) Uncontracted indices are represented by lines with free ends; upper (contravariant) indices may be represented by lines terminating at the top of the diagram ('arms') and lower (covariant) indices by lines terminating at the bottom ('legs'). Index permutation is represented by crossing over of lines. Addition (or subtraction) is generally represented in the normal way with a '+' (or '−') sign between diagrams. To see which index-line end-points correspond to one another in a sum, we imagine the diagrams for the different terms to be superimposed upon one another. An example is given in Fig. A-1.

A characteristic feature of the notation is the way in which Kronecker deltas and expressions built up from Kronecker deltas are to be represented. A single line (generally more or less vertical) which is unattached to any other tensor or spinor symbol represents a Kronecker delta. The laws

$$\delta^{\alpha}_{\beta}\delta^{\beta}_{\gamma} = \delta^{\alpha}_{\gamma}, \quad A_{\gamma}\delta^{\gamma}_{\rho} = A_{\rho}, \quad B^{\alpha}\delta^{\beta}_{\alpha} = B^{\beta}$$

$$P^{\alpha_1\cdots\alpha_6}_{\beta_1\cdots\beta_6} = \delta^{\alpha_1}_{\beta_2}\,\delta^{\alpha_2}_{\beta_1}\,\delta^{\alpha_3}_{\beta_3}\,\delta^{\alpha_4}_{\beta_6}\,\delta^{\alpha_5}_{\beta_4}\,\delta^{\alpha_6}_{\beta_5}$$

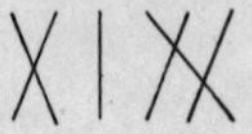

$$Q^{\alpha_1\cdots\alpha_6}_{\beta_1\cdots\beta_6} = \delta^{\alpha_1}_{\beta_4}\,\delta^{\alpha_2}_{\beta_2}\,\delta^{\alpha_3}_{\beta_1}\,\delta^{\alpha_4}_{\beta_5}\,\delta^{\alpha_5}_{\beta_6}\,\delta^{\alpha_6}_{\beta_3}$$

$$P^{\alpha_1\cdots\alpha_6}_{\beta_1\cdots\beta_6}\,Q^{\beta_1\cdots\beta_6}_{\gamma_1\cdots\gamma_6} = \delta^{\alpha_1}_{\gamma_2}\,\delta^{\alpha_2}_{\gamma_4}\,\delta^{\alpha_3}_{\gamma_1}\,\delta^{\alpha_4}_{\gamma_3}\,\delta^{\alpha_5}_{\gamma_5}\,\delta^{\alpha_6}_{\gamma_6}$$

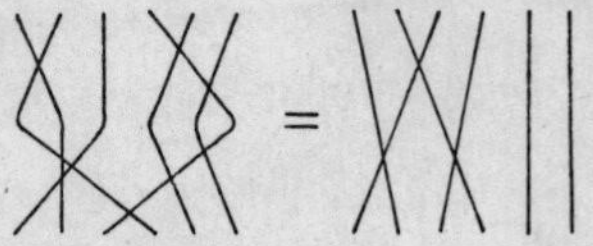

Fig. A-2. Permutation operators as products of Kronecker deltas.

are implicit in this notation, since to express the left-hand side in each case one need only extend an already existent line. Permutation operators now find a natural representation in terms of crossed lines ('Aitken's diagrams', Aitken 1958). To find the product of two permutations it is only necessary to place one diagram above the other, join the corresponding adjacent end-points and straighten out the lines. See Fig. A-2. This represents a product such as

$$\delta^{\alpha_1}_{\beta_{p(1)}} \ldots \delta^{\alpha_r}_{\beta_{p(r)}}\,\delta^{\beta_1}_{\gamma_{q(1)}} \ldots \delta^{\beta_r}_{\gamma_{q(r)}} = \delta^{\alpha_1}_{\gamma_{qp(1)}} \ldots \delta^{\alpha_r}_{\gamma_{qp(r)}}$$

where $p$ and $q$ are permutations of $1, 2, \ldots, r$ with $qp$ their product. A permutation is even or odd according as the number of crossing points (assumed to be simple intersections) is even or odd.

Symmetrization and anti-symmetrization operators are now sums and differences of such expressions. It is convenient to have special symbols to denote these operators, so that they can be used in products without our having to write out the sum explicitly. A set of $r$ vertical lines with a horizontal wavy line through them then denotes the sum of all the permutation symbols with the same end-points (Fig. A-3). This is $r! \times$ the

$$\begin{vmatrix} \delta^{\alpha_1}_{\beta_1} & \cdots & \delta^{\alpha_1}_{\beta_r} \\ \vdots & & \vdots \\ \delta^{\alpha_r}_{\beta_1} & \cdots & \delta^{\alpha_r}_{\beta_r} \end{vmatrix}^{+}, \qquad \delta^{\alpha_1\cdots\alpha_r}_{\beta_1\cdots\beta_r} = \begin{vmatrix} \delta^{\alpha_1}_{\beta_1} & \cdots & \delta^{\alpha_1}_{\beta_r} \\ \vdots & & \vdots \\ \delta^{\alpha_r}_{\beta_1} & \cdots & \delta^{\alpha_r}_{\beta_r} \end{vmatrix}$$

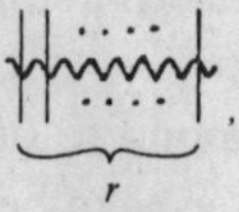

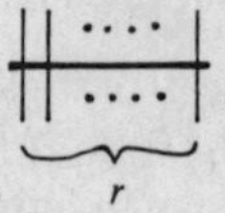

e.g.,

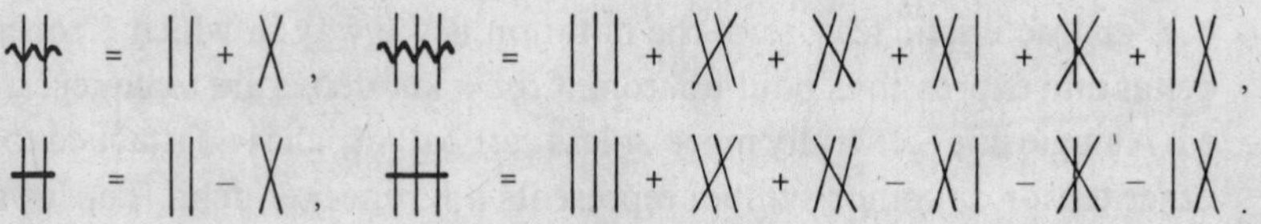

Fig. A-3. Symmetrizers and anti-symmetrizers.

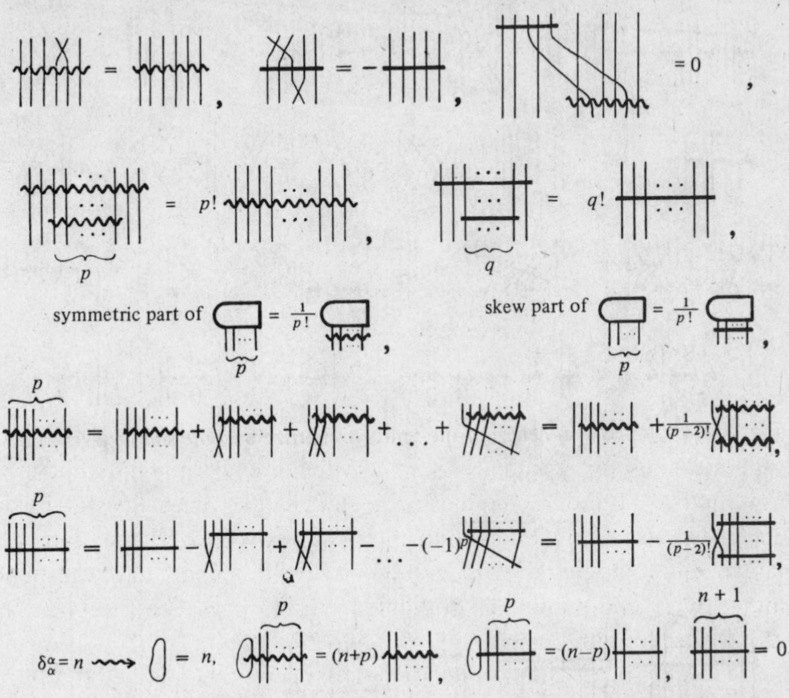

Fig. A-4. Some properties of the symmetrizer and anti-symmetrizer tensors.

symmetrization operator. Similarly, if the wavy line is replaced by a horizontal straight line, the symbol now represents $r! \times$ the anti-symmetrization operator. This is the sum of all the positive permutations minus the sum of the negatiye permutations and the tensor obtained is the 'generalized Kronecker delta'

$$\delta^{\alpha_1 \ldots \alpha_r}_{\beta_1 \ldots \beta_r}.$$

To represent the symmetric part of a set of $r$ indices, therefore, we draw a wavy line across the relevant index-lines and divide the expression by $r!$ Correspondingly, for the anti-symmetric part of a tensor we draw a straight line. (If preferred, the factor $(r!)^{-1}$ could be incorporated into the definitions of these symbols. The form chosen here has some computational advantages for the detailed expansion of diagrams into their constituent parts.) Some simple identities are given in Fig. A-4 which are of considerable help when diagrams are manipulated. The dimension of the space is $n$. Sometimes it is convenient to express symmetries or anti-symmetries in *groups* of indices (see p. 134). In this case the horizontal wavy line or straight line is drawn only *between* the groups without the

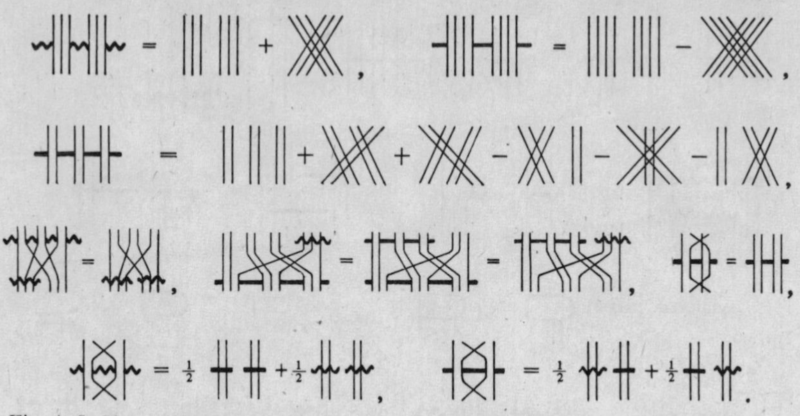

Fig. A-5. Clumped symmetrizers and anti-symmetrizers; some simple properties.

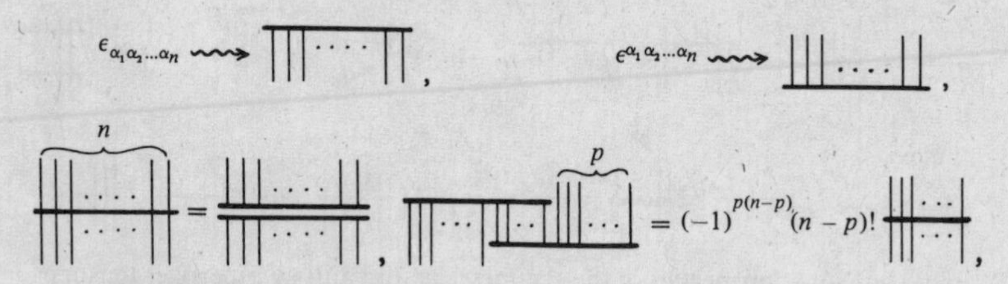


Fig. A-6. Alternating tensor and properties.

line continuing across the lines within a group (see Fig. A-5). A convenient notation for the alternating tensors, which brings out their relationship with the generalized Kronecker deltas, is a horizontal line at which the $n$ vertical index lines originate (Fig. A-6). In the covariant case the vertical lines point downwards and in the contravariant case, upwards.

When a (symmetric, non-degenerate) metric tensor $g_{\alpha\beta}$ is present it is convenient to represent this by a single 'hoop' as in Fig. A-7. This is the Kronecker delta line 'bent over', and the tensor $g^{\alpha\beta}$ may correspondingly

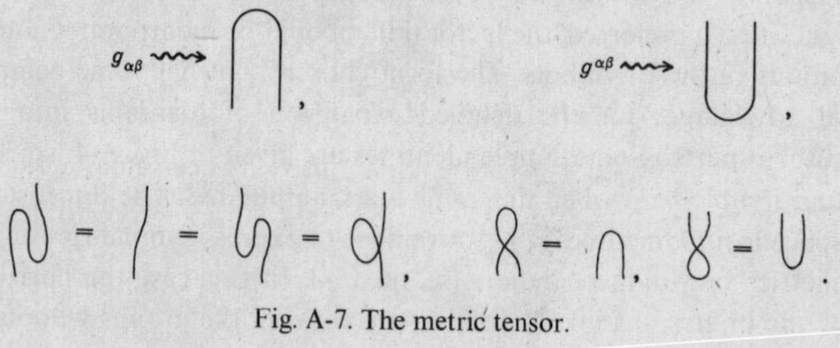

Fig. A-7. The metric tensor.

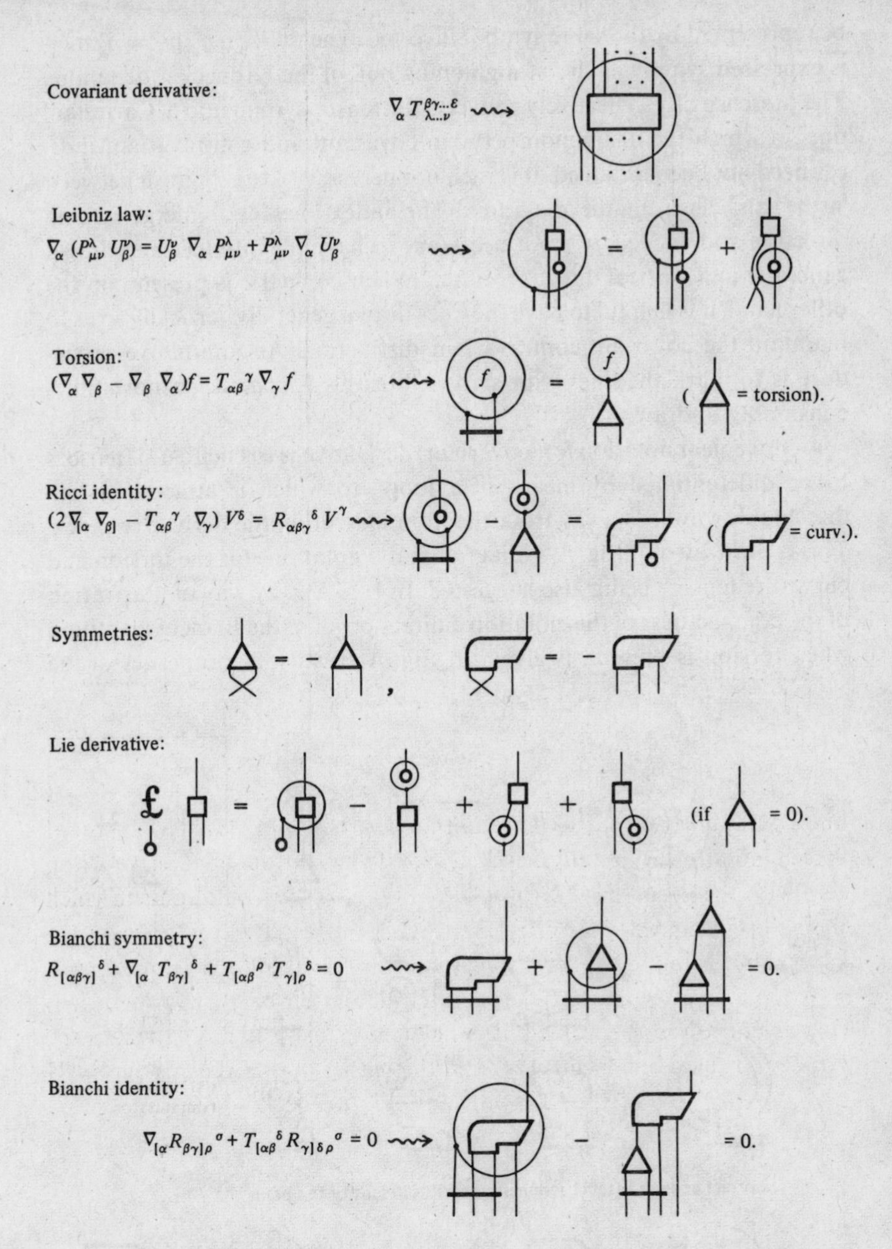


Fig. A-8. The 'loop' notation for (covariant) derivative; curvature, torsion.

be represented by the same symbol inverted. The relation $g_{\alpha\beta}g^{\gamma\beta} = \delta_\alpha^\gamma$ then is expressed symply as the 'straightening out' of the Kronecker delta line. The presence of $g_{\alpha\beta}$ effectively converts the tensor system into a 'Cartesian' one, in which the distinction between covariant and contravariant indices need not be maintained. It is then not necessary to distinguish between 'arms' and 'legs' in the diagrams. The index-lines can emerge in any direction and, indeed, it is not necessary to have the internal lines drawn generally in a vertical direction either. When *no* metric is present, on the other hand, it is helpful to have the lines drawn generally vertically so as to maintain the covariant/contravariant distinction. An alternative procedure is to mark the lines with arrows, but this is somewhat more time-consuming to draw.

A convenient notation for (covariant) derivative is to encircle all tensors to be differentiated by means of a loop – to which is attached a line directed downwards away from the loop indicating the derivative index. This is illustrated in Fig. A-8, diagrammatic notations for the torsion and curvature tensors being also suggested. In Fig. A-9, by way of illustration of the compactness of this notation a direct proof of the Bianchi identities, when torsion is present, is given. In Fig. A-10 various properties of the

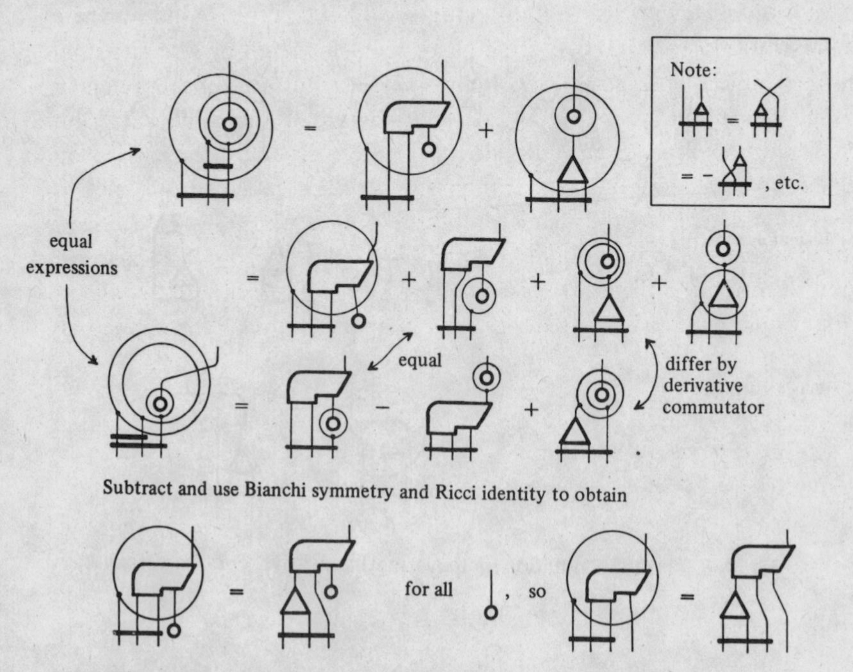


Fig. A-9. Proof, by diagrams, of the Bianchi identity with torsion.

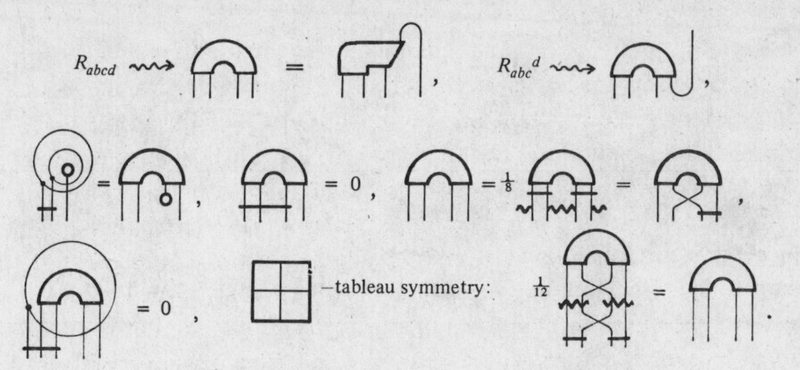


Fig. A-10. The Riemann tensor and its basic properties.

Riemann tensor are illustrated which hold in standard (pseudo-) Riemannian geometry (including the tableau property mentioned on p. 144).

For *spinors* various notations are possible which clearly distinguish the unprimed from the primed indices, such as having the corresponding index-lines at two different angles to the vertical (Fig. A-11), or else using two colours. In practice this is often unnecessary and it is frequently adequate to use the same type of index-line for both types of index, where generally unprimed indices would be grouped at the left of each term and primed indices at the right. World-tensor indices can always be represented as *pairs* of spinor index-lines. A convenient notation for *complex conjugation* is simply *reflection in the vertical*. This entails that the ordering on the page of the index lines representing $AB \ldots D$ must be opposite to that of $A'B' \ldots D'$ in the representation of a spinor $\phi_{AB\ldots DA'B'\ldots D''}$ (since, for example, this spinor might be a *real world-tensor* $\phi_{ab\ldots d}$ whose description, according to the above convention with regard to complex conjugation, would have to be left-right reflection-symmetric). For consistency we

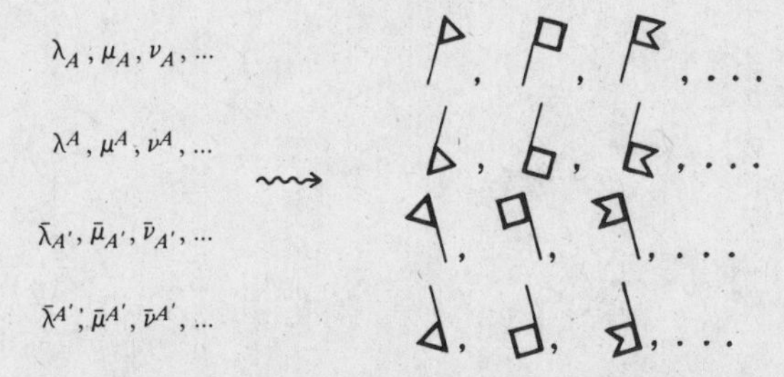


Fig. A-11. Spin-vectors; duals, complex conjugates.

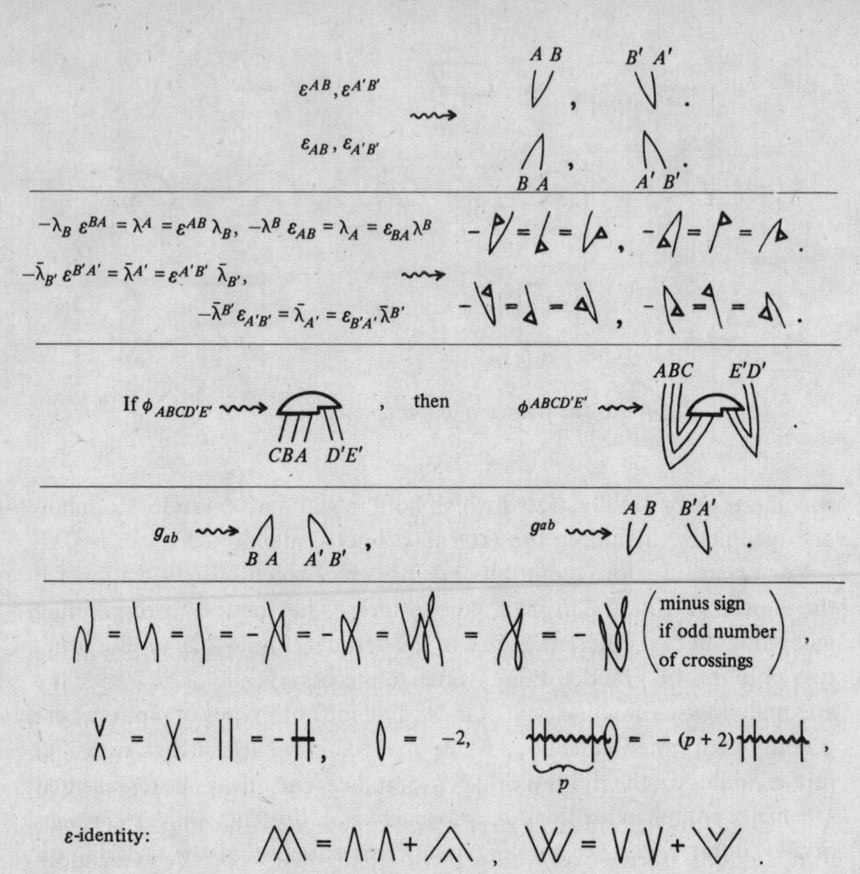


Fig. A-12. The ε-spinors and their basic properties. Note the ordering of index labels. As in the last block of diagrams, the sloping of lines need not be strictly adhered to if no confusion thereby arises. (The ε-identity may be used to convert expressions to a basis in which all lines are uncrossed).

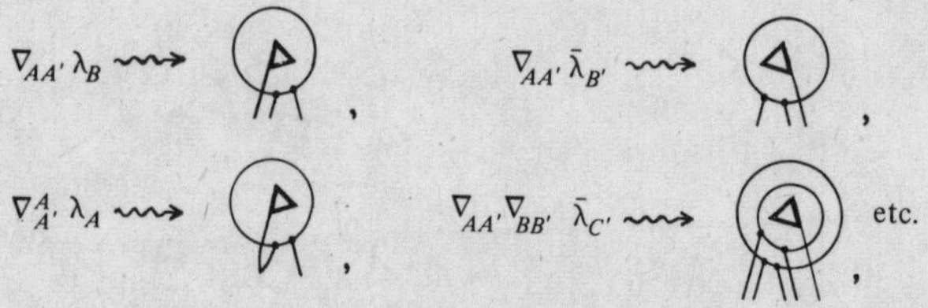

$\nabla_{AA'}\lambda_B$, $\nabla_{AA'}\bar{\lambda}_{B'}$, $\nabla^{A}_{A'}\lambda_A$, $\nabla_{AA'}\nabla_{BB'}\,\bar{\lambda}_{C'}$, etc.

Fig. A-13. Spinor covariant derivative.

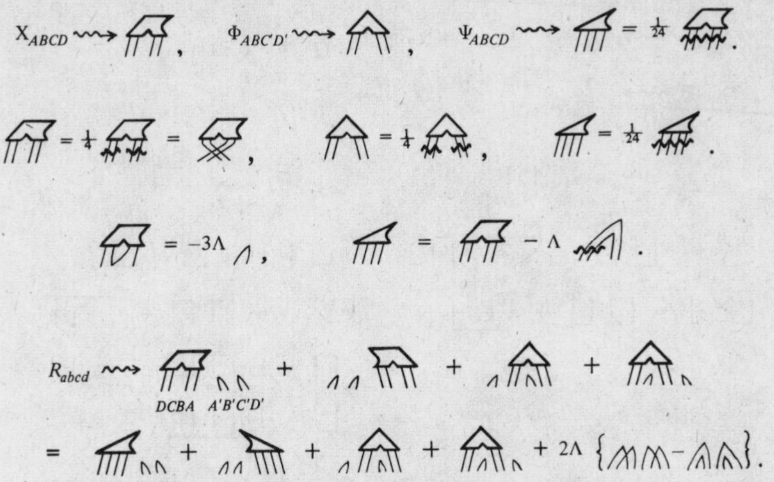

Fig. A-14. The spinor curvature quantities.

adopt the convention that the ordering of the end-points of the 'unprimed arms' and of the 'primed legs' is the *same* as that of the index letters in the ordinary spinor expression that is being represented, whereas for the 'primed arms' and 'unprimed legs' it is the *opposite* order. (*cf.* Fig. A-12.) The notation described in Fig. A-6 for alternating symbols generally can be used for the ε-spinors, but a more convenient notation is given in Fig. A-12 which has the advantage of compactness, and the somewhat awkward spinor sign rules for raising and lowering indices can be made more memorable.

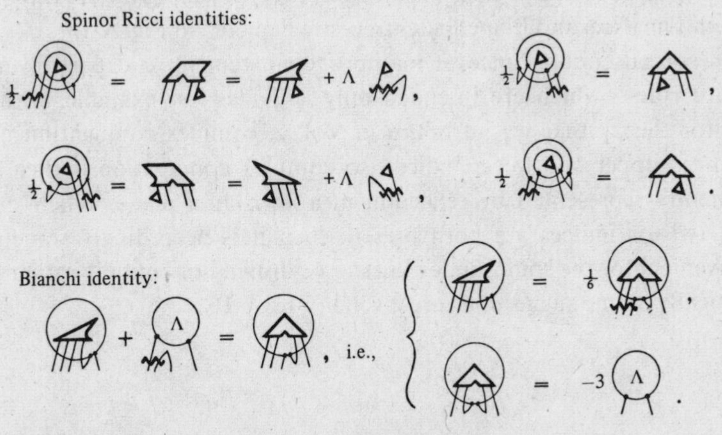


Fig. A-15. Spinor Ricci and Bianchi identities.

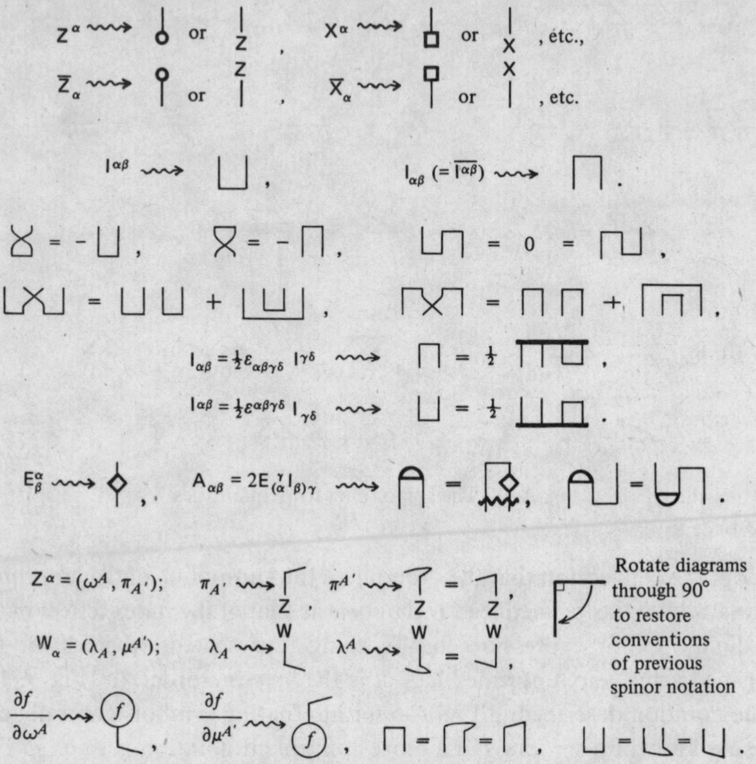

$Z^\alpha \rightsquigarrow$ or , $X^\alpha \rightsquigarrow$ or , etc.,

$\bar{Z}_\alpha \rightsquigarrow$ or , $\bar{X}_\alpha \rightsquigarrow$ or , etc.

$I^{\alpha\beta} \rightsquigarrow$ , $I_{\alpha\beta}\ (= \overline{I^{\alpha\beta}}) \rightsquigarrow$ .

$I_{\alpha\beta} = \frac{1}{2}\varepsilon_{\alpha\beta\gamma\delta}\, I^{\gamma\delta} \rightsquigarrow$

$I^{\alpha\beta} = \frac{1}{2}\varepsilon^{\alpha\beta\gamma\delta}\, I_{\gamma\delta} \rightsquigarrow$

$E^\alpha_\beta \rightsquigarrow$ , $A_{\alpha\beta} = 2E_{(\alpha}{}^{\gamma} I_{\beta)\gamma} \rightsquigarrow$

$Z^\alpha = (\omega^A, \pi_{A'})$; $\pi_{A'} \rightsquigarrow$ , $\pi^{A'} \rightsquigarrow$

$W_\alpha = (\lambda_A, \mu^{A'})$; $\lambda_A \rightsquigarrow$ , $\lambda^A \rightsquigarrow$

Rotate diagrams through 90° to restore conventions of previous spinor notation

$\frac{\partial f}{\partial \omega^A} \rightsquigarrow$ , $\frac{\partial f}{\partial \mu^{A'}} \rightsquigarrow$

Fig. A-16. A diagrammatic notation for twistor theory.

In Fig. A-13 the notation for spinor covariant derivatives is indicated and in Fig. A-14 suggestions for the spinors representing the curvature are given. The Ricci and Bianchi identities are depicted in Fig. A-15.

In special circumstances it may prove convenient to depart from the above rules – which are intended only a guide. For example, with the twistor theory that we introduce in Vol. 2, complex conjugation interchanges upper and lower indices, so complex conjugation is then conveniently represented by reflection in a *horizontal* plane. When spinor and twistor indices are both present then it is accordingly sometimes convenient for the spinor index lines to be drawn horizontally rather than vertically. Some suggestions are given in Fig. A-16.

# References

Agrawala, V.K. and Belinfante, J.G. (1968). Graphical formulation of recoupling theory for any compact group. *Ann. Phys (N.Y.)* **49**, 130–70.

Aharanov, Y. and Bohm, D. (1959). Significance of electromagnetic Potentials in the quantum theory. *Phys. Rev.* **115**, 485–91.

Aharanov, Y. and Susskind, L. (1967). Observability of the sign change of spinors under $2\pi$ rotations. *Phys. Rev.* **158**, 1237–8.

Ahlfors, L.V. and Sario, L. (1960). *Riemann Surfaces* (Princeton University Press, Princeton).

Aichelburg, P.C. and Urbantke, H.K. (1981). Necessary and sufficient conditions for trivial solutions in supergravity. *Gen. Rel. Grav.* **13**, 817–28.

Aitken, A.C. (1949). *Determinants and Matrices*, 6th edn (Oliver & Boyd, Edinburgh).

Arnol'd, V.I. (1978). *Mathematical Methods of Classical Mechanics* (Springer, New York).

Arnowitt, R., Deser, S. and Misner, C.W. (1962). The dynamics of general relativity, in *Gravitation: An Introduction to Current Research*, ed. L. Witten (Wiley, New York).

Ashtekar, A., Horowitz, G.T. and Magnon-Ashtekar, A. (1982). A generalization of tensor calculus and its applications to physics. *Gen. Rel. Grav.* **14**, 411–28.

Bade, W.L. and Jehle, H. (1953). An introduction to spinors. *Rev. Mod. Phys.* **25**, 714–28.

Bass, R.W. and Witten, L. (1957). Remarks on cosmological models. *Rev. Mod. Phys.* **29**, 452–3.

Bel, L. (1959). Introduction d'un tenseur du quatrième ordre. *Comptes Rendus* **248**, 1297–300.

Bell, P. and Szekeres, P. (1972). Some properties of higher spin rest-mass zero fields in general relativity. *Int. J. Theor. Phys.* **6**, 111–21.

Bergmann, P.G. (1957). Two-component spinors in general relativity. *Phys. Rev.* **107**, 624–9.

Bernstein, H.J. (1967). Spin precession during interferometry of fermions and the phase factor associated with rotations through $2\pi$ radians. *Phys. Rev. Lett.* **18**, 1102–3.

Bjorken, J.D. and Drell, S.D. (1964). *Relativistic Quantum Mechanics* (McGraw-Hill, New York).

Bondi, H., van der Burg, M.G.J. and Metzner, A.W.K. (1962). Gravitational waves in general relativity. VII. Waves from axi-symmetric isolated systems. *Proc. Roy. Soc. London* **A269**, 21–52.

Bott, R. and Mather, J. (1968). Topics in topology and differential geometry, in *Battelle Rencontres*, ed. C.M. DeWitt and J.A. Wheeler (Benjamin, New York).

Brauer, R. and Weyl, H. (1935). Spinors in *n* dimensions. *Am. J. Math.* **57**, 425–49.

Bruhat, Y. (1962). The Cauchy problem, in *Gravitation: An Introduction to Current Research*, ed. L. Witten (Wiley, New York).

Buchdahl, H.A. (1958). On the compatibility of relativistic wave equations for particles of higher spin in the presence of a gravitational field. *Nuovo Cim.* **10**, 96–103.

Buchdahl, H.A. (1959). On extended conformal transformations of spinors and spinor equations. *Nuovo Cim.* **11**, 496–506.

Buchdahl, H.A. (1962). On the compatibility of relativistic wave equations in Riemann spaces. *Nuovo Cim.* **25**, 486–96.

Buchdahl, H.A. (1982). Ditto II; III. *J. Phys.* **A15**, 1–5; 1057–62.

Campbell, S.J. and Wainwright, J. (1977). Algebraic computing and the Newman–Penrose formalism in general relativity. *Gen. Rel. Grav.* **8**, 987–1001.

Carmeli, M. (1977). *Group theory and general relativity* (McGraw-Hill, New York).

Cartan, É. (1913), Les groupes projectifs qui ne laissent invariante aucune multiplicité plane. *Bull. Soc. Math. France* **41**, 53–96.

Cartan, É. (1923, 1924, 1925). Sur les variétés à connexion affine et la théorie de la relativité généralisée I, I(suite), II. *Ann. Ec. Norm. Sup.* **40**, 325–412; **41**, 1–25; **42**, 17–88.

Cartan, É. (1966). *The Theory of Spinors* (Hermann, Paris).

Chandrasekhar, S. (1979). An introduction to the theory of the Kerr metric and its perturbations, in *General Relativity: An Einstein Centenary Survey*, ed. S.W. Hawking and W. Israel (Cambridge University Press, Cambridge).

Chevalley, C. (1946). *Theory of Lie Groups* (Princeton University Press, Princeton).

Chevalley, C. (1954). *The Algebraic Theory of Spinors* (Columbia University Press, New York).

Choquet-Bruhat, Y., DeWitt-Morette, C. and Dillard-Bleick, M. (1977). *Analysis, Manifolds and Physics* (North-Holland, Amsterdam).

Christenson, J.H., Cronin, J.W., Fitch, V.L. and Turlay, R. (1964). Evidence for the $2\pi$ decay of the $K^0$ meson. *Phys. Rev. Lett.* **13**, 138–40.

Corson, E.M. (1953). *Introduction to Tensors, Spinors and Relativistic Wave Equations* (Blackie, Glasgow).

Courant, R. and Hilbert, D. (1965). *Methods of Mathematical Physics*, Vol. 1 (Interscience Publishers, New York).

Curtis, G.E. (1975). *Twistor Theory and the Collision of Plane Fronted Impulsive Gravitational Waves*. D. Phil. thesis, University of Oxford.

Cvitanovic, P. (1976). Group theory for Feynman diagrams in non-abelian gauge theories. *Phys. Rev.* **D14**, 1536–53.

Cvitanović, P. and Kennedy, A.D. (1982). *Spinors in Negative Dimensions*, Phys. Scripta. **26**, 5–14.

d'Adhémar, R. (1905). Sur une équation aux dérivées partielles du type hyperbolique. Etude de l'intégrale près d'une frontière caractéristique. *Rend. Circ. Matem. Palermo* **20**, 142–59.

Darmois, G. (1927). *Les équations de la gravitation Einsteinienne*. Mémorial des science Mathématiques, Fascicule XXV (Gauthier-Villars, Paris).

Debever, R. (1958). La super-énergie en relativité générale. *Bull. Soc. Math. Belgique* **10**, 112–47.

Deser, S. and Zumino, B. (1976). Consistent supergravity. *Phys. Lett.* **62B**, 335–7.

Dirac, P.A.M. (1928). The quantum theory of the electron. *Proc. Roy. Soc.* **A117**, 610–24; *ditto*, Part II, *ibid.* **A118**, 351–61.

Dirac, P.A.M. (1936*a*) Relativistic wave equations *Proc. Roy. Soc. London* **A155**, 447–59.

Dirac, P.A.M. (1936*b*). Wave equations in conformal space. *Ann. Math.* **37**, 429–42.

Dirac, P.A.M. (1939). La théorie de l'électron et du champ électromagnétique. *Ann. de l'Inst. H. Poincare* **9**, 13–49.
Dirac, P.A.M. (1982). Pretty mathematics. *Int. J. Theor. Phys.* **21**, 603–5.
Dodson, C.T.J. and Poston, T. (1977). *Tensor Geometry* (Pitman, London).
Dowker, J.S. and Dowker, Y.P. (1966). Interaction of massless particles of arbitrary spin. *Proc. Roy. Soc.* **A294**, 175–94.
Duff, G.F.D. (1956). *Partial Differential Equations* (Oxford University Press, London).
Eastwood, M.G. and Tod, K.P. (1982). Edth – a differential operator on the sphere. *Math. Proc. Camb. Phil. Soc.* **92**, 317–30.
Eckmann, B. (1968). Continuous solutions of linear equations – some exceptional dimensions in topology, in *Battelle Rencontres*, ed. C.M. DeWitt and J.A. Wheeler (Benjamin, New York).
Ehlers, J. (1974). The geometry of the (modified) GH – formalism. *Commun. Math. Phys.* **37**, 327–9.
Ehlers, J., Rindler, W. and Robinson, I. (1966). Quaternions, bivectors, and the Lorentz group, in *Perspective in Geometry and Relativity*, ed. B. Hoffmann (Indiana Univ. Press, Bloomington, Ind.)
Einstein, A., (1916). Die Grundlage der allegemeinen Relativitätstheorie. *Ann. Phys.* **49**, 769–822, translated as The Foundation of the general theory of relativity, in Einstein *et al.* 1923.
Einstein, A., Lorentz, H.A., Weyl, H. and Minkowski, H. (1923). *The Principle of Relativity* (Methuen and Co., republished by Dover).
Einstein, A. and Mayer, W. (1932). Semivektoren und Spinoren. *Sitz. Ber. Preuss. Akad. Wiss. Berlin*, 522–50.
Engelking, R. (1968). *Outline of General Topology* (North-Holland & PWN, Amsterdam).
Exton, A.R., Newman, E.T. and Penrose, R. (1969). Conserved quantities in the Einstein–Maxwell Theory, *J. Math. Phys.* **10**, 1566–70.
Ferrara, S. and Zumino, B. (1974). Supergauge invariant Yang–Mills theories. *Nucl. Phys.* **B79**, 413–21.
Ferrara, S., Zumino, B. and Wess, J. (1974). Supergauge multiplets and superfields. *Phys. Lett.* **51B**, 239–41.
Feynman, R.P. and Gell-Mann, M. (1958). Theory of the Fermi interaction. *Phys. Rev.* (2) **109**, 193–8.
Feynman, R.P. and Hibbs, A.R. (1965). Quantum Mechanics and Path Integrals. pp. 34–6 (McGraw-Hill, New York).
Fierz, M. (1938). Uber die Relativitische Theorie Kräftefreier Teilchen mit beliebigem Spin. *Helv. Phys. Acta* **12**, 3–37.
Fierz, M. (1940). Uber den Drehimpuls von Teilchen mit Ruhemasse null und beliebigem Spin. *Helv. Phys. Acta* **13**, 45–60.
Fierz, M and Pauli, W. (1939). On relativistic wave equations for particles of arbitrary spin in an electromagnetic field. *Proc. Roy. Soc. London.* **A173**, 211–32.
Flanders, H. (1963). *Differential Forms* (Academic Press, New York).
Fock, V.A. (1937). Die Eigenzeit in der Klassischen und in der Quantenmechanik. *Physik. Z. Sowjetunion* **12**, 404–25.
Fordy, A.P. (1977). Zero-rest-mass fields in an algebraically special curved space–time. *Gen. Rel. Grav.* **8**, 227–43.
Frame, J.S., Robinson, G. de B. and Thrall, R.M. (1954). The hook graph of the symmetric group. *Can. J. Math.* **6**, 316–24.

Freedman, D.Z., van Nieuwenhuizen, P. and Ferrara, S. (1976). Progress toward a theory of supergravity. *Phys. Rev.* **D13**, 3214–18.
Friedlander, F.G. (1975). *The Wave Equation on a Curved Space–time.* Cambridge Monographs on Mathematical Physics 2 (Cambridge University Press, Cambridge).
Gardner, M. (1967). *The Ambidextrous Universe* (Allen Lane, The Penguin Press, London).
Gel'fand, I.M., Graev, M.I. and Vilenkin, N. Ya. (1966). *Generalized Functions, Vol. 5: Integral Geometry and Representation Theory* (Academic Press, New York).
Geroch, R. (1968). Spinor structure of space–times in general relativity I *J. Math. Phys.* **9**, 1739–44.
Geroch, R. (1970). Spinor structure of space–times in general relativity II. *J. Math. Phys.* **11**, **343**–8.
Geroch, R., Held, A. and Penrose, R. (1973). A space–time calculus based on pairs of null directions. *J. Math. Phys.* **14**, 874–81.
Goldberg, J.N., Macfarlane, A.J., Newman, E.T., Rohrlich, F. and Sudarshan, E.C.G. (1967). Spin-*s* Spherical Harmonics and eth *J. Math. Phys.* **8**, 2155–61.
Goldstein, H. (1980). *Classical Mechanics*, 2nd edn (Addison-Wesley, Reading, Mass)
Gunning, R.C. (1966). *Lectures on Riemann Surfaces* (Princeton University Press, Princeton).
Hadamard, J. (1952). *Lectures on Cauchy's Problem in Linear Partial Differential Equations* (Dover, New York).
Hansen, R.O., Janis, A.I., Newman, E.T., Porter, J.R. and Winicour, J. (1976). Tensors, spinors and functions on the unit sphere. *Gen. Rel. Grav.* **7**, 687–93.
Hawking, S.W. and Ellis, G.F.R. (1973). *The Large Scale Structure of Space–Time* (Cambridge University Press, Cambridge).
Hehl, F.W., von der Heyde, P., Kerlick, G.D. and Nester, J.M. (1976). General relativity with spin and torsion: foundations and prospects. *Rev. Mod. Phys.* **48**, 393–416.
Held, A. (1974). A formalism for the investigation of algebraically special metrics I. *Com. Math. Phys.* **37**, 311–26.
Held, A. (1975). *Ditto* II. *Com. Math. Phys.* **44**, 211–22.
Held, A. and Voorhees, B.H. (1974). Some zero rest mass test fields in general relativity. *Int. J. Theor. Phys.* **10**, 179–87.
Hermann, R. (1968). *Differential Geometry and the Calculus of Variations* (Academic Press, New York).
Herstein, I.N. (1964). *Topics in Algebra* (Blaisdell, New York).
Hicks, N.J. (1965). *Notes on Differential Geometry* (Van Nostrand, Princeton).
Hitchin, N. (1974). Compact four-dimensional Einstein Manifolds. *J. Diff. Geom.* **9**, 435–41.
Hochschild, G. (1965). *The Structure of Lie Groups* (Holden–Day, San Francisco).
Hughston, L.P. and Ward, R.S. (1979). (eds). *Advances in Twistor Theory.* Research Notes in Mathematics, 37, (Pitman, San Francisco).
Infeld, L. and van der Waerden, B.L. (1933). Die Wellengleichung des Elektrons in der allgemeinen Relativitätstheorie. *Sitz. Ber. Preuss. Akad. Wiss. Physik.-math.*, Kl., **9**, 380–401.
Jordan, P., Ehlers, J. and Sachs, R. (1961). Beiträge zur Theorie der reinen Gravitationsstrahlung. *Akad. Wiss. u. Lit. in Mainz, Math–Naturwiss. Kl.* 1960, No. 1.

Kelley, J.L. (1955). *General Topology* (Van Nostrand, New York).

Kibble, T.W.B. (1961). Lorentz invariance and the gravitational field. *J. Math. Phys.* **2**, 212–21.

Kirchhoff, G. (1882). Zur Theorie der Lichtstrahlen. *Sitz. Ber. K. Preuss. Akad. Wiss. Berlin*, 641–69.

Klein, A.G. and Opat, G.I. (1975). Observability of $2\pi$ rotations: A proposed experiment. *Phys. Rev.* **D11**, 523–8.

Klein, F. (1897). *The Mathematical Theory of the Top* (Scribners, New York) Chapter 5

Kobayashi, S. and Nomizu, K. (1963). *Foundations of Differential Geometry*, Vol. 1 (Interscience Publishers, London).

Kramer, D., Stephani, H., MacCallum, M. and Herlt, E. (1980). *Exact Solutions of Einstein's Field Equations* (VEB Deutscher Verlag der Wisenschaften, Berlin).

Lang, S. (1972). *Differentiable manifolds* (Addison Wesley, Reading, Mass.).

Laporte, O. and Uhlenbeck, G.E. (1931). Application of Spinor Analysis to the Maxwell and Dirac Equations. *Phys. Rev.* **37**, 1380–1552.

Lee, T.D., Oehme, R. and Yang, C.N. (1957). Remark on Possible Noninvariance under Time Reversal and Charge Conjugation. *Phys. Rev.* **106**, 340–5.

Lee, T.D. and Yang, C.N. (1956). Question of Parity Conservation in Weak Interactions. *Phys. Rev.* **104**, 254–8.

Lefschetz, S. (1949). *Introduction to Topology* (Princeton University Press, Princeton).

Lichnerowicz, A. (1968). Topics on Space–Time, in *Battelle Rencontres, 1967 Lectures in Mathematics and Physics*, ed. C.M. DeWitt and J.A. Wheeler (Benjamin, New York).

Littlewood, D.E. (1950). The Theory of Group Characters (Clarendon Press, Oxford).

Lorenz, L. (1867). On the identity of the vibrations of light with electrical currents. *Phil. Mag.* **34**, 287–301, translated from *Ann. Phys. Chem.* **131**, 243, translated from *Oversigt over det K. Danske Vidensk. Selsk. Forhandl.* **1**, 26.

Mac Lane, S. and Birkhoff, G. (1967). *Algebra* (Macmillan, New York).

McLenaghan, R.G., (1969). An explicit determination of the empty space–times on which the wave equation satisfies Huygens' principle. *Proc. Camb. Phil. Soc.* **65**, 139–55.

Milnor, J. (1963). Spin Structures on Manifolds. *Enseign. Math.* **9**, 198–203.

Milnor, J.W. and Stasheff, J.D. (1974). *Characteristic Classes*. Annals of Mathematics Studies No. 76 (Princeton University Press, Princeton).

Minkowski, H. (1908). *Space and Time*. Address delivered at the 80th Assembly of German Natural Scientists and Physicians, at Cologne, 21 September, 1908, translated in Einstein *et al.* 1923.

Misner, C.W. and Wheeler, J.A. (1957). Classical Physics as Geometry: Gravitation, Electromagnetism, Unquantized Charge, and Mass as Properties of Curved Empty Space. *Ann. of Phys.* **2**, 525–603.

Moore, E.H. (1920). On the Reciprocal of the General Algebraic Matrix (abstract). *Bull. Amer. Math. Soc.* **26**, 394–5.

Morrow J. and Kodaira, K. (1971). *Complex Manifolds* (Holt, Rinehart and Winston, New York).

Naimark, M.A. (1964). *Linear Representations of the Lorentz Group* (Pergamon, Oxford).

Nashed, M.Z. (1976) (ed.). *Generalized Inverses and Applications* (Academic Press, New York).

Nelson, E. (1967). *Tensor Analysis* (Princeton University Press and University of Tokyo Press, Princeton, New Jersey)
Newman, E.T. and Penrose, R. (1962). An approach to gravitational radiation by a method of spin coefficients. *J. Math. Phys.* **3**, 566–78; Errata: *ditto. Ibid.* **4**, 998.
Newman, E.T. and Penrose, R. (1966). Note on the Bondi–Metzner–Sachs group. *J. Math. Phys.* **7**, 863–70.
Newman, E.T. and Penrose, R. (1968). New conservation laws for zero rest–mass fields in asymptotically flat space–time. *Proc. Roy. Soc.* **A305**, 175–204.
Newman, E.T. and Unti, T.W.J. (1962). Behaviour of asymptotically flat empty spaces. *J. Math. Phys.* **3**, 891–901.
Newman, M.H.A. (1942). On a string problem of Dirac. *Journ. Lond. Math. Soc.* **17**, 173–7.
Nomizu, K. (1956). *Lie Groups and Differential Geometry* (The Mathematical Society of Japan, Tokyo).
Papapetrou, A. (1974). *Lectures on General Relativity* (D. Reidel Publishing Company, Dordrecht).
Pauli, W. (1927). Zur Quantenmechanik des magnetischen Elektrons. *Z. Phys.* **43**, 601–25.
Payne, W.T. (1952). Elementary spinor theory. *Am. J. Phys.* **20**, 253–62.
Penrose, R. (1955). A generalised inverse for matrices. *Proc. Camb. Phil. Soc.* **51**, 406–413.
Penrose, R. (1959). The apparent shape of a relativistically moving sphere. *Proc. Camb. Phil. Soc.* **55**, 137–9.
Penrose, R. (1960). A Spinor Approach to General Relativity. *Ann. Phys.* (New York) **10**, 171–201.
Penrose, R. (1962). General Relativity in Spinor Form, in *Les Théories Relativistes de la Gravitation*, eds A. Lichnerowicz and M.A. Tonnelat (CNRS, Paris).
Penrose, R. (1963). Null hypersurface initial data for classical fields of arbitrary spin and for general relativity, in *Aerospace Research Laboratories Report 63–56* (P.G. Bergmann) reprinted (1980) in *Gen. Rel. Grav.* **12**, 225–64.
Penrose, R. (1965). Zero rest–mass fields including gravitation: asymptotic behaviour. *Proc. Roy. Soc. London* **A284**, 159–203.
Penrose, R. (1966*a*). General–Relativistic Energy Flux and Elementary Optics, *ibid.* 259–274.
Penrose, R. (1966*b*). An Analysis of the Structure of Space–time, Adams Prize Essay, Cambridge University.
Penrose, R. (1968). Structure of space–time, in *Battelle Rencontres, 1967 Lectures in Mathematics and Physics*, eds C.M. DeWitt and J.A. Wheeler (Benjamin, New York).
Penrose, R. (1971). Application of Negative Dimensional Tensors, in *Combinatorial Mathematics and its Applications*, ed. D.J.A. Welch (Academic Press, London).
Penrose, R. (1972*a*). The Geometry of impulsive gravitational waves, in *General Relativity, Papers in Honour of J.L. Synge*, ed. L. O'Raifeartaigh (Clarendon Press, Oxford).
Penrose, R. (1972*b*). Spinor classification of energy tensors, *Gravitatsiya Nauk dumka*, Kiev, 203.
Penrose, R. (1974). Relativistic Symmetry Groups, in *Group Theory in Non-Linear Problems*, ed. A.O. Barut (Reidel, Dordrecht). pp. 1–58.

Penrose, R. (1975). Twistor Theory, its aims and achievements, in *Quantum Gravity: an Oxford Symposium*, eds. C.J. Isham, R. Penrose and D.W. Sciama (Clarendon Press, Oxford).

Penrose, R. (1983). Spinors and torsion in general relativity. *Found. Phys.* **13**, 325–40.

Penrose, R. and MacCallum, M.A.H. (1972). Twistor theory: An approach to the quantisation of fields and space–time. *Phys. Rep.* **6**, 241–316.

Penrose, R. and Ward, R.S. (1980). Twistors for flat and curved space–time, in *General Relativity and Gravitation, One Hundred Years after the Birth of Albert Einstein*, Vol. II, ed. A. Held (Plenum Press, New York).

Pirani, F.A.E. (1965). Introduction to gravitational radiation theory, in Trautman, Pirani and Bondi (1965), 249–373.

Pirani, F.A.E. and Schild, A. (1961). Geometrical and physical interpretation of the Weyl conformal curvature tensor. *Bull. Acad. Pol. Sci.*, Ser. Sci. Math. Astron. Phys. **9**, 543–7.

Plebański, J. (1965). The 'vectorial' optics of fields with arbitrary spin, rest–mass zero. *Acta Phys. Polon.* **27**, 361–93.

Proca, A. (1937). Sur un article de M.E. Whittaker, intitulé 'Les relations entre le calcul tensoriel et le calcul des spineurs'. *J. Phys. et radium* **8**, 363–5.

Rainich, G.Y. (1925). Electrodynamics in the general relativity theory. *Trans. Am. Math. Soc.* **27**, 106–36.

Rarita, W. and Schwinger, J. (1941). On a theory of particles with half–integral spin. *Phys. Rev.* **60**, 61.

Rauch, H. Zeilinger, A., Badurek, G., Wilfing, A., Bauspiess, W. and Bonse, U. (1975). Verification of coherent spinor rotation of fermions. *Phys. Lett.* **A54**, 425–7.

Riesz, M. (1949). L'Intégrale de Riemann–Liouville et le problème de Cauchy. *Acta Math.* **81**, 1–223.

Rindler, W. (1977). *Essential Relativity; Special, General and Cosmological* (Second edition) (Springer-Verlag, New York).

Rindler, W. (1982). *Introduction to Special Relativity* (Clarendon Press, Oxford).

Rutherford, D.E. (1948). *Substitutional Analysis* (Edinburgh University Press, Edinburgh).

Sachs, R. (1961). Gravitational waves in general relativity VI. The outgoing radiation condition. *Proc. Roy. Soc. London* **A264**, 309–38.

Sachs, R.K. (1962*a*). Gravitational waves in general relativity. VIII. Waves in asymptotically flat space–time. *Proc. Roy. Soc. London* **A270**, 103–26.

Sachs, R.K. (1962*b*). On the characteristic initial value problem in gravitational theory. *J. Math. Phys.* **3**, 908–14.

Sachs, R.K. and Bergmann, P.G. (1958). Structure of particles in linearized gravitational theory. *Phys. Rev.* **112**, 674–80.

Schiff, L.I. (1955). *Quantum Mechanics* (McGraw-Hill, New York).

Schild, A. (1967). Lectures on general relativity theory, in *Relativity Theory and Astrophysics, I. Relativity and Cosmology*, ed. J. Ehlers (Amer. Math. Soc. Providence).

Schrödinger, E. (1950). *Space–time structure* (Cambridge University Press, Cambridge).

Sciama, D.W. (1962). On the analogy between charge and spin in general relativity, in *Recent Developments in General Relativity* (Pergamon & PWN, Oxford).

Sommerfeld, A. (1936). Uber die Klein'schen Parameter $\alpha$, $\beta$, $\gamma$, $\delta$, und ihre

Bedeutung für die Dirac-Theorie. *Sitz. Ber. Akad. Wiss. Wien* **145**, 639–50.

Staruszkiewicz, A. (1976). On two kinds of spinors *Acta Physica Polonica* **B7**, 557–65.

Stellmacher, K.L. (1951). Geometrische Deutung konforminvarianter Eigenschaften des Riemannschen Raumes. *Math. Annalen* **123**, 34–52.

Stewart, J.M. (1979). Hertz–Bromwich–Debye–Whittaker–Penrose potentials in general relativity. *Proc. Roy. Soc. London* **A367**, 527–38.

Stewart, J.M. and Walker, M. (1974). Perturbations of space–times in general relativity. *Proc. Roy. Soc. London* **A341**, 49–74.

Synge, J.L. (1955). *Relativity: The Special Theory* (North-Holland Publishing Co., London).

Synge, J.L. (1957). *The Relativistic Gas* (North-Holland, Amsterdam).

Synge, J.L. (1960). *Relativity: The General Theory* (North-Holland, Amsterdam).

Szekeres, P. (1963). Spaces conformal to a class of spaces in general relativity. *Proc. Roy. Soc.* **A274**, 206–12.

Taub, A.H. and Veblen, O. (1934). Projective differentiation of spinors. *Nat. Acad. Sc. Proc.* **20**, 85–92.

Terrell, J. (1959). Invisibility of the Lorentz contraction. *Phys. Rev.* **116**, 1041–5.

't Hooft, G. and Veltman, M. (1973). *Diagrammar*, CERN preprint 73–9 (CERN, Geneva), unpublished.

Tod, K.P. (1983). The singularities of $\mathscr{H}$-space. *Math. Proc. Camb. Phil. Soc.*, to appear.

Trautman, A. (1962). Conservation laws in general relativity, in *Gravitation: An Introduction to Current Research*, ed. L. Witten (Wiley, New York).

Trautman, A. (1972, 1973). On the Einstein–Cartan equations I–IV. *Bull. Acad. Pol. Sci.*, Ser. Sci. Math. Astron. Phys. **20**, 185–90; 503–6; 895–6; **21**, 345–6.

Trautman, A., Pirani, F.A.E. and Bondi, H. (1965). *Lectures on General relativity* (Prentice-Hall, Englewood Cliffs)

Unruh, W.G. (1976). Self force on charged particles, *Proc. Roy. Soc. London* **A348** 447–65.

van der Waerden, B.L. (1929). Spinoranalyse. *Nachr. Akad. Wiss. Götting.* Math.-Phsyik Kl, 100–9.

Veblen, O. (1933*a*). Geometry of Two-Component Spinors. *Proc. Nat. Acad. Sci.* **19**, 462–74.

Veblen, O. (1933*b*). Geometry of Four-Component Spinors. *Proc. Nat. Acad. Sci.* **19**, 503–17.

Veblen, O. (1934). Spinors. *Science* **80**, 415–19.

Veblen, O. and von Neumann, J. (1936). *Geometry of Complex Domains.* mimeographed notes, issued by the Institute for Advanced Study Princeton (reissued 1955).

Wells, R.O. (1973). *Differential analysis in complex manifolds* (Prentice Hall, Englewood Cliffs).

Werner, S.A., Colella, R., Overhauser, A.W. and Eagen, C.F. (1975). Observation of the Phase Shift of a Neutron Due to Precession in a Magnetic Field. *Phys. Rev. Lett.* **35**, 1053–5.

Weyl, H. (1923). Gravitation and Electricity (translated from 'Gravitation und Elektricität', *Sitz. Ber. Preuss. Akad. Wiss.* 1918, 465–80), in *The Principle of Relativity*, by A. Einstein, H.A. Lorentz, H. Weyl and H. Minkowski (Methuen and Co., republished by Dover).

Weyl, H. (1929). Elektron und Gravitation I. *Z. Phys.* **56**, 330–52.

Weyl, H. (1931). *The Theory of Groups and Quantum Mechanics* (Methuen, London). p. 358.

Whittaker E.T. (1910). *The History of the Theories of Aether and Electricity* (Longman, London).

Whittaker, E.T. (1937). On the Relations of the Tensor-calculus to the Spinor-calculus. *Proc. Roy. Soc. London* **A158**, 38–46.

Witten, L. (1959). Invariants of General Relativity and the Classification of Spaces. *Phys. Rev.* **113**, 357–62.

Witten, L. (1962). A geometric theory of the electromagnetic and gravitational fields. In *Gravitation: An Introduction to Current Research*, ed. L. Witten (Wiley, New York).

Wu, C.S. Ambler, E., Hayward, R., Hoppes, D. and Hudson, R. (1957). Experimental Test of Parity Conservation in Beta Decay. *Phys. Rev.* **105**, 1413–15; Further Experiments on $\beta$ Decay of Polarized Nuclei. *Phys. Rev.* **106**, 1361–63.

Wu, T.T. and Yang, C.N. (1964). Phenomenological analysis of violation of *CP* invariance in decay of $K^0$ and $\bar{K}^0$. *Phys. Rev. Lett.* **13**, 380–5.

Wu, T.T. and Yang, C.N. (1976). Dirac monopole without strings: monopole harmonics. *Nucl. Phys.* **B107**, 365–80.

Yang, C.N. and Mills, R. (1954). Conservation of Isotopic Spin and Isotopic Gauge Invariance. *Phys. Rev.* **96**, 191–5.

Young, A. (1900). On quantitative substitutional analysis I. *Proc. Lond. Math. Soc.* **33**, 97–146.

# Subject and author index

*Italics denote main references; page numbers in parentheses refer to footnotes; a dash indicates continuation for more than two pages.*


aberration formula 28
abstract index notation viii, 68–, *76*–, 353
affine parameter 387
Aharonov, Y. (46), 318
Aichelburg, P.C. 370
Aitken, A.C. 426
Aitken's diagrams 426
alternating tensor 92, 137–, 166
  diagram for 428
Ambler, E. (4)
analytic 341, 346, 386–
  *see also* holomorphic
antipodal map 10, 30, 291
anti-self-dual *151*, 236, 240, 351
anti-sky mapping *10*, 11, 30
anti-symmetric (skew) tensor 91, 92, *134*–
  dual of 149–, 166
  simple (145), *165*–, 324
  *see also* bivector, differential form
anti-symmetrization 132
  diagram for 426–
Argand (–Wessel–Gauss) plane *10*, 19, 25, 60–, 274
Arnol'd, V.I. 19
Arnowitt, R. 390
Ashtekar, A. 335, 342

background fields 402–
Badurek, G. (47)
basis 1
  change of 97
  coordinate 122
  dual *94*, 102
  existence of 82, *91*–
  for local cross-sections of bundle 335
  *see also* spinor basis, spin-frame
Bass. R.W. 55
Bauspiess, W. (47)
Bel, L. 240
Bel–Robinson tensor *240*–, 246, 325–
  for Einstein–Maxwell theory 326
  positive-definiteness 328
Bell, P. 367
Bergmann, P.G. 231, 364
Bernstein H.J. (47)
Bessel function 414
Bianchi identity 194, 210, 385
  contracted 210
    spin-coefficient form of 260
    spinor form of 246
  diagram for 429
  diagrammatic proof of, with torsion 430
  spin-coefficient form of 259
  spinor form of 245, 246, 362
    diagram for 433
Bianchi symmetry (cyclic identity) 143, 194, *195*, 232, 233
  diagram for 429
Birkhoff, G. 69, (90)
bivector *149*–
  [anti-] self-dual 151
  complex null 131
  dual of 150
  duality rotation of 152
  simple 324
Bjorken, J.D. 368, 423
Bohm, D. (318)
Bondi, H. 399
Bondi–Sachs mass 278
Bonse, U. (47)
boost *18*, 20
boost-weight 253
Bott, R. 335
Bramson, B.D. (401)
Bruhat, Y. (Choquet–Bruhat, Y.) 207, 390
Buchdahl, H.A. 366, 367, 370, 385
bump function 100, *181*, 183
bundle: *see* vector bundle

Campbell, S.J. 224
canonical decomposition of symmetric spinors 162
Carmeli, M. 224, (294)
Cartan, É. vii, 205, 237, 262
Cartan calculus
  of differential forms *203*–, *262*–, 424

of moving frames vii, 262–
relation to spin-coefficients 265–
causal world-vector 3
causality 353
Cayley–Klein parameters 20
celestial sphere *9*, 26
Chandrasekhar, S. 224
characteristic hypersurface 374, 389
*see also* null hypersurface
charge 312
conservation of 322
elementary 314
charge-current vector *320*, 369, 371
Dirac 377
Schrödinger–Klein–Gordon 378
charged fields (charged modules) *313*–, 356, 357, 367
Chevalley, C. 180
Choquet–Bruhat, Y. 207, 390
Christenson, J.H. (4)
Christoffel symbols 208, *230*, 250, 262
*see also* covariant derivative
Clifford algebra 124
Clifford–Dirac equation (124) (221)
clumped (composite) indices *87*–, 134
Colella, R. (47)
colour spaces of particle physics 334
commutator
in compacted spin-coefficient formalism *258*, 270, 283
on null hypersurface 283
on two-sphere 292
on two-surface 270, 272
of covariant derivatives *193*–, 312–, 385
on bundle cross-section 340, 341
with charged fields 318, 320–
in ECSK theory 237
spinor form of *242*–, 320–, 350
with Yang–Mills-charged fields 349, 350
Lie bracket 206, 246
of Lie derivatives 207
in spin-coefficient formalism 247, 248
compacted spin-coefficient (GHP) formalism 250–
basic equations 258
Bianchi identity in 259
derivatives of symmetric spinors in 257
exterior calculus in 278–
massless field equation, twistor equation, wave equation in 260
complex analytic: *see* holomorphic
complex angle 32
complex conformal rescaling (353), (356), (360)
complex conjugation
of spinor components 114
of spinors 107–
of tensors, as spinors 117
complex curvature of 2-surface 277
complex null bivector 131
complex null direction 131
complex null vector 126, 129–
complex numbers
field of 56, 162
significance of viii, 250
complex projective line 14
*see also* Riemann sphere
complex scalars 188, 212
complex stereographic coordinate *11*, (12), 33–, 274, 309–
complex tensor 117, 188
complexified sphere 131
complexion of Maxwell field 330
components
of spin-vector 47
of spinor 113
of tensor 70–, *95*
of vector 35, *91*, 185
composite (clumped) indices *87*–, 134
conformal density 289, 354, *355*, 359, 366
conformal flatness 358, 367, 398–
conformal group (355)
conformal invariance *355*, 358–, 399–
conformal rescaling 352
change of covariant derivative under 355–
complex (353), (356), (360)
of curvature 357
of spin-coefficients 358, 359
conformal structure
of $\mathscr{S}^+$, $\mathscr{S}^-$, $S^+$, $S^-$ 38, 60, 129
of space–time 352
conformal transformation
of 2-surface 25, 289–, 300
*see also* conformal rescaling
conformal weight 289–, *355*, 360
conformally invariant equations 361, 366, 369
conformally invariant eth, thorn 360–
conformally invariant integral on 2-surface 278
connection 191
expressions independent of 202–
symbols *199*–, 313
*see also* covariant derivative
connection 1-form 264
consistency condition
Buchdahl 366, 367
Fierz–Pauli 366, 367
constant scalars 181
constant spinors 294
constraints 390
avoidance of 386, 402

continuous function 181
contraction 71, 74, 84, 85
contracted (inner) product *86*, 88, 89
contravariant 72
  classical definition of 185
  vector 72, 184
coordinate basis 122, 197
coordinate derivative 197–
coordinates
  Minkowskian 3
  normal 387
  of a vector 1
Corson, E.M. 312, 384, 390
cosmological constant 234, 235, 379
Courant, R. 297
covariant 72
  vector (covector) 72
covariant derivative 191–
  algebraic properties of 192
  change of *196*, 341
    under conformal rescalings 217, *355*–
  Christoffel (Levi–Civita) 209, 239, 341
    extended to spinor fields 211, 213–; (existence) 217–, 223–; (non-uniqueness) 216, 217; (uniqueness) 214–, 231
  diagram for 429–
  extended to bundles 340, 341
  extended to charged fields 315–
  extended to complex tensors 212, *218*
  exterior 263–
  spinor components of 225
  Taylor (–Gregory) series in 386–
  tensor components of 199
  in a two-surface 271
covector 72
covering 99, 335
covering space 50
  universal 45–
Cronin, J.W. (4)
cross-ratio 29–, *30*
  real, harmonic, equianharmonic 30, 31
cross-section of bundle (270), 334, *335*
Curtis, G.E. 365
curvature, Gaussian 272
curvature spinors 231–
  from Cartan formalism 263–
  change under conformal scaling 355
curvature tensor 194
  components of 200
  diagram for 429–
  Riemann tensor 146, 209, 231
curvature two-form 263
Cvitanović, P. 75, 424
cyclic identity (Bianchi symmetry) 143, 194, *195*, 232, 233
d'Adhémar, R. 389, 401
D'Alembert equation 375
  *see also* wave equation
Darmois, G. 393
de Donder gauge 364
degrees of freedom 374, 389, 390
derivation 184–
derivative, coordinate, covariant and directional: *see* coordinate derivative, covariant derivative *and* directional derivative
Deser, S. 370, 390
de-Sitter space 379
DeWitt-Morette C. 207
diagrammatic notation ix, 75, *424*–
differential 189
differential form ($p$-form) 203
  duals of 264
  spinor-valued 262
  suppression of indices on 203, 262
  tensor-valued 203, 262
  *see also* anti-symmetric tensor
Dillard-Bleik, M. 207
Dirac, P.A.M. vii, 220, 221, 362, 384, 403
Dirac delta-function 408, 413
  *see also* distribution
Dirac electron equation (field) 220–, 312, 377, 412–
  charge–current vector for 377
  degrees of freedom for 390
  energy tensor for 368
  Maxwell–Dirac equations 379, 417–
Dirac(–Fierz) higher spin fields 384
  degrees of freedom for 390
Dirac gamma matrices 124 (221)
Dirac scissors problem 43
Dirac spinors (4-spinors) vii, 143, (*221*), 370, 377
Dirac–Weyl neutrino equation (field) 220–, 362, 367, 375
directional derivative 180, 185, 190, 226, 227
  covariant 206
distributional solution
  of massless free-field equation 410, 412
  of wave equation 409
distributions 407–
divergence equation 322, 371
divergence theorem 205
  *see also* exterior calculus
division ring 69, 70
Dodson, C.T.J. 198
dominant energy condition 327
Doppler factor 20
Dowker, J.S. 370
Dowker, Y.P. 370

Drell, S.D. 368, 423
dual
  [anti-] self-dual *151–*, 320, 351
  of basis *94*, 102
  of bivector 150
  of dual of module 79
  left and right duals 232–
  of module 78, 79
  of *p*-form 264
duality rotation *152*, 232, 233, 242, 330
Duff, G.F.D. 389
Duffin–Kemmer equation 384
dummy index 85, 424
dyad *112*, 223–
  *see* spinor basis

Eagen, C.F. (47)
Eastwood, M.G. 291
Eckmann B. (93)
edth 255
  *see also* eth
Ehlers, J. (24)
eigenspinor 173
Einstein, A. 237
Einstein–Cartan–Sciama–Kibble (ECSK) theory 216, *237–*
Einstein field equations *234*, 235, 238, 246, 260, 379–
  degrees of freedom for 390
  weak-field limit of *362–*, 375
Einstein–Maxwell (electrovac) equations *325–*, 379, 382
Einstein static universe 55
Einstein tensor 234
electric and magnetic 3-vectors 323
electromagnetic (Maxwell) field tensor 312–, *318*
  complex 323
  complexion of 330
  components of 323
  components of dual of 324
  duality rotation of 330
  nullity of 164, *324*
  PNDs of 324, 328
  and potential 320, 373, 378
  purely electric [magnetic] 324, 330
  scalar invariants of 164, *324*
  simple 324
electromagnetic potential 312, *316–*
  and field tensor 320, 373, 378
electromagnetic spinor 312, *320–*, 362. 375
  components of 323
  PNDs of 324, 328
Ellis, G.F.R. 55, 180, (191), 207
energy(–momentum) tensor 235
  conformal invariance of, if trace-free 371
  Dirac 368
  Dirac–Weyl 368
  electromagnetic 173, *325–*
  invariance of, under duality rotation 325
  massless scalar 369
  non-symmetric 238
  positive-definiteness of 326–
  supergravity 370
  Yang–Mills 351
energy positivity condition 327
Engelking, R. 183
epsilon spinors: *see* 2-spinors
eth (ð) 255
  conformally invariant ($ð_{\mathscr{C}}$) 360–
  coordinate descriptions of 275, *308–*
  equations in 297–
  generalized inverse of 299
Euler homogeneity operators 287, 291
exact set (of interacting fields) 373–
  Dirac field as 377, 378
  Dirac–Fierz field as 384–
  Einstein field as 379–
  Einstein–Maxwell field as 382, 383
  invariant 375
  massless free field as 375
  Schrödinger–Klein–Gordon field as 378
  Yang–Mills field as 383
exponentiation 176–, 207, 346
extent (*36*), 39, 387, 394
exterior calculus 02–
  compacted spin-coefficient form of
    in conformal form 361, 362
    on null hypersurface *281–*, 362, 396–
    on 2-surface 278–
  fundamental theorem of *205*, 281–, 362, 396
exterior derivative *202–*, 208
  covariant 263–
  in spin-coefficient forms 278, 279, 284, 361
exterior product 203–, 262–
Exton, A.R. 299
extrinsic curvature quantity of 2-surface 277

*f*-equivalence 99
Ferrara, S. 370
Feynman, R.P. 383, 417
Feynman diagrams 423
Fierz, M. 363, 366, 367, 384, 385
Fitch, V.L. (4)
flag plane *37*, 39, 40, 53, 128
flagpole *37*, 39, 40, 53, 127
Fock, V.A. 383
Fock–Feynmann–Gell-Mann equation 383
form (145), 203
  *see also* differential form
Fordy, A.P. 252

four-screw *28*, 174
Frame, J.S. (144)
Freedman, D.Z. 370
Friedlander, F.G. 407, 408
fundamental group (first homotopy group) 44
fundamental theorem of exterior calculus 205
  *see also* exterior calculus
future-pointing world-vector *4*, 127

Gardner, M. (4)
gauge 251, 253, *316*, 345, 378
gauge transformation *317*, 345, 364
Gauss, C.F. 205
Gauss–Bonnet theorem 277, 278
Gaussian curvature 272, 276, 288
Gel'fand I.M. (301)
Gell-Mann, M. 383
general relativity: *see* Einstein field equations
generalized inverse of ð 299
generalized Kirchhoff–d'Adhémar formula 393
  *see also* Kirchhoff–d'Adhémar formula
generator of null hypersurface 283
genus 277
geodesic 283, *387*, 395
geodetic deviation 392, 393
geometrodynamics 329
geometry
  of intersecting null hypersurfaces 393
  of light cone (null cone) 392, 393
  of spinor operations 59: *see also* spinor operations
Geroch, R. 51, 55, 93, 252
Geroch–Held–Penrose (GHP) formalism: *see* compacted spin-coefficient formalism
Goldberg, J.N. 272, (297), 305, (306), (307)
Goldstein, H. 19, 20
gradient of scalar (72) *189*–
Graev, M.I. (301)
graph of a function 333
Grassmann calculus 424
  *see also* Cartan calculus
gravitational constant (Newton's) 235, 238
gravitational spinor 235
  *see also* Weyl (conformal) spinor
Green, G. 205
Grgin phenomenon (411)
Gunning, R.C. 298

Hadamard, J. 389
Hausdorff manifold 48, 98
Hawking, S.W. 55, 180, (191), 207
Hayward, R. (4)
Hehl, F.W. 237
Held, A. 252
helicity 308, 362
Herlt, E. 119, 224
Hermitian 16, 123, 124
Hermitian-null vector 131, 132
Hermitian scalar product on sphere 301
Herstein, I.N. 69, (90)
Hibbs, A.R. 417
Hicks, N.J. 180
Hilbert, D. 297
Hitchin, N. 51
Hochschild, G. (176)
holomorphic (15), (*25*), (287), 298
holomorphic coordinate for 2-surface 273–
holonomic basis (93), *199*
homotopy type (46)
Hoppes, D. (4)
Horowitz, G.T. 335, 342
Hudson, R. (4)
Huygens' principle 407

index permutation 71, 75, 84
index substitution 84
indices
  bold-face upright (numerical)*2*–, 76, 93–
  capital Greek (bundle) 335
  composite (clumped) *87*–, 134
  light-face sloping (abstract) *76*–, 93–
  lowering and raising of 104–, 118, 209, 225
  moving of, across each other 110
  staggering of 87, 105
  suppression of 203, 262
Infeld, L. 221
Infeld–van der Waerden symbols *123*–, 228–, 262, 266
infinitesimal transformation 175–, 364
initial value problem, initial data 374, *385*–
  explicit integrals for, in $\mathbb{M}$ 393–
  on intersecting null hypersurfaces 393
  on light cone (null data) 385–
  on spacelike hypersurface 389, 401, 402
inner product 2
  of spin-vector *56*–, 59–, 104, 137
  of tensors 86
intrinsic derivative 226
  *see also* directional derivative
irreducibility 141–
isotopic spin-space 179, 334

Jacobi identity 206

$K^0$-decay (4), 56
Kelley, J.L. 93, 183
Kelvin, Lord 205
Kennedy, A.D. 424

Kerlick, G.D. 237
Kibble, T.W.B. 237
Kirchhoff, G. 401
Kirchhoff–d'Adhémar formula (generalized) 394–
  in conformally space–time 398–
  defining background fields 402–
  derivatives of 406, 408
  proof of 396–
  satisfaction of field equations 405–
  use with spacelike data 401, 402
Klein, A.G. (47)
Klein–Gordon equation (field) 378
  degrees of freedom for 390
Kobayashi, S. 180, 197
Kramer, D. 119, 224
Kronecker delta
  abstract *89*, 118, 123, 353
    diagram for 425–
  mixed 93–
  numerical *73*, 94, 97, 112

labelling set 77, 78, 116
Lang, S. 180
Lee, T.D. (4)
left-handed massless particles 362
Leibniz law 191–, 214, 255
  in diagrams 429
Levi–Civita connection: *see* covariant derivative, Christoffel
Levi–Civita symbol 112, 113, 115, 137, 139
Lichnerowicz, A. 52
Lie algebra 346
Lie bracket (199), 202, *206*, 246, 247
Lie derivative *202*, 207, 208
  diagram for 429
  geometrical meaning of 207
Lie group 343
light cone 285–, 385, 388
  geometry of 392, 393
  initial data on 385–, *388*, 394–
  invariant volume 3-form on 415, 416
  *see also* null cone
linearized Einstein field 362–, 375
line bundle (253), 344, 355
  *see also* vector bundle
local coordinates 182
loop notation for (covariant) derivative 429–
loops in $SO(3)$, classes I and II of 41–
Lorentz, H.A. (321)
Lorentz force 403
Lorentz group 5, 6
  representations of 141, 146, 237, (294), 300, (301)
  *see also* rotations of 2-sphere
Lorentz norm 3
Lorentz transformations *5*–, 167–
  improper 167, 171, 172
  infinitesimal 175–
  involutory (reflections) 171–
  kinds of 27, 170
  proper *172*–
  and spin transformations 14, 121, 167–
Lorentzian metric
Lorenz, L. (321)
Lorenz gauge *321*, 322, 373, 378, 383, 417
luminosity parameter 399

MacCallum, M.A.H. 119, 224, 362
Macfarlane, A.J. 272, (297), 305, (306), (307)
Mac Lane, S. 69, (90)
Magnon-Ashtekar, A. 335, 342
Majorana 4-spinor-tensor 370
manifold 48, *179*–
  analytic 346, 386
  connected, Hausdorff, paracompact 48, 98, 180, 183, 184, 335
  covering of 99
  *see also* space–time manifold
massless (free-) field equations 246, 322, *362*, 394
  derivatives of 375, 376
  giving exact set 375
  initial data for 388, 395
  in spin-coefficient form 260
    conformally invariant 361
massless particles 362
  *see also* massless field equations
Mather, J. 335
Maxwell equations 319–
  conformal invariance of 372, 373
  degrees of freedom for 390
  with differential forms 322, 373
    Einstein–Maxwell equations 379
  first half 319
  free space 260, 319, 362, 367, 375
    Maxwell–Dirac equations 379, 417–
  second half 320
  in spinor form 322, 376
Maxwell field 150, (151), *318*
  *see also* electromagnetic field
Maxwell theory 312–, 333
  as Yang–Mills theory 344
metric 266, 275
  *see also* metric tensor
metric tensor 208, 266
  diagram for 428
  as spinor 117–
Metzner, A.W.K. 399
Mills, R. 342
Milnor, J. 52
Minkowski space–time *5*–, 285–, 318, (355), 366, 375, 393–

Minkowski tetrads 3–
global 55
orthochronous 4
proper and improper 2
restricted *4*, 58, 120
and spin-frames 58, (59), 120
Minkowski vector space *1*–, 48
orientation (space–time orientation) 2
space-orientation 4, 121
time-orientation 4, 121
Misner, C.W. 329, 330, 390
Möbius band 338, 339
module 69, 70, 76–
axioms for 78
dimension of 92
dual 78, 79
reflexive 80
𝔖-module 78
totally reflexive 80
Moore, E.H. 299
multilinear map 80

Naimark, N.A. (257), (294), (301)
Nashed, M.Z. 299
neighbourhood 99, 184
𝔗-neighbourhood 181–
Nester, J.M. 237
neutrino (220)
*see also* Dirac–Weyl equation
neutron (47), 179
Newman, E.T. 119, 224, 262, 272, 275, 291, 292, 293, (297), 299, 305, (306), (307), 307, 308, (399)
Newman–Penrose (NP)
constants 404
formalism 223–, 246–
Newman, M.H.A. (43)
Nomizu, K. 180, 197
normal coordinates 387
nucleon 179
null cone *8*–, 352, 388
conformal structure of cross-section of 38
*see also* light cone
null datum *388*–, 395
null-flag bundle 48
null flags 32–, 128
and orthonormal frames in $\mathbb{R}^4$ 53
*PP′* description of, on $\mathscr{S}^+$ 34–
and spin-vectors 39–, 352
null hypersurface (251), *281*
conservation law on 285
fundamental theorem of exterior calculus on 281–
generator of 283
geometry of 392, 393
initial data on 374, 385–, *388*–, 393–
invariant volume 3-form for light cone 415, 416
volume element 3-form 284
null rotation *28*, (29), 174, 178
null symmetric spinor 164, 324
null tetrad 119, 125
null two-planes 37
angle between 38
null vector: *see* world-vector, null
null zig-zag 415–
numerical indices *2*, 76

Oehme, R. (4)
Opat, G.I. (47)
orientation
of Minkowski vector space 2, 4, 121
of $\mathscr{S}^+$ [$\mathscr{S}^-$, $S^+$, $S^-$] 129, 274, 308
of space–time manifold 49, 129
of tetrads 2
orthochronous Minkowski tetrad 4
Ostrogradski (205)
Overhauser, A.W. (47)

Papapetrou, A. 224
paracompact 48, 98
parallelizability *92*, 93
partition of unity 98, *100*, 336
Pauli, W. 362, 366, 367, 385
Pauli spin matrices 125
Penrose, R. 26, 51, 55, 75, 119, 224, 239, 251, 252, 262, 275, 291, 292, 293, (297), 299, 305, 307, 308, 353, (353), 362, 374, 385, 392, 395, (399), (401), 402, (404)
photon 323
left- and right-handed (151)
Pirani, F.A.E. 392
Planck's constant 377
Plebanski, J. 366
PND 162–
*see also* principal null direction
Poincaré group 6, (355)
Poincaré lemma [converse of] 204
Poincaré transformation 5, 6, 7, 8
position vector 5, 362, *387*, 405
positive frequency 362
Poston, T. 198
potential
electromagnetic 312, *316*–, 320, 373, 378
Yang–Mills 347–
Poynting vector 327
prime operation 226–, 250, 256–, 268
principal null direction (PND) *162*–, 173, 367
*k*-fold 162, 163
of Maxwell field 328
principal null vector 162, 328

principal spinor 162–
product
  contracted (inner) *86*, 88, 89
  exterior 203–, 262–
  inner (scalar) 2
    of spin-vectors *56*–, 59–, 104
  outer (tensor) 71, 74, *83*
  scalar multiplication 1, 56, 83
projection operator 271
proper [improper] tetrad 2
proton 179

quantum mechanics viii, (47), 220
quaternions *21*–, (45), 52, 55, (92), (93)
  conjugate of 22
norm of 22

Rainich, G.Y. 329
Rainich conditions 328–,
  differential 331, 332
rank 72
rapidity *20*, 27, 32
Rarita, W. 384
Rarita–Schwinger equation 384
Rauch, H. (47)
reducibility 141–
reduction to symmetric spinors 139–
  of curvature 236
reflexive 80
  *see also* total reflexivity
restricted Lorentz group 5
  representations of 141–
restricted Lorentz transformation *5*, 17, 167–
restricted Minkowski tetrad 4
restricted Poincaré group 6
  representations of (141)
Ricci identity 194
  diagram for 429
  spinor form of 243, 244
    with charged fields 320–, 366
    diagram for 433
Ricci rotation coefficients 265
Ricci spinor 231–, *234*
  diagram for 433
  dyad components of 248
  weight types of 256
Ricci tensor 146, 210, 239
  non-symmetric 238
  trace-free 146
Riemann sphere 10–, *11*, 24–, 274, 289
Riemann (–Christoffel) tensor 146, *209*, 231, 239, 341
  and curvature 2-form 263
  diagram for 431
Riemannian connection: *see* covariant derivative, Christoffel

Riesz, M. 389
right-handed massless particles 362
Rindler, W. 24, 352
Robinson, G. de B. (144)
Robinson I. 24, (31), 153, 241
Robinson–Bel tensor 240–
  *see also* Bel–Robinson tensor
Rohrlich, F. 272, (297), 305, (306), (307)
rotations of 2-sphere 18–, 294–
  irreducible functions under 296–
    completeness of 297

Sachs, R.K. 199, 260, 364, 392, 393, 399
Sachs (asterisk-) operation 260, *261*, 269, 270
scalar curvature 210, 233
scalar fields 180
scalar multiplication 1, 56, 83
scalar product 1
  between module and its dual 79
  numerical expression for 94, 95
  *see also* product
scalars 73
  complex 188
  ring of 76, 77
scattering 419–
Schiff, L.I. 306
Schild, A. 392
Schrödinger, E. 371
Schrödinger–Klein–Gordon field (equation) 378
  degrees of freedom for 390
Schwinger, J. 384
Sciama, D.W. 237
see-saw 106, 110
self-dual 151, 351
signature
  Lorentzian *2*, 24, (121), 235
  positive-definite 24
simple skew tensors (145), *165*–
simply-connected 41
  non-simply-connected region of space-time 219, 318, 331
skew symmetry: *see* anti-symmetry
sky mapping *9*, 30
spacelike hypersurface, initial data on 389, 401, 402
spacelike 2-surface 252, *267*–
  orientation of 268
  surface element 2-form for *280*, 301, 394
spacelike world-vector 3
space-orientation of 𝕍 4
space-time manifold 48–, 103, *210*–
  Minkowskian 5
  space-orientation of 129
  space time orientation of *49*, 129
  spin structure on *49*–, 129

time-orientation of *49*, 129
Sparling, G.A.J. (401)
speed of light 235
spherical harmonics: *see* spin-weighted spherical harmonics
spin
of a massive particle 384
of a massless particle 362
spin-coefficient Bianchi identity (compacted form) 259
spin-coefficient commutator equations 247, 248
in compacted form 258
spin-coefficient equations 247–
in compacted form 258
spin-coefficients *223*–, 246–, 254–
conformal rescalings of 358, 359
relation to Cartan's moving frames 265–
spin density 237–
spin-frame *58*, *110*–, 119–, 224, 228, 353
conformal rescaling of 358
and Minkowski tetrad 58, (59), 120
spin group 16
representations of 141–, 237
spin matrix *15*–, 115
spin-space 56
spin structure *49*–, 129
ambiguity of 51–
obstructions to 50–
spin transformation *15*–, 58, 167–
infinitesimal 176–
positive-definite Hermitian 20
unitary 18–
spin-vector bundle (48), 50–
spin-vectors 32, 39–, *47*, 126–, 352
components of 56–, *58*, 112
diagrams for 431
fields of 69, 70, 103–
operations on: *see* spinor operations
spin-weight *253*, 273, 290–, 391
spin-weighted spherical harmonics *297*–, 391, 392, 399
basis (${}_sY_{j,m}$) for 305
conformal behaviour of 300–
coordinate descriptions of 308–
orthogonality of 302
reciprocity relations for 307
scalar product of 305
table of dimensions of 298
spinor basis *110*–, 223–
conformal rescaling of 358
global existence of 93, 111
normalized: *see* spin-frame
spinor operations
algebraic 56–
complex conjugation 107–
on fields 103–
on general spinor 108–
geometry of 59–
inner product *59*–, 137, 352
scalar multiplication 39, *40*, 59, 128, 129
sum 63
spinor structure 48–, *54*, 129
in non-compact space–times 55
spinorial object 41–, *46*,
in nature (46)
1-spinors 273
2-spinors vii, 70, 103–, *108*–, 423
components of 113–
curvature spinors 231–
diagrams for 431–
ε-spinors 104–
covariant constancy of 215
covariant non-constancy of 216, 217
diagrams for 432
identity in 106, 136
4-spinors (Dirac spinors) vii, 143, (*221*), 370, 377
splitting off skew indices 106, 137
squared interval 5
Staruszkiewicz, A. (32)
Stasheff, J.D. 52
Stephani, H. 119, 224
stereographic coordinate (complex) *11*, (12), 33–, 274, 309, 310
stereographic projection *10*–, 25, 60
Stewart, J.M. 252
Stieffel–Whitney classes 51, *52*
Stokes' theorem (205)
stretch band 339, 340
Sudarshan, E.C.G. 272, (297), 305, (306), (307)
summation convention *2*, 71–, 76
supersymmetry (supergravity) 370
surface: *see* null hypersurface, spacelike hypersurface, spacelike 2-surface
surface area 2-form 280, 301
Susskind, L. (46)
symmetric spinors 139–
canonical decomposition of 162
constant 294
derivatives of, in compacted formalism 257
null 164
number of components of 147
reduction of arbitrary spinor to 139
reduction of curvature to 236
symmetric tensors 134–
trace-free 146, 147, 240
trace-free part of 148
trace-reversed 149
symmetrization 132
diagram for 426–

symmetry operations 84, *132–*
diagrams for 426–
Synge, J.L. 28, 387, 416
Szekeres, P. 367

tangent space 179
at a point 187
tangent vectors 179
bundle of 333
field of 184
at a point 187
Taylor (Gregory) series (covariant) 386–
tensor translation of spinor algebra 147–
of contraction 158
of even spinor 155
of odd spinor 156
of outer product 156, 157
of spinor-index permutation 153–
of sum 157, 158
tensor translation of spinor differential equations 219–
of Dirac–Weyl equation 220, 221
of Dirac equation 222
tensors vii, 70–
abstract algebraic view of 73–
[anti-] symmetric: *see* [anti-] symmetric tensors
classical algebra of 70–
classical transformations of 71, 97, 101
components of 70–
definitions of
as formal-sum (type II) *81–*, 95–
as multilinear-map (type I) *80–*, 95–
as rule (type III) 72, 95–, *100–*
invariant under a linear transformation 168
operations on
contraction 71, 74, 84, 85
diagrams for 425–
index [substitution] permutation 71, 75, 84
outer product 71, 74, *83*
sum 70, 74, 82
real and complex 188
restricted to open set 98–
Terrell, J. 26
tetrad 1–
improper, proper 2
Minkowski 3–, 58, 120
null 119, 125
orientation of 2
t' Hooft, G. 75
thorn (þ) 255–
conformally invariant ($þ_{\mathscr{C}}$) 360
Thrall, R.M. (144)
three-sphere
left [right] translations of 55, (92), (93)
parallelizability (92), (93)
time-orientable space–time 49
time-orientation of $\mathbb{V}$ 4
timelike 2-surface 268–
*see also* spacelike 2-surface
timelike world-vector 3
Tod, K.P. 291
torsion-free 193, 200, 202–, 216–, 223–, 355–, 386
torsion tensor 193, 200, 215, 216, 237, (353), (356), 370
diagram for 429, 430
total reflexivity *80–*, 82
from existence of basis 95–
on a manifold *98–*, 183, 188
trace-reversal 149
transformation laws of components
for spinors 115
for tensors 71, 97, 101
transvection 86
Trautman, A. 119, 237, 364
trivialization (317), 345
Turlay, R. (4)
twistor equation 260
in spin-coefficient form 260
conformal 361
twistors (viii), 130, 143, 273, 348, (401), 423
diagrams for 434
types of weighted quantity 253–, 289–

unimodular matrix 15, 115
universal covering space 45–
Unruh, W.G. 403
unscrambler (Robinson's) 153, 241
Unti, T.W.J. 224
Urbantke, H.K. 370

valence 72, 78
van der Burg, M.G.J. 399
van der Waerden, B.L. vii, (107), 221
vector bundle 48–, (253), 269, *332–*
base space of 333
complex line bundle (253)
unitary 344
connection on 340
change of 341
cross-section of (270), 334, *335*
local basis for 335
curvature of 340, 341
explicit construction of 337–
fibre of 333
Hermitian structure on 344
Lie group structure on 343, 344
fields on 341, 342
null-flag [spin-vector] bundle 48–
tangent bundle 333–
vector field 69, *184–*,

components of 185
vector space 1
spin-space as 58
vectors, contravariant and covariant 72
Veltman, M. 75
Vilenkin, N. Ya. (301)
von der Heyde, P. 237
Voorhees, B.H. 252

Wainwright, J. 224
Walker, M. 252
Ward, R.S. 352
wave equation 260, 374
conformally invariant 260, 395
degrees of freedom for 390
energy tensor for 369
initial data for 390, 401
in spin-coefficient form 260
weak energy condition 327
weak-field limit of Einstein equations 362–, 375
weak interactions, non-invariance under space-reflections (4), 56
wedge (exterior) product 203–, 262–
weighted quantity 253–
type of 253–
Wells, R.O. Junior 274
Werner, S.A. (47)
Weyl, H. 42, (143), 220, 356
Weyl geometry 356, 357
Weyl neutrino equation 220–
*see* Dirac–Weyl equation
Weyl (conformal) spinor *236*, 239, 240–, 366, 367, 373
dyad components of 248
weight types of 256
Weyl (conformal) tensor 146, 236, *240*–
[anti-] self-dual parts of 236, 237, *240*–
Wheeler, J.A. 329, 330
Whittaker, E.T. 222, (321)
Wilfing, A. (47)
Witten, L. 55, 329
Woodhouse, N.M.J. 205
world-tensor, as spinor 116–
complex and real 117, 124
world-tensor calculus vii
world-vector *1*–, 116–
causal, spacelike, timelike 3
complex, and improper Lorentz transformations 171
complex null 126, 129
future-and past-pointing *4*, 127
Hermitian-null 131, 132
null *3*, 126
Wu, C.S. (4)
Wu, T.T. (4), (315)

Yang, C.N. (4), (315), 342
Yang–Mills (YM) theory 333, *342*–
bundle connection in 344–
current for 350
energy tensor in 351
exact sets with 383, 386
field equations in 350
spinor form of 351
field tensor in 349
[anti-] self-dual 351
spinor form of 350, 351
gauge in 345
gauge transformations in 345, *348*
Hermitian metric in 347
potentials for 347
change under gauge transformations 349
YM-charged fields 333, *342*–, 356
Young, A. (143)
Young tableau 143, (*143*–)
number of independent components (144)
tensor diagram for 431

Zeilinger, A. (47)
zero rest-mass field equations 362
*see also* massless field equations
zero spin-vector 47
zero tensor 73
Zumino, B. 370

# Index of symbols

*Symbols are listed in order of appearence in text; page numbers in italics denote main references; page numbers in parentheses refer to footnotes; a dash indicates continuation for more than two pages.*


$\mathbb{R}$ 1, 180–
$\mathbb{V}$ 1–
$\mathbf{g}_{\mathrm{i}}$ 1–
$\mathbf{a}, \mathbf{A}, \boldsymbol{\alpha}, \boldsymbol{\Gamma}$ (bold upright indices: numerical) *2–*, 76, 335–
$\| U \|$ 3
$\mathbb{M}$ 5–, 285–, 393–
$\overrightarrow{PQ} = \mathrm{vec}(P, Q)$ 5
$\mathbb{R}^4$ 5, 52–
$\mathscr{S}^+, \mathscr{S}^-, S^+, S^-$ *8–*, 24–, 26, 33–, 60–
$\zeta$ (complex stereographic coordinate) *10–*, 33–, 274, 309, 310
$\boldsymbol{K}$ 14, 32
$SL(2, \mathbb{C})$ *16–*, 45–, 141–, 237
$\mathbf{A}^*$ (conjugate transpose) 16
$w$ (Doppler factor) *20*, 27
$\phi = \log w$ (rapidity) 27
$PP'$ 34
$SO(3)$ *41–*, 344
$\pi_1$ (first homotopy group) 44
$\tilde{T}$, $\widetilde{SO(3)}$ (universal covering space) 45–
$SU(2)$, $O^{\uparrow}_{+}(1, 3)$ 45–
$\boldsymbol{\kappa}$ (spin-vector) 47
$\mathbf{0}$ (zero spin-vector, tensor) 47, 73
$\mathscr{M}$ 48–, 98–, *180–*, 210–
$C^\infty$ 48
$\mathscr{F}, \mathscr{F}', \mathscr{F}_P, \tilde{\mathscr{F}}_P$ 48–
$SO(4)$ 52
$S^3$ 55, 92
$\mathfrak{S}^{\cdot}$ 56, *76–*
$\mathbb{C}$ 56, 294
$\{\boldsymbol{\kappa}, \boldsymbol{\omega}\}$ 56–, 104, 352
$\boldsymbol{o}, \boldsymbol{\iota}$ 58
$[{}^{p}_{q}], [{}^{pq}_{rs}]$ 72, 108
$\boldsymbol{\delta}, \delta^{\boldsymbol{\beta}}_{\boldsymbol{\alpha}}$ 73
$a, A, \alpha, \Gamma$ (lightface sloping indices: abstract) *76–*, 335–
$\mathscr{L}$ (labelling set) 77, 103
$\mathfrak{S}^{\alpha\ldots}_{\lambda\ldots}$ 80
$\delta^{\beta}_{\alpha}$ *89*, 90, 425
$\alpha^*$ (reversed-position index) 90
$\mathscr{A}, \mathscr{B}, \ldots$ (composite indices) 90
$\mathfrak{S}^{\mathscr{B}\ldots}_{\mathscr{A}\ldots}$ 90
$\mathscr{L}'$ 91, 107
$\varepsilon_{\alpha_1 \cdots \alpha_n}$ 92
$\delta^{\alpha}_{\boldsymbol{\alpha}}, \delta^{\boldsymbol{\alpha}}_{\alpha}$ 93, 94
$A^{\boldsymbol{\alpha}\ldots}_{\boldsymbol{\lambda}\ldots}, A^{\alpha\ldots}_{\lambda\ldots}$ 95
$\delta^{\hat{\boldsymbol{\alpha}}}_{\boldsymbol{\alpha}}, \delta^{\boldsymbol{\alpha}}_{\hat{\boldsymbol{\alpha}}}$ 97
$\mathfrak{S}^{\alpha\ldots}_{\beta\ldots}(f)$ 99
$\{\mathscr{U}_{\mathrm{i}}\}$ 99, 335
$\mathfrak{S}^{B\ldots}_{A\ldots}$ 103, 212–
$\varepsilon_{AB}, \varepsilon^{AB}, \varepsilon_A{}^B$ 104, 105, 432
$\overline{\kappa^A} = \bar{\kappa}^{A'}$ 107
$\mathfrak{S}^{A\ldots P'\ldots}_{L\ldots U'\ldots}$ 108
$o_A, o^A, \iota_A, \iota^A$ 111
$\chi = o_A \iota^A$ 111, 224, 252, 254
$\varepsilon_{\mathbf{AB}}, \varepsilon_{\mathbf{A}}{}^{A}, \ldots$ 111–, 223, 432
$\mathscr{K}$ (world-tensor labelling set) 116
$a = AA'$ etc. 116
$\mathfrak{S}^{a\ldots d}_{p\ldots r}, \mathfrak{T}^{a\ldots d}_{p\ldots r}$ 117
$\mathfrak{T}$ 117, *180*
$g_{ab}, g_a{}^b, g^{ab}$ 117, 208
$l^a, m^a, \bar{m}^a, n^a$ 119, 226, 253
$g_{\mathbf{ab}}, g_{\mathbf{a}}{}^{\mathbf{b}}, g^{\mathbf{ab}}, g_{\mathbf{a}}{}^{a}, g_a{}^{\mathbf{a}}$ 121
$g_{\mathbf{a}}{}^{\mathbf{AA'}}, g_{\mathbf{AA'}}{}^{\mathbf{a}}$ (Infeld–van der Waerden symbols) *123*, 266
$\chi_{\ldots(\ldots)\ldots}, \chi_{\ldots[\ldots]\ldots}, \chi_{\ldots|\ldots|\ldots}$ 132–
$\mathfrak{S}_{\ldots(\ldots)\ldots}, \mathfrak{S}_{\ldots[\ldots]\ldots}$ 134
$e_{abcd}, e^{abcd}$ 137, *138*
$\oplus$ (direct sum) 141, 142, 180, 212, 343
$\dfrac{\partial}{\partial Y^{\alpha}}, \dfrac{\partial}{\partial o^{A}}, \dfrac{\partial}{\partial x^{a}}, \dfrac{\partial}{\partial y^{\Phi}}$ (abstract indices!) (145), *287*, 341
$GL(n, \mathbb{C})$ 146
$\hat{T}_{ab}$ (trace-reversed tensor) *149*, 364
$^*F_{ab}, {}^*G_{ab\mathscr{A}}$ *150*, 264

$^{\dagger}J_{abc\mathscr{A}}$ *151*, 264, 371
$^{\ddagger}K_{a\mathscr{B}}$ *151*, 264
$^{-}F_{ab}$, $^{+}F_{ab}$ *151*, 320
$^{(\theta)}F_{ab}$ 152
$U_{ab}{}^{cd}$ 153–
exp(...) *176*, 177, 207, 346, 387
$\mathfrak{R}$ 181
$B_{a,b}(x)$ *181*, 183
$(\mathscr{U}, x^{\mathbf{i}})$ 182–
$V(f)$ 184
$\mathfrak{T}^{\cdot}$ 184
$\partial/\partial x^{\mathbf{a}}$ (on a manifold) (184), 189
$V[P]$, $V[\mathscr{S}]$, $\mathfrak{T}^{\cdot}[P]$, $\mathfrak{T}[\mathscr{S}]$, $\mathfrak{T}^{\cdot}[\mathscr{S}]$ 187
$\mathfrak{T}^{\alpha\ldots\gamma}_{\lambda\ldots\nu}[P]$, $\mathfrak{S}^{\alpha\ldots\gamma}_{\lambda\ldots\nu}[\mathscr{S}]$ 188
d 189, 203–, 263–
$dx^{\alpha}$ 189
$\nabla_{\alpha}$ 190–, *192*, 312–, 340, 429
$\Delta_{\alpha\beta}$ *193*–, 242
$T_{\alpha\beta}{}^{\gamma}$ *193*–, 237, 429
$\Pi_{\alpha\beta}$ *193*–, 340
$R_{\alpha\beta\gamma}{}^{\delta}$ *194*, 209, 231, 429
$Q_{\alpha\beta}{}^{\gamma}$ *196*, 215, 341
$\partial_{\alpha}$ *198*, 348
$\Gamma_{\alpha\beta}{}^{\gamma}$ 199
$[U, V]$ (199), *206*, 246, 247
$\nabla_2 A_3$ 201
$\underset{V}{£}$ *202*, 207
$A := A_{1112\ldots1p}$ 203
$p$-form 203–
$\boldsymbol{A} \wedge \boldsymbol{C}$ 203–, 262–
$\int_{\mathscr{P}} \boldsymbol{A}$ 205
$\boldsymbol{U} \circ \boldsymbol{V}$ (composition) 206
$\underset{X}{\nabla}$ 206
$\Theta_{aB}{}^{C}$ 215–, 355
$\Upsilon_{a}$ 217, 356–
$\gamma_{\mathbf{AA'C}}{}^{\mathbf{B}}$, $\gamma_{\mathbf{AA'BC}}$ (spin-coefficients) 223–
$\rho, \sigma, \kappa, \tau, \ldots, \varepsilon'$ (spin-coefficients) *225*–, 248, 249, 254–, 270, 358, 359
$(\ldots)'$ 226–
$D, \delta, D', \delta'$ 227
$R_{abcd} = R_{AA'BB'CC'DD'}$ 231
$X_{ABCD}$, $\Phi_{ABC'D'}$ 231–, 357, 433
$*(\ldots)$, $(\ldots)*$, $^{(\theta)}(\ldots)$, $(\ldots)^{(\theta)}$ 232
$\Lambda$ 233–, 357
$\gamma$ (Newton's constant) 235
$\Psi_{ABCD}$ 235–, 312, 357, 433
$C_{abcd}$, $^{+}C_{abcd}, \ldots$ 236, 237
$\square_{AB}$, $\square_{A'B'}$ *242*–, 320–, 350
$\Psi_0, \Psi_1, \ldots$ 248
$\Phi_{00}, \Phi_{01}, \ldots$ 248
$\Pi$ 248
$\{r', r; t', t\}$ (type) 253
$\{p, q\}$ (type): $p = r' - r$, $q = t' - t$ 253
$[s, w]$ (type): $s = \frac{1}{2}(p - q)$, $w = \frac{1}{2}(p + q)$ 253, 273, 289–, *291*–
þ, ð, þ′, ð′ 255–, 270–, 308–
$\square$ *260*, 364, 409
$(\ldots)^*$ (Sachs operation) 260, 261
$\Omega_a{}^b$, $\Omega_A{}^B$ 263
$\theta^a$ 264, 265
$\omega_a{}^b$ 264, 265
$\boldsymbol{l}, \boldsymbol{m}, \bar{\boldsymbol{m}}, \boldsymbol{n}$ 266, 267
$ds^2$ 266, 275
$\delta\mathscr{S}$, $\mathscr{S}$ 268–
$\Delta_a$ 271–
$K$ (complex curvature of 2-surface) 272, 277
$\mathscr{S}$ (area 2-form) 280, 301, 394
$\mathscr{N}$ (null hypersurface) 281–, 388–
$\partial$ (boundary) 282
$\mathscr{N}$ (null volume 3-form) 284
$\mathscr{N}$ (invariant volume 3-form of light cone) 415
$f(\xi)$, $f(\xi, \bar{\xi})$ (287)
$\mathbb{S}\ldots$ *294*–, 396
$\langle h, f \rangle$ 301–
${}_sY_{j,m}$, ${}_sZ_{j,m}$ 305–
$\varphi_{AB}$ 312–, *320*
$\Phi_a$ 312–, *316*
$F_{ab}$ (Maxwell field) 312, *318*–
$\overset{e}{\mathfrak{S}}{}^{\cdots}_{\cdots}$, $\overset{e}{\mathfrak{S}}{}^{\cdots}_{\cdots}[P]$ 313–
$\varepsilon$ (elementary charge) 314–, 344
$K$ (complex Maxwell scalar) *324*, 330
$\mathfrak{T}^{\Phi}$, $\mathfrak{S}^{\Phi}$ 336, 343
$\mathscr{V}^{\cdot}$ (fibre) 338–
$K_{ab\Omega}{}^{\Phi}$ 340, 341
$\Phi', \ldots$ (conjugated bundle index) 343
$\alpha_{\Psi}{}^{\Psi}$ 345
$q_{\hat{\Psi}}{}^{\Psi}$ 345
$q_{\Psi}{}^{\Theta}$, $p_{\Psi}{}^{\Theta}$ 346
$\Phi_{a\Psi}{}^{\Theta}$ 347
$\hat{g}_{ab}$, $\hat{\phi}_{\ldots}$ (conformally rescaled) 352–
$\Omega$ 352–
$w_0, w_1, w$ (conformal weights) 358, 360
$þ_{\mathscr{C}}$, $ð_{\mathscr{C}}$, $þ'_{\mathscr{C}}$, $ð'_{\mathscr{C}}$ *360*, 395
$x^a$ (position vector) 362, *387*, 405
$h_{ab}$ 363
$\mu = 2^{-\frac{1}{2}}\hbar^{-1}m$ 377, 384, 412
$\Delta_j$ (distributional field) 408–